T3-ANT-659

UNIVERSITY CALCULUS

PART TWO

Joel Hass
University of California, Davis

Maurice D. Weir
Naval Postgraduate School

George B. Thomas, Jr.
Massachusetts Institute of Technology

PEARSON
Addison
Wesley

Boston San Francisco New York
London Toronto Sydney Tokyo Singapore Madrid
Mexico City Munich Paris Cape Town Hong Kong Montreal

Publisher:	Greg Tobin
Senior Acquisitions Editor:	William Hoffman
Senior Project Editor:	Rachel S. Reeve
Editorial Assistant:	Lianna Kong
Managing Editor:	Karen Wernholm
Senior Production Supervisor:	Jeffrey Holcomb
Senior Designer:	Barbara T. Atkinson
Supplements Supervisor:	Emily Portwood
Software Development:	Mary Durnwald and Bob Carroll
Marketing Manager:	Phyllis Hubbard
Marketing Assistant:	Celena Carr
Senior Author Support/ Technology Specialist:	Joe Vetere
Senior Prepress Supervisor:	Caroline Fell
Senior Manufacturing Buyer:	Evelyn Beaton
Production Coordinator:	Kathy Diamond
Composition:	Nesbitt Graphics, Inc.
Illustrations:	Techsetters, Inc.

Cover Image: "Flying Spinnakers," photographed by Morris and Stanley Rosenfeld on July 9, 1937 © Mystic Seaport, Rosenfeld Collection, image acquired in honor of Franz Schneider. The photograph is of the 70' Clinton Crane–designed, Henry B. Nevins–built 10-meter class sloop GLEAM and the 70' Sparkman & Stephens–designed, Henry B. Nevins–built 12-meter class sloop NORTHERN LIGHT, running down Long Island Sound on the NYYC Cruise. Visible in the image is the cropped port quarter view of GLEAM and the spinnaker and mainsail of NORTHERN LIGHT running before the wind under spinnakers and mainsails.

For permission to use copyrighted material, grateful acknowledgment is made to the copyright holders on page C-1, which is hereby made part of this copyright page.

Many of the designations used by manufacturers and sellers to distinguish their products are claimed as trademarks. Where those designations appear in this book, and Addison-Wesley was aware of a trademark claim, the designations have been printed in initial caps or all caps.

Library of Congress Cataloging-in-Publication Data

Hass, Joel.
    University calculus/Joel Hass, Maurice D. Weir, George B. Thomas, Jr.
      p. cm.
    ISBN 0-321-45421-9
    1. Calculus—Textbooks. I. Weir, Maurice D. II. Thomas, George Brinton, 1914– .
III. Title.

QA303.2.H37 2007
515—dc22   2005048720

Copyright © 2007 Pearson Education, Inc. All rights reserved. No part of this publication may be reproduced, stored in a retrieval system, or transmitted, in any form or by any means, electronic, mechanical, photocopying, recording, or otherwise, without the prior written permission of the publisher. Printed in the United States of America. For information on obtaining permission for use of material in this work, please submit a written request to Pearson Education, Inc., Rights and Contracts Department, 75 Arlington Street, Suite 300, Boston, MA 02116, fax your request to 617-848-7047, or e-mail at http://www.pearsoned.com/legal/permissions.htm.

2 3 4 5 6 7 8 9 10—VH—10 09 08 07

CONTENTS

14   Integration in Vector Fields   851

15   First-Order Differential Equations (online)

16   Second-Order Differential Equations (online)

Appendices   AP-1

Answers   A-1

Index   I-1

A Brief Table of Integrals   T-1

Credits   C-1

PREFACE

OVERVIEW *University Calculus* offers a more streamlined and faster-paced text that maintains the high standards and focused applications that have been the hallmark of *Thomas' Calculus*.

Condensing material from a well-established book is a difficult task. We have preserved the careful development of the main ideas in *Thomas' Calculus*, and we have resisted the temptation to lower its level of rigor. We think that having high standards inspires students to seek excellence. On the other hand, a solid grounding in the various types of functions is essential to understanding calculus. In this vein, we have retained a shortened first chapter reviewing the basic functions. We understand that some professors will prefer to skip this review, but we also believe that many students need to see this material again. The first chapter is not an introduction to calculus, but rather a helpful aid for the average student.

Today, an increasing number of students become familiar with the terminology and operational methods of calculus in high school. However, their conceptual understanding of calculus is often quite limited when they enter college. We have acknowledged this reality by concentrating on concepts and their applications throughout.

To achieve the goals of *University Calculus*, we consulted with many colleagues and reviewers across the country. They helped us decide which topics to retain and which to remove or condense. We are grateful for their thoughtful advice, from which we can now offer this new text.

Pedagogical Features

EXERCISES Exercises and examples play a crucial role in learning calculus. We have included many of the exercises that appeared in previous editions of *Thomas' Calculus*, and which constituted a great strength of those editions. Within each section we have organized and grouped the exercises by topic, progressing from computational problems to applied and theoretical problems. This arrangement gives students the opportunity to develop skills in using the methods of calculus and to deepen their understanding of its applications.

RIGOR The level of rigor is consistent throughout. We give both formal and informal discussions, making clear the distinction between the two, and we include precise definitions and accessible proofs for the students. The text is organized so the material can be covered informally, giving the instructor a degree of flexibility. For example, while we do not prove that a continuous function on a closed and bounded interval has a maximum there, we do state this theorem carefully and use it to prove several subsequent results.

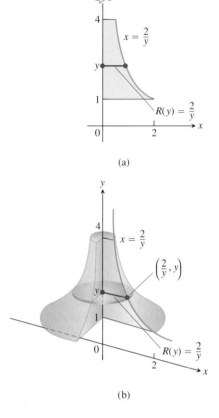

(a)

(b)

FIGURE 6.11, page 396
Finding the volume of the solid
generated by revolving the region (a)
about the *y*-axis.

ART   We realize that figures and illustrations are a critical component to learning calculus. We paid special attention to the clarity with which the figures illustrate their associated concepts. This is especially evident with the three-dimensional graphics, where we were able to better indicate depth, layering, and rotation (see figures on this page).

END-OF-CHAPTER REVIEWS AND PROJECTS   In addition to problems appearing after each section, each chapter culminates with review questions, practice exercises, and a series of Additional and Advanced Exercises. Student projects are available over the Internet at **www.aw-bc.com/thomas**.

WRITING EXERCISES   Writing exercises placed throughout the text ask students to explore a variety of calculus concepts and applications. In addition, each chapter contains questions asking students to summarize what they have learned. Many of these questions ask for written descriptions that test conceptual understanding.

ANSWERS   Answers are provided for all odd-numbered exercises when appropriate, and these have been carefully checked for correctness.

MATHEMATICAL CORRECTNESS   We have been careful to say only what is true and mathematically sound. Every definition, theorem, corollary, and proof has been reviewed for clarity and mathematical correctness.

WRITING AND APPLICATIONS   This text continues to be easy to read, conversational, and mathematically rich. Each new topic is motivated by clear, easy-to-understand examples and applications.

TECHNOLOGY   Technology can be incorporated according to the taste of the instructor. Each section contains exercises requiring the use of technology; these are marked with a ▨ if suitable for calculator or computer usage or are labeled **Computer Explorations** if a computer algebra system (CAS, such as *Maple* or *Mathematica*) is required.

FIGURE 6.13, page 397
The cross-sections of the
solid of revolution generated here are washers,
not disks.

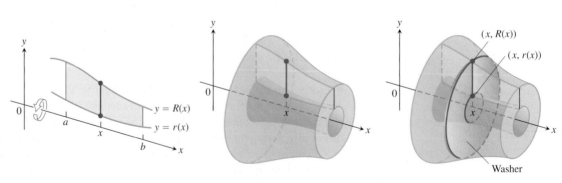

Print Supplements

STUDENT EDITION OF UNIVERSITY CALCULUS
ISBN 0-321-35014-6

INSTRUCTOR'S SOLUTIONS MANUAL
Part One (Chapters 1–9), ISBN 0-321-38848-8
Part Two (Chapters 10–14), ISBN 0-321-38698-1
The *Instructor's Solutions Manual* by William Ardis, et al., contains complete worked-out solutions to all of the exercises in the text.

ANSWER BOOK
ISBN 0-321-39423-2
The *Answer Book* by William Ardis, et al., contains short answers to most of the exercises in the text.

STUDENT OUTLINES
Part One (Chapters 1–9), ISBN 0-321-39551-4
Part Two (Chapters 8, 10–14), ISBN 0-321-39969-2
Organized to correspond to the text, the *Student Outlines* by Joseph Borzellino and Patricia Nelson reinforces important concepts and provides an outline of the important topics, theorems, and definitions, as well as study tips and additional practice problems.

STUDENT'S SOLUTIONS MANUAL
Part One (Chapters 1–9), ISBN 0-321-38849-6
Part Two (Chapters 8, 10–14), ISBN 0-321-38850-X
The *Student's Solutions Manual* by William Ardis, et al., is designed for the student and contains carefully worked-out solutions to all the odd-numbered exercises in the text.

JUST-IN-TIME ALGEBRA AND TRIGONOMETRY FOR EARLY TRANSCENDENTALS CALCULUS,
Third Edition
ISBN 0-321-32050-6
Sharp algebra and trigonometry skills are critical to mastering calculus, and *Just-in-Time Algebra and Trigonometry for Early Transcendentals Calculus,* Third Edition, by Guntram Mueller and Ronald I. Brent is designed to bolster these skills while students study calculus. As students make their way through calculus, this text is with them every step of the way, showing them the necessary algebra or trigonometry topics and pointing out potential problem spots. The easy-to-use contents has algebra and trigonometry topics arranged in the order in which students will need them as they study calculus.

ADDISON-WESLEY'S CALCULUS REVIEW CARD
The Calculus Review Card is a resource for students containing important formulas, functions, definitions, and theorems that correspond precisely to *University Calculus.* This card can work as a reference for completing homework assignments or as an aid in studying and is available bundled with a new text. Contact your Addison-Wesley sales representative for more information.

Media and Online Supplements

TECHNOLOGY RESOURCE MANUALS

Maple Manual by Donald Hartig, California Polytechnic State University

Mathematica Manual by Marie Vanisko, California State University Stanislaus, and Lyle Cochran, Whitworth College

TI-Graphing Calculator Manual by Luz DeAlba, Drake University

These manuals cover *Maple* 9, *Mathematica* 5, and the TI-83 Plus/TI-84 Plus, TI-85/TI-86, and TI-89/TI-92 Plus, respectively. Each manual provides detailed guidance for integrating a specific software package or graphing calculator throughout the course, including syntax and commands. These manuals are available to qualified instructors through **http://www.aw-bc.com/irc/**.

MYMATHLAB

MyMathLab is a series of text-specific, easily customizable online courses for Addison-Wesley textbooks in mathematics and statistics. Powered by CourseCompass™ (Pearson Education's online teaching and learning environment) and MathXL (our online homework, tutorial, and assessment system), MyMathLab gives you the tools you need to deliver all or a portion of your course online, whether your students are in a lab setting or working from home. MyMathLab provides a rich and flexible set of course materials, featuring free-response exercises that are algorithmically generated for unlimited practice and mastery. Students can also use online tools, such as video lectures, animations, multimedia textbook, and Maple/Mathematica projects, to independently improve their understanding and performance. Instructors can use MyMathLab's homework and test managers to select and assign online exercises correlated directly to the textbook, and they can also create and assign their own online exercises and import TestGen tests for added flexibility. MyMathLab's online gradebook—designed specifically for mathematics and statistics—automatically tracks students' homework and test results and gives the instructor control over how to calculate final grades. Instructors can also add offline (paper-and-pencil) grades to the gradebook. MyMathLab is available to qualified adopters. For more information, visit our Web site at **www.mymathlab.com** or contact your Addison-Wesley sales representative.

MATHXL®

MathXL is a powerful online homework, tutorial, and assessment system that accompanies Addison-Wesley textbooks in mathematics or statistics. With MathXL, instructors can create, edit, and assign online homework and tests using algorithmically generated exercises correlated at the objective level to the textbook. They can also create and assign their own online exercises and import TestGen tests for added flexibility. All student work is tracked in MathXL's online gradebook. Students can take chapter tests in MathXL and receive personalized study plans based on their test results. The study plan diagnoses weaknesses and links students directly to tutorial exercises for the objectives they need to study and retest. Students can also access supplemental animations and video clips directly from selected exercises. MathXL is available to qualified adopters. For more information, visit our Web site at **www.mathxl.com** or contact your Addison-Wesley sales representative.

TESTGEN

TestGen enables instructors to build, edit, print, and administer tests using a computerized bank of questions developed to cover all the objectives of the text. TestGen is algorithmically based, allowing instructors to create multiple but equivalent versions of the same question or test with the click of a button. Instructors can also modify test bank questions or add new questions. Tests can be printed or administered online. The software is available on a dual-platform Windows/Macintosh CD-ROM.

DIGITAL VIDEO TUTOR

The Digital Video Tutor features an engaging team of mathematics instructors who present comprehensive coverage of topics in the text. The lecturers' presentations include examples and exercises from the text and support an approach that emphasizes visualization and problem solving. The video lectures are available on CD-ROM, making it easy and convenient for students to watch the videos from a computer at home or on campus. The complete digitized video set, affordable and portable for students, is ideal for distance learning or supplemental instruction.

WEB SITE   www.aw-bc.com/thomas

The *University Calculus* Web site provides the expanded historical biographies and essays referenced in the text. Also available is a collection of *Maple* and *Mathematica* modules that can be used as projects by individual students or groups of students.

ADDISON-WESLEY MATH TUTOR CENTER

The Addison-Wesley Math Tutor Center is staffed by qualified mathematics and statistics instructors who provide students with tutoring on examples and odd-numbered exercises from the textbook. Tutoring is available via toll-free telephone, toll-free fax, e-mail, and the Internet. Interactive, Web-based technology allows tutors and students to view and work through problems together in real time over the Internet. The Addison-Wesley Math Tutor Center is available to qualified adopters. For more information, please visit our Web site at **www.aw-bc.com/tutorcenter** or call us at 1-888-777-0463.

Acknowledgments

For their careful checking of this text for accuracy, we extend our gratitude to Marie Vanisko and Thomas Wegleitner. We also extend our sincere thanks for the suggestions and contributions made by the following reviewers:

Harry Allen, *Ohio State University*
Edoh Amiran, *Western Washington University*
Anthony Bedenikovic, *Bradley University*
Deborah Brandon, *Carnegie Mellon University*
Said Fariabi, *San Antonio College*
Krystyna Kuperberg, *Auburn University*
Paul Sacks, *Iowa State University*
Stephen Summers, *University of Florida*
Blake Thornton, *Washington University in St. Louis*
Ilie Ugarcovici, *Rice University*

Finally, we are grateful for the advice, suggestions, and encouragment provided by our development editor, David Chelton.

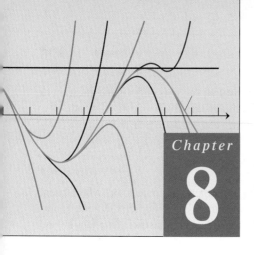

INFINITE SEQUENCES AND SERIES

OVERVIEW Everyone knows how to add two numbers together, or even several. But how do you add infinitely many numbers together? In this chapter we answer this question, which is part of the theory of infinite sequences and series.

An important application of this theory is a method for representing a known differentiable function $f(x)$ as an infinite sum of powers of x, so it looks like a "polynomial with infinitely many terms." Moreover, the method extends our knowledge of how to evaluate, differentiate, and integrate polynomials, so we can work with even more general functions than those encountered so far. These new functions often solve important problems in science and engineering.

8.1 Sequences

HISTORICAL ESSAY

Sequences and Series

A sequence is a list of numbers

$$a_1, a_2, a_3, \ldots, a_n, \ldots$$

in a given order. Each of a_1, a_2, a_3 and so on represents a number. These are the **terms** of the sequence. For example the sequence

$$2, 4, 6, 8, 10, 12, \ldots, 2n, \ldots$$

has first term $a_1 = 2$, second term $a_2 = 4$ and nth term $a_n = 2n$. The integer n is called the **index** of a_n, and indicates where a_n occurs in the list. Order is important. The sequence $2, 4, 6, 8 \ldots$ is not the same as the sequence $4, 2, 6, 8 \ldots$.

We can think of the sequence

$$a_1, a_2, a_3, \ldots, a_n, \ldots$$

as a function that sends 1 to a_1, 2 to a_2, 3 to a_3, and in general sends the positive integer n to the nth term a_n. This leads to the formal definition of a sequence.

> **DEFINITION** An **infinite sequence** of numbers is a function whose domain is the set of positive integers.

The function associated with the sequence

$$2, 4, 6, 8, 10, 12, \ldots, 2n, \ldots$$

sends 1 to $a_1 = 2$, 2 to $a_2 = 4$, and so on. The general behavior of this sequence is described by the formula $a_n = 2n$.

We can equally well make the domain the integers larger than a given number n_0, and we allow sequences of this type also. For example, the sequence

$$12, 14, 16, 18, 20, 22\ldots$$

is described by the formula $a_n = 10 + 2n$. It can also be described by the simpler formula $b_n = 2n$, where the index n starts at 6 and increases. To allow such simpler formulas, we let the first index of the sequence be any integer. In the sequence above, $\{a_n\}$ starts with a_1 while $\{b_n\}$ starts with b_6.

Sequences can be described by writing rules that specify their terms, such as

$$a_n = \sqrt{n}, \qquad b_n = (-1)^{n+1}\frac{1}{n}, \qquad c_n = \frac{n-1}{n}, \qquad d_n = (-1)^{n+1}$$

or by listing terms,

$$\{a_n\} = \left\{\sqrt{1}, \sqrt{2}, \sqrt{3}, \ldots, \sqrt{n}, \ldots\right\}$$

$$\{b_n\} = \left\{1, -\frac{1}{2}, \frac{1}{3}, -\frac{1}{4}, \ldots, (-1)^{n+1}\frac{1}{n}, \ldots\right\}$$

$$\{c_n\} = \left\{0, \frac{1}{2}, \frac{2}{3}, \frac{3}{4}, \frac{4}{5}, \ldots, \frac{n-1}{n}, \ldots\right\}$$

$$\{d_n\} = \left\{1, -1, 1, -1, 1, -1, \ldots, (-1)^{n+1}, \ldots\right\}.$$

We also sometimes write

$$\{a_n\} = \left\{\sqrt{n}\right\}_{n=1}^{\infty}.$$

Figure 8.1 shows two ways to represent sequences graphically. The first marks the first few points from $a_1, a_2, a_3, \ldots, a_n, \ldots$ on the real axis. The second method shows the graph of the function defining the sequence. The function is defined only on integer inputs, and the graph consists of some points in the xy-plane, located at $(1, a_1), (2, a_2), \ldots, (n, a_n), \ldots$.

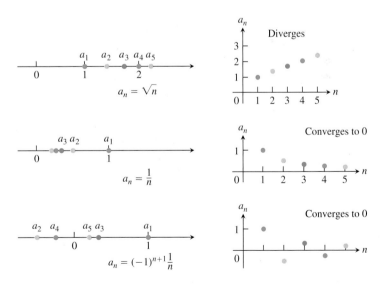

FIGURE 8.1 Sequences can be represented as points on the real line or as points in the plane where the horizontal axis n is the index number of the term and the vertical axis a_n is its value.

Convergence and Divergence

Sometimes the numbers in a sequence approach a single value as the index n increases. This happens in the sequence

$$\left\{ 1, \frac{1}{2}, \frac{1}{3}, \frac{1}{4}, \dots, \frac{1}{n}, \dots \right\}$$

whose terms approach 0 as n gets large, and in the sequence

$$\left\{ 0, \frac{1}{2}, \frac{2}{3}, \frac{3}{4}, \frac{4}{5}, \dots, 1 - \frac{1}{n}, \dots \right\}$$

whose terms approach 1. On the other hand, sequences like

$$\left\{ \sqrt{1}, \sqrt{2}, \sqrt{3}, \dots, \sqrt{n}, \dots \right\}$$

have terms that get larger than any number as n increases, and sequences like

$$\{ 1, -1, 1, -1, 1, -1, \dots, (-1)^{n+1}, \dots \}$$

bounce back and forth between 1 and −1, never converging to a single value. The following definition captures the meaning of having a sequence converge to a limiting value. It says that if we go far enough out in the sequence, by taking the index n to be larger than some value N, the difference between a_n and the limit of the sequence becomes less than any preselected number $\epsilon > 0$.

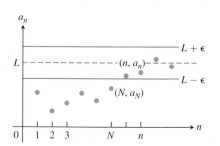

FIGURE 8.2 $a_n \to L$ if $y = L$ is a horizontal asymptote of the sequence of points $\{(n, a_n)\}$. In this figure, all the a_n's after a_N lie within ϵ of L.

> **DEFINITIONS** The sequence $\{a_n\}$ **converges** to the number L if for every positive number ϵ there corresponds an integer N such that for all n,
>
> $$n > N \quad \Rightarrow \quad |a_n - L| < \epsilon.$$
>
> If no such number L exists, we say that $\{a_n\}$ **diverges.**
> If $\{a_n\}$ converges to L, we write $\lim_{n \to \infty} a_n = L$, or simply $a_n \to L$, and call L the **limit** of the sequence (Figure 8.2).

HISTORICAL BIOGRAPHY

Nicole Oresme
(ca. 1320–1382)

The definition is very similar to the definition of the limit of a function $f(x)$ as x tends to ∞ ($\lim_{x \to \infty} f(x)$ in Section 2.4). We will exploit this connection to calculate limits of sequences.

EXAMPLE 1 Show that

(a) $\lim_{n \to \infty} \frac{1}{n} = 0$ **(b)** $\lim_{n \to \infty} k = k$ (any constant k)

Solution

(a) Let $\epsilon > 0$ be given. We must show that there exists an integer N such that for all n,

$$n > N \quad \Rightarrow \quad \left| \frac{1}{n} - 0 \right| < \epsilon.$$

This implication will hold if $(1/n) < \epsilon$ or $n > 1/\epsilon$. If N is any integer greater than $1/\epsilon$, the implication will hold for all $n > N$. This proves that $\lim_{n \to \infty} (1/n) = 0$.

(b) Let $\epsilon > 0$ be given. We must show that there exists an integer N such that for all n,

$$n > N \quad \Rightarrow \quad |k - k| < \epsilon.$$

Since $k - k = 0$, we can use any positive integer for N and the implication will hold. This proves that $\lim_{n \to \infty} k = k$ for any constant k. ∎

EXAMPLE 2   Show that the sequence $\{1, -1, 1, -1, 1, -1, \ldots, (-1)^{n+1}, \ldots\}$ diverges.

Solution   Suppose the sequence converges to some number L. By choosing $\epsilon = 1/2$ in the definition of the limit, all terms a_n of the sequence with index n larger than some N must lie within $\epsilon = 1/2$ of L. Since the number 1 appears repeatedly as every other term of the sequence, we must have that the number 1 lies within the distance $\epsilon = 1/2$ of L. It follows that $|L - 1| < 1/2$, or equivalently, $1/2 < L < 3/2$. Likewise, the number -1 appears repeatedly in the sequence with arbitrarily high index. So we must also have that $|L - (-1)| < 1/2$, or equivalently, $-3/2 < L < -1/2$. But the number L cannot lie in both of the intervals $(1/2, 3/2)$ and $(-3/2, -1/2)$ because they have no overlap. Therefore, no such limit L exists and so the sequence diverges.

Note that the same argument works for any positive number ϵ smaller than 1, not just $1/2$. ∎

The sequence $\{\sqrt{n}\}$ also diverges, but for a different reason. As n increases, its terms become larger than any fixed number. We describe the behavior of this sequence by writing

$$\lim_{n \to \infty} \sqrt{n} = \infty.$$

In writing infinity as the limit of a sequence, we are not saying that the differences between the terms a_n and ∞ become small as n increases. Nor are we asserting that there is some number infinity that the sequence approaches. We are merely using a notation that captures the idea that a_n eventually gets and stays larger than any fixed number as n gets large.

DEFINITION   The sequence $\{a_n\}$ **diverges to infinity** if for every number M there is an integer N such that for all n larger than N, $a_n > M$. If this condition holds we write

$$\lim_{n \to \infty} a_n = \infty \qquad \text{or} \qquad a_n \to \infty.$$

Similarly if for every number m there is an integer N such that for all $n > N$ we have $a_n < m$, then we say $\{a_n\}$ **diverges to negative infinity** and write

$$\lim_{n \to \infty} a_n = -\infty \qquad \text{or} \qquad a_n \to -\infty.$$

A sequence may diverge without diverging to infinity or negative infinity, as we saw in Example 2. The sequences $\{1, -2, 3, -4, 5, -6, 7, -8, \ldots\}$ and $\{1, 0, 2, 0, 3, 0, \ldots\}$ are also examples of such divergence.

Calculating Limits of Sequences

Since sequences are functions with domain restricted to the positive integers, it is not surprising that the theorems on limits of functions given in Chapter 2 have versions for sequences.

THEOREM 1   Let $\{a_n\}$ and $\{b_n\}$ be sequences of real numbers and let A and B be real numbers. The following rules hold if $\lim_{n\to\infty} a_n = A$ and $\lim_{n\to\infty} b_n = B$.

1. *Sum Rule:* $\lim_{n\to\infty}(a_n + b_n) = A + B$

2. *Difference Rule:* $\lim_{n\to\infty}(a_n - b_n) = A - B$

3. *Product Rule:* $\lim_{n\to\infty}(a_n \cdot b_n) = A \cdot B$

4. *Constant Multiple Rule:* $\lim_{n\to\infty}(k \cdot b_n) = k \cdot B$ (Any number k)

5. *Quotient Rule:* $\lim_{n\to\infty} \dfrac{a_n}{b_n} = \dfrac{A}{B}$ if $B \neq 0$

The proof is similar to that of Theorem 1 of Section 2.2, and is omitted.

EXAMPLE 3   By combining Theorem 1 with the limits of Example 1, we have:

(a) $\displaystyle \lim_{n\to\infty}\left(-\frac{1}{n}\right) = -1 \cdot \lim_{n\to\infty}\frac{1}{n} = -1 \cdot 0 = 0$ Constant Multiple Rule and Example 1a

(b) $\displaystyle \lim_{n\to\infty}\left(\frac{n-1}{n}\right) = \lim_{n\to\infty}\left(1 - \frac{1}{n}\right) = \lim_{n\to\infty}1 - \lim_{n\to\infty}\frac{1}{n} = 1 - 0 = 1$ Difference Rule and Example 1a

(c) $\displaystyle \lim_{n\to\infty}\frac{5}{n^2} = 5 \cdot \lim_{n\to\infty}\frac{1}{n} \cdot \lim_{n\to\infty}\frac{1}{n} = 5 \cdot 0 \cdot 0 = 0$ Product Rule

(d) $\displaystyle \lim_{n\to\infty}\frac{4 - 7n^6}{n^6 + 3} = \lim_{n\to\infty}\frac{(4/n^6) - 7}{1 + (3/n^6)} = \frac{0 - 7}{1 + 0} = -7.$ Sum and Quotient rules ■

Be cautious in applying Theorem 1. It does not say, for example, that each of the sequences $\{a_n\}$ and $\{b_n\}$ have limits if their sum $\{a_n + b_n\}$ has a limit. For instance, $\{a_n\} = \{1, 2, 3, \ldots\}$ and $\{b_n\} = \{-1, -2, -3, \ldots\}$ both diverge, but their sum $\{a_n + b_n\} = \{0, 0, 0, \ldots\}$ clearly converges to 0.

One consequence of Theorem 1 is that every nonzero multiple of a divergent sequence $\{a_n\}$ diverges. For suppose, to the contrary, that $\{ca_n\}$ converges for some number $c \neq 0$. Then, by taking $k = 1/c$ in the Constant Multiple Rule in Theorem 1, we see that the sequence

$$\left\{\frac{1}{c} \cdot ca_n\right\} = \{a_n\}$$

converges. Thus, $\{ca_n\}$ cannot converge unless $\{a_n\}$ also converges. If $\{a_n\}$ does not converge, then $\{ca_n\}$ does not converge.

The next theorem is the sequence version of the Sandwich Theorem in Section 2.2. You are asked to prove the theorem in Exercise 95.

THEOREM 2—The Sandwich Theorem for Sequences   Let $\{a_n\}, \{b_n\}$, and $\{c_n\}$ be sequences of real numbers. If $a_n \leq b_n \leq c_n$ holds for all n beyond some index N, and if $\lim_{n\to\infty} a_n = \lim_{n\to\infty} c_n = L$, then $\lim_{n\to\infty} b_n = L$ also.

An immediate consequence of Theorem 2 is that, if $|b_n| \leq c_n$ and $c_n \to 0$, then $b_n \to 0$ because $-c_n \leq b_n \leq c_n$. We use this fact in the next example.

EXAMPLE 4   Since $1/n \to 0$, we know that

(a) $\dfrac{\cos n}{n} \to 0$ because $-\dfrac{1}{n} \le \dfrac{\cos n}{n} \le \dfrac{1}{n};$

(b) $\dfrac{1}{2^n} \to 0$ because $0 \le \dfrac{1}{2^n} \le \dfrac{1}{n};$

(c) $(-1)^n \dfrac{1}{n} \to 0$ because $-\dfrac{1}{n} \le (-1)^n \dfrac{1}{n} \le \dfrac{1}{n}.$ ∎

The application of Theorems 1 and 2 is broadened by a theorem stating that applying a continuous function to a convergent sequence produces a convergent sequence. We state the theorem without proof (Exercise 96).

THEOREM 3—The Continuous Function Theorem for Sequences   Let $\{a_n\}$ be a sequence of real numbers. If $a_n \to L$ and if f is a function that is continuous at L and defined at all a_n, then $f(a_n) \to f(L)$.

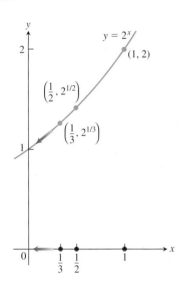

FIGURE 8.3   As $n \to \infty$, $1/n \to 0$ and $2^{1/n} \to 2^0$ (Example 6).

EXAMPLE 5   Show that $\sqrt{(n+1)/n} \to 1$.

Solution   We know that $(n+1)/n \to 1$. Taking $f(x) = \sqrt{x}$ and $L = 1$ in Theorem 3 gives $\sqrt{(n+1)/n} \to \sqrt{1} = 1.$ ∎

EXAMPLE 6   The sequence $\{1/n\}$ converges to 0. By taking $a_n = 1/n$, $f(x) = 2^x$, and $L = 0$ in Theorem 3, we see that $2^{1/n} = f(1/n) \to f(L) = 2^0 = 1$. The sequence $\{2^{1/n}\}$ converges to 1 (Figure 8.3).

Using l'Hôpital's Rule

The next theorem enables us to use l'Hôpital's Rule to find the limits of some sequences. It formalizes the connection between $\lim_{n \to \infty} a_n$ and $\lim_{x \to \infty} f(x)$.

THEOREM 4   Suppose that $f(x)$ is a function defined for all $x \ge n_0$ and that $\{a_n\}$ is a sequence of real numbers such that $a_n = f(n)$ for $n \ge n_0$. Then

$$\lim_{x \to \infty} f(x) = L \quad \Rightarrow \quad \lim_{n \to \infty} a_n = L.$$

Proof   Suppose that $\lim_{x \to \infty} f(x) = L$. Then for each positive number ϵ there is a number M such that for all x,

$$x > M \quad \Rightarrow \quad |f(x) - L| < \epsilon.$$

Let N be an integer greater than M and greater than or equal to n_0. Then

$$n > N \quad \Rightarrow \quad a_n = f(n) \quad \text{and} \quad |a_n - L| = |f(n) - L| < \epsilon.$$ ∎

EXAMPLE 7   Show that

$$\lim_{n \to \infty} \frac{\ln n}{n} = 0.$$

Solution   The function $(\ln x)/x$ is defined for all $x \geq 1$ and agrees with the given sequence at positive integers. Therefore, by Theorem 5, $\lim_{n \to \infty} (\ln n)/n$ will equal $\lim_{x \to \infty} (\ln x)/x$ if the latter exists. A single application of l'Hôpital's Rule shows that

$$\lim_{x \to \infty} \frac{\ln x}{x} = \lim_{x \to \infty} \frac{1/x}{1} = \frac{0}{1} = 0.$$

We conclude that $\lim_{n \to \infty} (\ln n)/n = 0$.   ∎

When we use l'Hôpital's Rule to find the limit of a sequence, we often treat n as a continuous real variable and differentiate directly with respect to n. This saves us from having to rewrite the formula for a_n as we did in Example 7.

EXAMPLE 8   Does the sequence whose nth term is

$$a_n = \left(\frac{n + 1}{n - 1}\right)^n$$

converge? If so, find $\lim_{n \to \infty} a_n$.

Solution   The limit leads to the indeterminate form 1^∞. We can apply l'Hôpital's Rule if we first change the form to $\infty \cdot 0$ by taking the natural logarithm of a_n:

$$\ln a_n = \ln\left(\frac{n + 1}{n - 1}\right)^n$$

$$= n \ln\left(\frac{n + 1}{n - 1}\right).$$

Then,

$$\lim_{n \to \infty} \ln a_n = \lim_{n \to \infty} n \ln\left(\frac{n + 1}{n - 1}\right) \qquad \infty \cdot 0$$

$$= \lim_{n \to \infty} \frac{\ln\left(\dfrac{n + 1}{n - 1}\right)}{1/n} \qquad \frac{0}{0}$$

$$= \lim_{n \to \infty} \frac{-2/(n^2 - 1)}{-1/n^2} \qquad \text{l'Hôpital's Rule}$$

$$= \lim_{n \to \infty} \frac{2n^2}{n^2 - 1} = 2.$$

Since $\ln a_n \to 2$ and $f(x) = e^x$ is continuous, Theorem 4 tells us that

$$a_n = e^{\ln a_n} \to e^2.$$

The sequence $\{a_n\}$ converges to e^2.   ∎

Commonly Occurring Limits

The next theorem gives some limits that arise frequently.

THEOREM 5   The following six sequences converge to the limits listed below:

1. $\lim\limits_{n \to \infty} \dfrac{\ln n}{n} = 0$

2. $\lim\limits_{n \to \infty} \sqrt[n]{n} = 1$

3. $\lim\limits_{n \to \infty} x^{1/n} = 1 \qquad (x > 0)$

4. $\lim\limits_{n \to \infty} x^n = 0 \qquad (|x| < 1)$

5. $\lim\limits_{n \to \infty} \left(1 + \dfrac{x}{n}\right)^n = e^x \qquad (\text{any } x)$

6. $\lim\limits_{n \to \infty} \dfrac{x^n}{n!} = 0 \qquad (\text{any } x)$

In Formulas (3) through (6), x remains fixed as $n \to \infty$.

Factorial Notation

The notation $n!$ ("n factorial") means the product $1 \cdot 2 \cdot 3 \cdots n$ of the integers from 1 to n. Notice that $(n + 1)! = (n + 1) \cdot n!$. Thus, $4! = 1 \cdot 2 \cdot 3 \cdot 4 = 24$ and $5! = 1 \cdot 2 \cdot 3 \cdot 4 \cdot 5 = 5 \cdot 4! = 120$. We define $0!$ to be 1. Factorials grow even faster than exponentials, as the table suggests.

n	e^n (rounded)	$n!$
1	3	1
5	148	120
10	22,026	3,628,800
20	4.9×10^8	2.4×10^{18}

Proof   The first limit was computed in Example 7. The next two can be proved by taking logarithms and applying Theorem 4 (Exercises 93 and 94). The remaining proofs are given in Appendix 6.   ∎

EXAMPLE 9

(a) $\dfrac{\ln (n^2)}{n} = \dfrac{2 \ln n}{n} \to 2 \cdot 0 = 0$ 　　　Formula 1

(b) $\sqrt[n]{n^2} = n^{2/n} = (n^{1/n})^2 \to (1)^2 = 1$ 　　　Formula 2

(c) $\sqrt[n]{3n} = 3^{1/n}(n^{1/n}) \to 1 \cdot 1 = 1$ 　　　Formula 3 with $x = 3$ and Formula 2

(d) $\left(-\dfrac{1}{2}\right)^n \to 0$ 　　　Formula 4 with $x = -\dfrac{1}{2}$

(e) $\left(\dfrac{n - 2}{n}\right)^n = \left(1 + \dfrac{-2}{n}\right)^n \to e^{-2}$ 　　　Formula 5 with $x = -2$

(f) $\dfrac{100^n}{n!} \to 0$ 　　　Formula 6 with $x = 100$ ∎

Recursive Definitions

So far, we have calculated each a_n directly from the value of n. But sequences are often defined **recursively** by giving

1. The value(s) of the initial term or terms, and

2. A rule, called a **recursion formula**, for calculating any later term from terms that precede it.

EXAMPLE 10

(a) The statements $a_1 = 1$ and $a_n = a_{n-1} + 1$ define the sequence $1, 2, 3, \ldots, n, \ldots$ of positive integers. With $a_1 = 1$, we have $a_2 = a_1 + 1 = 2$, $a_3 = a_2 + 1 = 3$, and so on.

(b) The statements $a_1 = 1$ and $a_n = n \cdot a_{n-1}$ define the sequence $1, 2, 6, 24, \ldots, n!, \ldots$ of factorials. With $a_1 = 1$, we have $a_2 = 2 \cdot a_1 = 2, a_3 = 3 \cdot a_2 = 6, a_4 = 4 \cdot a_3 = 24$, and so on.

(c) The statements $a_1 = 1, a_2 = 1$, and $a_{n+1} = a_n + a_{n-1}$ define the sequence $1, 1, 2, 3, 5, \ldots$ of **Fibonacci numbers**. With $a_1 = 1$ and $a_2 = 1$, we have $a_3 = 1 + 1 = 2, a_4 = 2 + 1 = 3, a_5 = 3 + 2 = 5$, and so on.

(d) As we can see by applying Newton's method, the statements $x_0 = 1$ and $x_{n+1} = x_n - [(\sin x_n - x_n^2)/(\cos x_n - 2x_n)]$ define a sequence that, when it converges, gives a solution to the equation $\sin x - x^2 = 0$. ∎

Bounded Nondecreasing Sequences

The terms of a general sequence can bounce around, sometimes getting larger, sometimes smaller. An important special kind of sequence is one for which each term is at least as large as its predecessor.

DEFINITION   A sequence $\{a_n\}$ with the property that $a_n \leq a_{n+1}$ for all n is called a **nondecreasing sequence**.

EXAMPLE 11   The following sequences are nondecreasing.

(a) The sequence $1, 2, 3, \ldots, n, \ldots$ of natural numbers

(b) The sequence $\dfrac{1}{2}, \dfrac{2}{3}, \dfrac{3}{4}, \ldots, \dfrac{n}{n+1}, \ldots$

(c) The constant sequence $\{3\}$ ∎

There are two kinds of nondecreasing sequences—those whose terms increase beyond any finite bound and those whose terms do not.

DEFINITIONS   A sequence $\{a_n\}$ is **bounded from above** if there exists a number M such that $a_n \leq M$ for all n. The number M is an **upper bound** for $\{a_n\}$. If M is an upper bound for $\{a_n\}$ but no number less than M is an upper bound for $\{a_n\}$, then M is the **least upper bound** for $\{a_n\}$.

EXAMPLE 12

(a) The sequence $1, 2, 3, \ldots, n, \ldots$ has no upper bound.

(b) The sequence $\dfrac{1}{2}, \dfrac{2}{3}, \dfrac{3}{4}, \ldots, \dfrac{n}{n+1}, \ldots$ is bounded above by $M = 1$.

No number less than 1 is an upper bound for the sequence, so 1 is the least upper bound (Exercise 113). ∎

A nondecreasing sequence that is bounded from above always has a least upper bound. This is the completeness property of the real numbers, discussed in Appendix 7. We will prove that if L is the least upper bound then the sequence converges to L.

Suppose we plot the points $(1, a_1), (2, a_2), \ldots, (n, a_n), \ldots$ in the xy-plane. If M is an upper bound of the sequence, all these points will lie on or below the line $y = M$

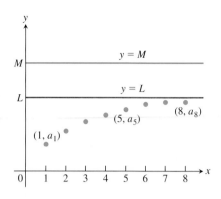

FIGURE 8.4 If the terms of a nondecreasing sequence have an upper bound M, they have a limit $L \leq M$.

(Figure 8.4). The line $y = L$ is the lowest such line. None of the points (n, a_n) lies above $y = L$, but some do lie above any lower line $y = L - \epsilon$, if ϵ is a positive number. The sequence converges to L because

(a) $a_n \leq L$ for *all* values of n and

(b) given any $\epsilon > 0$, there exists at least one integer N for which $a_N > L - \epsilon$.

The fact that $\{a_n\}$ is nondecreasing tells us further that

$$a_n \geq a_N > L - \epsilon \qquad \text{for all } n \geq N.$$

Thus, *all* the numbers a_n beyond the Nth number lie within ϵ of L. This is precisely the condition for L to be the limit of the sequence $\{a_n\}$.

The facts for nondecreasing sequences are summarized in the following theorem. A similar result holds for nonincreasing sequences (Exercise 107).

THEOREM 6—The Nondecreasing Sequence Theorem   A nondecreasing sequence of real numbers converges if and only if it is bounded from above. If a nondecreasing sequence converges, it converges to its least upper bound.

Theorem 6 implies that a nondecreasing sequence converges when it is bounded from above. It diverges to infinity if it is not bounded from above.

EXERCISES 8.1

Each of Exercises 1–6 gives a formula for the nth term a_n of a sequence $\{a_n\}$. Find the values of a_1, a_2, a_3, and a_4.

1. $a_n = \dfrac{1 - n}{n^2}$

2. $a_n = \dfrac{1}{n!}$

3. $a_n = \dfrac{(-1)^{n+1}}{2n - 1}$

4. $a_n = 2 + (-1)^n$

5. $a_n = \dfrac{2^n}{2^{n+1}}$

6. $a_n = \dfrac{2^n - 1}{2^n}$

Each of Exercises 7–12 gives the first term or two of a sequence along with a recursion formula for the remaining terms. Write out the first ten terms of the sequence.

7. $a_1 = 1, \quad a_{n+1} = a_n + (1/2^n)$

8. $a_1 = 1, \quad a_{n+1} = a_n/(n + 1)$

9. $a_1 = 2, \quad a_{n+1} = (-1)^{n+1}a_n/2$

10. $a_1 = -2, \quad a_{n+1} = na_n/(n + 1)$

11. $a_1 = a_2 = 1, \quad a_{n+2} = a_{n+1} + a_n$

12. $a_1 = 2, \quad a_2 = -1, \quad a_{n+2} = a_{n+1}/a_n$

In Exercises 13–22, find a formula for the nth term of the sequence.

13. The sequence $1, -1, 1, -1, 1, \ldots$ 1's with alternating signs

14. The sequence $-1, 1, -1, 1, -1, \ldots$ 1's with alternating signs

15. The sequence $1, -4, 9, -16, 25, \ldots$ Squares of the positive integers, with alternating signs

16. The sequence $1, -\dfrac{1}{4}, \dfrac{1}{9}, -\dfrac{1}{16}, \dfrac{1}{25}, \ldots$ Reciprocals of squares of the positive integers, with alternating signs

17. The sequence $0, 3, 8, 15, 24, \ldots$ Squares of the positive integers diminished by 1

18. The sequence $-3, -2, -1, 0, 1, \ldots$ Integers beginning with -3

19. The sequence $1, 5, 9, 13, 17, \ldots$ Every other odd positive integer

20. The sequence $2, 6, 10, 14, 18, \ldots$ Every other even positive integer

21. The sequence $1, 0, 1, 0, 1, \ldots$ Alternating 1's and 0's

22. The sequence $0, 1, 1, 2, 2, 3, 3, 4, \ldots$ Each positive integer repeated

Which of the sequences $\{a_n\}$ in Exercises 23–84 converge, and which diverge? Find the limit of each convergent sequence.

23. $a_n = 2 + (0.1)^n$

24. $a_n = \dfrac{n + (-1)^n}{n}$

25. $a_n = \dfrac{1 - 2n}{1 + 2n}$

26. $a_n = \dfrac{2n + 1}{1 - 3\sqrt{n}}$

27. $a_n = \dfrac{1 - 5n^4}{n^4 + 8n^3}$

28. $a_n = \dfrac{n + 3}{n^2 + 5n + 6}$

29. $a_n = \dfrac{n^2 - 2n + 1}{n - 1}$

30. $a_n = \dfrac{1 - n^3}{70 - 4n^2}$

31. $a_n = 1 + (-1)^n$

32. $a_n = (-1)^n \left(1 - \dfrac{1}{n}\right)$

33. $a_n = \left(\dfrac{n + 1}{2n}\right)\left(1 - \dfrac{1}{n}\right)$

34. $a_n = \left(2 - \dfrac{1}{2^n}\right)\left(3 + \dfrac{1}{2^n}\right)$

35. $a_n = \dfrac{(-1)^{n+1}}{2n - 1}$

36. $a_n = \left(-\dfrac{1}{2}\right)^n$

37. $a_n = \sqrt{\dfrac{2n}{n + 1}}$

38. $a_n = \dfrac{1}{(0.9)^n}$

39. $a_n = \sin\left(\dfrac{\pi}{2} + \dfrac{1}{n}\right)$

40. $a_n = n\pi \cos(n\pi)$

41. $a_n = \dfrac{\sin n}{n}$

42. $a_n = \dfrac{\sin^2 n}{2^n}$

43. $a_n = \dfrac{n}{2^n}$

44. $a_n = \dfrac{3^n}{n^3}$

45. $a_n = \dfrac{\ln(n + 1)}{\sqrt{n}}$

46. $a_n = \dfrac{\ln n}{\ln 2n}$

47. $a_n = 8^{1/n}$

48. $a_n = (0.03)^{1/n}$

49. $a_n = \left(1 + \dfrac{7}{n}\right)^n$

50. $a_n = \left(1 - \dfrac{1}{n}\right)^n$

51. $a_n = \sqrt[n]{10n}$

52. $a_n = \sqrt[n]{n^2}$

53. $a_n = \left(\dfrac{3}{n}\right)^{1/n}$

54. $a_n = (n + 4)^{1/(n+4)}$

55. $a_n = \dfrac{\ln n}{n^{1/n}}$

56. $a_n = \ln n - \ln(n + 1)$

57. $a_n = \sqrt[n]{4^n n}$

58. $a_n = \sqrt[n]{3^{2n+1}}$

59. $a_n = \dfrac{n!}{n^n}$ (*Hint:* Compare with $1/n$.)

60. $a_n = \dfrac{(-4)^n}{n!}$

61. $a_n = \dfrac{n!}{10^{6n}}$

62. $a_n = \dfrac{n!}{2^n \cdot 3^n}$

63. $a_n = \left(\dfrac{1}{n}\right)^{1/(\ln n)}$

64. $a_n = \ln\left(1 + \dfrac{1}{n}\right)^n$

65. $a_n = \left(\dfrac{3n + 1}{3n - 1}\right)^n$

66. $a_n = \left(\dfrac{n}{n + 1}\right)^n$

67. $a_n = \left(\dfrac{x^n}{2n + 1}\right)^{1/n}, \quad x > 0$

68. $a_n = \left(1 - \dfrac{1}{n^2}\right)^n$

69. $a_n = \dfrac{3^n \cdot 6^n}{2^{-n} \cdot n!}$

70. $a_n = \dfrac{(10/11)^n}{(9/10)^n + (11/12)^n}$

71. $a_n = \tanh n$

72. $a_n = \sinh(\ln n)$

73. $a_n = \dfrac{n^2}{2n - 1} \sin \dfrac{1}{n}$

74. $a_n = n\left(1 - \cos \dfrac{1}{n}\right)$

75. $a_n = \tan^{-1} n$

76. $a_n = \dfrac{1}{\sqrt{n}} \tan^{-1} n$

77. $a_n = \left(\dfrac{1}{3}\right)^n + \dfrac{1}{\sqrt{2^n}}$

78. $a_n = \sqrt[n]{n^2 + n}$

79. $a_n = \dfrac{(\ln n)^{200}}{n}$

80. $a_n = \dfrac{(\ln n)^5}{\sqrt{n}}$

81. $a_n = n - \sqrt{n^2 - n}$

82. $a_n = \dfrac{1}{\sqrt{n^2 - 1} - \sqrt{n^2 + n}}$

83. $a_n = \dfrac{1}{n} \displaystyle\int_1^n \dfrac{1}{x}\, dx$

84. $a_n = \displaystyle\int_1^n \dfrac{1}{x^p}\, dx, \quad p > 1$

85. The first term of a sequence is $x_1 = 1$. Each succeeding term is the sum of all those that come before it:

$$x_{n+1} = x_1 + x_2 + \cdots + x_n.$$

Write out enough early terms of the sequence to deduce a general formula for x_n that holds for $n \geq 2$.

86. A sequence of rational numbers is described as follows:

$$\frac{1}{1}, \frac{3}{2}, \frac{7}{5}, \frac{17}{12}, \ldots, \frac{a}{b}, \frac{a + 2b}{a + b}, \ldots$$

Here the numerators form one sequence, the denominators form a second sequence, and their ratios form a third sequence. Let x_n and y_n be, respectively, the numerator and the denominator of the nth fraction $r_n = x_n/y_n$.

a. Verify that $x_1^2 - 2y_1^2 = -1$, $x_2^2 - 2y_2^2 = +1$ and, more generally, that if $a^2 - 2b^2 = -1$ or $+1$, then

$$(a + 2b)^2 - 2(a + b)^2 = +1 \quad \text{or} \quad -1,$$

respectively.

b. The fractions $r_n = x_n/y_n$ approach a limit as n increases. What is that limit? (*Hint:* Use part (a) to show that $r_n^2 - 2 = \pm(1/y_n)^2$ and that y_n is not less than n.)

87. **Newton's method** The following sequences come from the recursion formula for Newton's method,

$$x_{n+1} = x_n - \dfrac{f(x_n)}{f'(x_n)}.$$

Do the sequences converge? If so, to what value? In each case, begin by identifying the function f that generates the sequence.

a. $x_0 = 1, \quad x_{n+1} = x_n - \dfrac{x_n^2 - 2}{2x_n} = \dfrac{x_n}{2} + \dfrac{1}{x_n}$

b. $x_0 = 1, \quad x_{n+1} = x_n - \dfrac{\tan x_n - 1}{\sec^2 x_n}$

c. $x_0 = 1, \quad x_{n+1} = x_n - 1$

88. a. Suppose that $f(x)$ is differentiable for all x in $[0, 1]$ and that $f(0) = 0$. Define the sequence $\{a_n\}$ by the rule $a_n = nf(1/n)$. Show that $\lim_{n \to \infty} a_n = f'(0)$.

Use the result in part (a) to find the limits of the following sequences $\{a_n\}$.

b. $a_n = n \tan^{-1} \dfrac{1}{n}$ **c.** $a_n = n(e^{1/n} - 1)$

d. $a_n = n \ln\left(1 + \dfrac{2}{n}\right)$

89. Pythagorean triples A triple of positive integers a, b, and c is called a **Pythagorean triple** if $a^2 + b^2 = c^2$. Let a be an odd positive integer and let

$$b = \left\lfloor \frac{a^2}{2} \right\rfloor \quad \text{and} \quad c = \left\lceil \frac{a^2}{2} \right\rceil$$

be, respectively, the integer floor and ceiling for $a^2/2$.

a. Show that $a^2 + b^2 = c^2$. (*Hint:* Let $a = 2n + 1$ and express b and c in terms of n.)

b. By direct calculation, or by appealing to the figure here, find

$$\lim_{a \to \infty} \frac{\left\lfloor \dfrac{a^2}{2} \right\rfloor}{\left\lceil \dfrac{a^2}{2} \right\rceil}.$$

90. The nth root of $n!$

a. Show that $\lim_{n \to \infty} (2n\pi)^{1/(2n)} = 1$ and hence, using Stirling's approximation (Chapter 7, Additional Exercise 32a), that

$$\sqrt[n]{n!} \approx \frac{n}{e} \quad \text{for large values of } n.$$

T **b.** Test the approximation in part (a) for $n = 40, 50, 60, \ldots,$ as far as your calculator will allow.

91. a. Assuming that $\lim_{n \to \infty} (1/n^c) = 0$ if c is any positive constant, show that

$$\lim_{n \to \infty} \frac{\ln n}{n^c} = 0$$

if c is any positive constant.

b. Prove that $\lim_{n \to \infty} (1/n^c) = 0$ if c is any positive constant. (*Hint:* If $\epsilon = 0.001$ and $c = 0.04$, how large should N be to ensure that $|1/n^c - 0| < \epsilon$ if $n > N$?)

92. The zipper theorem Prove the "zipper theorem" for sequences: If $\{a_n\}$ and $\{b_n\}$ both converge to L, then the sequence

$$a_1, b_1, a_2, b_2, \ldots, a_n, b_n, \ldots$$

converges to L.

93. Prove that $\lim_{n \to \infty} \sqrt[n]{n} = 1$.

94. Prove that $\lim_{n \to \infty} x^{1/n} = 1, (x > 0)$.

95. Prove Theorem 2.   **96.** Prove Theorem 3.

In Exercises 97–100, determine if the sequence is nondecreasing and if it is bounded from above.

97. $a_n = \dfrac{3n + 1}{n + 1}$ **98.** $a_n = \dfrac{(2n + 3)!}{(n + 1)!}$

99. $a_n = \dfrac{2^n 3^n}{n!}$ **100.** $a_n = 2 - \dfrac{2}{n} - \dfrac{1}{2^n}$

Which of the sequences in Exercises 101–106 converge, and which diverge? Give reasons for your answers.

101. $a_n = 1 - \dfrac{1}{n}$ **102.** $a_n = n - \dfrac{1}{n}$

103. $a_n = \dfrac{2^n - 1}{2^n}$ **104.** $a_n = \dfrac{2^n - 1}{3^n}$

105. $a_n = ((-1)^n + 1)\left(\dfrac{n + 1}{n}\right)$

106. The first term of a sequence is $x_1 = \cos(1)$. The next terms are $x_2 = x_1$ or $\cos(2)$, whichever is larger; and $x_3 = x_2$ or $\cos(3)$, whichever is larger (farther to the right). In general,

$$x_{n+1} = \max\{x_n, \cos(n + 1)\}.$$

107. Nonincreasing sequences A sequence of numbers $\{a_n\}$ in which $a_n \geq a_{n+1}$ for every n is called a **nonincreasing sequence**. A sequence $\{a_n\}$ is **bounded from below** if there is a number M with $M \leq a_n$ for every n. Such a number M is called a **lower bound** for the sequence. Deduce from Theorem 6 that a nonincreasing sequence that is bounded from below converges and that a nonincreasing sequence that is not bounded from below diverges.

(*Continuation of Exercise 107.*) Using the conclusion of Exercise 107, determine which of the sequences in Exercises 108–112 converge and which diverge.

108. $a_n = \dfrac{n + 1}{n}$ **109.** $a_n = \dfrac{1 + \sqrt{2n}}{\sqrt{n}}$

110. $a_n = \dfrac{1 - 4^n}{2^n}$ **111.** $a_n = \dfrac{4^{n+1} + 3^n}{4^n}$

112. $a_1 = 1, \quad a_{n+1} = 2a_n - 3$

113. The sequence $\{n/(n + 1)\}$ has a least upper bound of 1 Show that if M is a number less than 1, then the terms of $\{n/(n + 1)\}$ eventually exceed M. That is, if $M < 1$ there is an integer N such that $n/(n + 1) > M$ whenever $n > N$. Since $n/(n + 1) < 1$ for every n, this proves that 1 is a least upper bound for $\{n/(n + 1)\}$.

114. Uniqueness of least upper bounds Show that if M_1 and M_2 are least upper bounds for the sequence $\{a_n\}$, then $M_1 = M_2$. That is, a sequence cannot have two different least upper bounds.

115. Is it true that a sequence $\{a_n\}$ of positive numbers must converge if it is bounded from above? Give reasons for your answer.

116. Prove that if $\{a_n\}$ is a convergent sequence, then to every positive number ϵ there corresponds an integer N such that for all m and n,

$$m > N \quad \text{and} \quad n > N \implies |a_m - a_n| < \epsilon.$$

117. Uniqueness of limits Prove that limits of sequences are unique. That is, show that if L_1 and L_2 are numbers such that $a_n \to L_1$ and $a_n \to L_2$, then $L_1 = L_2$.

118. Limits and subsequences If the terms of one sequence appear in another sequence in their given order, we call the first sequence a **subsequence** of the second. Prove that if two subsequences of a sequence $\{a_n\}$ have different limits $L_1 \neq L_2$, then $\{a_n\}$ diverges.

119. For a sequence $\{a_n\}$ the terms of even index are denoted by a_{2k} and the terms of odd index by a_{2k+1}. Prove that if $a_{2k} \to L$ and $a_{2k+1} \to L$, then $a_n \to L$.

120. Prove that a sequence $\{a_n\}$ converges to 0 if and only if the sequence of absolute values $\{|a_n|\}$ converges to 0.

T In Exercises 121–124, experiment with a calculator to find a value of N that will make the inequality hold for all $n > N$. Assuming that the inequality is the one from the formal definition of the limit of a sequence, what sequence is being considered in each case and what is its limit?

121. $|\sqrt[n]{0.5} - 1| < 10^{-3}$ **122.** $|\sqrt[n]{n} - 1| < 10^{-3}$

123. $(0.9)^n < 10^{-3}$ **124.** $2^n/n! < 10^{-7}$

125. Sequences generated by Newton's method Newton's method, applied to a differentiable function $f(x)$, begins with a starting value x_0 and constructs from it a sequence of numbers $\{x_n\}$ that under favorable circumstances converges to a zero of f. The recursion formula for the sequence is

$$x_{n+1} = x_n - \frac{f(x_n)}{f'(x_n)}.$$

a. Show that the recursion formula for $f(x) = x^2 - a, a > 0$, can be written as $x_{n+1} = (x_n + a/x_n)/2$.

b. Starting with $x_0 = 1$ and $a = 3$, calculate successive terms of the sequence until the display begins to repeat. What number is being approximated? Explain.

126. (*Continuation of Exercise 125.*) Repeat part (b) of Exercise 125 with $a = 2$ in place of $a = 3$.

127. A recursive definition of $\pi/2$ If you start with $x_1 = 1$ and define the subsequent terms of $\{x_n\}$ by the rule $x_n = x_{n-1} + \cos x_{n-1}$, you generate a sequence that converges rapidly to $\pi/2$. **a.** Try it. **b.** Use the accompanying figure to explain why the convergence is so rapid.

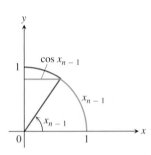

128. According to a front-page article in the December 15, 1992, issue of the *Wall Street Journal*, Ford Motor Company used about $7\frac{1}{4}$ hours of labor to produce stampings for the average vehicle, down from an estimated 15 hours in 1980. The Japanese automobile companies needed only about $3\frac{1}{2}$ hours.

Ford's improvement since 1980 represents an average decrease of 6% per year. If that rate continues, then n years from 1992 Ford will use about

$$S_n = 7.25(0.94)^n$$

hours of labor to produce stampings for the average vehicle. Assuming that the Japanese continue to spend $3\frac{1}{2}$ hours per vehicle, how many more years will it take Ford to catch up? Find out two ways:

a. Find the first term of the sequence $\{S_n\}$ that is less than or equal to 3.5.

T **b.** Graph $f(x) = 7.25(0.94)^x$ and use Trace to find where the graph crosses the line $y = 3.5$.

COMPUTER EXPLORATIONS

Use a CAS to perform the following steps for the sequences in Exercises 129–140.

a. Calculate and then plot the first 25 terms of the sequence. Does the sequence appear to be bounded from above or below? Does it appear to converge or diverge? If it does converge, what is the limit L?

b. If the sequence converges, find an integer N such that $|a_n - L| \leq 0.01$ for $n \geq N$. How far in the sequence do you have to get for the terms to lie within 0.0001 of L?

129. $a_n = \sqrt[n]{n}$ **130.** $a_n = \left(1 + \dfrac{0.5}{n}\right)^n$

131. $a_1 = 1, \quad a_{n+1} = a_n + \dfrac{1}{5^n}$

132. $a_1 = 1, \quad a_{n+1} = a_n + (-2)^n$

133. $a_n = \sin n$ **134.** $a_n = n \sin \dfrac{1}{n}$

135. $a_n = \dfrac{\sin n}{n}$ **136.** $a_n = \dfrac{\ln n}{n}$

137. $a_n = (0.9999)^n$ **138.** $a_n = 123456^{1/n}$

139. $a_n = \dfrac{8^n}{n!}$ **140.** $a_n = \dfrac{n^{41}}{19^n}$

8.2 Infinite Series

An *infinite series* is the sum of an infinite sequence of numbers

$$a_1 + a_2 + a_3 + \cdots + a_n + \cdots$$

The goal of this section is to understand the meaning of such an infinite sum and to develop methods to calculate it. Since there are infinitely many terms to add in an infinite series, we cannot just keep adding to see what comes out. Instead we look at the result of summing the first n terms of the sequence and stopping. The sum of the first n terms

$$s_n = a_1 + a_2 + a_3 + \cdots + a_n$$

is an ordinary finite sum and can be calculated by normal addition. It is called the *nth partial sum*. As n gets larger, we expect the partial sums to get closer and closer to a limiting value in the same sense that the terms of a sequence approach a limit, as discussed in Section 8.1.

For example, to assign meaning to an expression like

$$1 + \frac{1}{2} + \frac{1}{4} + \frac{1}{8} + \frac{1}{16} + \cdots$$

we add the terms one at a time from the beginning and look for a pattern in how these partial sums grow.

Partial sum		Suggestive expression for partial sum	Value
First:	$s_1 = 1$	$2 - 1$	1
Second:	$s_2 = 1 + \dfrac{1}{2}$	$2 - \dfrac{1}{2}$	$\dfrac{3}{2}$
Third:	$s_3 = 1 + \dfrac{1}{2} + \dfrac{1}{4}$	$2 - \dfrac{1}{4}$	$\dfrac{7}{4}$
$\vdots$	$\vdots$	$\vdots$	$\vdots$
nth:	$s_n = 1 + \dfrac{1}{2} + \dfrac{1}{4} + \cdots + \dfrac{1}{2^{n-1}}$	$2 - \dfrac{1}{2^{n-1}}$	$\dfrac{2^n - 1}{2^{n-1}}$

Indeed there is a pattern. The partial sums form a sequence whose nth term is

$$s_n = 2 - \frac{1}{2^{n-1}}.$$

This sequence of partial sums converges to 2 because $\lim_{n\to\infty}(1/2^{n-1}) = 0$. We say

"the sum of the infinite series $1 + \dfrac{1}{2} + \dfrac{1}{4} + \cdots + \dfrac{1}{2^{n-1}} + \cdots$ is 2."

Is the sum of any finite number of terms in this series equal to 2? No. Can we actually add an infinite number of terms one by one? No. But we can still define their sum by defining it to be the limit of the sequence of partial sums as $n \to \infty$, in this case 2 (Figure 8.5). Our knowledge of sequences and limits enables us to break away from the confines of finite sums.

FIGURE 8.5 As the lengths $1, \frac{1}{2}, \frac{1}{4}, \frac{1}{8}, \dots$ are added one by one, the sum approaches 2.

HISTORICAL BIOGRAPHY

Blaise Pascal
(1623–1662)

DEFINITIONS   Given a sequence of numbers $\{a_n\}$, an expression of the form

$$a_1 + a_2 + a_3 + \cdots + a_n + \cdots$$

is an **infinite series**. The number a_n is the **nth term** of the series. The sequence $\{s_n\}$ defined by

$$s_1 = a_1$$
$$s_2 = a_1 + a_2$$
$$\vdots$$
$$s_n = a_1 + a_2 + \cdots + a_n = \sum_{k=1}^{n} a_k$$
$$\vdots$$

is the **sequence of partial sums** of the series, the number s_n being the **nth partial sum**. If the sequence of partial sums converges to a limit L, we say that the series **converges** and that its **sum** is L. In this case, we also write

$$a_1 + a_2 + \cdots + a_n + \cdots = \sum_{n=1}^{\infty} a_n = L.$$

If the sequence of partial sums of the series does not converge, we say that the series **diverges**.

When we begin to study a given series $a_1 + a_2 + \cdots + a_n + \cdots$, we might not know whether it converges or diverges. In either case, it is convenient to use sigma notation to write the series as

$$\sum_{n=1}^{\infty} a_n, \qquad \sum_{k=1}^{\infty} a_k, \qquad \text{or} \qquad \sum a_n$$

A useful shorthand when summation from 1 to ∞ is understood

Geometric Series

Geometric series are series of the form

$$a + ar + ar^2 + \cdots + ar^{n-1} + \cdots = \sum_{n=1}^{\infty} ar^{n-1}$$

in which a and r are fixed real numbers and $a \neq 0$. The series can also be written as $\sum_{n=0}^{\infty} ar^n$. The **ratio** r can be positive, as in

$$1 + \frac{1}{2} + \frac{1}{4} + \cdots + \left(\frac{1}{2}\right)^{n-1} + \cdots, \qquad r = 1/2, a = 1$$

or negative, as in

$$1 - \frac{1}{3} + \frac{1}{9} - \cdots + \left(-\frac{1}{3}\right)^{n-1} + \cdots. \qquad r = -1/3, a = 1$$

If $r = 1$, the nth partial sum of the geometric series is

$$s_n = a + a(1) + a(1)^2 + \cdots + a(1)^{n-1} = na,$$

and the series diverges because $\lim_{n \to \infty} s_n = \pm\infty$, depending on the sign of a. If $r = -1$, the series diverges because the nth partial sums alternate between a and 0. If $|r| \neq 1$, we can determine the convergence or divergence of the series in the following way:

$$s_n = a + ar + ar^2 + \cdots + ar^{n-1}$$

$$rs_n = ar + ar^2 + \cdots + ar^{n-1} + ar^n \qquad \text{Multiply } s_n \text{ by } r.$$

$$s_n - rs_n = a - ar^n \qquad \begin{array}{l}\text{Subtract } rs_n \text{ from } s_n. \text{ Most of}\\ \text{the terms on the right cancel.}\end{array}$$

$$s_n(1 - r) = a(1 - r^n) \qquad \text{Factor.}$$

$$s_n = \frac{a(1 - r^n)}{1 - r}, \qquad (r \neq 1). \qquad \text{We can solve for } s_n \text{ if } r \neq 1.$$

If $|r| < 1$, then $r^n \to 0$ as $n \to \infty$ (as in Section 8.1) and $s_n \to a/(1 - r)$. If $|r| > 1$, then $|r^n| \to \infty$ and the series diverges.

If $|r| < 1$, the geometric series $a + ar + ar^2 + \cdots + ar^{n-1} + \cdots$ converges to $a/(1 - r)$:

$$\sum_{n=1}^{\infty} ar^{n-1} = \frac{a}{1 - r}, \qquad |r| < 1.$$

If $|r| \geq 1$, the series diverges.

We have determined when a geometric series converges or diverges, and to what value. Often we can determine that a series converges without knowing the value to which it converges, as we will see in the next several sections. The formula $a/(1 - r)$ for the sum of a geometric series applies *only* when the summation index begins with $n = 1$ in the expression $\sum_{n=1}^{\infty} ar^{n-1}$ (or with the index $n = 0$ if we write the series as $\sum_{n=0}^{\infty} ar^n$).

EXAMPLE 1   The geometric series with $a = 1/9$ and $r = 1/3$ is

$$\frac{1}{9} + \frac{1}{27} + \frac{1}{81} + \cdots = \sum_{n=1}^{\infty} \frac{1}{9}\left(\frac{1}{3}\right)^{n-1} = \frac{1/9}{1 - (1/3)} = \frac{1}{6}. \qquad \blacksquare$$

EXAMPLE 2   The series

$$\sum_{n=0}^{\infty} \frac{(-1)^n 5}{4^n} = 5 - \frac{5}{4} + \frac{5}{16} - \frac{5}{64} + \cdots$$

is a geometric series with $a = 5$ and $r = -1/4$. It converges to

$$\frac{a}{1 - r} = \frac{5}{1 + (1/4)} = 4. \qquad \blacksquare$$

EXAMPLE 3   You drop a ball from a meters above a flat surface. Each time the ball hits the surface after falling a distance h, it rebounds a distance rh, where r is positive but less than 1. Find the total distance the ball travels up and down (Figure 8.6).

(a)

(b)

FIGURE 8.6   (a) Example 3 shows how to use a geometric series to calculate the total vertical distance traveled by a bouncing ball if the height of each rebound is reduced by the factor r. (b) A stroboscopic photo of a bouncing ball.

Solution   The total distance is

$$\underbrace{s = a + 2ar + 2ar^2 + 2ar^3 + \cdots}_{\text{This sum is } 2ar/(1-r).} = a + \frac{2ar}{1-r} = a\frac{1+r}{1-r}.$$

If $a = 6$ m and $r = 2/3$, for instance, the distance is

$$s = 6\frac{1 + (2/3)}{1 - (2/3)} = 6\left(\frac{5/3}{1/3}\right) = 30 \text{ m}. \qquad \blacksquare$$

EXAMPLE 4   Express the repeating decimal $5.232323\ldots$ as the ratio of two integers.

Solution

$$5.232323\ldots = 5 + \frac{23}{100} + \frac{23}{(100)^2} + \frac{23}{(100)^3} + \cdots$$

$$= 5 + \frac{23}{100}\underbrace{\left(1 + \frac{1}{100} + \left(\frac{1}{100}\right)^2 + \cdots\right)}_{1/(1-0.01)} \qquad \begin{array}{l} a = 1, \\ r = 1/100 \end{array}$$

$$= 5 + \frac{23}{100}\left(\frac{1}{0.99}\right) = 5 + \frac{23}{99} = \frac{518}{99} \qquad \blacksquare$$

Unfortunately, formulas like the one for the sum of a convergent geometric series are rare and we usually have to settle for an estimate of a series' sum (more about this later). The next example, however, is another case in which we can find the sum exactly.

EXAMPLE 5   Find the sum of the "telescoping" series $\displaystyle\sum_{n=1}^{\infty} \frac{1}{n(n+1)}$.

Solution   We look for a pattern in the sequence of partial sums that might lead to a formula for s_k. The key observation is the partial fraction decomposition

$$\frac{1}{n(n+1)} = \frac{1}{n} - \frac{1}{n+1},$$

so

$$\sum_{n=1}^{k} \frac{1}{n(n+1)} = \sum_{n=1}^{k}\left(\frac{1}{n} - \frac{1}{n+1}\right)$$

and

$$s_k = \left(\frac{1}{1} - \frac{1}{2}\right) + \left(\frac{1}{2} - \frac{1}{3}\right) + \left(\frac{1}{3} - \frac{1}{4}\right) + \cdots + \left(\frac{1}{k} - \frac{1}{k+1}\right).$$

Removing parentheses and canceling adjacent terms of opposite sign collapses the sum to

$$s_k = 1 - \frac{1}{k+1}.$$

We now see that $s_k \to 1$ as $k \to \infty$. The series converges, and its sum is 1:

$$\sum_{n=1}^{\infty} \frac{1}{n(n+1)} = 1. \qquad \blacksquare$$

Divergent Series

One reason that a series may fail to converge is that its terms don't become small.

EXAMPLE 6   The series

$$\sum_{n=1}^{\infty} \frac{n+1}{n} = \frac{2}{1} + \frac{3}{2} + \frac{4}{3} + \cdots + \frac{n+1}{n} + \cdots$$

diverges because the partial sums eventually outgrow every preassigned number. Each term is greater than 1, so the sum of n terms is greater than n. ∎

The nth-Term Test for Divergence

Notice that $\lim_{n\to\infty} a_n$ must equal zero if the series $\sum_{n=1}^{\infty} a_n$ converges. To see why, let S represent the series' sum and $s_n = a_1 + a_2 + \cdots + a_n$ the nth partial sum. When n is large, both s_n and s_{n-1} are close to S, so their difference, a_n, is close to zero. More formally,

$$a_n = s_n - s_{n-1} \quad \rightarrow \quad S - S = 0. \qquad \text{Difference Rule for sequences}$$

This establishes the following theorem.

> **Caution**
> Theorem 7 *does not say* that $\sum_{n=1}^{\infty} a_n$ converges if $a_n \to 0$. It is possible for a series to diverge when $a_n \to 0$.

THEOREM 7   If $\sum_{n=1}^{\infty} a_n$ converges, then $a_n \to 0$.

Theorem 7 leads to a test for detecting the kind of divergence that occurred in Example 6.

The nth-Term Test for Divergence

$\sum_{n=1}^{\infty} a_n$ diverges if $\lim_{n\to\infty} a_n$ fails to exist or is different from zero.

EXAMPLE 7

(a) $\sum_{n=1}^{\infty} n^2$ diverges because $n^2 \to \infty$

(b) $\sum_{n=1}^{\infty} \frac{n+1}{n}$ diverges because $\frac{n+1}{n} \to 1$ $\qquad \lim_{n\to\infty} a_n \neq 0$

(c) $\sum_{n=1}^{\infty} (-1)^{n+1}$ diverges because $\lim_{n\to\infty} (-1)^{n+1}$ does not exist

(d) $\sum_{n=1}^{\infty} \frac{-n}{2n+5}$ diverges because $\lim_{n\to\infty} \frac{-n}{2n+5} = -\frac{1}{2} \neq 0$. ∎

EXAMPLE 8   The series

$$1 + \underbrace{\frac{1}{2} + \frac{1}{2}}_{2 \text{ terms}} + \underbrace{\frac{1}{4} + \frac{1}{4} + \frac{1}{4} + \frac{1}{4}}_{4 \text{ terms}} + \cdots + \underbrace{\frac{1}{2^n} + \frac{1}{2^n} + \cdots + \frac{1}{2^n}}_{2^n \text{ terms}} + \cdots$$

diverges because the terms can be grouped into clusters that add to 1, so the partial sums increase without bound. However, the terms of the series form a sequence that converges to 0. Example 1 of Section 8.3 shows that the harmonic series also behaves in this manner. ∎

Combining Series

Whenever we have two convergent series, we can add them term by term, subtract them term by term, or multiply them by constants to make new convergent series.

THEOREM 8   If $\sum a_n = A$ and $\sum b_n = B$ are convergent series, then

1. *Sum Rule:* $\sum(a_n + b_n) = \sum a_n + \sum b_n = A + B$
2. *Difference Rule:* $\sum(a_n - b_n) = \sum a_n - \sum b_n = A - B$
3. *Constant Multiple Rule:* $\sum ka_n = k\sum a_n = kA$ (Any number k).

Proof   The three rules for series follow from the analogous rules for sequences in Theorem 1, Section 8.1. To prove the Sum Rule for series, let

$$A_n = a_1 + a_2 + \cdots + a_n, \quad B_n = b_1 + b_2 + \cdots + b_n.$$

Then the partial sums of $\sum(a_n + b_n)$ are

$$\begin{aligned} s_n &= (a_1 + b_1) + (a_2 + b_2) + \cdots + (a_n + b_n) \\ &= (a_1 + \cdots + a_n) + (b_1 + \cdots + b_n) \\ &= A_n + B_n. \end{aligned}$$

Since $A_n \to A$ and $B_n \to B$, we have $s_n \to A + B$ by the Sum Rule for sequences. The proof of the Difference Rule is similar.

To prove the Constant Multiple Rule for series, observe that the partial sums of $\sum ka_n$ form the sequence

$$s_n = ka_1 + ka_2 + \cdots + ka_n = k(a_1 + a_2 + \cdots + a_n) = kA_n,$$

which converges to kA by the Constant Multiple Rule for sequences.   ∎

As corollaries of Theorem 8, we have the following results. We omit the proofs.

---

1. Every nonzero constant multiple of a divergent series diverges.

2. If $\sum a_n$ converges and $\sum b_n$ diverges, then $\sum(a_n + b_n)$ and $\sum(a_n - b_n)$ both diverge.

Caution   Remember that $\sum(a_n + b_n)$ can converge when $\sum a_n$ and $\sum b_n$ both diverge. For example, $\sum a_n = 1 + 1 + 1 + \cdots$ and $\sum b_n = (-1) + (-1) + (-1) + \cdots$ diverge, whereas $\sum(a_n + b_n) = 0 + 0 + 0 + \cdots$ converges to 0.

EXAMPLE 9   Find the sums of the following series.

(a) $\displaystyle\sum_{n=1}^{\infty} \frac{3^{n-1} - 1}{6^{n-1}} = \sum_{n=1}^{\infty} \left(\frac{1}{2^{n-1}} - \frac{1}{6^{n-1}} \right)$

$\displaystyle = \sum_{n=1}^{\infty} \frac{1}{2^{n-1}} - \sum_{n=1}^{\infty} \frac{1}{6^{n-1}}$ Difference Rule

$$= \frac{1}{1 - (1/2)} - \frac{1}{1 - (1/6)} \qquad \text{Geometric series with } a = 1 \text{ and } r = 1/2, 1/6$$

$$= 2 - \frac{6}{5} = \frac{4}{5}$$

(b) $\displaystyle\sum_{n=0}^{\infty} \frac{4}{2^n} = 4 \sum_{n=0}^{\infty} \frac{1}{2^n}$ Constant Multiple Rule

$$= 4\left(\frac{1}{1 - (1/2)}\right) \qquad \text{Geometric series with } a = 1, r = 1/2$$

$$= 8 \qquad\qquad\qquad\qquad\qquad\qquad \blacksquare$$

Adding or Deleting Terms

We can add a finite number of terms to a series or delete a finite number of terms without altering the series' convergence or divergence, although in the case of convergence this will usually change the sum. If $\sum_{n=1}^{\infty} a_n$ converges, then $\sum_{n=k}^{\infty} a_n$ converges for any $k > 1$ and

$$\sum_{n=1}^{\infty} a_n = a_1 + a_2 + \cdots + a_{k-1} + \sum_{n=k}^{\infty} a_n.$$

Conversely, if $\sum_{n=k}^{\infty} a_n$ converges for any $k > 1$, then $\sum_{n=1}^{\infty} a_n$ converges. Thus,

$$\sum_{n=1}^{\infty} \frac{1}{5^n} = \frac{1}{5} + \frac{1}{25} + \frac{1}{125} + \sum_{n=4}^{\infty} \frac{1}{5^n}$$

and

$$\sum_{n=4}^{\infty} \frac{1}{5^n} = \left(\sum_{n=1}^{\infty} \frac{1}{5^n}\right) - \frac{1}{5} - \frac{1}{25} - \frac{1}{125}.$$

Reindexing

HISTORICAL BIOGRAPHY

Richard Dedekind
(1831–1916)

As long as we preserve the order of its terms, we can reindex any series without altering its convergence. To raise the starting value of the index h units, replace the n in the formula for a_n by $n - h$:

$$\sum_{n=1}^{\infty} a_n = \sum_{n=1+h}^{\infty} a_{n-h} = a_1 + a_2 + a_3 + \cdots.$$

To lower the starting value of the index h units, replace the n in the formula for a_n by $n + h$:

$$\sum_{n=1}^{\infty} a_n = \sum_{n=1-h}^{\infty} a_{n+h} = a_1 + a_2 + a_3 + \cdots.$$

It works like a horizontal shift. We saw this in starting a geometric series with the index $n = 0$ instead of the index $n = 1$, but we can use any other starting index value as well. We usually give preference to indexings that lead to simple expressions.

EXAMPLE 10 We can write the geometric series

$$\sum_{n=1}^{\infty} \frac{1}{2^{n-1}} = 1 + \frac{1}{2} + \frac{1}{4} + \cdots$$

as

$$\sum_{n=0}^{\infty} \frac{1}{2^n}, \qquad \sum_{n=5}^{\infty} \frac{1}{2^{n-5}}, \qquad \text{or even} \qquad \sum_{n=-4}^{\infty} \frac{1}{2^{n+4}}.$$

The partial sums remain the same no matter what indexing we choose. $\blacksquare$

EXERCISES 8.2

In Exercises 1–6, find a formula for the nth partial sum of each series and use it to find the series' sum if the series converges.

1. $2 + \dfrac{2}{3} + \dfrac{2}{9} + \dfrac{2}{27} + \cdots + \dfrac{2}{3^{n-1}} + \cdots$

2. $\dfrac{9}{100} + \dfrac{9}{100^2} + \dfrac{9}{100^3} + \cdots + \dfrac{9}{100^n} + \cdots$

3. $1 - \dfrac{1}{2} + \dfrac{1}{4} - \dfrac{1}{8} + \cdots + (-1)^{n-1}\dfrac{1}{2^{n-1}} + \cdots$

4. $1 - 2 + 4 - 8 + \cdots + (-1)^{n-1}2^{n-1} + \cdots$

5. $\dfrac{1}{2 \cdot 3} + \dfrac{1}{3 \cdot 4} + \dfrac{1}{4 \cdot 5} + \cdots + \dfrac{1}{(n+1)(n+2)} + \cdots$

6. $\dfrac{5}{1 \cdot 2} + \dfrac{5}{2 \cdot 3} + \dfrac{5}{3 \cdot 4} + \cdots + \dfrac{5}{n(n+1)} + \cdots$

In Exercises 7–14, write out the first few terms of each series to show how the series starts. Then find the sum of the series.

7. $\displaystyle\sum_{n=0}^{\infty} \dfrac{(-1)^n}{4^n}$

8. $\displaystyle\sum_{n=2}^{\infty} \dfrac{1}{4^n}$

9. $\displaystyle\sum_{n=1}^{\infty} \dfrac{7}{4^n}$

10. $\displaystyle\sum_{n=0}^{\infty} (-1)^n \dfrac{5}{4^n}$

11. $\displaystyle\sum_{n=0}^{\infty} \left(\dfrac{5}{2^n} + \dfrac{1}{3^n}\right)$

12. $\displaystyle\sum_{n=0}^{\infty} \left(\dfrac{5}{2^n} - \dfrac{1}{3^n}\right)$

13. $\displaystyle\sum_{n=0}^{\infty} \left(\dfrac{1}{2^n} + \dfrac{(-1)^n}{5^n}\right)$

14. $\displaystyle\sum_{n=0}^{\infty} \left(\dfrac{2^{n+1}}{5^n}\right)$

Find the sum of each series in Exercises 15–22.

15. $\displaystyle\sum_{n=1}^{\infty} \dfrac{4}{(4n-3)(4n+1)}$

16. $\displaystyle\sum_{n=1}^{\infty} \dfrac{6}{(2n-1)(2n+1)}$

17. $\displaystyle\sum_{n=1}^{\infty} \dfrac{40n}{(2n-1)^2(2n+1)^2}$

18. $\displaystyle\sum_{n=1}^{\infty} \dfrac{2n+1}{n^2(n+1)^2}$

19. $\displaystyle\sum_{n=1}^{\infty} \left(\dfrac{1}{\sqrt{n}} - \dfrac{1}{\sqrt{n+1}}\right)$

20. $\displaystyle\sum_{n=1}^{\infty} \left(\dfrac{1}{2^{1/n}} - \dfrac{1}{2^{1/(n+1)}}\right)$

21. $\displaystyle\sum_{n=1}^{\infty} \left(\dfrac{1}{\ln(n+2)} - \dfrac{1}{\ln(n+1)}\right)$

22. $\displaystyle\sum_{n=1}^{\infty} (\tan^{-1}(n) - \tan^{-1}(n+1))$

Which series in Exercises 23–40 converge, and which diverge? Give reasons for your answers. If a series converges, find its sum.

23. $\displaystyle\sum_{n=0}^{\infty} \left(\dfrac{1}{\sqrt{2}}\right)^n$

24. $\displaystyle\sum_{n=0}^{\infty} \left(\sqrt{2}\right)^n$

25. $\displaystyle\sum_{n=1}^{\infty} (-1)^{n+1} \dfrac{3}{2^n}$

26. $\displaystyle\sum_{n=1}^{\infty} (-1)^{n+1} n$

27. $\displaystyle\sum_{n=0}^{\infty} \cos n\pi$

28. $\displaystyle\sum_{n=0}^{\infty} \dfrac{\cos n\pi}{5^n}$

29. $\displaystyle\sum_{n=0}^{\infty} e^{-2n}$

30. $\displaystyle\sum_{n=1}^{\infty} \ln \dfrac{1}{n}$

31. $\displaystyle\sum_{n=1}^{\infty} \dfrac{2}{10^n}$

32. $\displaystyle\sum_{n=0}^{\infty} \dfrac{1}{x^n}, \quad |x| > 1$

33. $\displaystyle\sum_{n=0}^{\infty} \dfrac{2^n - 1}{3^n}$

34. $\displaystyle\sum_{n=1}^{\infty} \left(1 - \dfrac{1}{n}\right)^n$

35. $\displaystyle\sum_{n=0}^{\infty} \dfrac{n!}{1000^n}$

36. $\displaystyle\sum_{n=1}^{\infty} \dfrac{n^n}{n!}$

37. $\displaystyle\sum_{n=1}^{\infty} \ln\left(\dfrac{n}{n+1}\right)$

38. $\displaystyle\sum_{n=1}^{\infty} \ln\left(\dfrac{n}{2n+1}\right)$

39. $\displaystyle\sum_{n=0}^{\infty} \left(\dfrac{e}{\pi}\right)^n$

40. $\displaystyle\sum_{n=0}^{\infty} \dfrac{e^{n\pi}}{\pi^{ne}}$

In each of the geometric series in Exercises 41–44, write out the first few terms of the series to find a and r, and find the sum of the series. Then express the inequality $|r| < 1$ in terms of x and find the values of x for which the inequality holds and the series converges.

41. $\displaystyle\sum_{n=0}^{\infty} (-1)^n x^n$

42. $\displaystyle\sum_{n=0}^{\infty} (-1)^n x^{2n}$

43. $\displaystyle\sum_{n=0}^{\infty} 3\left(\dfrac{x-1}{2}\right)^n$

44. $\displaystyle\sum_{n=0}^{\infty} \dfrac{(-1)^n}{2} \left(\dfrac{1}{3+\sin x}\right)^n$

In Exercises 45–50, find the values of x for which the given geometric series converges. Also, find the sum of the series (as a function of x) for those values of x.

45. $\displaystyle\sum_{n=0}^{\infty} 2^n x^n$

46. $\displaystyle\sum_{n=0}^{\infty} (-1)^n x^{-2n}$

47. $\displaystyle\sum_{n=0}^{\infty} (-1)^n (x+1)^n$

48. $\displaystyle\sum_{n=0}^{\infty} \left(-\dfrac{1}{2}\right)^n (x-3)^n$

49. $\displaystyle\sum_{n=0}^{\infty} \sin^n x$

50. $\displaystyle\sum_{n=0}^{\infty} (\ln x)^n$

Express each of the numbers in Exercises 51–58 as the ratio of two integers.

51. $0.\overline{23} = 0.23\ 23\ 23\ldots$

52. $0.\overline{234} = 0.234\ 234\ 234\ldots$

53. $0.\overline{7} = 0.7777\ldots$

54. $0.\overline{d} = 0.dddd\ldots, \quad$ where d is a digit

55. $0.0\overline{6} = 0.06666\ldots$

56. $1.\overline{414} = 1.414\ 414\ 414\ldots$

57. $1.24\overline{123} = 1.24\ 123\ 123\ 123\ldots$

58. $3.\overline{142857} = 3.142857\ 142857\ldots$

59. The series in Exercise 5 can also be written as

$$\sum_{n=1}^{\infty} \frac{1}{(n+1)(n+2)} \quad \text{and} \quad \sum_{n=-1}^{\infty} \frac{1}{(n+3)(n+4)}.$$

Write it as a sum beginning with **(a)** $n = -2$, **(b)** $n = 0$, **(c)** $n = 5$.

60. The series in Exercise 6 can also be written as

$$\sum_{n=1}^{\infty} \frac{5}{n(n+1)} \quad \text{and} \quad \sum_{n=0}^{\infty} \frac{5}{(n+1)(n+2)}.$$

Write it as a sum beginning with **(a)** $n = -1$, **(b)** $n = 3$, **(c)** $n = 20$.

61. Make up an infinite series of nonzero terms whose sum is

   a. 1   **b.** -3 **c.** 0.

62. (*Continuation of Exercise 61.*) Can you make an infinite series of nonzero terms that converges to any number you want? Explain.

63. Show by example that $\Sigma(a_n/b_n)$ may diverge even though Σa_n and Σb_n converge and no b_n equals 0.

64. Find convergent geometric series $A = \Sigma a_n$ and $B = \Sigma b_n$ that illustrate the fact that $\Sigma a_n b_n$ may converge without being equal to AB.

65. Show by example that $\Sigma(a_n/b_n)$ may converge to something other than A/B even when $A = \Sigma a_n$, $B = \Sigma b_n \neq 0$, and no b_n equals 0.

66. If Σa_n converges and $a_n > 0$ for all n, can anything be said about $\Sigma(1/a_n)$? Give reasons for your answer.

67. What happens if you add a finite number of terms to a divergent series or delete a finite number of terms from a divergent series? Give reasons for your answer.

68. If Σa_n converges and Σb_n diverges, can anything be said about their term-by-term sum $\Sigma(a_n + b_n)$? Give reasons for your answer.

69. Make up a geometric series Σar^{n-1} that converges to the number 5 if

   a. $a = 2$ **b.** $a = 13/2$.

70. Find the value of b for which

$$1 + e^b + e^{2b} + e^{3b} + \cdots = 9.$$

71. For what values of r does the infinite series

$$1 + 2r + r^2 + 2r^3 + r^4 + 2r^5 + r^6 + \cdots$$

converge? Find the sum of the series when it converges.

72. Show that the error $(L - s_n)$ obtained by replacing a convergent geometric series with one of its partial sums s_n is $ar^n/(1 - r)$.

8.3 The Integral Test

Given a series, we want to know whether it converges or not. In this section and the next two, we study series with nonnegative terms. Such a series converges if its sequence of partial sums is bounded. If we establish that a given series does converge, we generally do not have a formula available for its sum, so we investigate methods to approximate the sum instead.

Nondecreasing Partial Sums

Suppose that $\sum_{n=1}^{\infty} a_n$ is an infinite series with $a_n \geq 0$ for all n. Then each partial sum is greater than or equal to its predecessor because $s_{n+1} = s_n + a_n$:

$$s_1 \leq s_2 \leq s_3 \leq \cdots \leq s_n \leq s_{n+1} \leq \cdots.$$

Since the partial sums form a nondecreasing sequence, the Nondecreasing Sequence Theorem (Theorem 6, Section 8.1) tells us that the series will converge if and only if the partial sums are bounded from above.

> **Corollary of Theorem 6**   A series $\sum_{n=1}^{\infty} a_n$ of nonnegative terms converges if and only if its partial sums are bounded from above.

EXAMPLE 1 The series

$$\sum_{n=1}^{\infty} \frac{1}{n} = 1 + \frac{1}{2} + \frac{1}{3} + \cdots + \frac{1}{n} + \cdots$$

is called the **harmonic series**. The harmonic series is divergent, but this doesn't follow from the nth-Term Test. The nth term $1/n$ does go to zero, but the series still diverges. The reason it diverges is because there is no upper bound for its partial sums. To see why, group the terms of the series in the following way:

$$1 + \frac{1}{2} + \underbrace{\left(\frac{1}{3} + \frac{1}{4}\right)}_{> \frac{2}{4} = \frac{1}{2}} + \underbrace{\left(\frac{1}{5} + \frac{1}{6} + \frac{1}{7} + \frac{1}{8}\right)}_{> \frac{4}{8} = \frac{1}{2}} + \underbrace{\left(\frac{1}{9} + \frac{1}{10} + \cdots + \frac{1}{16}\right)}_{> \frac{8}{16} = \frac{1}{2}} + \cdots.$$

The sum of the first two terms is 1.5. The sum of the next two terms is $1/3 + 1/4$, which is greater than $1/4 + 1/4 = 1/2$. The sum of the next four terms is $1/5 + 1/6 + 1/7 + 1/8$, which is greater than $1/8 + 1/8 + 1/8 + 1/8 = 1/2$. The sum of the next eight terms is $1/9 + 1/10 + 1/11 + 1/12 + 1/13 + 1/14 + 1/15 + 1/16$, which is greater than $8/16 = 1/2$. The sum of the next 16 terms is greater than $16/32 = 1/2$, and so on. In general, the sum of 2^n terms ending with $1/2^{n+1}$ is greater than $2^n/2^{n+1} = 1/2$. The sequence of partial sums is not bounded from above: If $n = 2^k$, the partial sum s_n is greater than $k/2$. The harmonic series diverges. ∎

The Integral Test

We introduce the Integral Test with a series that is related to the harmonic series, but whose nth term is $1/n^2$ instead of $1/n$.

EXAMPLE 2 Does the following series converge?

$$\sum_{n=1}^{\infty} \frac{1}{n^2} = 1 + \frac{1}{4} + \frac{1}{9} + \frac{1}{16} + \cdots + \frac{1}{n^2} + \cdots$$

Solution We determine the convergence of $\sum_{n=1}^{\infty}(1/n^2)$ by comparing it with $\int_1^{\infty}(1/x^2)\, dx$. To carry out the comparison, we think of the terms of the series as values of the function $f(x) = 1/x^2$ and interpret these values as the areas of rectangles under the curve $y = 1/x^2$.

As Figure 8.7 shows,

$$s_n = \frac{1}{1^2} + \frac{1}{2^2} + \frac{1}{3^2} + \cdots + \frac{1}{n^2}$$

$$= f(1) + f(2) + f(3) + \cdots + f(n)$$

$$< f(1) + \int_1^n \frac{1}{x^2}\, dx \qquad \text{Rectangle areas sum to less than area under graph.}$$

$$< 1 + \int_1^{\infty} \frac{1}{x^2}\, dx \qquad \int_1^n(1/x^2)\, dx < \int_1^{\infty}(1/x^2)\, dx$$

$$< 1 + 1 = 2. \qquad \text{As in Section 7.7, Example 3,} \int_1^{\infty}(1/x^2)\, dx = 1.$$

Thus the partial sums of $\sum_{n=1}^{\infty} 1/n^2$ are bounded from above (by 2) and the series converges. The sum of the series is known to be $\pi^2/6 \approx 1.64493$. ∎

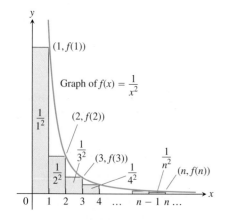

FIGURE 8.7 The sum of the areas of the rectangles under the graph of $f(x) = 1/x^2$ is less than the area under the graph (Example 2).

Caution
The series and integral need not have the same value in the convergent case. As we noted in Example 2, $\sum_{n=1}^{\infty} (1/n^2) = \pi^2/6$ while $\int_1^{\infty} (1/x^2)\, dx = 1$.

> **THEOREM 9—The Integral Test**   Let $\{a_n\}$ be a sequence of positive terms. Suppose that $a_n = f(n)$, where f is a continuous, positive, decreasing function of x for all $x \geq N$ (N a positive integer). Then the series $\sum_{n=N}^{\infty} a_n$ and the integral $\int_N^{\infty} f(x)\, dx$ both converge or both diverge.

Proof   We establish the test for the case $N = 1$. The proof for general N is similar.

We start with the assumption that f is a decreasing function with $f(n) = a_n$ for every n. This leads us to observe that the rectangles in Figure 8.8a, which have areas $a_1, a_2, \ldots, a_n$, collectively enclose more area than that under the curve $y = f(x)$ from $x = 1$ to $x = n + 1$. That is,

$$\int_1^{n+1} f(x)\, dx \leq a_1 + a_2 + \cdots + a_n.$$

In Figure 8.8b the rectangles have been faced to the left instead of to the right. If we momentarily disregard the first rectangle, of area a_1, we see that

$$a_2 + a_3 + \cdots + a_n \leq \int_1^n f(x)\, dx.$$

If we include a_1, we have

$$a_1 + a_2 + \cdots + a_n \leq a_1 + \int_1^n f(x)\, dx.$$

Combining these results gives

$$\int_1^{n+1} f(x)\, dx \leq a_1 + a_2 + \cdots + a_n \leq a_1 + \int_1^n f(x)\, dx.$$

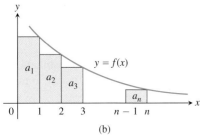

FIGURE 8.8   Subject to the conditions of the Integral Test, the series $\sum_{n=1}^{\infty} a_n$ and the integral $\int_1^{\infty} (x)\, dx$ both converge or both diverge.

These inequalities hold for each n, and continue to hold as $n \to \infty$.

If $\int_1^{\infty} f(x)\, dx$ is finite, the right-hand inequality shows that $\sum a_n$ is finite. If $\int_1^{\infty} f(x)\, dx$ is infinite, the left-hand inequality shows that $\sum a_n$ is infinite. Hence the series and the integral are both finite or both infinite. ∎

EXAMPLE 3   Show that the *p-series*

$$\sum_{n=1}^{\infty} \frac{1}{n^p} = \frac{1}{1^p} + \frac{1}{2^p} + \frac{1}{3^p} + \cdots + \frac{1}{n^p} + \cdots$$

(p a real constant) converges if $p > 1$, and diverges if $p \leq 1$.

Solution   If $p > 1$, then $f(x) = 1/x^p$ is a positive decreasing function of x. Since

$$\int_1^{\infty} \frac{1}{x^p}\, dx = \int_1^{\infty} x^{-p}\, dx = \lim_{b \to \infty} \left[\frac{x^{-p+1}}{-p + 1} \right]_1^b$$

$$= \frac{1}{1 - p} \lim_{b \to \infty} \left(\frac{1}{b^{p-1}} - 1 \right)$$

$$= \frac{1}{1 - p} (0 - 1) = \frac{1}{p - 1},$$

$b^{p-1} \to \infty$ as $b \to \infty$ because $p - 1 > 0$.

the series converges by the Integral Test. We emphasize that the sum of the p-series is *not* $1/(p - 1)$. The series converges, but we don't know the value it converges to.

If $p < 1$, then $1 - p > 0$ and

$$\int_1^\infty \frac{1}{x^p}\, dx = \frac{1}{1 - p} \lim_{b \to \infty} (b^{1-p} - 1) = \infty.$$

The series diverges by the Integral Test.

If $p = 1$, we have the (divergent) harmonic series

$$1 + \frac{1}{2} + \frac{1}{3} + \cdots + \frac{1}{n} + \cdots.$$

We have convergence for $p > 1$ but divergence for every other value of p. ∎

The p-series with $p = 1$ is the **harmonic series** (Example 1). The p-Series Test shows that the harmonic series is just *barely* divergent; if we increase p to 1.000000001, for instance, the series converges!

The slowness with which the partial sums of the harmonic series approaches infinity is impressive. For instance, it takes more than 178 million terms of the harmonic series to move the partial sums beyond 20. (See also Exercise 33b.)

Error Estimation

If a series Σa_n is shown to be convergent by the integral test, we may want to estimate the size of the **remainder** R_n between the total sum S of the series and its nth partial sum s_n. That is, we wish to estimate

$$R_n = S - s_n = a_{n+1} + a_{n+2} + a_{n+3} + \cdots.$$

To get a lower bound for the remainder, we compare the sum of the areas of the rectangles with the area under the curve $y = f(x)$ for $x \geq n$ (see Figure 8.8a). We see that

$$R_n = a_{n+1} + a_{n+2} + a_{n+3} + \cdots \geq \int_{n+1}^\infty f(x)\, dx.$$

Similarly, from Figure 8.8b, we find an upper bound with

$$R_n = a_{n+1} + a_{n+2} + a_{n+3} + \cdots \leq \int_n^\infty f(x)\, dx.$$

These comparisons prove the following result giving bounds on the size of the remainder.

Bounds for the Remainder in the Integral Test

Suppose $\{a_k\}$ is a sequence of positive terms with $a_k = f(k)$, where f is a continuous positive decreasing function of x for all $x \geq n$, and that Σa_n converges to S. Then the remainder $R_n = S - s_n$ satisfies the inequalities

$$\int_{n+1}^\infty f(x)\, dx \leq R_n \leq \int_n^\infty f(x)\, dx. \tag{1}$$

If we add the partial sum s_n to each side of the inequalities in (1), we get

$$s_n + \int_{n+1}^{\infty} f(x)\,dx \le S \le s_n + \int_{n}^{\infty} f(x)\,dx \tag{2}$$

since $s_n + R_n = S$. The inequalities in (2) are useful for estimating the error in approximating the sum of a convergent series. The error can be no larger than the length of the interval containing S, as given by (2).

EXAMPLE 4   Estimate the sum of the series $\Sigma\left(1/n^2\right)$ using the inequalities in (2) and $n = 10$.

Solution   We have that

$$\int_{n}^{\infty} \frac{1}{x^2}\,dx = \lim_{b\to\infty} \left[-\frac{1}{x}\right]_{n}^{b} = \lim_{b\to\infty}\left(-\frac{1}{b} + \frac{1}{n}\right) = \frac{1}{n}.$$

Using this result with the inequalities in (2), we get

$$s_{10} + \frac{1}{11} \le S \le s_{10} + \frac{1}{10}.$$

Taking $s_{10} = 1 + (1/4) + (1/9) + (1/16) + \cdots + (1/100) \approx 1.54977$, these last inequalities give

$$1.64068 \le S \le 1.64997.$$

If we approximate the sum S by the midpoint of this interval, we find that

$$\sum_{n=1}^{\infty} \frac{1}{n^2} \approx 1.6453.$$

The error in this approximation is less than half the length of the interval, so the error is less than 0.005.   ∎

EXERCISES 8.3

Which of the series in Exercises 1–30 converge, and which diverge? Give reasons for your answers. (When you check an answer, remember that there may be more than one way to determine the series' convergence or divergence.)

1. $\displaystyle\sum_{n=1}^{\infty} \frac{1}{10^n}$

2. $\displaystyle\sum_{n=1}^{\infty} e^{-n}$

3. $\displaystyle\sum_{n=1}^{\infty} \frac{n}{n+1}$

4. $\displaystyle\sum_{n=1}^{\infty} \frac{5}{n+1}$

5. $\displaystyle\sum_{n=1}^{\infty} \frac{3}{\sqrt{n}}$

6. $\displaystyle\sum_{n=1}^{\infty} \frac{-2}{n\sqrt{n}}$

7. $\displaystyle\sum_{n=1}^{\infty} -\frac{1}{8^n}$

8. $\displaystyle\sum_{n=1}^{\infty} \frac{-8}{n}$

9. $\displaystyle\sum_{n=2}^{\infty} \frac{\ln n}{n}$

10. $\displaystyle\sum_{n=2}^{\infty} \frac{\ln n}{\sqrt{n}}$

11. $\displaystyle\sum_{n=1}^{\infty} \frac{2^n}{3^n}$

12. $\displaystyle\sum_{n=1}^{\infty} \frac{5^n}{4^n + 3}$

13. $\displaystyle\sum_{n=0}^{\infty} \frac{-2}{n+1}$

14. $\displaystyle\sum_{n=1}^{\infty} \frac{1}{2n-1}$

15. $\displaystyle\sum_{n=1}^{\infty} \frac{2^n}{n+1}$

16. $\displaystyle\sum_{n=1}^{\infty} \frac{1}{\sqrt{n}(\sqrt{n}+1)}$

17. $\displaystyle\sum_{n=2}^{\infty} \frac{\sqrt{n}}{\ln n}$

18. $\displaystyle\sum_{n=1}^{\infty} \left(1 + \frac{1}{n}\right)^n$

19. $\displaystyle\sum_{n=1}^{\infty} \frac{1}{(\ln 2)^n}$

20. $\displaystyle\sum_{n=1}^{\infty} \frac{1}{(\ln 3)^n}$

21. $\displaystyle\sum_{n=3}^{\infty} \frac{(1/n)}{(\ln n)\sqrt{\ln^2 n - 1}}$

22. $\displaystyle\sum_{n=1}^{\infty} \frac{1}{n(1 + \ln^2 n)}$

23. $\displaystyle\sum_{n=1}^{\infty} n \sin\frac{1}{n}$

24. $\displaystyle\sum_{n=1}^{\infty} n \tan\frac{1}{n}$

25. $\displaystyle\sum_{n=1}^{\infty} \frac{e^n}{1 + e^{2n}}$

26. $\displaystyle\sum_{n=1}^{\infty} \frac{2}{1 + e^n}$

27. $\displaystyle\sum_{n=1}^{\infty} \frac{8 \tan^{-1} n}{1 + n^2}$

28. $\displaystyle\sum_{n=1}^{\infty} \frac{n}{n^2 + 1}$

29. $\displaystyle\sum_{n=1}^{\infty} \operatorname{sech} n$

30. $\displaystyle\sum_{n=1}^{\infty} \operatorname{sech}^2 n$

For what values of a, if any, do the series in Exercises 31 and 32 converge?

31. $\displaystyle\sum_{n=1}^{\infty} \left(\frac{a}{n+2} - \frac{1}{n+4} \right)$ **32.** $\displaystyle\sum_{n=3}^{\infty} \left(\frac{1}{n-1} - \frac{2a}{n+1} \right)$

33. a. Draw illustrations like those in Figures 8.7 and 8.8 to show that the partial sums of the harmonic series satisfy the inequalities

$$\ln (n+1) = \int_1^{n+1} \frac{1}{x}\,dx \le 1 + \frac{1}{2} + \cdots + \frac{1}{n}$$

$$\le 1 + \int_1^n \frac{1}{x}\,dx = 1 + \ln n.$$

T **b.** There is absolutely no empirical evidence for the divergence of the harmonic series even though we know it diverges. The partial sums just grow too slowly. To see what we mean, suppose you had started with $s_1 = 1$ the day the universe was formed, 13 billion years ago, and added a new term every *second*. About how large would the partial sum s_n be today, assuming a 365-day year?

34. Are there any values of x for which $\sum_{n=1}^{\infty}(1/(nx))$ converges? Give reasons for your answer.

35. Is it true that if $\sum_{n=1}^{\infty} a_n$ is a divergent series of positive numbers then there is also a divergent series $\sum_{n=1}^{\infty} b_n$ of positive numbers with $b_n < a_n$ for every n? Is there a "smallest" divergent series of positive numbers? Give reasons for your answers.

36. (*Continuation of Exercise 35.*) Is there a "largest" convergent series of positive numbers? Explain.

37. The Cauchy condensation test The Cauchy condensation test says: Let $\{a_n\}$ be a nonincreasing sequence ($a_n \ge a_{n+1}$ for all n) of positive terms that converges to 0. Then $\sum a_n$ converges if and only if $\sum 2^n a_{2^n}$ converges. For example, $\sum(1/n)$ diverges because $\sum 2^n \cdot (1/2^n) = \sum 1$ diverges. Show why the test works.

38. Use the Cauchy condensation test from Exercise 37 to show that

a. $\displaystyle\sum_{n=2}^{\infty} \frac{1}{n \ln n}$ diverges;

b. $\displaystyle\sum_{n=1}^{\infty} \frac{1}{n^p}$ converges if $p > 1$ and diverges if $p \le 1$.

39. Logarithmic p-series

a. Show that

$$\int_2^{\infty} \frac{dx}{x(\ln x)^p} \quad (p \text{ a positive constant})$$

converges if and only if $p > 1$.

b. What implications does the fact in part (a) have for the convergence of the series

$$\sum_{n=2}^{\infty} \frac{1}{n(\ln n)^p} ?$$

Give reasons for your answer.

40. (*Continuation of Exercise 39.*) Use the result in Exercise 39 to determine which of the following series converge and which diverge. Support your answer in each case.

a. $\displaystyle\sum_{n=2}^{\infty} \frac{1}{n(\ln n)}$ **b.** $\displaystyle\sum_{n=2}^{\infty} \frac{1}{n(\ln n)^{1.01}}$

c. $\displaystyle\sum_{n=2}^{\infty} \frac{1}{n \ln(n^3)}$ **d.** $\displaystyle\sum_{n=2}^{\infty} \frac{1}{n(\ln n)^3}$

41. Euler's constant Graphs like those in Figure 8.8 suggest that as n increases there is little change in the difference between the sum

$$1 + \frac{1}{2} + \cdots + \frac{1}{n}$$

and the integral

$$\ln n = \int_1^n \frac{1}{x}\,dx.$$

To explore this idea, carry out the following steps.

a. By taking $f(x) = 1/x$ in the proof of Theorem 9, show that

$$\ln (n+1) \le 1 + \frac{1}{2} + \cdots + \frac{1}{n} \le 1 + \ln n$$

or

$$0 < \ln (n+1) - \ln n \le 1 + \frac{1}{2} + \cdots + \frac{1}{n} - \ln n \le 1.$$

Thus, the sequence

$$a_n = 1 + \frac{1}{2} + \cdots + \frac{1}{n} - \ln n$$

is bounded from below and from above.

b. Show that

$$\frac{1}{n+1} < \int_n^{n+1} \frac{1}{x}\,dx = \ln (n+1) - \ln n,$$

and use this result to show that the sequence $\{a_n\}$ in part (a) is decreasing.

Since a decreasing sequence that is bounded from below converges (Exercise 107 in Section 8.1), the numbers a_n defined in part (a) converge:

$$1 + \frac{1}{2} + \cdots + \frac{1}{n} - \ln n \to \gamma.$$

The number γ, whose value is $0.5772\ldots$, is called *Euler's constant*. In contrast to other special numbers like π and e, no other expression with a simple law of formulation has ever been found for γ.

42. Use the integral test to show that

$$\sum_{n=0}^{\infty} e^{-n^2}$$

converges.

43. a. For the series $\sum(1/n^3)$, use the inequalities in (2) with $n = 10$ to find an interval containing the sum S.

b. As in Example 4, use the midpoint of the interval found in part (a) to approximate the sum of the series. What is the maximum error for your approximation?

44. Repeat Exercise 43 using the series $\sum(1/n^4)$.

8.4 Comparison Tests

We have seen how to determine the convergence of geometric series, *p*-series, and a few others. We can test the convergence of many more series by comparing their terms to those of a series whose convergence is known.

> **THEOREM 10—The Comparison Test** Let $\sum a_n$ be a series with nonnegative terms.
>
> (a) $\sum a_n$ converges if there is a convergent series $\sum c_n$ with $a_n \leq c_n$ for all $n > N$, for some integer N.
>
> (b) $\sum a_n$ diverges if there is a divergent series of nonnegative terms $\sum d_n$ with $a_n \geq d_n$ for all $n > N$, for some integer N.

HISTORICAL BIOGRAPHY

Albert of Saxony
(ca. 1316–1390)

Proof In Part (a), the partial sums of $\sum a_n$ are bounded above by

$$M = a_1 + a_2 + \cdots + a_N + \sum_{n=N+1}^{\infty} c_n.$$

They therefore form a nondecreasing sequence with a limit $L \leq M$.

In Part (b), the partial sums of $\sum a_n$ are not bounded from above. If they were, the partial sums for $\sum d_n$ would be bounded by

$$M^* = d_1 + d_2 + \cdots + d_N + \sum_{n=N+1}^{\infty} a_n$$

and $\sum d_n$ would have to converge instead of diverge. ∎

EXAMPLE 1

(a) The series

$$\sum_{n=1}^{\infty} \frac{5}{5n - 1}$$

diverges because its *n*th term

$$\frac{5}{5n - 1} = \frac{1}{n - \frac{1}{5}} > \frac{1}{n}$$

is greater than the *n*th term of the divergent harmonic series.

(b) The series

$$\sum_{n=0}^{\infty} \frac{1}{n!} = 1 + \frac{1}{1!} + \frac{1}{2!} + \frac{1}{3!} + \cdots$$

converges because its terms are all positive and less than or equal to the corresponding terms of

$$1 + \sum_{n=0}^{\infty} \frac{1}{2^n} = 1 + 1 + \frac{1}{2} + \frac{1}{2^2} + \cdots.$$

The geometric series on the left converges and we have

$$1 + \sum_{n=0}^{\infty} \frac{1}{2^n} = 1 + \frac{1}{1 - (1/2)} = 3.$$

The fact that 3 is an upper bound for the partial sums of $\sum_{n=0}^{\infty}(1/n!)$ does not mean that the series converges to 3. As we will see in Section 8.9, the series converges to e.

(c) The series

$$5 + \frac{2}{3} + \frac{1}{7} + 1 + \frac{1}{2 + \sqrt{1}} + \frac{1}{4 + \sqrt{2}} + \frac{1}{8 + \sqrt{3}} + \cdots + \frac{1}{2^n + \sqrt{n}} + \cdots$$

converges. To see this, we ignore the first three terms and compare the remaining terms with those of the convergent geometric series $\sum_{n=0}^{\infty}(1/2^n)$. The term $1/(2^n + \sqrt{n})$ of the truncated sequence is less than the corresponding term $1/2^n$ of the geometric series. We see that term by term we have the comparison,

$$1 + \frac{1}{2 + \sqrt{1}} + \frac{1}{4 + \sqrt{2}} + \frac{1}{8 + \sqrt{3}} + \cdots \leq 1 + \frac{1}{2} + \frac{1}{4} + \frac{1}{8} + \cdots$$

So the truncated series and the original series converge by an application of the Comparison Test. ∎

The Limit Comparison Test

We now introduce a comparison test that is particularly useful for series in which a_n is a rational function of n.

THEOREM 11—Limit Comparison Test   Suppose that $a_n > 0$ and $b_n > 0$ for all $n \geq N$ (N an integer).

1.   If $\lim\limits_{n\to\infty} \dfrac{a_n}{b_n} = c > 0$, then $\sum a_n$ and $\sum b_n$ both converge or both diverge.

2.   If $\lim\limits_{n\to\infty} \dfrac{a_n}{b_n} = 0$ and $\sum b_n$ converges, then $\sum a_n$ converges.

3.   If $\lim\limits_{n\to\infty} \dfrac{a_n}{b_n} = \infty$ and $\sum b_n$ diverges, then $\sum a_n$ diverges.

Proof   We will prove Part 1. Parts 2 and 3 are left as Exercises 37(a) and (b).

Since $c/2 > 0$, there exists an integer N such that for all n

$$n > N \;\Rightarrow\; \left| \frac{a_n}{b_n} - c \right| < \frac{c}{2}.$$

Limit definition with $\epsilon = c/2$, $L = c$, and a_n replaced by a_n/b_n

Thus, for $n > N$,

$$-\frac{c}{2} < \frac{a_n}{b_n} - c < \frac{c}{2},$$

$$\frac{c}{2} < \frac{a_n}{b_n} < \frac{3c}{2},$$

$$\left(\frac{c}{2}\right)b_n < a_n < \left(\frac{3c}{2}\right)b_n.$$

If $\sum b_n$ converges, then $\sum(3c/2)b_n$ converges and $\sum a_n$ converges by the Direct Comparison Test. If $\sum b_n$ diverges, then $\sum(c/2)b_n$ diverges and $\sum a_n$ diverges by the Direct Comparison Test. ∎

EXAMPLE 2 Which of the following series converge, and which diverge?

(a) $\dfrac{3}{4} + \dfrac{5}{9} + \dfrac{7}{16} + \dfrac{9}{25} + \cdots = \displaystyle\sum_{n=1}^{\infty} \dfrac{2n+1}{(n+1)^2} = \sum_{n=1}^{\infty} \dfrac{2n+1}{n^2+2n+1}$

(b) $\dfrac{1}{1} + \dfrac{1}{3} + \dfrac{1}{7} + \dfrac{1}{15} + \cdots = \displaystyle\sum_{n=1}^{\infty} \dfrac{1}{2^n - 1}$

(c) $\dfrac{1+2\ln 2}{9} + \dfrac{1+3\ln 3}{14} + \dfrac{1+4\ln 4}{21} + \cdots = \displaystyle\sum_{n=2}^{\infty} \dfrac{1+n\ln n}{n^2+5}$

Solution

(a) Let $a_n = (2n+1)/(n^2+2n+1)$. For large n, we expect a_n to behave like $2n/n^2 = 2/n$ since the leading terms dominate for large n, so we let $b_n = 1/n$. Since

$$\sum_{n=1}^{\infty} b_n = \sum_{n=1}^{\infty} \frac{1}{n} \text{ diverges}$$

and

$$\lim_{n\to\infty} \frac{a_n}{b_n} = \lim_{n\to\infty} \frac{2n^2+n}{n^2+2n+1} = 2,$$

$\sum a_n$ diverges by Part 1 of the Limit Comparison Test. We could just as well have taken $b_n = 2/n$, but $1/n$ is simpler.

(b) Let $a_n = 1/(2^n - 1)$. For large n, we expect a_n to behave like $1/2^n$, so we let $b_n = 1/2^n$. Since

$$\sum_{n=1}^{\infty} b_n = \sum_{n=1}^{\infty} \frac{1}{2^n} \text{ converges}$$

and

$$\begin{aligned}
\lim_{n\to\infty} \frac{a_n}{b_n} &= \lim_{n\to\infty} \frac{2^n}{2^n - 1} \\
&= \lim_{n\to\infty} \frac{1}{1-(1/2^n)} \\
&= 1,
\end{aligned}$$

$\sum a_n$ converges by Part 1 of the Limit Comparison Test.

(c) Let $a_n = (1+n\ln n)/(n^2+5)$. For large n, we expect a_n to behave like $(n\ln n)/n^2 = (\ln n)/n$, which is greater than $1/n$ for $n \geq 3$, so we take $b_n = 1/n$. Since

$$\sum_{n=2}^{\infty} b_n = \sum_{n=2}^{\infty} \frac{1}{n} \text{ diverges}$$

and

$$\begin{aligned}
\lim_{n\to\infty} \frac{a_n}{b_n} &= \lim_{n\to\infty} \frac{n+n^2\ln n}{n^2+5} \\
&= \infty,
\end{aligned}$$

$\sum a_n$ diverges by Part 3 of the Limit Comparison Test. ∎

EXAMPLE 3 Does $\displaystyle\sum_{n=1}^{\infty} \dfrac{\ln n}{n^{3/2}}$ converge?

Solution Because $\ln n$ grows more slowly than n^c for any positive constant c (Section 8.1, Exercise 91), we would expect to have

$$\frac{\ln n}{n^{3/2}} < \frac{n^{1/4}}{n^{3/2}} = \frac{1}{n^{5/4}}$$

for n sufficiently large. Indeed, taking $a_n = (\ln n)/n^{3/2}$ and $b_n = 1/n^{5/4}$, we have

$$\lim_{n \to \infty} \frac{a_n}{b_n} = \lim_{n \to \infty} \frac{\ln n}{n^{1/4}}$$

$$= \lim_{n \to \infty} \frac{1/n}{(1/4)n^{-3/4}} \qquad \text{l'Hôpital's Rule}$$

$$= \lim_{n \to \infty} \frac{4}{n^{1/4}} = 0.$$

Since $\sum b_n = \sum (1/n^{5/4})$ (a p-series with $p > 1$) converges, $\sum a_n$ converges by Part 2 of the Limit Comparison Test. ∎

EXERCISES 8.4

Which of the series in Exercises 1–36 converge, and which diverge? Give reasons for your answers.

1. $\displaystyle\sum_{n=1}^{\infty} \frac{1}{2\sqrt{n} + \sqrt[3]{n}}$

2. $\displaystyle\sum_{n=1}^{\infty} \frac{3}{n + \sqrt{n}}$

3. $\displaystyle\sum_{n=1}^{\infty} \frac{\sin^2 n}{2^n}$

4. $\displaystyle\sum_{n=1}^{\infty} \frac{1 + \cos n}{n^2}$

5. $\displaystyle\sum_{n=1}^{\infty} \frac{2n}{3n - 1}$

6. $\displaystyle\sum_{n=1}^{\infty} \frac{n + 1}{n^2 \sqrt{n}}$

7. $\displaystyle\sum_{n=1}^{\infty} \left(\frac{n}{3n + 1}\right)^n$

8. $\displaystyle\sum_{n=1}^{\infty} \frac{1}{\sqrt{n^3 + 2}}$

9. $\displaystyle\sum_{n=3}^{\infty} \frac{1}{\ln(\ln n)}$

10. $\displaystyle\sum_{n=2}^{\infty} \frac{1}{(\ln n)^2}$

11. $\displaystyle\sum_{n=1}^{\infty} \frac{(\ln n)^2}{n^3}$

12. $\displaystyle\sum_{n=1}^{\infty} \frac{(\ln n)^3}{n^3}$

13. $\displaystyle\sum_{n=2}^{\infty} \frac{1}{\sqrt{n}\ln n}$

14. $\displaystyle\sum_{n=1}^{\infty} \frac{(\ln n)^2}{n^{3/2}}$

15. $\displaystyle\sum_{n=1}^{\infty} \frac{1}{1 + \ln n}$

16. $\displaystyle\sum_{n=1}^{\infty} \frac{1}{(1 + \ln n)^2}$

17. $\displaystyle\sum_{n=2}^{\infty} \frac{\ln(n + 1)}{n + 1}$

18. $\displaystyle\sum_{n=1}^{\infty} \frac{1}{(1 + \ln^2 n)}$

19. $\displaystyle\sum_{n=2}^{\infty} \frac{1}{n\sqrt{n^2 - 1}}$

20. $\displaystyle\sum_{n=1}^{\infty} \frac{\sqrt{n}}{n^2 + 1}$

21. $\displaystyle\sum_{n=1}^{\infty} \frac{1 - n}{n2^n}$

22. $\displaystyle\sum_{n=1}^{\infty} \frac{n + 2^n}{n^2 2^n}$

23. $\displaystyle\sum_{n=1}^{\infty} \frac{1}{3^{n-1} + 1}$

24. $\displaystyle\sum_{n=1}^{\infty} \frac{3^{n-1} + 1}{3^n}$

25. $\displaystyle\sum_{n=1}^{\infty} \sin \frac{1}{n}$

26. $\displaystyle\sum_{n=1}^{\infty} \tan \frac{1}{n}$

27. $\displaystyle\sum_{n=1}^{\infty} \frac{10n + 1}{n(n + 1)(n + 2)}$

28. $\displaystyle\sum_{n=3}^{\infty} \frac{5n^3 - 3n}{n^2(n - 2)(n^2 + 5)}$

29. $\displaystyle\sum_{n=1}^{\infty} \frac{\tan^{-1} n}{n^{1.1}}$

30. $\displaystyle\sum_{n=1}^{\infty} \frac{\sec^{-1} n}{n^{1.3}}$

31. $\displaystyle\sum_{n=1}^{\infty} \frac{\coth n}{n^2}$

32. $\displaystyle\sum_{n=1}^{\infty} \frac{\tanh n}{n^2}$

33. $\displaystyle\sum_{n=1}^{\infty} \frac{1}{n\sqrt[n]{n}}$

34. $\displaystyle\sum_{n=1}^{\infty} \frac{\sqrt[n]{n}}{n^2}$

35. $\displaystyle\sum_{n=1}^{\infty} \frac{1}{1 + 2 + 3 + \cdots + n}$

36. $\displaystyle\sum_{n=1}^{\infty} \frac{1}{1 + 2^2 + 3^2 + \cdots + n^2}$

37. Prove **(a)** Part 2 and **(b)** Part 3 of the Limit Comparison Test.

38. If $\sum_{n=1}^{\infty} a_n$ is a convergent series of nonnegative numbers, can anything be said about $\sum_{n=1}^{\infty} (a_n/n)$? Explain.

39. Suppose that $a_n > 0$ and $b_n > 0$ for $n \geq N$ (N an integer). If $\lim_{n \to \infty} (a_n/b_n) = \infty$ and $\sum a_n$ converges, can anything be said about $\sum b_n$? Give reasons for your answer.

40. Prove that if $\sum a_n$ is a convergent series of nonnegative terms, then $\sum a_n^2$ converges.

COMPUTER EXPLORATION

41. It is not yet known whether the series

$$\sum_{n=1}^{\infty} \frac{1}{n^3 \sin^2 n}$$

converges or diverges. Use a CAS to explore the behavior of the series by performing the following steps.

a. Define the sequence of partial sums

$$s_k = \sum_{n=1}^{k} \frac{1}{n^3 \sin^2 n}.$$

What happens when you try to find the limit of s_k as $k \to \infty$? Does your CAS find a closed form answer for this limit?

b. Plot the first 100 points (k, s_k) for the sequence of partial sums. Do they appear to converge? What would you estimate the limit to be?

c. Next plot the first 200 points (k, s_k). Discuss the behavior in your own words.

d. Plot the first 400 points (k, s_k). What happens when $k = 355$? Calculate the number $355/113$. Explain from your calculation what happened at $k = 355$. For what values of k would you guess this behavior might occur again?

You will find an interesting discussion of this series in Chapter 72 of *Mazes for the Mind* by Clifford A. Pickover, St. Martin's Press, Inc., New York, 1992.

8.5   The Ratio and Root Tests

The Ratio Test measures the rate of growth (or decline) of a series by examining the ratio a_{n+1}/a_n. For a geometric series $\sum ar^n$, this rate is a constant $((ar^{n+1})/(ar^n) = r)$, and the series converges if and only if its ratio is less than 1 in absolute value. The Ratio Test is a powerful rule extending that result.

THEOREM 12—The Ratio Test   Let $\sum a_n$ be a series with positive terms and suppose that

$$\lim_{n \to \infty} \frac{a_{n+1}}{a_n} = \rho.$$

Then **(a)** the series *converges* if $\rho < 1$, **(b)** the series *diverges* if $\rho > 1$ or ρ is infinite, **(c)** the test is *inconclusive* if $\rho = 1$.

Proof

(a) $\boldsymbol{\rho < 1.}$ Let r be a number between ρ and 1. Then the number $\epsilon = r - \rho$ is positive. Since

$$\frac{a_{n+1}}{a_n} \to \rho,$$

a_{n+1}/a_n must lie within ϵ of ρ when n is large enough, say for all $n \geq N$. In particular

$$\frac{a_{n+1}}{a_n} < \rho + \epsilon = r, \qquad \text{when } n \geq N.$$

That is,

$$a_{N+1} < ra_N,$$
$$a_{N+2} < ra_{N+1} < r^2 a_N,$$
$$a_{N+3} < ra_{N+2} < r^3 a_N,$$
$$\vdots$$
$$a_{N+m} < ra_{N+m-1} < r^m a_N.$$

These inequalities show that the terms of our series, after the Nth term, approach zero more rapidly than the terms in a geometric series with ratio $r < 1$. More precisely, consider the series $\sum c_n$, where $c_n = a_n$ for $n = 1, 2, \ldots, N$ and $c_{N+1} = ra_N, c_{N+2} = r^2 a_N, \ldots, c_{N+m} = r^m a_N, \ldots$. Now $a_n \leq c_n$ for all n, and

$$\sum_{n=1}^{\infty} c_n = a_1 + a_2 + \cdots + a_{N-1} + a_N + ra_N + r^2 a_N + \cdots$$
$$= a_1 + a_2 + \cdots + a_{N-1} + a_N(1 + r + r^2 + \cdots).$$

The geometric series $1 + r + r^2 + \cdots$ converges because $|r| < 1$, so $\sum c_n$ converges. Since $a_n \leq c_n$, $\sum a_n$ also converges.

(b) $\boldsymbol{1 < \rho \leq \infty.}$ From some index M on,

$$\frac{a_{n+1}}{a_n} > 1 \qquad \text{and} \qquad a_M < a_{M+1} < a_{M+2} < \cdots.$$

The terms of the series do not approach zero as n becomes infinite, and the series diverges by the nth-Term Test.

(c) $\rho = 1.$ The two series

$$\sum_{n=1}^{\infty} \frac{1}{n} \quad \text{and} \quad \sum_{n=1}^{\infty} \frac{1}{n^2}$$

show that some other test for convergence must be used when $\rho = 1$.

$$\text{For } \sum_{n=1}^{\infty} \frac{1}{n}: \quad \frac{a_{n+1}}{a_n} = \frac{1/(n+1)}{1/n} = \frac{n}{n+1} \to 1.$$

$$\text{For } \sum_{n=1}^{\infty} \frac{1}{n^2}: \quad \frac{a_{n+1}}{a_n} = \frac{1/(n+1)^2}{1/n^2} = \left(\frac{n}{n+1}\right)^2 \to 1^2 = 1.$$

In both cases, $\rho = 1$, yet the first series diverges, whereas the second converges. ∎

The Ratio Test is often effective when the terms of a series contain factorials of expressions involving n or expressions raised to a power involving n.

EXAMPLE 1   Investigate the convergence of the following series.

(a) $\sum_{n=0}^{\infty} \frac{2^n + 5}{3^n}$ **(b)** $\sum_{n=1}^{\infty} \frac{(2n)!}{n!n!}$ **(c)** $\sum_{n=1}^{\infty} \frac{4^n n! n!}{(2n)!}$

Solution

(a) For the series $\sum_{n=0}^{\infty}(2^n + 5)/3^n$,

$$\frac{a_{n+1}}{a_n} = \frac{(2^{n+1} + 5)/3^{n+1}}{(2^n + 5)/3^n} = \frac{1}{3} \cdot \frac{2^{n+1} + 5}{2^n + 5} = \frac{1}{3} \cdot \left(\frac{2 + 5 \cdot 2^{-n}}{1 + 5 \cdot 2^{-n}}\right) \to \frac{1}{3} \cdot \frac{2}{1} = \frac{2}{3}.$$

The series converges because $\rho = 2/3$ is less than 1. This does *not* mean that $2/3$ is the sum of the series. In fact,

$$\sum_{n=0}^{\infty} \frac{2^n + 5}{3^n} = \sum_{n=0}^{\infty} \left(\frac{2}{3}\right)^n + \sum_{n=0}^{\infty} \frac{5}{3^n} = \frac{1}{1 - (2/3)} + \frac{5}{1 - (1/3)} = \frac{21}{2}.$$

(b) If $a_n = \dfrac{(2n)!}{n!n!}$, then $a_{n+1} = \dfrac{(2n + 2)!}{(n + 1)!(n + 1)!}$ and

$$\frac{a_{n+1}}{a_n} = \frac{n!n!(2n + 2)(2n + 1)(2n)!}{(n + 1)!(n + 1)!(2n)!}$$

$$= \frac{(2n + 2)(2n + 1)}{(n + 1)(n + 1)} = \frac{4n + 2}{n + 1} \to 4.$$

The series diverges because $\rho = 4$ is greater than 1.

(c) If $a_n = 4^n n! n!/(2n)!$, then

$$\frac{a_{n+1}}{a_n} = \frac{4^{n+1}(n + 1)!(n + 1)!}{(2n + 2)(2n + 1)(2n)!} \cdot \frac{(2n)!}{4^n n! n!}$$

$$= \frac{4(n + 1)(n + 1)}{(2n + 2)(2n + 1)} = \frac{2(n + 1)}{2n + 1} \to 1.$$

Because the limit is $\rho = 1$, we cannot decide from the Ratio Test whether the series converges. When we notice that $a_{n+1}/a_n = (2n + 2)/(2n + 1)$, we conclude that a_{n+1} is always greater than a_n because $(2n + 2)/(2n + 1)$ is always greater than 1.

Therefore, all terms are greater than or equal to $a_1 = 2$, and the nth term does not approach zero as $n \to \infty$. The series diverges.   ∎

The Root Test

The convergence tests we have so far for Σa_n work best when the formula for a_n is relatively simple. However, consider the following series.

EXAMPLE 2   Let $a_n = \begin{cases} n/2^n, & n \text{ odd} \\ 1/2^n, & n \text{ even.} \end{cases}$ Does Σa_n converge?

Solution   We write out several terms of the series:

$$\sum_{n=1}^{\infty} a_n = \frac{1}{2^1} + \frac{1}{2^2} + \frac{3}{2^3} + \frac{1}{2^4} + \frac{5}{2^5} + \frac{1}{2^6} + \frac{7}{2^7} + \cdots$$

$$= \frac{1}{2} + \frac{1}{4} + \frac{3}{8} + \frac{1}{16} + \frac{5}{32} + \frac{1}{64} + \frac{7}{128} + \cdots.$$

Clearly, this is not a geometric series. The nth term approaches zero as $n \to \infty$, so we do not know if the series diverges. The Integral Test does not look promising. The Ratio Test produces

$$\frac{a_{n+1}}{a_n} = \begin{cases} \dfrac{1}{2n}, & n \text{ odd} \\[2ex] \dfrac{n+1}{2}, & n \text{ even.} \end{cases}$$

As $n \to \infty$, the ratio is alternately small and large and has no limit.

A test that will answer the question (the series converges) is the Root Test.   ∎

THEOREM 13—The Root Test   Let Σa_n be a series with $a_n \geq 0$ for $n \geq N$, and suppose that

$$\lim_{n \to \infty} \sqrt[n]{a_n} = \rho.$$

Then **(a)** the series *converges* if $\rho < 1$, **(b)** the series *diverges* if $\rho > 1$ or ρ is infinite, **(c)** the test is *inconclusive* if $\rho = 1$.

Proof

(a) $\rho < 1$. Choose an $\epsilon > 0$ so small that $\rho + \epsilon < 1$. Since $\sqrt[n]{a_n} \to \rho$, the terms $\sqrt[n]{a_n}$ eventually get closer than ϵ to ρ. In other words, there exists an index $M \geq N$ such that

$$\sqrt[n]{a_n} < \rho + \epsilon \qquad \text{when } n \geq M.$$

Then it is also true that

$$a_n < (\rho + \epsilon)^n \qquad \text{for } n \geq M.$$

Now, $\sum_{n=M}^{\infty} (\rho + \epsilon)^n$, a geometric series with ratio $(\rho + \epsilon) < 1$, converges. By comparison, $\sum_{n=M}^{\infty} a_n$ converges, from which it follows that

$$\sum_{n=1}^{\infty} a_n = a_1 + \cdots + a_{M-1} + \sum_{n=M}^{\infty} a_n$$

converges.

(b) $1 < \rho \leq \infty$. For all indices beyond some integer M, we have $\sqrt[n]{a_n} > 1$, so that $a_n > 1$ for $n > M$. The terms of the series do not converge to zero. The series diverges by the nth-Term Test.

(c) $\rho = 1$. The series $\sum_{n=1}^{\infty}(1/n)$ and $\sum_{n=1}^{\infty}(1/n^2)$ show that the test is not conclusive when $\rho = 1$. The first series diverges and the second converges, but in both cases $\sqrt[n]{a_n} \to 1$. ∎

EXAMPLE 3   Which of the following series converge, and which diverge?

(a) $\sum_{n=1}^{\infty} \dfrac{n^2}{2^n}$ **(b)** $\sum_{n=1}^{\infty} \dfrac{2^n}{n^2}$ **(c)** $\sum_{n=1}^{\infty} \left(\dfrac{1}{1+n}\right)^n$

Solution

(a) $\sum_{n=1}^{\infty} \dfrac{n^2}{2^n}$ converges because $\sqrt[n]{\dfrac{n^2}{2^n}} = \dfrac{\sqrt[n]{n^2}}{\sqrt[n]{2^n}} = \dfrac{\left(\sqrt[n]{n}\right)^2}{2} \to \dfrac{1}{2} < 1$.

(b) $\sum_{n=1}^{\infty} \dfrac{2^n}{n^2}$ diverges because $\sqrt[n]{\dfrac{2^n}{n^2}} = \dfrac{2}{\left(\sqrt[n]{n}\right)^2} \to \dfrac{2}{1} > 1$.

(c) $\sum_{n=1}^{\infty} \left(\dfrac{1}{1+n}\right)^n$ converges because $\sqrt[n]{\left(\dfrac{1}{1+n}\right)^n} = \dfrac{1}{1+n} \to 0 < 1$. ∎

EXAMPLE 4   (Example 2 Revisited)

Let $a_n = \begin{cases} n/2^n, & n \text{ odd} \\ 1/2^n, & n \text{ even.} \end{cases}$ Does $\sum a_n$ converge?

Solution   We apply the Root Test, finding that

$$\sqrt[n]{a_n} = \begin{cases} \sqrt[n]{n}/2, & n \text{ odd} \\ 1/2, & n \text{ even.} \end{cases}$$

Therefore,

$$\frac{1}{2} \leq \sqrt[n]{a_n} \leq \frac{\sqrt[n]{n}}{2}.$$

Since $\sqrt[n]{n} \to 1$ (Section 8.1, Theorem 5), we have $\lim_{n\to\infty} \sqrt[n]{a_n} = 1/2$ by the Sandwich Theorem. The limit is less than 1, so the series converges by the Root Test. ∎

EXERCISES 8.5

Which of the series in Exercises 1–26 converge, and which diverge? Give reasons for your answers. (When checking your answers, remember there may be more than one way to determine a series' convergence or divergence.)

1. $\sum_{n=1}^{\infty} \dfrac{n^{\sqrt{2}}}{2^n}$

2. $\sum_{n=1}^{\infty} n^2 e^{-n}$

3. $\sum_{n=1}^{\infty} n! e^{-n}$

4. $\sum_{n=1}^{\infty} \dfrac{n!}{10^n}$

5. $\sum_{n=1}^{\infty} \dfrac{n^{10}}{10^n}$

6. $\sum_{n=1}^{\infty} \left(\dfrac{n-2}{n}\right)^n$

7. $\sum_{n=1}^{\infty} \dfrac{2 + (-1)^n}{1.25^n}$

8. $\sum_{n=1}^{\infty} \dfrac{(-2)^n}{3^n}$

9. $\sum_{n=1}^{\infty} \left(1 - \dfrac{3}{n}\right)^n$

10. $\sum_{n=1}^{\infty} \left(1 - \dfrac{1}{3n}\right)^n$

11. $\sum_{n=1}^{\infty} \dfrac{\ln n}{n^3}$

12. $\sum_{n=1}^{\infty} \dfrac{(\ln n)^n}{n^n}$

13. $\displaystyle\sum_{n=1}^{\infty} \left(\frac{1}{n} - \frac{1}{n^2} \right)$

14. $\displaystyle\sum_{n=1}^{\infty} \left(\frac{1}{n} - \frac{1}{n^2} \right)^n$

15. $\displaystyle\sum_{n=1}^{\infty} \frac{\ln n}{n}$

16. $\displaystyle\sum_{n=1}^{\infty} \frac{n \ln n}{2^n}$

17. $\displaystyle\sum_{n=1}^{\infty} \frac{(n+1)(n+2)}{n!}$

18. $\displaystyle\sum_{n=1}^{\infty} e^{-n}(n^3)$

19. $\displaystyle\sum_{n=1}^{\infty} \frac{(n+3)!}{3!n!3^n}$

20. $\displaystyle\sum_{n=1}^{\infty} \frac{n 2^n (n+1)!}{3^n n!}$

21. $\displaystyle\sum_{n=1}^{\infty} \frac{n!}{(2n+1)!}$

22. $\displaystyle\sum_{n=1}^{\infty} \frac{n!}{n^n}$

23. $\displaystyle\sum_{n=2}^{\infty} \frac{n}{(\ln n)^n}$

24. $\displaystyle\sum_{n=2}^{\infty} \frac{n}{(\ln n)^{(n/2)}}$

25. $\displaystyle\sum_{n=1}^{\infty} \frac{n! \ln n}{n(n+2)!}$

26. $\displaystyle\sum_{n=1}^{\infty} \frac{3^n}{n^3 2^n}$

Which of the series $\sum_{n=1}^{\infty} a_n$ defined by the formulas in Exercises 27–38 converge, and which diverge? Give reasons for your answers.

27. $a_1 = 2, \quad a_{n+1} = \dfrac{1 + \sin n}{n} a_n$

28. $a_1 = 1, \quad a_{n+1} = \dfrac{1 + \tan^{-1} n}{n} a_n$

29. $a_1 = \dfrac{1}{3}, \quad a_{n+1} = \dfrac{3n-1}{2n+5} a_n$

30. $a_1 = 3, \quad a_{n+1} = \dfrac{n}{n+1} a_n$

31. $a_1 = 2, \quad a_{n+1} = \dfrac{2}{n} a_n$

32. $a_1 = 5, \quad a_{n+1} = \dfrac{\sqrt[n]{n}}{2} a_n$

33. $a_1 = 1, \quad a_{n+1} = \dfrac{1 + \ln n}{n} a_n$

34. $a_1 = \dfrac{1}{2}, \quad a_{n+1} = \dfrac{n + \ln n}{n + 10} a_n$

35. $a_1 = \dfrac{1}{3}, \quad a_{n+1} = \sqrt[n]{a_n}$

36. $a_1 = \dfrac{1}{2}, \quad a_{n+1} = (a_n)^{n+1}$

37. $a_n = \dfrac{2^n n! n!}{(2n)!}$

38. $a_n = \dfrac{(3n)!}{n!(n+1)!(n+2)!}$

Which of the series in Exercises 39–44 converge, and which diverge? Give reasons for your answers.

39. $\displaystyle\sum_{n=1}^{\infty} \frac{(n!)^n}{(n^n)^2}$

40. $\displaystyle\sum_{n=1}^{\infty} \frac{(n!)^n}{n^{(n^2)}}$

41. $\displaystyle\sum_{n=1}^{\infty} \frac{n^n}{2^{(n^2)}}$

42. $\displaystyle\sum_{n=1}^{\infty} \frac{n^n}{(2^n)^2}$

43. $\displaystyle\sum_{n=1}^{\infty} \frac{1 \cdot 3 \cdot \cdots \cdot (2n-1)}{4^n 2^n n!}$

44. $\displaystyle\sum_{n=1}^{\infty} \frac{1 \cdot 3 \cdot \cdots \cdot (2n-1)}{[2 \cdot 4 \cdot \cdots \cdot (2n)](3^n + 1)}$

45. Neither the Ratio nor the Root Test helps with p-series. Try them on

$$\sum_{n=1}^{\infty} \frac{1}{n^p}$$

and show that both tests fail to provide information about convergence.

46. Show that neither the Ratio Test nor the Root Test provides information about the convergence of

$$\sum_{n=2}^{\infty} \frac{1}{(\ln n)^p} \qquad (p \text{ constant}).$$

47. Let $a_n = \begin{cases} n/2^n, & \text{if } n \text{ is a prime number} \\ 1/2^n, & \text{otherwise.} \end{cases}$

Does $\sum a_n$ converge? Give reasons for your answer.

8.6   Alternating Series, Absolute and Conditional Convergence

A series in which the terms are alternately positive and negative is an **alternating series**. Here are three examples:

$$1 - \frac{1}{2} + \frac{1}{3} - \frac{1}{4} + \frac{1}{5} - \cdots + \frac{(-1)^{n+1}}{n} + \cdots \qquad (1)$$

$$-2 + 1 - \frac{1}{2} + \frac{1}{4} - \frac{1}{8} + \cdots + \frac{(-1)^n 4}{2^n} + \cdots \qquad (2)$$

$$1 - 2 + 3 - 4 + 5 - 6 + \cdots + (-1)^{n+1} n + \cdots \qquad (3)$$

Series (1), called the **alternating harmonic series**, converges, as we will see in a moment. Series (2), a geometric series with ratio $r = -1/2$, converges to $-2/[1 + (1/2)] = -4/3$. Series (3) diverges because the nth term does not approach zero.

We prove the convergence of the alternating harmonic series by applying the Alternating Series Test.

THEOREM 14—The Alternating Series Test (Leibniz's Theorem)   The series

$$\sum_{n=1}^{\infty} (-1)^{n+1} u_n = u_1 - u_2 + u_3 - u_4 + \cdots$$

converges if all three of the following conditions are satisfied:

1. The u_n's are all positive.
2. $u_n \geq u_{n+1}$ for all $n \geq N$, for some integer N.
3. $u_n \to 0$.

Proof   Assume $N = 1$. If n is an even integer, say $n = 2m$, then the sum of the first n terms is

$$s_{2m} = (u_1 - u_2) + (u_3 - u_4) + \cdots + (u_{2m-1} - u_{2m})$$
$$= u_1 - (u_2 - u_3) - (u_4 - u_5) - \cdots - (u_{2m-2} - u_{2m-1}) - u_{2m}.$$

The first equality shows that s_{2m} is the sum of m nonnegative terms, since each term in parentheses is positive or zero. Hence $s_{2m+2} \geq s_{2m}$, and the sequence $\{s_{2m}\}$ is nondecreasing. The second equality shows that $s_{2m} \leq u_1$. Since $\{s_{2m}\}$ is nondecreasing and bounded from above, it has a limit, say

$$\lim_{m \to \infty} s_{2m} = L. \tag{4}$$

If n is an odd integer, say $n = 2m + 1$, then the sum of the first n terms is $s_{2m+1} = s_{2m} + u_{2m+1}$. Since $u_n \to 0$,

$$\lim_{m \to \infty} u_{2m+1} = 0$$

and, as $m \to \infty$,

$$s_{2m+1} = s_{2m} + u_{2m+1} \to L + 0 = L. \tag{5}$$

Combining the results of Equations (4) and (5) gives $\lim_{n \to \infty} s_n = L$ (Section 8.1, Exercise 119). ∎

EXAMPLE 1   The alternating harmonic series

$$\sum_{n=1}^{\infty} (-1)^{n+1} \frac{1}{n} = 1 - \frac{1}{2} + \frac{1}{3} - \frac{1}{4} + \cdots$$

satisfies the three requirements of Theorem 14 with $N = 1$; it therefore converges. ∎

FIGURE 8.9 The partial sums of an alternating series that satisfies the hypotheses of Theorem 14 for $N = 1$ straddle the limit from the beginning.

A graphical interpretation of the partial sums (Figure 8.9) shows how an alternating series converges to its limit L when the three conditions of Theorem 14 are satisfied with $N = 1$. Starting from the origin of the x-axis, we lay off the positive distance $s_1 = u_1$. To find the point corresponding to $s_2 = u_1 - u_2$, we back up a distance equal to u_2. Since

$u_2 \leq u_1$, we do not back up any farther than the origin. We continue in this seesaw fashion, backing up or going forward as the signs in the series demand. But for $n \geq N$, each forward or backward step is shorter than (or at most the same size as) the preceding step, because $u_{n+1} \leq u_n$. And since the nth term approaches zero as n increases, the size of step we take forward or backward gets smaller and smaller. We oscillate across the limit L, and the amplitude of oscillation approaches zero. The limit L lies between any two successive sums s_n and s_{n+1} and hence differs from s_n by an amount less than u_{n+1}.

Because

$$|L - s_n| < u_{n+1} \qquad \text{for } n \geq N,$$

we can make useful estimates of the sums of convergent alternating series.

THEOREM 15—The Alternating Series Estimation Theorem   If the alternating series $\sum_{n=1}^{\infty} (-1)^{n+1} u_n$ satisfies the three conditions of Theorem 14, then for $n \geq N$,

$$s_n = u_1 - u_2 + \cdots + (-1)^{n+1} u_n$$

approximates the sum L of the series with an error whose absolute value is less than u_{n+1}, the absolute value of the first unused term. Furthermore, the remainder, $L - s_n$, has the same sign as the first unused term.

We leave the verification of the sign of the remainder for Exercise 53.

EXAMPLE 2   We try Theorem 15 on a series whose sum we know:

$$\sum_{n=0}^{\infty} (-1)^n \frac{1}{2^n} = 1 - \frac{1}{2} + \frac{1}{4} - \frac{1}{8} + \frac{1}{16} - \frac{1}{32} + \frac{1}{64} - \frac{1}{128} \,\Big|\, + \frac{1}{256} - \cdots.$$

The theorem says that if we truncate the series after the eighth term, we throw away a total that is positive and less than $1/256$. The sum of the first eight terms is 0.6640625. The sum of the series is

$$\frac{1}{1 - (-1/2)} = \frac{1}{3/2} = \frac{2}{3}.$$

The difference, $(2/3) - 0.6640625 = 0.0026041666\ldots$, is positive and less than $(1/256) = 0.00390625$.   ∎

Absolute and Conditional Convergence

DEFINITION   A series $\sum a_n$ **converges absolutely** (is **absolutely convergent**) if the corresponding series of absolute values, $\sum |a_n|$, converges.

The geometric series in Example 2 converges absolutely because the corresponding series of absolute values

$$1 + \frac{1}{2} + \frac{1}{4} + \frac{1}{8} + \cdots$$

converges. The alternating harmonic series does not converge absolutely. The corresponding series of absolute values is the (divergent) harmonic series.

DEFINITION   A series that converges but does not converge absolutely **converges conditionally**.

The alternating harmonic series converges conditionally.

Absolute convergence is important for two reasons. First, we have good tests for convergence of series of positive terms. Second, if a series converges absolutely, then it converges, as we now prove.

THEOREM 16—The Absolute Convergence Test   If $\displaystyle\sum_{n=1}^{\infty} |a_n|$ converges, then $\displaystyle\sum_{n=1}^{\infty} a_n$ converges.

Proof   For each n,

$$-|a_n| \le a_n \le |a_n|, \qquad \text{so} \qquad 0 \le a_n + |a_n| \le 2|a_n|.$$

If $\sum_{n=1}^{\infty} |a_n|$ converges, then $\sum_{n=1}^{\infty} 2|a_n|$ converges and, by the Direct Comparison Test, the nonnegative series $\sum_{n=1}^{\infty} (a_n + |a_n|)$ converges. The equality $a_n = (a_n + |a_n|) - |a_n|$ now lets us express $\sum_{n=1}^{\infty} a_n$ as the difference of two convergent series:

$$\sum_{n=1}^{\infty} a_n = \sum_{n=1}^{\infty} (a_n + |a_n| - |a_n|) = \sum_{n=1}^{\infty} (a_n + |a_n|) - \sum_{n=1}^{\infty} |a_n|.$$

Therefore, $\sum_{n=1}^{\infty} a_n$ converges.   ∎

Caution   We can rephrase Theorem 16 to say that every absolutely convergent series converges. However, the converse statement is false: Many convergent series do not converge absolutely (such as the alternating harmonic series in Example 1).

EXAMPLE 3

(a) For $\displaystyle\sum_{n=1}^{\infty} (-1)^{n+1} \frac{1}{n^2} = 1 - \frac{1}{4} + \frac{1}{9} - \frac{1}{16} + \cdots$, the corresponding series of absolute values is the convergent series

$$\sum_{n=1}^{\infty} \frac{1}{n^2} = 1 + \frac{1}{4} + \frac{1}{9} + \frac{1}{16} + \cdots.$$

The original series converges because it converges absolutely.

(b) For $\displaystyle\sum_{n=1}^{\infty} \frac{\sin n}{n^2} = \frac{\sin 1}{1} + \frac{\sin 2}{4} + \frac{\sin 3}{9} + \cdots$, the corresponding series of absolute values is

$$\sum_{n=1}^{\infty} \left| \frac{\sin n}{n^2} \right| = \frac{|\sin 1|}{1} + \frac{|\sin 2|}{4} + \cdots,$$

which converges by comparison with $\sum_{n=1}^{\infty} (1/n^2)$ because $|\sin n| \leq 1$ for every n. The original series converges absolutely; therefore it converges. ■

EXAMPLE 4 If p is a positive constant, the sequence $\{1/n^p\}$ is a decreasing sequence with limit zero. Therefore the alternating p-series

$$\sum_{n=1}^{\infty} \frac{(-1)^{n-1}}{n^p} = 1 - \frac{1}{2^p} + \frac{1}{3^p} - \frac{1}{4^p} + \cdots, \qquad p > 0$$

converges.

If $p > 1$, the series converges absolutely. If $0 < p \leq 1$, the series converges conditionally.

Conditional convergence: $\qquad 1 - \dfrac{1}{\sqrt{2}} + \dfrac{1}{\sqrt{3}} - \dfrac{1}{\sqrt{4}} + \cdots$

Absolute convergence: $\qquad 1 - \dfrac{1}{2^{3/2}} + \dfrac{1}{3^{3/2}} - \dfrac{1}{4^{3/2}} + \cdots$ ■

Rearranging Series

We can always rearrange the terms of a *finite* sum. The same result is true for an infinite series that is absolutely convergent (see Exercise 60 for an outline of the proof).

THEOREM 17—The Rearrangement Theorem for Absolutely Convergent Series If $\sum_{n=1}^{\infty} a_n$ converges absolutely, and $b_1, b_2, \ldots, b_n, \ldots$ is any arrangement of the sequence $\{a_n\}$, then $\sum b_n$ converges absolutely and

$$\sum_{n=1}^{\infty} b_n = \sum_{n=1}^{\infty} a_n.$$

If we rearrange the terms of a conditionally convergent series, we get different results. In fact, it can be proved that for any real number r, a given conditionally convergent series can be rearranged so its sum is equal to r. (We omit the proof of this fact.) Therefore we must always add the terms of a conditionally convergent series in the order given.

We have now developed several tests for convergence and divergence of series. In summary:

---

1. **The nth-Term Test:** Unless $a_n \to 0$, the series diverges.

2. **Geometric series:** $\sum ar^n$ converges if $|r| < 1$; otherwise it diverges.

3. **p-series:** $\sum 1/n^p$ converges if $p > 1$; otherwise it diverges.

4. **Series with nonnegative terms:** Try the Integral Test, Ratio Test, or Root Test. Try comparing to a known series with the Comparison Test.

5. **Series with some negative terms:** Does $\sum |a_n|$ converge? If yes, so does $\sum a_n$, since absolute convergence implies convergence.

6. **Alternating series:** $\sum a_n$ converges if the series satisfies the conditions of the Alternating Series Test.

EXERCISES 8.6

Which of the alternating series in Exercises 1–10 converge, and which diverge? Give reasons for your answers.

1. $\displaystyle\sum_{n=1}^{\infty}(-1)^{n+1}\frac{1}{n^2}$

2. $\displaystyle\sum_{n=1}^{\infty}(-1)^{n+1}\frac{1}{n^{3/2}}$

3. $\displaystyle\sum_{n=1}^{\infty}(-1)^{n+1}\left(\frac{n}{10}\right)^n$

4. $\displaystyle\sum_{n=1}^{\infty}(-1)^{n+1}\frac{10^n}{n^{10}}$

5. $\displaystyle\sum_{n=2}^{\infty}(-1)^{n+1}\frac{1}{\ln n}$

6. $\displaystyle\sum_{n=1}^{\infty}(-1)^{n+1}\frac{\ln n}{n}$

7. $\displaystyle\sum_{n=2}^{\infty}(-1)^{n+1}\frac{\ln n}{\ln n^2}$

8. $\displaystyle\sum_{n=1}^{\infty}(-1)^n \ln\left(1+\frac{1}{n}\right)$

9. $\displaystyle\sum_{n=1}^{\infty}(-1)^{n+1}\frac{\sqrt{n}+1}{n+1}$

10. $\displaystyle\sum_{n=1}^{\infty}(-1)^{n+1}\frac{3\sqrt{n}+1}{\sqrt{n}+1}$

Which of the series in Exercises 11–44 converge absolutely, which converge, and which diverge? Give reasons for your answers.

11. $\displaystyle\sum_{n=1}^{\infty}(-1)^{n+1}(0.1)^n$

12. $\displaystyle\sum_{n=1}^{\infty}(-1)^{n+1}\frac{(0.1)^n}{n}$

13. $\displaystyle\sum_{n=1}^{\infty}(-1)^n\frac{1}{\sqrt{n}}$

14. $\displaystyle\sum_{n=1}^{\infty}\frac{(-1)^n}{1+\sqrt{n}}$

15. $\displaystyle\sum_{n=1}^{\infty}(-1)^{n+1}\frac{n}{n^3+1}$

16. $\displaystyle\sum_{n=1}^{\infty}(-1)^{n+1}\frac{n!}{2^n}$

17. $\displaystyle\sum_{n=1}^{\infty}(-1)^n\frac{1}{n+3}$

18. $\displaystyle\sum_{n=1}^{\infty}(-1)^n\frac{\sin n}{n^2}$

19. $\displaystyle\sum_{n=1}^{\infty}(-1)^{n+1}\frac{3+n}{5+n}$

20. $\displaystyle\sum_{n=2}^{\infty}(-1)^n\frac{1}{\ln(n^3)}$

21. $\displaystyle\sum_{n=1}^{\infty}(-1)^{n+1}\frac{1+n}{n^2}$

22. $\displaystyle\sum_{n=1}^{\infty}\frac{(-2)^{n+1}}{n+5^n}$

23. $\displaystyle\sum_{n=1}^{\infty}(-1)^n n^2(2/3)^n$

24. $\displaystyle\sum_{n=1}^{\infty}(-1)^{n+1}\left(\sqrt[n]{10}\right)$

25. $\displaystyle\sum_{n=1}^{\infty}(-1)^n\frac{\tan^{-1}n}{n^2+1}$

26. $\displaystyle\sum_{n=2}^{\infty}(-1)^{n+1}\frac{1}{n\ln n}$

27. $\displaystyle\sum_{n=1}^{\infty}(-1)^n\frac{n}{n+1}$

28. $\displaystyle\sum_{n=1}^{\infty}(-1)^n\frac{\ln n}{n-\ln n}$

29. $\displaystyle\sum_{n=1}^{\infty}\frac{(-100)^n}{n!}$

30. $\displaystyle\sum_{n=1}^{\infty}(-5)^{-n}$

31. $\displaystyle\sum_{n=1}^{\infty}\frac{(-1)^{n-1}}{n^2+2n+1}$

32. $\displaystyle\sum_{n=2}^{\infty}(-1)^n\left(\frac{\ln n}{\ln n^2}\right)^n$

33. $\displaystyle\sum_{n=1}^{\infty}\frac{\cos n\pi}{n\sqrt{n}}$

34. $\displaystyle\sum_{n=1}^{\infty}\frac{\cos n\pi}{n}$

35. $\displaystyle\sum_{n=1}^{\infty}\frac{(-1)^n(n+1)^n}{(2n)^n}$

36. $\displaystyle\sum_{n=1}^{\infty}\frac{(-1)^{n+1}(n!)^2}{(2n)!}$

37. $\displaystyle\sum_{n=1}^{\infty}(-1)^n\frac{(2n)!}{2^n n! n}$

38. $\displaystyle\sum_{n=1}^{\infty}(-1)^n\frac{(n!)^2 3^n}{(2n+1)!}$

39. $\displaystyle\sum_{n=1}^{\infty}(-1)^n\left(\sqrt{n+1}-\sqrt{n}\right)$ **40.** $\displaystyle\sum_{n=1}^{\infty}(-1)^n\left(\sqrt{n^2+n}-n\right)$

41. $\displaystyle\sum_{n=1}^{\infty}(-1)^n\left(\sqrt{n+\sqrt{n}}-\sqrt{n}\right)$

42. $\displaystyle\sum_{n=1}^{\infty}\frac{(-1)^n}{\sqrt{n}+\sqrt{n+1}}$

43. $\displaystyle\sum_{n=1}^{\infty}(-1)^n \text{ sech } n$

44. $\displaystyle\sum_{n=1}^{\infty}(-1)^n \text{ csch } n$

In Exercises 45–48, estimate the magnitude of the error involved in using the sum of the first four terms to approximate the sum of the entire series.

45. $\displaystyle\sum_{n=1}^{\infty}(-1)^{n+1}\frac{1}{n}$ It can be shown that the sum is $\ln 2$.

46. $\displaystyle\sum_{n=1}^{\infty}(-1)^{n+1}\frac{1}{10^n}$

47. $\displaystyle\sum_{n=1}^{\infty}(-1)^{n+1}\frac{(0.01)^n}{n}$ As you will see in Section 8.7, the sum is $\ln(1.01)$.

48. $\displaystyle\frac{1}{1+t}=\sum_{n=0}^{\infty}(-1)^n t^n,\quad 0<t<1$

T Approximate the sums in Exercises 49 and 50 with an error of magnitude less than 5×10^{-6}.

49. $\displaystyle\sum_{n=0}^{\infty}(-1)^n\frac{1}{(2n)!}$ As you will see in Section 8.9, the sum is $\cos 1$, the cosine of 1 radian.

50. $\displaystyle\sum_{n=0}^{\infty}(-1)^n\frac{1}{n!}$ As you will see in Section 8.9, the sum is e^{-1}.

51. a. The series

$$\frac{1}{3}-\frac{1}{2}+\frac{1}{9}-\frac{1}{4}+\frac{1}{27}-\frac{1}{8}+\cdots+\frac{1}{3^n}-\frac{1}{2^n}+\cdots$$

does not meet one of the conditions of Theorem 14. Which one?

b. Find the sum of the series in part (a).

T **52.** The limit L of an alternating series that satisfies the conditions of Theorem 14 lies between the values of any two consecutive partial sums. This suggests using the average

$$\frac{s_n+s_{n+1}}{2}=s_n+\frac{1}{2}(-1)^{n+2}a_{n+1}$$

to estimate L. Compute

$$s_{20} + \frac{1}{2} \cdot \frac{1}{21}$$

as an approximation to the sum of the alternating harmonic series. The exact sum is $\ln 2 = 0.6931\ldots$.

53. The sign of the remainder of an alternating series that satisfies the conditions of Theorem 14 Prove the assertion in Theorem 15 that whenever an alternating series satisfying the conditions of Theorem 14 is approximated with one of its partial sums, then the remainder (sum of the unused terms) has the same sign as the first unused term. (*Hint:* Group the remainder's terms in consecutive pairs.)

54. Show that the sum of the first $2n$ terms of the series

$$1 - \frac{1}{2} + \frac{1}{2} - \frac{1}{3} + \frac{1}{3} - \frac{1}{4} + \frac{1}{4} - \frac{1}{5} + \frac{1}{5} - \frac{1}{6} + \cdots$$

is the same as the sum of the first n terms of the series

$$\frac{1}{1 \cdot 2} + \frac{1}{2 \cdot 3} + \frac{1}{3 \cdot 4} + \frac{1}{4 \cdot 5} + \frac{1}{5 \cdot 6} + \cdots.$$

Do these series converge? What is the sum of the first $2n + 1$ terms of the first series? If the series converge, what is their sum?

55. Show that if $\sum_{n=1}^{\infty} a_n$ diverges, then $\sum_{n=1}^{\infty} |a_n|$ diverges.

56. Show that if $\sum_{n=1}^{\infty} a_n$ converges absolutely, then

$$\left| \sum_{n=1}^{\infty} a_n \right| \leq \sum_{n=1}^{\infty} |a_n|.$$

57. Show that if $\sum_{n=1}^{\infty} a_n$ and $\sum_{n=1}^{\infty} b_n$ both converge absolutely, then so does

a. $\displaystyle\sum_{n=1}^{\infty} (a_n + b_n)$ **b.** $\displaystyle\sum_{n=1}^{\infty} (a_n - b_n)$

c. $\displaystyle\sum_{n=1}^{\infty} ka_n$ (k any number)

58. Show by example that $\sum_{n=1}^{\infty} a_n b_n$ may diverge even if $\sum_{n=1}^{\infty} a_n$ and $\sum_{n=1}^{\infty} b_n$ both converge.

T 59. In the alternating harmonic series, suppose the goal is to arrange the terms to get a new series that converges to $-1/2$. Start the new arrangement with the first negative term, which is $-1/2$. Whenever you have a sum that is less than or equal to $-1/2$, start introducing positive terms, taken in order, until the new total is greater than $-1/2$. Then add negative terms until the total is less than or equal to $-1/2$ again. Continue this process until your partial sums have been above the target at least three times and finish at or below it. If s_n is the sum of the first n terms of your new series, plot the points (n, s_n) to illustrate how the sums are behaving.

60. Outline of the proof of the Rearrangement Theorem (Theorem 17)

a. Let ϵ be a positive real number, let $L = \sum_{n=1}^{\infty} a_n$, and let $s_k = \sum_{n=1}^{k} a_n$. Show that for some index N_1 and for some index $N_2 \geq N_1$,

$$\sum_{n=N_1}^{\infty} |a_n| < \frac{\epsilon}{2} \quad \text{and} \quad |s_{N_2} - L| < \frac{\epsilon}{2}.$$

Since all the terms $a_1, a_2, \ldots, a_{N_2}$ appear somewhere in the sequence $\{b_n\}$, there is an index $N_3 \geq N_2$ such that if $n \geq N_3$, then $\left(\sum_{k=1}^{n} b_k\right) - s_{N_2}$ is at most a sum of terms a_m with $m \geq N_1$. Therefore, if $n \geq N_3$,

$$\left| \sum_{k=1}^{n} b_k - L \right| \leq \left| \sum_{k=1}^{n} b_k - s_{N_2} \right| + |s_{N_2} - L|$$

$$\leq \sum_{k=N_1}^{\infty} |a_k| + |s_{N_2} - L| < \epsilon.$$

b. The argument in part (a) shows that if $\sum_{n=1}^{\infty} a_n$ converges absolutely then $\sum_{n=1}^{\infty} b_n$ converges and $\sum_{n=1}^{\infty} b_n = \sum_{n=1}^{\infty} a_n$. Now show that because $\sum_{n=1}^{\infty} |a_n|$ converges, $\sum_{n=1}^{\infty} |b_n|$ converges to $\sum_{n=1}^{\infty} |a_n|$.

8.7 Power Series

Now that we can test infinite series for convergence, we can study sums that look like "infinite polynomials." We call these sums *power series* because they are defined as infinite series of powers of some variable, in our case x. Like polynomials, power series can be added, subtracted, multiplied, differentiated, and integrated to give new power series.

Power Series and Convergence

We begin with the formal definition.

DEFINITIONS A **power series about $x = 0$** is a series of the form

$$\sum_{n=0}^{\infty} c_n x^n = c_0 + c_1 x + c_2 x^2 + \cdots + c_n x^n + \cdots. \qquad (1)$$

A **power series about $x = a$** is a series of the form

$$\sum_{n=0}^{\infty} c_n (x - a)^n = c_0 + c_1(x - a) + c_2(x - a)^2 + \cdots + c_n(x - a)^n + \cdots \qquad (2)$$

in which the **center** a and the **coefficients** $c_0, c_1, c_2, \ldots, c_n, \ldots$ are constants.

Equation (1) is the special case obtained by taking $a = 0$ in Equation (2).

EXAMPLE 1 Taking all the coefficients to be 1 in Equation (1) gives the geometric power series

$$\sum_{n=0}^{\infty} x^n = 1 + x + x^2 + \cdots + x^n + \cdots.$$

This is the geometric series with first term 1 and ratio x. It converges to $1/(1 - x)$ for $|x| < 1$. We express this fact by writing

$$\frac{1}{1 - x} = 1 + x + x^2 + \cdots + x^n + \cdots, \qquad -1 < x < 1. \qquad (3)$$

■

Up to now, we have used Equation (3) as a formula for the sum of the series on the right. We now change the focus: We think of the partial sums of the series on the right as polynomials $P_n(x)$ that approximate the function on the left. For values of x near zero, we need take only a few terms of the series to get a good approximation. As we move toward $x = 1$, or -1, we must take more terms. Figure 8.10 shows the graphs of $f(x) = 1/(1 - x)$, and the approximating polynomials $y_n = P_n(x)$ for $n = 0, 1, 2,$ and 8. The function $f(x) = 1/(1 - x)$ is not continuous on intervals containing $x = 1$, where it has a vertical asymptote. The approximations do not apply when $x \geq 1$.

EXAMPLE 2 The power series

$$1 - \frac{1}{2}(x - 2) + \frac{1}{4}(x - 2)^2 + \cdots + \left(-\frac{1}{2}\right)^n (x - 2)^n + \cdots \qquad (4)$$

matches Equation (2) with $a = 2$, $c_0 = 1$, $c_1 = -1/2$, $c_2 = 1/4, \ldots, c_n = (-1/2)^n$. This is a geometric series with first term 1 and ratio $r = -\dfrac{x - 2}{2}$. The series converges for

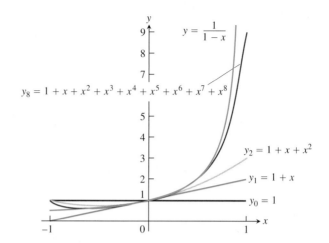

FIGURE 8.10 The graphs of $f(x) = 1/(1 - x)$ and four of its polynomial approximations (Example 1).

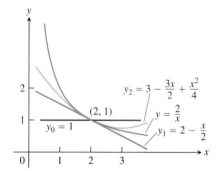

FIGURE 8.11 The graphs of $f(x) = 2/x$ and its first three polynomial approximations (Example 2).

$\left| \dfrac{x - 2}{2} \right| < 1$ or $0 < x < 4$. The sum is

$$\frac{1}{1 - r} = \frac{1}{1 + \dfrac{x - 2}{2}} = \frac{2}{x},$$

so

$$\frac{2}{x} = 1 - \frac{(x - 2)}{2} + \frac{(x - 2)^2}{4} - \cdots + \left(-\frac{1}{2}\right)^n (x - 2)^n + \cdots, \qquad 0 < x < 4.$$

Series (4) generates useful polynomial approximations of $f(x) = 2/x$ for values of x near 2:

$$P_0(x) = 1$$

$$P_1(x) = 1 - \frac{1}{2}(x - 2) = 2 - \frac{x}{2}$$

$$P_2(x) = 1 - \frac{1}{2}(x - 2) + \frac{1}{4}(x - 2)^2 = 3 - \frac{3x}{2} + \frac{x^2}{4},$$

and so on (Figure 8.11). ∎

The following example illustrates how we usually test a power series for convergence and the possible results.

EXAMPLE 3 For what values of x do the following power series converge?

(a) $\displaystyle\sum_{n=1}^{\infty} (-1)^{n-1} \frac{x^n}{n} = x - \frac{x^2}{2} + \frac{x^3}{3} - \cdots$

(b) $\displaystyle\sum_{n=1}^{\infty} (-1)^{n-1} \frac{x^{2n-1}}{2n - 1} = x - \frac{x^3}{3} + \frac{x^5}{5} - \cdots$

(c) $\displaystyle\sum_{n=0}^{\infty} \frac{x^n}{n!} = 1 + x + \frac{x^2}{2!} + \frac{x^3}{3!} + \cdots$

(d) $\displaystyle\sum_{n=0}^{\infty} n! x^n = 1 + x + 2! x^2 + 3! x^3 + \cdots$

Solution Apply the Ratio Test to the series $\sum |u_n|$, where u_n is the nth term of the series in question.

(a) $\left| \dfrac{u_{n+1}}{u_n} \right| = \dfrac{n}{n+1} |x| \rightarrow |x|.$

The series converges absolutely for $|x| < 1$. It diverges if $|x| > 1$ because the nth term does not converge to zero. At $x = 1$, we get the alternating harmonic series $1 - 1/2 + 1/3 - 1/4 + \cdots$, which converges. At $x = -1$ we get $-1 - 1/2 - 1/3 - 1/4 - \cdots$, the negative of the harmonic series; it diverges. Series (a) converges for $-1 < x \le 1$ and diverges elsewhere.

(b) $\left| \dfrac{u_{n+1}}{u_n} \right| = \dfrac{2n-1}{2n+1} x^2 \rightarrow x^2.$

The series converges absolutely for $x^2 < 1$. It diverges for $x^2 > 1$ because the nth term does not converge to zero. At $x = 1$ the series becomes $1 - 1/3 + 1/5 - 1/7 + \cdots$, which converges by the Alternating Series Theorem. It also converges at $x = -1$ because it is again an alternating series that satisfies the conditions for convergence. The value at $x = -1$ is the negative of the value at $x = 1$. Series (b) converges for $-1 \le x \le 1$ and diverges elsewhere.

(c) $\left| \dfrac{u_{n+1}}{u_n} \right| = \left| \dfrac{x^{n+1}}{(n+1)!} \cdot \dfrac{n!}{x^n} \right| = \dfrac{|x|}{n+1} \rightarrow 0$ for every x.

The series converges absolutely for all x.

(d) $\left| \dfrac{u_{n+1}}{u_n} \right| = \left| \dfrac{(n+1)! x^{n+1}}{n! x^n} \right| = (n+1)|x| \rightarrow \infty$ unless $x = 0$.

The series diverges for all values of x except $x = 0$.

The previous example illustrated how a power series might converge. The next result shows that if a power series converges at more than one value, then it converges over an entire interval of values.

> **THEOREM 18—The Convergence Theorem for Power Series**   If the power series
> $$\sum_{n=0}^{\infty} a_n x^n = a_0 + a_1 x + a_2 x^2 + \cdots$$ converges for $x = c \neq 0$, then it converges
> absolutely for all x with $|x| < |c|$. If the series diverges for $x = d$, then it
> diverges for all x with $|x| > |d|$.

Proof   Suppose the series $\sum_{n=0}^{\infty} a_n c^n$ converges. Then $\lim_{n\to\infty} a_n c^n = 0$. Hence, there is
an integer N such that $|a_n c^n| < 1$ for all $n \geq N$. That is,

$$|a_n| < \frac{1}{|c|^n} \qquad \text{for } n \geq N. \tag{5}$$

Now take any x such that $|x| < |c|$ and consider

$$|a_0| + |a_1 x| + \cdots + |a_{N-1} x^{N-1}| + |a_N x^N| + |a_{N+1} x^{N+1}| + \cdots .$$

There are only a finite number of terms prior to $|a_N x^N|$, and their sum is finite. Starting
with $|a_N x^N|$ and beyond, the terms are less than

$$\left|\frac{x}{c}\right|^N + \left|\frac{x}{c}\right|^{N+1} + \left|\frac{x}{c}\right|^{N+2} + \cdots \tag{6}$$

because of Inequality (5). But Series (6) is a geometric series with ratio $r = |x/c|$, which
is less than 1, since $|x| < |c|$. Hence Series (6) converges, so the original series converges
absolutely. This proves the first half of the theorem.

The second half of the theorem follows from the first. If the series diverges at $x = d$
and converges at a value x_0 with $|x_0| > |d|$, we may take $c = x_0$ in the first half of the the-
orem and conclude that the series converges absolutely at d. But the series cannot converge
absolutely and diverge at one and the same time. Hence, if it diverges at d, it diverges for
all x with $|x| > |d|$. ∎

To simplify the notation, Theorem 18 deals with the convergence of series of the form
$\sum a_n x^n$. For series of the form $\sum a_n(x - a)^n$ we can replace $x - a$ by x' and apply the results to the series $\sum a_n(x')^n$.

The Radius of Convergence of a Power Series

The theorem we have just proved and the examples we have studied lead to the conclusion
that a power series $\sum c_n(x - a)^n$ behaves in one of three possible ways. It might converge
only at $x = a$, or converge everywhere, or converge on some interval of radius R centered
at $x = a$. We prove this as a Corollary to Theorem 18.

COROLLARY TO THEOREM 18 The convergence of the series $\sum c_n(x - a)^n$ is described by one of the following three possibilities:

1. There is a positive number R such that the series diverges for x with $|x - a| > R$ but converges absolutely for x with $|x - a| < R$. The series may or may not converge at either of the endpoints $x = a - R$ and $x = a + R$.

2. The series converges absolutely for every x ($R = \infty$).

3. The series converges at $x = a$ and diverges elsewhere ($R = 0$).

Proof We assume first that $a = 0$, so that the power series is centered at 0. If the series converges everywhere we are in Case 2. If it converges only at $x = 0$ we are in Case 3. Otherwise there is a nonzero number d such that $\sum c_n d^n$ diverges. The set S of values of x for which the series $\sum c_n x^n$ converges is nonempty because it contains 0 and a positive number p as well. By Theorem 18, the series diverges for all x with $|x| > |d|$, so $|x| \leq |d|$ for all $x \in S$, and S is a bounded set. By the Completeness Property of the real numbers (see Appendix 7) a nonempty, bounded set has a least upper bound R. (The least upper bound is the smallest number with the property that the elements $x \in S$ satisfy $x \leq R$.) If $|x| > R \geq p$, then $x \notin S$ so the series $\sum c_n x^n$ diverges. If $|x| < R$, then $|x|$ is not an upper bound for S (because it's smaller than the least upper bound) so there is a number $b \in S$ such that $b > |x|$. Since $b \in S$, the series $\sum c_n b^n$ converges and therefore the series $\sum c_n |x|^n$ converges by Theorem 18. This proves the Corollary for power series centered at $a = 0$.

For a power series centered at $a \neq 0$, we set $x' = (x - a)$ and repeat the argument with x'. Since $x' = 0$ when $x = a$, a radius R interval of convergence for $\sum c_n(x')^n$ centered at $x' = 0$ is the same as a radius R interval of convergence for $\sum c_n(x - a)^n$ centered at $x = a$. This establishes the Corollary for the general case. ∎

R is called the **radius of convergence** of the power series and the interval of radius R centered at $x = a$ is called the **interval of convergence**. The interval of convergence may be open, closed, or half-open, depending on the particular series. At points x with $|x - a| < R$, the series converges absolutely. If the series converges for all values of x, we say its radius of convergence is infinite. If it converges only at $x = a$, we say its radius of convergence is zero.

How to Test a Power Series for Convergence

1. *Use the Ratio Test (or nth-Root Test) to find the interval where the series converges absolutely*. Ordinarily, this is an open interval

$$|x - a| < R \qquad \text{or} \qquad a - R < x < a + R.$$

2. *If the interval of absolute convergence is finite, test for convergence or divergence at each endpoint*, as in Examples 3a and b. Use a Comparison Test, the Integral Test, or the Alternating Series Test.

3. *If the interval of absolute convergence is $a - R < x < a + R$, the series diverges for $|x - a| > R$* (it does not even converge conditionally), because the nth term does not approach zero for those values of x.

Term-by-Term Differentiation

A theorem from advanced calculus says that a power series can be differentiated term by term at each interior point of its interval of convergence.

THEOREM 19—The Term-by-Term Differentiation Theorem If $\sum c_n(x - a)^n$ converges for $a - R < x < a + R$ for some $R > 0$, it defines a function f:

$$f(x) = \sum_{n=0}^{\infty} c_n(x - a)^n, \qquad a - R < x < a + R.$$

Such a function f has derivatives of all orders inside the interval of convergence. We can obtain the derivatives by differentiating the original series term by term:

$$f'(x) = \sum_{n=1}^{\infty} nc_n(x - a)^{n-1}$$

$$f''(x) = \sum_{n=2}^{\infty} n(n - 1)c_n(x - a)^{n-2},$$

and so on. Each of these derived series converges at every interior point of the interval of convergence of the original series.

EXAMPLE 4 Find series for $f'(x)$ and $f''(x)$ if

$$f(x) = \frac{1}{1 - x} = 1 + x + x^2 + x^3 + x^4 + \cdots + x^n + \cdots$$

$$= \sum_{n=0}^{\infty} x^n, \qquad -1 < x < 1$$

Solution

$$f'(x) = \frac{1}{(1 - x)^2} = 1 + 2x + 3x^2 + 4x^3 + \cdots + nx^{n-1} + \cdots$$

$$= \sum_{n=1}^{\infty} nx^{n-1}, \qquad -1 < x < 1$$

$$f''(x) = \frac{2}{(1 - x)^3} = 2 + 6x + 12x^2 + \cdots + n(n - 1)x^{n-2} + \cdots$$

$$= \sum_{n=2}^{\infty} n(n - 1)x^{n-2}, \qquad -1 < x < 1 \qquad \blacksquare$$

Caution   Term-by-term differentiation might not work for other kinds of series. For example, the trigonometric series

$$\sum_{n=1}^{\infty} \frac{\sin (n!x)}{n^2}$$

converges for all x. But if we differentiate term by term we get the series

$$\sum_{n=1}^{\infty} \frac{n!\cos(n!x)}{n^2},$$

which diverges for all x. This is not a power series, since it is not a sum of positive integer powers of x.

Term-by-Term Integration

It is also true that a power series can be integrated term by term throughout its interval of convergence. This result is proved in a more advanced course.

THEOREM 20—The Term-by-Term Integration Theorem   Suppose that

$$f(x) = \sum_{n=0}^{\infty} c_n (x - a)^n$$

converges for $a - R < x < a + R$ $(R > 0)$. Then

$$\sum_{n=0}^{\infty} c_n \frac{(x - a)^{n+1}}{n + 1}$$

converges for $a - R < x < a + R$ and

$$\int f(x)\, dx = \sum_{n=0}^{\infty} c_n \frac{(x - a)^{n+1}}{n + 1} + C$$

for $a - R < x < a + R$.

EXAMPLE 5   Identify the function

$$f(x) = x - \frac{x^3}{3} + \frac{x^5}{5} - \cdots, \qquad -1 \le x \le 1.$$

Solution   We differentiate the original series term by term and get

$$f'(x) = 1 - x^2 + x^4 - x^6 + \cdots, \qquad -1 < x < 1.$$

This is a geometric series with first term 1 and ratio $-x^2$, so

$$f'(x) = \frac{1}{1 - (-x^2)} = \frac{1}{1 + x^2}.$$

We can now integrate $f'(x) = 1/(1 + x^2)$ to get

$$\int f'(x)\, dx = \int \frac{dx}{1 + x^2} = \tan^{-1} x + C.$$

The series for $f(x)$ is zero when $x = 0$, so $C = 0$. Hence

$$f(x) = x - \frac{x^3}{3} + \frac{x^5}{5} - \frac{x^7}{7} + \cdots = \tan^{-1} x, \qquad -1 < x < 1. \tag{7}$$

It can be shown that the series also converges to $\tan^{-1} x$ at $x = \pm 1$, but we omit the proof. ∎

Notice that the original series in Example 5 converges at both endpoints of the original interval of convergence, but Theorem 20 can guarantee the convergence of the differentiated series only inside the interval.

EXAMPLE 6 The series

$$\frac{1}{1+t} = 1 - t + t^2 - t^3 + \cdots$$

converges on the open interval $-1 < t < 1$. Therefore,

$$\ln(1+x) = \int_0^x \frac{1}{1+t}\,dt = t - \frac{t^2}{2} + \frac{t^3}{3} - \frac{t^4}{4} + \cdots \Big]_0^x \qquad \text{Theorem 20}$$

$$= x - \frac{x^2}{2} + \frac{x^3}{3} - \frac{x^4}{4} + \cdots, \qquad -1 < x < 1.$$

It can also be shown that the series converges at $x = 1$ to the number $\ln 2$, but that was not guaranteed by the theorem. ∎

Multiplication of Power Series

Another theorem states that absolutely converging power series can be multiplied the way we multiply polynomials. We again omit the proof.

> **THEOREM 21—The Series Multiplication Theorem for Power Series** If $A(x) = \sum_{n=0}^{\infty} a_n x^n$ and $B(x) = \sum_{n=0}^{\infty} b_n x^n$ converge absolutely for $|x| < R$, and
>
> $$c_n = a_0 b_n + a_1 b_{n-1} + a_2 b_{n-2} + \cdots + a_{n-1}b_1 + a_n b_0 = \sum_{k=0}^{n} a_k b_{n-k},$$
>
> then $\sum_{n=0}^{\infty} c_n x^n$ converges absolutely to $A(x)B(x)$ for $|x| < R$:
>
> $$\left(\sum_{n=0}^{\infty} a_n x^n\right) \cdot \left(\sum_{n=0}^{\infty} b_n x^n\right) = \sum_{n=0}^{\infty} c_n x^n.$$

Finding the general coefficient c_n in the product of two given power series can be very tedious. In practice, we usually find just the first few terms instead. For instance, from Examples 1 and 6, we have

$$\frac{1}{1-x}\ln(1+x) = \left(1 + x + x^2 + \cdots\right)\left(x - \frac{x^2}{2} + \frac{x^3}{3} - \cdots\right)$$

$$= x + \frac{x^2}{2} + \frac{5x^3}{6} - \frac{x^4}{6}\cdots.$$

EXERCISES 8.7

In Exercises 1–32, **(a)** find the series' radius and interval of convergence. For what values of x does the series converge **(b)** absolutely, **(c)** conditionally?

1. $\sum_{n=0}^{\infty} x^n$

2. $\sum_{n=0}^{\infty} (x + 5)^n$

3. $\sum_{n=0}^{\infty} (-1)^n (4x + 1)^n$

4. $\sum_{n=1}^{\infty} \dfrac{(3x - 2)^n}{n}$

5. $\sum_{n=0}^{\infty} \dfrac{(x - 2)^n}{10^n}$

6. $\sum_{n=0}^{\infty} (2x)^n$

7. $\sum_{n=0}^{\infty} \dfrac{nx^n}{n + 2}$

8. $\sum_{n=1}^{\infty} \dfrac{(-1)^n (x + 2)^n}{n}$

9. $\sum_{n=1}^{\infty} \dfrac{x^n}{n\sqrt{n}\, 3^n}$

10. $\sum_{n=1}^{\infty} \dfrac{(x - 1)^n}{\sqrt{n}}$

11. $\sum_{n=0}^{\infty} \dfrac{(-1)^n x^n}{n!}$

12. $\sum_{n=0}^{\infty} \dfrac{3^n x^n}{n!}$

13. $\sum_{n=0}^{\infty} \dfrac{x^{2n+1}}{n!}$

14. $\sum_{n=0}^{\infty} \dfrac{(2x + 3)^{2n+1}}{n!}$

15. $\sum_{n=0}^{\infty} \dfrac{x^n}{\sqrt{n^2 + 3}}$

16. $\sum_{n=0}^{\infty} \dfrac{(-1)^n x^n}{\sqrt{n^2 + 3}}$

17. $\sum_{n=0}^{\infty} \dfrac{n(x + 3)^n}{5^n}$

18. $\sum_{n=0}^{\infty} \dfrac{nx^n}{4^n(n^2 + 1)}$

19. $\sum_{n=0}^{\infty} \dfrac{\sqrt{n}\, x^n}{3^n}$

20. $\sum_{n=1}^{\infty} \sqrt[n]{n}\,(2x + 5)^n$

21. $\sum_{n=1}^{\infty} \left(1 + \dfrac{1}{n}\right)^n x^n$

22. $\sum_{n=1}^{\infty} (\ln n) x^n$

23. $\sum_{n=1}^{\infty} n^n x^n$

24. $\sum_{n=0}^{\infty} n!(x - 4)^n$

25. $\sum_{n=1}^{\infty} \dfrac{(-1)^{n+1}(x + 2)^n}{n 2^n}$

26. $\sum_{n=0}^{\infty} (-2)^n (n + 1)(x - 1)^n$

27. $\sum_{n=2}^{\infty} \dfrac{x^n}{n(\ln n)^2}$ Get the information you need about $\sum 1/(n(\ln n)^2)$ from Section 8.3, Exercise 39.

28. $\sum_{n=2}^{\infty} \dfrac{x^n}{n \ln n}$ Get the information you need about $\sum 1/(n \ln n)$ from Section 8.3, Exercise 38.

29. $\sum_{n=1}^{\infty} \dfrac{(4x - 5)^{2n+1}}{n^{3/2}}$

30. $\sum_{n=1}^{\infty} \dfrac{(3x + 1)^{n+1}}{2n + 2}$

31. $\sum_{n=1}^{\infty} \dfrac{(x + \pi)^n}{\sqrt{n}}$

32. $\sum_{n=0}^{\infty} \dfrac{\left(x - \sqrt{2}\right)^{2n+1}}{2^n}$

In Exercises 33–38, find the series' interval of convergence and, within this interval, the sum of the series as a function of x.

33. $\sum_{n=0}^{\infty} \dfrac{(x - 1)^{2n}}{4^n}$

34. $\sum_{n=0}^{\infty} \dfrac{(x + 1)^{2n}}{9^n}$

35. $\sum_{n=0}^{\infty} \left(\dfrac{\sqrt{x}}{2} - 1\right)^n$

36. $\sum_{n=0}^{\infty} (\ln x)^n$

37. $\sum_{n=0}^{\infty} \left(\dfrac{x^2 + 1}{3}\right)^n$

38. $\sum_{n=0}^{\infty} \left(\dfrac{x^2 - 1}{2}\right)^n$

39. For what values of x does the series

$$1 - \frac{1}{2}(x - 3) + \frac{1}{4}(x - 3)^2 + \cdots + \left(-\frac{1}{2}\right)^n (x - 3)^n + \cdots$$

converge? What is its sum? What series do you get if you differentiate the given series term by term? For what values of x does the new series converge? What is its sum?

40. If you integrate the series in Exercise 39 term by term, what new series do you get? For what values of x does the new series converge, and what is another name for its sum?

41. The series

$$\sin x = x - \frac{x^3}{3!} + \frac{x^5}{5!} - \frac{x^7}{7!} + \frac{x^9}{9!} - \frac{x^{11}}{11!} + \cdots$$

converges to $\sin x$ for all x.

a. Find the first six terms of a series for $\cos x$. For what values of x should the series converge?

b. By replacing x by $2x$ in the series for $\sin x$, find a series that converges to $\sin 2x$ for all x.

c. Using the result in part (a) and series multiplication, calculate the first six terms of a series for $2 \sin x \cos x$. Compare your answer with the answer in part (b).

42. The series

$$e^x = 1 + x + \frac{x^2}{2!} + \frac{x^3}{3!} + \frac{x^4}{4!} + \frac{x^5}{5!} + \cdots$$

converges to e^x for all x.

a. Find a series for $(d/dx)e^x$. Do you get the series for e^x? Explain your answer.

b. Find a series for $\int e^x \, dx$. Do you get the series for e^x? Explain your answer.

c. Replace x by $-x$ in the series for e^x to find a series that converges to e^{-x} for all x. Then multiply the series for e^x and e^{-x} to find the first six terms of a series for $e^{-x} \cdot e^x$.

43. The series

$$\tan x = x + \frac{x^3}{3} + \frac{2x^5}{15} + \frac{17x^7}{315} + \frac{62x^9}{2835} + \cdots$$

converges to $\tan x$ for $-\pi/2 < x < \pi/2$.

a. Find the first five terms of the series for $\ln|\sec x|$. For what values of x should the series converge?

b. Find the first five terms of the series for $\sec^2 x$. For what values of x should this series converge?

c. Check your result in part (b) by squaring the series given for $\sec x$ in Exercise 44.

44. The series

$$\sec x = 1 + \frac{x^2}{2} + \frac{5}{24}x^4 + \frac{61}{720}x^6 + \frac{277}{8064}x^8 + \cdots$$

converges to $\sec x$ for $-\pi/2 < x < \pi/2$.

a. Find the first five terms of a power series for the function $\ln|\sec x + \tan x|$. For what values of x should the series converge?

b. Find the first four terms of a series for $\sec x \tan x$. For what values of x should the series converge?

c. Check your result in part (b) by multiplying the series for $\sec x$ by the series given for $\tan x$ in Exercise 43.

45. Uniqueness of convergent power series

a. Show that if two power series $\sum_{n=0}^{\infty} a_n x^n$ and $\sum_{n=0}^{\infty} b_n x^n$ are convergent and equal for all values of x in an open interval $(-c, c)$, then $a_n = b_n$ for every n. (*Hint:* Let $f(x) = \sum_{n=0}^{\infty} a_n x^n = \sum_{n=0}^{\infty} b_n x^n$. Differentiate term by term to show that a_n and b_n both equal $f^{(n)}(0)/(n!)$.

b. Show that if $\sum_{n=0}^{\infty} a_n x^n = 0$ for all x in an open interval $(-c, c)$, then $a_n = 0$ for every n.

46. The sum of the series $\sum_{n=0}^{\infty} (n^2/2^n)$ To find the sum of this series, express $1/(1 - x)$ as a geometric series, differentiate both sides of the resulting equation with respect to x, multiply both sides of the result by x, differentiate again, multiply by x again, and set x equal to 1/2. What do you get? (*Source:* David E. Dobbs' letter to the editor, *Illinois Mathematics Teacher*, Vol. 33, Issue 4, 1982, p. 27.)

47. Convergence at endpoints Show by examples that the convergence of a power series at an endpoint of its interval of convergence may be either conditional or absolute.

48. Make up a power series whose interval of convergence is

a. $(-3, 3)$ **b.** $(-2, 0)$ **c.** $(1, 5)$.

8.8 Taylor and Maclaurin Series

This section shows how functions that are infinitely differentiable generate power series called Taylor series. In many cases, these series can provide useful polynomial approximations of the generating functions.

Series Representations

We know from Theorem 19 that within its interval of convergence the sum of a power series is a continuous function with derivatives of all orders. But what about the other way around? If a function $f(x)$ has derivatives of all orders on an interval I, can it be expressed as a power series on I? And if it can, what will its coefficients be?

We can answer the last question readily if we assume that $f(x)$ is the sum of a power series

$$f(x) = \sum_{n=0}^{\infty} a_n(x - a)^n$$
$$= a_0 + a_1(x - a) + a_2(x - a)^2 + \cdots + a_n(x - a)^n + \cdots$$

with a positive radius of convergence. By repeated term-by-term differentiation within the interval of convergence I we obtain

$$f'(x) = a_1 + 2a_2(x - a) + 3a_3(x - a)^2 + \cdots + na_n(x - a)^{n-1} + \cdots$$

$$f''(x) = 1 \cdot 2a_2 + 2 \cdot 3a_3(x - a) + 3 \cdot 4a_4(x - a)^2 + \cdots$$

$$f'''(x) = 1 \cdot 2 \cdot 3a_3 + 2 \cdot 3 \cdot 4a_4(x - a) + 3 \cdot 4 \cdot 5a_5(x - a)^2 + \cdots,$$

with the nth derivative, for all n, being

$$f^{(n)}(x) = n!a_n + \text{a sum of terms with } (x - a) \text{ as a factor}.$$

Since these equations all hold at $x = a$, we have

$$f'(a) = a_1,$$
$$f''(a) = 1 \cdot 2a_2,$$
$$f'''(a) = 1 \cdot 2 \cdot 3a_3,$$

and, in general,

$$f^{(n)}(a) = n!a_n.$$

These formulas reveal a pattern in the coefficients of any power series $\sum_{n=0}^{\infty} a_n(x - a)^n$ that converges to the values of f on I ("represents f on I"). If there *is* such a series (still an open question), then there is only one such series and its nth coefficient is

$$a_n = \frac{f^{(n)}(a)}{n!}.$$

If f has a series representation, then the series must be

$$f(x) = f(a) + f'(a)(x - a) + \frac{f''(a)}{2!}(x - a)^2$$

$$+ \cdots + \frac{f^{(n)}(a)}{n!}(x - a)^n + \cdots. \tag{1}$$

But if we start with an arbitrary function f that is infinitely differentiable on an interval I centered at $x = a$ and use it to generate the series in Equation (1), will the series then converge to $f(x)$ at each x in the interior of I? The answer is maybe—for some functions it will but for other functions it will not, as we will see.

Taylor and Maclaurin Series

HISTORICAL BIOGRAPHIES

Brook Taylor
(1685–1731)

Colin Maclaurin
(1698–1746)

> **DEFINITIONS**   Let f be a function with derivatives of all orders throughout some interval containing a as an interior point. Then the **Taylor series generated by f at $x = a$** is
>
> $$\sum_{k=0}^{\infty} \frac{f^{(k)}(a)}{k!}(x - a)^k = f(a) + f'(a)(x - a) + \frac{f''(a)}{2!}(x - a)^2$$
>
> $$+ \cdots + \frac{f^{(n)}(a)}{n!}(x - a)^n + \cdots.$$
>
> The **Maclaurin series generated by f is**
>
> $$\sum_{k=0}^{\infty} \frac{f^{(k)}(0)}{k!}x^k = f(0) + f'(0)x + \frac{f''(0)}{2!}x^2 + \cdots + \frac{f^{(n)}(0)}{n!}x^n + \cdots,$$
>
> the Taylor series generated by f at $x = 0$.

The Maclaurin series generated by f is often just called the Taylor series of f.

EXAMPLE 1   Find the Taylor series generated by $f(x) = 1/x$ at $a = 2$. Where, if anywhere, does the series converge to $1/x$?

Solution   We need to find $f(2), f'(2), f''(2), \dots$. Taking derivatives we get

$$f(x) = x^{-1}, \qquad\qquad f(2) = 2^{-1} = \frac{1}{2},$$

$$f'(x) = -x^{-2}, \qquad\qquad f'(2) = -\frac{1}{2^2},$$

$$f''(x) = 2!x^{-3}, \qquad\qquad \frac{f''(2)}{2!} = 2^{-3} = \frac{1}{2^3},$$

$$f'''(x) = -3!x^{-4}, \qquad\qquad \frac{f'''(2)}{3!} = -\frac{1}{2^4},$$

$$\vdots \qquad\qquad\qquad \vdots$$

$$f^{(n)}(x) = (-1)^n n! x^{-(n+1)}, \qquad \frac{f^{(n)}(2)}{n!} = \frac{(-1)^n}{2^{n+1}}.$$

The Taylor series is

$$f(2) + f'(2)(x - 2) + \frac{f''(2)}{2!}(x - 2)^2 + \cdots + \frac{f^{(n)}(2)}{n!}(x - 2)^n + \cdots$$

$$= \frac{1}{2} - \frac{(x - 2)}{2^2} + \frac{(x - 2)^2}{2^3} - \cdots + (-1)^n \frac{(x - 2)^n}{2^{n+1}} + \cdots.$$

This is a geometric series with first term $1/2$ and ratio $r = -(x - 2)/2$. It converges absolutely for $|x - 2| < 2$ and its sum is

$$\frac{1/2}{1 + (x - 2)/2} = \frac{1}{2 + (x - 2)} = \frac{1}{x}.$$

In this example the Taylor series generated by $f(x) = 1/x$ at $a = 2$ converges to $1/x$ for $|x - 2| < 2$ or $0 < x < 4$.   ∎

Taylor Polynomials

The linearization of a differentiable function f at a point a is the polynomial of degree one given by

$$P_1(x) = f(a) + f'(a)(x - a).$$

In Section 3.10 we used this linearization to approximate $f(x)$ at values of x near a. If f has derivatives of higher order at a, then it has higher-order polynomial approximations as well, one for each available derivative. These polynomials are called the Taylor polynomials of f.

DEFINITION   Let f be a function with derivatives of order k for $k = 1, 2, \ldots, N$ in some interval containing a as an interior point. Then for any integer n from 0 through N, the **Taylor polynomial of order n** generated by f at $x = a$ is the polynomial

$$P_n(x) = f(a) + f'(a)(x - a) + \frac{f''(a)}{2!}(x - a)^2 + \cdots$$

$$+ \frac{f^{(k)}(a)}{k!}(x - a)^k + \cdots + \frac{f^{(n)}(a)}{n!}(x - a)^n.$$

We speak of a Taylor polynomial of *order n* rather than *degree n* because $f^{(n)}(a)$ may be zero. The first two Taylor polynomials of $f(x) = \cos x$ at $x = 0$, for example, are $P_0(x) = 1$ and $P_1(x) = 1$. The first-order Taylor polynomial has degree zero, not one.

Just as the linearization of f at $x = a$ provides the best linear approximation of f in the neighborhood of a, the higher-order Taylor polynomials provide the "best" polynomial approximations of their respective degrees. (See Exercise 32.)

EXAMPLE 2   Find the Taylor series and the Taylor polynomials generated by $f(x) = e^x$ at $x = 0$.

Solution   Since

$$f(x) = e^x, \qquad f'(x) = e^x, \qquad \ldots, \qquad f^{(n)}(x) = e^x, \qquad \ldots,$$

we have

$$f(0) = e^0 = 1, \qquad f'(0) = 1, \qquad \ldots, \qquad f^{(n)}(0) = 1, \qquad \ldots.$$

The Taylor series generated by f at $x = 0$ is

$$f(0) + f'(0)x + \frac{f''(0)}{2!}x^2 + \cdots + \frac{f^{(n)}(0)}{n!}x^n + \cdots$$

$$= 1 + x + \frac{x^2}{2} + \cdots + \frac{x^n}{n!} + \cdots$$

$$= \sum_{k=0}^{\infty} \frac{x^k}{k!}.$$

This is also the Maclaurin series for e^x. In Section 8.9 we will see that the series converges to e^x at every x.

The Taylor polynomial of order n at $x = 0$ is

$$P_n(x) = 1 + x + \frac{x^2}{2} + \cdots + \frac{x^n}{n!}.$$

See Figure 8.12.   ∎

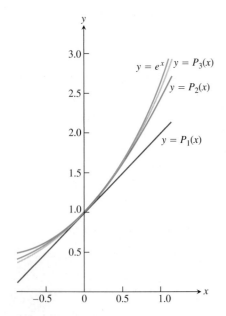

FIGURE 8.12   The graph of $f(x) = e^x$ and its Taylor polynomials

$P_1(x) = 1 + x$

$P_2(x) = 1 + x + (x^2/2!)$

$P_3(x) = 1 + x + (x^2/2!) + (x^3/3!)$.

Notice the very close agreement near the center $x = 0$ (Example 2).

EXAMPLE 3   Find the Taylor series and Taylor polynomials generated by $f(x) = \cos x$ at $x = 0$.

Solution   The cosine and its derivatives are

$$f(x) = \cos x, \qquad f'(x) = -\sin x,$$
$$f''(x) = -\cos x, \qquad f^{(3)}(x) = \sin x,$$
$$\vdots \qquad\qquad\qquad \vdots$$
$$f^{(2n)}(x) = (-1)^n \cos x, \qquad f^{(2n+1)}(x) = (-1)^{n+1} \sin x.$$

At $x = 0$, the cosines are 1 and the sines are 0, so

$$f^{(2n)}(0) = (-1)^n, \qquad f^{(2n+1)}(0) = 0.$$

The Taylor series generated by f at 0 is

$$f(0) + f'(0)x + \frac{f''(0)}{2!}x^2 + \frac{f'''(0)}{3!}x^3 + \cdots + \frac{f^{(n)}(0)}{n!}x^n + \cdots$$

$$= 1 + 0 \cdot x - \frac{x^2}{2!} + 0 \cdot x^3 + \frac{x^4}{4!} + \cdots + (-1)^n \frac{x^{2n}}{(2n)!} + \cdots$$

$$= \sum_{k=0}^{\infty} \frac{(-1)^k x^{2k}}{(2k)!}.$$

This is also the Maclaurin series for $\cos x$. In Section 8.9, we will see that the series converges to $\cos x$ at every x.

Because $f^{(2n+1)}(0) = 0$, the Taylor polynomials of orders $2n$ and $2n + 1$ are identical:

$$P_{2n}(x) = P_{2n+1}(x) = 1 - \frac{x^2}{2!} + \frac{x^4}{4!} - \cdots + (-1)^n \frac{x^{2n}}{(2n)!}.$$

Figure 8.13 shows how well these polynomials approximate $f(x) = \cos x$ near $x = 0$. Only the right-hand portions of the graphs are given because the graphs are symmetric about the y-axis. ∎

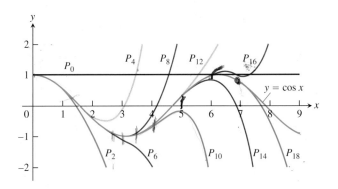

FIGURE 8.13   The polynomials

$$P_{2n}(x) = \sum_{k=0}^{n} \frac{(-1)^k x^{2k}}{(2k)!}$$

converge to $\cos x$ as $n \to \infty$. We can deduce the behavior of $\cos x$ arbitrarily far away solely from knowing the values of the cosine and its derivatives at $x = 0$ (Example 3).

EXAMPLE 4   It can be shown (though not easily) that

$$f(x) = \begin{cases} 0, & x = 0 \\ e^{-1/x^2}, & x \neq 0 \end{cases}$$

(Figure 8.14) has derivatives of all orders at $x = 0$ and that $f^{(n)}(0) = 0$ for all n. This means that the Taylor series generated by f at $x = 0$ is

$$f(0) + f'(0)x + \frac{f''(0)}{2!}x^2 + \cdots + \frac{f^{(n)}(0)}{n!}x^n + \cdots$$

$$= 0 + 0 \cdot x + 0 \cdot x^2 + \cdots + 0 \cdot x^n + \cdots$$

$$= 0 + 0 + \cdots + 0 + \cdots.$$

The series converges for every x (its sum is 0) but converges to $f(x)$ only at $x = 0$.   ■

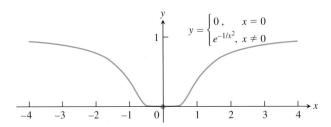

FIGURE 8.14   The graph of the continuous extension of $y = e^{-1/x^2}$ is so flat at the origin that all of its derivatives there are zero (Example 4).

Two questions still remain.

1. For what values of x can we normally expect a Taylor series to converge to its generating function?
2. How accurately do a function's Taylor polynomials approximate the function on a given interval?

The answers are provided by a theorem of Taylor in the next section.

EXERCISES 8.8

In Exercises 1–8, find the Taylor polynomials of orders 0, 1, 2, and 3 generated by f at a.

1. $f(x) = \ln x, \quad a = 1$
2. $f(x) = \ln(1 + x), \quad a = 0$
3. $f(x) = 1/x, \quad a = 2$
4. $f(x) = 1/(x + 2), \quad a = 0$
5. $f(x) = \sin x, \quad a = \pi/4$
6. $f(x) = \cos x, \quad a = \pi/4$
7. $f(x) = \sqrt{x}, \quad a = 4$
8. $f(x) = \sqrt{x + 4}, \quad a = 0$

Find the Maclaurin series for the functions in Exercises 9–20.

9. e^{-x}
10. $e^{x/2}$
11. $\dfrac{1}{1 + x}$
12. $\dfrac{1}{1 - x}$

13. $\sin 3x$
14. $\sin \dfrac{x}{2}$
15. $7 \cos(-x)$
16. $5 \cos \pi x$
17. $\cosh x = \dfrac{e^x + e^{-x}}{2}$
18. $\sinh x = \dfrac{e^x - e^{-x}}{2}$
19. $x^4 - 2x^3 - 5x + 4$
20. $(x + 1)^2$

In Exercises 21–28, find the Taylor series generated by f at $x = a$.

21. $f(x) = x^3 - 2x + 4, \quad a = 2$
22. $f(x) = 2x^3 + x^2 + 3x - 8, \quad a = 1$
23. $f(x) = x^4 + x^2 + 1, \quad a = -2$

24. $f(x) = 3x^5 - x^4 + 2x^3 + x^2 - 2, \quad a = -1$

25. $f(x) = 1/x^2, \quad a = 1$

26. $f(x) = x/(1 - x), \quad a = 0$

27. $f(x) = e^x, \quad a = 2$

28. $f(x) = 2^x, \quad a = 1$

29. Use the Taylor series generated by e^x at $x = a$ to show that

$$e^x = e^a \left[1 + (x - a) + \frac{(x - a)^2}{2!} + \cdots \right].$$

30. (*Continuation of Exercise 29.*) Find the Taylor series generated by e^x at $x = 1$. Compare your answer with the formula in Exercise 29.

31. Let $f(x)$ have derivatives through order n at $x = a$. Show that the Taylor polynomial of order n and its first n derivatives have the same values that f and its first n derivatives have at $x = a$.

32. Approximation properties of Taylor polynomials Suppose that $f(x)$ is differentiable on an interval centered at $x = a$ and that $g(x) = b_0 + b_1(x - a) + \cdots + b_n(x - a)^n$ is a polynomial of degree n with constant coefficients $b_0, \ldots, b_n$. Let $E(x) = f(x) - g(x)$. Show that if we impose on g the conditions

a. $E(a) = 0$ The approximation error is zero at $x = a$.

b. $\displaystyle \lim_{x \to a} \frac{E(x)}{(x - a)^n} = 0,$ The error is negligible when compared to $(x - a)^n$.

then

$$g(x) = f(a) + f'(a)(x - a) + \frac{f''(a)}{2!}(x - a)^2 + \cdots$$
$$+ \frac{f^{(n)}(a)}{n!}(x - a)^n.$$

Thus, the Taylor polynomial $P_n(x)$ is the only polynomial of degree less than or equal to n whose error is both zero at $x = a$ and negligible when compared with $(x - a)^n$.

Quadratic Approximations The Taylor polynomial of order 2 generated by a twice-differentiable function $f(x)$ at $x = a$ is called the *quadratic approximation* of f at $x = a$. In Exercises 33–38, find the **(a)** linearization (Taylor polynomial of order 1) and **(b)** quadratic approximation of f at $x = 0$.

33. $f(x) = \ln(\cos x)$ **34.** $f(x) = e^{\sin x}$

35. $f(x) = 1/\sqrt{1 - x^2}$ **36.** $f(x) = \cosh x$

37. $f(x) = \sin x$ **38.** $f(x) = \tan x$

8.9   Convergence of Taylor Series

In the last section, we asked when a Taylor series for a function can be expected to converge to that (generating) function. We answer the question in this section with the following theorem.

> **THEOREM 22—Taylor's Theorem**   If f and its first n derivatives $f', f'', \ldots, f^{(n)}$ are continuous on the closed interval between a and b, and $f^{(n)}$ is differentiable on the open interval between a and b, then there exists a number c between a and b such that
>
> $$f(b) = f(a) + f'(a)(b - a) + \frac{f''(a)}{2!}(b - a)^2 + \cdots$$
> $$+ \frac{f^{(n)}(a)}{n!}(b - a)^n + \frac{f^{(n+1)}(c)}{(n + 1)!}(b - a)^{n+1}.$$

Taylor's Theorem is a generalization of the Mean Value Theorem (Exercise 39). There is a proof of Taylor's Theorem at the end of this section.

When we apply Taylor's Theorem, we usually want to hold a fixed and treat b as an independent variable. Taylor's formula is easier to use in circumstances like these if we change b to x. Here is a version of the theorem with this change.

Taylor's Formula

If f has derivatives of all orders in an open interval I containing a, then for each positive integer n and for each x in I,

$$f(x) = f(a) + f'(a)(x - a) + \frac{f''(a)}{2!}(x - a)^2 + \cdots$$

$$+ \frac{f^{(n)}(a)}{n!}(x - a)^n + R_n(x), \tag{1}$$

where

$$R_n(x) = \frac{f^{(n+1)}(c)}{(n+1)!}(x - a)^{n+1} \qquad \text{for some } c \text{ between } a \text{ and } x. \tag{2}$$

When we state Taylor's theorem this way, it says that for each $x \in I$,

$$f(x) = P_n(x) + R_n(x).$$

The function $R_n(x)$ is determined by the value of the $(n + 1)$st derivative $f^{(n+1)}$ at a point c that depends on both a and x, and which lies somewhere between them. For any value of n we want, the equation gives both a polynomial approximation of f of that order and a formula for the error involved in using that approximation over the interval I.

Equation (1) is called **Taylor's formula**. The function $R_n(x)$ is called the **remainder of order n** or the **error term** for the approximation of f by $P_n(x)$ over I. If $R_n(x) \to 0$ as $n \to \infty$ for all $x \in I$, we say that the Taylor series generated by f at $x = a$ **converges** to f on I, and we write

$$f(x) = \sum_{k=0}^{\infty} \frac{f^{(k)}(a)}{k!}(x - a)^k.$$

Often we can estimate R_n without knowing the value of c, as the following example illustrates.

EXAMPLE 1   Show that the Taylor series generated by $f(x) = e^x$ at $x = 0$ converges to $f(x)$ for every real value of x.

Solution The function has derivatives of all orders throughout the interval $I = (-\infty, \infty)$. Equations (1) and (2) with $f(x) = e^x$ and $a = 0$ give

$$e^x = 1 + x + \frac{x^2}{2!} + \cdots + \frac{x^n}{n!} + R_n(x) \qquad \text{Polynomial from Section 8.8, Example 2}$$

and

$$R_n(x) = \frac{e^c}{(n+1)!} x^{n+1} \qquad \text{for some } c \text{ between } 0 \text{ and } x.$$

Since e^x is an increasing function of x, e^c lies between $e^0 = 1$ and e^x. When x is negative, so is c, and $e^c < 1$. When x is zero, $e^x = 1$ and $R_n(x) = 0$. When x is positive, so is c, and $e^c < e^x$. Thus, for $R_n(x)$ given as above,

$$|R_n(x)| \le \frac{|x|^{n+1}}{(n+1)!} \qquad \text{when } x \le 0, \qquad e^c < 1$$

and

$$|R_n(x)| < e^x \frac{x^{n+1}}{(n+1)!} \qquad \text{when } x > 0. \qquad {\scriptstyle e^c < e^x}$$

Finally, because

$$\lim_{n \to \infty} \frac{x^{n+1}}{(n+1)!} = 0 \qquad \text{for every } x, \qquad {\scriptstyle \text{Section 8.1}}$$

$\lim_{n \to \infty} R_n(x) = 0$, and the series converges to e^x for every x. Thus,

$$e^x = \sum_{k=0}^{\infty} \frac{x^k}{k!} = 1 + x + \frac{x^2}{2!} + \cdots + \frac{x^k}{k!} + \cdots. \tag{3}$$

∎

We can use the result of Example 1 with $x = 1$ to write

$$e = 1 + 1 + \frac{1}{2!} + \cdots + \frac{1}{n!} + R_n(1),$$

where for some c between 0 and 1,

$$R_n(1) = e^c \frac{1}{(n+1)!} < \frac{e}{(n+1)!}. \qquad {\scriptstyle e^c < e^1}$$

Estimating the Remainder

It is often possible to estimate $R_n(x)$ as we did in Example 1. This method of estimation is so convenient that we state it as a theorem for future reference.

> **THEOREM 23—The Remainder Estimation Theorem** If there is a positive constant M such that $|f^{(n+1)}(t)| \le M$ for all t between x and a, inclusive, then the remainder term $R_n(x)$ in Taylor's Theorem satisfies the inequality
>
> $$|R_n(x)| \le M \frac{|x - a|^{n+1}}{(n+1)!}.$$
>
> If this condition holds for every n and the other conditions of Taylor's Theorem are satisfied by f, then the series converges to $f(x)$.

Here's an application of Theorem 23.

EXAMPLE 2 Show that the Taylor series for $\sin x$ at $x = 0$ converges for all x.

Solution The function and its derivatives are

$$
\begin{array}{llll}
f(x) = & \sin x, & f'(x) = & \cos x, \\
f''(x) = & -\sin x, & f'''(x) = & -\cos x, \\
\;\;\vdots & & \;\;\vdots & \\
f^{(2k)}(x) = & (-1)^k \sin x, & f^{(2k+1)}(x) = & (-1)^k \cos x,
\end{array}
$$

so

$$f^{(2k)}(0) = 0 \quad \text{and} \quad f^{(2k+1)}(0) = (-1)^k.$$

The series has only odd-powered terms and, for $n = 2k + 1$, Taylor's Theorem gives

$$\sin x = x - \frac{x^3}{3!} + \frac{x^5}{5!} - \cdots + \frac{(-1)^k x^{2k+1}}{(2k + 1)!} + R_{2k+1}(x).$$

All the derivatives of $\sin x$ have absolute values less than or equal to 1, so we can apply the Remainder Estimation Theorem with $M = 1$ to obtain

$$\left| R_{2k+1}(x) \right| \leq 1 \cdot \frac{|x|^{2k+2}}{(2k + 2)!}.$$

Since $(|x|^{2k+2}/(2k + 2)!) \to 0$ as $k \to \infty$, whatever the value of x, $R_{2k+1}(x) \to 0$, and the Maclaurin series for $\sin x$ converges to $\sin x$ for every x. Thus,

$$\sin x = \sum_{k=0}^{\infty} \frac{(-1)^k x^{2k+1}}{(2k + 1)!} = x - \frac{x^3}{3!} + \frac{x^5}{5!} - \frac{x^7}{7!} + \cdots. \tag{4}$$

∎

EXAMPLE 3 Show that the Taylor series for $\cos x$ at $x = 0$ converges to $\cos x$ for every value of x.

Solution We add the remainder term to the Taylor polynomial for $\cos x$ (Section 8.8, Example 3) to obtain Taylor's formula for $\cos x$ with $n = 2k$:

$$\cos x = 1 - \frac{x^2}{2!} + \frac{x^4}{4!} - \cdots + (-1)^k \frac{x^{2k}}{(2k)!} + R_{2k}(x).$$

Because the derivatives of the cosine have absolute value less than or equal to 1, the Remainder Estimation Theorem with $M = 1$ gives

$$\left| R_{2k}(x) \right| \leq 1 \cdot \frac{|x|^{2k+1}}{(2k + 1)!}.$$

For every value of x, $R_{2k}(x) \to 0$ as $k \to \infty$. Therefore, the series converges to $\cos x$ for every value of x. Thus,

$$\cos x = \sum_{k=0}^{\infty} \frac{(-1)^k x^{2k}}{(2k)!} = 1 - \frac{x^2}{2!} + \frac{x^4}{4!} - \frac{x^6}{6!} + \cdots. \tag{5}$$

∎

We can find the Taylor series for $\cos 2x$ by substituting $2x$ for x in the Taylor series for $\cos x$:

$$\cos 2x = \sum_{k=0}^{\infty} \frac{(-1)^k (2x)^{2k}}{(2k)!} = 1 - \frac{(2x)^2}{2!} + \frac{(2x)^4}{4!} - \frac{(2x)^6}{6!} + \cdots \qquad \text{Equation (5)} \atop \text{with } 2x \text{ for } x$$

$$= 1 - \frac{2^2 x^2}{2!} + \frac{2^4 x^4}{4!} - \frac{2^6 x^6}{6!} + \cdots$$

$$= \sum_{k=0}^{\infty} (-1)^k \frac{2^{2k} x^{2k}}{(2k)!}.$$

EXAMPLE 4   For what values of x can we replace $\sin x$ by $x - (x^3/3!)$ with an error of magnitude no greater than 3×10^{-4}?

Solution   Here we can take advantage of the fact that the Taylor series for $\sin x$ is an alternating series for every nonzero value of x. According to the Alternating Series Estimation Theorem (Section 8.6), the error in truncating

$$\sin x = x - \frac{x^3}{3!} \;\Big|\; + \frac{x^5}{5!} - \cdots$$

after $(x^3/3!)$ is no greater than

$$\left| \frac{x^5}{5!} \right| = \frac{|x|^5}{120}.$$

Therefore the error will be less than or equal to 3×10^{-4} if

$$\frac{|x|^5}{120} < 3 \times 10^{-4} \qquad \text{or} \qquad |x| < \sqrt[5]{360 \times 10^{-4}} \approx 0.514. \qquad \text{Rounded down, to be safe}$$

The Alternating Series Estimation Theorem tells us something that the Remainder Estimation Theorem does not: namely, that the estimate $x - (x^3/3!)$ for $\sin x$ is an underestimate when x is positive because then $x^5/120$ is positive.

Figure 8.15 shows the graph of $\sin x$, along with the graphs of a number of its approximating Taylor polynomials. The graph of $P_3(x) = x - (x^3/3!)$ is almost indistinguishable from the sine curve when $0 \le x \le 1$.   ∎

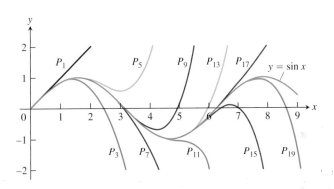

FIGURE 8.15   The polynomials

$$P_{2n+1}(x) = \sum_{k=0}^{n} \frac{(-1)^k x^{2k+1}}{(2k+1)!}$$

converge to $\sin x$ as $n \to \infty$. Notice how closely $P_3(x)$ approximates the sine curve for $x \le 1$ (Example 4).

Applying Taylor Series

On the intersection of their intervals of convergence, Taylor series can be added, subtracted, and multiplied by constants, and the results are once again Taylor series. The Taylor series for $f(x) + g(x)$ is the sum of the Taylor series for $f(x)$ and $g(x)$ because the nth derivative of $f + g$ is $f^{(n)} + g^{(n)}$, and so on. Thus we obtain the Taylor series for

$(1 + \cos 2x)/2$ by adding 1 to the Taylor series for $\cos 2x$ and dividing the combined results by 2, and the Taylor series for $\sin x + \cos x$ is the term-by-term sum of the Taylor series for $\sin x$ and $\cos x$.

Taylor series can be used to express nonelementary integrals in terms of series. For example, integrals like $\int \sin x^2 \, dx$ arise in the study of the diffraction of light. However, there is no simple formula for an antiderivative of $\sin x^2$.

EXAMPLE 5   Express $\int \sin x^2 \, dx$ as a power series.

Solution   From the series for $\sin x$ we obtain

$$\sin x^2 = x^2 - \frac{x^6}{3!} + \frac{x^{10}}{5!} - \frac{x^{14}}{7!} + \frac{x^{18}}{9!} - \cdots.$$

Therefore,

$$\int \sin x^2 \, dx = C + \frac{x^3}{3} - \frac{x^7}{7 \cdot 3!} + \frac{x^{11}}{11 \cdot 5!} - \frac{x^{15}}{15 \cdot 7!} + \frac{x^{10}}{19 \cdot 9!} - \cdots. \quad \blacksquare$$

EXAMPLE 6   Estimate $\int_0^1 \sin x^2 \, dx$ with an error of less than 0.001.

Solution   From the indefinite integral in Example 5,

$$\int_0^1 \sin x^2 \, dx = \frac{1}{3} - \frac{1}{7 \cdot 3!} + \frac{1}{11 \cdot 5!} - \frac{1}{15 \cdot 7!} + \frac{1}{19 \cdot 9!} - \cdots.$$

The series alternates, and we find by experiment that

$$\frac{1}{11 \cdot 5!} \approx 0.00076$$

is the first term to be numerically less than 0.001. The sum of the preceding two terms gives

$$\int_0^1 \sin x^2 \, dx \approx \frac{1}{3} - \frac{1}{42} \approx 0.310. \quad \blacksquare$$

We can sometimes evaluate indeterminate forms by expressing the functions involved as Taylor series.

EXAMPLE 7   Evaluate

$$\lim_{x \to 0} \frac{\sin x - \tan x}{x^3}.$$

Solution   The Taylor series for $\sin x$ and $\tan x$, to terms in x^5, are

$$\sin x = x - \frac{x^3}{3!} + \frac{x^5}{5!} - \cdots, \qquad \tan x = x + \frac{x^3}{3} + \frac{2x^5}{15} + \cdots.$$

Hence,

$$\sin x - \tan x = -\frac{x^3}{2} - \frac{x^5}{8} - \cdots = x^3 \left(-\frac{1}{2} - \frac{x^2}{8} - \cdots \right)$$

and

$$\lim_{x \to 0} \frac{\sin x - \tan x}{x^3} = \lim_{x \to 0} \left(-\frac{1}{2} - \frac{x^2}{8} - \cdots \right)$$

$$= -\frac{1}{2}. \qquad \blacksquare$$

Euler's Identity

As you may recall, a complex number is a number of the form $a + bi$, where a and b are real numbers and $i = \sqrt{-1}$. If we substitute $x = i\theta$ (θ real) in the Taylor series for e^x and use the relations

$$i^2 = -1, \qquad i^3 = i^2 i = -i, \qquad i^4 = i^2 i^2 = 1, \qquad i^5 = i^4 i = i,$$

and so on, to simplify the result, we obtain

$$e^{i\theta} = 1 + \frac{i\theta}{1!} + \frac{i^2 \theta^2}{2!} + \frac{i^3 \theta^3}{3!} + \frac{i^4 \theta^4}{4!} + \frac{i^5 \theta^5}{5!} + \frac{i^6 \theta^6}{6!} + \cdots$$

$$= \left(1 - \frac{\theta^2}{2!} + \frac{\theta^4}{4!} - \frac{\theta^6}{6!} + \cdots \right) + i \left(\theta - \frac{\theta^3}{3!} + \frac{\theta^5}{5!} - \cdots \right) = \cos \theta + i \sin \theta.$$

This does not *prove* that $e^{i\theta} = \cos \theta + i \sin \theta$ because we have not yet defined what it means to raise e to an imaginary power. Rather, it says how to define $e^{i\theta}$ to be consistent with other things we know.

DEFINITION

For any real number θ, $e^{i\theta} = \cos \theta + i \sin \theta$. (6)

Equation (6), called **Euler's identity**, enables us to define e^{a+bi} to be $e^a \cdot e^{bi}$ for any complex number $a + bi$. One consequence of the identity is the equation

$$e^{i\pi} = -1.$$

When written in the form $e^{i\pi} + 1 = 0$, this equation combines five of the most important constants in mathematics.

A Proof of Taylor's Theorem

We prove Taylor's theorem assuming $a < b$. The proof for $a > b$ is nearly the same.

The Taylor polynomial

$$P_n(x) = f(a) + f'(a)(x - a) + \frac{f''(a)}{2!}(x - a)^2 + \cdots + \frac{f^{(n)}(a)}{n!}(x - a)^n$$

and its first n derivatives match the function f and its first n derivatives at $x = a$. We do not disturb that matching if we add another term of the form $K(x - a)^{n+1}$, where K is any constant, because such a term and its first n derivatives are all equal to zero at $x = a$. The new function

$$\phi_n(x) = P_n(x) + K(x - a)^{n+1}$$

and its first n derivatives still agree with f and its first n derivatives at $x = a$.

We now choose the particular value of K that makes the curve $y = \phi_n(x)$ agree with the original curve $y = f(x)$ at $x = b$. In symbols,

$$f(b) = P_n(b) + K(b - a)^{n+1}, \quad \text{or} \quad K = \frac{f(b) - P_n(b)}{(b - a)^{n+1}}. \tag{7}$$

With K defined by Equation (7), the function

$$F(x) = f(x) - \phi_n(x)$$

measures the difference between the original function f and the approximating function ϕ_n for each x in $[a, b]$.

We now use Rolle's Theorem (Section 4.2). First, because $F(a) = F(b) = 0$ and both F and F' are continuous on $[a, b]$, we know that

$$F'(c_1) = 0 \qquad \text{for some } c_1 \text{ in } (a, b).$$

Next, because $F'(a) = F'(c_1) = 0$ and both F' and F'' are continuous on $[a, c_1]$, we know that

$$F''(c_2) = 0 \qquad \text{for some } c_2 \text{ in } (a, c_1).$$

Rolle's Theorem, applied successively to $F'', F''', \ldots, F^{(n-1)}$ implies the existence of

$$c_3 \quad \text{in } (a, c_2) \qquad \text{such that } F'''(c_3) = 0,$$

$$c_4 \quad \text{in } (a, c_3) \qquad \text{such that } F^{(4)}(c_4) = 0,$$

$$\vdots$$

$$c_n \quad \text{in } (a, c_{n-1}) \qquad \text{such that } F^{(n)}(c_n) = 0.$$

Finally, because $F^{(n)}$ is continuous on $[a, c_n]$ and differentiable on (a, c_n), and $F^{(n)}(a) = F^{(n)}(c_n) = 0$, Rolle's Theorem implies that there is a number c_{n+1} in (a, c_n) such that

$$F^{(n+1)}(c_{n+1}) = 0. \tag{8}$$

If we differentiate $F(x) = f(x) - P_n(x) - K(x - a)^{n+1}$ a total of $n + 1$ times, we get

$$F^{(n+1)}(x) = f^{(n+1)}(x) - 0 - (n + 1)!K. \tag{9}$$

Equations (8) and (9) together give

$$K = \frac{f^{(n+1)}(c)}{(n + 1)!} \qquad \text{for some number } c = c_{n+1} \text{ in } (a, b). \tag{10}$$

Equations (7) and (10) give

$$f(b) = P_n(b) + \frac{f^{(n+1)}(c)}{(n + 1)!}(b - a)^{n+1}.$$

This concludes the proof. $\blacksquare$

EXERCISES 8.9

Use substitution (as in Example 4) to find the Taylor series at $x = 0$ of the functions in Exercises 1–6.

1. e^{-5x} **2.** $e^{-x/2}$ **3.** $5 \sin (-x)$

4. $\sin \left(\dfrac{\pi x}{2} \right)$ **5.** $\cos \sqrt{x + 1}$ **6.** $\cos \left(x^{3/2} / \sqrt{2} \right)$

Find Taylor series at $x = 0$ for the functions in Exercises 7–18.

7. xe^x **8.** $x^2 \sin x$ **9.** $\dfrac{x^2}{2} - 1 + \cos x$

10. $\sin x - x + \dfrac{x^3}{3!}$ **11.** $x \cos \pi x$ **12.** $x^2 \cos (x^2)$

13. $\cos^2 x$ (*Hint:* $\cos^2 x = (1 + \cos 2x)/2$.)

14. $\sin^2 x$ **15.** $\dfrac{x^2}{1 - 2x}$ **16.** $x \ln (1 + 2x)$

17. $\dfrac{1}{(1 - x)^2}$ **18.** $\dfrac{2}{(1 - x)^3}$

19. For approximately what values of x can you replace $\sin x$ by $x - (x^3/6)$ with an error of magnitude no greater than 5×10^{-4}? Give reasons for your answer.

20. If $\cos x$ is replaced by $1 - (x^2/2)$ and $|x| < 0.5$, what estimate can be made of the error? Does $1 - (x^2/2)$ tend to be too large, or too small? Give reasons for your answer.

21. How close is the approximation $\sin x = x$ when $|x| < 10^{-3}$? For which of these values of x is $x < \sin x$?

22. The estimate $\sqrt{1 + x} = 1 + (x/2)$ is used when x is small. Estimate the error when $|x| < 0.01$.

23. The approximation $e^x = 1 + x + (x^2/2)$ is used when x is small. Use the Remainder Estimation Theorem to estimate the error when $|x| < 0.1$.

24. (*Continuation of Exercise 23.*) When $x < 0$, the series for e^x is an alternating series. Use the Alternating Series Estimation Theorem to estimate the error that results from replacing e^x by $1 + x + (x^2/2)$ when $-0.1 < x < 0$. Compare your estimate with the one you obtained in Exercise 23.

In Exercises 25–28, find a polynomial that will approximate $F(x)$ throughout the given interval with an error of magnitude less than 10^{-3}.

25. $F(x) = \displaystyle\int_0^x \sin t^2 \, dt,$ $[0, 1]$

26. $F(x) = \displaystyle\int_0^x t^2 e^{-t^2} \, dt,$ $[0, 1]$

27. $F(x) = \displaystyle\int_0^x \tan^{-1} t \, dt,$ **(a)** $[0, 0.5]$ **(b)** $[0, 1]$

28. $F(x) = \displaystyle\int_0^x \dfrac{\ln (1 + t)}{t} \, dt,$ **(a)** $[0, 0.5]$ **(b)** $[0, 1]$

Use series to evaluate the limits in Exercises 29–34.

29. $\lim\limits_{x \to 0} \dfrac{e^x - (1 + x)}{x^2}$ **30.** $\lim\limits_{x \to 0} \dfrac{e^x - e^{-x}}{x}$

31. $\lim\limits_{t \to 0} \dfrac{1 - \cos t - (t^2/2)}{t^4}$ **32.** $\lim\limits_{\theta \to 0} \dfrac{\sin \theta - \theta + (\theta^3/6)}{\theta^5}$

33. $\lim\limits_{y \to 0} \dfrac{y - \tan^{-1} y}{y^3}$ **34.** $\lim\limits_{y \to 0} \dfrac{\tan^{-1} y - \sin y}{y^3 \cos y}$

35. Multiply the Maclaurin series for e^x and $\sin x$ together to find the first five nonzero terms of the Maclaurin series for $e^x \sin x$.

36. Multiply the Maclaurin series for e^x and $\cos x$ together to find the first five nonzero terms of the Maclaurin series for $e^x \cos x$.

37. Use the identity $\sin^2 x = (1 - \cos 2x)/2$ to obtain the Maclaurin series for $\sin^2 x$. Then differentiate this series to obtain the Maclaurin series for $2 \sin x \cos x$. Check that this is the series for $\sin 2x$.

38. (*Continuation of Exercise 37.*) Use the identity $\cos^2 x = \cos 2x + \sin^2 x$ to obtain a power series for $\cos^2 x$.

39. Taylor's Theorem and the Mean Value Theorem Explain how the Mean Value Theorem (Section 4.2, Theorem 4) is a special case of Taylor's Theorem.

40. Linearizations at inflection points Show that if the graph of a twice-differentiable function $f(x)$ has an inflection point at $x = a$, then the linearization of f at $x = a$ is also the quadratic approximation of f at $x = a$. This explains why tangent lines fit so well at inflection points.

41. The (second) second derivative test Use the equation

$$f(x) = f(a) + f'(a)(x - a) + \dfrac{f''(c_2)}{2}(x - a)^2$$

to establish the following test.

Let f have continuous first and second derivatives and suppose that $f'(a) = 0$. Then

a. f has a local maximum at a if $f'' \leq 0$ throughout an interval whose interior contains a;

b. f has a local minimum at a if $f'' \geq 0$ throughout an interval whose interior contains a.

42. A cubic approximation Use Taylor's formula with $a = 0$ and $n = 3$ to find the standard cubic approximation of $f(x) = 1/(1 - x)$ at $x = 0$. Give an upper bound for the magnitude of the error in the approximation when $|x| \leq 0.1$.

43. a. Use Taylor's formula with $n = 2$ to find the quadratic approximation of $f(x) = (1 + x)^k$ at $x = 0$ (k a constant).

b. If $k = 3$, for approximately what values of x in the interval $[0, 1]$ will the error in the quadratic approximation be less than $1/100$?

44. Improving approximations to π

  a. Let P be an approximation of π accurate to n decimals. Show that $P + \sin P$ gives an approximation correct to $3n$ decimals. (*Hint:* Let $P = \pi + x$.)

  T b. Try it with a calculator.

45. The Taylor series generated by $f(x) = \sum_{n=0}^{\infty} a_n x^n$ **is** $\sum_{n=0}^{\infty} a_n x^n$ A function defined by a power series $\sum_{n=0}^{\infty} a_n x^n$ with a radius of convergence $c > 0$ has a Taylor series that converges to the function at every point of $(-c, c)$. Show this by showing that the Taylor series generated by $f(x) = \sum_{n=0}^{\infty} a_n x^n$ is the series $\sum_{n=0}^{\infty} a_n x^n$ itself.

  An immediate consequence of this is that series like

$$x \sin x = x^2 - \frac{x^4}{3!} + \frac{x^6}{5!} - \frac{x^8}{7!} + \cdots$$

and

$$x^2 e^x = x^2 + x^3 + \frac{x^4}{2!} + \frac{x^5}{3!} + \cdots,$$

obtained by multiplying Taylor series by powers of x, as well as series obtained by integration and differentiation of convergent power series, are themselves the Taylor series generated by the functions they represent.

46. Taylor series for even functions and odd functions (*Continuation of Section 8.7, Exercise 45.*) Suppose that $f(x) = \sum_{n=0}^{\infty} a_n x^n$ converges for all x in an open interval $(-c, c)$. Show that

  a. If f is even, then $a_1 = a_3 = a_5 = \cdots = 0$, i.e., the Taylor series for f at $x = 0$ contains only even powers of x.

  b. If f is odd, then $a_0 = a_2 = a_4 = \cdots = 0$, i.e., the Taylor series for f at $x = 0$ contains only odd powers of x.

47. Taylor polynomials of periodic functions

  a. Show that every continuous periodic function $f(x)$, $-\infty < x < \infty$, is bounded in magnitude by showing that there exists a positive constant M such that $|f(x)| \leq M$ for all x.

  b. Show that the graph of every Taylor polynomial of positive degree generated by $f(x) = \cos x$ must eventually move away from the graph of $\cos x$ as $|x|$ increases. You can see this in Figure 8.13. The Taylor polynomials of $\sin x$ behave in a similar way (Figure 8.15).

T 48. a. Graph the curves $y = (1/3) - (x^2)/5$ and $y = (x - \tan^{-1} x)/x^3$ together with the line $y = 1/3$.

  b. Use a Taylor series to explain what you see. What is

$$\lim_{x \to 0} \frac{x - \tan^{-1} x}{x^3} ?$$

49. Use Equation (6) to write the following powers of e in the form $a + bi$.

  a. $e^{-i\pi}$

  b. $e^{i\pi/4}$

  c. $e^{-i\pi/2}$

50. Use Equation (6) to show that

$$\cos \theta = \frac{e^{i\theta} + e^{-i\theta}}{2} \quad \text{and} \quad \sin \theta = \frac{e^{i\theta} - e^{-i\theta}}{2i}.$$

51. Establish the equations in Exercise 50 by combining the formal Taylor series for $e^{i\theta}$ and $e^{-i\theta}$.

52. Show that

  a. $\cosh i\theta = \cos \theta$, **b.** $\sinh i\theta = i \sin \theta$.

53. By multiplying the Taylor series for e^x and $\sin x$, find the terms through x^5 of the Taylor series for $e^x \sin x$. This series is the imaginary part of the series for

$$e^x \cdot e^{ix} = e^{(1+i)x}.$$

Use this fact to check your answer. For what values of x should the series for $e^x \sin x$ converge?

54. When a and b are real, we define $e^{(a+ib)x}$ with the equation

$$e^{(a+ib)x} = e^{ax} \cdot e^{ibx} = e^{ax}(\cos bx + i \sin bx).$$

Differentiate the right-hand side of this equation to show that

$$\frac{d}{dx} e^{(a+ib)x} = (a + ib) e^{(a+ib)x}.$$

Thus the familiar rule $(d/dx)e^{kx} = k e^{kx}$ holds for k complex as well as real.

55. Use the definition of $e^{i\theta}$ to show that for any real numbers θ, θ_1, and θ_2,

  a. $e^{i\theta_1} e^{i\theta_2} = e^{i(\theta_1 + \theta_2)}$, **b.** $e^{-i\theta} = 1/e^{i\theta}$.

56. Two complex numbers $a + ib$ and $c + id$ are equal if and only if $a = c$ and $b = d$. Use this fact to evaluate

$$\int e^{ax} \cos bx \, dx \quad \text{and} \quad \int e^{ax} \sin bx \, dx$$

from

$$\int e^{(a+ib)x} \, dx = \frac{a - ib}{a^2 + b^2} e^{(a+ib)x} + C,$$

where $C = C_1 + iC_2$ is a complex constant of integration.

COMPUTER EXPLORATIONS

Taylor's formula with $n = 1$ and $a = 0$ gives the linearization of a function at $x = 0$. With $n = 2$ and $n = 3$ we obtain the standard quadratic and cubic approximations. In these exercises we explore the errors associated with these approximations. We seek answers to two questions:

  a. For what values of x can the function be replaced by each approximation with an error less than 10^{-2}?

  b. What is the maximum error we could expect if we replace the function by each approximation over the specified interval?

  Using a CAS, perform the following steps to aid in answering questions (a) and (b) for the functions and intervals in Exercises 57–62.

  Step 1: Plot the function over the specified interval.

Step 2: Find the Taylor polynomials $P_1(x)$, $P_2(x)$, and $P_3(x)$ at $x = 0$.

Step 3: Calculate the $(n + 1)$st derivative $f^{(n+1)}(c)$ associated with the remainder term for each Taylor polynomial. Plot the derivative as a function of c over the specified interval and estimate its maximum absolute value, M.

Step 4: Calculate the remainder $R_n(x)$ for each polynomial. Using the estimate M from Step 3 in place of $f^{(n+1)}(c)$, plot $R_n(x)$ over the specified interval. Then estimate the values of x that answer question (a).

Step 5: Compare your estimated error with the actual error $E_n(x) = |f(x) - P_n(x)|$ by plotting $E_n(x)$ over the specified interval. This will help answer question (b).

Step 6: Graph the function and its three Taylor approximations together. Discuss the graphs in relation to the information discovered in Steps 4 and 5.

57. $f(x) = \dfrac{1}{\sqrt{1 + x}}, \quad |x| \le \dfrac{3}{4}$

58. $f(x) = (1 + x)^{3/2}, \quad -\dfrac{1}{2} \le x \le 2$

59. $f(x) = \dfrac{x}{x^2 + 1}, \quad |x| \le 2$

60. $f(x) = (\cos x)(\sin 2x), \quad |x| \le 2$

61. $f(x) = e^{-x} \cos 2x, \quad |x| \le 1$

62. $f(x) = e^{x/3} \sin 2x, \quad |x| \le 2$

8.10 The Binomial Series

This section introduces the binomial series for estimating powers and roots of binomial expressions $(1 + x)^m$. This section concludes with a reference table of frequently used series.

The Binomial Series for Powers and Roots

The Taylor series generated by $f(x) = (1 + x)^m$, when m is constant, is

$$1 + mx + \frac{m(m - 1)}{2!}x^2 + \frac{m(m - 1)(m - 2)}{3!}x^3 + \cdots$$

$$+ \frac{m(m - 1)(m - 2)\cdots(m - k + 1)}{k!}x^k + \cdots. \tag{1}$$

This series, called the **binomial series**, converges absolutely for $|x| < 1$. To derive the series, we first list the function and its derivatives:

$$f(x) = (1 + x)^m$$
$$f'(x) = m(1 + x)^{m-1}$$
$$f''(x) = m(m - 1)(1 + x)^{m-2}$$
$$f'''(x) = m(m - 1)(m - 2)(1 + x)^{m-3}$$
$$\vdots$$
$$f^{(k)}(x) = m(m - 1)(m - 2)\cdots(m - k + 1)(1 + x)^{m-k}.$$

We then evaluate these at $x = 0$ and substitute into the Taylor series formula to obtain Series (1).

If m is an integer greater than or equal to zero, the series stops after $(m + 1)$ terms because the coefficients from $k = m + 1$ on are zero.

If m is not a positive integer or zero, the series is infinite and converges for $|x| < 1$. To see why, let u_k be the term involving x^k. Then apply the Ratio Test for absolute convergence to see that

$$\left|\frac{u_{k+1}}{u_k}\right| = \left|\frac{m - k}{k + 1}x\right| \to |x| \qquad \text{as } k \to \infty.$$

Our derivation of the binomial series shows only that it is generated by $(1 + x)^m$ and converges for $|x| < 1$. The derivation does not show that the series converges to $(1 + x)^m$. It does, but we omit the proof. (See Exercise 20.)

The Binomial Series

For $-1 < x < 1$,

$$(1 + x)^m = 1 + \sum_{k=1}^{\infty} \binom{m}{k} x^k,$$

where we define

$$\binom{m}{1} = m, \qquad \binom{m}{2} = \frac{m(m - 1)}{2!},$$

and

$$\binom{m}{k} = \frac{m(m - 1)(m - 2)\cdots(m - k + 1)}{k!} \qquad \text{for } k \geq 3.$$

EXAMPLE 1   If $m = -1$,

$$\binom{-1}{1} = -1, \qquad \binom{-1}{2} = \frac{-1(-2)}{2!} = 1,$$

and

$$\binom{-1}{k} = \frac{-1(-2)(-3)\cdots(-1 - k + 1)}{k!} = (-1)^k \left(\frac{k!}{k!}\right) = (-1)^k.$$

With these coefficient values and with x replaced by $-x$, the binomial series formula gives the familiar geometric series

$$(1 + x)^{-1} = 1 + \sum_{k=1}^{\infty} (-1)^k x^k = 1 - x + x^2 - x^3 + \cdots + (-1)^k x^k + \cdots. \quad\blacksquare$$

EXAMPLE 2   We know from Section 3.10, Example 1, that $\sqrt{1 + x} \approx 1 + (x/2)$ for $|x|$ small. With $m = 1/2$, the binomial series gives quadratic and higher-order approximations as well, along with error estimates that come from the Alternating Series Estimation Theorem:

$$(1 + x)^{1/2} = 1 + \frac{x}{2} + \frac{\left(\frac{1}{2}\right)\left(-\frac{1}{2}\right)}{2!} x^2 + \frac{\left(\frac{1}{2}\right)\left(-\frac{1}{2}\right)\left(-\frac{3}{2}\right)}{3!} x^3$$

$$+ \frac{\left(\frac{1}{2}\right)\left(-\frac{1}{2}\right)\left(-\frac{3}{2}\right)\left(-\frac{5}{2}\right)}{4!} x^4 + \cdots$$

$$= 1 + \frac{x}{2} - \frac{x^2}{8} + \frac{x^3}{16} - \frac{5x^4}{128} + \cdots.$$

Substitution for x gives still other approximations. For example,

$$\sqrt{1 - x^2} \approx 1 - \frac{x^2}{2} - \frac{x^4}{8} \qquad \text{for } |x^2| \text{ small}$$

$$\sqrt{1 - \frac{1}{x}} \approx 1 - \frac{1}{2x} - \frac{1}{8x^2} \qquad \text{for } \left|\frac{1}{x}\right| \text{ small, that is, } |x| \text{ large.} \qquad \blacksquare$$

TABLE 8.1 Frequently used Taylor series

$$\frac{1}{1 - x} = 1 + x + x^2 + \cdots + x^n + \cdots = \sum_{n=0}^{\infty} x^n, \qquad |x| < 1$$

$$\frac{1}{1 + x} = 1 - x + x^2 - \cdots + (-x)^n + \cdots = \sum_{n=0}^{\infty} (-1)^n x^n, \qquad |x| < 1$$

$$e^x = 1 + x + \frac{x^2}{2!} + \cdots + \frac{x^n}{n!} + \cdots = \sum_{n=0}^{\infty} \frac{x^n}{n!}, \qquad |x| < \infty$$

$$\sin x = x - \frac{x^3}{3!} + \frac{x^5}{5!} - \cdots + (-1)^n \frac{x^{2n+1}}{(2n+1)!} + \cdots = \sum_{n=0}^{\infty} \frac{(-1)^n x^{2n+1}}{(2n+1)!}, \qquad |x| < \infty$$

$$\cos x = 1 - \frac{x^2}{2!} + \frac{x^4}{4!} - \cdots + (-1)^n \frac{x^{2n}}{(2n)!} + \cdots = \sum_{n=0}^{\infty} \frac{(-1)^n x^{2n}}{(2n)!}, \qquad |x| < \infty$$

$$\ln(1 + x) = x - \frac{x^2}{2} + \frac{x^3}{3} - \cdots + (-1)^{n-1} \frac{x^n}{n} + \cdots = \sum_{n=1}^{\infty} \frac{(-1)^{n-1} x^n}{n}, \qquad -1 < x \le 1$$

$$\ln \frac{1 + x}{1 - x} = 2 \tanh^{-1} x = 2\left(x + \frac{x^3}{3} + \frac{x^5}{5} + \cdots + \frac{x^{2n+1}}{2n+1} + \cdots\right) = 2 \sum_{n=0}^{\infty} \frac{x^{2n+1}}{2n+1}, \qquad |x| < 1$$

$$\tan^{-1} x = x - \frac{x^3}{3} + \frac{x^5}{5} - \cdots + (-1)^n \frac{x^{2n+1}}{2n+1} + \cdots = \sum_{n=0}^{\infty} \frac{(-1)^n x^{2n+1}}{2n+1}, \qquad |x| \le 1$$

Binomial Series

$$(1 + x)^m = 1 + mx + \frac{m(m-1)x^2}{2!} + \frac{m(m-1)(m-2)x^3}{3!} + \cdots + \frac{m(m-1)(m-2)\cdots(m-k+1)x^k}{k!} + \cdots$$

$$= 1 + \sum_{k=1}^{\infty} \binom{m}{k} x^k, \qquad |x| < 1,$$

where

$$\binom{m}{1} = m, \qquad \binom{m}{2} = \frac{m(m-1)}{2!}, \qquad \binom{m}{k} = \frac{m(m-1)\cdots(m-k+1)}{k!} \qquad \text{for } k \ge 3.$$

Note: To write the binomial series compactly, it is customary to define $\binom{m}{0}$ to be 1 and to take $x^0 = 1$ (even in the usually excluded case where $x = 0$), yielding $(1 + x)^m = \sum_{k=0}^{\infty} \binom{m}{k} x^k$. If m is a *positive integer*, the series terminates at x^m and the result converges for all x.

EXERCISES 8.10

Find the first four terms of the binomial series for the functions in Exercises 1–10.

1. $(1 + x)^{1/2}$ **2.** $(1 + x)^{1/3}$ **3.** $(1 - x)^{-1/2}$

4. $(1 - 2x)^{1/2}$ **5.** $\left(1 + \dfrac{x}{2}\right)^{-2}$ **6.** $\left(1 - \dfrac{x}{2}\right)^{-2}$

7. $(1 + x^3)^{-1/2}$ **8.** $(1 + x^2)^{-1/3}$

9. $\left(1 + \dfrac{1}{x}\right)^{1/2}$ **10.** $\left(1 - \dfrac{2}{x}\right)^{1/3}$

Find the binomial series for the functions in Exercises 11–14.

11. $(1 + x)^4$ **12.** $(1 + x^2)^3$

13. $(1 - 2x)^3$ **14.** $\left(1 - \dfrac{x}{2}\right)^4$

15. a. Use the binomial series and the fact that

$$\frac{d}{dx}\sin^{-1}x = (1 - x^2)^{-1/2}$$

to generate the first four nonzero terms of the Taylor series for $\sin^{-1}x$. What is the radius of convergence?

b. Series for $\cos^{-1}x$ Use your result in part (a) to find the first five nonzero terms of the Taylor series for $\cos^{-1}x$.

16 a. Series for $\sinh^{-1}x$ Find the first four nonzero terms of the Taylor series for

$$\sinh^{-1}x = \int_0^x \frac{dt}{\sqrt{1 + t^2}}.$$

T   **b.** Use the first *three* terms of the series in part (a) to estimate $\sinh^{-1} 0.25$. Give an upper bound for the magnitude of the estimation error.

17. Obtain the Taylor series for $1/(1 + x)^2$ from the binomial series for $-1/(1 + x)$.

18. Use the binomial series for $1/(1 - x^2)$ to obtain a power series for $2x/(1 - x^2)^2$.

19. Series for $\sin^{-1}x$ Integrate the binomial series for $(1 - x^2)^{-1/2}$ to show that for $|x| < 1$,

$$\sin^{-1}x = x + \sum_{n=1}^{\infty} \frac{1 \cdot 3 \cdot 5 \cdot \cdots \cdot (2n - 1)}{2 \cdot 4 \cdot 6 \cdot \cdots \cdot (2n)} \frac{x^{2n+1}}{2n + 1}.$$

20. Use the following steps to prove Equation (1).

a. Differentiate the series

$$f(x) = 1 + \sum_{k=1}^{\infty} \binom{m}{k} x^k$$

to show that

$$f'(x) = \frac{mf(x)}{1 + x}, \quad -1 < x < 1.$$

b. Define $g(x) = (1 + x)^{-m} f(x)$ and show that $g'(x) = 0$.

c. From part (b), show that

$$f(x) = (1 + x)^m.$$

Chapter 8   Questions to Guide Your Review

1. What is an infinite sequence? What does it mean for such a sequence to converge? To diverge? Give examples.

2. What is a nondecreasing sequence? Under what circumstances does such a sequence have a limit? Give examples.

3. What theorems are available for calculating limits of sequences? Give examples.

4. What theorem sometimes enables us to use l'Hôpital's Rule to calculate the limit of a sequence? Give an example.

5. What six sequence limits are likely to arise when you work with sequences and series?

6. What is an infinite series? What does it mean for such a series to converge? To diverge? Give examples.

7. What is a geometric series? When does such a series converge? Diverge? When it does converge, what is its sum? Give examples.

8. Besides geometric series, what other convergent and divergent series do you know?

9. What is the *n*th-Term Test for Divergence? What is the idea behind the test?

10. What can be said about term-by-term sums and differences of convergent series? About constant multiples of convergent and divergent series?

11. What happens if you add a finite number of terms to a convergent series? A divergent series? What happens if you delete a finite number of terms from a convergent series? A divergent series?

12. How do you reindex a series? Why might you want to do this?

13. Under what circumstances will an infinite series of nonnegative terms converge? Diverge? Why study series of nonnegative terms?

14. What is the Integral Test? What is the reasoning behind it? Give an example of its use.

15. When do p-series converge? Diverge? How do you know? Give examples of convergent and divergent p-series.

16. What are the Direct Comparison Test and the Limit Comparison Test? What is the reasoning behind these tests? Give examples of their use.

17. What are the Ratio and Root Tests? Do they always give you the information you need to determine convergence or divergence? Give examples.

18. What is an alternating series? What theorem is available for determining the convergence of such a series?

19. How can you estimate the error involved in approximating the sum of an alternating series with one of the series' partial sums? What is the reasoning behind the estimate?

20. What is absolute convergence? Conditional convergence? How are the two related?

21. What do you know about rearranging the terms of an absolutely convergent series? Of a conditionally convergent series?

22. What is a power series? How do you test a power series for convergence? What are the possible outcomes?

23. What are the basic facts about

   a. term-by-term differentiation of power series?

   b. term-by-term integration of power series?

   c. multiplication of power series?

   Give examples.

24. What is the Taylor series generated by a function $f(x)$ at a point $x = a$? What information do you need about f to construct the series? Give an example.

25. What is a Maclaurin series?

26. Does a Taylor series always converge to its generating function? Explain.

27. What are Taylor polynomials? Of what use are they?

28. What is Taylor's formula? What does it say about the errors involved in using Taylor polynomials to approximate functions? In particular, what does Taylor's formula say about the error in a linearization? A quadratic approximation?

29. What is the binomial series? On what interval does it converge? How is it used?

30. How can you sometimes use power series to estimate the values of nonelementary definite integrals?

31. What are the Taylor series for $1/(1 - x)$, $1/(1 + x)$, e^x, $\sin x$, $\cos x$, $\ln (1 + x)$, $\ln [(1 + x)/(1 - x)]$, and $\tan^{-1} x$? How do you estimate the errors involved in replacing these series with their partial sums?

Chapter 8   Practice Exercises

Which of the sequences whose nth terms appear in Exercises 1–18 converge, and which diverge? Find the limit of each convergent sequence.

1. $a_n = 1 + \dfrac{(-1)^n}{n}$

2. $a_n = \dfrac{1 - (-1)^n}{\sqrt{n}}$

3. $a_n = \dfrac{1 - 2^n}{2^n}$

4. $a_n = 1 + (0.9)^n$

5. $a_n = \sin \dfrac{n\pi}{2}$

6. $a_n = \sin n\pi$

7. $a_n = \dfrac{\ln (n^2)}{n}$

8. $a_n = \dfrac{\ln (2n + 1)}{n}$

9. $a_n = \dfrac{n + \ln n}{n}$

10. $a_n = \dfrac{\ln (2n^3 + 1)}{n}$

11. $a_n = \left(\dfrac{n - 5}{n}\right)^n$

12. $a_n = \left(1 + \dfrac{1}{n}\right)^{-n}$

13. $a_n = \sqrt[n]{\dfrac{3^n}{n}}$

14. $a_n = \left(\dfrac{3}{n}\right)^{1/n}$

15. $a_n = n(2^{1/n} - 1)$

16. $a_n = \sqrt[n]{2n + 1}$

17. $a_n = \dfrac{(n + 1)!}{n!}$

18. $a_n = \dfrac{(-4)^n}{n!}$

Find the sums of the series in Exercises 19–24.

19. $\displaystyle\sum_{n=3}^{\infty} \dfrac{1}{(2n - 3)(2n - 1)}$

20. $\displaystyle\sum_{n=2}^{\infty} \dfrac{-2}{n(n + 1)}$

21. $\displaystyle\sum_{n=1}^{\infty} \dfrac{9}{(3n - 1)(3n + 2)}$

22. $\displaystyle\sum_{n=3}^{\infty} \dfrac{-8}{(4n - 3)(4n + 1)}$

23. $\displaystyle\sum_{n=0}^{\infty} e^{-n}$

24. $\displaystyle\sum_{n=1}^{\infty} (-1)^n \dfrac{3}{4^n}$

Which of the series in Exercises 25–40 converge absolutely, which converge conditionally, and which diverge? Give reasons for your answers.

25. $\displaystyle\sum_{n=1}^{\infty} \dfrac{1}{\sqrt{n}}$

26. $\displaystyle\sum_{n=1}^{\infty} \dfrac{-5}{n}$

27. $\displaystyle\sum_{n=1}^{\infty} \dfrac{(-1)^n}{\sqrt{n}}$

28. $\displaystyle\sum_{n=1}^{\infty} \dfrac{1}{2n^3}$

29. $\displaystyle\sum_{n=1}^{\infty} \dfrac{(-1)^n}{\ln (n + 1)}$

30. $\displaystyle\sum_{n=2}^{\infty} \dfrac{1}{n (\ln n)^2}$

31. $\displaystyle\sum_{n=1}^{\infty} \dfrac{\ln n}{n^3}$

32. $\displaystyle\sum_{n=3}^{\infty} \dfrac{\ln n}{\ln (\ln n)}$

33. $\displaystyle\sum_{n=1}^{\infty} \frac{(-1)^n}{n\sqrt{n^2+1}}$

34. $\displaystyle\sum_{n=1}^{\infty} \frac{(-1)^n 3n^2}{n^3+1}$

35. $\displaystyle\sum_{n=1}^{\infty} \frac{n+1}{n!}$

36. $\displaystyle\sum_{n=1}^{\infty} \frac{(-1)^n(n^2+1)}{2n^2+n-1}$

37. $\displaystyle\sum_{n=1}^{\infty} \frac{(-3)^n}{n!}$

38. $\displaystyle\sum_{n=1}^{\infty} \frac{2^n 3^n}{n^n}$

39. $\displaystyle\sum_{n=1}^{\infty} \frac{1}{\sqrt{n(n+1)(n+2)}}$

40. $\displaystyle\sum_{n=2}^{\infty} \frac{1}{n\sqrt{n^2-1}}$

In Exercises 41–50, **(a)** find the series' radius and interval of convergence. Then identify the values of x for which the series converges **(b)** absolutely and **(c)** conditionally.

41. $\displaystyle\sum_{n=1}^{\infty} \frac{(x+4)^n}{n3^n}$

42. $\displaystyle\sum_{n=1}^{\infty} \frac{(x-1)^{2n-2}}{(2n-1)!}$

43. $\displaystyle\sum_{n=1}^{\infty} \frac{(-1)^{n-1}(3x-1)^n}{n^2}$

44. $\displaystyle\sum_{n=0}^{\infty} \frac{(n+1)(2x+1)^n}{(2n+1)2^n}$

45. $\displaystyle\sum_{n=1}^{\infty} \frac{x^n}{n^n}$

46. $\displaystyle\sum_{n=1}^{\infty} \frac{x^n}{\sqrt{n}}$

47. $\displaystyle\sum_{n=0}^{\infty} \frac{(n+1)x^{2n-1}}{3^n}$

48. $\displaystyle\sum_{n=0}^{\infty} \frac{(-1)^n(x-1)^{2n+1}}{2n+1}$

49. $\displaystyle\sum_{n=1}^{\infty} (\operatorname{csch} n)x^n$

50. $\displaystyle\sum_{n=1}^{\infty} (\coth n)x^n$

Each of the series in Exercises 51–56 is the value of the Taylor series at $x = 0$ of a function $f(x)$ at a particular point. What function and what point? What is the sum of the series?

51. $1 - \dfrac{1}{4} + \dfrac{1}{16} - \cdots + (-1)^n \dfrac{1}{4^n} + \cdots$

52. $\dfrac{2}{3} - \dfrac{4}{18} + \dfrac{8}{81} - \cdots + (-1)^{n-1} \dfrac{2^n}{n3^n} + \cdots$

53. $\pi - \dfrac{\pi^3}{3!} + \dfrac{\pi^5}{5!} - \cdots + (-1)^n \dfrac{\pi^{2n+1}}{(2n+1)!} + \cdots$

54. $1 - \dfrac{\pi^2}{9\cdot 2!} + \dfrac{\pi^4}{81\cdot 4!} - \cdots + (-1)^n \dfrac{\pi^{2n}}{3^{2n}(2n)!} + \cdots$

55. $1 + \ln 2 + \dfrac{(\ln 2)^2}{2!} + \cdots + \dfrac{(\ln 2)^n}{n!} + \cdots$

56. $\dfrac{1}{\sqrt{3}} - \dfrac{1}{9\sqrt{3}} + \dfrac{1}{45\sqrt{3}} - \cdots$

$\qquad + (-1)^{n-1}\dfrac{1}{(2n-1)\left(\sqrt{3}\right)^{2n-1}} + \cdots$

Find Taylor series at $x = 0$ for the functions in Exercises 57–64.

57. $\dfrac{1}{1-2x}$

58. $\dfrac{1}{1+x^3}$

59. $\sin \pi x$

60. $\sin \dfrac{2x}{3}$

61. $\cos (x^{5/2})$

62. $\cos \sqrt{5x}$

63. $e^{(\pi x/2)}$

64. e^{-x^2}

In Exercises 65–68, find the first four nonzero terms of the Taylor series generated by f at $x = a$.

65. $f(x) = \sqrt{3+x^2}$ at $x = -1$

66. $f(x) = 1/(1-x)$ at $x = 2$

67. $f(x) = 1/(x+1)$ at $x = 3$

68. $f(x) = 1/x$ at $x = a > 0$

Use series to approximate the values of the integrals in Exercises 69–72 with an error of magnitude less than 10^{-8}. (The answer section gives the integrals' values rounded to 10 decimal places.)

69. $\displaystyle\int_0^{1/2} e^{-x^3}\, dx$

70. $\displaystyle\int_0^1 x\sin(x^3)\, dx$

71. $\displaystyle\int_0^{1/2} \frac{\tan^{-1}x}{x}\, dx$

72. $\displaystyle\int_0^{1/64} \frac{\tan^{-1}x}{\sqrt{x}}\, dx$

In Exercises 73–78:

   a. Use power series to evaluate the limit.

T **b.** Then use a grapher to support your calculation.

73. $\displaystyle\lim_{x\to 0} \frac{7\sin x}{e^{2x}-1}$

74. $\displaystyle\lim_{\theta\to 0} \frac{e^{\theta}-e^{-\theta}-2\theta}{\theta-\sin\theta}$

75. $\displaystyle\lim_{t\to 0} \left(\frac{1}{2-2\cos t} - \frac{1}{t^2}\right)$

76. $\displaystyle\lim_{h\to 0} \frac{(\sin h)/h - \cos h}{h^2}$

77. $\displaystyle\lim_{z\to 0} \frac{1-\cos^2 z}{\ln(1-z)+\sin z}$

78. $\displaystyle\lim_{y\to 0} \frac{y^2}{\cos y - \cosh y}$

79. Use a series representation of $\sin 3x$ to find values of r and s for which

$$\lim_{x\to 0}\left(\frac{\sin 3x}{x^3} + \frac{r}{x^2} + s\right) = 0.$$

T 80. Compare the accuracies of the approximations $\sin x \approx x$ and $\sin x \approx 6x/(6+x^2)$ by comparing the graphs of $f(x) = \sin x - x$ and $g(x) = \sin x - (6x/(6+x^2))$. Describe what you find.

81. Find the radius of convergence of the series

$$\sum_{n=1}^{\infty} \frac{2\cdot 5\cdot 8\cdot \cdots \cdot (3n-1)}{2\cdot 4\cdot 6\cdot \cdots \cdot (2n)}x^n.$$

82. Find the radius of convergence of the series

$$\sum_{n=1}^{\infty} \frac{3\cdot 5\cdot 7\cdot \cdots \cdot (2n+1)}{4\cdot 9\cdot 14\cdot \cdots \cdot (5n-1)}(x-1)^n.$$

83. Find a closed-form formula for the nth partial sum of the series $\sum_{n=2}^{\infty} \ln(1-(1/n^2))$ and use it to determine the convergence or divergence of the series.

84. Evaluate $\sum_{k=2}^{\infty} (1/(k^2-1))$ by finding the limits as $n\to\infty$ of the series' nth partial sum.

85. **a.** Find the interval of convergence of the series

$$y = 1 + \frac{1}{6}x^3 + \frac{1}{180}x^6 + \cdots$$

$$+ \frac{1\cdot 4\cdot 7\cdot \cdots \cdot (3n-2)}{(3n)!}x^{3n} + \cdots.$$

b. Show that the function defined by the series satisfies a differential equation of the form

$$\frac{d^2y}{dx^2} = x^a y + b$$

and find the values of the constants a and b.

86. a. Find the Maclaurin series for the function $x^2/(1 + x)$.

  b. Does the series converge at $x = 1$? Explain.

87. If $\sum_{n=1}^{\infty} a_n$ and $\sum_{n=1}^{\infty} b_n$ are convergent series of nonnegative numbers, can anything be said about $\sum_{n=1}^{\infty} a_n b_n$? Give reasons for your answer.

88. If $\sum_{n=1}^{\infty} a_n$ and $\sum_{n=1}^{\infty} b_n$ are divergent series of nonnegative numbers, can anything be said about $\sum_{n=1}^{\infty} a_n b_n$? Give reasons for your answer.

89. Prove that the sequence $\{x_n\}$ and the series $\sum_{k=1}^{\infty} (x_{k+1} - x_k)$ both converge or both diverge.

90. Prove that $\sum_{n=1}^{\infty} (a_n/(1 + a_n))$ converges if $a_n > 0$ for all n and $\sum_{n=1}^{\infty} a_n$ converges.

91. Suppose that $a_1, a_2, a_3, \ldots, a_n$ are positive numbers satisfying the following conditions:

  i. $a_1 \geq a_2 \geq a_3 \geq \cdots$;

  ii. the series $a_2 + a_4 + a_8 + a_{16} + \cdots$ diverges.

  Show that the series

$$\frac{a_1}{1} + \frac{a_2}{2} + \frac{a_3}{3} + \cdots$$

  diverges.

92. Use the result in Exercise 91 to show that

$$1 + \sum_{n=2}^{\infty} \frac{1}{n \ln n}$$

diverges.

Chapter 8 Additional and Advanced Exercises

Which of the series $\sum_{n=1}^{\infty} a_n$ defined by the formulas in Exercises 1–4 converge, and which diverge? Give reasons for your answers.

1. $\sum_{n=1}^{\infty} \frac{1}{(3n - 2)^{n+(1/2)}}$

2. $\sum_{n=1}^{\infty} \frac{(\tan^{-1} n)^2}{n^2 + 1}$

3. $\sum_{n=1}^{\infty} (-1)^n \tanh n$

4. $\sum_{n=2}^{\infty} \frac{\log_n (n!)}{n^3}$

Which of the series $\sum_{n=1}^{\infty} a_n$ defined by the formulas in Exercises 5–8 converge, and which diverge? Give reasons for your answers.

5. $a_1 = 1, \quad a_{n+1} = \dfrac{n(n + 1)}{(n + 2)(n + 3)} a_n$

  (*Hint:* Write out several terms, see which factors cancel, and then generalize.)

6. $a_1 = a_2 = 7, \quad a_{n+1} = \dfrac{n}{(n - 1)(n + 1)} a_n$ if $n \geq 2$

7. $a_1 = a_2 = 1, \quad a_{n+1} = \dfrac{1}{1 + a_n}$ if $n \geq 2$

8. $a_n = 1/3^n$ if n is odd, $a_n = n/3^n$ if n is even

Taylor's formula

$$f(x) = f(a) + f'(a)(x - a) + \frac{f''(a)}{2!}(x - a)^2 + \cdots$$

$$+ \frac{f^{(n)}(a)}{n!}(x - a)^n + \frac{f^{(n+1)}(c)}{(n + 1)!}(x - a)^{n+1}$$

expresses the value of f at x in terms of the values of f and its derivatives at $x = a$. In numerical computations, we therefore need a to be a point where we know the values of f and its derivatives. We also need a to be close enough to the values of f we are interested in to make $(x - a)^{n+1}$ so small we can neglect the remainder.

In Exercises 9–14, what Taylor series would you choose to represent the function near the given value of x? (There may be more than one good answer.) Write out the first four nonzero terms of the series you choose.

9. $\cos x$ near $x = 1$

10. $\sin x$ near $x = 6.3$

11. e^x near $x = 0.4$

12. $\ln x$ near $x = 1.3$

13. $\cos x$ near $x = 69$

14. $\tan^{-1} x$ near $x = 2$

15. Let a and b be constants with $0 < a < b$. Does the sequence $\{(a^n + b^n)^{1/n}\}$ converge? If it does converge, what is the limit?

16. Find the sum of the infinite series

$$1 + \frac{2}{10} + \frac{3}{10^2} + \frac{7}{10^3} + \frac{2}{10^4} + \frac{3}{10^5} + \frac{7}{10^6} + \frac{2}{10^7}$$

$$+ \frac{3}{10^8} + \frac{7}{10^9} + \cdots.$$

17. Evaluate

$$\sum_{n=0}^{\infty} \int_n^{n+1} \frac{1}{1 + x^2} \, dx.$$

18. Find all values of x for which

$$\sum_{n=1}^{\infty} \frac{nx^n}{(n + 1)(2x + 1)^n}$$

converges absolutely.

T 19. a. Does the value of

$$\lim_{n \to \infty} \left(1 - \frac{\cos (a/n)}{n}\right)^n, \quad a \text{ constant},$$

appear to depend on the value of a? If so, how?

b. Does the value of

$$\lim_{n \to \infty} \left(1 - \frac{\cos{(a/n)}}{bn} \right)^n, \quad a \text{ and } b \text{ constant, } b \neq 0,$$

appear to depend on the value of b? If so, how?

c. Use calculus to confirm your findings in parts (a) and (b).

20. Show that if $\sum_{n=1}^{\infty} a_n$ converges, then

$$\sum_{n=1}^{\infty} \left(\frac{1 + \sin{(a_n)}}{2} \right)^n$$

converges.

21. Find a value for the constant b that will make the radius of convergence of the power series

$$\sum_{n=2}^{\infty} \frac{b^n x^n}{\ln n}$$

equal to 5.

22. How do you know that the functions $\sin x$, $\ln x$, and e^x are not polynomials? Give reasons for your answer.

23. Find the value of a for which the limit

$$\lim_{x \to 0} \frac{\sin{(ax)} - \sin x - x}{x^3}$$

is finite and evaluate the limit.

24. Find values of a and b for which

$$\lim_{x \to 0} \frac{\cos{(ax)} - b}{2x^2} = -1.$$

25. Raabe's (or Gauss's) test   The following test, which we state without proof, is an extension of the Ratio Test.

Raabe's test: If $\sum_{n=1}^{\infty} u_n$ is a series of positive constants and there exist constants C, K, and N such that

$$\frac{u_n}{u_{n+1}} = 1 + \frac{C}{n} + \frac{f(n)}{n^2}, \tag{1}$$

where $|f(n)| < K$ for $n \geq N$, then $\sum_{n=1}^{\infty} u_n$ converges if $C > 1$ and diverges if $C \leq 1$.

Show that the results of Raabe's test agree with what you know about the series $\sum_{n=1}^{\infty} (1/n^2)$ and $\sum_{n=1}^{\infty} (1/n)$.

26. (*Continuation of Exercise 25.*) Suppose that the terms of $\sum_{n=1}^{\infty} u_n$ are defined recursively by the formulas

$$u_1 = 1, \quad u_{n+1} = \frac{(2n - 1)^2}{(2n)(2n + 1)} u_n.$$

Apply Raabe's test to determine whether the series converges.

27. If $\sum_{n=1}^{\infty} a_n$ converges, and if $a_n \neq 1$ and $a_n > 0$ for all n,

a. Show that $\sum_{n=1}^{\infty} a_n^2$ converges.

b. Does $\sum_{n=1}^{\infty} a_n/(1 - a_n)$ converge? Explain.

28. (*Continuation of Exercise 27.*) If $\sum_{n=1}^{\infty} a_n$ converges, and if $1 > a_n > 0$ for all n, show that $\sum_{n=1}^{\infty} \ln{(1 - a_n)}$ converges.

(*Hint:* First show that $|\ln{(1 - a_n)}| \leq a_n/(1 - a_n)$.)

29. Nicole Oresme's Theorem   Prove Nicole Oresme's Theorem that

$$1 + \frac{1}{2} \cdot 2 + \frac{1}{4} \cdot 3 + \cdots + \frac{n}{2^{n-1}} + \cdots = 4.$$

(*Hint:* Differentiate both sides of the equation $1/(1 - x) = 1 + \sum_{n=1}^{\infty} x^n$.)

30. a. Show that

$$\sum_{n=1}^{\infty} \frac{n(n + 1)}{x^n} = \frac{2x^2}{(x - 1)^3}$$

for $|x| > 1$ by differentiating the identity

$$\sum_{n=1}^{\infty} x^{n+1} = \frac{x^2}{1 - x}$$

twice, multiplying the result by x, and then replacing x by $1/x$.

b. Use part (a) to find the real solution greater than 1 of the equation

$$x = \sum_{n=1}^{\infty} \frac{n(n + 1)}{x^n}.$$

31. Quality control

a. Differentiate the series

$$\frac{1}{1 - x} = 1 + x + x^2 + \cdots + x^n + \cdots$$

to obtain a series for $1/(1 - x)^2$.

b. In one throw of two dice, the probability of getting a roll of 7 is $p = 1/6$. If you throw the dice repeatedly, the probability that a 7 will appear for the first time at the nth throw is $q^{n-1}p$, where $q = 1 - p = 5/6$. The expected number of throws until a 7 first appears is $\sum_{n=1}^{\infty} nq^{n-1}p$. Find the sum of this series.

c. As an engineer applying statistical control to an industrial operation, you inspect items taken at random from the assembly line. You classify each sampled item as either "good" or "bad." If the probability of an item's being good is p and of an item's being bad is $q = 1 - p$, the probability that the first bad item found is the nth one inspected is $p^{n-1}q$. The average number inspected up to and including the first bad item found is $\sum_{n=1}^{\infty} np^{n-1}q$. Evaluate this sum, assuming $0 < p < 1$.

32. Expected value   Suppose that a random variable X may assume the values 1, 2, 3, ..., with probabilities $p_1, p_2, p_3, \ldots$, where p_k is the probability that X equals k ($k = 1, 2, 3, \ldots$). Suppose also that $p_k \geq 0$ and that $\sum_{k=1}^{\infty} p_k = 1$. The **expected value** of X, denoted by $E(X)$, is the number $\sum_{k=1}^{\infty} k p_k$, provided the series converges. In each of the following cases, show that $\sum_{k=1}^{\infty} p_k = 1$ and find $E(X)$ if it exists. (*Hint:* See Exercise 31.)

a. $p_k = 2^{-k}$ **b.** $p_k = \dfrac{5^{k-1}}{6^k}$

c. $p_k = \dfrac{1}{k(k + 1)} = \dfrac{1}{k} - \dfrac{1}{k + 1}$

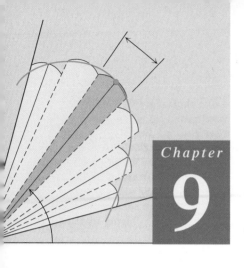

POLAR COORDINATES AND CONICS

OVERVIEW   In this chapter we study curves, derivatives, and integrals in polar coordinates. Polar coordinates are useful for describing rotational motion, as in planetary and satellite motion, and they can also simplify the calculations of multiple integrals (Chapter 13). Additionally, we review the geometric definitions of parabolas, ellipses, and hyperbolas and derive their standard equations. These curves are called *conic sections*, or *conics*, and model the paths traveled by projectiles, planets, and any other object moving under the sole influence of a gravitational or electromagnetic force.

9.1   Polar Coordinates

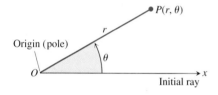

FIGURE 9.1   To define polar coordinates for the plane, we start with an origin, called the pole, and an initial ray.

In this section, we study polar coordinates and their relation to Cartesian coordinates. While a point in the plane has just one pair of Cartesian coordinates, it has infinitely many pairs of polar coordinates.

Definition of Polar Coordinates

To define polar coordinates, we first fix an **origin** O (called the **pole**) and an **initial ray** from O (Figure 9.1). Then each point P can be located by assigning to it a **polar coordinate pair** (r, θ) in which r gives the directed distance from O to P and θ gives the directed angle from the initial ray to ray OP.

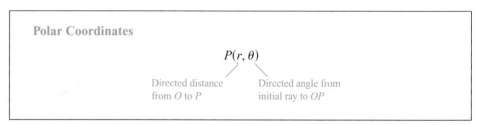

As in trigonometry, θ is positive when measured counterclockwise and negative when measured clockwise. The angle associated with a given point is not unique. For instance, the point 2 units from the origin along the ray $\theta = \pi/6$ has polar coordinates $r = 2$, $\theta = \pi/6$. It also has coordinates $r = 2$, $\theta = -11\pi/6$ (Figure 9.2). In some situations we allow r to be negative. That is why we use directed distance in defining $P(r, \theta)$. The point $P(2, 7\pi/6)$ can be reached by turning $7\pi/6$ radians counterclockwise from the initial ray and going forward 2 units (Figure 9.3). It can also be reached by turning $\pi/6$ radians

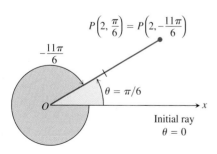

FIGURE 9.2   Polar coordinates are not unique.

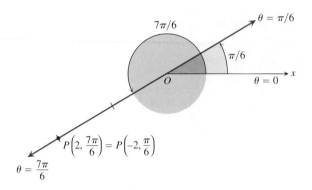

FIGURE 9.3 Polar coordinates can have negative r-values.

counterclockwise from the initial ray and going *backward* 2 units. So the point also has polar coordinates $r = -2, \theta = \pi/6$.

EXAMPLE 1 Find all the polar coordinates of the point $P(2, \pi/6)$.

Solution We sketch the initial ray of the coordinate system, draw the ray from the origin that makes an angle of $\pi/6$ radians with the initial ray, and mark the point $(2, \pi/6)$ (Figure 9.4). We then find the angles for the other coordinate pairs of P in which $r = 2$ and $r = -2$.

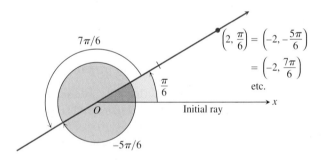

FIGURE 9.4 The point $P(2, \pi/6)$ has infinitely many polar coordinate pairs (Example 1).

For $r = 2$, the complete list of angles is

$$\frac{\pi}{6}, \quad \frac{\pi}{6} \pm 2\pi, \quad \frac{\pi}{6} \pm 4\pi, \quad \frac{\pi}{6} \pm 6\pi, \quad \dots.$$

For $r = -2$, the angles are

$$-\frac{5\pi}{6}, \quad -\frac{5\pi}{6} \pm 2\pi, \quad -\frac{5\pi}{6} \pm 4\pi, \quad -\frac{5\pi}{6} \pm 6\pi, \quad \dots.$$

The corresponding coordinate pairs of P are

$$\left(2, \frac{\pi}{6} + 2n\pi\right), \qquad n = 0, \pm 1, \pm 2, \dots$$

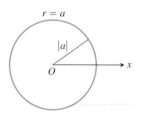

FIGURE 9.5 The polar equation for a circle is $r = a$.

and

$$\left(-2, -\frac{5\pi}{6} + 2n\pi\right), \qquad n = 0, \pm1, \pm2, \dots.$$

When $n = 0$, the formulas give $(2, \pi/6)$ and $(-2, -5\pi/6)$. When $n = 1$, they give $(2, 13\pi/6)$ and $(-2, 7\pi/6)$, and so on. ∎

Polar Equations and Graphs

If we hold r fixed at a constant value $r = a \neq 0$, the point $P(r, \theta)$ will lie $|a|$ units from the origin O. As θ varies over any interval of length 2π, P then traces a circle of radius $|a|$ centered at O (Figure 9.5).

If we hold θ fixed at a constant value $\theta = \theta_0$ and let r vary between $-\infty$ and ∞, the point $P(r, \theta)$ traces the line through O that makes an angle of measure θ_0 with the initial ray.

Equation	Graph
$r = a$	Circle of radius $|a|$ centered at O
$\theta = \theta_0$	Line through O making an angle θ_0 with the initial ray

EXAMPLE 2

(a) $r = 1$ and $r = -1$ are equations for the circle of radius 1 centered at O.

(b) $\theta = \pi/6, \theta = 7\pi/6$, and $\theta = -5\pi/6$ are equations for the line in Figure 9.4. ∎

Equations of the form $r = a$ and $\theta = \theta_0$ can be combined to define regions, segments, and rays.

EXAMPLE 3   Graph the sets of points whose polar coordinates satisfy the following conditions.

(a) $1 \leq r \leq 2 \qquad$ and $\qquad 0 \leq \theta \leq \dfrac{\pi}{2}$

(b) $-3 \leq r \leq 2 \qquad$ and $\qquad \theta = \dfrac{\pi}{4}$

(c) $r \leq 0 \qquad$ and $\qquad \theta = \dfrac{\pi}{4}$

(d) $\dfrac{2\pi}{3} \leq \theta \leq \dfrac{5\pi}{6} \qquad$ (no restriction on r)

Solution   The graphs are shown in Figure 9.6. ∎

Relating Polar and Cartesian Coordinates

When we use both polar and Cartesian coordinates in a plane, we place the two origins together and take the initial polar ray as the positive x-axis. The ray $\theta = \pi/2, r > 0$,

(a)

(b)

(c)

(d)

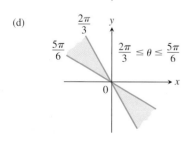

FIGURE 9.6 The graphs of typical inequalities in r and θ (Example 3).

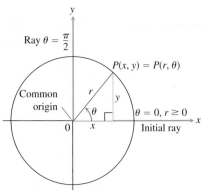

FIGURE 9.7 The usual way to relate polar and Cartesian coordinates.

becomes the positive y-axis (Figure 9.7). The two coordinate systems are then related by the following equations.

Equations Relating Polar and Cartesian Coordinates

$$x = r \cos \theta, \qquad y = r \sin \theta, \qquad x^2 + y^2 = r^2$$

The first two of these equations uniquely determine the Cartesian coordinates x and y given the polar coordinates r and θ. On the other hand, if x and y are given, the third equation gives two possible choices for r (a positive and a negative value). For each selection, there is a unique $\theta \in [0, 2\pi)$ satisfying the first two equations, each then giving a polar coordinate representation of the Cartesian point (x, y). The other polar coordinate representations for the point can be determined from these two, as in Example 1.

EXAMPLE 4   Here are some equivalent equations.

Polar equation	Cartesian equivalent
$r \cos \theta = 2$	$x = 2$
$r^2 \cos \theta \sin \theta = 4$	$xy = 4$
$r^2 \cos^2 \theta - r^2 \sin^2 \theta = 1$	$x^2 - y^2 = 1$
$r = 1 + 2r \cos \theta$	$y^2 - 3x^2 - 4x - 1 = 0$
$r = 1 - \cos \theta$	$x^4 + y^4 + 2x^2y^2 + 2x^3 + 2xy^2 - y^2 = 0$

With some curves, we are better off with polar coordinates; with others, we aren't. ∎

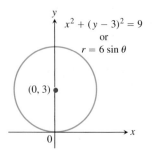

FIGURE 9.8 The circle in Example 5.

EXAMPLE 5   Find a polar equation for the circle $x^2 + (y - 3)^2 = 9$ (Figure 9.8).

Solution

$$\begin{aligned}
x^2 + y^2 - 6y + 9 &= 9 && \text{Expand } (y - 3)^2. \\
x^2 + y^2 - 6y &= 0 && \text{The 9's cancel.} \\
r^2 - 6r \sin \theta &= 0 && x^2 + y^2 = r^2 \\
r = 0 \quad \text{or} \quad r - 6 \sin \theta &= 0 \\
r &= 6 \sin \theta && \text{Includes both possibilities}
\end{aligned}$$

∎

EXAMPLE 6   Replace the following polar equations by equivalent Cartesian equations, and identify their graphs.

(a) $r \cos \theta = -4$

(b) $r^2 = 4r \cos \theta$

(c) $r = \dfrac{4}{2 \cos \theta - \sin \theta}$

Solution   We use the substitutions $r \cos \theta = x$, $r \sin \theta = y$, $r^2 = x^2 + y^2$.

(a) $r \cos \theta = -4$

The Cartesian equation: $r \cos \theta = -4$
$$x = -4$$

The graph: Vertical line through $x = -4$ on the x-axis

(b) $r^2 = 4r \cos \theta$

The Cartesian equation:
$$r^2 = 4r \cos \theta$$
$$x^2 + y^2 = 4x$$
$$x^2 - 4x + y^2 = 0$$
$$x^2 - 4x + 4 + y^2 = 4 \qquad \text{Completing the square}$$
$$(x - 2)^2 + y^2 = 4$$

The graph: Circle, radius 2, center $(h, k) = (2, 0)$

(c) $r = \dfrac{4}{2 \cos \theta - \sin \theta}$

The Cartesian equation:
$$r(2 \cos \theta - \sin \theta) = 4$$
$$2r \cos \theta - r \sin \theta = 4$$
$$2x - y = 4$$
$$y = 2x - 4$$

The graph: Line, slope $m = 2$, y-intercept $b = -4$ ■

EXERCISES 9.1

1. Which polar coordinate pairs label the same point?

a. $(3, 0)$ **b.** $(-3, 0)$ **c.** $(2, 2\pi/3)$

d. $(2, 7\pi/3)$ **e.** $(-3, \pi)$ **f.** $(2, \pi/3)$

g. $(-3, 2\pi)$ **h.** $(-2, -\pi/3)$

2. Which polar coordinate pairs label the same point?

a. $(-2, \pi/3)$ **b.** $(2, -\pi/3)$ **c.** (r, θ)

d. $(r, \theta + \pi)$ **e.** $(-r, \theta)$ **f.** $(2, -2\pi/3)$

g. $(-r, \theta + \pi)$ **h.** $(-2, 2\pi/3)$

3. Plot the following points (given in polar coordinates). Then find all the polar coordinates of each point.

a. $(2, \pi/2)$ **b.** $(2, 0)$

c. $(-2, \pi/2)$ **d.** $(-2, 0)$

4. Plot the following points (given in polar coordinates). Then find all the polar coordinates of each point.

a. $(3, \pi/4)$ **b.** $(-3, \pi/4)$

c. $(3, -\pi/4)$ **d.** $(-3, -\pi/4)$

5. Find the Cartesian coordinates of the points in Exercise 1.

6. Find the Cartesian coordinates of the following points (given in polar coordinates).

a. $\left(\sqrt{2}, \pi/4 \right)$ **b.** $(1, 0)$

c. $(0, \pi/2)$ **d.** $\left(-\sqrt{2}, \pi/4 \right)$

e. $(-3, 5\pi/6)$ **f.** $(5, \tan^{-1}(4/3))$

g. $(-1, 7\pi)$ **h.** $\left(2\sqrt{3}, 2\pi/3 \right)$

Graph the sets of points whose polar coordinates satisfy the equations and inequalities in Exercises 7–22.

7. $r = 2$ **8.** $0 \le r \le 2$

9. $r \ge 1$ **10.** $1 \le r \le 2$

11. $0 \le \theta \le \pi/6, \quad r \ge 0$ **12.** $\theta = 2\pi/3, \quad r \le -2$

13. $\theta = \pi/3, \quad -1 \le r \le 3$ **14.** $\theta = 11\pi/4, \quad r \ge -1$

15. $\theta = \pi/2, \quad r \ge 0$ **16.** $\theta = \pi/2, \quad r \le 0$

17. $0 \le \theta \le \pi, \quad r = 1$ **18.** $0 \le \theta \le \pi, \quad r = -1$

19. $\pi/4 \le \theta \le 3\pi/4, \quad 0 \le r \le 1$

20. $-\pi/4 \le \theta \le \pi/4, \quad -1 \le r \le 1$

21. $-\pi/2 \le \theta \le \pi/2, \quad 1 \le r \le 2$

22. $0 \le \theta \le \pi/2, \quad 1 \le |r| \le 2$

Replace the polar equations in Exercises 23–48 by equivalent Cartesian equations. Then describe or identify the graph.

23. $r \cos \theta = 2$ **24.** $r \sin \theta = -1$

25. $r \sin \theta = 0$ **26.** $r \cos \theta = 0$

27. $r = 4 \csc \theta$

28. $r = -3 \sec \theta$

29. $r \cos \theta + r \sin \theta = 1$

30. $r \sin \theta = r \cos \theta$

31. $r^2 = 1$

32. $r^2 = 4r \sin \theta$

33. $r = \dfrac{5}{\sin \theta - 2 \cos \theta}$

34. $r^2 \sin 2\theta = 2$

35. $r = \cot \theta \csc \theta$

36. $r = 4 \tan \theta \sec \theta$

37. $r = \csc \theta \, e^{r \cos \theta}$

38. $r \sin \theta = \ln r + \ln \cos \theta$

39. $r^2 + 2r^2 \cos \theta \sin \theta = 1$

40. $\cos^2 \theta = \sin^2 \theta$

41. $r^2 = -4r \cos \theta$

42. $r^2 = -6r \sin \theta$

43. $r = 8 \sin \theta$

44. $r = 3 \cos \theta$

45. $r = 2 \cos \theta + 2 \sin \theta$

46. $r = 2 \cos \theta - \sin \theta$

47. $r \sin \left(\theta + \dfrac{\pi}{6} \right) = 2$

48. $r \sin \left(\dfrac{2\pi}{3} - \theta \right) = 5$

Replace the Cartesian equations in Exercises 49–62 by equivalent polar equations.

49. $x = 7$ **50.** $y = 1$ **51.** $x = y$

52. $x - y = 3$ **53.** $x^2 + y^2 = 4$ **54.** $x^2 - y^2 = 1$

55. $\dfrac{x^2}{9} + \dfrac{y^2}{4} = 1$ **56.** $xy = 2$

57. $y^2 = 4x$ **58.** $x^2 + xy + y^2 = 1$

59. $x^2 + (y - 2)^2 = 4$ **60.** $(x - 5)^2 + y^2 = 25$

61. $(x - 3)^2 + (y + 1)^2 = 4$ **62.** $(x + 2)^2 + (y - 5)^2 = 16$

63. Find all polar coordinates of the origin.

64. **Vertical and horizontal lines**

  a. Show that every vertical line in the *xy*-plane has a polar equation of the form $r = a \sec \theta$.

  b. Find the analogous polar equation for horizontal lines in the *xy*-plane.

9.2   Graphing in Polar Coordinates

This section describes techniques for graphing equations in polar coordinates.

Symmetry

Figure 9.9 illustrates the standard polar coordinate tests for symmetry.

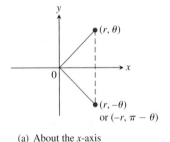

(a) About the *x*-axis   (b) About the *y*-axis   (c) About the origin

FIGURE 9.9   Three tests for symmetry in polar coordinates.

Symmetry Tests for Polar Graphs

1. *Symmetry about the x-axis:* If the point (r, θ) lies on the graph, the point $(r, -\theta)$ or $(-r, \pi - \theta)$ lies on the graph (Figure 9.9a).

2. *Symmetry about the y-axis:* If the point (r, θ) lies on the graph, the point $(r, \pi - \theta)$ or $(-r, -\theta)$ lies on the graph (Figure 9.9b).

3. *Symmetry about the origin:* If the point (r, θ) lies on the graph, the point $(-r, \theta)$ or $(r, \theta + \pi)$ lies on the graph (Figure 9.9c).

Slope

The slope of a polar curve $r = f(\theta)$ is given by dy/dx, not by $r' = df/d\theta$. To see why, think of the graph of f as the graph of the parametric equations

$$x = r\cos\theta = f(\theta)\cos\theta, \qquad y = r\sin\theta = f(\theta)\sin\theta.$$

If f is a differentiable function of θ, then so are x and y and, when $dx/d\theta \neq 0$, we can calculate dy/dx from the parametric formula

$$\frac{dy}{dx} = \frac{dy/d\theta}{dx/d\theta}$$
Section 3.5, Equation (2) with $t = \theta$

$$= \frac{\dfrac{d}{d\theta}\,(f(\theta)\cdot\sin\theta)}{\dfrac{d}{d\theta}\,(f(\theta)\cdot\cos\theta)}$$

$$= \frac{\dfrac{df}{d\theta}\sin\theta + f(\theta)\cos\theta}{\dfrac{df}{d\theta}\cos\theta - f(\theta)\sin\theta}$$
Product Rule for derivatives

Slope of the Curve $r = f(\theta)$

$$\left.\frac{dy}{dx}\right|_{(r,\theta)} = \frac{f'(\theta)\sin\theta + f(\theta)\cos\theta}{f'(\theta)\cos\theta - f(\theta)\sin\theta},$$

provided $dx/d\theta \neq 0$ at (r,θ).

If the curve $r = f(\theta)$ passes through the origin at $\theta = \theta_0$, then $f(\theta_0) = 0$, and the slope equation gives

$$\left.\frac{dy}{dx}\right|_{(0,\theta_0)} = \frac{f'(\theta_0)\sin\theta_0}{f'(\theta_0)\cos\theta_0} = \tan\theta_0.$$

If the graph of $r = f(\theta)$ passes through the origin at the value $\theta = \theta_0$, the slope of the curve there is $\tan\theta_0$. The reason we say "slope at $(0, \theta_0)$" and not just "slope at the origin" is that a polar curve may pass through the origin (or any point) more than once, with different slopes at different θ-values. This is not the case in our first example, however.

EXAMPLE 1   Graph the curve $r = 1 - \cos\theta$.

Solution   The curve is symmetric about the x-axis because

$$(r, \theta) \text{ on the graph} \Rightarrow r = 1 - \cos\theta$$

$$\Rightarrow r = 1 - \cos(-\theta) \qquad \cos\theta = \cos(-\theta)$$

$$\Rightarrow (r, -\theta) \text{ on the graph}.$$

θ	$r = 1 - \cos\theta$
0	0
$\dfrac{\pi}{3}$	$\dfrac{1}{2}$
$\dfrac{\pi}{2}$	1
$\dfrac{2\pi}{3}$	$\dfrac{3}{2}$
π	2

(a)

(b)

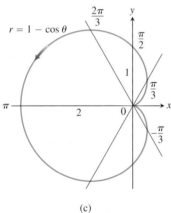

(c)

FIGURE 9.10 The steps in graphing the cardioid $r = 1 - \cos\theta$ (Example 1). The arrow shows the direction of increasing θ.

As θ increases from 0 to π, $\cos\theta$ decreases from 1 to -1, and $r = 1 - \cos\theta$ increases from a minimum value of 0 to a maximum value of 2. As θ continues on from π to 2π, $\cos\theta$ increases from -1 back to 1 and r decreases from 2 back to 0. The curve starts to repeat when $\theta = 2\pi$ because the cosine has period 2π.

The curve leaves the origin with slope $\tan(0) = 0$ and returns to the origin with slope $\tan(2\pi) = 0$.

We make a table of values from $\theta = 0$ to $\theta = \pi$, plot the points, draw a smooth curve through them with a horizontal tangent at the origin, and reflect the curve across the x-axis to complete the graph (Figure 9.10). The curve is called a *cardioid* because of its heart shape. ∎

EXAMPLE 2   Graph the curve $r^2 = 4\cos\theta$.

Solution   The equation $r^2 = 4\cos\theta$ requires $\cos\theta \geq 0$, so we get the entire graph by running θ from $-\pi/2$ to $\pi/2$. The curve is symmetric about the x-axis because

$$(r, \theta) \text{ on the graph} \Rightarrow r^2 = 4\cos\theta$$

$$\Rightarrow r^2 = 4\cos(-\theta) \qquad \cos\theta = \cos(-\theta)$$

$$\Rightarrow (r, -\theta) \text{ on the graph.}$$

The curve is also symmetric about the origin because

$$(r, \theta) \text{ on the graph} \Rightarrow r^2 = 4\cos\theta$$

$$\Rightarrow (-r)^2 = 4\cos\theta$$

$$\Rightarrow (-r, \theta) \text{ on the graph.}$$

Together, these two symmetries imply symmetry about the y-axis.

The curve passes through the origin when $\theta = -\pi/2$ and $\theta = \pi/2$. It has a vertical tangent both times because $\tan\theta$ is infinite.

For each value of θ in the interval between $-\pi/2$ and $\pi/2$, the formula $r^2 = 4\cos\theta$ gives two values of r:

$$r = \pm 2\sqrt{\cos\theta}.$$

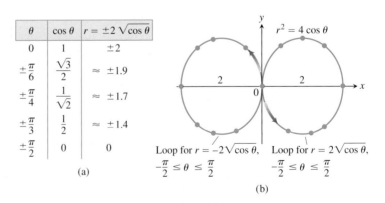

(a)

(b)

FIGURE 9.11 The graph of $r^2 = 4\cos\theta$. The arrows show the direction of increasing θ. The values of r in the table are rounded (Example 2).

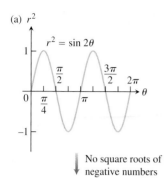

(a) r^2

$r^2 = \sin 2\theta$

No square roots of negative numbers

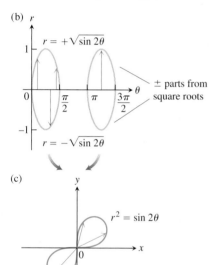

(b) r

$r = +\sqrt{\sin 2\theta}$

$\pm$ parts from square roots

$r = -\sqrt{\sin 2\theta}$

(c)

$r^2 = \sin 2\theta$

FIGURE 9.12 To plot $r = f(\theta)$ in the Cartesian $r\theta$-plane in (b), we first plot $r^2 = \sin 2\theta$ in the $r^2\theta$-plane in (a) and then ignore the values of θ for which $\sin 2\theta$ is negative. The radii from the sketch in (b) cover the polar graph of the lemniscate in (c) twice (Example 3).

We make a short table of values, plot the corresponding points, and use information about symmetry and tangents to guide us in connecting the points with a smooth curve (Figure 9.11). ∎

A Technique for Graphing

One way to graph a polar equation $r = f(\theta)$ is to make a table of (r, θ)-values, plot the corresponding points, and connect them in order of increasing θ. This can work well if enough points have been plotted to reveal all the loops and dimples in the graph. Another method of graphing that is usually quicker and more reliable is to

1. first graph $r = f(\theta)$ in the *Cartesian $r\theta$-plane*,

2. then use the Cartesian graph as a "table" and guide to sketch the *polar* coordinate graph.

This method is better than simple point plotting because the first Cartesian graph, even when hastily drawn, shows at a glance where r is positive, negative, and nonexistent, as well as where r is increasing and decreasing. Here's an example.

EXAMPLE 3   Graph the *lemniscate* curve

$$r^2 = \sin 2\theta.$$

Solution   Here we begin by plotting r^2 (not r) as a function of θ in the Cartesian $r^2\theta$-plane. See Figure 9.12a. We pass from there to the graph of $r = \pm\sqrt{\sin 2\theta}$ in the $r\theta$-plane (Figure 9.12b), and then draw the polar graph (Figure 9.12c). The graph in Figure 9.12b "covers" the final polar graph in Figure 9.12c twice. We could have managed with either loop alone, with the two upper halves, or with the two lower halves. The double covering does no harm, however, and we actually learn a little more about the behavior of the function this way. ∎

USING TECHNOLOGY   Graphing Polar Curves Parametrically

For complicated polar curves we may need to use a graphing calculator or computer to graph the curve. If the device does not plot polar graphs directly, we can convert $r = f(\theta)$ into parametric form using the equations

$$x = r\cos\theta = f(\theta)\cos\theta, \qquad y = r\sin\theta = f(\theta)\sin\theta.$$

Then we use the device to draw a parametrized curve in the Cartesian xy-plane. It may be necessary to use the parameter t rather than θ for the graphing device.

EXERCISES 9.2

Identify the symmetries of the curves in Exercises 1–12. Then sketch the curves.

1. $r = 1 + \cos\theta$

2. $r = 2 - 2\cos\theta$

3. $r = 1 - \sin\theta$

4. $r = 1 + \sin\theta$

5. $r = 2 + \sin\theta$

6. $r = 1 + 2\sin\theta$

7. $r = \sin(\theta/2)$

8. $r = \cos(\theta/2)$

9. $r^2 = \cos\theta$

10. $r^2 = \sin\theta$

11. $r^2 = -\sin\theta$

12. $r^2 = -\cos\theta$

Graph the lemniscates in Exercises 13–16. What symmetries do these curves have?

13. $r^2 = 4\cos 2\theta$

14. $r^2 = 4\sin 2\theta$

15. $r^2 = -\sin 2\theta$

16. $r^2 = -\cos 2\theta$

Find the slopes of the curves in Exercises 17–20 at the given points. Sketch the curves along with their tangents at these points.

17. Cardioid $r = -1 + \cos \theta$; $\theta = \pm \pi/2$

18. Cardioid $r = -1 + \sin \theta$; $\theta = 0, \pi$

19. Four-leaved rose $r = \sin 2\theta$; $\theta = \pm \pi/4, \pm 3\pi/4$

20. Four-leaved rose $r = \cos 2\theta$; $\theta = 0, \pm \pi/2, \pi$

Graph the limaçons in Exercises 21–24. Limaçon ("*lee*-ma-sahn") is Old French for "snail." You will understand the name when you graph the limaçons in Exercise 21. Equations for limaçons have the form $r = a \pm b \cos \theta$ or $r = a \pm b \sin \theta$. There are four basic shapes.

21. Limaçons with an inner loop

a. $r = \dfrac{1}{2} + \cos \theta$ b. $r = \dfrac{1}{2} + \sin \theta$

22. Cardioids

a. $r = 1 - \cos \theta$ b. $r = -1 + \sin \theta$

23. Dimpled limaçons

a. $r = \dfrac{3}{2} + \cos \theta$ b. $r = \dfrac{3}{2} - \sin \theta$

24. Oval limaçons

a. $r = 2 + \cos \theta$ b. $r = -2 + \sin \theta$

25. Sketch the region defined by the inequalities $-1 \leq r \leq 2$ and $-\pi/2 \leq \theta \leq \pi/2$.

26. Sketch the region defined by the inequalities $0 \leq r \leq 2 \sec \theta$ and $-\pi/4 \leq \theta \leq \pi/4$.

In Exercises 27 and 28, sketch the region defined by the inequality.

27. $0 \leq r \leq 2 - 2 \cos \theta$ **28.** $0 \leq r^2 \leq \cos \theta$

T 29. Which of the following has the same graph as $r = 1 - \cos \theta$?

a. $r = -1 - \cos \theta$

b. $r = 1 + \cos \theta$

Confirm your answer with algebra.

T 30. Which of the following has the same graph as $r = \cos 2\theta$?

a. $r = -\sin (2\theta + \pi/2)$

b. $r = -\cos (\theta/2)$

Confirm your answer with algebra.

T 31. A rose within a rose   Graph the equation $r = 1 - 2 \sin 3\theta$.

T 32. The nephroid of Freeth   Graph the nephroid of Freeth:

$$r = 1 + 2 \sin \frac{\theta}{2}.$$

T 33. Roses   Graph the roses $r = \cos m\theta$ for $m = 1/3, 2, 3$, and 7.

T 34. Spirals   Polar coordinates are just the thing for defining spirals. Graph the following spirals.

a. $r = \theta$

b. $r = -\theta$

c. *A logarithmic spiral:* $r = e^{\theta/10}$

d. *A hyperbolic spiral:* $r = 8/\theta$

e. *An equilateral hyperbola:* $r = \pm 10/\sqrt{\theta}$

(Use different colors for the two branches.)

9.3   Areas and Lengths in Polar Coordinates

This section shows how to calculate areas of plane regions and lengths of curves in polar coordinates.

Area in the Plane

The region *OTS* in Figure 9.13 is bounded by the rays $\theta = \alpha$ and $\theta = \beta$ and the curve $r = f(\theta)$. We approximate the region with n nonoverlapping fan-shaped circular sectors based on a partition P of angle *TOS*. The typical sector has radius $r_k = f(\theta_k)$ and central angle of radian measure $\Delta \theta_k$. Its area is $\Delta \theta_k / 2\pi$ times the area of a circle of radius r_k, or

$$A_k = \frac{1}{2} r_k^2 \, \Delta \theta_k = \frac{1}{2} \big(f(\theta_k) \big)^2 \, \Delta \theta_k.$$

The area of region *OTS* is approximately

$$\sum_{k=1}^{n} A_k = \sum_{k=1}^{n} \frac{1}{2} \big(f(\theta_k) \big)^2 \, \Delta \theta_k.$$

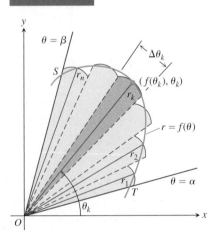

FIGURE 9.13   To derive a formula for the area of region *OTS*, we approximate the region with fan-shaped circular sectors.

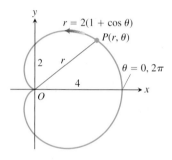

FIGURE 9.14 The area differential dA for the curve $r = f(\theta)$.

If f is continuous, we expect the approximations to improve as the norm of the partition P goes to zero, where the norm of P is the largest value of $\Delta\theta_k$. We are then led to the following formula for the region's area:

$$A = \lim_{\|P\|\to 0} \sum_{k=1}^{n} \frac{1}{2}\left(f(\theta_k)\right)^2 \Delta\theta_k$$

$$= \int_{\alpha}^{\beta} \frac{1}{2}\left(f(\theta)\right)^2 d\theta.$$

Area of the Fan-Shaped Region Between the Origin and the Curve $r = f(\theta),\ \alpha \le \theta \le \beta$

$$A = \int_{\alpha}^{\beta} \frac{1}{2} r^2 \, d\theta.$$

This is the integral of the **area differential** (Figure 9.14)

$$dA = \frac{1}{2} r^2 \, d\theta = \frac{1}{2}\left(f(\theta)\right)^2 d\theta.$$

EXAMPLE 1   Find the area of the region in the plane enclosed by the cardioid $r = 2(1 + \cos\theta)$.

Solution   We graph the cardioid (Figure 9.15) and determine that the radius OP sweeps out the region exactly once as θ runs from 0 to 2π. The area is therefore

$$\int_{\theta=0}^{\theta=2\pi} \frac{1}{2} r^2 \, d\theta = \int_{0}^{2\pi} \frac{1}{2} \cdot 4(1 + \cos\theta)^2 \, d\theta$$

$$= \int_{0}^{2\pi} 2(1 + 2\cos\theta + \cos^2\theta) \, d\theta$$

$$= \int_{0}^{2\pi} \left(2 + 4\cos\theta + 2\,\frac{1 + \cos 2\theta}{2}\right) d\theta$$

$$= \int_{0}^{2\pi} (3 + 4\cos\theta + \cos 2\theta) \, d\theta$$

$$= \left[3\theta + 4\sin\theta + \frac{\sin 2\theta}{2}\right]_{0}^{2\pi} = 6\pi - 0 = 6\pi.$$ ∎

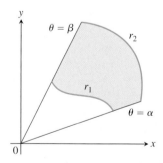

FIGURE 9.15 The cardioid in Example 1.

To find the area of a region like the one in Figure 9.16, which lies between two polar curves $r_1 = r_1(\theta)$ and $r_2 = r_2(\theta)$ from $\theta = \alpha$ to $\theta = \beta$, we subtract the integral of $(1/2)r_1^2 \, d\theta$ from the integral of $(1/2)r_2^2 \, d\theta$. This leads to the following formula.

Area of the Region $0 \le r_1(\theta) \le r \le r_2(\theta),\quad \alpha \le \theta \le \beta$

$$A = \int_{\alpha}^{\beta} \frac{1}{2} r_2^2 \, d\theta - \int_{\alpha}^{\beta} \frac{1}{2} r_1^2 \, d\theta = \int_{\alpha}^{\beta} \frac{1}{2}\left(r_2^2 - r_1^2\right) d\theta \qquad (1)$$

FIGURE 9.16 The area of the shaded region is calculated by subtracting the area of the region between r_1 and the origin from the area of the region between r_2 and the origin.

EXAMPLE 2   Find the area of the region that lies inside the circle $r = 1$ and outside the cardioid $r = 1 - \cos\theta$.

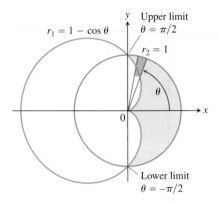

FIGURE 9.17 The region and limits of integration in Example 2.

Solution We sketch the region to determine its boundaries and find the limits of integration (Figure 9.17). The outer curve is $r_2 = 1$, the inner curve is $r_1 = 1 - \cos\theta$, and θ runs from $-\pi/2$ to $\pi/2$. The area, from Equation (1), is

$$A = \int_{-\pi/2}^{\pi/2} \frac{1}{2}\left(r_2{}^2 - r_1{}^2\right) d\theta$$

$$= 2\int_{0}^{\pi/2} \frac{1}{2}\left(r_2{}^2 - r_1{}^2\right) d\theta \qquad \text{Symmetry}$$

$$= \int_{0}^{\pi/2} \left(1 - \left(1 - 2\cos\theta + \cos^2\theta\right)\right) d\theta$$

$$= \int_{0}^{\pi/2} \left(2\cos\theta - \cos^2\theta\right) d\theta = \int_{0}^{\pi/2} \left(2\cos\theta - \frac{1 + \cos 2\theta}{2}\right) d\theta$$

$$= \left[2\sin\theta - \frac{\theta}{2} - \frac{\sin 2\theta}{4}\right]_{0}^{\pi/2} = 2 - \frac{\pi}{4}. \qquad \blacksquare$$

The fact that we can represent a point in different ways in polar coordinates requires extra care in deciding when a point lies on the graph of a polar equation and in determining the points in which polar graphs intersect. (We needed intersection points in Example 2.) In Cartesian coordinates, we can always find the points where two curves cross by solving their equations simultaneously. In polar coordinates, the story is different. Simultaneous solution may reveal some intersection points without revealing others. One way to identify all the points of intersection is to graph the equations.

Length of a Polar Curve

We can obtain a polar coordinate formula for the length of a curve $r = f(\theta)$, $\alpha \le \theta \le \beta$, by parametrizing the curve as

$$x = r\cos\theta = f(\theta)\cos\theta, \qquad y = r\sin\theta = f(\theta)\sin\theta, \qquad \alpha \le \theta \le \beta. \qquad (2)$$

The parametric length formula, Equation (1) from Section 6.3, then gives the length as

$$L = \int_{\alpha}^{\beta} \sqrt{\left(\frac{dx}{d\theta}\right)^2 + \left(\frac{dy}{d\theta}\right)^2}\, d\theta.$$

This equation becomes

$$L = \int_{\alpha}^{\beta} \sqrt{r^2 + \left(\frac{dr}{d\theta}\right)^2}\, d\theta$$

when Equations (2) are substituted for x and y (Exercise 27).

Length of a Polar Curve

If $r = f(\theta)$ has a continuous first derivative for $\alpha \le \theta \le \beta$ and if the point $P(r, \theta)$ traces the curve $r = f(\theta)$ exactly once as θ runs from α to β, then the length of the curve is

$$L = \int_{\alpha}^{\beta} \sqrt{r^2 + \left(\frac{dr}{d\theta}\right)^2}\, d\theta. \qquad (3)$$

EXAMPLE 3 Find the length of the cardioid $r = 1 - \cos\theta$.

Solution We sketch the cardioid to determine the limits of integration (Figure 9.18). The point $P(r, \theta)$ traces the curve once, counterclockwise as θ runs from 0 to 2π, so these are the values we take for α and β.

With

$$r = 1 - \cos\theta, \qquad \frac{dr}{d\theta} = \sin\theta,$$

we have

$$r^2 + \left(\frac{dr}{d\theta}\right)^2 = (1 - \cos\theta)^2 + (\sin\theta)^2$$

$$= 1 - 2\cos\theta + \underbrace{\cos^2\theta + \sin^2\theta}_{1} = 2 - 2\cos\theta$$

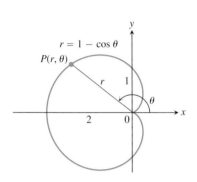

$r = 1 - \cos\theta$

$P(r, \theta)$

and

$$L = \int_\alpha^\beta \sqrt{r^2 + \left(\frac{dr}{d\theta}\right)^2}\, d\theta = \int_0^{2\pi} \sqrt{2 - 2\cos\theta}\, d\theta$$

$$= \int_0^{2\pi} \sqrt{4\sin^2\frac{\theta}{2}}\, d\theta \qquad 1 - \cos\theta = 2\sin^2\frac{\theta}{2}$$

$$= \int_0^{2\pi} 2\left|\sin\frac{\theta}{2}\right| d\theta$$

$$= \int_0^{2\pi} 2\sin\frac{\theta}{2}\, d\theta \qquad \sin\frac{\theta}{2} \geq 0 \quad \text{for} \quad 0 \leq \theta \leq 2\pi$$

$$= \left[-4\cos\frac{\theta}{2}\right]_0^{2\pi} = 4 + 4 = 8. \qquad \blacksquare$$

FIGURE 9.18 Calculating the length of a cardioid (Example 3).

EXERCISES 9.3

Find the areas of the regions in Exercises 1–6.

1. Inside the oval limaçon $r = 4 + 2\cos\theta$

2. Inside the cardioid $r = a(1 + \cos\theta), \quad a > 0$

3. Inside one leaf of the four-leaved rose $r = \cos 2\theta$

4. Inside the lemniscate $r^2 = 2a^2\cos 2\theta, \quad a > 0$

5. Inside one loop of the lemniscate $r^2 = 4\sin 2\theta$

6. Inside the six-leaved rose $r^2 = 2\sin 3\theta$

Find the areas of the regions in Exercises 7–16.

7. Shared by the circles $r = 2\cos\theta$ and $r = 2\sin\theta$

8. Shared by the circles $r = 1$ and $r = 2\sin\theta$

9. Shared by the circle $r = 2$ and the cardioid $r = 2(1 - \cos\theta)$

10. Shared by the cardioids $r = 2(1 + \cos\theta)$ and $r = 2(1 - \cos\theta)$

11. Inside the lemniscate $r^2 = 6\cos 2\theta$ and outside the circle $r = \sqrt{3}$

12. Inside the circle $r = 3a\cos\theta$ and outside the cardioid $r = a(1 + \cos\theta), a > 0$

13. Inside the circle $r = -2\cos\theta$ and outside the circle $r = 1$

14. Inside the circle $r = 6$ above the line $r = 3\csc\theta$

15. a. Find the area of the shaded region in the accompanying figure.

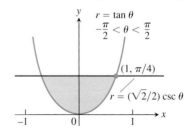

y $r = \tan\theta$

$-\frac{\pi}{2} < \theta < \frac{\pi}{2}$

$(1, \pi/4)$

$r = (\sqrt{2}/2)\csc\theta$

b. It looks as if the graph of $r = \tan\theta, -\pi/2 < \theta < \pi/2$, could be asymptotic to the lines $x = 1$ and $x = -1$. Is it? Give reasons for your answer.

16. The area of the region that lies inside the cardioid curve $r = \cos\theta + 1$ and outside the circle $r = \cos\theta$ is not

$$\frac{1}{2}\int_0^{2\pi} [(\cos\theta + 1)^2 - \cos^2\theta]\, d\theta = \pi.$$

Why not? What *is* the area? Give reasons for your answers.

Find the lengths of the curves in Exercises 17–25.

17. The spiral $r = \theta^2$, $\quad 0 \le \theta \le \sqrt{5}$

18. The spiral $r = e^\theta/\sqrt{2}$, $\quad 0 \le \theta \le \pi$

19. The cardioid $r = 1 + \cos\theta$

20. The curve $r = a\sin^2(\theta/2)$, $\quad 0 \le \theta \le \pi$, $\quad a > 0$

21. The parabolic segment $r = 6/(1 + \cos\theta)$, $\quad 0 \le \theta \le \pi/2$

22. The parabolic segment $r = 2/(1 - \cos\theta)$, $\quad \pi/2 \le \theta \le \pi$

23. The curve $r = \cos^3(\theta/3)$, $\quad 0 \le \theta \le \pi/4$

24. The curve $r = \sqrt{1 + \sin 2\theta}$, $\quad 0 \le \theta \le \pi\sqrt{2}$

25. The curve $r = \sqrt{1 + \cos 2\theta}$, $\quad 0 \le \theta \le \pi\sqrt{2}$

26. **Circumferences of circles** As usual, when faced with a new formula, it is a good idea to try it on familiar objects to be sure it gives results consistent with past experience. Use the length formula in Equation (3) to calculate the circumferences of the following circles ($a > 0$):

    a. $r = a$ **b.** $r = a\cos\theta$ **c.** $r = a\sin\theta$

27. **The length of the curve $r = f(\theta)$, $\alpha \le \theta \le \beta$** Assuming that the necessary derivatives are continuous, show how the substitutions

    $$x = f(\theta)\cos\theta, \quad y = f(\theta)\sin\theta$$

(Equations 2 in the text) transform

$$L = \int_\alpha^\beta \sqrt{\left(\frac{dx}{d\theta}\right)^2 + \left(\frac{dy}{d\theta}\right)^2}\, d\theta$$

into

$$L = \int_\alpha^\beta \sqrt{r^2 + \left(\frac{dr}{d\theta}\right)^2}\, d\theta.$$

28. **Average value** If f is continuous, the average value of the polar coordinate r over the curve $r = f(\theta)$, $\alpha \le \theta \le \beta$, with respect to θ is given by the formula

    $$r_{\text{av}} = \frac{1}{\beta - \alpha}\int_\alpha^\beta f(\theta)\, d\theta.$$

    Use this formula to find the average value of r with respect to θ over the following curves ($a > 0$).

    a. The cardioid $r = a(1 - \cos\theta)$

    b. The circle $r = a$

    c. The circle $r = a\cos\theta$, $\quad -\pi/2 \le \theta \le \pi/2$

29. $r = f(\theta)$ **vs.** $r = 2f(\theta)$ Can anything be said about the relative lengths of the curves $r = f(\theta)$, $\alpha \le \theta \le \beta$, and $r = 2f(\theta)$, $\alpha \le \theta \le \beta$? Give reasons for your answer.

30. $r = f(\theta)$ **vs.** $r = 2f(\theta)$ The curves $r = f(\theta)$, $\alpha \le \theta \le \beta$, and $r = 2f(\theta)$, $\alpha \le \theta \le \beta$, are revolved about the x-axis to generate surfaces. Can anything be said about the relative areas of these surfaces? Give reasons for your answer.

9.4   Conic Sections

In this section we define parabolas, ellipses, and hyperbolas geometrically and derive their standard equations. These curves are called *conic sections* or *conics* because they are formed by cutting a double cone with a plane (Figure 9.19), which is the only way they could be described by Greek mathematicians who did not have our tools of Cartesian or polar coordinates.

Parabolas

> **DEFINITIONS**   A set that consists of all the points in a plane equidistant from a given fixed point and a given fixed line in the plane is a **parabola**. The fixed point is the **focus** of the parabola. The fixed line is the **directrix**.

If the focus F lies on the directrix L, the parabola is the line through F perpendicular to L. We consider this to be a degenerate case and assume henceforth that F does not lie on L.

A parabola has its simplest equation when its focus and directrix straddle one of the coordinate axes. For example, suppose that the focus lies at the point $F(0, p)$ on the positive

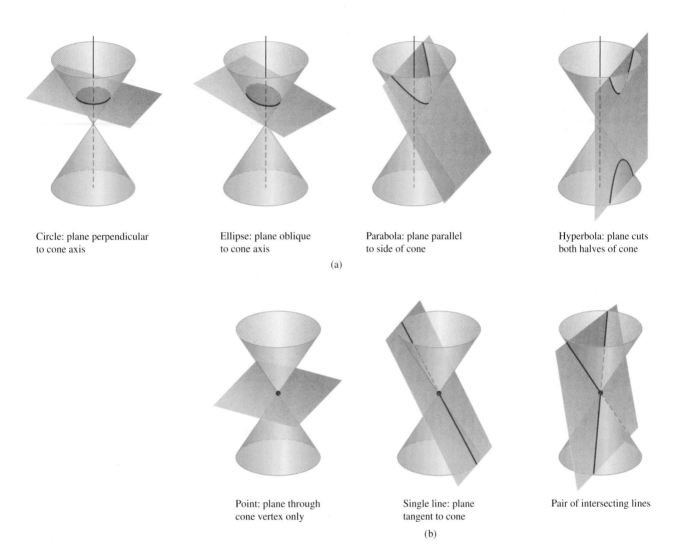

Circle: plane perpendicular
to cone axis

Ellipse: plane oblique
to cone axis

Parabola: plane parallel
to side of cone

Hyperbola: plane cuts
both halves of cone

(a)

Point: plane through
cone vertex only

Single line: plane
tangent to cone

Pair of intersecting lines

(b)

FIGURE 9.19 The standard conic sections (a) are the curves in which a plane cuts a *double* cone. Hyperbolas come in two parts, called *branches*. The point and lines obtained by passing the plane through the cone's vertex (b) are *degenerate* conic sections.

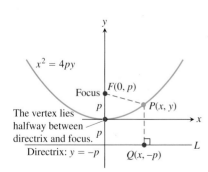

FIGURE 9.20 The standard form of the parabola $x^2 = 4py, p > 0$.

y-axis and that the directrix is the line $y = -p$ (Figure 9.20). In the notation of the figure, a point $P(x, y)$ lies on the parabola if and only if $PF = PQ$. From the distance formula,

$$PF = \sqrt{(x - 0)^2 + (y - p)^2} = \sqrt{x^2 + (y - p)^2}$$

$$PQ = \sqrt{(x - x)^2 + (y - (-p))^2} = \sqrt{(y + p)^2}.$$

When we equate these expressions, square, and simplify, we get

$$y = \frac{x^2}{4p} \quad \text{or} \quad x^2 = 4py. \quad \text{Standard form} \quad (1)$$

These equations reveal the parabola's symmetry about the y-axis. We call the y-axis the **axis** of the parabola (short for "axis of symmetry").

The point where a parabola crosses its axis is the **vertex**. The vertex of the parabola $x^2 = 4py$ lies at the origin (Figure 9.20). The positive number p is the parabola's **focal length**.

If the parabola opens downward, with its focus at $(0, -p)$ and its directrix the line $y = p$, then Equations (1) become

$$y = -\frac{x^2}{4p} \quad \text{and} \quad x^2 = -4py.$$

By interchanging the variables x and y, we obtain similar equations for parabolas opening to the right or to the left (Figure 9.21).

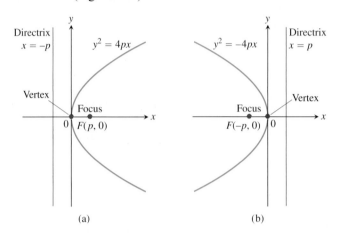

(a)   (b)

FIGURE 9.21   (a) The parabola $y^2 = 4px$. (b) The parabola $y^2 = -4px$.

EXAMPLE 1   Find the focus and directrix of the parabola $y^2 = 10x$.

Solution   We find the value of p in the standard equation $y^2 = 4px$:

$$4p = 10, \quad \text{so} \quad p = \frac{10}{4} = \frac{5}{2}.$$

Then we find the focus and directrix for this value of p:

$$\text{Focus:} \quad (p, 0) = \left(\frac{5}{2}, 0\right)$$

$$\text{Directrix:} \quad x = -p \quad \text{or} \quad x = -\frac{5}{2}. \quad \blacksquare$$

Ellipses

FIGURE 9.22   Points on the focal axis of an ellipse.

> **DEFINITIONS**   An **ellipse** is the set of points in a plane whose distances from two fixed points in the plane have a constant sum. The two fixed points are the **foci** of the ellipse.
> The line through the foci of an ellipse is the ellipse's **focal axis**. The point on the axis halfway between the foci is the **center**. The points where the focal axis and ellipse cross are the ellipse's **vertices** (Figure 9.22).

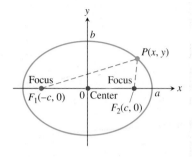

FIGURE 9.23 The ellipse defined by the equation $PF_1 + PF_2 = 2a$ is the graph of the equation $(x^2/a^2) + (y^2/b^2) = 1$, where $b^2 = a^2 - c^2$.

If the foci are $F_1(-c, 0)$ and $F_2(c, 0)$ (Figure 9.23), and $PF_1 + PF_2$ is denoted by $2a$, then the coordinates of a point P on the ellipse satisfy the equation

$$\sqrt{(x + c)^2 + y^2} + \sqrt{(x - c)^2 + y^2} = 2a.$$

To simplify this equation, we move the second radical to the right-hand side, square, isolate the remaining radical, and square again, obtaining

$$\frac{x^2}{a^2} + \frac{y^2}{a^2 - c^2} = 1. \tag{2}$$

Since $PF_1 + PF_2$ is greater than the length F_1F_2 (by the triangle inequality for triangle PF_1F_2), the number $2a$ is greater than $2c$. Accordingly, $a > c$ and the number $a^2 - c^2$ in Equation (2) is positive.

The algebraic steps leading to Equation (2) can be reversed to show that every point P whose coordinates satisfy an equation of this form with $0 < c < a$ also satisfies the equation $PF_1 + PF_2 = 2a$. A point therefore lies on the ellipse if and only if its coordinates satisfy Equation (2).

If

$$b = \sqrt{a^2 - c^2}, \tag{3}$$

then $a^2 - c^2 = b^2$ and Equation (2) takes the form

$$\frac{x^2}{a^2} + \frac{y^2}{b^2} = 1. \tag{4}$$

Equation (4) reveals that this ellipse is symmetric with respect to the origin and both coordinate axes. It lies inside the rectangle bounded by the lines $x = \pm a$ and $y = \pm b$. It crosses the axes at the points $(\pm a, 0)$ and $(0, \pm b)$. The tangents at these points are perpendicular to the axes because

$$\frac{dy}{dx} = -\frac{b^2 x}{a^2 y} \qquad \text{Obtained from Equation (4) by implicit differentiation}$$

is zero if $x = 0$ and infinite if $y = 0$.

The **major axis** of the ellipse in Equation (4) is the line segment of length $2a$ joining the points $(\pm a, 0)$. The **minor axis** is the line segment of length $2b$ joining the points $(0, \pm b)$. The number a itself is the **semimajor axis**, the number b the **semiminor axis**. The number c, found from Equation (3) as

$$c = \sqrt{a^2 - b^2},$$

is the **center-to-focus distance** of the ellipse. If $a = b$, the ellipse is a circle.

EXAMPLE 2 The ellipse

$$\frac{x^2}{16} + \frac{y^2}{9} = 1 \tag{5}$$

(Figure 9.24) has

Semimajor axis: $a = \sqrt{16} = 4,$ Semiminor axis: $b = \sqrt{9} = 3$
Center-to-focus distance: $c = \sqrt{16 - 9} = \sqrt{7}$
Foci: $(\pm c, 0) = (\pm\sqrt{7}, 0)$
Vertices: $(\pm a, 0) = (\pm 4, 0)$
Center: $(0, 0)$. ∎

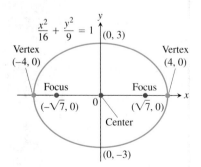

FIGURE 9.24 An ellipse with its major axis horizontal (Example 2).

If we interchange x and y in Equation (5), we have the equation

$$\frac{x^2}{9} + \frac{y^2}{16} = 1 \tag{6}$$

The major axis of this ellipse is now vertical instead of horizontal, with the foci and vertices on the y-axis. There is no confusion in analyzing Equations (5) and (6). If we find the intercepts on the coordinate axes, we will know which way the major axis runs because it is the longer of the two axes.

Standard-Form Equations for Ellipses Centered at the Origin

Foci on the x-axis: $\quad \dfrac{x^2}{a^2} + \dfrac{y^2}{b^2} = 1 \quad (a > b)$

$\qquad\qquad$ Center-to-focus distance: $\quad c = \sqrt{a^2 - b^2}$

$\qquad\qquad$ Foci: $\quad (\pm c, 0)$

$\qquad\qquad$ Vertices: $\quad (\pm a, 0)$

Foci on the y-axis: $\quad \dfrac{x^2}{b^2} + \dfrac{y^2}{a^2} = 1 \quad (a > b)$

$\qquad\qquad$ Center-to-focus distance: $\quad c = \sqrt{a^2 - b^2}$

$\qquad\qquad$ Foci: $\quad (0, \pm c)$

$\qquad\qquad$ Vertices: $\quad (0, \pm a)$

In each case, a is the semimajor axis and b is the semiminor axis.

Hyperbolas

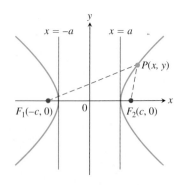

FIGURE 9.25 Points on the focal axis of a hyperbola.

FIGURE 9.26 Hyperbolas have two branches. For points on the right-hand branch of the hyperbola shown here, $PF_1 - PF_2 = 2a$. For points on the left-hand branch, $PF_2 - PF_1 = 2a$. We then let $b = \sqrt{c^2 - a^2}$.

DEFINITIONS A **hyperbola** is the set of points in a plane whose distances from two fixed points in the plane have a constant difference. The two fixed points are the **foci** of the hyperbola.

The line through the foci of a hyperbola is the **focal axis**. The point on the axis halfway between the foci is the hyperbola's **center**. The points where the focal axis and hyperbola cross are the **vertices** (Figure 9.25).

If the foci are $F_1(-c, 0)$ and $F_2(c, 0)$ (Figure 9.26) and the constant difference is $2a$, then a point (x, y) lies on the hyperbola if and only if

$$\sqrt{(x + c)^2 + y^2} - \sqrt{(x - c)^2 + y^2} = \pm 2a. \tag{7}$$

To simplify this equation, we move the second radical to the right-hand side, square, isolate the remaining radical, and square again, obtaining

$$\frac{x^2}{a^2} + \frac{y^2}{a^2 - c^2} = 1. \tag{8}$$

So far, this looks just like the equation for an ellipse. But now $a^2 - c^2$ is negative because $2a$, being the difference of two sides of triangle PF_1F_2, is less than $2c$, the third side.

The algebraic steps leading to Equation (8) can be reversed to show that every point P whose coordinates satisfy an equation of this form with $0 < a < c$ also satisfies Equation (7). A point therefore lies on the hyperbola if and only if its coordinates satisfy Equation (8).

If we let b denote the positive square root of $c^2 - a^2$,

$$b = \sqrt{c^2 - a^2}, \qquad (9)$$

then $a^2 - c^2 = -b^2$ and Equation (8) takes the more compact form

$$\frac{x^2}{a^2} - \frac{y^2}{b^2} = 1. \qquad (10)$$

The differences between Equation (10) and the equation for an ellipse (Equation 4) are the minus sign and the new relation

$$c^2 = a^2 + b^2. \qquad \text{From Equation (9)}$$

Like the ellipse, the hyperbola is symmetric with respect to the origin and coordinate axes. It crosses the x-axis at the points $(\pm a, 0)$. The tangents at these points are vertical because

$$\frac{dy}{dx} = \frac{b^2 x}{a^2 y} \qquad \begin{array}{l}\text{Obtained from Equation (10)}\\ \text{by implicit differentiation}\end{array}$$

is infinite when $y = 0$. The hyperbola has no y-intercepts; in fact, no part of the curve lies between the lines $x = -a$ and $x = a$.

The lines

$$y = \pm \frac{b}{a} x$$

are the two **asymptotes** of the hyperbola defined by Equation (10). The fastest way to find the equations of the asymptotes is to replace the 1 in Equation (10) by 0 and solve the new equation for y:

$$\underbrace{\frac{x^2}{a^2} - \frac{y^2}{b^2} = 1}_{\text{hyperbola}} \rightarrow \underbrace{\frac{x^2}{a^2} - \frac{y^2}{b^2} = 0}_{\text{0 for 1}} \rightarrow \underbrace{y = \pm \frac{b}{a} x.}_{\text{asymptotes}}$$

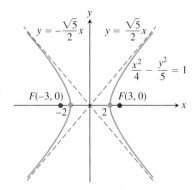

FIGURE 9.27 The hyperbola and its asymptotes in Example 3.

EXAMPLE 3 The equation

$$\frac{x^2}{4} - \frac{y^2}{5} = 1 \qquad (11)$$

is Equation (10) with $a^2 = 4$ and $b^2 = 5$ (Figure 9.27). We have

Center-to-focus distance: $c = \sqrt{a^2 + b^2} = \sqrt{4 + 5} = 3$

Foci: $(\pm c, 0) = (\pm 3, 0)$, Vertices: $(\pm a, 0) = (\pm 2, 0)$

Center: $(0, 0)$

Asymptotes: $\dfrac{x^2}{4} - \dfrac{y^2}{5} = 0$ or $y = \pm \dfrac{\sqrt{5}}{2} x.$ ∎

If we interchange x and y in Equation (11), the foci and vertices of the resulting hyperbola will lie along the y-axis. We still find the asymptotes in the same way as before, but now their equations will be $y = \pm 2x/\sqrt{5}$.

Standard-Form Equations for Hyperbolas Centered at the Origin

Foci on the x-axis: $\dfrac{x^2}{a^2} - \dfrac{y^2}{b^2} = 1$ *Foci on the y-axis:* $\dfrac{y^2}{a^2} - \dfrac{x^2}{b^2} = 1$

   Center-to-focus distance: $c = \sqrt{a^2 + b^2}$ Center-to-focus distance: $c = \sqrt{a^2 + b^2}$

   Foci: $(\pm c, 0)$ Foci: $(0, \pm c)$

   Vertices: $(\pm a, 0)$ Vertices: $(0, \pm a)$

   Asymptotes: $\dfrac{x^2}{a^2} - \dfrac{y^2}{b^2} = 0$ or $y = \pm\dfrac{b}{a}x$ Asymptotes: $\dfrac{y^2}{a^2} - \dfrac{x^2}{b^2} = 0$ or $y = \pm\dfrac{a}{b}x$

Notice the difference in the asymptote equations (b/a in the first, a/b in the second).

We shift conics using the principles reviewed in Section 1.2 replacing x by $x + h$ and y by $y + k$.

EXAMPLE 4   Show that the equation $x^2 - 4y^2 + 2x + 8y - 7 = 0$ represents a hyperbola. Find its center, asymptotes, and foci.

Solution   We reduce the equation to standard form by completing the square in x and y as follows:

$$(x^2 + 2x) - 4(y^2 - 2y) = 7$$

$$(x^2 + 2x + 1) - 4(y^2 - 2y + 1) = 7 + 1 - 4$$

$$\frac{(x + 1)^2}{4} - (y - 1)^2 = 1.$$

This is the standard form Equation (10) of a hyperbola with x replaced by $x + 1$ and y replaced by $y - 1$. The hyperbola is shifted one unit to the left and one unit upward, and it has center $x + 1 = 0$ and $y - 1 = 0$, or $x = -1$ and $y = 1$. Moreover,

$$a^2 = 4, \quad b^2 = 1, \quad c^2 = a^2 + b^2 = 5$$

so the asymptotes are the two lines

$$\frac{x + 1}{2} - (y - 1) = 0 \quad \text{and} \quad \frac{x + 1}{2} + (y - 1) = 0.$$

The shifted foci have coordinates $\left(-1 \pm \sqrt{5}, 1\right)$.   ∎

EXERCISES 9.4

Match the parabolas in Exercises 1–4 with the following equations:

$$x^2 = 2y, \quad x^2 = -6y, \quad y^2 = 8x, \quad y^2 = -4x.$$

Then find the parabola's focus and directrix.

1.

2.

3.

4.

Match each conic section in Exercises 5–8 with one of these equations:

$$\frac{x^2}{4} + \frac{y^2}{9} = 1, \qquad \frac{x^2}{2} + y^2 = 1,$$

$$\frac{y^2}{4} - x^2 = 1, \qquad \frac{x^2}{4} - \frac{y^2}{9} = 1.$$

Then find the conic section's foci and vertices. If the conic section is a hyperbola, find its asymptotes as well.

5.

6.

7.

8.
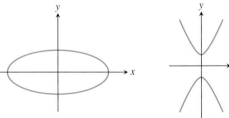

Exercises 9–16 give equations of parabolas. Find each parabola's focus and directrix. Then sketch the parabola. Include the focus and directrix in your sketch.

9. $y^2 = 12x$ **10.** $x^2 = 6y$ **11.** $x^2 = -8y$

12. $y^2 = -2x$ **13.** $y = 4x^2$ **14.** $y = -8x^2$

15. $x = -3y^2$ **16.** $x = 2y^2$

Exercises 17–24 give equations for ellipses. Put each equation in standard form. Then sketch the ellipse. Include the foci in your sketch.

17. $16x^2 + 25y^2 = 400$ **18.** $7x^2 + 16y^2 = 112$

19. $2x^2 + y^2 = 2$ **20.** $2x^2 + y^2 = 4$

21. $3x^2 + 2y^2 = 6$ **22.** $9x^2 + 10y^2 = 90$

23. $6x^2 + 9y^2 = 54$ **24.** $169x^2 + 25y^2 = 4225$

Exercises 25 and 26 give information about the foci and vertices of ellipses centered at the origin of the xy-plane. In each case, find the ellipse's standard-form equation from the given information.

25. Foci: $\left(\pm\sqrt{2}, 0\right)$ Vertices: $(\pm 2, 0)$

26. Foci: $(0, \pm 4)$ Vertices: $(0, \pm 5)$

Exercises 27–34 give equations for hyperbolas. Put each equation in standard form and find the hyperbola's asymptotes. Then sketch the hyperbola. Include the asymptotes and foci in your sketch.

27. $x^2 - y^2 = 1$ **28.** $9x^2 - 16y^2 = 144$

29. $y^2 - x^2 = 8$ **30.** $y^2 - x^2 = 4$

31. $8x^2 - 2y^2 = 16$ **32.** $y^2 - 3x^2 = 3$

33. $8y^2 - 2x^2 = 16$ **34.** $64x^2 - 36y^2 = 2304$

Exercises 35–38 give information about the foci, vertices, and asymptotes of hyperbolas centered at the origin of the xy-plane. In each case, find the hyperbola's standard-form equation from the information given.

35. Foci: $\left(0, \pm\sqrt{2}\right)$ **36.** Foci: $(\pm 2, 0)$

    Asymptotes: $y = \pm x$ Asymptotes: $y = \pm\dfrac{1}{\sqrt{3}}x$

37. Vertices: $(\pm 3, 0)$ **38.** Vertices: $(0, \pm 2)$

    Asymptotes: $y = \pm\dfrac{4}{3}x$ Asymptotes: $y = \pm\dfrac{1}{2}x$

You may wish to review Section 1.2 before solving Exercises 39–56.

39. The parabola $y^2 = 8x$ is shifted down 2 units and right 1 unit to generate the parabola $(y + 2)^2 = 8(x - 1)$.

    a. Find the new parabola's vertex, focus, and directrix.

    b. Plot the new vertex, focus, and directrix, and sketch in the parabola.

40. The parabola $x^2 = -4y$ is shifted left 1 unit and up 3 units to generate the parabola $(x + 1)^2 = -4(y - 3)$.

    a. Find the new parabola's vertex, focus, and directrix.

    b. Plot the new vertex, focus, and directrix, and sketch in the parabola.

41. The ellipse $(x^2/16) + (y^2/9) = 1$ is shifted 4 units to the right and 3 units up to generate the ellipse
$$\frac{(x-4)^2}{16} + \frac{(y-3)^2}{9} = 1.$$

    a. Find the foci, vertices, and center of the new ellipse.

    b. Plot the new foci, vertices, and center, and sketch in the new ellipse.

42. The ellipse $(x^2/9) + (y^2/25) = 1$ is shifted 3 units to the left and 2 units down to generate the ellipse
$$\frac{(x+3)^2}{9} + \frac{(y+2)^2}{25} = 1.$$

    a. Find the foci, vertices, and center of the new ellipse.

    b. Plot the new foci, vertices, and center, and sketch in the new ellipse.

43. The hyperbola $(x^2/16) - (y^2/9) = 1$ is shifted 2 units to the right to generate the hyperbola
$$\frac{(x-2)^2}{16} - \frac{y^2}{9} = 1.$$

    a. Find the center, foci, vertices, and asymptotes of the new hyperbola.

    b. Plot the new center, foci, vertices, and asymptotes, and sketch in the hyperbola.

44. The hyperbola $(y^2/4) - (x^2/5) = 1$ is shifted 2 units down to generate the hyperbola
$$\frac{(y+2)^2}{4} - \frac{x^2}{5} = 1.$$

    a. Find the center, foci, vertices, and asymptotes of the new hyperbola.

    b. Plot the new center, foci, vertices, and asymptotes, and sketch in the hyperbola.

Exercises 45–48 give equations for parabolas and tell how many units up or down and to the right or left each parabola is to be shifted. Find an equation for the new parabola, and find the new vertex, focus, and directrix.

45. $y^2 = 4x$,   left 2, down 3   46. $y^2 = -12x$,   right 4, up 3

47. $x^2 = 8y$,   right 1, down 7   48. $x^2 = 6y$,   left 3, down 2

Exercises 49–52 give equations for ellipses and tell how many units up or down and to the right or left each ellipse is to be shifted. Find an equation for the new ellipse, and find the new foci, vertices, and center.

49. $\dfrac{x^2}{6} + \dfrac{y^2}{9} = 1$,   left 2, down 1

50. $\dfrac{x^2}{2} + y^2 = 1$,   right 3, up 4

51. $\dfrac{x^2}{3} + \dfrac{y^2}{2} = 1$,   right 2, up 3

52. $\dfrac{x^2}{16} + \dfrac{y^2}{25} = 1$,   left 4, down 5

Exercises 53–56 give equations for hyperbolas and tell how many units up or down and to the right or left each hyperbola is to be shifted. Find an equation for the new hyperbola, and find the new center, foci, vertices, and asymptotes.

53. $\dfrac{x^2}{4} - \dfrac{y^2}{5} = 1$,   right 2, up 2

54. $\dfrac{x^2}{16} - \dfrac{y^2}{9} = 1$,   left 2, down 1

55. $y^2 - x^2 = 1$,   left 1, down 1

56. $\dfrac{y^2}{3} - x^2 = 1$,   right 1, up 3

Find the center, foci, vertices, asymptotes, and radius, as appropriate, of the conic sections in Exercises 57–68.

57. $x^2 + 4x + y^2 = 12$

58. $2x^2 + 2y^2 - 28x + 12y + 114 = 0$

59. $x^2 + 2x + 4y - 3 = 0$ 60. $y^2 - 4y - 8x - 12 = 0$

61. $x^2 + 5y^2 + 4x = 1$ 62. $9x^2 + 6y^2 + 36y = 0$

63. $x^2 + 2y^2 - 2x - 4y = -1$

64. $4x^2 + y^2 + 8x - 2y = -1$

65. $x^2 - y^2 - 2x + 4y = 4$ 66. $x^2 - y^2 + 4x - 6y = 6$

67. $2x^2 - y^2 + 6y = 3$ 68. $y^2 - 4x^2 + 16x = 24$

69. If lines are drawn parallel to the coordinate axes through a point P on the parabola $y^2 = kx, k > 0$, the parabola partitions the rectangular region bounded by these lines and the coordinate axes into two smaller regions, A and B.

    a. If the two smaller regions are revolved about the y-axis, show that they generate solids whose volumes have the ratio 4:1.

    b. What is the ratio of the volumes generated by revolving the regions about the x-axis?

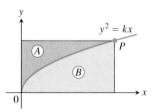

70. **Suspension bridge cables hang in parabolas** The suspension bridge cable shown on the next page supports a uniform load of w pounds per horizontal foot. It can be shown that if H is the horizontal tension of the cable at the origin, then the curve of the cable satisfies the equation
$$\frac{dy}{dx} = \frac{w}{H}x.$$

Show that the cable hangs in a parabola by solving this differential equation subject to the initial condition that $y = 0$ when $x = 0$.

71. The width of a parabola at the focus Show that the number $4p$ is the *width* of the parabola $x^2 = 4py$ $(p > 0)$ at the focus by showing that the line $y = p$ cuts the parabola at points that are $4p$ units apart.

72. The asymptotes of $(x^2/a^2) - (y^2/b^2) = 1$ Show that the vertical distance between the line $y = (b/a)x$ and the upper half of the right-hand branch $y = (b/a)\sqrt{x^2 - a^2}$ of the hyperbola $(x^2/a^2) - (y^2/b^2) = 1$ approaches 0 by showing that

$$\lim_{x \to \infty} \left(\frac{b}{a}x - \frac{b}{a}\sqrt{x^2 - a^2} \right) = \frac{b}{a} \lim_{x \to \infty} \left(x - \sqrt{x^2 - a^2} \right) = 0.$$

Similar results hold for the remaining portions of the hyperbola and the lines $y = \pm(b/a)x$.

73. Find the dimensions of the rectangle of largest area that can be inscribed in the ellipse $x^2 + 4y^2 = 4$ with its sides parallel to the coordinate axes. What is the area of the rectangle?

74. Find the volume of the solid generated by revolving the region enclosed by the ellipse $9x^2 + 4y^2 = 36$ about the **(a)** x-axis, **(b)** y-axis.

75. The "triangular" region in the first quadrant bounded by the x-axis, the line $x = 4$, and the hyperbola $9x^2 - 4y^2 = 36$ is revolved about the x-axis to generate a solid. Find the volume of the solid.

76. Show that the tangents to the curve $y^2 = 4px$ from any point on the line $x = -p$ are perpendicular.

9.5 Conics in Polar Coordinates

Polar coordinates are important in astronomy and astronautical engineering because satellites, moons, planets, and comets all move approximately along ellipses, parabolas, and hyperbolas that can be described with a single relatively simple polar coordinate equation. We develop that equation here after first introducing the idea of a conic section's *eccentricity*. The eccentricity reveals the conic section's type (circle, ellipse, parabola, or hyperbola), and the degree to which it is "squashed" or flattened.

Eccentricity

Although the center-to-focus distance c does not appear in the equation

$$\frac{x^2}{a^2} + \frac{y^2}{b^2} = 1, \quad (a > b)$$

for an ellipse, we can still determine c from the equation $c = \sqrt{a^2 - b^2}$. If we fix a and vary c over the interval $0 \leq c \leq a$, the resulting ellipses will vary in shape. They are circles if $c = 0$ (so that $a = b$) and flatten as c increases. If $c = a$, the foci and vertices overlap and the ellipse degenerates into a line segment. Thus we are led to consider the ratio $e = c/a$. We use this ratio for hyperbolas as well, only in this case c equals $\sqrt{a^2 + b^2}$ instead of $\sqrt{a^2 - b^2}$, and define these ratios with the somewhat familiar term *eccentricity*.

> **DEFINITION**
> The **eccentricity** of the ellipse $(x^2/a^2) + (y^2/b^2) = 1$ $(a > b)$ is
> $$e = \frac{c}{a} = \frac{\sqrt{a^2 - b^2}}{a}.$$
> The **eccentricity** of the hyperbola $(x^2/a^2) - (y^2/b^2) = 1$ is
> $$e = \frac{c}{a} = \frac{\sqrt{a^2 + b^2}}{a}.$$
> The **eccentricity** of a parabola is $e = 1$.

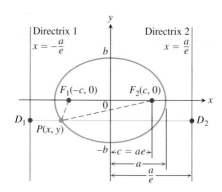

FIGURE 9.28 The foci and directrices of the ellipse $(x^2/a^2) + (y^2/b^2) = 1$. Directrix 1 corresponds to focus F_1, and directrix 2 to focus F_2.

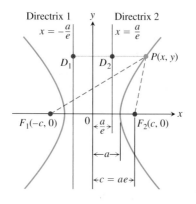

FIGURE 9.29 The foci and directrices of the hyperbola $(x^2/a^2) - (y^2/b^2) = 1$. No matter where P lies on the hyperbola, $PF_1 = e \cdot PD_1$ and $PF_2 = e \cdot PD_2$.

Whereas a parabola has one focus and one directrix, each **ellipse** has two foci and two **directrices**. These are the lines perpendicular to the major axis at distances $\pm a/e$ from the center. The parabola has the property that

$$PF = 1 \cdot PD \tag{1}$$

for any point P on it, where F is the focus and D is the point nearest P on the directrix. For an ellipse, it can be shown that the equations that replace Equation (1) are

$$PF_1 = e \cdot PD_1, \qquad PF_2 = e \cdot PD_2. \tag{2}$$

Here, e is the eccentricity, P is any point on the ellipse, F_1 and F_2 are the foci, and D_1 and D_2 are the points on the directrices nearest P (Figure 9.28).

In both Equations (2) the directrix and focus must correspond; that is, if we use the distance from P to F_1, we must also use the distance from P to the directrix at the same end of the ellipse. The directrix $x = -a/e$ corresponds to $F_1(-c, 0)$, and the directrix $x = a/e$ corresponds to $F_2(c, 0)$.

As with the ellipse, it can be shown that the lines $x = \pm a/e$ act as **directrices** for the **hyperbola** and that

$$PF_1 = e \cdot PD_1 \qquad \text{and} \qquad PF_2 = e \cdot PD_2. \tag{3}$$

Here P is any point on the hyperbola, F_1 and F_2 are the foci, and D_1 and D_2 are the points nearest P on the directrices (Figure 9.29).

In both the ellipse and the hyperbola, the eccentricity is the ratio of the distance between the foci to the distance between the vertices (because $c/a = 2c/2a$).

$$\text{Eccentricity} = \frac{\text{distance between foci}}{\text{distance between vertices}}$$

In an ellipse, the foci are closer together than the vertices and the ratio is less than 1. In a hyperbola, the foci are farther apart than the vertices and the ratio is greater than 1.

The "focus–directrix" equation $PF = e \cdot PD$ unites the parabola, ellipse, and hyperbola in the following way. Suppose that the distance PF of a point P from a fixed point F (the focus) is a constant multiple of its distance from a fixed line (the directrix). That is, suppose

$$PF = e \cdot PD, \tag{4}$$

where e is the constant of proportionality. Then the path traced by P is

(a) a *parabola* if $e = 1$,

(b) an *ellipse* of eccentricity e if $e < 1$, and

(c) a *hyperbola* of eccentricity e if $e > 1$.

There are no coordinates in Equation (4) and when we try to translate it into coordinate form it translates in different ways, depending on the size of e. At least, that is what happens in Cartesian coordinates. However, as we will see, in polar coordinates the equation $PF = e \cdot PD$ translates into a single equation regardless of the value of e.

Given the focus and corresponding directrix of a hyperbola centered at the origin and with foci on the x-axis, we can use the dimensions shown in Figure 9.29 to find e. Knowing

e, we can derive a Cartesian equation for the hyperbola from the equation $PF = e \cdot PD$, as in the next example. We can find equations for ellipses centered at the origin and with foci on the x-axis in a similar way, using the dimensions shown in Figure 9.28.

EXAMPLE 1 Find a Cartesian equation for the hyperbola centered at the origin that has a focus at $(3, 0)$ and the line $x = 1$ as the corresponding directrix.

Solution We first use the dimensions shown in Figure 9.29 to find the hyperbola's eccentricity. The focus is

$$(c, 0) = (3, 0) \qquad \text{so} \qquad c = 3.$$

The directrix is the line

$$x = \frac{a}{e} = 1, \qquad \text{so} \qquad a = e.$$

When combined with the equation $e = c/a$ that defines eccentricity, these results give

$$e = \frac{c}{a} = \frac{3}{e}, \qquad \text{so} \qquad e^2 = 3 \quad \text{and} \quad e = \sqrt{3}.$$

Knowing e, we can now derive the equation we want from the equation $PF = e \cdot PD$. In the notation of Figure 9.30, we have

$$PF = e \cdot PD \qquad\qquad \text{Equation (4)}$$
$$\sqrt{(x - 3)^2 + (y - 0)^2} = \sqrt{3}\,|x - 1| \qquad e = \sqrt{3}$$
$$x^2 - 6x + 9 + y^2 = 3(x^2 - 2x + 1)$$
$$2x^2 - y^2 = 6$$
$$\frac{x^2}{3} - \frac{y^2}{6} = 1.$$ ∎

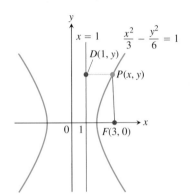

FIGURE 9.30 The hyperbola and directrix in Example 1.

Polar Equations

To find polar equations for ellipses, parabolas, and hyperbolas, we place one focus at the origin and the corresponding directrix to the right of the origin along the vertical line $x = k$ (Figure 9.31). This makes

$$PF = r$$

and

$$PD = k - FB = k - r\cos\theta.$$

The conic's focus–directrix equation $PF = e \cdot PD$ then becomes

$$r = e(k - r\cos\theta),$$

which can be solved for r to obtain the following expression.

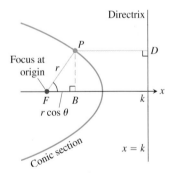

FIGURE 9.31 If a conic section is put in the position with its focus placed at the origin and a directrix perpendicular to the initial ray and right of the origin, we can find its polar equation from the conic's focus–directrix equation.

> **Polar Equation for a Conic with Eccentricity e**
> $$r = \frac{ke}{1 + e\cos\theta}, \qquad\qquad (5)$$
> where $x = k > 0$ is the vertical directrix.

EXAMPLE 2 Here are polar equations for three conics.

$$e = \frac{1}{2}: \quad \text{ellipse} \qquad r = \frac{k}{2 + \cos\theta}$$

$$e = 1: \quad \text{parabola} \qquad r = \frac{k}{1 + \cos\theta}$$

$$e = 2: \quad \text{hyperbola} \qquad r = \frac{2k}{1 + 2\cos\theta} \qquad \blacksquare$$

You may see variations of Equation (5), depending on the location of the directrix. If the directrix is the line $x = -k$ to the left of the origin (the origin is still a focus), we replace Equation (5) by

$$r = \frac{ke}{1 - e\cos\theta}.$$

The denominator now has a $(-)$ instead of a $(+)$. If the directrix is either of the lines $y = k$ or $y = -k$, the equations have sines in them instead of cosines, as shown in Figure 9.32.

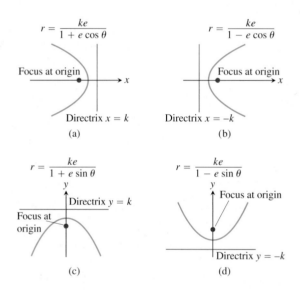

FIGURE 9.32 Equations for conic sections with eccentricity $e > 0$, but different locations of the directrix. The graphs here show a parabola, so $e = 1$.

EXAMPLE 3 Find an equation for the hyperbola with eccentricity 3/2 and directrix $x = 2$.

Solution We use Equation (5) with $k = 2$ and $e = 3/2$:

$$r = \frac{2(3/2)}{1 + (3/2)\cos\theta} \qquad \text{or} \qquad r = \frac{6}{2 + 3\cos\theta}. \qquad \blacksquare$$

EXAMPLE 4 Find the directrix of the parabola

$$r = \frac{25}{10 + 10\cos\theta}.$$

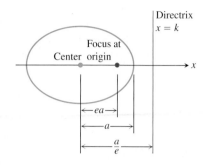

FIGURE 9.33 In an ellipse with semimajor axis a, the focus–directrix distance is $k = (a/e) - ea$, so $ke = a(1 - e^2)$.

Solution We divide the numerator and denominator by 10 to put the equation in standard polar form:

$$r = \frac{5/2}{1 + \cos \theta}.$$

This is the equation

$$r = \frac{ke}{1 + e \cos \theta}$$

with $k = 5/2$ and $e = 1$. The equation of the directrix is $x = 5/2$. ∎

From the ellipse diagram in Figure 9.33, we see that k is related to the eccentricity e and the semimajor axis a by the equation

$$k = \frac{a}{e} - ea.$$

From this, we find that $ke = a(1 - e^2)$. Replacing ke in Equation (5) by $a(1 - e^2)$ gives the standard polar equation for an ellipse.

Polar Equation for the Ellipse with Eccentricity e and Semimajor Axis a

$$r = \frac{a(1 - e^2)}{1 + e \cos \theta} \tag{6}$$

Notice that when $e = 0$, Equation (6) becomes $r = a$, which represents a circle.

Lines

Suppose the perpendicular from the origin to line L meets L at the point $P_0(r_0, \theta_0)$, with $r_0 \geq 0$ (Figure 9.34). Then, if $P(r, \theta)$ is any other point on L, the points P, P_0, and O are the vertices of a right triangle, from which we can read the relation

$$r_0 = r \cos (\theta - \theta_0).$$

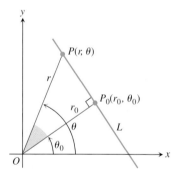

FIGURE 9.34 We can obtain a polar equation for line L by reading the relation $r_0 = r \cos (\theta - \theta_0)$ from the right triangle OP_0P.

The Standard Polar Equation for Lines

If the point $P_0(r_0, \theta_0)$ is the foot of the perpendicular from the origin to the line L, and $r_0 \geq 0$, then an equation for L is

$$r \cos (\theta - \theta_0) = r_0. \tag{7}$$

For example, if $\theta_0 = \pi/3$, we find that

$$r \cos \left(\theta - \frac{\pi}{3} \right) = 2$$

$$r \left(\cos \theta \cos \frac{\pi}{3} + \sin \theta \sin \frac{\pi}{3} \right) = 2$$

$$\frac{1}{2} r \cos \theta + \frac{\sqrt{3}}{2} r \sin \theta = 2, \quad \text{or} \quad x + \sqrt{3} \, y = 4.$$

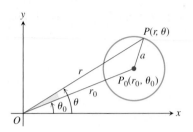

FIGURE 9.35 We can get a polar equation for this circle by applying the Law of Cosines to triangle OP_0P.

Circles

To find a polar equation for the circle of radius a centered at $P_0(r_0, \theta_0)$, we let $P(r, \theta)$ be a point on the circle and apply the Law of Cosines to triangle OP_0P (Figure 9.35). This gives

$$a^2 = r_0^2 + r^2 - 2r_0r\cos(\theta - \theta_0).$$

If the circle passes through the origin, then $r_0 = a$ and this equation simplifies to

$$a^2 = a^2 + r^2 - 2ar\cos(\theta - \theta_0)$$
$$r^2 = 2ar\cos(\theta - \theta_0)$$
$$r = 2a\cos(\theta - \theta_0).$$

If the circle's center lies on the positive x-axis, $\theta_0 = 0$ and we get the further simplification

$$r = 2a\cos\theta. \tag{8}$$

If the center lies on the positive y-axis, $\theta = \pi/2$, $\cos(\theta - \pi/2) = \sin\theta$, and the equation $r = 2a\cos(\theta - \theta_0)$ becomes

$$r = 2a\sin\theta. \tag{9}$$

Equations for circles through the origin centered on the negative x- and y-axes can be obtained by replacing r with $-r$ in the above equations.

EXAMPLE 5   Here are several polar equations for circles through the origin.

Radius	Center (polar coordinates)	Polar equation
3	$(3, 0)$	$r = 6\cos\theta$
2	$(2, \pi/2)$	$r = 4\sin\theta$
1/2	$(-1/2, 0)$	$r = -\cos\theta$
1	$(-1, \pi/2)$	$r = -2\sin\theta$

EXERCISES 9.5

In Exercises 1–8, find the eccentricity of the ellipse. Then find and graph the ellipse's foci and directrices.

1. $16x^2 + 25y^2 = 400$ **2.** $7x^2 + 16y^2 = 112$

3. $2x^2 + y^2 = 2$ **4.** $2x^2 + y^2 = 4$

5. $3x^2 + 2y^2 = 6$ **6.** $9x^2 + 10y^2 = 90$

7. $6x^2 + 9y^2 = 54$ **8.** $169x^2 + 25y^2 = 4225$

Exercises 9–12 give the foci or vertices and the eccentricities of ellipses centered at the origin of the xy-plane. In each case, find the ellipse's standard-form equation in Cartesian coordinates.

9. Foci: $(0, \pm 3)$
   Eccentricity: 0.5

10. Foci: $(\pm 8, 0)$
   Eccentricity: 0.2

11. Vertices: $(0, \pm 70)$
   Eccentricity: 0.1

12. Vertices: $(\pm 10, 0)$
   Eccentricity: 0.24

Exercises 13–16 give foci and corresponding directrices of ellipses centered at the origin of the xy-plane. In each case, use the dimensions in Figure 9.28 to find the eccentricity of the ellipse. Then find the ellipse's standard-form equation in Cartesian coordinates.

13. Focus: $\left(\sqrt{5}, 0\right)$
   Directrix: $x = \dfrac{9}{\sqrt{5}}$

14. Focus: $(4, 0)$
   Directrix: $x = \dfrac{16}{3}$

15. Focus: $(-4, 0)$
   Directrix: $x = -16$

16. Focus: $\left(-\sqrt{2}, 0\right)$
   Directrix: $x = -2\sqrt{2}$

In Exercises 17–24, find the eccentricity of the hyperbola. Then find and graph the hyperbola's foci and directrices.

17. $x^2 - y^2 = 1$ **18.** $9x^2 - 16y^2 = 144$

19. $y^2 - x^2 = 8$ **20.** $y^2 - x^2 = 4$

21. $8x^2 - 2y^2 = 16$ **22.** $y^2 - 3x^2 = 3$

23. $8y^2 - 2x^2 = 16$ **24.** $64x^2 - 36y^2 = 2304$

Exercises 25–28 give the eccentricities and the vertices or foci of hyperbolas centered at the origin of the xy-plane. In each case, find the hyperbola's standard-form equation in Cartesian coordinates.

25. Eccentricity: 3
Vertices: $(0, \pm 1)$

26. Eccentricity: 2
Vertices: $(\pm 2, 0)$

27. Eccentricity: 3
Foci: $(\pm 3, 0)$

28. Eccentricity: 1.25
Foci: $(0, \pm 5)$

Exercises 29–36 give the eccentricities of conic sections with one focus at the origin, along with the directrix corresponding to that focus. Find a polar equation for each conic section.

29. $e = 1, \quad x = 2$ **30.** $e = 1, \quad y = 2$

31. $e = 5, \quad y = -6$ **32.** $e = 2, \quad x = 4$

33. $e = 1/2, \quad x = 1$ **34.** $e = 1/4, \quad x = -2$

35. $e = 1/5, \quad y = -10$ **36.** $e = 1/3, \quad y = 6$

Sketch the parabolas and ellipses in Exercises 37–44. Include the directrix that corresponds to the focus at the origin. Label the vertices with appropriate polar coordinates. Label the centers of the ellipses as well.

37. $r = \dfrac{1}{1 + \cos\theta}$ **38.** $r = \dfrac{6}{2 + \cos\theta}$

39. $r = \dfrac{25}{10 - 5\cos\theta}$ **40.** $r = \dfrac{4}{2 - 2\cos\theta}$

41. $r = \dfrac{400}{16 + 8\sin\theta}$ **42.** $r = \dfrac{12}{3 + 3\sin\theta}$

43. $r = \dfrac{8}{2 - 2\sin\theta}$ **44.** $r = \dfrac{4}{2 - \sin\theta}$

Sketch the lines in Exercises 45–48 and find Cartesian equations for them.

45. $r\cos\left(\theta - \dfrac{\pi}{4}\right) = \sqrt{2}$ **46.** $r\cos\left(\theta + \dfrac{3\pi}{4}\right) = 1$

47. $r\cos\left(\theta - \dfrac{2\pi}{3}\right) = 3$ **48.** $r\cos\left(\theta + \dfrac{\pi}{3}\right) = 2$

Find a polar equation in the form $r\cos(\theta - \theta_0) = r_0$ for each of the lines in Exercises 49–52.

49. $\sqrt{2}\,x + \sqrt{2}\,y = 6$ **50.** $\sqrt{3}\,x - y = 1$

51. $y = -5$ **52.** $x = -4$

Sketch the circles in Exercises 53–56. Give polar coordinates for their centers and identify their radii.

53. $r = 4\cos\theta$ **54.** $r = 6\sin\theta$

55. $r = -2\cos\theta$ **56.** $r = -8\sin\theta$

Find polar equations for the circles in Exercises 57–64. Sketch each circle in the coordinate plane and label it with both its Cartesian and polar equations.

57. $(x - 6)^2 + y^2 = 36$ **58.** $(x + 2)^2 + y^2 = 4$

59. $x^2 + (y - 5)^2 = 25$ **60.** $x^2 + (y + 7)^2 = 49$

61. $x^2 + 2x + y^2 = 0$ **62.** $x^2 - 16x + y^2 = 0$

63. $x^2 + y^2 + y = 0$ **64.** $x^2 + y^2 - \dfrac{4}{3}y = 0$

T Graph the lines and conic sections in Exercises 65–74.

65. $r = 3\sec(\theta - \pi/3)$ **66.** $r = 4\sec(\theta + \pi/6)$

67. $r = 4\sin\theta$ **68.** $r = -2\cos\theta$

69. $r = 8/(4 + \cos\theta)$ **70.** $r = 8/(4 + \sin\theta)$

71. $r = 1/(1 - \sin\theta)$ **72.** $r = 1/(1 + \cos\theta)$

73. $r = 1/(1 + 2\sin\theta)$ **74.** $r = 1/(1 + 2\cos\theta)$

75. Perihelion and aphelion A planet travels about its sun in an ellipse whose semimajor axis has length a. (See accompanying figure.)

    a. Show that $r = a(1 - e)$ when the planet is closest to the sun and that $r = a(1 + e)$ when the planet is farthest from the sun.

    b. Use the data in the table in Exercise 76 to find how close each planet in our solar system comes to the sun and how far away each planet gets from the sun.

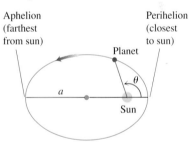

76. Planetary orbits Use the data in the table below and Equation (6) to find polar equations for the orbits of the planets.

Planet	Semimajor axis (astronomical units)	Eccentricity
Mercury	0.3871	0.2056
Venus	0.7233	0.0068
Earth	1.000	0.0167
Mars	1.524	0.0934
Jupiter	5.203	0.0484
Saturn	9.539	0.0543
Uranus	19.18	0.0460
Neptune	30.06	0.0082
Pluto	39.44	0.2481

For example, the orbit of Pluto is

$$r = \frac{39.44(1 - (0.25)^2)}{1 + 0.25\cos\theta} = \frac{147.9}{4 + \cos\theta}$$

In Exercises 77–80, find a polar equation for the given curve. In each case, sketch a typical curve.

77. $x^2 + y^2 - 2ay = 0$ **78.** $y^2 = 4ax + 4a^2$

79. $x \cos \alpha + y \sin \alpha = p$ $(\alpha, p \text{ constant})$

80. $(x^2 + y^2)^2 + 2ax(x^2 + y^2) - a^2y^2 = 0$

COMPUTER EXPLORATIONS

81. Use a CAS to plot the polar equation

$$r = \frac{ke}{1 + e \cos \theta}$$

for various values of k and e, $-\pi \le \theta \le \pi$. Answer the following questions.

a. Take $k = -2$. Describe what happens to the plots as you take e to be $3/4$, 1, and $5/4$. Repeat for $k = 2$.

b. Take $k = -1$. Describe what happens to the plots as you take e to be $7/6$, $5/4$, $4/3$, $3/2$, 2, 3, 5, 10, and 20. Repeat for $e = 1/2$, $1/3$, $1/4$, $1/10$, and $1/20$.

c. Now keep $e > 0$ fixed and describe what happens as you take k to be -1, -2, -3, -4, and -5. Be sure to look at graphs for parabolas, ellipses, and hyperbolas.

82. Use a CAS to plot the polar ellipse

$$r = \frac{a(1 - e^2)}{1 + e \cos \theta}$$

for various values of $a > 0$ and $0 < e < 1$, $-\pi \le \theta \le \pi$.

a. Take $e = 9/10$. Describe what happens to the plots as you let a equal 1, $3/2$, 2, 3, 5, and 10. Repeat with $e = 1/4$.

b. Take $a = 2$. Describe what happens as you take e to be $9/10$, $8/10$, $7/10$, $\ldots$, $1/10$, $1/20$, and $1/50$.

9.6 Conics and Parametric Equations; The Cycloid

Curves in the Cartesian plane defined by parametric equations, and the calculation of their derivatives, were introduced in Section 3.5. There we studied parametrizations of lines, circles, and ellipses. In this section we discuss parametrization of other plane curves, including parabolas, hyperbolas, cycloids, brachistochrones, and tautocrones.

Parabolas and Hyperbolas

In Section 3.5 we used the parametrization

$$x = \sqrt{t}, \qquad y = t, \qquad t \ge 0$$

to describe the motion of a particle moving along the right branch of the parabola $y = x^2$. In the following example we obtain a parametrization of the entire parabola, not just its right branch.

EXAMPLE 1   The position $P(x, y)$ of a particle moving in the xy-plane is given by the equations and parameter interval

$$x = t, \qquad y = t^2, \qquad -\infty < t < \infty.$$

Identify the particle's path and describe the motion.

Solution   We identify the path by eliminating t between the equations $x = t$ and $y = t^2$, obtaining

$$y = (t)^2 = x^2.$$

The particle's position coordinates satisfy the equation $y = x^2$, so the particle moves along this curve.

In contrast to Example 10 in Section 3.5, the particle now traverses the entire parabola. As t increases from $-\infty$ to ∞, the particle comes down the left-hand side, passes through the origin, and moves up the right-hand side (Figure 9.36).   ∎

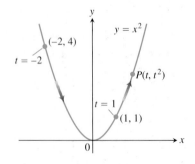

FIGURE 9.36   The path defined by $x = t, y = t^2, -\infty < t < \infty$ is the entire parabola $y = x^2$ (Example 1).

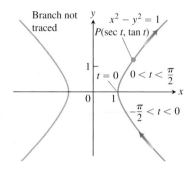

FIGURE 9.37 The equations $x = \sec t, y = \tan t$ and interval $-\pi/2 < t < \pi/2$ describe the right-hand branch of the hyperbola $x^2 - y^2 = 1$ (Example 2).

HISTORICAL BIOGRAPHY

Christiaan Huygens
(1629–1695)

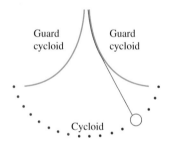

FIGURE 9.38 In Huygens' pendulum clock, the bob swings in a cycloid, so the frequency is independent of the amplitude.

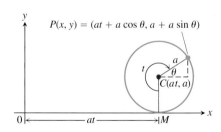

FIGURE 9.39 The position of $P(x, y)$ on the rolling wheel at angle t (Example 3).

As Example 1 illustrates, any curve $y = f(x)$ has the parametrization $x = t, y = f(t)$. This is so simple we usually do not use it, but the point of view is occasionally helpful.

EXAMPLE 2 Describe the motion of the particle whose position $P(x, y)$ at time t is given by

$$x = \sec t, \qquad y = \tan t, \qquad -\frac{\pi}{2} < t < \frac{\pi}{2}.$$

Solution We find a Cartesian equation for the coordinates of P by eliminating t between the equations

$$\sec t = x, \qquad \tan t = y.$$

We accomplish this with the identity $\sec^2 t - \tan^2 t = 1$, which yields

$$x^2 - y^2 = 1.$$

Since the particle's coordinates (x, y) satisfy the equation $x^2 - y^2 = 1$, the motion takes place somewhere on this hyperbola. As t runs between $-\pi/2$ and $\pi/2$, $x = \sec t$ remains positive and $y = \tan t$ runs between $-\infty$ and ∞, so P traverses the hyperbola's right-hand branch. It comes in along the branch's lower half as $t \to 0^-$, reaches $(1, 0)$ at $t = 0$, and moves out into the first quadrant as t increases toward $\pi/2$ (Figure 9.37). ∎

Cycloids

The problem with a pendulum clock whose bob swings in a circular arc is that the frequency of the swing depends on the amplitude of the swing. The wider the swing, the longer it takes the bob to return to center (its lowest position).

This does not happen if the bob can be made to swing in a *cycloid*. In 1673, Christiaan Huygens designed a pendulum clock whose bob would swing in a cycloid, a curve we define in Example 3. He hung the bob from a fine wire constrained by guards that caused it to draw up as it swung away from center (Figure 9.38).

EXAMPLE 3 A wheel of radius a rolls along a horizontal straight line. Find parametric equations for the path traced by a point P on the wheel's circumference. The path is called a **cycloid**.

Solution We take the line to be the x-axis, mark a point P on the wheel, start the wheel with P at the origin, and roll the wheel to the right. As parameter, we use the angle t through which the wheel turns, measured in radians. Figure 9.39 shows the wheel a short while later, when its base lies at units from the origin. The wheel's center C lies at (at, a) and the coordinates of P are

$$x = at + a \cos \theta, \qquad y = a + a \sin \theta.$$

To express θ in terms of t, we observe that $t + \theta = 3\pi/2$ in the figure, so that

$$\theta = \frac{3\pi}{2} - t.$$

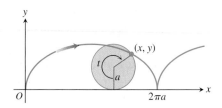

FIGURE 9.40 The cycloid $x = a(t - \sin t), y = a(1 - \cos t)$, for $t \geq 0$.

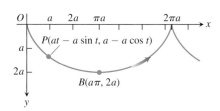

FIGURE 9.41 To study motion along an upside-down cycloid under the influence of gravity, we turn Figure 9.40 upside down. This points the y-axis in the direction of the gravitational force and makes the downward y-coordinates positive. The equations and parameter interval for the cycloid are still

$$x = a(t - \sin t),$$
$$y = a(1 - \cos t), \quad t \geq 0.$$

The arrow shows the direction of increasing t.

This makes

$$\cos \theta = \cos\left(\frac{3\pi}{2} - t\right) = -\sin t, \qquad \sin \theta = \sin\left(\frac{3\pi}{2} - t\right) = -\cos t.$$

The equations we seek are

$$x = at - a \sin t, \qquad y = a - a \cos t.$$

These are usually written with the a factored out:

$$x = a(t - \sin t), \qquad y = a(1 - \cos t). \tag{1}$$

Figure 9.40 shows the first arch of the cycloid and part of the next. ∎

Brachistochrones and Tautochrones

If we turn Figure 9.40 upside down, Equations (1) still apply and the resulting curve (Figure 9.41) has two interesting physical properties. The first relates to the origin O and the point B at the bottom of the first arch. Among all smooth curves joining these points, the cycloid is the curve along which a frictionless bead, subject only to the force of gravity, will slide from O to B the fastest. This makes the cycloid a **brachistochrone** ("*brah-kiss-*toe-krone"), or shortest time curve for these points. The second property is that even if you start the bead partway down the curve toward B, it will still take the bead the same amount of time to reach B. This makes the cycloid a **tautochrone** ("*taw-*toe-krone"), or same-time curve for O and B. It can be shown that the cycloid from O to B is the one and only brachistochrone for O and B. We omit the argument here.

EXERCISE 9.6

Exercises 1–12 give parametric equations and parameter intervals for the motion of a particle in the xy-plane. Identify the particle's path by finding a Cartesian equation for it. Graph the Cartesian equation. (The graphs will vary with the equation used.) Indicate the portion of the graph traced by the particle and the direction of motion.

1. $x = \cos t, \quad y = \sin t, \quad 0 \leq t \leq \pi$

2. $x = \sin(2\pi(1 - t)), \quad y = \cos(2\pi(1 - t)); \quad 0 \leq t \leq 1$

3. $x = 4\cos t, \quad y = 5\sin t; \quad 0 \leq t \leq \pi$

4. $x = 4\sin t, \quad y = 5\cos t; \quad 0 \leq t \leq 2\pi$

5. $x = t, \quad y = \sqrt{t}; \quad t \geq 0$

6. $x = \sec^2 t - 1, \quad y = \tan t; \quad -\pi/2 < t < \pi/2$

7. $x = -\sec t, \quad y = \tan t; \quad -\pi/2 < t < \pi/2$

8. $x = \csc t, \quad y = \cot t; \quad 0 < t < \pi$

9. $x = t, \quad y = \sqrt{4 - t^2}; \quad 0 \leq t \leq 2$

10. $x = t^2, \quad y = \sqrt{t^4 + 1}; \quad t \geq 0$

11. $x = -\cosh t, \quad y = \sinh t; \quad -\infty < t < \infty$

12. $x = 2\sinh t, \quad y = 2\cosh t; \quad -\infty < t < \infty$

13. Hypocycloids When a circle rolls on the inside of a fixed circle, any point P on the circumference of the rolling circle describes a *hypocycloid*. Let the fixed circle be $x^2 + y^2 = a^2$, let the radius of the rolling circle be b, and let the initial position of the tracing point P be $A(a, 0)$. Find parametric equations for the hypocycloid, using as the parameter the angle θ from the positive x-axis to the line joining the circles' centers. In particular, if $b = a/4$, as in the accompanying figure, show that the hypocycloid is the astroid

$$x = a \cos^3 \theta, \quad y = a \sin^3 \theta.$$

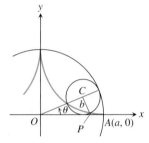

14. More about hypocycloids The accompanying figure shows a circle of radius a tangent to the inside of a circle of radius $2a$. The point P, shown as the point of tangency in the figure, is attached to the smaller circle. What path does P trace as the smaller circle rolls around the inside of the larger circle?

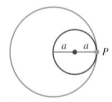

15. As the point N moves along the line $y = a$ in the accompanying figure, P moves in such a way that $OP = MN$. Find parametric equations for the coordinates of P as functions of the angle t that the line ON makes with the positive y-axis.

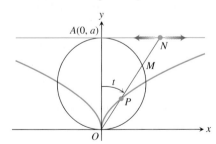

16. Trochoids A wheel of radius a rolls along a horizontal straight line without slipping. Find parametric equations for the curve traced out by a point P on a spoke of the wheel b units from its center. As parameter, use the angle θ through which the wheel turns. The curve is called a *trochoid*, which is a cycloid when $b = a$.

17. Find the point on the parabola $x = t, y = t^2, -\infty < t < \infty$, closest to the point $(2, 1/2)$. (*Hint:* Minimize the square of the distance as a function of t.)

18. Find the point on the ellipse $x = 2 \cos t, y = \sin t, 0 \le t \le 2\pi$ closest to the point $(3/4, 0)$. (*Hint:* Minimize the square of the distance as a function of t.)

T If you have a parametric equation grapher, graph the following equations over the given intervals.

19. Ellipse $x = 4 \cos t, \quad y = 2 \sin t, \quad$ over
   a. $0 \le t \le 2\pi$ **b.** $0 \le t \le \pi$
   c. $-\pi/2 \le t \le \pi/2$.

20. Hyperbola branch $x = \sec t$ (enter as $1/\cos (t)$), $\quad y = \tan t$ (enter as $\sin (t)/\cos (t)$), over
   a. $-1.5 \le t \le 1.5$ **b.** $-0.5 \le t \le 0.5$
   c. $-0.1 \le t \le 0.1$.

21. Parabola $x = 2t + 3, \quad y = t^2 - 1, \quad -2 \le t \le 2$

22. Cycloid $x = t - \sin t, \quad y = 1 - \cos t, \quad$ over
   a. $0 \le t \le 2\pi$ **b.** $0 \le t \le 4\pi$
   c. $\pi \le t \le 3\pi$.

23. A nice curve (a deltoid)

$$x = 2 \cos t + \cos 2t, \quad y = 2 \sin t - \sin 2t; \quad 0 \le t \le 2\pi$$

What happens if you replace 2 with -2 in the equations for x and y? Graph the new equations and find out.

24. An even nicer curve

$$x = 3 \cos t + \cos 3t, \quad y = 3 \sin t - \sin 3t; \quad 0 \le t \le 2\pi$$

What happens if you replace 3 with -3 in the equations for x and y? Graph the new equations and find out.

25. Three beautiful curves
   a. *Epicycloid:*
   $$x = 9 \cos t - \cos 9t, \quad y = 9 \sin t - \sin 9t; \quad 0 \le t \le 2\pi$$
   b. *Hypocycloid:*
   $$x = 8 \cos t + 2 \cos 4t, \quad y = 8 \sin t - 2 \sin 4t; \quad 0 \le t \le 2\pi$$
   c. *Hypotrochoid:*
   $$x = \cos t + 5 \cos 3t, \quad y = 6 \cos t - 5 \sin 3t; \quad 0 \le t \le 2\pi$$

26. More beautiful curves
   a. $x = 6 \cos t + 5 \cos 3t, \quad y = 6 \sin t - 5 \sin 3t; \quad 0 \le t \le 2\pi$
   b. $x = 6 \cos 2t + 5 \cos 6t, \quad y = 6 \sin 2t - 5 \sin 6t; \quad 0 \le t \le \pi$
   c. $x = 6 \cos t + 5 \cos 3t, \quad y = 6 \sin 2t - 5 \sin 3t; \quad 0 \le t \le 2\pi$
   d. $x = 6 \cos 2t + 5 \cos 6t, \quad y = 6 \sin 4t - 5 \sin 6t; \quad 0 \le t \le \pi$

Chapter **9** Questions to Guide Your Review

1. What are polar coordinates? What equations relate polar coordinates to Cartesian coordinates? Why might you want to change from one coordinate system to the other?

2. What consequence does the lack of uniqueness of polar coordinates have for graphing? Give an example.

3. How do you graph equations in polar coordinates? Include in your discussion symmetry, slope, behavior at the origin, and the use of Cartesian graphs. Give examples.

4. How do you find the area of a region $0 \le r_1(\theta) \le r \le r_2(\theta)$, $\alpha \le \theta \le \beta$, in the polar coordinate plane? Give examples.

5. Under what conditions can you find the length of a curve $r = f(\theta)$, $\alpha \le \theta \le \beta$, in the polar coordinate plane? Give an example of a typical calculation.

6. What is a parabola? What are the Cartesian equations for parabolas whose vertices lie at the origin and whose foci lie on the coordinate axes? How can you find the focus and directrix of such a parabola from its equation?

7. What is an ellipse? What are the Cartesian equations for ellipses centered at the origin with foci on one of the coordinate axes? How can you find the foci, vertices, and directrices of such an ellipse from its equation?

8. What is a hyperbola? What are the Cartesian equations for hyperbolas centered at the origin with foci on one of the coordinate axes? How can you find the foci, vertices, and directrices of such an ellipse from its equation?

9. What is the eccentricity of a conic section? How can you classify conic sections by eccentricity? How are an ellipse's shape and eccentricity related?

10. Explain the equation $PF = e \cdot PD$.

11. What are the standard equations for lines and conic sections in polar coordinates? Give examples.

12. What are some typical parametrizations for conic sections?

13. What is a cycloid? What are typical parametric equations for cycloids? What physical properties account for the importance of cycloids?

Chapter **9** Practice Exercises

Sketch the lines in Exercises 1–6. Also, find a Cartesian equation for each line.

1. $r \cos \left(\theta + \dfrac{\pi}{3} \right) = 2\sqrt{3}$ **2.** $r \cos \left(\theta - \dfrac{3\pi}{4} \right) = \dfrac{\sqrt{2}}{2}$

3. $r = 2 \sec \theta$ **4.** $r = -\sqrt{2} \sec \theta$

5. $r = -(3/2) \csc \theta$ **6.** $r = \left(3\sqrt{3} \right) \csc \theta$

Find Cartesian equations for the circles in Exercises 7–10. Sketch each circle in the coordinate plane and label it with both its Cartesian and polar equations.

7. $r = -4 \sin \theta$ **8.** $r = 3\sqrt{3} \sin \theta$

9. $r = 2\sqrt{2} \cos \theta$ **10.** $r = -6 \cos \theta$

Find polar equations for the circles in Exercises 11–14. Sketch each circle in the coordinate plane and label it with both its Cartesian and polar equations.

11. $x^2 + y^2 + 5y = 0$ **12.** $x^2 + y^2 - 2y = 0$

13. $x^2 + y^2 - 3x = 0$ **14.** $x^2 + y^2 + 4x = 0$

Sketch the regions defined by the polar coordinate inequalities in Exercises 15 and 16.

15. $0 \le r \le 6 \cos \theta$ **16.** $-4 \sin \theta \le r \le 0$

Match each graph in Exercises 17–24 with the appropriate equation (a)–(l). There are more equations than graphs, so some equations will not be matched.

a. $r = \cos 2\theta$ **b.** $r \cos \theta = 1$

c. $r = \dfrac{6}{1 - 2 \cos \theta}$ **d.** $r = \sin 2\theta$

e. $r = \theta$ **f.** $r^2 = \cos 2\theta$

g. $r = 1 + \cos \theta$ **h.** $r = 1 - \sin \theta$

i. $r = \dfrac{2}{1 - \cos \theta}$ **j.** $r^2 = \sin 2\theta$

k. $r = -\sin \theta$ **l.** $r = 2 \cos \theta + 1$

17. Four-leaved rose   **18.** Spiral

19. Limaçon   **20.** Lemniscate

21. Circle

22. Cardioid

23. Parabola

24. Lemniscate

Find the areas of the regions in the polar coordinate plane described in Exercises 25–28.

25. Enclosed by the limaçon $r = 2 - \cos\theta$

26. Enclosed by one leaf of the three-leaved rose $r = \sin 3\theta$

27. Inside the "figure eight" $r = 1 + \cos 2\theta$ and outside the circle $r = 1$

28. Inside the cardioid $r = 2(1 + \sin\theta)$ and outside the circle $r = 2\sin\theta$

Find the lengths of the curves given by the polar coordinate equations in Exercises 29–32.

29. $r = -1 + \cos\theta$

30. $r = 2\sin\theta + 2\cos\theta, \quad 0 \le \theta \le \pi/2$

31. $r = 8\sin^3(\theta/3), \quad 0 \le \theta \le \pi/4$

32. $r = \sqrt{1 + \cos 2\theta}, \quad -\pi/2 \le \theta \le \pi/2$

Sketch the parabolas in Exercises 33–36. Include the focus and directrix in each sketch.

33. $x^2 = -4y$

34. $x^2 = 2y$

35. $y^2 = 3x$

36. $y^2 = -(8/3)x$

Find the eccentricities of the ellipses and hyperbolas in Exercises 37–40. Sketch each conic section. Include the foci, vertices, and asymptotes (as appropriate) in your sketch.

37. $16x^2 + 7y^2 = 112$

38. $x^2 + 2y^2 = 4$

39. $3x^2 - y^2 = 3$

40. $5y^2 - 4x^2 = 20$

Exercises 41–46 give equations for conic sections and tell how many units up or down and to the right or left each curve is to be shifted. Find an equation for the new conic section and find the new foci, vertices, centers, and asymptotes, as appropriate. If the curve is a parabola, find the new directrix as well.

41. $x^2 = -12y, \quad$ right 2, up 3

42. $y^2 = 10x, \quad$ left 1/2, down 1

43. $\dfrac{x^2}{9} + \dfrac{y^2}{25} = 1, \quad$ left 3, down 5

44. $\dfrac{x^2}{169} + \dfrac{y^2}{144} = 1, \quad$ right 5, up 12

45. $\dfrac{y^2}{8} - \dfrac{x^2}{2} = 1, \quad$ right 2, up $2\sqrt{2}$

46. $\dfrac{x^2}{36} - \dfrac{y^2}{64} = 1, \quad$ left 10, down 3

Complete the squares to identify the conic sections in Exercises 47–54. Find their foci, vertices, centers, and asymptotes (as appropriate). If the curve is a parabola, find its directrix as well.

47. $x^2 - 4x - 4y^2 = 0$

48. $4x^2 - y^2 + 4y = 8$

49. $y^2 - 2y + 16x = -49$

50. $x^2 - 2x + 8y = -17$

51. $9x^2 + 16y^2 + 54x - 64y = -1$

52. $25x^2 + 9y^2 - 100x + 54y = 44$

53. $x^2 + y^2 - 2x - 2y = 0$

54. $x^2 + y^2 + 4x + 2y = 1$

Sketch the conic sections whose polar coordinate equations are given in Exercises 55–58. Give polar coordinates for the vertices and, in the case of ellipses, for the centers as well.

55. $r = \dfrac{2}{1 + \cos\theta}$

56. $r = \dfrac{8}{2 + \cos\theta}$

57. $r = \dfrac{6}{1 - 2\cos\theta}$

58. $r = \dfrac{12}{3 + \sin\theta}$

Exercises 59–62 give the eccentricities of conic sections with one focus at the origin of the polar coordinate plane, along with the directrix for that focus. Find a polar equation for each conic section.

59. $e = 2, \quad r\cos\theta = 2$

60. $e = 1, \quad r\cos\theta = -4$

61. $e = 1/2, \quad r\sin\theta = 2$

62. $e = 1/3, \quad r\sin\theta = -6$

Exercises 63–66 give parametric equations and parameter intervals for the motion of a particle in the xy-plane. Identify the particle's path by finding a Cartesian equation for it. Graph the Cartesian equation and indicate the direction of motion and the portion traced by the particle.

63. $x = (1/2)\tan t, \quad y = (1/2)\sec t; \quad -\pi/2 < t < \pi/2$

64. $x = -2\cos t, \quad y = 2\sin t; \quad 0 \le t \le \pi$

65. $x = -\cos t, \quad y = \cos^2 t; \quad 0 \le t \le \pi$

66. $x = 4\cos t, \quad y = 9\sin t; \quad 0 \le t \le 2\pi$

67. Find the volume of the solid generated by revolving the region enclosed by the ellipse $9x^2 + 4y^2 = 36$ about **(a)** the x-axis, **(b)** the y-axis.

68. The "triangular" region in the first quadrant bounded by the x-axis, the line $x = 4$, and the hyperbola $9x^2 - 4y^2 = 36$ is revolved about the x-axis to generate a solid. Find the volume of the solid.

69. Show that the equations $x = r \cos \theta$, $y = r \sin \theta$ transform the polar equation

$$r = \frac{k}{1 + e \cos \theta}$$

into the Cartesian equation

$$(1 - e^2)x^2 + y^2 + 2kex - k^2 = 0.$$

70. Archimedes spirals The graph of an equation of the form $r = a\theta$, where a is a nonzero constant, is called an *Archimedes spiral*. Is there anything special about the widths between the successive turns of such a spiral?

Chapter 9   Additional and Advanced Exercises

1. Find an equation for the parabola with focus $(4, 0)$ and directrix $x = 3$. Sketch the parabola together with its vertex, focus, and directrix.

2. Find the vertex, focus, and directrix of the parabola

$$x^2 - 6x - 12y + 9 = 0.$$

3. Find an equation for the curve traced by the point $P(x, y)$ if the distance from P to the vertex of the parabola $x^2 = 4y$ is twice the distance from P to the focus. Identify the curve.

4. A line segment of length $a + b$ runs from the x-axis to the y-axis. The point P on the segment lies a units from one end and b units from the other end. Show that P traces an ellipse as the ends of the segment slide along the axes.

5. The vertices of an ellipse of eccentricity 0.5 lie at the points $(0, \pm 2)$. Where do the foci lie?

6. Find an equation for the ellipse of eccentricity 2/3 that has the line $x = 2$ as a directrix and the point $(4, 0)$ as the corresponding focus.

7. One focus of a hyperbola lies at the point $(0, -7)$ and the corresponding directrix is the line $y = -1$. Find an equation for the hyperbola if its eccentricity is **(a)** 2, **(b)** 5.

8. Find an equation for the hyperbola with foci $(0, -2)$ and $(0, 2)$ that passes through the point $(12, 7)$.

9. Show that the line

$$b^2xx_1 + a^2yy_1 - a^2b^2 = 0$$

is tangent to the ellipse $b^2x^2 + a^2y^2 - a^2b^2 = 0$ at the point (x_1, y_1) on the ellipse.

10. Show that the line

$$b^2xx_1 - a^2yy_1 - a^2b^2 = 0$$

is tangent to the hyperbola $b^2x^2 - a^2y^2 - a^2b^2 = 0$ at the point (x_1, y_1) on the hyperbola.

What points in the xy-plane satisfy the equations and inequalities in Exercises 11–16? Draw a figure for each exercise.

11. $(x^2 - y^2 - 1)(x^2 + y^2 - 25)(x^2 + 4y^2 - 4) = 0$

12. $(x + y)(x^2 + y^2 - 1) = 0$

13. $(x^2/9) + (y^2/16) \leq 1$

14. $(x^2/9) - (y^2/16) \leq 1$

15. $(9x^2 + 4y^2 - 36)(4x^2 + 9y^2 - 16) \leq 0$

16. $(9x^2 + 4y^2 - 36)(4x^2 + 9y^2 - 16) > 0$

17. a. Find an equation in polar coordinates for the curve

$$x = e^{2t} \cos t, \quad y = e^{2t} \sin t; \quad -\infty < t < \infty.$$

b. Find the length of the curve from $t = 0$ to $t = 2\pi$.

18. Find the length of the curve $r = 2 \sin^3 (\theta/3)$, $0 \leq \theta \leq 3\pi$, in the polar coordinate plane.

Exercises 19–22 give the eccentricities of conic sections with one focus at the origin of the polar coordinate plane, along with the directrix for that focus. Find a polar equation for each conic section.

19. $e = 2$, $r \cos \theta = 2$ **20.** $e = 1$, $r \cos \theta = -4$

21. $e = 1/2$, $r \sin \theta = 2$ **22.** $e = 1/3$, $r \sin \theta = -6$

23. Epicycloids When a circle rolls externally along the circumference of a second, fixed circle, any point P on the circumference of the rolling circle describes an *epicycloid*, as shown here. Let the fixed circle have its center at the origin O and have radius a.

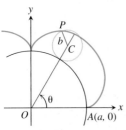

Let the radius of the rolling circle be b and let the initial position of the tracing point P be $A(a, 0)$. Find parametric equations for the epicycloid, using as the parameter the angle θ from the positive x-axis to the line through the circles' centers.

24. Find the centroid of the region enclosed by the x-axis and the cycloid arch

$$x = a(t - \sin t), \quad y = a(1 - \cos t); \quad 0 \leq t \leq 2\pi.$$

The Angle Between the Radius Vector and the Tangent Line to a Polar Coordinate Curve In Cartesian coordinates, when we want to discuss the direction of a curve at a point, we use the angle ϕ measured counterclockwise from the positive x-axis to the tangent line. In polar coordinates, it is more convenient to calculate the angle ψ from the *radius vector* to the tangent line (see the accompanying figure). The angle ϕ can then be calculated from the relation

$$\phi = \theta + \psi, \qquad (1)$$

which comes from applying the Exterior Angle Theorem to the triangle in the accompanying figure.

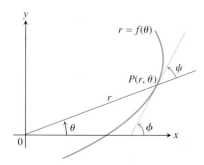

Suppose the equation of the curve is given in the form $r = f(\theta)$, where $f(\theta)$ is a differentiable function of θ. Then

$$x = r \cos \theta \quad \text{and} \quad y = r \sin \theta \qquad (2)$$

are differentiable functions of θ with

$$\frac{dx}{d\theta} = -r \sin \theta + \cos \theta \frac{dr}{d\theta},$$
$$\frac{dy}{d\theta} = r \cos \theta + \sin \theta \frac{dr}{d\theta}. \qquad (3)$$

Since $\psi = \phi - \theta$ from (1),

$$\tan \psi = \tan(\phi - \theta) = \frac{\tan \phi - \tan \theta}{1 + \tan \phi \tan \theta}.$$

Furthermore,

$$\tan \phi = \frac{dy}{dx} = \frac{dy/d\theta}{dx/d\theta}$$

because $\tan \phi$ is the slope of the curve at P. Also,

$$\tan \theta = \frac{y}{x}.$$

Hence

$$\tan \psi = \frac{\dfrac{dy/d\theta}{dx/d\theta} - \dfrac{y}{x}}{1 + \dfrac{y}{x} \dfrac{dy/d\theta}{dx/d\theta}} = \frac{x \dfrac{dy}{d\theta} - y \dfrac{dx}{d\theta}}{x \dfrac{dx}{d\theta} + y \dfrac{dy}{d\theta}}. \qquad (4)$$

The numerator in the last expression in Equation (4) is found from Equations (2) and (3) to be

$$x \frac{dy}{d\theta} - y \frac{dx}{d\theta} = r^2.$$

Similarly, the denominator is

$$x \frac{dx}{d\theta} + y \frac{dy}{d\theta} = r \frac{dr}{d\theta}.$$

When we substitute these into Equation (4), we obtain

$$\tan \psi = \frac{r}{dr/d\theta}. \qquad (5)$$

This is the equation we use for finding ψ as a function of θ.

25. Show, by reference to a figure, that the angle β between the tangents to two curves at a point of intersection may be found from the formula

$$\tan \beta = \frac{\tan \psi_2 - \tan \psi_1}{1 + \tan \psi_2 \tan \psi_1}. \qquad (6)$$

When will the two curves intersect at right angles?

26. Find the value of $\tan \psi$ for the curve $r = \sin^4(\theta/4)$.

27. Find the angle between the radius vector to the curve $r = 2a \sin 3\theta$ and its tangent when $\theta = \pi/6$.

T **28. a.** Graph the hyperbolic spiral $r\theta = 1$. What appears to happen to ψ as the spiral winds in around the origin?

  b. Confirm your finding in part (a) analytically.

29. The circles $r = \sqrt{3} \cos \theta$ and $r = \sin \theta$ intersect at the point $(\sqrt{3}/2, \pi/3)$. Show that their tangents are perpendicular there.

30. Find the angle at which the cardioid $r = a(1 - \cos \theta)$ crosses the ray $\theta = \pi/2$.

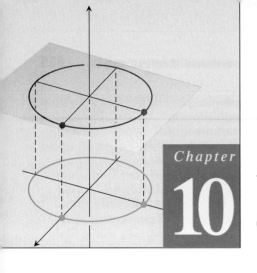

10

VECTORS AND THE GEOMETRY OF SPACE

OVERVIEW To apply calculus in many real-world situations and in higher mathematics, we need a mathematical description of three-dimensional space. In this chapter we introduce three-dimensional coordinate systems and vectors. Building on what we already know about coordinates in the xy-plane, we establish coordinates in space by adding a third axis that measures distance above and below the xy-plane. Vectors are used to study the analytic geometry of space, where they give simple ways to describe lines, planes, surfaces, and curves in space. We use these geometric ideas later in the book to study motion in space and the calculus of functions of several variables, with their many important applications in science, engineering, economics, and higher mathematics.

10.1   Three-Dimensional Coordinate Systems

To locate a point in space, we use three mutually perpendicular coordinate axes, arranged as in Figure 10.1. The axes shown there make a *right-handed* coordinate frame. When you hold your right hand so that the fingers curl from the positive x-axis toward the positive y-axis, your thumb points along the positive z-axis. So when you look down on the xy-plane from the positive direction of the z-axis, positive angles in the plane are measured counterclockwise from the positive x-axis and around the positive z-axis. (In a *left-handed* coordinate frame, the z-axis would point downward in Figure 10.1 and angles in the plane would be positive when measured clockwise from the positive x-axis. Right-handed and left-handed coordinate frames are not equivalent.)

The Cartesian coordinates (x, y, z) of a point P in space are the values at which the planes through P perpendicular to the axes cut the axes. Cartesian coordinates for space are also called **rectangular coordinates** because the axes that define them meet at right angles. Points on the x-axis have y- and z-coordinates equal to zero. That is, they have coordinates of the form $(x, 0, 0)$. Similarly, points on the y-axis have coordinates of the form $(0, y, 0)$, and points on the z-axis have coordinates of the form $(0, 0, z)$.

The planes determined by the coordinates axes are the **xy-plane**, whose standard equation is $z = 0$; the **yz-plane**, whose standard equation is $x = 0$; and the **xz-plane**, whose standard equation is $y = 0$. They meet at the **origin** $(0, 0, 0)$ (Figure 10.2). The origin is also identified by simply 0 or sometimes the letter O.

The three **coordinate planes** $x = 0$, $y = 0$, and $z = 0$ divide space into eight cells called **octants**. The octant in which the point coordinates are all positive is called the **first octant**; there is no convention for numbering the other seven octants.

The points in a plane perpendicular to the x-axis all have the same x-coordinate, this being the number at which that plane cuts the x-axis. The y- and z-coordinates can be any numbers.

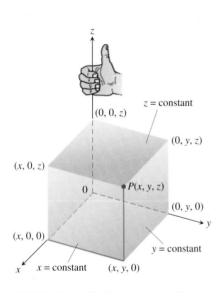

FIGURE 10.1 The Cartesian coordinate system is right-handed.

Similarly, the points in a plane perpendicular to the y-axis have a common y-coordinate and the points in a plane perpendicular to the z-axis have a common z-coordinate. To write equations for these planes, we name the common coordinate's value. The plane $x = 2$ is the plane perpendicular to the x-axis at $x = 2$. The plane $y = 3$ is the plane perpendicular to the y-axis at $y = 3$. The plane $z = 5$ is the plane perpendicular to the z-axis at $z = 5$. Figure 10.3 shows the planes $x = 2$, $y = 3$, and $z = 5$, together with their intersection point (2, 3, 5).

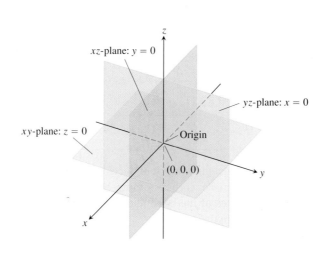

FIGURE 10.2 The planes $x = 0$, $y = 0$, and $z = 0$ divide space into eight octants.

FIGURE 10.3 The planes $x = 2$, $y = 3$, and $z = 5$ determine three lines through the point (2, 3, 5).

The planes $x = 2$ and $y = 3$ in Figure 10.3 intersect in a line parallel to the z-axis. This line is described by the *pair* of equations $x = 2$, $y = 3$. A point (x, y, z) lies on the line if and only if $x = 2$ and $y = 3$. Similarly, the line of intersection of the planes $y = 3$ and $z = 5$ is described by the equation pair $y = 3$, $z = 5$. This line runs parallel to the x-axis. The line of intersection of the planes $x = 2$ and $z = 5$, parallel to the y-axis, is described by the equation pair $x = 2$, $z = 5$.

In the following examples, we match coordinate equations and inequalities with the sets of points they define in space.

EXAMPLE 1 We interpret these equations and inequalities geometrically.

(a) $z \geq 0$ The half-space consisting of the points on and above the xy-plane.

(b) $x = -3$ The plane perpendicular to the x-axis at $x = -3$. This plane lies parallel to the yz-plane and 3 units behind it.

(c) $z = 0, x \leq 0, y \geq 0$ The second quadrant of the xy-plane.

(d) $x \geq 0, y \geq 0, z \geq 0$ The first octant.

(e) $-1 \leq y \leq 1$ The slab between the planes $y = -1$ and $y = 1$ (planes included).

(f) $y = -2, z = 2$ The line in which the planes $y = -2$ and $z = 2$ intersect. Alternatively, the line through the point $(0, -2, 2)$ parallel to the x-axis. ∎

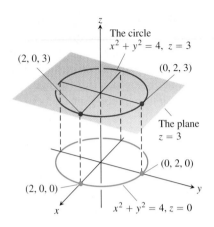

FIGURE 10.4 The circle $x^2 + y^2 = 4$ in the plane $z = 3$ (Example 2).

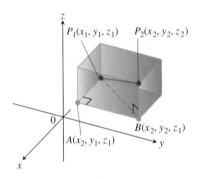

FIGURE 10.5 We find the distance between P_1 and P_2 by applying the Pythagorean theorem to the right triangles P_1AB and P_1BP_2.

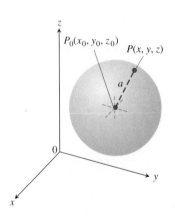

FIGURE 10.6 The sphere of radius a centered at the point (x_0, y_0, z_0).

EXAMPLE 2 What points $P(x, y, z)$ satisfy the equations

$$x^2 + y^2 = 4 \quad \text{and} \quad z = 3?$$

Solution The points lie in the horizontal plane $z = 3$ and, in this plane, make up the circle $x^2 + y^2 = 4$. We call this set of points "the circle $x^2 + y^2 = 4$ in the plane $z = 3$" or, more simply, "the circle $x^2 + y^2 = 4, z = 3$" (Figure 10.4). ∎

Distance and Spheres in Space

The formula for the distance between two points in the xy-plane extends to points in space.

The Distance Between $P_1(x_1, y_1, z_1)$ and $P_2(x_2, y_2, z_2)$ is

$$|P_1P_2| = \sqrt{(x_2 - x_1)^2 + (y_2 - y_1)^2 + (z_2 - z_1)^2}$$

Proof We construct a rectangular box with faces parallel to the coordinate planes and the points P_1 and P_2 at opposite corners of the box (Figure 10.5). If $A(x_2, y_1, z_1)$ and $B(x_2, y_2, z_1)$ are the vertices of the box indicated in the figure, then the three box edges P_1A, AB, and BP_2 have lengths

$$|P_1A| = |x_2 - x_1|, \qquad |AB| = |y_2 - y_1|, \qquad |BP_2| = |z_2 - z_1|.$$

Because triangles P_1BP_2 and P_1AB are both right-angled, two applications of the Pythagorean theorem give

$$|P_1P_2|^2 = |P_1B|^2 + |BP_2|^2 \quad \text{and} \quad |P_1B|^2 = |P_1A|^2 + |AB|^2$$

(see Figure 10.5).
So

$$
\begin{aligned}
|P_1P_2|^2 &= |P_1B|^2 + |BP_2|^2 \\
&= |P_1A|^2 + |AB|^2 + |BP_2|^2 \qquad \substack{\text{Substitute} \\ |P_1B|^2 = |P_1A|^2 + |AB|^2.} \\
&= |x_2 - x_1|^2 + |y_2 - y_1|^2 + |z_2 - z_1|^2 \\
&= (x_2 - x_1)^2 + (y_2 - y_1)^2 + (z_2 - z_1)^2
\end{aligned}
$$

Therefore

$$|P_1P_2| = \sqrt{(x_2 - x_1)^2 + (y_2 - y_1)^2 + (z_2 - z_1)^2} \qquad ∎$$

EXAMPLE 3 The distance between $P_1(2, 1, 5)$ and $P_2(-2, 3, 0)$ is

$$
\begin{aligned}
|P_1P_2| &= \sqrt{(-2 - 2)^2 + (3 - 1)^2 + (0 - 5)^2} \\
&= \sqrt{16 + 4 + 25} \\
&= \sqrt{45} \approx 6.708.
\end{aligned}
$$
∎

We can use the distance formula to write equations for spheres in space (Figure 10.6). A point $P(x, y, z)$ lies on the sphere of radius a centered at $P_0(x_0, y_0, z_0)$ precisely when $|P_0P| = a$ or

$$(x - x_0)^2 + (y - y_0)^2 + (z - z_0)^2 = a^2.$$

The Standard Equation for the Sphere of Radius a and Center (x_0, y_0, z_0)

$$(x - x_0)^2 + (y - y_0)^2 + (z - z_0)^2 = a^2$$

EXAMPLE 4   Find the center and radius of the sphere

$$x^2 + y^2 + z^2 + 3x - 4z + 1 = 0.$$

Solution   We find the center and radius of a sphere the way we find the center and radius of a circle: Complete the squares on the x-, y-, and z-terms as necessary and write each quadratic as a squared linear expression. Then, from the equation in standard form, read off the center and radius. For the sphere here, we have

$$x^2 + y^2 + z^2 + 3x - 4z + 1 = 0$$
$$(x^2 + 3x) + y^2 + (z^2 - 4z) = -1$$
$$\left(x^2 + 3x + \left(\frac{3}{2}\right)^2\right) + y^2 + \left(z^2 - 4z + \left(\frac{-4}{2}\right)^2\right) = -1 + \left(\frac{3}{2}\right)^2 + \left(\frac{-4}{2}\right)^2$$
$$\left(x + \frac{3}{2}\right)^2 + y^2 + (z - 2)^2 = -1 + \frac{9}{4} + 4 = \frac{21}{4}.$$

From this standard form, we read that $x_0 = -3/2$, $y_0 = 0$, $z_0 = 2$, and $a = \sqrt{21}/2$. The center is $(-3/2, 0, 2)$. The radius is $\sqrt{21}/2$.   ∎

EXAMPLE 5   Here are some geometric interpretations.

(a) $x^2 + y^2 + z^2 < 4$ The interior of the sphere $x^2 + y^2 + z^2 = 4$.

(b) $x^2 + y^2 + z^2 \leq 4$ The solid ball bounded by the sphere $x^2 + y^2 + z^2 = 4$. Alternatively, the sphere $x^2 + y^2 + z^2 = 4$ together with its interior.

(c) $x^2 + y^2 + z^2 > 4$ The exterior of the sphere $x^2 + y^2 + z^2 = 4$.

(d) $x^2 + y^2 + z^2 = 4, z \leq 0$ The lower hemisphere cut from the sphere $x^2 + y^2 + z^2 = 4$ by the xy-plane (the plane $z = 0$).   ∎

Just as polar coordinates give another way to locate points in the xy-plane (Section 9.1), alternative coordinate systems, different from the Cartesian coordinate system developed here, exist for three-dimensional space. We examine two of these coordinate systems in Section 13.7.

EXERCISES 10.1

In Exercises 1–12, give a geometric description of the set of points in space whose coordinates satisfy the given pairs of equations.

1. $x = 2$, $y = 3$

2. $x = -1$, $z = 0$

3. $y = 0$, $z = 0$

4. $x = 1$, $y = 0$

5. $x^2 + y^2 = 4$, $z = 0$

6. $x^2 + y^2 = 4$, $z = -2$

7. $x^2 + z^2 = 4$, $y = 0$ **8.** $y^2 + z^2 = 1$, $x = 0$

9. $x^2 + y^2 + z^2 = 1$, $x = 0$

10. $x^2 + y^2 + z^2 = 25$, $y = -4$

11. $x^2 + y^2 + (z + 3)^2 = 25$, $z = 0$

12. $x^2 + (y - 1)^2 + z^2 = 4$, $y = 0$

In Exercises 13–18, describe the sets of points in space whose coordinates satisfy the given inequalities or combinations of equations and inequalities.

13. **a.** $x \geq 0$, $y \geq 0$, $z = 0$ **b.** $x \geq 0$, $y \leq 0$, $z = 0$

14. **a.** $0 \leq x \leq 1$ **b.** $0 \leq x \leq 1$, $0 \leq y \leq 1$

 c. $0 \leq x \leq 1$, $0 \leq y \leq 1$, $0 \leq z \leq 1$

15. **a.** $x^2 + y^2 + z^2 \leq 1$ **b.** $x^2 + y^2 + z^2 > 1$

16. **a.** $x^2 + y^2 \leq 1$, $z = 0$ **b.** $x^2 + y^2 \leq 1$, $z = 3$

 c. $x^2 + y^2 \leq 1$, no restriction on z

17. **a.** $x^2 + y^2 + z^2 = 1$, $z \geq 0$

 b. $x^2 + y^2 + z^2 \leq 1$, $z \geq 0$

18. **a.** $x = y$, $z = 0$ **b.** $x = y$, no restriction on z

In Exercises 19–28, describe the given set with a single equation or with a pair of equations.

19. The plane perpendicular to the

 a. x-axis at $(3, 0, 0)$ **b.** y-axis at $(0, -1, 0)$

 c. z-axis at $(0, 0, -2)$

20. The plane through the point $(3, -1, 2)$ perpendicular to the

 a. x-axis **b.** y-axis **c.** z-axis

21. The plane through the point $(3, -1, 1)$ parallel to the

 a. xy-plane **b.** yz-plane **c.** xz-plane

22. The circle of radius 2 centered at $(0, 0, 0)$ and lying in the

 a. xy-plane **b.** yz-plane **c.** xz-plane

23. The circle of radius 2 centered at $(0, 2, 0)$ and lying in the

 a. xy-plane **b.** yz-plane **c.** plane $y = 2$

24. The circle of radius 1 centered at $(-3, 4, 1)$ and lying in a plane parallel to the

 a. xy-plane **b.** yz-plane **c.** xz-plane

25. The line through the point $(1, 3, -1)$ parallel to the

 a. x-axis **b.** y-axis **c.** z-axis

26. The set of points in space equidistant from the origin and the point $(0, 2, 0)$

27. The circle in which the plane through the point $(1, 1, 3)$ perpendicular to the z-axis meets the sphere of radius 5 centered at the origin

28. The set of points in space that lie 2 units from the point $(0, 0, 1)$ and, at the same time, 2 units from the point $(0, 0, -1)$

Write inequalities to describe the sets in Exercises 29–34.

29. The slab bounded by the planes $z = 0$ and $z = 1$ (planes included)

30. The solid cube in the first octant bounded by the coordinate planes and the planes $x = 2$, $y = 2$, and $z = 2$

31. The half-space consisting of the points on and below the xy-plane

32. The upper hemisphere of the sphere of radius 1 centered at the origin

33. The **(a)** interior and **(b)** exterior of the sphere of radius 1 centered at the point $(1, 1, 1)$

34. The closed region bounded by the spheres of radius 1 and radius 2 centered at the origin. (*Closed* means the spheres are to be included. Had we wanted the spheres left out, we would have asked for the *open* region bounded by the spheres. This is analogous to the way we use *closed* and *open* to describe intervals: *closed* means endpoints included, *open* means endpoints left out. Closed sets include boundaries; open sets leave them out.)

In Exercises 35–40, find the distance between points P_1 and P_2.

35. $P_1(1, 1, 1)$, $P_2(3, 3, 0)$

36. $P_1(-1, 1, 5)$, $P_2(2, 5, 0)$

37. $P_1(1, 4, 5)$, $P_2(4, -2, 7)$

38. $P_1(3, 4, 5)$, $P_2(2, 3, 4)$

39. $P_1(0, 0, 0)$, $P_2(2, -2, -2)$

40. $P_1(5, 3, -2)$, $P_2(0, 0, 0)$

Find the centers and radii of the spheres in Exercises 41–44.

41. $(x + 2)^2 + y^2 + (z - 2)^2 = 8$

42. $\left(x + \frac{1}{2}\right)^2 + \left(y + \frac{1}{2}\right)^2 + \left(z + \frac{1}{2}\right)^2 = \frac{21}{4}$

43. $(x - \sqrt{2})^2 + (y - \sqrt{2})^2 + (z + \sqrt{2})^2 = 2$

44. $x^2 + \left(y + \frac{1}{3}\right)^2 + \left(z - \frac{1}{3}\right)^2 = \frac{29}{9}$

Find equations for the spheres whose centers and radii are given in Exercises 45–48.

Center	Radius
45. $(1, 2, 3)$	$\sqrt{14}$
46. $(0, -1, 5)$	2
47. $(-2, 0, 0)$	$\sqrt{3}$
48. $(0, -7, 0)$	7

Find the centers and radii of the spheres in Exercises 49–52.

49. $x^2 + y^2 + z^2 + 4x - 4z = 0$

50. $x^2 + y^2 + z^2 - 6y + 8z = 0$

51. $2x^2 + 2y^2 + 2z^2 + x + y + z = 9$

52. $3x^2 + 3y^2 + 3z^2 + 2y - 2z = 9$

53. Find a formula for the distance from the point $P(x, y, z)$ to the

 a. x-axis **b.** y-axis **c.** z-axis

54. Find a formula for the distance from the point $P(x, y, z)$ to the

 a. xy-plane **b.** yz-plane **c.** xz-plane

55. Find the perimeter of the triangle with vertices $A(-1, 2, 1)$, $B(1, -1, 3)$, and $C(3, 4, 5)$.

56. Show that the point $P(3, 1, 2)$ is equidistant from the points $A(2, -1, 3)$ and $B(4, 3, 1)$.

Some of the things we measure are determined simply by their magnitudes. To record mass, length, or time, for example, we need only write down a number and name an appropriate unit of measure. We need more information to describe a force, displacement, or velocity. To describe a force, we need to record the direction in which it acts as well as how large it is. To describe a body's displacement, we have to say in what direction it moved as well as how far. To describe a body's velocity, we have to know where the body is headed as well as how fast it is going. In this section we show how to represent things that have both magnitude and direction in the plane or in space.

Component Form

A quantity such as force, displacement, or velocity is called a *vector* and is represented by a **directed line segment** (Figure 10.7). The arrow points in the direction of the action and its length gives the magnitude of the action in terms of a suitably chosen unit. For example, a force vector points in the direction in which the force acts; its length is a measure of the force's strength; a velocity vector points in the direction of motion and its length is the speed of the moving object. Figure 10.8 displays the velocity vector $\mathbf{v}$ at a specific location for a particle moving along a path in the plane or in space. (This application of vectors is studied in Chapter 11.)

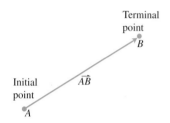

FIGURE 10.7 The directed line segment $\overrightarrow{AB}$.

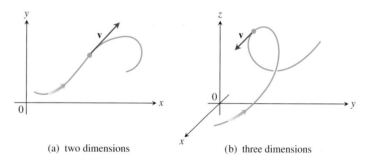

(a) two dimensions   (b) three dimensions

FIGURE 10.8 The velocity vector of a particle moving along a path (a) in the plane (b) in space. The arrowhead on the path indicates the direction of motion of the particle.

DEFINITIONS   A **vector** is a directed line segment. The directed line segment $\overrightarrow{AB}$ has **initial point** A and **terminal point** B; its **length** is denoted by $|\overrightarrow{AB}|$. Two vectors are **equal** if they have the same length and direction.

The arrows we use when we draw vectors are understood to represent the same vector if they have the same length, are parallel, and point in the same direction (Figure 10.9) regardless of the initial point.

In textbooks, vectors are usually written in lowercase, boldface letters, for example $\mathbf{u}$, $\mathbf{v}$, and $\mathbf{w}$. Sometimes we use uppercase boldface letters, such as $\mathbf{F}$, to denote a force vector. In handwritten form, it is customary to draw small arrows above the letters, for example $\vec{u}$, $\vec{v}$, $\vec{w}$, and $\vec{F}$.

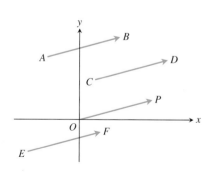

FIGURE 10.9 The four arrows in the plane (directed line segments) shown here have the same length and direction. They therefore represent the same vector, and we write $\overrightarrow{AB} = \overrightarrow{CD} = \overrightarrow{OP} = \overrightarrow{EF}$.

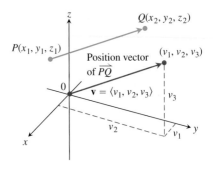

FIGURE 10.10 A vector $\overrightarrow{PQ}$ in standard position has its initial point at the origin. The directed line segments $\overrightarrow{PQ}$ and $\mathbf{v}$ are parallel and have the same length.

We need a way to represent vectors algebraically so that we can be more precise about the direction of a vector. Let $\mathbf{v} = \overrightarrow{PQ}$. There is one directed line segment equal to $\overrightarrow{PQ}$ whose initial point is the origin (Figure 10.10). It is the representative of $\mathbf{v}$ in **standard position** and is the vector we normally use to represent $\mathbf{v}$. We can specify $\mathbf{v}$ by writing the coordinates of its terminal point (v_1, v_2, v_3) when $\mathbf{v}$ is in standard position. If $\mathbf{v}$ is a vector in the plane its terminal point (v_1, v_2) has two coordinates.

DEFINITION

If $\mathbf{v}$ is a **two-dimensional** vector in the plane equal to the vector with initial point at the origin and terminal point (v_1, v_2), then the **component form** of $\mathbf{v}$ is

$$\mathbf{v} = \langle v_1, v_2 \rangle.$$

If $\mathbf{v}$ is a **three-dimensional** vector equal to the vector with initial point at the origin and terminal point (v_1, v_2, v_3), then the **component form** of $\mathbf{v}$ is

$$\mathbf{v} = \langle v_1, v_2, v_3 \rangle.$$

So a two-dimensional vector is an ordered pair $\mathbf{v} = \langle v_1, v_2 \rangle$ of real numbers, and a three-dimensional vector is an ordered triple $\mathbf{v} = \langle v_1, v_2, v_3 \rangle$ of real numbers. The numbers v_1, v_2, and v_3 are the **components** of $\mathbf{v}$.

If $\mathbf{v} = \langle v_1, v_2, v_3 \rangle$ is represented by the directed line segment $\overrightarrow{PQ}$, where the initial point is $P(x_1, y_1, z_1)$ and the terminal point is $Q(x_2, y_2, z_2)$, then $x_1 + v_1 = x_2, y_1 + v_2 = y_2$, and $z_1 + v_3 = z_2$ (see Figure 10.10). Thus, $v_1 = x_2 - x_1, v_2 = y_2 - y_1$, and $v_3 = z_2 - z_1$ are the components of $\overrightarrow{PQ}$.

In summary, given the points $P(x_1, y_1, z_1)$ and $Q(x_2, y_2, z_2)$, the standard position vector $\mathbf{v} = \langle v_1, v_2, v_3 \rangle$ equal to $\overrightarrow{PQ}$ is

$$\mathbf{v} = \langle x_2 - x_1, y_2 - y_1, z_2 - z_1 \rangle.$$

If $\mathbf{v}$ is two-dimensional with $P(x_1, y_1)$ and $Q(x_2, y_2)$ as points in the plane, then $\mathbf{v} = \langle x_2 - x_1, y_2 - y_1 \rangle$. There is no third component for planar vectors. With this understanding, we will develop the algebra of three-dimensional vectors and simply drop the third component when the vector is two-dimensional (a planar vector).

Two vectors are equal if and only if their standard position vectors are identical. Thus $\langle u_1, u_2, u_3 \rangle$ and $\langle v_1, v_2, v_3 \rangle$ are equal if and only if $u_1 = v_1, u_2 = v_2$, and $u_3 = v_3$.

The **magnitude** or **length** of the vector $\overrightarrow{PQ}$ is the length of any of its equivalent directed line segment representations. In particular, if $\mathbf{v} = \langle x_2 - x_1, y_2 - y_1, z_2 - z_1 \rangle$ is the standard position vector for $\overrightarrow{PQ}$, then the distance formula gives the magnitude or length of $\mathbf{v}$, denoted by the symbol $|\mathbf{v}|$ or $\|\mathbf{v}\|$.

The **magnitude** or **length** of the vector $\mathbf{v} = \overrightarrow{PQ}$ is the nonnegative number

$$|\mathbf{v}| = \sqrt{v_1^2 + v_2^2 + v_3^2} = \sqrt{(x_2 - x_1)^2 + (y_2 - y_1)^2 + (z_2 - z_1)^2}$$

(See Figure 10.10.)

The only vector with length 0 is the **zero vector** $\mathbf{0} = \langle 0, 0 \rangle$ or $\mathbf{0} = \langle 0, 0, 0 \rangle$. This vector is also the only vector with no specific direction.

EXAMPLE 1 Find the **(a)** component form and **(b)** length of the vector with initial point $P(-3, 4, 1)$ and terminal point $Q(-5, 2, 2)$.

Solution

(a) The standard position vector **v** representing $\overrightarrow{PQ}$ has components

$$v_1 = x_2 - x_1 = -5 - (-3) = -2, \qquad v_2 = y_2 - y_1 = 2 - 4 = -2,$$

and

$$v_3 = z_2 - z_1 = 2 - 1 = 1.$$

The component form of $\overrightarrow{PQ}$ is

$$\mathbf{v} = \langle -2, -2, 1 \rangle.$$

(b) The length or magnitude of $\mathbf{v} = \overrightarrow{PQ}$ is

$$|\mathbf{v}| = \sqrt{(-2)^2 + (-2)^2 + (1)^2} = \sqrt{9} = 3.$$ ∎

EXAMPLE 2 A small cart is being pulled along a smooth horizontal floor with a 20-lb force **F** making a 45° angle to the floor (Figure 10.11). What is the *effective* force moving the cart forward?

Solution The effective force is the horizontal component of $\mathbf{F} = \langle a, b \rangle$, given by

$$a = |\mathbf{F}| \cos 45° = (20)\left(\frac{\sqrt{2}}{2}\right) \approx 14.14 \text{ lb}.$$

Notice that **F** is a two-dimensional vector. ∎

FIGURE 10.11 The force pulling the cart forward is represented by the vector **F** of magnitude 20 (pounds) making an angle of 45° with the horizontal ground (positive *x*-axis) (Example 2).

Vector Algebra Operations

Two principal operations involving vectors are *vector addition* and *scalar multiplication*. A **scalar** is simply a real number, and is called such when we want to draw attention to its differences from vectors. Scalars can be positive, negative, or zero and are used to "scale" a vector by multiplication.

DEFINITIONS Let $\mathbf{u} = \langle u_1, u_2, u_3 \rangle$ and $\mathbf{v} = \langle v_1, v_2, v_3 \rangle$ be vectors with k a scalar.

> **Addition:** $\mathbf{u} + \mathbf{v} = \langle u_1 + v_1, u_2 + v_2, u_3 + v_3 \rangle$
>
> **Scalar multiplication:** $k\mathbf{u} = \langle ku_1, ku_2, ku_3 \rangle$

We add vectors by adding the corresponding components of the vectors. We multiply a vector by a scalar by multiplying each component by the scalar. The definitions apply to planar vectors except there are only two components, $\langle u_1, u_2 \rangle$ and $\langle v_1, v_2 \rangle$.

The definition of vector addition is illustrated geometrically for planar vectors in Figure 10.12a, where the initial point of one vector is placed at the terminal point of the other. Another interpretation is shown in Figure 10.12b (called the **parallelogram law** of addition), where the sum, called the **resultant vector**, is the diagonal of the parallelogram. In physics, forces add vectorially as do velocities, accelerations, and so on. So the force acting on a particle subject to two gravitational forces, for example, is obtained by adding the two force vectors.

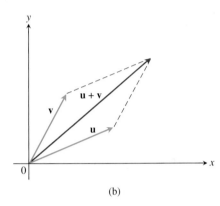

(a)   (b)

FIGURE 10.12   (a) Geometric interpretation of the vector sum. (b) The parallelogram law of vector addition.

Figure 10.13 displays a geometric interpretation of the product $k\mathbf{u}$ of the scalar k and vector $\mathbf{u}$. If $k > 0$, then $k\mathbf{u}$ has the same direction as $\mathbf{u}$; if $k < 0$, then the direction of $k\mathbf{u}$ is opposite to that of $\mathbf{u}$. Comparing the lengths of $\mathbf{u}$ and $k\mathbf{u}$, we see that

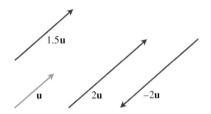

FIGURE 10.13   Scalar multiples of $\mathbf{u}$.

$$|k\mathbf{u}| = \sqrt{(ku_1)^2 + (ku_2)^2 + (ku_3)^2} = \sqrt{k^2(u_1^2 + u_2^2 + u_3^2)}$$
$$= \sqrt{k^2}\sqrt{u_1^2 + u_2^2 + u_3^2} = |k||\mathbf{u}|.$$

The length of $k\mathbf{u}$ is the absolute value of the scalar k times the length of $\mathbf{u}$. The vector $(-1)\mathbf{u} = -\mathbf{u}$ has the same length as $\mathbf{u}$ but points in the opposite direction.

The **difference** $\mathbf{u} - \mathbf{v}$ of two vectors is defined by

$$\mathbf{u} - \mathbf{v} = \mathbf{u} + (-\mathbf{v}).$$

If $\mathbf{u} = \langle u_1, u_2, u_3 \rangle$ and $\mathbf{v} = \langle v_1, v_2, v_3 \rangle$, then

$$\mathbf{u} - \mathbf{v} = \langle u_1 - v_1, u_2 - v_2, u_3 - v_3 \rangle.$$

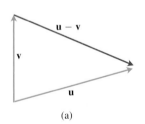

(a)

Note that $(\mathbf{u} - \mathbf{v}) + \mathbf{v} = \mathbf{u}$, so adding the vector $(\mathbf{u} - \mathbf{v})$ to $\mathbf{v}$ gives $\mathbf{u}$ (Figure 10.14a). Figure 10.14b shows the difference $\mathbf{u} - \mathbf{v}$ as the sum $\mathbf{u} + (-\mathbf{v})$.

EXAMPLE 3   Let $\mathbf{u} = \langle -1, 3, 1 \rangle$ and $\mathbf{v} = \langle 4, 7, 0 \rangle$. Find

(a) $2\mathbf{u} + 3\mathbf{v}$ **(b)** $\mathbf{u} - \mathbf{v}$ **(c)** $\left|\dfrac{1}{2}\mathbf{u}\right|$.

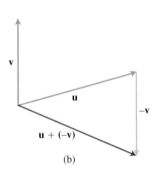

FIGURE 10.14   (a) The vector $\mathbf{u} - \mathbf{v}$, when added to $\mathbf{v}$, gives $\mathbf{u}$.
(b) $\mathbf{u} - \mathbf{v} = \mathbf{u} + (-\mathbf{v})$.

Solution

(a) $2\mathbf{u} + 3\mathbf{v} = 2\langle -1, 3, 1 \rangle + 3\langle 4, 7, 0 \rangle = \langle -2, 6, 2 \rangle + \langle 12, 21, 0 \rangle = \langle 10, 27, 2 \rangle$

(b) $\mathbf{u} - \mathbf{v} = \langle -1, 3, 1 \rangle - \langle 4, 7, 0 \rangle = \langle -1 - 4, 3 - 7, 1 - 0 \rangle = \langle -5, -4, 1 \rangle$

(c) $\left|\dfrac{1}{2}\mathbf{u}\right| = \left|\left\langle -\dfrac{1}{2}, \dfrac{3}{2}, \dfrac{1}{2} \right\rangle\right| = \sqrt{\left(-\dfrac{1}{2}\right)^2 + \left(\dfrac{3}{2}\right)^2 + \left(\dfrac{1}{2}\right)^2} = \dfrac{1}{2}\sqrt{11}.$ ∎

Vector operations have many of the properties of ordinary arithmetic. These properties are readily verified using the definitions of vector addition and multiplication by a scalar.

Properties of Vector Operations
Let **u**, **v**, **w** be vectors and a, b be scalars.

1. $\mathbf{u} + \mathbf{v} = \mathbf{v} + \mathbf{u}$
2. $(\mathbf{u} + \mathbf{v}) + \mathbf{w} = \mathbf{u} + (\mathbf{v} + \mathbf{w})$
3. $\mathbf{u} + \mathbf{0} = \mathbf{u}$
4. $\mathbf{u} + (-\mathbf{u}) = \mathbf{0}$
5. $0\mathbf{u} = \mathbf{0}$
6. $1\mathbf{u} = \mathbf{u}$
7. $a(b\mathbf{u}) = (ab)\mathbf{u}$
8. $a(\mathbf{u} + \mathbf{v}) = a\mathbf{u} + a\mathbf{v}$
9. $(a + b)\mathbf{u} = a\mathbf{u} + b\mathbf{u}$

An important application of vectors occurs in navigation.

EXAMPLE 4   A jet airliner, flying due east at 500 mph in still air, encounters a 70-mph tailwind blowing in the direction 60° north of east. The airplane holds its compass heading due east but, because of the wind, acquires a new ground speed and direction. What are they?

Solution   If **u** = the velocity of the airplane alone and **v** = the velocity of the tailwind, then $|\mathbf{u}| = 500$ and $|\mathbf{v}| = 70$ (Figure 10.15). The velocity of the airplane with respect to the ground is given by the magnitude and direction of the *resultant vector* **u** + **v**. If we let the positive x-axis represent east and the positive y-axis represent north, then the component forms of **u** and **v** are

$$\mathbf{u} = \langle 500, 0 \rangle \qquad \text{and} \qquad \mathbf{v} = \langle 70\cos 60°, 70\sin 60° \rangle = \langle 35, 35\sqrt{3} \rangle.$$

Therefore,

$$\mathbf{u} + \mathbf{v} = \langle 535, 35\sqrt{3} \rangle$$

$$|\mathbf{u} + \mathbf{v}| = \sqrt{535^2 + (35\sqrt{3})^2} \approx 538.4$$

and

$$\theta = \tan^{-1}\frac{35\sqrt{3}}{535} \approx 6.5°. \qquad \text{Figure 10.15}$$

The new ground speed of the airplane is about 538.4 mph, and its new direction is about 6.5° north of east. ∎

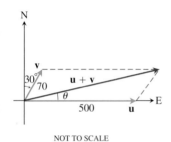

FIGURE 10.15   Vectors representing the velocities of the airplane **u** and tailwind **v** in Example 4.

Unit Vectors

A vector **v** of length 1 is called a **unit vector**. The **standard unit vectors** are

$$\mathbf{i} = \langle 1, 0, 0 \rangle, \qquad \mathbf{j} = \langle 0, 1, 0 \rangle, \qquad \text{and} \quad \mathbf{k} = \langle 0, 0, 1 \rangle.$$

Any vector $\mathbf{v} = \langle v_1, v_2, v_3 \rangle$ can be written as a *linear combination* of the standard unit vectors as follows:

$$\begin{aligned}
\mathbf{v} = \langle v_1, v_2, v_3 \rangle &= \langle v_1, 0, 0 \rangle + \langle 0, v_2, 0 \rangle + \langle 0, 0, v_3 \rangle \\
&= v_1 \langle 1, 0, 0 \rangle + v_2 \langle 0, 1, 0 \rangle + v_3 \langle 0, 0, 1 \rangle \\
&= v_1 \mathbf{i} + v_2 \mathbf{j} + v_3 \mathbf{k}.
\end{aligned}$$

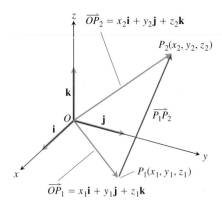

FIGURE 10.16 The vector from P_1 to P_2 is $\overrightarrow{P_1P_2} = (x_2 - x_1)\mathbf{i} + (y_2 - y_1)\mathbf{j} + (z_2 - z_1)\mathbf{k}$.

HISTORICAL BIOGRAPHY

Hermann Grassmann
(1809–1877)

We call the scalar (or number) v_1 the **i-component** of the vector $\mathbf{v}$, v_2 the **j-component**, and v_3 the **k-component**. In component form, the vector from $P_1(x_1, y_1, z_1)$ to $P_2(x_2, y_2, z_2)$ is

$$\overrightarrow{P_1P_2} = (x_2 - x_1)\mathbf{i} + (y_2 - y_1)\mathbf{j} + (z_2 - z_1)\mathbf{k}$$

(Figure 10.16).

Whenever $\mathbf{v} \neq \mathbf{0}$, its length $|\mathbf{v}|$ is not zero and

$$\left| \frac{1}{|\mathbf{v}|} \mathbf{v} \right| = \frac{1}{|\mathbf{v}|} |\mathbf{v}| = 1.$$

That is, $\mathbf{v}/|\mathbf{v}|$ is a unit vector in the direction of $\mathbf{v}$, called **the direction** of the nonzero vector $\mathbf{v}$.

EXAMPLE 5 Find a unit vector $\mathbf{u}$ in the direction of the vector from $P_1(1, 0, 1)$ to $P_2(3, 2, 0)$.

Solution We divide $\overrightarrow{P_1P_2}$ by its length:

$$\overrightarrow{P_1P_2} = (3 - 1)\mathbf{i} + (2 - 0)\mathbf{j} + (0 - 1)\mathbf{k} = 2\mathbf{i} + 2\mathbf{j} - \mathbf{k}$$

$$|\overrightarrow{P_1P_2}| = \sqrt{(2)^2 + (2)^2 + (-1)^2} = \sqrt{4 + 4 + 1} = \sqrt{9} = 3$$

$$\mathbf{u} = \frac{\overrightarrow{P_1P_2}}{|\overrightarrow{P_1P_2}|} = \frac{2\mathbf{i} + 2\mathbf{j} - \mathbf{k}}{3} = \frac{2}{3}\mathbf{i} + \frac{2}{3}\mathbf{j} - \frac{1}{3}\mathbf{k}.$$

The unit vector $\mathbf{u}$ is the direction of $\overrightarrow{P_1P_2}$. ∎

EXAMPLE 6 If $\mathbf{v} = 3\mathbf{i} - 4\mathbf{j}$ is a velocity vector, express $\mathbf{v}$ as a product of its speed times a unit vector in the direction of motion.

Solution Speed is the magnitude (length) of $\mathbf{v}$:

$$|\mathbf{v}| = \sqrt{(3)^2 + (-4)^2} = \sqrt{9 + 16} = 5.$$

The unit vector $\mathbf{v}/|\mathbf{v}|$ has the same direction as $\mathbf{v}$:

$$\frac{\mathbf{v}}{|\mathbf{v}|} = \frac{3\mathbf{i} - 4\mathbf{j}}{5} = \frac{3}{5}\mathbf{i} - \frac{4}{5}\mathbf{j}.$$

So

$$\mathbf{v} = 3\mathbf{i} - 4\mathbf{j} = 5\left(\frac{3}{5}\mathbf{i} - \frac{4}{5}\mathbf{j} \right).$$

Length
(speed)   Direction of motion

∎

In summary, we can express any nonzero vector $\mathbf{v}$ in terms of its two important features, length and direction, by writing $\mathbf{v} = |\mathbf{v}| \dfrac{\mathbf{v}}{|\mathbf{v}|}$.

> If $\mathbf{v} \neq \mathbf{0}$, then
>
> 1. $\dfrac{\mathbf{v}}{|\mathbf{v}|}$ is a unit vector in the direction of $\mathbf{v}$;
> 2. the equation $\mathbf{v} = |\mathbf{v}|\dfrac{\mathbf{v}}{|\mathbf{v}|}$ expresses $\mathbf{v}$ in terms of its length and direction.

EXAMPLE 7   A force of 6 newtons is applied in the direction of the vector $\mathbf{v} = 2\mathbf{i} + 2\mathbf{j} - \mathbf{k}$. Express the force $\mathbf{F}$ as a product of its magnitude and direction.

Solution   The force vector has magnitude 6 and direction $\dfrac{\mathbf{v}}{|\mathbf{v}|}$, so

$$\mathbf{F} = 6\frac{\mathbf{v}}{|\mathbf{v}|} = 6\frac{2\mathbf{i} + 2\mathbf{j} - \mathbf{k}}{\sqrt{2^2 + 2^2 + (-1)^2}} = 6\frac{2\mathbf{i} + 2\mathbf{j} - \mathbf{k}}{3}$$

$$= 6\left(\frac{2}{3}\mathbf{i} + \frac{2}{3}\mathbf{j} - \frac{1}{3}\mathbf{k}\right). \qquad\blacksquare$$

Midpoint of a Line Segment

Vectors are often useful in geometry. For example, the coordinates of the midpoint of a line segment are found by averaging.

> The **midpoint** M of the line segment joining points $P_1(x_1, y_1, z_1)$ and $P_2(x_2, y_2, z_2)$ is the point
>
> $$\left(\frac{x_1 + x_2}{2}, \frac{y_1 + y_2}{2}, \frac{z_1 + z_2}{2}\right).$$

$P_1(x_1, y_1, z_1)$

$M\left(\dfrac{x_1 + x_2}{2}, \dfrac{y_1 + y_2}{2}, \dfrac{z_1 + z_2}{2}\right)$

$P_2(x_2, y_2, z_2)$

O

FIGURE 10.17   The coordinates of the midpoint are the averages of the coordinates of P_1 and P_2.

To see why, observe (Figure 10.17) that

$$\overrightarrow{OM} = \overrightarrow{OP_1} + \frac{1}{2}(\overrightarrow{P_1 P_2}) = \overrightarrow{OP_1} + \frac{1}{2}(\overrightarrow{OP_2} - \overrightarrow{OP_1})$$

$$= \frac{1}{2}(\overrightarrow{OP_1} + \overrightarrow{OP_2})$$

$$= \frac{x_1 + x_2}{2}\mathbf{i} + \frac{y_1 + y_2}{2}\mathbf{j} + \frac{z_1 + z_2}{2}\mathbf{k}.$$

EXAMPLE 8   The midpoint of the segment joining $P_1(3, -2, 0)$ and $P_2(7, 4, 4)$ is

$$\left(\frac{3 + 7}{2}, \frac{-2 + 4}{2}, \frac{0 + 4}{2}\right) = (5, 1, 2). \qquad\blacksquare$$

EXERCISES 10.2

In Exercises 1–8, let $\mathbf{u} = \langle 3, -2 \rangle$ and $\mathbf{v} = \langle -2, 5 \rangle$. Find the **(a)** component form and **(b)** magnitude (length) of the vector.

1. $3\mathbf{u}$

2. $-2\mathbf{v}$

3. $\mathbf{u} + \mathbf{v}$

4. $\mathbf{u} - \mathbf{v}$

5. $2\mathbf{u} - 3\mathbf{v}$

6. $-2\mathbf{u} + 5\mathbf{v}$

7. $\dfrac{3}{5}\mathbf{u} + \dfrac{4}{5}\mathbf{v}$

8. $-\dfrac{5}{13}\mathbf{u} + \dfrac{12}{13}\mathbf{v}$

In Exercises 9–16, find the component form of the vector.

9. The vector $\overrightarrow{PQ}$, where $P = (1, 3)$ and $Q = (2, -1)$

10. The vector $\overrightarrow{OP}$ where O is the origin and P is the midpoint of segment RS, where $R = (2, -1)$ and $S = (-4, 3)$

11. The vector from the point $A = (2, 3)$ to the origin

12. The sum of $\overrightarrow{AB}$ and $\overrightarrow{CD}$, where $A = (1, -1), B = (2, 0)$, $C = (-1, 3)$, and $D = (-2, 2)$

13. The unit vector that makes an angle $\theta = 2\pi/3$ with the positive x-axis

14. The unit vector that makes an angle $\theta = -3\pi/4$ with the positive x-axis

15. The unit vector obtained by rotating the vector $\langle 0, 1 \rangle$ $120°$ counterclockwise about the origin

16. The unit vector obtained by rotating the vector $\langle 1, 0 \rangle$ $135°$ counterclockwise about the origin

In Exercises 17–22, express each vector in the form $\mathbf{v} = v_1\mathbf{i} + v_2\mathbf{j} + v_3\mathbf{k}$.

17. $\overrightarrow{P_1P_2}$ if P_1 is the point $(5, 7, -1)$ and P_2 is the point $(2, 9, -2)$

18. $\overrightarrow{P_1P_2}$ if P_1 is the point $(1, 2, 0)$ and P_2 is the point $(-3, 0, 5)$

19. $\overrightarrow{AB}$ if A is the point $(-7, -8, 1)$ and B is the point $(-10, 8, 1)$

20. $\overrightarrow{AB}$ if A is the point $(1, 0, 3)$ and B is the point $(-1, 4, 5)$

21. $5\mathbf{u} - \mathbf{v}$ if $\mathbf{u} = \langle 1, 1, -1 \rangle$ and $\mathbf{v} = \langle 2, 0, 3 \rangle$

22. $-2\mathbf{u} + 3\mathbf{v}$ if $\mathbf{u} = \langle -1, 0, 2 \rangle$ and $\mathbf{v} = \langle 1, 1, 1 \rangle$

In Exercises 23 and 24, copy vectors $\mathbf{u}$, $\mathbf{v}$, and $\mathbf{w}$ head to tail as needed to sketch the indicated vector.

23.

a. $\mathbf{u} + \mathbf{v}$ **b.** $\mathbf{u} + \mathbf{v} + \mathbf{w}$

c. $\mathbf{u} - \mathbf{v}$ **d.** $\mathbf{u} - \mathbf{w}$

24.

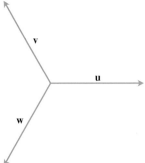

a. $\mathbf{u} - \mathbf{v}$ **b.** $\mathbf{u} - \mathbf{v} + \mathbf{w}$

c. $2\mathbf{u} - \mathbf{v}$ **d.** $\mathbf{u} + \mathbf{v} + \mathbf{w}$

In Exercises 25–30, express each vector as a product of its length and direction.

25. $2\mathbf{i} + \mathbf{j} - 2\mathbf{k}$

26. $9\mathbf{i} - 2\mathbf{j} + 6\mathbf{k}$

27. $5\mathbf{k}$

28. $\dfrac{3}{5}\mathbf{i} + \dfrac{4}{5}\mathbf{k}$

29. $\dfrac{1}{\sqrt{6}}\mathbf{i} - \dfrac{1}{\sqrt{6}}\mathbf{j} - \dfrac{1}{\sqrt{6}}\mathbf{k}$ **30.** $\dfrac{\mathbf{i}}{\sqrt{3}} + \dfrac{\mathbf{j}}{\sqrt{3}} + \dfrac{\mathbf{k}}{\sqrt{3}}$

31. Find the vectors whose lengths and directions are given. Try to do the calculations without writing.

	Length	Direction
a.	2	$\mathbf{i}$
b.	$\sqrt{3}$	$-\mathbf{k}$
c.	$\dfrac{1}{2}$	$\dfrac{3}{5}\mathbf{j} + \dfrac{4}{5}\mathbf{k}$
d.	7	$\dfrac{6}{7}\mathbf{i} - \dfrac{2}{7}\mathbf{j} + \dfrac{3}{7}\mathbf{k}$

32. Find the vectors whose lengths and directions are given. Try to do the calculations without writing.

	Length	Direction
a.	7	$-\mathbf{j}$
b.	$\sqrt{2}$	$-\dfrac{3}{5}\mathbf{i} - \dfrac{4}{5}\mathbf{k}$
c.	$\dfrac{13}{12}$	$\dfrac{3}{13}\mathbf{i} - \dfrac{4}{13}\mathbf{j} - \dfrac{12}{13}\mathbf{k}$
d.	$a > 0$	$\dfrac{1}{\sqrt{2}}\mathbf{i} + \dfrac{1}{\sqrt{3}}\mathbf{j} - \dfrac{1}{\sqrt{6}}\mathbf{k}$

33. Find a vector of magnitude 7 in the direction of $\mathbf{v} = 12\mathbf{i} - 5\mathbf{k}$.

34. Find a vector of magnitude 3 in the direction opposite to the direction of $\mathbf{v} = (1/2)\mathbf{i} - (1/2)\mathbf{j} - (1/2)\mathbf{k}$.

In Exercises 35–38, find

   a. the direction of $\overrightarrow{P_1P_2}$ and

   b. the midpoint of line segment P_1P_2.

35. $P_1(-1, 1, 5)$ $P_2(2, 5, 0)$

36. $P_1(1, 4, 5)$ $P_2(4, -2, 7)$

37. $P_1(3, 4, 5)$ $P_2(2, 3, 4)$

38. $P_1(0, 0, 0)$ $P_2(2, -2, -2)$

39. If $\overrightarrow{AB} = \mathbf{i} + 4\mathbf{j} - 2\mathbf{k}$ and B is the point $(5, 1, 3)$, find A.

40. If $\overrightarrow{AB} = -7\mathbf{i} + 3\mathbf{j} + 8\mathbf{k}$ and A is the point $(-2, -3, 6)$, find B.

41. Linear combination   Let $\mathbf{u} = 2\mathbf{i} + \mathbf{j}, \mathbf{v} = \mathbf{i} + \mathbf{j}$, and $\mathbf{w} = \mathbf{i} - \mathbf{j}$. Find scalars a and b such that $\mathbf{u} = a\mathbf{v} + b\mathbf{w}$.

42. Linear combination   Let $\mathbf{u} = \mathbf{i} - 2\mathbf{j}, \mathbf{v} = 2\mathbf{i} + 3\mathbf{j}$, and $\mathbf{w} = \mathbf{i} + \mathbf{j}$. Write $\mathbf{u} = \mathbf{u}_1 + \mathbf{u}_2$, where $\mathbf{u}_1$ is parallel to $\mathbf{v}$ and $\mathbf{u}_2$ is parallel to $\mathbf{w}$. (See Exercise 41.)

43. Force vector   You are pulling on a suitcase with a force $\mathbf{F}$ (pictured here) whose magnitude is $|\mathbf{F}| = 10$ lb. Find the $\mathbf{i}$- and $\mathbf{j}$-components of $\mathbf{F}$.

44. Force vector   A kite string exerts a 12-lb pull ($|\mathbf{F}| = 12$) on a kite and makes a $45°$ angle with the horizontal. Find the horizontal and vertical components of $\mathbf{F}$.

45. Velocity   An airplane is flying in the direction $25°$ west of north at 800 km/h. Find the component form of the velocity of the airplane, assuming that the positive x-axis represents due east and the positive y-axis represents due north.

46. Velocity   An airplane is flying in the direction $10°$ east of south at 600 km/h. Find the component form of the velocity of the airplane, assuming that the positive x-axis represents due east and the positive y-axis represents due north.

47. Location   A bird flies from its nest 5 km in the direction $60°$ north of east, where it stops to rest on a tree. It then flies 10 km in the direction due southeast and lands atop a telephone pole. Place an xy-coordinate system so that the origin is the bird's nest, the x-axis points east, and the y-axis points north.

   a. At what point is the tree located?

   b. At what point is the telephone pole?

48. Use similar triangles to find the coordinates of the point Q that divides the segment from $P_1(x_1, y_1, z_1)$ to $P_2(x_2, y_2, z_2)$ into two lengths whose ratio is $p/q = r$.

49. Medians of a triangle   Suppose that A, B, and C are the corner points of the thin triangular plate of constant density shown here.

   a. Find the vector from C to the midpoint M of side AB.

   b. Find the vector from C to the point that lies two-thirds of the way from C to M on the median CM.

   c. Find the coordinates of the point in which the medians of $\triangle ABC$ intersect. According to Exercise 21, Section 6.7, this point is the plate's center of mass.

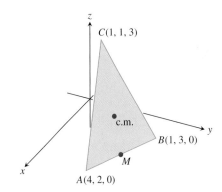

50. Find the vector from the origin to the point of intersection of the medians of the triangle whose vertices are

$$A(1, -1, 2), \quad B(2, 1, 3), \quad \text{and} \quad C(-1, 2, -1).$$

51. Let $ABCD$ be a general, not necessarily planar, quadrilateral in space. Show that the two segments joining the midpoints of opposite sides of $ABCD$ bisect each other. (*Hint:* Show that the segments have the same midpoint.)

52. Vectors are drawn from the center of a regular n-sided polygon in the plane to the vertices of the polygon. Show that the sum of the vectors is zero. (*Hint:* What happens to the sum if you rotate the polygon about its center?)

53. Suppose that A, B, and C are vertices of a triangle and that a, b, and c are, respectively, the midpoints of the opposite sides. Show that $\overrightarrow{Aa} + \overrightarrow{Bb} + \overrightarrow{Cc} = 0$.

54. Unit vectors in the plane   Show that a unit vector in the plane can be expressed as $\mathbf{u} = (\cos \theta)\mathbf{i} + (\sin \theta)\mathbf{j}$, obtained by rotating $\mathbf{i}$ through an angle θ in the counterclockwise direction. Explain why this form gives *every* unit vector in the plane.

Length = $|\mathbf{F}|\cos\theta$

FIGURE 10.18 The magnitude of the force $\mathbf{F}$ in the direction of vector $\mathbf{v}$ is the length $|\mathbf{F}|\cos\theta$ of the projection of $\mathbf{F}$ onto $\mathbf{v}$.

If a force $\mathbf{F}$ is applied to a particle moving along a path, we often need to know the magnitude of the force in the direction of motion. If $\mathbf{v}$ is parallel to the tangent line to the path at the point where $\mathbf{F}$ is applied, then we want the magnitude of $\mathbf{F}$ in the direction of $\mathbf{v}$. Figure 10.18 shows that the scalar quantity we seek is the length $|\mathbf{F}|\cos\theta$, where θ is the angle between the two vectors $\mathbf{F}$ and $\mathbf{v}$.

In this section, we show how to calculate easily the angle between two vectors directly from their components. A key part of the calculation is an expression called the *dot product*. Dot products are also called *inner* or *scalar* products because the product results in a scalar, not a vector. After investigating the dot product, we apply it to finding the projection of one vector onto another (as displayed in Figure 10.18) and to finding the work done by a constant force acting through a displacement.

Angle Between Vectors

When two nonzero vectors $\mathbf{u}$ and $\mathbf{v}$ are placed so their initial points coincide, they form an angle θ of measure $0 \leq \theta \leq \pi$ (Figure 10.19). If the vectors do not lie along the same line, the angle θ is measured in the plane containing both of them. If they do lie along the same line, the angle between them is 0 if they point in the same direction, and π if they point in opposite directions. The angle θ is the **angle between $\mathbf{u}$ and $\mathbf{v}$**. Theorem 1 gives a formula to determine this angle.

> **THEOREM 1—Angle Between Two Vectors**   The angle θ between two nonzero vectors $\mathbf{u} = \langle u_1, u_2, u_3 \rangle$ and $\mathbf{v} = \langle v_1, v_2, v_3 \rangle$ is given by
>
> $$\theta = \cos^{-1}\left(\frac{u_1 v_1 + u_2 v_2 + u_3 v_3}{|\mathbf{u}||\mathbf{v}|}\right).$$

FIGURE 10.19 The angle between $\mathbf{u}$ and $\mathbf{v}$.

Before proving Theorem 1, we focus attention on the expression $u_1 v_1 + u_2 v_2 + u_3 v_3$ in the calculation for θ.

> **DEFINITION**   The **dot product $\mathbf{u} \cdot \mathbf{v}$** ("$\mathbf{u}$ dot $\mathbf{v}$") of vectors $\mathbf{u} = \langle u_1, u_2, u_3 \rangle$ and $\mathbf{v} = \langle v_1, v_2, v_3 \rangle$ is
>
> $$\mathbf{u} \cdot \mathbf{v} = u_1 v_1 + u_2 v_2 + u_3 v_3.$$

EXAMPLE 1

(a) $\langle 1, -2, -1 \rangle \cdot \langle -6, 2, -3 \rangle = (1)(-6) + (-2)(2) + (-1)(-3)$
$$= -6 - 4 + 3 = -7$$

(b) $\left(\frac{1}{2}\mathbf{i} + 3\mathbf{j} + \mathbf{k}\right) \cdot (4\mathbf{i} - \mathbf{j} + 2\mathbf{k}) = \left(\frac{1}{2}\right)(4) + (3)(-1) + (1)(2) = 1$ ∎

The dot product of a pair of two-dimensional vectors is defined in a similar fashion:

$$\langle u_1, u_2 \rangle \cdot \langle v_1, v_2 \rangle = u_1 v_1 + u_2 v_2.$$

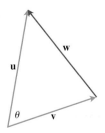

FIGURE 10.20   The parallelogram law of addition of vectors gives $\mathbf{w} = \mathbf{u} - \mathbf{v}$.

Proof of Theorem 1   Applying the law of cosines (Equation (8), Section 1.3) to the triangle in Figure 10.20, we find that

$$|\mathbf{w}|^2 = |\mathbf{u}|^2 + |\mathbf{v}|^2 - 2|\mathbf{u}||\mathbf{v}|\cos\theta \qquad \text{Law of cosines}$$

$$2|\mathbf{u}||\mathbf{v}|\cos\theta = |\mathbf{u}|^2 + |\mathbf{v}|^2 - |\mathbf{w}|^2.$$

Because $\mathbf{w} = \mathbf{u} - \mathbf{v}$, the component form of $\mathbf{w}$ is $\langle u_1 - v_1, u_2 - v_2, u_3 - v_3 \rangle$. So

$$|\mathbf{u}|^2 = \left(\sqrt{u_1^2 + u_2^2 + u_3^2}\right)^2 = u_1^2 + u_2^2 + u_3^2$$

$$|\mathbf{v}|^2 = \left(\sqrt{v_1^2 + v_2^2 + v_3^2}\right)^2 = v_1^2 + v_2^2 + v_3^2$$

$$|\mathbf{w}|^2 = \left(\sqrt{(u_1 - v_1)^2 + (u_2 - v_2)^2 + (u_3 - v_3)^2}\right)^2$$

$$= (u_1 - v_1)^2 + (u_2 - v_2)^2 + (u_3 - v_3)^2$$

$$= u_1^2 - 2u_1v_1 + v_1^2 + u_2^2 - 2u_2v_2 + v_2^2 + u_3^2 - 2u_3v_3 + v_3^2$$

and

$$|\mathbf{u}|^2 + |\mathbf{v}|^2 - |\mathbf{w}|^2 = 2(u_1v_1 + u_2v_2 + u_3v_3).$$

Therefore,

$$2|\mathbf{u}||\mathbf{v}|\cos\theta = |\mathbf{u}|^2 + |\mathbf{v}|^2 - |\mathbf{w}|^2 = 2(u_1v_1 + u_2v_2 + u_3v_3)$$

$$|\mathbf{u}||\mathbf{v}|\cos\theta = u_1v_1 + u_2v_2 + u_3v_3$$

$$\cos\theta = \frac{u_1v_1 + u_2v_2 + u_3v_3}{|\mathbf{u}||\mathbf{v}|}.$$

Since $0 \le \theta < \pi$, we have

$$\theta = \cos^{-1}\left(\frac{u_1v_1 + u_2v_2 + u_3v_3}{|\mathbf{u}||\mathbf{v}|}\right). \qquad \blacksquare$$

With the notation of the dot product, the angle between two vectors $\mathbf{u}$ and $\mathbf{v}$ can be written as

$$\theta = \cos^{-1}\left(\frac{\mathbf{u}\cdot\mathbf{v}}{|\mathbf{u}||\mathbf{v}|}\right).$$

EXAMPLE 2   Find the angle between $\mathbf{u} = \mathbf{i} - 2\mathbf{j} - 2\mathbf{k}$ and $\mathbf{v} = 6\mathbf{i} + 3\mathbf{j} + 2\mathbf{k}$.

Solution   We use the formula above:

$$\mathbf{u}\cdot\mathbf{v} = (1)(6) + (-2)(3) + (-2)(2) = 6 - 6 - 4 = -4$$

$$|\mathbf{u}| = \sqrt{(1)^2 + (-2)^2 + (-2)^2} = \sqrt{9} = 3$$

$$|\mathbf{v}| = \sqrt{(6)^2 + (3)^2 + (2)^2} = \sqrt{49} = 7$$

$$\theta = \cos^{-1}\left(\frac{\mathbf{u}\cdot\mathbf{v}}{|\mathbf{u}||\mathbf{v}|}\right)$$

$$= \cos^{-1}\left(\frac{-4}{(3)(7)}\right) \approx 1.76 \text{ radians}. \qquad \blacksquare$$

The angle formula applies to two-dimensional vectors as well.

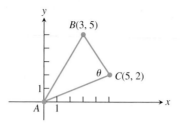

FIGURE 10.21 The triangle in Example 3.

EXAMPLE 3 Find the angle θ in the triangle ABC determined by the vertices $A = (0, 0)$, $B = (3, 5)$, and $C = (5, 2)$ (Figure 10.21).

Solution The angle θ is the angle between the vectors $\vec{CA}$ and $\vec{CB}$. The component forms of these two vectors are

$$\vec{CA} = \langle -5, -2 \rangle \quad \text{and} \quad \vec{CB} = \langle -2, 3 \rangle.$$

First we calculate the dot product and magnitudes of these two vectors.

$$\vec{CA} \cdot \vec{CB} = (-5)(-2) + (-2)(3) = 4$$

$$|\vec{CA}| = \sqrt{(-5)^2 + (-2)^2} = \sqrt{29}$$

$$|\vec{CB}| = \sqrt{(-2)^2 + (3)^2} = \sqrt{13}$$

Then applying the angle formula, we have

$$\theta = \cos^{-1}\left(\frac{\vec{CA} \cdot \vec{CB}}{|\vec{CA}||\vec{CB}|}\right)$$

$$= \cos^{-1}\left(\frac{4}{(\sqrt{29})(\sqrt{13})}\right)$$

$$\approx 78.1° \quad \text{or} \quad 1.36 \text{ radians}. \quad \blacksquare$$

Perpendicular (Orthogonal) Vectors

Two nonzero vectors $\mathbf{u}$ and $\mathbf{v}$ are perpendicular or **orthogonal** if the angle between them is $\pi/2$. For such vectors, we have $\mathbf{u} \cdot \mathbf{v} = 0$ because $\cos(\pi/2) = 0$. The converse is also true. If $\mathbf{u}$ and $\mathbf{v}$ are nonzero vectors with $\mathbf{u} \cdot \mathbf{v} = |\mathbf{u}||\mathbf{v}| \cos \theta = 0$, then $\cos \theta = 0$ and $\theta = \cos^{-1} 0 = \pi/2$.

DEFINITION Vectors $\mathbf{u}$ and $\mathbf{v}$ are **orthogonal** (or **perpendicular**) if and only if $\mathbf{u} \cdot \mathbf{v} = 0$.

EXAMPLE 4

(a) $\mathbf{u} = \langle 3, -2 \rangle$ and $\mathbf{v} = \langle 4, 6 \rangle$ are orthogonal because $\mathbf{u} \cdot \mathbf{v} = (3)(4) + (-2)(6) = 0$.

(b) $\mathbf{u} = 3\mathbf{i} - 2\mathbf{j} + \mathbf{k}$ and $\mathbf{v} = 2\mathbf{j} + 4\mathbf{k}$ are orthogonal because $\mathbf{u} \cdot \mathbf{v} = (3)(0) + (-2)(2) + (1)(4) = 0$.

(c) $\mathbf{0}$ is orthogonal to every vector $\mathbf{u}$ since

$$\mathbf{0} \cdot \mathbf{u} = \langle 0, 0, 0 \rangle \cdot \langle u_1, u_2, u_3 \rangle$$
$$= (0)(u_1) + (0)(u_2) + (0)(u_3)$$
$$= 0. \quad \blacksquare$$

Dot Product Properties and Vector Projections

The dot product obeys many of the laws that hold for ordinary products of real numbers (scalars).

Properties of the Dot Product

If $\mathbf{u}$, $\mathbf{v}$, and $\mathbf{w}$ are any vectors and c is a scalar, then

1. $\mathbf{u} \cdot \mathbf{v} = \mathbf{v} \cdot \mathbf{u}$
2. $(c\mathbf{u}) \cdot \mathbf{v} = \mathbf{u} \cdot (c\mathbf{v}) = c(\mathbf{u} \cdot \mathbf{v})$
3. $\mathbf{u} \cdot (\mathbf{v} + \mathbf{w}) = \mathbf{u} \cdot \mathbf{v} + \mathbf{u} \cdot \mathbf{w}$
4. $\mathbf{u} \cdot \mathbf{u} = |\mathbf{u}|^2$
5. $\mathbf{0} \cdot \mathbf{u} = 0$.

HISTORICAL BIOGRAPHY

Carl Friedrich Gauss
(1777–1855)

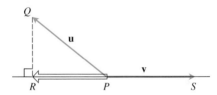

FIGURE 10.22 The vector projection of $\mathbf{u}$ onto $\mathbf{v}$.

FIGURE 10.23 If we pull on the box with force $\mathbf{u}$, the effective force moving the box forward in the direction $\mathbf{v}$ is the projection of $\mathbf{u}$ onto $\mathbf{v}$.

Proofs of Properties 1 and 3 The properties are easy to prove using the definition. For instance, here are the proofs of Properties 1 and 3.

1. $\mathbf{u} \cdot \mathbf{v} = u_1 v_1 + u_2 v_2 + u_3 v_3 = v_1 u_1 + v_2 u_2 + v_3 u_3 = \mathbf{v} \cdot \mathbf{u}$

3. $\mathbf{u} \cdot (\mathbf{v} + \mathbf{w}) = \langle u_1, u_2, u_3 \rangle \cdot \langle v_1 + w_1, v_2 + w_2, v_3 + w_3 \rangle$

$$= u_1(v_1 + w_1) + u_2(v_2 + w_2) + u_3(v_3 + w_3)$$

$$= u_1 v_1 + u_1 w_1 + u_2 v_2 + u_2 w_2 + u_3 v_3 + u_3 w_3$$

$$= (u_1 v_1 + u_2 v_2 + u_3 v_3) + (u_1 w_1 + u_2 w_2 + u_3 w_3)$$

$$= \mathbf{u} \cdot \mathbf{v} + \mathbf{u} \cdot \mathbf{w} \qquad \blacksquare$$

We now return to the problem of projecting one vector onto another, posed in the opening to this section. The **vector projection** of $\mathbf{u} = \overrightarrow{PQ}$ onto a nonzero vector $\mathbf{v} = \overrightarrow{PS}$ (Figure 10.22) is the vector $\overrightarrow{PR}$ determined by dropping a perpendicular from Q to the line PS. The notation for this vector is

$$\text{proj}_{\mathbf{v}}\,\mathbf{u} \qquad (\text{"the vector projection of } \mathbf{u} \text{ onto } \mathbf{v}\text{"}).$$

If $\mathbf{u}$ represents a force, then $\text{proj}_{\mathbf{v}}\,\mathbf{u}$ represents the effective force in the direction of $\mathbf{v}$ (Figure 10.23).

If the angle θ between $\mathbf{u}$ and $\mathbf{v}$ is acute, $\text{proj}_{\mathbf{v}}\,\mathbf{u}$ has length $|\mathbf{u}| \cos\theta$ and direction $\mathbf{v}/|\mathbf{v}|$ (Figure 10.24). If θ is obtuse, $\cos\theta < 0$ and $\text{proj}_{\mathbf{v}}\,\mathbf{u}$ has length $-|\mathbf{u}| \cos\theta$ and direction $-\mathbf{v}/|\mathbf{v}|$. In both cases,

$$\text{proj}_{\mathbf{v}}\,\mathbf{u} = (|\mathbf{u}| \cos\theta)\frac{\mathbf{v}}{|\mathbf{v}|}$$

$$= \left(\frac{\mathbf{u} \cdot \mathbf{v}}{|\mathbf{v}|}\right)\frac{\mathbf{v}}{|\mathbf{v}|} \qquad |\mathbf{u}|\cos\theta = \frac{|\mathbf{u}||\mathbf{v}|\cos\theta}{|\mathbf{v}|} = \frac{\mathbf{u} \cdot \mathbf{v}}{|\mathbf{v}|}$$

$$= \left(\frac{\mathbf{u} \cdot \mathbf{v}}{|\mathbf{v}|^2}\right)\mathbf{v}.$$

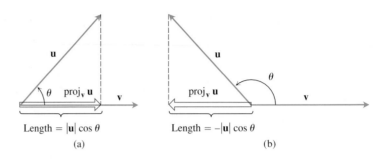

FIGURE 10.24 The length of $\text{proj}_\mathbf{v}\mathbf{u}$ is (a) $|\mathbf{u}|\cos\theta$ if $\cos\theta \geq 0$ and (b) $-|\mathbf{u}|\cos\theta$ if $\cos\theta < 0$.

The number $|\mathbf{u}|\cos\theta$ is called the **scalar component of u in the direction of v**. To summarize,

---

Vector projection of **u** onto **v**:

$$\text{proj}_\mathbf{v}\mathbf{u} = \left(\frac{\mathbf{u}\cdot\mathbf{v}}{|\mathbf{v}|^2}\right)\mathbf{v} \qquad (1)$$

Scalar component of **u** in the direction of **v**:

$$|\mathbf{u}|\cos\theta = \frac{\mathbf{u}\cdot\mathbf{v}}{|\mathbf{v}|} = \mathbf{u}\cdot\frac{\mathbf{v}}{|\mathbf{v}|} \qquad (2)$$

Note that both the vector projection of **u** onto **v** and the scalar component of **u** onto **v** depend only on the direction of the vector **v** and not its length (because we dot **u** with $\mathbf{v}/|\mathbf{v}|$, which is the direction of **v**).

EXAMPLE 5   Find the vector projection of $\mathbf{u} = 6\mathbf{i} + 3\mathbf{j} + 2\mathbf{k}$ onto $\mathbf{v} = \mathbf{i} - 2\mathbf{j} - 2\mathbf{k}$ and the scalar component of **u** in the direction of **v**.

Solution   We find $\text{proj}_\mathbf{v}\mathbf{u}$ from Equation (1):

$$\text{proj}_\mathbf{v}\mathbf{u} = \frac{\mathbf{u}\cdot\mathbf{v}}{\mathbf{v}\cdot\mathbf{v}}\mathbf{v} = \frac{6-6-4}{1+4+4}(\mathbf{i}-2\mathbf{j}-2\mathbf{k})$$

$$= -\frac{4}{9}(\mathbf{i}-2\mathbf{j}-2\mathbf{k}) = -\frac{4}{9}\mathbf{i} + \frac{8}{9}\mathbf{j} + \frac{8}{9}\mathbf{k}.$$

We find the scalar component of **u** in the direction of **v** from Equation (2):

$$|\mathbf{u}|\cos\theta = \mathbf{u}\cdot\frac{\mathbf{v}}{|\mathbf{v}|} = (6\mathbf{i}+3\mathbf{j}+2\mathbf{k})\cdot\left(\frac{1}{3}\mathbf{i} - \frac{2}{3}\mathbf{j} - \frac{2}{3}\mathbf{k}\right)$$

$$= 2 - 2 - \frac{4}{3} = -\frac{4}{3}. \qquad \blacksquare$$

Equations (1) and (2) also apply to two-dimensional vectors.

EXAMPLE 6   Find the vector projection of a force $\mathbf{F} = 5\mathbf{i} + 2\mathbf{j}$ onto $\mathbf{v} = \mathbf{i} - 3\mathbf{j}$ and the scalar component of $\mathbf{F}$ in the direction of $\mathbf{v}$.

Solution   The vector projection is

$$
\text{proj}_{\mathbf{v}}\,\mathbf{F} = \left(\frac{\mathbf{F} \cdot \mathbf{v}}{|\mathbf{v}|^2}\right)\mathbf{v}
$$

$$
= \frac{5 - 6}{1 + 9}(\mathbf{i} - 3\mathbf{j}) = -\frac{1}{10}(\mathbf{i} - 3\mathbf{j})
$$

$$
= -\frac{1}{10}\mathbf{i} + \frac{3}{10}\mathbf{j}.
$$

The scalar component of $\mathbf{F}$ in the direction of $\mathbf{v}$ is

$$
|\mathbf{F}|\cos\theta = \frac{\mathbf{F} \cdot \mathbf{v}}{|\mathbf{v}|} = \frac{5 - 6}{\sqrt{1 + 9}} = -\frac{1}{\sqrt{10}}.
$$ ∎

Work

In Chapter 6, we calculated the work done by a constant force of magnitude F in moving an object through a distance d as $W = Fd$. That formula holds only if the force is directed along the line of motion. If a force $\mathbf{F}$ moving an object through a displacement $\mathbf{D} = \vec{PQ}$ has some other direction, the work is performed by the component of $\mathbf{F}$ in the direction of $\mathbf{D}$. If θ is the angle between $\mathbf{F}$ and $\mathbf{D}$ (Figure 10.25), then

$$
\text{Work} = \begin{pmatrix}\text{scalar component of }\mathbf{F} \\ \text{in the direction of }\mathbf{D}\end{pmatrix}(\text{length of }\mathbf{D})
$$

$$
= (|\mathbf{F}|\cos\theta)|\mathbf{D}|
$$

$$
= \mathbf{F} \cdot \mathbf{D}.
$$

FIGURE 10.25   The work done by a constant force $\mathbf{F}$ during a displacement $\mathbf{D}$ is $(|\mathbf{F}|\cos\theta)|\mathbf{D}|$.

> **DEFINITION**   The **work** done by a constant force $\mathbf{F}$ acting through a displacement $\mathbf{D} = \vec{PQ}$ is
>
> $$W = \mathbf{F} \cdot \mathbf{D} = |\mathbf{F}||\mathbf{D}|\cos\theta,$$
>
> where θ is the angle between $\mathbf{F}$ and $\mathbf{D}$.

EXAMPLE 7   If $|\mathbf{F}| = 40$ N (newtons), $|\mathbf{D}| = 3$ m, and $\theta = 60°$, the work done by $\mathbf{F}$ in acting from P to Q is

$$
\begin{aligned}
\text{Work} &= |\mathbf{F}||\mathbf{D}|\cos\theta &&\text{Definition}\\
&= (40)(3)\cos 60° &&\text{Given values}\\
&= (120)(1/2)\\
&= 60\text{ J (joules)}.
\end{aligned}
$$ ∎

We encounter more challenging work problems in Chapter 14 when we learn to find the work done by a variable force along a *path* in space.

EXERCISES 10.3

In Exercises 1–8, find

    a. $\mathbf{v} \cdot \mathbf{u}, |\mathbf{v}|, |\mathbf{u}|$

    b. the cosine of the angle between $\mathbf{v}$ and $\mathbf{u}$

    c. the scalar component of $\mathbf{u}$ in the direction of $\mathbf{v}$

    d. the vector $\text{proj}_{\mathbf{v}}\,\mathbf{u}$.

1. $\mathbf{v} = 2\mathbf{i} - 4\mathbf{j} + \sqrt{5}\mathbf{k}, \quad \mathbf{u} = -2\mathbf{i} + 4\mathbf{j} - \sqrt{5}\mathbf{k}$
2. $\mathbf{v} = (3/5)\mathbf{i} + (4/5)\mathbf{k}, \quad \mathbf{u} = 5\mathbf{i} + 12\mathbf{j}$
3. $\mathbf{v} = 10\mathbf{i} + 11\mathbf{j} - 2\mathbf{k}, \quad \mathbf{u} = 3\mathbf{j} + 4\mathbf{k}$
4. $\mathbf{v} = 2\mathbf{i} + 10\mathbf{j} - 11\mathbf{k}, \quad \mathbf{u} = 2\mathbf{i} + 2\mathbf{j} + \mathbf{k}$
5. $\mathbf{v} = 5\mathbf{j} - 3\mathbf{k}, \quad \mathbf{u} = \mathbf{i} + \mathbf{j} + \mathbf{k}$
6. $\mathbf{v} = -\mathbf{i} + \mathbf{j}, \quad \mathbf{u} = \sqrt{2}\mathbf{i} + \sqrt{3}\mathbf{j} + 2\mathbf{k}$
7. $\mathbf{v} = 5\mathbf{i} + \mathbf{j}, \quad \mathbf{u} = 2\mathbf{i} + \sqrt{17}\mathbf{j}$
8. $\mathbf{v} = \left\langle \dfrac{1}{\sqrt{2}}, \dfrac{1}{\sqrt{3}} \right\rangle, \quad \mathbf{u} = \left\langle \dfrac{1}{\sqrt{2}}, -\dfrac{1}{\sqrt{3}} \right\rangle$

T Find the angles between the vectors in Exercises 9–12 to the nearest hundredth of a radian.

9. $\mathbf{u} = 2\mathbf{i} + \mathbf{j}, \quad \mathbf{v} = \mathbf{i} + 2\mathbf{j} - \mathbf{k}$
10. $\mathbf{u} = 2\mathbf{i} - 2\mathbf{j} + \mathbf{k}, \quad \mathbf{v} = 3\mathbf{i} + 4\mathbf{k}$
11. $\mathbf{u} = \sqrt{3}\mathbf{i} - 7\mathbf{j}, \quad \mathbf{v} = \sqrt{3}\mathbf{i} + \mathbf{j} - 2\mathbf{k}$
12. $\mathbf{u} = \mathbf{i} + \sqrt{2}\mathbf{j} - \sqrt{2}\mathbf{k}, \quad \mathbf{v} = -\mathbf{i} + \mathbf{j} + \mathbf{k}$

13. **Triangle** Find the measures of the angles of the triangle whose vertices are $A = (-1, 0)$, $B = (2, 1)$, and $C = (1, -2)$.

14. **Rectangle** Find the measures of the angles between the diagonals of the rectangle whose vertices are $A = (1, 0)$, $B = (0, 3)$, $C = (3, 4)$, and $D = (4, 1)$.

15. **Direction angles and direction cosines** The *direction angles* α, β, and γ of a vector $\mathbf{v} = a\mathbf{i} + b\mathbf{j} + c\mathbf{k}$ are defined as follows:

    α is the angle between $\mathbf{v}$ and the positive x-axis ($0 \le \alpha \le \pi$)

    β is the angle between $\mathbf{v}$ and the positive y-axis ($0 \le \beta \le \pi$)

    γ is the angle between $\mathbf{v}$ and the positive z-axis ($0 \le \gamma \le \pi$).

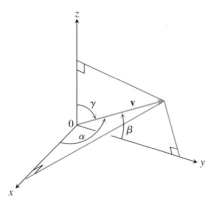

    a. Show that

$$\cos\alpha = \frac{a}{|\mathbf{v}|}, \qquad \cos\beta = \frac{b}{|\mathbf{v}|}, \qquad \cos\gamma = \frac{c}{|\mathbf{v}|},$$

    and $\cos^2\alpha + \cos^2\beta + \cos^2\gamma = 1$. These cosines are called the *direction cosines* of $\mathbf{v}$.

    b. Unit vectors are built from direction cosines Show that if $\mathbf{v} = a\mathbf{i} + b\mathbf{j} + c\mathbf{k}$ is a unit vector, then a, b, and c are the direction cosines of $\mathbf{v}$.

16. **Water main construction** A water main is to be constructed with a 20% grade in the north direction and a 10% grade in the east direction. Determine the angle θ required in the water main for the turn from north to east.

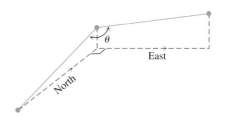

17. **Sums and differences** In the accompanying figure, it looks as if $\mathbf{v}_1 + \mathbf{v}_2$ and $\mathbf{v}_1 - \mathbf{v}_2$ are orthogonal. Is this mere coincidence, or are there circumstances under which we may expect the sum of two vectors to be orthogonal to their difference? Give reasons for your answer.

18. **Orthogonality on a circle** Suppose that AB is the diameter of a circle with center O and that C is a point on one of the two arcs joining A and B. Show that $\vec{CA}$ and $\vec{CB}$ are orthogonal.

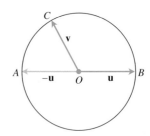

19. Diagonals of a rhombus Show that the diagonals of a rhombus (parallelogram with sides of equal length) are perpendicular.

20. Perpendicular diagonals Show that squares are the only rectangles with perpendicular diagonals.

21. When parallelograms are rectangles Prove that a parallelogram is a rectangle if and only if its diagonals are equal in length. (This fact is often exploited by carpenters.)

22. Diagonal of parallelogram Show that the indicated diagonal of the parallelogram determined by vectors **u** and **v** bisects the angle between **u** and **v** if $|\mathbf{u}| = |\mathbf{v}|$.

23. Projectile motion A gun with muzzle velocity of 1200 ft/sec is fired at an angle of 8° above the horizontal. Find the horizontal and vertical components of the velocity.

24. Inclined plane Suppose that a box is being towed up an inclined plane as shown in the figure. Find the force **w** needed to make the component of the force parallel to the inclined plane equal to 2.5 lb.

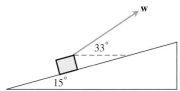

25. a. Cauchy-Schwartz inequality Use the fact that $\mathbf{u} \cdot \mathbf{v} = |\mathbf{u}||\mathbf{v}| \cos \theta$ to show that the inequality $|\mathbf{u} \cdot \mathbf{v}| \leq |\mathbf{u}||\mathbf{v}|$ holds for any vectors **u** and **v**.

b. Under what circumstances, if any, does $|\mathbf{u} \cdot \mathbf{v}|$ equal $|\mathbf{u}||\mathbf{v}|$? Give reasons for your answer.

26. Copy the axes and vector shown here. Then shade in the points (x, y) for which $(x\mathbf{i} + y\mathbf{j}) \cdot \mathbf{v} \leq 0$. Justify your answer.

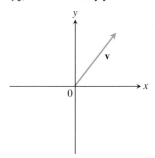

27. Orthogonal unit vectors If $\mathbf{u}_1$ and $\mathbf{u}_2$ are orthogonal unit vectors and $\mathbf{v} = a\mathbf{u}_1 + b\mathbf{u}_2$, find $\mathbf{v} \cdot \mathbf{u}_1$.

28. Cancellation in dot products In real-number multiplication, if $uv_1 = uv_2$ and $u \neq 0$, we can cancel the u and conclude that $v_1 = v_2$. Does the same rule hold for the dot product: If $\mathbf{u} \cdot \mathbf{v}_1 = \mathbf{u} \cdot \mathbf{v}_2$ and $\mathbf{u} \neq \mathbf{0}$, can you conclude that $\mathbf{v}_1 = \mathbf{v}_2$? Give reasons for your answer.

29. Line perpendicular to a vector Show that the vector $\mathbf{v} = a\mathbf{i} + b\mathbf{j}$ is perpendicular to the line $ax + by = c$ by establishing that the slope of **v** is the negative reciprocal of the slope of the given line.

30. Line parallel to a vector Show that the vector $\mathbf{v} = a\mathbf{i} + b\mathbf{j}$ is parallel to the line $bx - ay = c$ by establishing that the slope of the line segment representing **v** is the same as the slope of the given line.

In Exercises 31–34, use the result of Exercise 29 to find an equation for the line through P perpendicular to **v**. Then sketch the line. Include **v** in your sketch *as a vector starting at the origin*.

31. $P(2, 1)$, $\mathbf{v} = \mathbf{i} + 2\mathbf{j}$ **32.** $P(-1, 2)$, $\mathbf{v} = -2\mathbf{i} - \mathbf{j}$

33. $P(-2, -7)$, $\mathbf{v} = -2\mathbf{i} + \mathbf{j}$ **34.** $P(11, 10)$, $\mathbf{v} = 2\mathbf{i} - 3\mathbf{j}$

In Exercises 35–38, use the result of Exercise 30 to find an equation for the line through P parallel to **v**. Then sketch the line. Include **v** in your sketch *as a vector starting at the origin*.

35. $P(-2, 1)$, $\mathbf{v} = \mathbf{i} - \mathbf{j}$ **36.** $P(0, -2)$, $\mathbf{v} = 2\mathbf{i} + 3\mathbf{j}$

37. $P(1, 2)$, $\mathbf{v} = -\mathbf{i} - 2\mathbf{j}$ **38.** $P(1, 3)$, $\mathbf{v} = 3\mathbf{i} - 2\mathbf{j}$

39. Work along a line Find the work done by a force $\mathbf{F} = 5\mathbf{i}$ (magnitude 5 N) in moving an object along the line from the origin to the point $(1, 1)$ (distance in meters).

40. Locomotive The Union Pacific's *Big Boy* locomotive could pull 6000-ton trains with a tractive effort (pull) of 602,148 N (135,375 lb). At this level of effort, about how much work did *Big Boy* do on the (approximately straight) 605-km journey from San Francisco to Los Angeles?

41. Inclined plane How much work does it take to slide a crate 20 m along a loading dock by pulling on it with a 200 N force at an angle of 30° from the horizontal?

42. Sailboat The wind passing over a boat's sail exerted a 1000-lb magnitude force **F** as shown here. How much work did the wind perform in moving the boat forward 1 mi? Answer in foot-pounds.

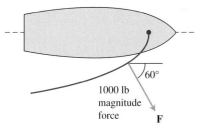

The **acute angle between intersecting lines** that do not cross at right angles is the same as the angle determined by vectors normal to the lines or by the vectors parallel to the lines.

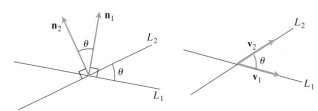

Use this fact and the results of Exercise 29 or 30 to find the acute angles between the lines in Exercises 43–48.

43. $3x + y = 5$, $\quad 2x - y = 4$

44. $y = \sqrt{3}x - 1$, $\quad y = -\sqrt{3}x + 2$

45. $\sqrt{3}x - y = -2$, $\quad x - \sqrt{3}y = 1$

46. $x + \sqrt{3}y = 1$, $\quad \left(1 - \sqrt{3}\right)x + \left(1 + \sqrt{3}\right)y = 8$

47. $3x - 4y = 3$, $\quad x - y = 7$

48. $12x + 5y = 1$, $\quad 2x - 2y = 3$

10.4 The Cross Product

In studying lines in the plane, when we needed to describe how a line was tilting, we used the notions of slope and angle of inclination. In space, we want a way to describe how a *plane* is tilting. We accomplish this by multiplying two vectors in the plane together to get a third vector perpendicular to the plane. The direction of this third vector tells us the "inclination" of the plane. The product we use to multiply the vectors together is the *vector* or *cross product*, the second of the two vector multiplication methods. We study the cross product in this section.

The Cross Product of Two Vectors in Space

We start with two nonzero vectors **u** and **v** in space. If **u** and **v** are not parallel, they determine a plane. We select a unit vector **n** perpendicular to the plane by the **right-hand rule**. This means that we choose **n** to be the unit (normal) vector that points the way your right thumb points when your fingers curl through the angle θ from **u** to **v** (Figure 10.26). Then the **cross product u $\times$ v** ("u cross v") is the *vector* defined as follows.

FIGURE 10.26 The construction of **u $\times$ v**.

DEFINITION

$$\mathbf{u} \times \mathbf{v} = \left(\,|\mathbf{u}|\,|\mathbf{v}|\, \sin \theta\right) \mathbf{n}$$

Unlike the dot product, the cross product is a vector. For this reason it's also called the **vector product** of **u** and **v**, and applies *only* to vectors in space. The vector **u $\times$ v** is orthogonal to both **u** and **v** because it is a scalar multiple of **n**.

There is a straightforward way to calculate the cross product of two vectors from their components. The method does not require that we know the angle between them (as suggested by the definition), but we postpone that calculation momentarily so we can focus first on the properties of the cross product.

Since the sines of 0 and π are both zero, it makes sense to define the cross product of two parallel nonzero vectors to be **0**. If one or both of **u** and **v** are zero, we also define **u $\times$ v** to be zero. This way, the cross product of two vectors **u** and **v** is zero if and only if **u** and **v** are parallel or one or both of them are zero.

Parallel Vectors

Nonzero vectors **u** and **v** are parallel if and only if **u $\times$ v** = **0**.

The cross product obeys the following laws.

FIGURE 10.27 The construction of $\mathbf{v} \times \mathbf{u}$.

> **Properties of the Cross Product**
> If $\mathbf{u}$, $\mathbf{v}$, and $\mathbf{w}$ are any vectors and r, s are scalars, then
> 1. $(r\mathbf{u}) \times (s\mathbf{v}) = (rs)(\mathbf{u} \times \mathbf{v})$
> 2. $\mathbf{u} \times (\mathbf{v} + \mathbf{w}) = \mathbf{u} \times \mathbf{v} + \mathbf{u} \times \mathbf{w}$
> 3. $(\mathbf{v} + \mathbf{w}) \times \mathbf{u} = \mathbf{v} \times \mathbf{u} + \mathbf{w} \times \mathbf{u}$
> 4. $\mathbf{v} \times \mathbf{u} = -(\mathbf{u} \times \mathbf{v})$
> 5. $\mathbf{0} \times \mathbf{u} = \mathbf{0}$

To visualize Property 4, for example, notice that when the fingers of a right hand curl through the angle θ from $\mathbf{v}$ to $\mathbf{u}$, the thumb points the opposite way and the unit vector we choose in forming $\mathbf{v} \times \mathbf{u}$ is the negative of the one we choose in forming $\mathbf{u} \times \mathbf{v}$ (Figure 10.27).

Property 1 can be verified by applying the definition of cross product to both sides of the equation and comparing the results. Property 2 is proved in Appendix 8. Property 3 follows by multiplying both sides of the equation in Property 2 by -1 and reversing the order of the products using Property 4. Property 5 is a definition. As a rule, cross product multiplication is *not associative* so $(\mathbf{u} \times \mathbf{v}) \times \mathbf{w}$ does not generally equal $\mathbf{u} \times (\mathbf{v} \times \mathbf{w})$. (See Additional Exercise 15.)

When we apply the definition to calculate the pairwise cross products of $\mathbf{i}$, $\mathbf{j}$, and $\mathbf{k}$, we find (Figure 10.28)

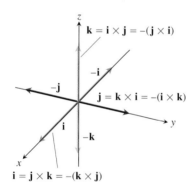

FIGURE 10.28 The pairwise cross products of $\mathbf{i}$, $\mathbf{j}$, and $\mathbf{k}$.

and

$$\mathbf{i} \times \mathbf{j} = -(\mathbf{j} \times \mathbf{i}) = \mathbf{k}$$

$$\mathbf{j} \times \mathbf{k} = -(\mathbf{k} \times \mathbf{j}) = \mathbf{i}$$

$$\mathbf{k} \times \mathbf{i} = -(\mathbf{i} \times \mathbf{k}) = \mathbf{j}$$

Diagram for recalling these products

$$\mathbf{i} \times \mathbf{i} = \mathbf{j} \times \mathbf{j} = \mathbf{k} \times \mathbf{k} = \mathbf{0}.$$

$|\mathbf{u} \times \mathbf{v}|$ Is the Area of a Parallelogram

Because $\mathbf{n}$ is a unit vector, the magnitude of $\mathbf{u} \times \mathbf{v}$ is

$$|\mathbf{u} \times \mathbf{v}| = |\mathbf{u}||\mathbf{v}|\,|\sin\theta||\mathbf{n}| = |\mathbf{u}||\mathbf{v}|\sin\theta.$$

This is the area of the parallelogram determined by $\mathbf{u}$ and $\mathbf{v}$ (Figure 10.29), $|\mathbf{u}|$ being the base of the parallelogram and $|\mathbf{v}||\sin\theta|$ the height.

Determinant Formula for $\mathbf{u} \times \mathbf{v}$

Our next objective is to calculate $\mathbf{u} \times \mathbf{v}$ from the components of $\mathbf{u}$ and $\mathbf{v}$ relative to a Cartesian coordinate system.

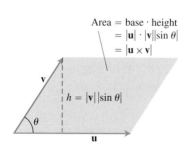

FIGURE 10.29 The parallelogram determined by $\mathbf{u}$ and $\mathbf{v}$.

Determinants

2×2 and 3×3 determinants are evaluated as follows:

$$\begin{vmatrix} a & b \\ c & d \end{vmatrix} = ad - bc$$

EXAMPLE

$$\begin{vmatrix} 2 & 1 \\ -4 & 3 \end{vmatrix} = (2)(3) - (1)(-4)$$
$$= 6 + 4 = 10$$

$$\begin{vmatrix} a_1 & a_2 & a_3 \\ b_1 & b_2 & b_3 \\ c_1 & c_2 & c_3 \end{vmatrix} = a_1 \begin{vmatrix} b_2 & b_3 \\ c_2 & c_3 \end{vmatrix}$$

$$- a_2 \begin{vmatrix} b_1 & b_3 \\ c_1 & c_3 \end{vmatrix} + a_3 \begin{vmatrix} b_1 & b_2 \\ c_1 & c_2 \end{vmatrix}$$

EXAMPLE

$$\begin{vmatrix} -5 & 3 & 1 \\ 2 & 1 & 1 \\ -4 & 3 & 1 \end{vmatrix} = (-5) \begin{vmatrix} 1 & 1 \\ 3 & 1 \end{vmatrix}$$

$$- (3) \begin{vmatrix} 2 & 1 \\ -4 & 1 \end{vmatrix} + (1) \begin{vmatrix} 2 & 1 \\ -4 & 3 \end{vmatrix}$$

$$= -5(1 - 3) - 3(2 + 4)$$
$$+ 1(6 + 4)$$
$$= 10 - 18 + 10 = 2$$

(For more information, see the Web site at **www.aw-bc.com/thomas**.)

Suppose that

$$\mathbf{u} = u_1 \mathbf{i} + u_2 \mathbf{j} + u_3 \mathbf{k}, \qquad \mathbf{v} = v_1 \mathbf{i} + v_2 \mathbf{j} + v_3 \mathbf{k}.$$

Then the distributive laws and the rules for multiplying $\mathbf{i}$, $\mathbf{j}$, and $\mathbf{k}$ tell us that

$$\mathbf{u} \times \mathbf{v} = (u_1 \mathbf{i} + u_2 \mathbf{j} + u_3 \mathbf{k}) \times (v_1 \mathbf{i} + v_2 \mathbf{j} + v_3 \mathbf{k})$$

$$= u_1 v_1 \mathbf{i} \times \mathbf{i} + u_1 v_2 \mathbf{i} \times \mathbf{j} + u_1 v_3 \mathbf{i} \times \mathbf{k}$$

$$+ u_2 v_1 \mathbf{j} \times \mathbf{i} + u_2 v_2 \mathbf{j} \times \mathbf{j} + u_2 v_3 \mathbf{j} \times \mathbf{k}$$

$$+ u_3 v_1 \mathbf{k} \times \mathbf{i} + u_3 v_2 \mathbf{k} \times \mathbf{j} + u_3 v_3 \mathbf{k} \times \mathbf{k}$$

$$= (u_2 v_3 - u_3 v_2) \mathbf{i} - (u_1 v_3 - u_3 v_1) \mathbf{j} + (u_1 v_2 - u_2 v_1) \mathbf{k}.$$

The terms in the last line are the same as the terms in the expansion of the symbolic determinant

$$\begin{vmatrix} \mathbf{i} & \mathbf{j} & \mathbf{k} \\ u_1 & u_2 & u_3 \\ v_1 & v_2 & v_3 \end{vmatrix}.$$

We therefore have the following rule.

Calculating Cross Products Using Determinants

If $\mathbf{u} = u_1 \mathbf{i} + u_2 \mathbf{j} + u_3 \mathbf{k}$ and $\mathbf{v} = v_1 \mathbf{i} + v_2 \mathbf{j} + v_3 \mathbf{k}$, then

$$\mathbf{u} \times \mathbf{v} = \begin{vmatrix} \mathbf{i} & \mathbf{j} & \mathbf{k} \\ u_1 & u_2 & u_3 \\ v_1 & v_2 & v_3 \end{vmatrix}.$$

EXAMPLE 1 Find $\mathbf{u} \times \mathbf{v}$ and $\mathbf{v} \times \mathbf{u}$ if $\mathbf{u} = 2\mathbf{i} + \mathbf{j} + \mathbf{k}$ and $\mathbf{v} = -4\mathbf{i} + 3\mathbf{j} + \mathbf{k}$.

Solution

$$\mathbf{u} \times \mathbf{v} = \begin{vmatrix} \mathbf{i} & \mathbf{j} & \mathbf{k} \\ 2 & 1 & 1 \\ -4 & 3 & 1 \end{vmatrix} = \begin{vmatrix} 1 & 1 \\ 3 & 1 \end{vmatrix} \mathbf{i} - \begin{vmatrix} 2 & 1 \\ -4 & 1 \end{vmatrix} \mathbf{j} + \begin{vmatrix} 2 & 1 \\ -4 & 3 \end{vmatrix} \mathbf{k}$$

$$= -2\mathbf{i} - 6\mathbf{j} + 10\mathbf{k}$$

$$\mathbf{v} \times \mathbf{u} = -(\mathbf{u} \times \mathbf{v}) = 2\mathbf{i} + 6\mathbf{j} - 10\mathbf{k} \qquad \blacksquare$$

EXAMPLE 2 Find a vector perpendicular to the plane of $P(1, -1, 0)$, $Q(2, 1, -1)$, and $R(-1, 1, 2)$ (Figure 10.30).

Solution The vector $\vec{PQ} \times \vec{PR}$ is perpendicular to the plane because it is perpendicular to both vectors. In terms of components,

$$\vec{PQ} = (2 - 1)\mathbf{i} + (1 + 1)\mathbf{j} + (-1 - 0)\mathbf{k} = \mathbf{i} + 2\mathbf{j} - \mathbf{k}$$

$$\vec{PR} = (-1 - 1)\mathbf{i} + (1 + 1)\mathbf{j} + (2 - 0)\mathbf{k} = -2\mathbf{i} + 2\mathbf{j} + 2\mathbf{k}$$

FIGURE 10.30 The area of triangle PQR is half of $|\vec{PQ} \times \vec{PR}|$ (Example 2).

$$\vec{PQ} \times \vec{PR} = \begin{vmatrix} \mathbf{i} & \mathbf{j} & \mathbf{k} \\ 1 & 2 & -1 \\ -2 & 2 & 2 \end{vmatrix} = \begin{vmatrix} 2 & -1 \\ 2 & 2 \end{vmatrix}\mathbf{i} - \begin{vmatrix} 1 & -1 \\ -2 & 2 \end{vmatrix}\mathbf{j} + \begin{vmatrix} 1 & 2 \\ -2 & 2 \end{vmatrix}\mathbf{k}$$

$$= 6\mathbf{i} + 6\mathbf{k}.$$

■

EXAMPLE 3   Find the area of the triangle with vertices $P(1, -1, 0)$, $Q(2, 1, -1)$, and $R(-1, 1, 2)$ (Figure 10.30).

Solution   The area of the parallelogram determined by P, Q, and R is

$$|\vec{PQ} \times \vec{PR}| = |6\mathbf{i} + 6\mathbf{k}| \qquad\qquad \text{Values from Example 2.}$$

$$= \sqrt{(6)^2 + (6)^2} = \sqrt{2 \cdot 36} = 6\sqrt{2}.$$

The triangle's area is half of this, or $3\sqrt{2}$.

■

EXAMPLE 4   Find a unit vector perpendicular to the plane of $P(1, -1, 0)$, $Q(2, 1, -1)$, and $R(-1, 1, 2)$.

Solution   Since $\vec{PQ} \times \vec{PR}$ is perpendicular to the plane, its direction $\mathbf{n}$ is a unit vector perpendicular to the plane. Taking values from Examples 2 and 3, we have

$$\mathbf{n} = \frac{\vec{PQ} \times \vec{PR}}{|\vec{PQ} \times \vec{PR}|} = \frac{6\mathbf{i} + 6\mathbf{k}}{6\sqrt{2}} = \frac{1}{\sqrt{2}}\mathbf{i} + \frac{1}{\sqrt{2}}\mathbf{k}.$$

■

For ease in calculating the cross product using determinants, we usually write vectors in the form $\mathbf{v} = v_1\mathbf{i} + v_2\mathbf{j} + v_3\mathbf{k}$ rather than as ordered triples $\mathbf{v} = \langle v_1, v_2, v_3 \rangle$.

Torque

When we turn a bolt by applying a force $\mathbf{F}$ to a wrench (Figure 10.31), the torque we produce acts along the axis of the bolt to drive the bolt forward. The magnitude of the torque depends on how far out on the wrench the force is applied and on how much of the force is perpendicular to the wrench at the point of application. The number we use to measure the torque's magnitude is the product of the length of the lever arm $\mathbf{r}$ and the scalar component of $\mathbf{F}$ perpendicular to $\mathbf{r}$. In the notation of Figure 10.31,

$$\text{Magnitude of torque vector} = |\mathbf{r}||\mathbf{F}|\sin\theta,$$

or $|\mathbf{r} \times \mathbf{F}|$. If we let $\mathbf{n}$ be a unit vector along the axis of the bolt in the direction of the torque, then a complete description of the torque vector is $\mathbf{r} \times \mathbf{F}$, or

$$\text{Torque vector} = (|\mathbf{r}||\mathbf{F}|\sin\theta)\mathbf{n}.$$

Recall that we defined $\mathbf{u} \times \mathbf{v}$ to be $\mathbf{0}$ when $\mathbf{u}$ and $\mathbf{v}$ are parallel. This is consistent with the torque interpretation as well. If the force $\mathbf{F}$ in Figure 10.31 is parallel to the wrench, meaning that we are trying to turn the bolt by pushing or pulling along the line of the wrench's handle, the torque produced is zero.

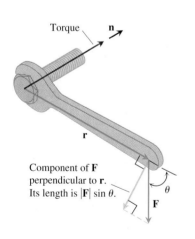

Torque $\mathbf{n}$

$\mathbf{r}$

Component of $\mathbf{F}$
perpendicular to $\mathbf{r}$.
Its length is $|\mathbf{F}|\sin\theta$.

θ

$\mathbf{F}$

FIGURE 10.31   The torque vector describes the tendency of the force $\mathbf{F}$ to drive the bolt forward.

FIGURE 10.32 The magnitude of the torque exerted by **F** at P is about 56.4 ft-lb (Example 5).

EXAMPLE 5   The magnitude of the torque generated by force **F** at the pivot point P in Figure 10.32 is

$$|\vec{PQ} \times \mathbf{F}| = |\vec{PQ}||\mathbf{F}|\sin 70°$$

$$\approx (3)(20)(0.94)$$

$$\approx 56.4 \text{ ft-lb.} \qquad \blacksquare$$

Triple Scalar or Box Product

The product $(\mathbf{u} \times \mathbf{v}) \cdot \mathbf{w}$ is called the **triple scalar product** of **u**, **v**, and **w** (in that order). As you can see from the formula

$$|(\mathbf{u} \times \mathbf{v}) \cdot \mathbf{w}| = |\mathbf{u} \times \mathbf{v}||\mathbf{w}||\cos\theta|,$$

the absolute value of the product is the volume of the parallelepiped (parallelogram-sided box) determined by **u**, **v**, and **w** (Figure 10.33). The number $|\mathbf{u} \times \mathbf{v}|$ is the area of the base parallelogram. The number $|\mathbf{w}||\cos\theta|$ is the parallelepiped's height. Because of this geometry, $(\mathbf{u} \times \mathbf{v}) \cdot \mathbf{w}$ is also called the **box product** of **u**, **v**, and **w**.

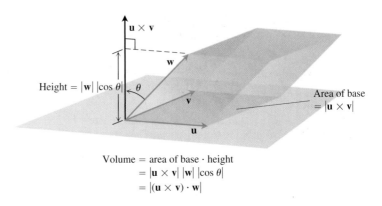

Volume = area of base · height
= $|\mathbf{u} \times \mathbf{v}| |\mathbf{w}| |\cos\theta|$
= $|(\mathbf{u} \times \mathbf{v}) \cdot \mathbf{w}|$

FIGURE 10.33 The number $|(\mathbf{u} \times \mathbf{v}) \cdot \mathbf{w}|$ is the volume of a parallelepiped.

By treating the planes of **v** and **w** and of **w** and **u** as the base planes of the parallelepiped determined by **u**, **v**, and **w**, we see that

$$(\mathbf{u} \times \mathbf{v}) \cdot \mathbf{w} = (\mathbf{v} \times \mathbf{w}) \cdot \mathbf{u} = (\mathbf{w} \times \mathbf{u}) \cdot \mathbf{v}.$$

Since the dot product is commutative, we also have

$$(\mathbf{u} \times \mathbf{v}) \cdot \mathbf{w} = \mathbf{u} \cdot (\mathbf{v} \times \mathbf{w}).$$

The dot and cross may be interchanged in a triple scalar product without altering its value.

The triple scalar product can be evaluated as a determinant:

$$(\mathbf{u} \times \mathbf{v}) \cdot \mathbf{w} = \left[\begin{vmatrix} u_2 & u_3 \\ v_2 & v_3 \end{vmatrix}\mathbf{i} - \begin{vmatrix} u_1 & u_3 \\ v_1 & v_3 \end{vmatrix}\mathbf{j} + \begin{vmatrix} u_1 & u_2 \\ v_1 & v_2 \end{vmatrix}\mathbf{k}\right] \cdot \mathbf{w}$$

$$= w_1\begin{vmatrix} u_2 & u_3 \\ v_2 & v_3 \end{vmatrix} - w_2\begin{vmatrix} u_1 & u_3 \\ v_1 & v_3 \end{vmatrix} + w_3\begin{vmatrix} u_1 & u_2 \\ v_1 & v_2 \end{vmatrix}$$

$$= \begin{vmatrix} u_1 & u_2 & u_3 \\ v_1 & v_2 & v_3 \\ w_1 & w_2 & w_3 \end{vmatrix}.$$

Calculating the Triple Scalar Product

$$(\mathbf{u} \times \mathbf{v}) \cdot \mathbf{w} = \begin{vmatrix} u_1 & u_2 & u_3 \\ v_1 & v_2 & v_3 \\ w_1 & w_2 & w_3 \end{vmatrix}$$

EXAMPLE 6 Find the volume of the box (parallelepiped) determined by $\mathbf{u} = \mathbf{i} + 2\mathbf{j} - \mathbf{k}$, $\mathbf{v} = -2\mathbf{i} + 3\mathbf{k}$, and $\mathbf{w} = 7\mathbf{j} - 4\mathbf{k}$.

Solution Using the rule for calculating determinants, we find

$$(\mathbf{u} \times \mathbf{v}) \cdot \mathbf{w} = \begin{vmatrix} 1 & 2 & -1 \\ -2 & 0 & 3 \\ 0 & 7 & -4 \end{vmatrix} = -23.$$

The volume is $|(\mathbf{u} \times \mathbf{v}) \cdot \mathbf{w}| = 23$ units cubed. ∎

EXERCISES 10.4

In Exercises 1–8, find the length and direction (when defined) of $\mathbf{u} \times \mathbf{v}$ and $\mathbf{v} \times \mathbf{u}$.

1. $\mathbf{u} = 2\mathbf{i} - 2\mathbf{j} - \mathbf{k}, \quad \mathbf{v} = \mathbf{i} - \mathbf{k}$

2. $\mathbf{u} = 2\mathbf{i} + 3\mathbf{j}, \quad \mathbf{v} = -\mathbf{i} + \mathbf{j}$

3. $\mathbf{u} = 2\mathbf{i} - 2\mathbf{j} + 4\mathbf{k}, \quad \mathbf{v} = -\mathbf{i} + \mathbf{j} - 2\mathbf{k}$

4. $\mathbf{u} = \mathbf{i} + \mathbf{j} - \mathbf{k}, \quad \mathbf{v} = 0$

5. $\mathbf{u} = 2\mathbf{i}, \quad \mathbf{v} = -3\mathbf{j}$

6. $\mathbf{u} = \mathbf{i} \times \mathbf{j}, \quad \mathbf{v} = \mathbf{j} \times \mathbf{k}$

7. $\mathbf{u} = -8\mathbf{i} - 2\mathbf{j} - 4\mathbf{k}, \quad \mathbf{v} = 2\mathbf{i} + 2\mathbf{j} + \mathbf{k}$

8. $\mathbf{u} = \frac{3}{2}\mathbf{i} - \frac{1}{2}\mathbf{j} + \mathbf{k}, \quad \mathbf{v} = \mathbf{i} + \mathbf{j} + 2\mathbf{k}$

In Exercises 9–14, sketch the coordinate axes and then include the vectors $\mathbf{u}$, $\mathbf{v}$ and $\mathbf{u} \times \mathbf{v}$ as vectors starting at the origin.

9. $\mathbf{u} = \mathbf{i}, \quad \mathbf{v} = \mathbf{j}$ 10. $\mathbf{u} = \mathbf{i} - \mathbf{k}, \quad \mathbf{v} = \mathbf{j}$

11. $\mathbf{u} = \mathbf{i} - \mathbf{k}, \quad \mathbf{v} = \mathbf{j} + \mathbf{k}$ 12. $\mathbf{u} = 2\mathbf{i} - \mathbf{j}, \quad \mathbf{v} = \mathbf{i} + 2\mathbf{j}$

13. $\mathbf{u} = \mathbf{i} + \mathbf{j}, \quad \mathbf{v} = \mathbf{i} - \mathbf{j}$ 14. $\mathbf{u} = \mathbf{j} + 2\mathbf{k}, \quad \mathbf{v} = \mathbf{i}$

In Exercises 15–18,

  a. Find the area of the triangle determined by the points P, Q, and R.

  b. Find a unit vector perpendicular to plane PQR.

15. $P(1, -1, 2), \quad Q(2, 0, -1), \quad R(0, 2, 1)$

16. $P(1, 1, 1), \quad Q(2, 1, 3), \quad R(3, -1, 1)$

17. $P(2, -2, 1), \quad Q(3, -1, 2), \quad R(3, -1, 1)$

18. $P(-2, 2, 0), \quad Q(0, 1, -1), \quad R(-1, 2, -2)$

In Exercises 19–22, verify that $(\mathbf{u} \times \mathbf{v}) \cdot \mathbf{w} = (\mathbf{v} \times \mathbf{w}) \cdot \mathbf{u} = (\mathbf{w} \times \mathbf{u}) \cdot \mathbf{v}$ and find the volume of the parallelepiped (box) determined by $\mathbf{u}$, $\mathbf{v}$, and $\mathbf{w}$.

	u	**v**	**w**
19.	$2\mathbf{i}$	$2\mathbf{j}$	$2\mathbf{k}$
20.	$\mathbf{i} - \mathbf{j} + \mathbf{k}$	$2\mathbf{i} + \mathbf{j} - 2\mathbf{k}$	$-\mathbf{i} + 2\mathbf{j} - \mathbf{k}$
21.	$2\mathbf{i} + \mathbf{j}$	$2\mathbf{i} - \mathbf{j} + \mathbf{k}$	$\mathbf{i} + 2\mathbf{k}$
22.	$\mathbf{i} + \mathbf{j} - 2\mathbf{k}$	$-\mathbf{i} - \mathbf{k}$	$2\mathbf{i} + 4\mathbf{j} - 2\mathbf{k}$

23. **Parallel and perpendicular vectors** Let $\mathbf{u} = 5\mathbf{i} - \mathbf{j} + \mathbf{k}$, $\mathbf{v} = \mathbf{j} - 5\mathbf{k}$, $\mathbf{w} = -15\mathbf{i} + 3\mathbf{j} - 3\mathbf{k}$. Which vectors, if any, are **(a)** perpendicular? **(b)** Parallel? Give reasons for your answers.

24. **Parallel and perpendicular vectors** Let $\mathbf{u} = \mathbf{i} + 2\mathbf{j} - \mathbf{k}$, $\mathbf{v} = -\mathbf{i} + \mathbf{j} + \mathbf{k}$, $\mathbf{w} = \mathbf{i} + \mathbf{k}$, $\mathbf{r} = -(\pi/2)\mathbf{i} - \pi\mathbf{j} + (\pi/2)\mathbf{k}$. Which vectors, if any, are **(a)** perpendicular? **(b)** Parallel? Give reasons for your answers.

In Exercises 25 and 26, find the magnitude of the torque exerted by **F** on the bolt at P if $|\vec{PQ}| = 8$ in. and $|\mathbf{F}| = 30$ lb. Answer in foot-pounds.

25.   **26.**

27. Which of the following are *always true*, and which are *not always true*? Give reasons for your answers.

   a. $|\mathbf{u}| = \sqrt{\mathbf{u} \cdot \mathbf{u}}$

   b. $\mathbf{u} \cdot \mathbf{u} = |\mathbf{u}|$

   c. $\mathbf{u} \times \mathbf{0} = \mathbf{0} \times \mathbf{u} = \mathbf{0}$

   d. $\mathbf{u} \times (-\mathbf{u}) = \mathbf{0}$

   e. $\mathbf{u} \times \mathbf{v} = \mathbf{v} \times \mathbf{u}$

   f. $\mathbf{u} \times (\mathbf{v} + \mathbf{w}) = \mathbf{u} \times \mathbf{v} + \mathbf{u} \times \mathbf{w}$

   g. $(\mathbf{u} \times \mathbf{v}) \cdot \mathbf{v} = 0$

   h. $(\mathbf{u} \times \mathbf{v}) \cdot \mathbf{w} = \mathbf{u} \cdot (\mathbf{v} \times \mathbf{w})$

28. Which of the following are *always true*, and which are *not always true*? Give reasons for your answers.

   a. $\mathbf{u} \cdot \mathbf{v} = \mathbf{v} \cdot \mathbf{u}$ **b.** $\mathbf{u} \times \mathbf{v} = -(\mathbf{v} \times \mathbf{u})$

   c. $(-\mathbf{u}) \times \mathbf{v} = -(\mathbf{u} \times \mathbf{v})$

   d. $(c\mathbf{u}) \cdot \mathbf{v} = \mathbf{u} \cdot (c\mathbf{v}) = c(\mathbf{u} \cdot \mathbf{v})$ (any number c)

   e. $c(\mathbf{u} \times \mathbf{v}) = (c\mathbf{u}) \times \mathbf{v} = \mathbf{u} \times (c\mathbf{v})$ (any number c)

   f. $\mathbf{u} \cdot \mathbf{u} = |\mathbf{u}|^2$ **g.** $(\mathbf{u} \times \mathbf{u}) \cdot \mathbf{u} = 0$

   h. $(\mathbf{u} \times \mathbf{v}) \cdot \mathbf{u} = \mathbf{v} \cdot (\mathbf{u} \times \mathbf{v})$

29. Given nonzero vectors **u**, **v**, and **w**, use dot product and cross product notation, as appropriate, to describe the following.

   a. The vector projection of **u** onto **v**

   b. A vector orthogonal to **u** and **v**

   c. A vector orthogonal to $\mathbf{u} \times \mathbf{v}$ and **w**

   d. The volume of the parallelepiped determined by **u**, **v**, and **w**

30. Given nonzero vectors **u**, **v**, and **w**, use dot product and cross product notation to describe the following.

   a. A vector orthogonal to $\mathbf{u} \times \mathbf{v}$ and $\mathbf{u} \times \mathbf{w}$

   b. A vector orthogonal to $\mathbf{u} + \mathbf{v}$ and $\mathbf{u} - \mathbf{v}$

   c. A vector of length $|\mathbf{u}|$ in the direction of **v**

   d. The area of the parallelogram determined by **u** and **w**

31. Let **u**, **v**, and **w** be vectors. Which of the following make sense, and which do not? Give reasons for your answers.

   a. $(\mathbf{u} \times \mathbf{v}) \cdot \mathbf{w}$

   b. $\mathbf{u} \times (\mathbf{v} \cdot \mathbf{w})$

   c. $\mathbf{u} \times (\mathbf{v} \times \mathbf{w})$

   d. $\mathbf{u} \cdot (\mathbf{v} \cdot \mathbf{w})$

32. Cross products of three vectors Show that except in degenerate cases, $(\mathbf{u} \times \mathbf{v}) \times \mathbf{w}$ lies in the plane of **u** and **v**, whereas $\mathbf{u} \times (\mathbf{v} \times \mathbf{w})$ lies in the plane of **v** and **w**. What *are* the degenerate cases?

33. Cancellation in cross products If $\mathbf{u} \times \mathbf{v} = \mathbf{u} \times \mathbf{w}$ and $\mathbf{u} \neq \mathbf{0}$, then does $\mathbf{v} = \mathbf{w}$? Give reasons for your answer.

34. Double cancellation If $\mathbf{u} \neq \mathbf{0}$ and if $\mathbf{u} \times \mathbf{v} = \mathbf{u} \times \mathbf{w}$ and $\mathbf{u} \cdot \mathbf{v} = \mathbf{u} \cdot \mathbf{w}$, then does $\mathbf{v} = \mathbf{w}$? Give reasons for your answer.

Find the areas of the parallelograms whose vertices are given in Exercises 35–38.

35. $A(1, 0)$, $B(0, 1)$, $C(-1, 0)$, $D(0, -1)$

36. $A(0, 0)$, $B(7, 3)$, $C(9, 8)$, $D(2, 5)$

37. $A(-1, 2)$, $B(2, 0)$, $C(7, 1)$, $D(4, 3)$

38. $A(-6, 0)$, $B(1, -4)$, $C(3, 1)$, $D(-4, 5)$

Find the areas of the triangles whose vertices are given in Exercises 39–42.

39. $A(0, 0)$, $B(-2, 3)$, $C(3, 1)$

40. $A(-1, -1)$, $B(3, 3)$, $C(2, 1)$

41. $A(-5, 3)$, $B(1, -2)$, $C(6, -2)$

42. $A(-6, 0)$, $B(10, -5)$, $C(-2, 4)$

43. Triangle area Find a formula for the area of the triangle in the xy-plane with vertices at $(0, 0), (a_1, a_2)$, and (b_1, b_2). Explain your work.

44. Triangle area Find a concise formula for the area of a triangle with vertices $(a_1, a_2), (b_1, b_2)$, and (c_1, c_2).

10.5   Lines and Planes in Space

This section shows how to use scalar and vector products to write equations for lines, line segments, and planes in space. We will use these representations throughout the rest of the book.

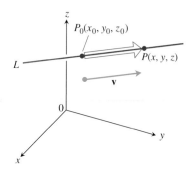

FIGURE 10.34 A point P lies on L through P_0 parallel to $\mathbf{v}$ if and only if $\overrightarrow{P_0P}$ is a scalar multiple of $\mathbf{v}$.

Lines and Line Segments in Space

In the plane, a line is determined by a point and a number giving the slope of the line. In space a line is determined by a point and a *vector* giving the direction of the line.

Suppose that L is a line in space passing through a point $P_0(x_0, y_0, z_0)$ parallel to a vector $\mathbf{v} = v_1\mathbf{i} + v_2\mathbf{j} + v_3\mathbf{k}$. Then L is the set of all points $P(x, y, z)$ for which $\overrightarrow{P_0P}$ is parallel to $\mathbf{v}$ (Figure 10.34). Thus, $\overrightarrow{P_0P} = t\mathbf{v}$ for some scalar parameter t. The value of t depends on the location of the point P along the line, and the domain of t is $(-\infty, \infty)$. The expanded form of the equation $\overrightarrow{P_0P} = t\mathbf{v}$ is

$$(x - x_0)\mathbf{i} + (y - y_0)\mathbf{j} + (z - z_0)\mathbf{k} = t(v_1\mathbf{i} + v_2\mathbf{j} + v_3\mathbf{k}),$$

which can be rewritten as

$$x\mathbf{i} + y\mathbf{j} + z\mathbf{k} = x_0\mathbf{i} + y_0\mathbf{j} + z_0\mathbf{k} + t(v_1\mathbf{i} + v_2\mathbf{j} + v_3\mathbf{k}). \tag{1}$$

If $\mathbf{r}(t)$ is the position vector of a point $P(x, y, z)$ on the line and $\mathbf{r}_0$ is the position vector of the point $P_0(x_0, y_0, z_0)$, then Equation (1) gives the following vector form for the equation of a line in space.

Vector Equation for a Line

A vector equation for the line L through $P_0(x_0, y_0, z_0)$ parallel to $\mathbf{v}$ is

$$\mathbf{r}(t) = \mathbf{r}_0 + t\mathbf{v}, \qquad -\infty < t < \infty, \tag{2}$$

where $\mathbf{r}$ is the position vector of a point $P(x, y, z)$ on L and $\mathbf{r}_0$ is the position vector of $P_0(x_0, y_0, z_0)$.

Equating the corresponding components of the two sides of Equation (1) gives three scalar equations involving the parameter t:

$$x = x_0 + tv_1, \qquad y = y_0 + tv_2, \qquad z = z_0 + tv_3.$$

These equations give us the standard parametrization of the line for the parameter interval $-\infty < t < \infty$.

Parametric Equations for a Line

The standard parametrization of the line through $P_0(x_0, y_0, z_0)$ parallel to $\mathbf{v} = v_1\mathbf{i} + v_2\mathbf{j} + v_3\mathbf{k}$ is

$$x = x_0 + tv_1, \quad y = y_0 + tv_2, \quad z = z_0 + tv_3, \quad -\infty < t < \infty \tag{3}$$

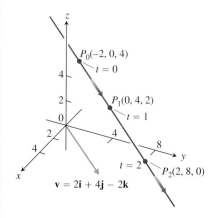

FIGURE 10.35 Selected points and parameter values on the line in Example 1. The arrows show the direction of increasing t.

EXAMPLE 1 Find parametric equations for the line through $(-2, 0, 4)$ parallel to $\mathbf{v} = 2\mathbf{i} + 4\mathbf{j} - 2\mathbf{k}$ (Figure 10.35).

Solution With $P_0(x_0, y_0, z_0)$ equal to $(-2, 0, 4)$ and $v_1\mathbf{i} + v_2\mathbf{j} + v_3\mathbf{k}$ equal to $2\mathbf{i} + 4\mathbf{j} - 2\mathbf{k}$, Equations (3) become

$$x = -2 + 2t, \qquad y = 4t, \qquad z = 4 - 2t. \qquad \blacksquare$$

EXAMPLE 2 Find parametric equations for the line through $P(-3, 2, -3)$ and $Q(1, -1, 4)$.

Solution The vector

$$\vec{PQ} = (1 - (-3))\mathbf{i} + (-1 - 2)\mathbf{j} + (4 - (-3))\mathbf{k}$$
$$= 4\mathbf{i} - 3\mathbf{j} + 7\mathbf{k}$$

is parallel to the line, and Equations (3) with $(x_0, y_0, z_0) = (-3, 2, -3)$ give

$$x = -3 + 4t, \qquad y = 2 - 3t, \qquad z = -3 + 7t.$$

We could have chosen $Q(1, -1, 4)$ as the "base point" and written

$$x = 1 + 4t, \quad y = -1 - 3t, \quad z = 4 + 7t.$$

These equations serve as well as the first; they simply place you at a different point on the line for a given value of t. ∎

Notice that parametrizations are not unique. Not only can the "base point" change, but so can the parameter. The equations $x = -3 + 4t^3$, $y = 2 - 3t^3$, and $z = -3 + 7t^3$ also parametrize the line in Example 2.

To parametrize a line segment joining two points, we first parametrize the line through the points. We then find the t-values for the endpoints and restrict t to lie in the closed interval bounded by these values. The line equations together with this added restriction parametrize the segment.

EXAMPLE 3 Parametrize the line segment joining the points $P(-3, 2, -3)$ and $Q(1, -1, 4)$ (Figure 10.36).

Solution We begin with equations for the line through P and Q, taking them, in this case, from Example 2:

$$x = -3 + 4t, \qquad y = 2 - 3t, \qquad z = -3 + 7t.$$

We observe that the point

$$(x, y, z) = (-3 + 4t, 2 - 3t, -3 + 7t)$$

on the line passes through $P(-3, 2, -3)$ at $t = 0$ and $Q(1, -1, 4)$ at $t = 1$. We add the restriction $0 \leq t \leq 1$ to parametrize the segment:

$$x = -3 + 4t, \qquad y = 2 - 3t, \qquad z = -3 + 7t, \qquad 0 \leq t \leq 1. \quad ∎$$

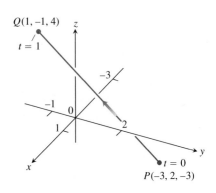

FIGURE 10.36 Example 3 derives a parametrization of line segment PQ. The arrow shows the direction of increasing t.

The vector form (Equation (2)) for a line in space is more revealing if we think of a line as the path of a particle starting at position $P_0(x_0, y_0, z_0)$ and moving in the direction of vector $\mathbf{v}$. Rewriting Equation (2), we have

$$\mathbf{r}(t) = \mathbf{r}_0 + t\mathbf{v}$$

$$= \mathbf{r}_0 + t|\mathbf{v}| \frac{\mathbf{v}}{|\mathbf{v}|}. \tag{4}$$

Initial position   Time   Speed   Direction

In other words, the position of the particle at time t is its initial position plus its distance moved (speed $\times$ time) in the direction $\mathbf{v}/|\mathbf{v}|$ of its straight-line motion.

EXAMPLE 4   A helicopter is to fly directly from a helipad at the origin in the direction of the point $(1, 1, 1)$ at a speed of 60 ft/sec. What is the position of the helicopter after 10 sec?

Solution   We place the origin at the starting position (helipad) of the helicopter. Then the unit vector

$$\mathbf{u} = \frac{1}{\sqrt{3}}\mathbf{i} + \frac{1}{\sqrt{3}}\mathbf{j} + \frac{1}{\sqrt{3}}\mathbf{k}$$

gives the flight direction of the helicopter. From Equation (4), the position of the helicopter at any time t is

$$\mathbf{r}(t) = \mathbf{r}_0 + t(\text{speed})\mathbf{u}$$

$$= \mathbf{0} + t(60)\left(\frac{1}{\sqrt{3}}\mathbf{i} + \frac{1}{\sqrt{3}}\mathbf{j} + \frac{1}{\sqrt{3}}\mathbf{k}\right)$$

$$= 20\sqrt{3}\,t(\mathbf{i} + \mathbf{j} + \mathbf{k}).$$

When $t = 10$ sec,

$$\mathbf{r}(10) = 200\sqrt{3}\,(\mathbf{i} + \mathbf{j} + \mathbf{k})$$

$$= \left\langle 200\sqrt{3}, 200\sqrt{3}, 200\sqrt{3} \right\rangle.$$

After 10 sec of flight from the origin toward $(1, 1, 1)$, the helicopter is located at the point $(200\sqrt{3}, 200\sqrt{3}, 200\sqrt{3})$ in space. It has traveled a distance of $(60\ \text{ft/sec})(10\ \text{sec}) = 600$ ft, which is the length of the vector $\mathbf{r}(10)$. ∎

The Distance from a Point to a Line in Space

To find the distance from a point S to a line that passes through a point P parallel to a vector $\mathbf{v}$, we find the absolute value of the scalar component of $\overrightarrow{PS}$ in the direction of a vector normal to the line (Figure 10.37). In the notation of the figure, the absolute value of the scalar component is $|\overrightarrow{PS}|\sin\theta$, which is $\dfrac{|\overrightarrow{PS} \times \mathbf{v}|}{|\mathbf{v}|}$.

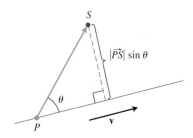

FIGURE 10.37   The distance from S to the line through P parallel to $\mathbf{v}$ is $|\overrightarrow{PS}|\sin\theta$, where θ is the angle between $\overrightarrow{PS}$ and $\mathbf{v}$.

Distance from a Point S to a Line Through P Parallel to v

$$d = \frac{|\overrightarrow{PS} \times \mathbf{v}|}{|\mathbf{v}|} \tag{5}$$

EXAMPLE 5 Find the distance from the point $S(1, 1, 5)$ to the line

$$L: \quad x = 1 + t, \quad y = 3 - t, \quad z = 2t.$$

Solution We see from the equations for L that L passes through $P(1, 3, 0)$ parallel to $\mathbf{v} = \mathbf{i} - \mathbf{j} + 2\mathbf{k}$. With

$$\vec{PS} = (1 - 1)\mathbf{i} + (1 - 3)\mathbf{j} + (5 - 0)\mathbf{k} = -2\mathbf{j} + 5\mathbf{k}$$

and

$$\vec{PS} \times \mathbf{v} = \begin{vmatrix} \mathbf{i} & \mathbf{j} & \mathbf{k} \\ 0 & -2 & 5 \\ 1 & -1 & 2 \end{vmatrix} = \mathbf{i} + 5\mathbf{j} + 2\mathbf{k},$$

Equation (5) gives

$$d = \frac{|\vec{PS} \times \mathbf{v}|}{|\mathbf{v}|} = \frac{\sqrt{1 + 25 + 4}}{\sqrt{1 + 1 + 4}} = \frac{\sqrt{30}}{\sqrt{6}} = \sqrt{5}. \qquad \blacksquare$$

An Equation for a Plane in Space

A plane in space is determined by knowing a point on the plane and its "tilt" or orientation. This "tilt" is defined by specifying a vector that is perpendicular or normal to the plane.

Suppose that plane M passes through a point $P_0(x_0, y_0, z_0)$ and is normal to the nonzero vector $\mathbf{n} = A\mathbf{i} + B\mathbf{j} + C\mathbf{k}$. Then M is the set of all points $P(x, y, z)$ for which $\vec{P_0P}$ is orthogonal to $\mathbf{n}$ (Figure 10.38). Thus, the dot product $\mathbf{n} \cdot \vec{P_0P} = 0$. This equation is equivalent to

$$(A\mathbf{i} + B\mathbf{j} + C\mathbf{k}) \cdot [(x - x_0)\mathbf{i} + (y - y_0)\mathbf{j} + (z - z_0)\mathbf{k}] = 0$$

or

$$A(x - x_0) + B(y - y_0) + C(z - z_0) = 0.$$

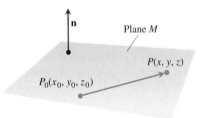

FIGURE 10.38 The standard equation for a plane in space is defined in terms of a vector normal to the plane: A point P lies in the plane through P_0 normal to $\mathbf{n}$ if and only if $\mathbf{n} \cdot \vec{P_0P} = 0$.

Equation for a Plane
The plane through $P_0(x_0, y_0, z_0)$ normal to $\mathbf{n} = A\mathbf{i} + B\mathbf{j} + C\mathbf{k}$ has

Vector equation:	$\mathbf{n} \cdot \vec{P_0P} = 0$
Component equation:	$A(x - x_0) + B(y - y_0) + C(z - z_0) = 0$
Component equation simplified:	$Ax + By + Cz = D,$ where
	$D = Ax_0 + By_0 + Cz_0$

EXAMPLE 6 Find an equation for the plane through $P_0(-3, 0, 7)$ perpendicular to $\mathbf{n} = 5\mathbf{i} + 2\mathbf{j} - \mathbf{k}$.

Solution The component equation is

$$5(x - (-3)) + 2(y - 0) + (-1)(z - 7) = 0.$$

Simplifying, we obtain

$$5x + 15 + 2y - z + 7 = 0$$

$$5x + 2y - z = -22.$$ ■

Notice in Example 6 how the components of $\mathbf{n} = 5\mathbf{i} + 2\mathbf{j} - \mathbf{k}$ became the coefficients of x, y, and z in the equation $5x + 2y - z = -22$. The vector $\mathbf{n} = A\mathbf{i} + B\mathbf{j} + C\mathbf{k}$ is normal to the plane $Ax + By + Cz = D$.

EXAMPLE 7 Find an equation for the plane through $A(0, 0, 1)$, $B(2, 0, 0)$, and $C(0, 3, 0)$.

Solution We find a vector normal to the plane and use it with one of the points (it does not matter which) to write an equation for the plane.

The cross product

$$\overrightarrow{AB} \times \overrightarrow{AC} = \begin{vmatrix} \mathbf{i} & \mathbf{j} & \mathbf{k} \\ 2 & 0 & -1 \\ 0 & 3 & -1 \end{vmatrix} = 3\mathbf{i} + 2\mathbf{j} + 6\mathbf{k}$$

is normal to the plane. We substitute the components of this vector and the coordinates of $A(0, 0, 1)$ into the component form of the equation to obtain

$$3(x - 0) + 2(y - 0) + 6(z - 1) = 0$$

$$3x + 2y + 6z = 6.$$ ■

Lines of Intersection

Just as lines are parallel if and only if they have the same direction, two planes are **parallel** if and only if their normals are parallel, or $\mathbf{n}_1 = k\mathbf{n}_2$ for some scalar k. Two planes that are not parallel intersect in a line.

EXAMPLE 8 Find a vector parallel to the line of intersection of the planes $3x - 6y - 2z = 15$ and $2x + y - 2z = 5$.

Solution The line of intersection of two planes is perpendicular to both planes' normal vectors $\mathbf{n}_1$ and $\mathbf{n}_2$ (Figure 10.39) and therefore parallel to $\mathbf{n}_1 \times \mathbf{n}_2$. Turning this around, $\mathbf{n}_1 \times \mathbf{n}_2$ is a vector parallel to the planes' line of intersection. In our case,

$$\mathbf{n}_1 \times \mathbf{n}_2 = \begin{vmatrix} \mathbf{i} & \mathbf{j} & \mathbf{k} \\ 3 & -6 & -2 \\ 2 & 1 & -2 \end{vmatrix} = 14\mathbf{i} + 2\mathbf{j} + 15\mathbf{k}.$$

Any nonzero scalar multiple of $\mathbf{n}_1 \times \mathbf{n}_2$ will do as well. ■

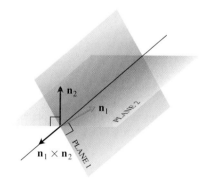

FIGURE 10.39 How the line of intersection of two planes is related to the planes' normal vectors (Example 8).

EXAMPLE 9 Find parametric equations for the line in which the planes $3x - 6y - 2z = 15$ and $2x + y - 2z = 5$ intersect.

Solution We find a vector parallel to the line and a point on the line and use Equations (3).

Example 8 identifies $\mathbf{v} = 14\mathbf{i} + 2\mathbf{j} + 15\mathbf{k}$ as a vector parallel to the line. To find a point on the line, we can take any point common to the two planes. Substituting $z = 0$ in the plane equations and solving for x and y simultaneously identifies one of these points as $(3, -1, 0)$. The line is

$$x = 3 + 14t, \qquad y = -1 + 2t, \qquad z = 15t.$$

The choice $z = 0$ is arbitrary and we could have chosen $z = 1$ or $z = -1$ just as well. Or we could have let $x = 0$ and solved for y and z. The different choices would simply give different parametrizations of the same line. ∎

Sometimes we want to know where a line and a plane intersect. For example, if we are looking at a flat plate and a line segment passes through it, we may be interested in knowing what portion of the line segment is hidden from our view by the plate. This application is used in computer graphics (Exercise 74).

EXAMPLE 10   Find the point where the line

$$x = \frac{8}{3} + 2t, \qquad y = -2t, \qquad z = 1 + t$$

intersects the plane $3x + 2y + 6z = 6$.

Solution   The point

$$\left(\frac{8}{3} + 2t, -2t, 1 + t \right)$$

lies in the plane if its coordinates satisfy the equation of the plane, that is, if

$$3\left(\frac{8}{3} + 2t \right) + 2(-2t) + 6(1 + t) = 6$$

$$8 + 6t - 4t + 6 + 6t = 6$$

$$8t = -8$$

$$t = -1.$$

The point of intersection is

$$(x, y, z)\big|_{t=-1} = \left(\frac{8}{3} - 2, 2, 1 - 1 \right) = \left(\frac{2}{3}, 2, 0 \right).$$ ∎

The Distance from a Point to a Plane

If P is a point on a plane with normal $\mathbf{n}$, then the distance from any point S to the plane is the length of the vector projection of $\vec{PS}$ onto $\mathbf{n}$. That is, the distance from S to the plane is

$$d = \left| \vec{PS} \cdot \frac{\mathbf{n}}{|\mathbf{n}|} \right| \tag{6}$$

where $\mathbf{n} = A\mathbf{i} + B\mathbf{j} + C\mathbf{k}$ is normal to the plane.

EXAMPLE 11   Find the distance from $S(1, 1, 3)$ to the plane $3x + 2y + 6z = 6$.

Solution   We find a point P in the plane and calculate the length of the vector projection of $\vec{PS}$ onto a vector $\mathbf{n}$ normal to the plane (Figure 10.40). The coefficients in the equation $3x + 2y + 6z = 6$ give

$$\mathbf{n} = 3\mathbf{i} + 2\mathbf{j} + 6\mathbf{k}.$$

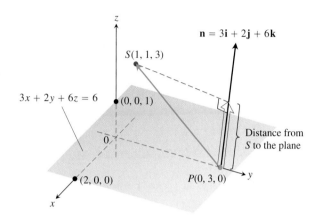

FIGURE 10.40 The distance from S to the plane is the length of the vector projection of $\vec{PS}$ onto $\mathbf{n}$ (Example 11).

The points on the plane easiest to find from the plane's equation are the intercepts. If we take P to be the y-intercept $(0, 3, 0)$, then

$$\vec{PS} = (1 - 0)\mathbf{i} + (1 - 3)\mathbf{j} + (3 - 0)\mathbf{k}$$
$$= \mathbf{i} - 2\mathbf{j} + 3\mathbf{k},$$
$$|\mathbf{n}| = \sqrt{(3)^2 + (2)^2 + (6)^2} = \sqrt{49} = 7.$$

The distance from S to the plane is

$$d = \left| \vec{PS} \cdot \frac{\mathbf{n}}{|\mathbf{n}|} \right| \qquad \text{length of proj}_\mathbf{n} \vec{PS}$$
$$= \left| (\mathbf{i} - 2\mathbf{j} + 3\mathbf{k}) \cdot \left(\frac{3}{7}\mathbf{i} + \frac{2}{7}\mathbf{j} + \frac{6}{7}\mathbf{k} \right) \right|$$
$$= \left| \frac{3}{7} - \frac{4}{7} + \frac{18}{7} \right| = \frac{17}{7}. \qquad \blacksquare$$

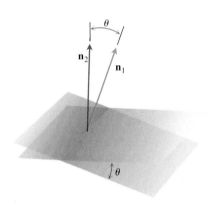

FIGURE 10.41 The angle between two planes is obtained from the angle between their normals.

Angles Between Planes

The angle between two intersecting planes is defined to be the (acute) angle between their normal vectors (Figure 10.41).

EXAMPLE 12 Find the angle between the planes $3x - 6y - 2z = 15$ and $2x + y - 2z = 5$.

Solution The vectors

$$\mathbf{n}_1 = 3\mathbf{i} - 6\mathbf{j} - 2\mathbf{k}, \qquad \mathbf{n}_2 = 2\mathbf{i} + \mathbf{j} - 2\mathbf{k}$$

are normals to the planes. The angle between them is

$$\theta = \cos^{-1}\left(\frac{\mathbf{n}_1 \cdot \mathbf{n}_2}{|\mathbf{n}_1||\mathbf{n}_2|} \right)$$
$$= \cos^{-1}\left(\frac{4}{21} \right)$$
$$\approx 1.38 \text{ radians.} \qquad \text{About 79 deg} \qquad \blacksquare$$

EXERCISES 10.5

Find parametric equations for the lines in Exercises 1–12.

1. The line through the point $P(3, -4, -1)$ parallel to the vector $\mathbf{i} + \mathbf{j} + \mathbf{k}$

2. The line through $P(1, 2, -1)$ and $Q(-1, 0, 1)$

3. The line through $P(-2, 0, 3)$ and $Q(3, 5, -2)$

4. The line through $P(1, 2, 0)$ and $Q(1, 1, -1)$

5. The line through the origin parallel to the vector $2\mathbf{j} + \mathbf{k}$

6. The line through the point $(3, -2, 1)$ parallel to the line $x = 1 + 2t, y = 2 - t, z = 3t$

7. The line through $(1, 1, 1)$ parallel to the z-axis

8. The line through $(2, 4, 5)$ perpendicular to the plane $3x + 7y - 5z = 21$

9. The line through $(0, -7, 0)$ perpendicular to the plane $x + 2y + 2z = 13$

10. The line through $(2, 3, 0)$ perpendicular to the vectors $\mathbf{u} = \mathbf{i} + 2\mathbf{j} + 3\mathbf{k}$ and $\mathbf{v} = 3\mathbf{i} + 4\mathbf{j} + 5\mathbf{k}$

11. The x-axis   12. The z-axis

Find parametrizations for the line segments joining the points in Exercises 13–20. Draw coordinate axes and sketch each segment, indicating the direction of increasing t for your parametrization.

13. $(0, 0, 0)$, $(1, 1, 3/2)$ 14. $(0, 0, 0)$, $(1, 0, 0)$

15. $(1, 0, 0)$, $(1, 1, 0)$ 16. $(1, 1, 0)$, $(1, 1, 1)$

17. $(0, 1, 1)$, $(0, -1, 1)$ 18. $(0, 2, 0)$, $(3, 0, 0)$

19. $(2, 0, 2)$, $(0, 2, 0)$ 20. $(1, 0, -1)$, $(0, 3, 0)$

Find equations for the planes in Exercises 21–26.

21. The plane through $P_0(0, 2, -1)$ normal to $\mathbf{n} = 3\mathbf{i} - 2\mathbf{j} - \mathbf{k}$

22. The plane through $(1, -1, 3)$ parallel to the plane

$$3x + y + z = 7$$

23. The plane through $(1, 1, -1)$, $(2, 0, 2)$, and $(0, -2, 1)$

24. The plane through $(2, 4, 5)$, $(1, 5, 7)$, and $(-1, 6, 8)$

25. The plane through $P_0(2, 4, 5)$ perpendicular to the line

$$x = 5 + t, \quad y = 1 + 3t, \quad z = 4t$$

26. The plane through $A(1, -2, 1)$ perpendicular to the vector from the origin to A

27. Find the point of intersection of the lines $x = 2t + 1, y = 3t + 2, z = 4t + 3$, and $x = s + 2, y = 2s + 4, z = -4s - 1$, and then find the plane determined by these lines.

28. Find the point of intersection of the lines $x = t, y = -t + 2, z = t + 1$, and $x = 2s + 2, y = s + 3, z = 5s + 6$, and then find the plane determined by these lines.

In Exercises 29 and 30, find the plane determined by the intersecting lines.

29. $L1$: $x = -1 + t$, $y = 2 + t$, $z = 1 - t$; $-\infty < t < \infty$

    $L2$: $x = 1 - 4s$, $y = 1 + 2s$, $z = 2 - 2s$; $-\infty < s < \infty$

30. $L1$: $x = t$, $y = 3 - 3t$, $z = -2 - t$; $-\infty < t < \infty$

    $L2$: $x = 1 + s$, $y = 4 + s$, $z = -1 + s$; $-\infty < s < \infty$

31. Find a plane through $P_0(2, 1, -1)$ and perpendicular to the line of intersection of the planes $2x + y - z = 3, x + 2y + z = 2$.

32. Find a plane through the points $P_1(1, 2, 3)$, $P_2(3, 2, 1)$ and perpendicular to the plane $4x - y + 2z = 7$.

In Exercises 33–38, find the distance from the point to the line.

33. $(0, 0, 12)$; $x = 4t$, $y = -2t$, $z = 2t$

34. $(0, 0, 0)$; $x = 5 + 3t$, $y = 5 + 4t$, $z = -3 - 5t$

35. $(2, 1, 3)$; $x = 2 + 2t$, $y = 1 + 6t$, $z = 3$

36. $(2, 1, -1)$; $x = 2t$, $y = 1 + 2t$, $z = 2t$

37. $(3, -1, 4)$; $x = 4 - t$, $y = 3 + 2t$, $z = -5 + 3t$

38. $(-1, 4, 3)$; $x = 10 + 4t$, $y = -3$, $z = 4t$

In Exercises 39–44, find the distance from the point to the plane.

39. $(2, -3, 4)$, $x + 2y + 2z = 13$

40. $(0, 0, 0)$, $3x + 2y + 6z = 6$

41. $(0, 1, 1)$, $4y + 3z = -12$

42. $(2, 2, 3)$, $2x + y + 2z = 4$

43. $(0, -1, 0)$, $2x + y + 2z = 4$

44. $(1, 0, -1)$, $-4x + y + z = 4$

45. Find the distance from the plane $x + 2y + 6z = 1$ to the plane $x + 2y + 6z = 10$.

46. Find the distance from the line $x = 2 + t, y = 1 + t, z = -(1/2) - (1/2)t$ to the plane $x + 2y + 6z = 10$.

Find the angles between the planes in Exercises 47 and 48.

47. $x + y = 1$, $2x + y - 2z = 2$

48. $5x + y - z = 10$, $x - 2y + 3z = -1$

T Use a calculator to find the acute angles between the planes in Exercises 49–52 to the nearest hundredth of a radian.

49. $2x + 2y + 2z = 3$, $2x - 2y - z = 5$

50. $x + y + z = 1$, $z = 0$ (the xy-plane)

51. $2x + 2y - z = 3$, $x + 2y + z = 2$

52. $4y + 3z = -12$, $3x + 2y + 6z = 6$

In Exercises 53–56, find the point in which the line meets the plane.

53. $x = 1 - t$, $y = 3t$, $z = 1 + t$; $2x - y + 3z = 6$

54. $x = 2, \quad y = 3 + 2t, \quad z = -2 - 2t; \quad 6x + 3y - 4z = -12$

55. $x = 1 + 2t, \quad y = 1 + 5t, \quad z = 3t; \quad x + y + z = 2$

56. $x = -1 + 3t, \quad y = -2, \quad z = 5t; \quad 2x - 3z = 7$

Find parametrizations for the lines in which the planes in Exercises 57–60 intersect.

57. $x + y + z = 1, \quad x + y = 2$

58. $3x - 6y - 2z = 3, \quad 2x + y - 2z = 2$

59. $x - 2y + 4z = 2, \quad x + y - 2z = 5$

60. $5x - 2y = 11, \quad 4y - 5z = -17$

Given two lines in space, either they are parallel, or they intersect, or they are skew (imagine, for example, the flight paths of two planes in the sky). Exercises 61 and 62 each give three lines. In each exercise, determine whether the lines, taken two at a time, are parallel, intersect, or are skew. If they intersect, find the point of intersection.

61. $L1: x = 3 + 2t, y = -1 + 4t, z = 2 - t; \quad -\infty < t < \infty$

$L2: x = 1 + 4s, y = 1 + 2s, z = -3 + 4s; \quad -\infty < s < \infty$

$L3: x = 3 + 2r, y = 2 + r, z = -2 + 2r; \quad -\infty < r < \infty$

62. $L1: x = 1 + 2t, \quad y = -1 - t, \quad z = 3t; \quad -\infty < t < \infty$

$L2: x = 2 - s, \quad y = 3s, \quad z = 1 + s; \quad -\infty < s < \infty$

$L3: x = 5 + 2r, \quad y = 1 - r, \quad z = 8 + 3r; \quad -\infty < r < \infty$

63. Use Equations (3) to generate a parametrization of the line through $P(2, -4, 7)$ parallel to $\mathbf{v}_1 = 2\mathbf{i} - \mathbf{j} + 3\mathbf{k}$. Then generate another parametrization of the line using the point $P_2(-2, -2, 1)$ and the vector $\mathbf{v}_2 = -\mathbf{i} + (1/2)\mathbf{j} - (3/2)\mathbf{k}$.

64. Use the component form to generate an equation for the plane through $P_1(4, 1, 5)$ normal to $\mathbf{n}_1 = \mathbf{i} - 2\mathbf{j} + \mathbf{k}$. Then generate another equation for the same plane using the point $P_2(3, -2, 0)$ and the normal vector $\mathbf{n}_2 = -\sqrt{2}\mathbf{i} + 2\sqrt{2}\mathbf{j} - \sqrt{2}\mathbf{k}$.

65. Find the points in which the line $x = 1 + 2t, y = -1 - t, z = 3t$ meets the coordinate planes. Describe the reasoning behind your answer.

66. Find equations for the line in the plane $z = 3$ that makes an angle of $\pi/6$ rad with $\mathbf{i}$ and an angle of $\pi/3$ rad with $\mathbf{j}$. Describe the reasoning behind your answer.

67. Is the line $x = 1 - 2t, y = 2 + 5t, z = -3t$ parallel to the plane $2x + y - z = 8$? Give reasons for your answer.

68. How can you tell when two planes $A_1x + B_1y + C_1z = D_1$ and $A_2x + B_2y + C_2z = D_2$ are parallel? Perpendicular? Give reasons for your answer.

69. Find two different planes whose intersection is the line $x = 1 + t, y = 2 - t, z = 3 + 2t$. Write equations for each plane in the form $Ax + By + Cz = D$.

70. Find a plane through the origin that meets the plane $M: 2x + 3y + z = 12$ in a right angle. How do you know that your plane is perpendicular to M?

71. For any nonzero numbers a, b, and c, the graph of $(x/a) + (y/b) + (z/c) = 1$ is a plane. Which planes have an equation of this form?

72. Suppose L_1 and L_2 are disjoint (nonintersecting) nonparallel lines. Is it possible for a nonzero vector to be perpendicular to both L_1 and L_2? Give reasons for your answer.

73. Perspective in computer graphics In computer graphics and perspective drawing, we need to represent objects seen by the eye in space as images on a two-dimensional plane. Suppose that the eye is at $E(x_0, 0, 0)$ as shown here and that we want to represent a point $P_1(x_1, y_1, z_1)$ as a point on the yz-plane. We do this by projecting P_1 onto the plane with a ray from E. The point P_1 will be portrayed as the point $P(0, y, z)$. The problem for us as graphics designers is to find y and z given E and P_1.

a. Write a vector equation that holds between $\vec{EP}$ and $\vec{EP_1}$. Use the equation to express y and z in terms of x_0, x_1, y_1, and z_1.

b. Test the formulas obtained for y and z in part (a) by investigating their behavior at $x_1 = 0$ and $x_1 = x_0$ and by seeing what happens as $x_0 \to \infty$. What do you find?

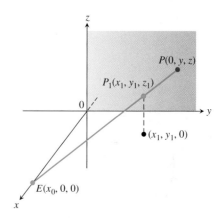

74. Hidden lines Here is another typical problem in computer graphics. Your eye is at $(4, 0, 0)$. You are looking at a triangular plate whose vertices are at $(1, 0, 1)$, $(1, 1, 0)$, and $(-2, 2, 2)$. The line segment from $(1, 0, 0)$ to $(0, 2, 2)$ passes through the plate. What portion of the line segment is hidden from your view by the plate? (This is an exercise in finding intersections of lines and planes.)

10.6 Cylinders and Quadric Surfaces

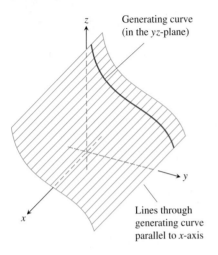

FIGURE 10.42 A cylinder and generating curve.

Up to now, we have studied two special types of surfaces: spheres and planes. In this section, we extend our inventory to include a variety of cylinders and quadric surfaces. Quadric surfaces are surfaces defined by second-degree equations in x, y, and z. Spheres are quadric surfaces, but there are others of equal interest.

Cylinders

A **cylinder** is a surface that is generated by moving a straight line along a given planar curve while holding the line parallel to a given fixed line. The curve is called a **generating curve** for the cylinder (Figure 10.42). In solid geometry, where *cylinder* means *circular cylinder*, the generating curves are circles, but now we allow generating curves of any kind. The cylinder in our first example is generated by a parabola.

EXAMPLE 1 Find an equation for the cylinder made by the lines parallel to the z-axis that pass through the parabola $y = x^2, z = 0$ (Figure 10.43).

Solution The point $P_0(x_0, x_0^2, 0)$ lies on the parabola $y = x^2$ in the xy-plane. Then, for any value of z, the point $Q(x_0, x_0^2, z)$ lies on the cylinder because it lies on the line $x = x_0, y = x_0^2$ through P_0 parallel to the z-axis. Conversely, any point $Q(x_0, x_0^2, z)$ whose y-coordinate is the square of its x-coordinate lies on the cylinder because it lies on the line $x = x_0, y = x_0^2$ through P_0 parallel to the z-axis (Figure 10.43).

Regardless of the value of z, therefore, the points on the surface are the points whose coordinates satisfy the equation $y = x^2$. This makes $y = x^2$ an equation for the cylinder. Because of this, we call the cylinder "the cylinder $y = x^2$." ∎

As Example 1 suggests, any curve $f(x, y) = c$ in the xy-plane defines a cylinder parallel to the z-axis whose equation is also $f(x, y) = c$. For instance, the equation $x^2 + y^2 = 1$ defines the circular cylinder made by the lines parallel to the z-axis that pass through the circle $x^2 + y^2 = 1$ in the xy-plane.

In a similar way, any curve $g(x, z) = c$ in the xz-plane defines a cylinder parallel to the y-axis whose space equation is also $g(x, z) = c$. Any curve $h(y, z) = c$ defines a cylinder parallel to the x-axis whose space equation is also $h(y, z) = c$. The axis of a cylinder need not be parallel to a coordinate axis, however.

Quadric Surfaces

A **quadric surface** is the graph in space of a second-degree equation in x, y, and z. We focus on the special equation

$$Ax^2 + By^2 + Cz^2 + Dz = E,$$

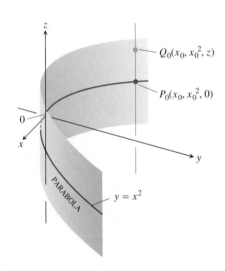

FIGURE 10.43 Every point of the cylinder in Example 1 has coordinates of the form (x_0, x_0^2, z). We call it "the cylinder $y = x^2$."

where A, B, C, D, and E are constants. The basic quadric surfaces are **ellipsoids**, **paraboloids**, **elliptical cones**, and **hyperboloids**. Spheres are special cases of ellipsoids. We present a few examples illustrating how to sketch a quadric surface, and then give a summary table of graphs of the basic types.

EXAMPLE 2 The **ellipsoid**

$$\frac{x^2}{a^2} + \frac{y^2}{b^2} + \frac{z^2}{c^2} = 1$$

(Figure 10.44) cuts the coordinate axes at $(\pm a, 0, 0)$, $(0, \pm b, 0)$, and $(0, 0, \pm c)$. It lies within the rectangular box defined by the inequalities $|x| \le a$, $|y| \le b$, and $|z| \le c$. The surface is symmetric with respect to each of the coordinate planes because each variable in the defining equation is squared.

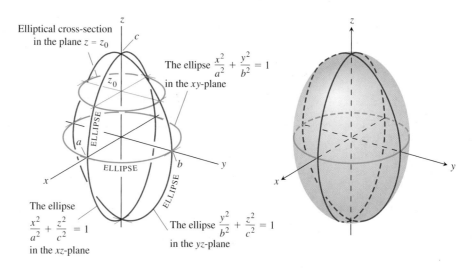

FIGURE 10.44 The ellipsoid

$$\frac{x^2}{a^2} + \frac{y^2}{b^2} + \frac{z^2}{c^2} = 1$$

in Example 2 has elliptical cross-sections in each of the three coordinate planes.

The curves in which the three coordinate planes cut the surface are ellipses. For example,

$$\frac{x^2}{a^2} + \frac{y^2}{b^2} = 1 \qquad \text{when} \qquad z = 0.$$

The curve cut from the surface by the plane $z = z_0$, $|z_0| < c$, is the ellipse

$$\frac{x^2}{a^2(1 - (z_0/c)^2)} + \frac{y^2}{b^2(1 - (z_0/c)^2)} = 1.$$

If any two of the semiaxes a, b, and c are equal, the surface is an **ellipsoid of revolution**. If all three are equal, the surface is a sphere. ∎

EXAMPLE 3   The **hyperbolic paraboloid**

$$\frac{y^2}{b^2} - \frac{x^2}{a^2} = \frac{z}{c}, \qquad c > 0$$

has symmetry with respect to the planes $x = 0$ and $y = 0$ (Figure 10.45). The cross-sections in these planes are

$$x = 0: \quad \text{the parabola } z = \frac{c}{b^2} y^2. \tag{1}$$

$$y = 0: \quad \text{the parabola } z = -\frac{c}{a^2} x^2. \tag{2}$$

In the plane $x = 0$, the parabola opens upward from the origin. The parabola in the plane $y = 0$ opens downward.

If we cut the surface by a plane $z = z_0 > 0$, the cross-section is a hyperbola,

$$\frac{y^2}{b^2} - \frac{x^2}{a^2} = \frac{z_0}{c},$$

with its focal axis parallel to the y-axis and its vertices on the parabola in Equation (1). If z_0 is negative, the focal axis is parallel to the x-axis and the vertices lie on the parabola in Equation (2).

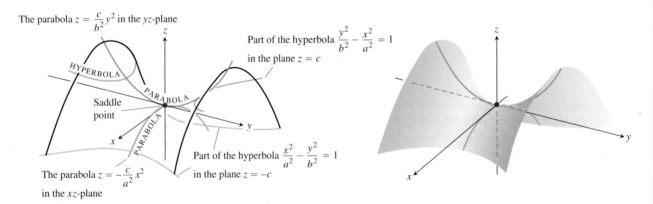

FIGURE 10.45   The hyperbolic paraboloid $(y^2/b^2) - (x^2/a^2) = z/c$, $c > 0$. The cross-sections in planes perpendicular to the z-axis above and below the xy-plane are hyperbolas. The cross-sections in planes perpendicular to the other axes are parabolas.

Near the origin, the surface is shaped like a saddle or mountain pass. To a person traveling along the surface in the yz-plane the origin looks like a minimum. To person traveling the xz-plane the origin looks like a maximum. Such a point is called a **saddle point** of a surface. We will say more about saddle points in Section 12.7.   ∎

Table 10.1 shows graphs of the six basic types of quadric surfaces. Each surface shown is symmetric with respect to the z-axis, but other coordinate axes can serve as well (with appropriate changes to the equation).

TABLE 10.1 Graphs of Quadric Surfaces

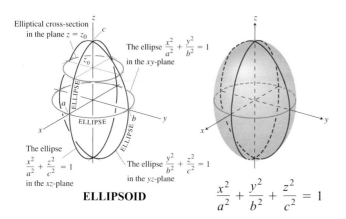

The ellipse $\frac{x^2}{a^2} + \frac{y^2}{b^2} = 1$ in the xy-plane

Elliptical cross-section in the plane $z = z_0$

The ellipse $\frac{x^2}{a^2} + \frac{z^2}{c^2} = 1$ in the xz-plane

The ellipse $\frac{y^2}{b^2} + \frac{z^2}{c^2} = 1$ in the yz-plane

ELLIPSOID
$$\frac{x^2}{a^2} + \frac{y^2}{b^2} + \frac{z^2}{c^2} = 1$$

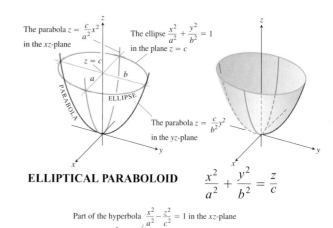

The parabola $z = \frac{c}{a^2}x^2$ in the xz-plane

The ellipse $\frac{x^2}{a^2} + \frac{y^2}{b^2} = 1$ in the plane $z = c$

The parabola $z = \frac{c}{b^2}y^2$ in the yz-plane

ELLIPTICAL PARABOLOID
$$\frac{x^2}{a^2} + \frac{y^2}{b^2} = \frac{z}{c}$$

The line $z = -\frac{c}{b}y$ in the yz-plane

The ellipse $\frac{x^2}{a^2} + \frac{y^2}{b^2} = 1$ in the plane $z = c$

The line $z = \frac{c}{a}x$ in the xz-plane

ELLIPTICAL CONE
$$\frac{x^2}{a^2} + \frac{y^2}{b^2} = \frac{z^2}{c^2}$$

Part of the hyperbola $\frac{x^2}{a^2} - \frac{z^2}{c^2} = 1$ in the xz-plane

The ellipse $\frac{x^2}{a^2} + \frac{y^2}{b^2} = 2$ in the plane $z = c$

The ellipse $\frac{x^2}{a^2} + \frac{y^2}{b^2} = 1$ in the xy-plane

Part of the hyperbola $\frac{y^2}{b^2} - \frac{z^2}{c^2} = 1$ in the yz-plane

HYPERBOLOID OF ONE SHEET
$$\frac{x^2}{a^2} + \frac{y^2}{b^2} - \frac{z^2}{c^2} = 1$$

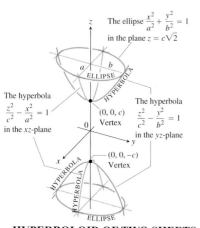

The ellipse $\frac{x^2}{a^2} + \frac{y^2}{b^2} = 1$ in the plane $z = c\sqrt{2}$

The hyperbola $\frac{z^2}{c^2} - \frac{x^2}{a^2} = 1$ in the xz-plane

$(0, 0, c)$ Vertex

The hyperbola $\frac{z^2}{c^2} - \frac{y^2}{b^2} = 1$ in the yz-plane

$(0, 0, -c)$ Vertex

HYPERBOLOID OF TWO SHEETS
$$\frac{z^2}{c^2} - \frac{x^2}{a^2} - \frac{y^2}{b^2} = 1$$

The parabola $z = \frac{c}{b^2}y^2$ in the yz-plane

Part of the hyperbola $\frac{y^2}{b^2} - \frac{x^2}{a^2} = 1$ in the plane $z = c$

Saddle point

The parabola $z = -\frac{c}{a^2}x^2$ in the xz-plane

Part of the hyperbola $\frac{x^2}{a^2} - \frac{y^2}{b^2} = 1$ in the plane $z = -c$

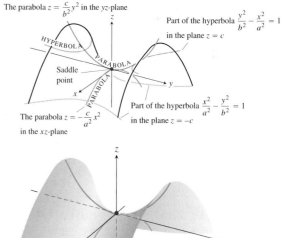

HYPERBOLIC PARABOLOID
$$\frac{y^2}{b^2} - \frac{x^2}{a^2} = \frac{z}{c}, \quad c > 0$$

EXERCISES 10.6

In Exercises 1–12, match the equation with the surface it defines. Also, identify each surface by type (paraboloid, ellipsoid, etc.) The surfaces are labeled (a)–(l).

1. $x^2 + y^2 + 4z^2 = 10$

2. $z^2 + 4y^2 - 4x^2 = 4$

3. $9y^2 + z^2 = 16$

4. $y^2 + z^2 = x^2$

5. $x = y^2 - z^2$

6. $x = -y^2 - z^2$

7. $x^2 + 2z^2 = 8$

8. $z^2 + x^2 - y^2 = 1$

9. $x = z^2 - y^2$

10. $z = -4x^2 - y^2$

11. $x^2 + 4z^2 = y^2$

12. $9x^2 + 4y^2 + 2z^2 = 36$

a.

b.

c.

d.

e.

f.

g.

h.

i.

j.

k.

l.

Sketch the surfaces in Exercises 13–44.

CYLINDERS

13. $x^2 + y^2 = 4$

14. $z = y^2 - 1$

15. $x^2 + 4z^2 = 16$

16. $4x^2 + y^2 = 36$

ELLIPSOIDS

17. $9x^2 + y^2 + z^2 = 9$

18. $4x^2 + 4y^2 + z^2 = 16$

19. $4x^2 + 9y^2 + 4z^2 = 36$

20. $9x^2 + 4y^2 + 36z^2 = 36$

PARABOLOIDS AND CONES

21. $z = x^2 + 4y^2$

22. $z = 8 - x^2 - y^2$

23. $x = 4 - 4y^2 - z^2$

24. $y = 1 - x^2 - z^2$

25. $x^2 + y^2 = z^2$

26. $4x^2 + 9z^2 = 9y^2$

HYPERBOLOIDS

27. $x^2 + y^2 - z^2 = 1$

28. $y^2 + z^2 - x^2 = 1$

29. $z^2 - x^2 - y^2 = 1$

30. $(y^2/4) - (x^2/4) - z^2 = 1$

HYPERBOLIC PARABOLOIDS

31. $y^2 - x^2 = z$

32. $x^2 - y^2 = z$

ASSORTED

33. $z = 1 + y^2 - x^2$

34. $4x^2 + 4y^2 = z^2$

35. $y = -(x^2 + z^2)$

36. $16x^2 + 4y^2 = 1$

37. $x^2 + y^2 - z^2 = 4$

38. $x^2 + z^2 = y$

39. $x^2 + z^2 = 1$

40. $16y^2 + 9z^2 = 4x^2$

41. $z = -(x^2 + y^2)$

42. $y^2 - x^2 - z^2 = 1$

43. $4y^2 + z^2 - 4x^2 = 4$

44. $x^2 + y^2 = z$

45. a. Express the area A of the cross-section cut from the ellipsoid

$$x^2 + \frac{y^2}{4} + \frac{z^2}{9} = 1$$

by the plane $z = c$ as a function of c. (The area of an ellipse with semiaxes a and b is πab.)

b. Use slices perpendicular to the z-axis to find the volume of the ellipsoid in part (a).

c. Now find the volume of the ellipsoid

$$\frac{x^2}{a^2} + \frac{y^2}{b^2} + \frac{z^2}{c^2} = 1.$$

Does your formula give the volume of a sphere of radius a if $a = b = c$?

46. The barrel shown here is shaped like an ellipsoid with equal pieces cut from the ends by planes perpendicular to the z-axis. The cross-sections perpendicular to the z-axis are circular. The barrel is $2h$ units high, its midsection radius is R, and its end radii are both r. Find a formula for the barrel's volume. Then check two things. First, suppose the sides of the barrel are straightened to turn the barrel into a cylinder of radius R and height $2h$. Does your formula give the cylinder's volume? Second, suppose $r = 0$ and $h = R$ so the barrel is a sphere. Does your formula give the sphere's volume?

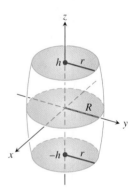

47. Show that the volume of the segment cut from the paraboloid

$$\frac{x^2}{a^2} + \frac{y^2}{b^2} = \frac{z}{c}$$

by the plane $z = h$ equals half the segment's base times its altitude.

48. a. Find the volume of the solid bounded by the hyperboloid

$$\frac{x^2}{a^2} + \frac{y^2}{b^2} - \frac{z^2}{c^2} = 1$$

and the planes $z = 0$ and $z = h, h > 0$.

b. Express your answer in part (a) in terms of h and the areas A_0 and A_h of the regions cut by the hyperboloid from the planes $z = 0$ and $z = h$.

c. Show that the volume in part (a) is also given by the formula

$$V = \frac{h}{6}(A_0 + 4A_m + A_h),$$

where A_m is the area of the region cut by the hyperboloid from the plane $z = h/2$.

T Plot the surfaces in Exercises 49–52 over the indicated domains. If you can, rotate the surface into different viewing positions.

49. $z = y^2$, $\quad -2 \le x \le 2$, $\quad -0.5 \le y \le 2$

50. $z = 1 - y^2$, $\quad -2 \le x \le 2$, $\quad -2 \le y \le 2$

51. $z = x^2 + y^2$, $\quad -3 \le x \le 3$, $\quad -3 \le y \le 3$

52. $z = x^2 + 2y^2$ over

a. $-3 \le x \le 3$, $\quad -3 \le y \le 3$

b. $-1 \le x \le 1$, $\quad -2 \le y \le 3$

c. $-2 \le x \le 2$, $\quad -2 \le y \le 2$

d. $-2 \le x \le 2$, $\quad -1 \le y \le 1$

COMPUTER EXPLORATIONS

Use a CAS to plot the surfaces in Exercises 53–58. Identify the type of quadric surface from your graph.

53. $\dfrac{x^2}{9} + \dfrac{y^2}{36} = 1 - \dfrac{z^2}{25}$

54. $\dfrac{x^2}{9} - \dfrac{z^2}{9} = 1 - \dfrac{y^2}{16}$

55. $5x^2 = z^2 - 3y^2$

56. $\dfrac{y^2}{16} = 1 - \dfrac{x^2}{9} + z$

57. $\dfrac{x^2}{9} - 1 = \dfrac{y^2}{16} + \dfrac{z^2}{2}$

58. $y - \sqrt{4 - z^2} = 0$

Chapter 10 Questions to Guide Your Review

1. When do directed line segments in the plane represent the same vector?

2. How are vectors added and subtracted geometrically? Algebraically?

3. How do you find a vector's magnitude and direction?

4. If a vector is multiplied by a positive scalar, how is the result related to the original vector? What if the scalar is zero? Negative?

5. Define the *dot product* (*scalar product*) of two vectors. Which algebraic laws are satisfied by dot products? Give examples. When is the dot product of two vectors equal to zero?

6. What geometric interpretation does the dot product have? Give examples.

7. What is the vector projection of a vector **u** onto a vector **v**? Give an example of a useful application of a vector projection.

8. Define the *cross product* (*vector product*) of two vectors. Which algebraic laws are satisfied by cross products, and which are not? Give examples. When is the cross product of two vectors equal to zero?

9. What geometric or physical interpretations do cross products have? Give examples.

10. What is the determinant formula for calculating the cross product of two vectors relative to the Cartesian **i**, **j**, **k**-coordinate system? Use it in an example.

11. How do you find equations for lines, line segments, and planes in space? Give examples. Can you express a line in space by a single equation? A plane?

12. How do you find the distance from a point to a line in space? From a point to a plane? Give examples.

13. What are box products? What significance do they have? How are they evaluated? Give an example.

14. How do you find equations for spheres in space? Give examples.

15. How do you find the intersection of two lines in space? A line and a plane? Two planes? Give examples.

16. What is a cylinder? Give examples of equations that define cylinders in Cartesian coordinates.

17. What are quadric surfaces? Give examples of different kinds of ellipsoids, paraboloids, cones, and hyperboloids (equations and sketches).

Chapter 10   Practice Exercises

In Exercises 1–4, let $\mathbf{u} = \langle -3, 4 \rangle$ and $\mathbf{v} = \langle 2, -5 \rangle$. Find **(a)** the component form of the vector and **(b)** its magnitude.

1. $3\mathbf{u} - 4\mathbf{v}$ **2.** $\mathbf{u} + \mathbf{v}$

3. $-2\mathbf{u}$ **4.** $5\mathbf{v}$

In Exercises 5–8, find the component form of the vector.

5. The vector obtained by rotating $\langle 0, 1 \rangle$ through an angle of $2\pi/3$ radians

6. The unit vector that makes an angle of $\pi/6$ radian with the positive x-axis

7. The vector 2 units long in the direction $4\mathbf{i} - \mathbf{j}$

8. The vector 5 units long in the direction opposite to the direction of $(3/5)\mathbf{i} + (4/5)\mathbf{j}$

Express the vectors in Exercises 9–12 in terms of their lengths and directions.

9. $\sqrt{2}\mathbf{i} + \sqrt{2}\mathbf{j}$ **10.** $-\mathbf{i} - \mathbf{j}$

11. Velocity vector $\mathbf{v} = (-2 \sin t)\mathbf{i} + (2 \cos t)\mathbf{j}$ when $t = \pi/2$.

12. Velocity vector $\mathbf{v} = (e^t \cos t - e^t \sin t)\mathbf{i} + (e^t \sin t + e^t \cos t)\mathbf{j}$ when $t = \ln 2$.

Express the vectors in Exercises 13 and 14 in terms of their lengths and directions.

13. $2\mathbf{i} - 3\mathbf{j} + 6\mathbf{k}$ **14.** $\mathbf{i} + 2\mathbf{j} - \mathbf{k}$

15. Find a vector 2 units long in the direction of $\mathbf{v} = 4\mathbf{i} - \mathbf{j} + 4\mathbf{k}$.

16. Find a vector 5 units long in the direction opposite to the direction of $\mathbf{v} = (3/5)\,\mathbf{i} + (4/5)\mathbf{k}$.

In Exercises 17 and 18, find $|\mathbf{v}|$, $|\mathbf{u}|$, $\mathbf{v} \cdot \mathbf{u}$, $\mathbf{u} \cdot \mathbf{v}$, $\mathbf{v} \times \mathbf{u}$, $\mathbf{u} \times \mathbf{v}$, $|\mathbf{v} \times \mathbf{u}|$, the angle between $\mathbf{v}$ and $\mathbf{u}$, the scalar component of $\mathbf{u}$ in the direction of $\mathbf{v}$, and the vector projection of $\mathbf{u}$ onto $\mathbf{v}$.

17. $\mathbf{v} = \mathbf{i} + \mathbf{j}$ **18.** $\mathbf{v} = \mathbf{i} + \mathbf{j} + 2\mathbf{k}$
    $\mathbf{u} = 2\mathbf{i} + \mathbf{j} - 2\mathbf{k}$ $\mathbf{u} = -\mathbf{i} - \mathbf{k}$

In Exercises 19 and 20, find $\text{proj}_{\mathbf{v}}\,\mathbf{u}$.

19. $\mathbf{v} = 2\mathbf{i} + \mathbf{j} - \mathbf{k}$ **20.** $\mathbf{u} = \mathbf{i} - 2\mathbf{j}$
    $\mathbf{u} = \mathbf{i} + \mathbf{j} - 5\mathbf{k}$ $\mathbf{v} = \mathbf{i} + \mathbf{j} + \mathbf{k}$

In Exercises 21 and 22, draw coordinate axes and then sketch $\mathbf{u}$, $\mathbf{v}$, and $\mathbf{u} \times \mathbf{v}$ as vectors at the origin.

21. $\mathbf{u} = \mathbf{i}$, $\mathbf{v} = \mathbf{i} + \mathbf{j}$ **22.** $\mathbf{u} = \mathbf{i} - \mathbf{j}$, $\mathbf{v} = \mathbf{i} + \mathbf{j}$

23. If $|\mathbf{v}| = 2$, $|\mathbf{w}| = 3$, and the angle between $\mathbf{v}$ and $\mathbf{w}$ is $\pi/3$, find $|\mathbf{v} - 2\mathbf{w}|$.

24. For what value or values of a will the vectors $\mathbf{u} = 2\mathbf{i} + 4\mathbf{j} - 5\mathbf{k}$ and $\mathbf{v} = -4\mathbf{i} - 8\mathbf{j} + a\mathbf{k}$ be parallel?

In Exercises 25 and 26, find **(a)** the area of the parallelogram determined by vectors $\mathbf{u}$ and $\mathbf{v}$ and **(b)** the volume of the parallelepiped determined by the vectors $\mathbf{u}$, $\mathbf{v}$, and $\mathbf{w}$.

25. $\mathbf{u} = \mathbf{i} + \mathbf{j} - \mathbf{k}$, $\mathbf{v} = 2\mathbf{i} + \mathbf{j} + \mathbf{k}$, $\mathbf{w} = -\mathbf{i} - 2\mathbf{j} + 3\mathbf{k}$

26. $\mathbf{u} = \mathbf{i} + \mathbf{j}$, $\mathbf{v} = \mathbf{j}$, $\mathbf{w} = \mathbf{i} + \mathbf{j} + \mathbf{k}$

27. Suppose that $\mathbf{n}$ is normal to a plane and that $\mathbf{v}$ is parallel to the plane. Describe how you would find a vector $\mathbf{n}$ that is both perpendicular to $\mathbf{v}$ and parallel to the plane.

28. Find a vector in the plane parallel to the line $ax + by = c$.

In Exercises 29 and 30, find the distance from the point to the line.

29. $(2, 2, 0)$; $x = -t$, $y = t$, $z = -1 + t$

30. $(0, 4, 1)$; $x = 2 + t$, $y = 2 + t$, $z = t$

31. Parametrize the line that passes through the point $(1, 2, 3)$ parallel to the vector $\mathbf{v} = -3\mathbf{i} + 7\mathbf{k}$.

32. Parametrize the line segment joining the points $P(1, 2, 0)$ and $Q(1, 3, -1)$.

In Exercises 33 and 34, find the distance from the point to the plane.

33. $(6, 0, -6)$, $x - y = 4$

34. $(3, 0, 10)$, $2x + 3y + z = 2$

35. Find an equation for the plane that passes through the point $(3, -2, 1)$ normal to the vector $\mathbf{n} = 2\mathbf{i} + \mathbf{j} + \mathbf{k}$.

36. Find an equation for the plane that passes through the point $(-1, 6, 0)$ perpendicular to the line $x = -1 + t, y = 6 - 2t,$ $z = 3t$.

In Exercises 37 and 38, find an equation for the plane through points $P, Q,$ and R.

37. $P(1, -1, 2), \quad Q(2, 1, 3), \quad R(-1, 2, -1)$

38. $P(1, 0, 0), \quad Q(0, 1, 0), \quad R(0, 0, 1)$

39. Find the points in which the line $x = 1 + 2t, y = -1 - t,$ $z = 3t$ meets the three coordinate planes.

40. Find the point in which the line through the origin perpendicular to the plane $2x - y - z = 4$ meets the plane $3x - 5y + 2z = 6$.

41. Find the acute angle between the planes $x = 7$ and $x + y + \sqrt{2}z = -3$.

42. Find the acute angle between the planes $x + y = 1$ and $y + z = 1$.

43. Find parametric equations for the line in which the planes $x + 2y + z = 1$ and $x - y + 2z = -8$ intersect.

44. Show that the line in which the planes

$$x + 2y - 2z = 5 \quad \text{and} \quad 5x - 2y - z = 0$$

intersect is parallel to the line

$$x = -3 + 2t, \quad y = 3t, \quad z = 1 + 4t.$$

45. The planes $3x + 6z = 1$ and $2x + 2y - z = 3$ intersect in a line.

  a. Show that the planes are orthogonal.

  b. Find equations for the line of intersection.

46. Find an equation for the plane that passes through the point $(1, 2, 3)$ parallel to $\mathbf{u} = 2\mathbf{i} + 3\mathbf{j} + \mathbf{k}$ and $\mathbf{v} = \mathbf{i} - \mathbf{j} + 2\mathbf{k}$.

47. Is $\mathbf{v} = 2\mathbf{i} - 4\mathbf{j} + \mathbf{k}$ related in any special way to the plane $2x + y = 5$? Give reasons for your answer.

48. The equation $\mathbf{n} \cdot \overrightarrow{P_0P} = 0$ represents the plane through P_0 normal to $\mathbf{n}$. What set does the inequality $\mathbf{n} \cdot \overrightarrow{P_0P} > 0$ represent?

49. Find the distance from the point $P(1, 4, 0)$ to the plane through $A(0, 0, 0), B(2, 0, -1)$ and $C(2, -1, 0)$.

50. Find the distance from the point $(2, 2, 3)$ to the plane $2x + 3y + 5z = 0$.

51. Find a vector parallel to the plane $2x - y - z = 4$ and orthogonal to $\mathbf{i} + \mathbf{j} + \mathbf{k}$.

52. Find a unit vector orthogonal to $\mathbf{A}$ in the plane of $\mathbf{B}$ and $\mathbf{C}$ if $\mathbf{A} = 2\mathbf{i} - \mathbf{j} + \mathbf{k}, \mathbf{B} = \mathbf{i} + 2\mathbf{j} + \mathbf{k}$, and $\mathbf{C} = \mathbf{i} + \mathbf{j} - 2\mathbf{k}$.

53. Find a vector of magnitude 2 parallel to the line of intersection of the planes $x + 2y + z - 1 = 0$ and $x - y + 2z + 7 = 0$.

54. Find the point in which the line through the origin perpendicular to the plane $2x - y - z = 4$ meets the plane $3x - 5y + 2z = 6$.

55. Find the point in which the line through $P(3, 2, 1)$ normal to the plane $2x - y + 2z = -2$ meets the plane.

56. What angle does the line of intersection of the planes $2x + y - z = 0$ and $x + y + 2z = 0$ make with the positive x-axis?

57. The line

$$L: \quad x = 3 + 2t, \quad y = 2t, \quad z = t$$

intersects the plane $x + 3y - z = -4$ in a point P. Find the co-ordinates of P and find equations for the line in the plane through P perpendicular to L.

58. Show that for every real number k the plane

$$x - 2y + z + 3 + k(2x - y - z + 1) = 0$$

contains the line of intersection of the planes

$$x - 2y + z + 3 = 0 \quad \text{and} \quad 2x - y - z + 1 = 0.$$

59. Find an equation for the plane through $A(-2, 0, -3)$ and $B(1, -2, 1)$ that lies parallel to the line through $C(-2, -13/5, 26/5)$ and $D(16/5, -13/5, 0)$.

60. Is the line $x = 1 + 2t, y = -2 + 3t, z = -5t$ related in any way to the plane $-4x - 6y + 10z = 9$? Give reasons for your answer.

61. Which of the following are equations for the plane through the points $P(1, 1, -1), Q(3, 0, 2)$, and $R(-2, 1, 0)$?

  a. $(2\mathbf{i} - 3\mathbf{j} + 3\mathbf{k}) \cdot ((x + 2)\mathbf{i} + (y - 1)\mathbf{j} + z\mathbf{k}) = 0$

  b. $x = 3 - t, \quad y = -11t, \quad z = 2 - 3t$

  c. $(x + 2) + 11(y - 1) = 3z$

  d. $(2\mathbf{i} - 3\mathbf{j} + 3\mathbf{k}) \times ((x + 2)\mathbf{i} + (y - 1)\mathbf{j} + z\mathbf{k}) = \mathbf{0}$

  e. $(2\mathbf{i} - \mathbf{j} + 3\mathbf{k}) \times (-3\mathbf{i} + \mathbf{k}) \cdot ((x + 2)\mathbf{i} + (y - 1)\mathbf{j} + z\mathbf{k}) = 0$

62. The parallelogram shown here has vertices at $A(2, -1, 4),$ $B(1, 0, -1), C(1, 2, 3)$, and D. Find

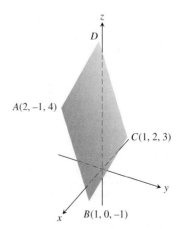

a. the coordinates of D,

b. the cosine of the interior angle at B,

c. the vector projection of $\vec{BA}$ onto $\vec{BC}$,

d. the area of the parallelogram,

e. an equation for the plane of the parallelogram,

f. the areas of the orthogonal projections of the parallelogram on the three coordinate planes.

63. Distance between lines Find the distance between the line L_1 through the points $A(1, 0, -1)$ and $B(-1, 1, 0)$ and the line L_2 through the points $C(3, 1, -1)$ and $D(4, 5, -2)$. The distance is to be measured along the line perpendicular to the two lines. First find a vector **n** perpendicular to both lines. Then project $\vec{AC}$ onto **n**.

64. (*Continuation of Exercise 63.*) Find the distance between the line through $A(4, 0, 2)$ and $B(2, 4, 1)$ and the line through $C(1, 3, 2)$ and $D(2, 2, 4)$.

Identify and sketch the surfaces in Exercises 65–76.

65. $x^2 + y^2 + z^2 = 4$

66. $x^2 + (y - 1)^2 + z^2 = 1$

67. $4x^2 + 4y^2 + z^2 = 4$

68. $36x^2 + 9y^2 + 4z^2 = 36$

69. $z = -(x^2 + y^2)$

70. $y = -(x^2 + z^2)$

71. $x^2 + y^2 = z^2$

72. $x^2 + z^2 = y^2$

73. $x^2 + y^2 - z^2 = 4$

74. $4y^2 + z^2 - 4x^2 = 4$

75. $y^2 - x^2 - z^2 = 1$

76. $z^2 - x^2 - y^2 = 1$

Chapter **10**   Additional and Advanced Exercises

1. Submarine hunting Two surface ships on maneuvers are trying to determine a submarine's course and speed to prepare for an aircraft intercept. As shown here, ship A is located at $(4, 0, 0)$, whereas ship B is located at $(0, 5, 0)$. All coordinates are given in thousands of feet. Ship A locates the submarine in the direction of the vector $2\mathbf{i} + 3\mathbf{j} - (1/3)\mathbf{k}$, and ship B locates it in the direction of the vector $18\mathbf{i} - 6\mathbf{j} - \mathbf{k}$. Four minutes ago, the submarine was located at $(2, -1, -1/3)$. The aircraft is due in 20 min. Assuming that the submarine moves in a straight line at a constant speed, to what position should the surface ships direct the aircraft?

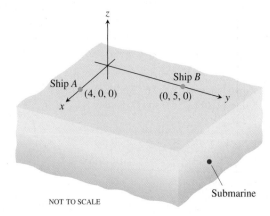

NOT TO SCALE

2. A helicopter rescue Two helicopters, H_1 and H_2, are traveling together. At time $t = 0$, they separate and follow different straight-line paths given by

H_1: $x = 6 + 40t$, $y = -3 + 10t$, $z = -3 + 2t$

H_2: $x = 6 + 110t$, $y = -3 + 4t$, $z = -3 + t$.

Time t is measured in hours and all coordinates are measured in miles. Due to system malfunctions, H_2 stops its flight at $(446, 13, 1)$ and, in a negligible amount of time, lands at $(446, 13, 0)$.

Two hours later, H_1 is advised of this fact and heads toward H_2 at 150 mph. How long will it take H_1 to reach H_2?

3. Torque The operator's manual for the Toro® 21 in. lawnmower says "tighten the spark plug to 15 ft-lb (20.4 N·m)." If you are installing the plug with a 10.5-in. socket wrench that places the center of your hand 9 in. from the axis of the spark plug, about how hard should you pull? Answer in pounds.

9 in.

4. Rotating body The line through the origin and the point $A(1, 1, 1)$ is the axis of rotation of a right body rotating with a constant angular speed of $3/2$ rad/sec. The rotation appears to be clockwise when we look toward the origin from A. Find the velocity **v** of the point of the body that is at the position $B(1, 3, 2)$.

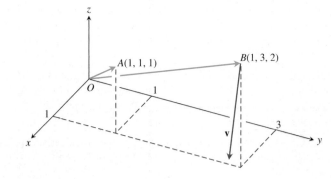

5. Determinants and planes

  a. Show that

$$\begin{vmatrix} x_1 - x & y_1 - y & z_1 - z \\ x_2 - x & y_2 - y & z_2 - z \\ x_3 - x & y_3 - y & z_3 - z \end{vmatrix} = 0$$

is an equation for the plane through the three noncollinear points $P_1(x_1, y_1, z_1)$, $P_2(x_2, y_2, z_2)$, and $P_3(x_3, y_3, z_3)$.

  b. What set of points in space is described by the equation

$$\begin{vmatrix} x & y & z & 1 \\ x_1 & y_1 & z_1 & 1 \\ x_2 & y_2 & z_2 & 1 \\ x_3 & y_3 & z_3 & 1 \end{vmatrix} = 0?$$

6. Determinants and lines   Show that the lines

$$x = a_1 s + b_1, y = a_2 s + b_2, z = a_3 s + b_3, \quad -\infty < s < \infty,$$

and

$$x = c_1 t + d_1, y = c_2 t + d_2, z = c_3 t + d_3, \quad -\infty < t < \infty,$$

intersect or are parallel if and only if

$$\begin{vmatrix} a_1 & c_1 & b_1 - d_1 \\ a_2 & c_2 & b_2 - d_2 \\ a_3 & c_3 & b_3 - d_3 \end{vmatrix} = 0.$$

7. Parallelogram   The accompanying figure shows parallelogram $ABCD$ and the midpoint P of diagonal BD.

  a. Express $\vec{BD}$ in terms of $\vec{AB}$ and $\vec{AD}$.

  b. Express $\vec{AP}$ in terms of $\vec{AB}$ and $\vec{AD}$.

  c. Prove that P is also the midpoint of diagonal AC.

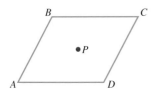

8. In the figure here, D is the midpoint of side AB of triangle ABC, and E is one-third of the way between C and B. Use vectors to prove that F is the midpoint of line segment CD.

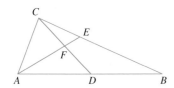

9. Use vectors to show that the distance from $P_1(x_1, y_1)$ to the line $ax + by = c$ is

$$d = \frac{|ax_1 + by_1 - c|}{\sqrt{a^2 + b^2}}.$$

10. a. Use vectors to show that the distance from $P_1(x_1, y_1, z_1)$ to the plane $Ax + By + Cz = D$ is

$$d = \frac{|Ax_1 + By_1 + Cz_1 - D|}{\sqrt{A^2 + B^2 + C^2}}.$$

  b. Find an equation for the sphere that is tangent to the planes $x + y + z = 3$ and $x + y + z = 9$ if the planes $2x - y = 0$ and $3x - z = 0$ pass through the center of the sphere.

11. a. Show that the distance between the parallel planes $Ax + By + Cz = D_1$ and $Ax + By + Cz = D_2$ is

$$d = \frac{|D_1 - D_2|}{|A\mathbf{i} + B\mathbf{j} + C\mathbf{k}|}.$$

  b. Find the distance between the planes $2x + 3y - z = 6$ and $2x + 3y - z = 12$.

  c. Find an equation for the plane parallel to the plane $2x - y + 2z = -4$ if the point $(3, 2, -1)$ is equidistant from the two planes.

  d. Write equations for the planes that lie parallel to and 5 units away from the plane $x - 2y + z = 3$.

12. Prove that four points A, B, C, and D are coplanar (lie in a common plane) if and only if $\vec{AD} \cdot (\vec{AB} \times \vec{BC}) = 0$.

13. The projection of a vector on a plane   Let P be a plane in space and let $\mathbf{v}$ be a vector. The vector projection of $\mathbf{v}$ onto the plane P, $\text{proj}_P\mathbf{v}$, can be defined informally as follows. Suppose the sun is shining so that its rays are normal to the plane P. Then $\text{proj}_P\mathbf{v}$ is the "shadow" of $\mathbf{v}$ onto P. If P is the plane $x + 2y + 6z = 6$ and $\mathbf{v} = \mathbf{i} + \mathbf{j} + \mathbf{k}$, find $\text{proj}_P\mathbf{v}$.

14. The accompanying figure shows nonzero vectors $\mathbf{v}$, $\mathbf{w}$, and $\mathbf{z}$, with $\mathbf{z}$ orthogonal to the line L, and $\mathbf{v}$ and $\mathbf{w}$ making equal angles β with L. Assuming $|\mathbf{v}| = |\mathbf{w}|$, find $\mathbf{w}$ in terms of $\mathbf{v}$ and $\mathbf{z}$.

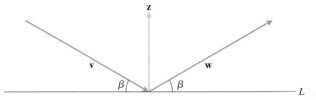

15. Triple vector products   The *triple vector products* $(\mathbf{u} \times \mathbf{v}) \times \mathbf{w}$ and $\mathbf{u} \times (\mathbf{v} \times \mathbf{w})$ are usually not equal, although the formulas for evaluating them from components are similar:

$$(\mathbf{u} \times \mathbf{v}) \times \mathbf{w} = (\mathbf{u} \cdot \mathbf{w})\mathbf{v} - (\mathbf{v} \cdot \mathbf{w})\mathbf{u}.$$

$$\mathbf{u} \times (\mathbf{v} \times \mathbf{w}) = (\mathbf{u} \cdot \mathbf{w})\mathbf{v} - (\mathbf{u} \cdot \mathbf{v})\mathbf{w}.$$

Verify each formula for the following vectors by evaluating its two sides and comparing the results.

	u	**v**	**w**
a.	$2\mathbf{i}$	$2\mathbf{j}$	$2\mathbf{k}$
b.	$\mathbf{i} - \mathbf{j} + \mathbf{k}$	$2\mathbf{i} + \mathbf{j} - 2\mathbf{k}$	$-\mathbf{i} + 2\mathbf{j} - \mathbf{k}$
c.	$2\mathbf{i} + \mathbf{j}$	$2\mathbf{i} - \mathbf{j} + \mathbf{k}$	$\mathbf{i} + 2\mathbf{k}$
d.	$\mathbf{i} + \mathbf{j} - 2\mathbf{k}$	$-\mathbf{i} - \mathbf{k}$	$2\mathbf{i} + 4\mathbf{j} - 2\mathbf{k}$

16. Cross and dot products Show that if **u, v, w,** and **r** are any vectors, then

a. $\mathbf{u} \times (\mathbf{v} \times \mathbf{w}) + \mathbf{v} \times (\mathbf{w} \times \mathbf{u}) + \mathbf{w} \times (\mathbf{u} \times \mathbf{v}) = \mathbf{0}$

b. $\mathbf{u} \times \mathbf{v} = (\mathbf{u} \cdot \mathbf{v} \times \mathbf{i})\mathbf{i} + (\mathbf{u} \cdot \mathbf{v} \times \mathbf{j})\mathbf{j} + (\mathbf{u} \cdot \mathbf{v} \times \mathbf{k})\mathbf{k}$

c. $(\mathbf{u} \times \mathbf{v}) \cdot (\mathbf{w} \times \mathbf{r}) = \begin{vmatrix} \mathbf{u} \cdot \mathbf{w} & \mathbf{v} \cdot \mathbf{w} \\ \mathbf{u} \cdot \mathbf{r} & \mathbf{v} \cdot \mathbf{r} \end{vmatrix}.$

17. Cross and dot products Prove or disprove the formula

$$\mathbf{u} \times (\mathbf{u} \times (\mathbf{u} \times \mathbf{v})) \cdot \mathbf{w} = -|\mathbf{u}|^2 \mathbf{u} \cdot \mathbf{v} \times \mathbf{w}.$$

18. By forming the cross product of two appropriate vectors, derive the trigonometric identity

$$\sin(A - B) = \sin A \cos B - \cos A \sin B.$$

19. Use vectors to prove that

$$(a^2 + b^2)(c^2 + d^2) \geq (ac + bd)^2$$

for any four numbers a, b, c, and d. (*Hint:* Let $\mathbf{u} = a\mathbf{i} + b\mathbf{j}$ and $\mathbf{v} = c\mathbf{i} + d\mathbf{j}$.)

20. Prove that the vector $\mathbf{u} - \text{proj}_\mathbf{v} \, \mathbf{u}$ is orthogonal to the projection vector $\text{proj}_\mathbf{v} \, \mathbf{u}$ by showing their dot product is zero.

21. Show that $|\mathbf{u} + \mathbf{v}| \leq |\mathbf{u}| + |\mathbf{v}|$ for any vectors **u** and **v**.

22. Show that $\mathbf{w} = |\mathbf{v}|\mathbf{u} + |\mathbf{u}|\mathbf{v}$ bisects the angle between **u** and **v**.

23. Show that $|\mathbf{v}|\mathbf{u} + |\mathbf{u}|\mathbf{v}$ and $|\mathbf{v}|\mathbf{u} - |\mathbf{u}|\mathbf{v}$ are orthogonal.

24. Dot multiplication is positive definite Show that dot multiplication of vectors is *positive definite*; that is, show that $\mathbf{u} \cdot \mathbf{u} \geq 0$ for every vector **u** and that $\mathbf{u} \cdot \mathbf{u} = 0$ if and only if $\mathbf{u} = \mathbf{0}$.

Chapter 11

Vector-Valued Functions and Motion in Space

OVERVIEW In this chapter we introduce the calculus of vector-valued functions, including their derivatives and integrals. Here we use vector calculus to study the paths, velocities, and accelerations of moving bodies. As we proceed, we will see how questions are answered about the paths and motions of projectiles and other objects moving in a plane or in space.

11.1 Vector Functions and Their Derivatives

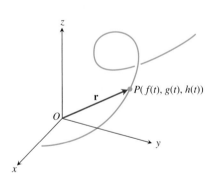

FIGURE 11.1 The position vector $\mathbf{r} = \overrightarrow{OP}$ of a particle moving through space is a function of time.

When a particle moves through space during a time interval I, we think of the particle's coordinates as functions defined on I:

$$x = f(t), \qquad y = g(t), \qquad z = h(t), \qquad t \in I. \tag{1}$$

The points $(x, y, z) = (f(t), g(t), h(t))$, $t \in I$, make up the **curve** in space that we call the particle's **path**. The equations and interval in Equation (1) **parametrize** the curve.

A curve in space can also be represented in vector form. The vector

$$\mathbf{r}(t) = \overrightarrow{OP} = f(t)\mathbf{i} + g(t)\mathbf{j} + h(t)\mathbf{k} \tag{2}$$

from the origin to the particle's **position** $P(f(t), g(t), h(t))$ at time t is the particle's **position vector** (Figure 11.1). The functions f, g, and h are the **component functions (components)** of the position vector. We think of the particle's path as the **curve traced by r** during the time interval I. Figure 11.2 displays several space curves generated by a computer graphing program. It would not be easy to plot these curves by hand.

Equation (2) defines $\mathbf{r}$ as a vector function of the real variable t on the interval I. More generally, a **vector-valued function** or **vector function** on a domain set D is a rule that assigns a vector in space to each element in D. For now, the domains will be intervals of real numbers resulting in a space curve. Later, in Chapter 14, the domains will be regions in the plane. Vector functions will then represent surfaces in space. Vector functions on a domain in the plane or space also give rise to "vector fields," which are important to the study of the flow of a fluid, gravitational fields, and electromagnetic phenomena. We investigate vector fields and their applications in Chapter 14.

Real-valued functions are called **scalar functions** to distinguish them from vector functions. The components of $\mathbf{r}$ are scalar functions of t. The domain of a vector-valued function is the common domain of its components.

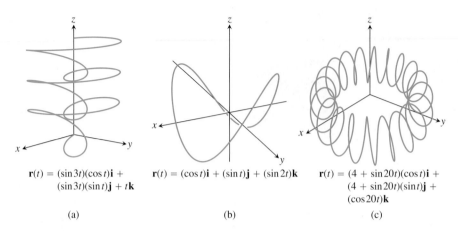

$$\mathbf{r}(t) = (\sin 3t)(\cos t)\mathbf{i} + \qquad \mathbf{r}(t) = (\cos t)\mathbf{i} + (\sin t)\mathbf{j} + (\sin 2t)\mathbf{k} \qquad \mathbf{r}(t) = (4 + \sin 20t)(\cos t)\mathbf{i} + (\sin 3t)(\sin t)\mathbf{j} + t\mathbf{k} \qquad\qquad\qquad\qquad\qquad\qquad (4 + \sin 20t)(\sin t)\mathbf{j} + (\cos 20t)\mathbf{k}$$

(a)   (b)   (c)

FIGURE 11.2 Computer-generated space curves are defined by the position vectors $\mathbf{r}(t)$.

EXAMPLE 1   Graph the vector function

$$\mathbf{r}(t) = (\cos t)\mathbf{i} + (\sin t)\mathbf{j} + t\mathbf{k}.$$

Solution   The vector function

$$\mathbf{r}(t) = (\cos t)\mathbf{i} + (\sin t)\mathbf{j} + t\mathbf{k}$$

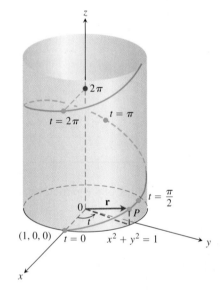

FIGURE 11.3 The upper half of the helix $\mathbf{r}(t) = (\cos t)\mathbf{i} + (\sin t)\mathbf{j} + t\mathbf{k}$ (Example 1).

is defined for all real values of t. The curve traced by $\mathbf{r}$ winds around the circular cylinder $x^2 + y^2 = 1$ (Figure 11.3). The curve lies on the cylinder because the $\mathbf{i}$- and $\mathbf{j}$-components of $\mathbf{r}$, being the x- and y-coordinates of the tip of $\mathbf{r}$, satisfy the cylinder's equation:

$$x^2 + y^2 = (\cos t)^2 + (\sin t)^2 = 1.$$

The curve rises as the $\mathbf{k}$-component $z = t$ increases. Each time t increases by 2π, the curve completes one turn around the cylinder. The curve is called a **helix** (from an old Greek word for "spiral"). The equations

$$x = \cos t, \qquad y = \sin t, \qquad z = t$$

parametrize the helix, the interval $-\infty < t < \infty$ being understood. You will find more helices in Figure 11.4.   ∎

$$\mathbf{r}(t) = (\cos t)\mathbf{i} + (\sin t)\mathbf{j} + t\mathbf{k} \qquad\qquad \mathbf{r}(t) = (\cos t)\mathbf{i} + (\sin t)\mathbf{j} + 0.3t\mathbf{k} \qquad\qquad \mathbf{r}(t) = (\cos 5t)\mathbf{i} + (\sin 5t)\mathbf{j} + t\mathbf{k}$$

FIGURE 11.4   Helices drawn by computer.

Limits and Continuity

The way we define limits of vector-valued functions is similar to the way we define limits of real-valued functions.

DEFINITION   Let $\mathbf{r}(t) = f(t)\mathbf{i} + g(t)\mathbf{j} + h(t)\mathbf{k}$ be a vector function and $\mathbf{L}$ a vector. We say that $\mathbf{r}$ has **limit L** as t approaches t_0 and write

$$\lim_{t \to t_0} \mathbf{r}(t) = \mathbf{L}$$

if, for every number $\epsilon > 0$, there exists a corresponding number $\delta > 0$ such that for all t

$$0 < |t - t_0| < \delta \quad \Rightarrow \quad |\mathbf{r}(t) - \mathbf{L}| < \epsilon.$$

If $\mathbf{L} = L_1\mathbf{i} + L_2\mathbf{j} + L_3\mathbf{k}$, then it can be shown that $\lim_{t \to t_0} \mathbf{r}(t) = \mathbf{L}$ precisely when

$$\lim_{t \to t_0} f(t) = L_1, \qquad \lim_{t \to t_0} g(t) = L_2, \qquad \text{and} \qquad \lim_{t \to t_0} h(t) = L_3.$$

We omit the proof. The equation

$$\lim_{t \to t_0} \mathbf{r}(t) = \left(\lim_{t \to t_0} f(t) \right)\mathbf{i} + \left(\lim_{t \to t_0} g(t) \right)\mathbf{j} + \left(\lim_{t \to t_0} h(t) \right)\mathbf{k} \tag{3}$$

provides a practical way to calculate limits of vector functions.

EXAMPLE 2   If $\mathbf{r}(t) = (\cos t)\mathbf{i} + (\sin t)\mathbf{j} + t\mathbf{k}$, then

$$\lim_{t \to \pi/4} \mathbf{r}(t) = \left(\lim_{t \to \pi/4} \cos t \right)\mathbf{i} + \left(\lim_{t \to \pi/4} \sin t \right)\mathbf{j} + \left(\lim_{t \to \pi/4} t \right)\mathbf{k}$$

$$= \frac{\sqrt{2}}{2}\mathbf{i} + \frac{\sqrt{2}}{2}\mathbf{j} + \frac{\pi}{4}\mathbf{k}. \qquad \blacksquare$$

We define continuity for vector functions the same way we define continuity for scalar functions.

DEFINITION   A vector function $\mathbf{r}(t)$ is **continuous at a point** $t = t_0$ in its domain if $\lim_{t \to t_0} \mathbf{r}(t) = \mathbf{r}(t_0)$. The function is **continuous** if it is continuous at every point in its domain.

From Equation (3), we see that $\mathbf{r}(t)$ is continuous at $t = t_0$ if and only if each component function is continuous there (Exercise 31).

EXAMPLE 3

(a) All the space curves shown in Figures 11.2 and 11.4 are continuous because their component functions are continuous at every value of t in $(-\infty, \infty)$.

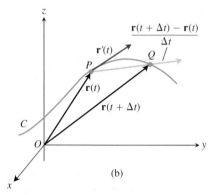

FIGURE 11.5 As $\Delta t \rightarrow 0$, the point Q approaches the point P along the curve C. In the limit, the vector $\vec{PQ}/\Delta t$ becomes the tangent vector $\mathbf{r}'(t)$.

(b) The function

$$\mathbf{g}(t) = (\cos t)\mathbf{i} + (\sin t)\mathbf{j} + \lfloor t \rfloor \mathbf{k}$$

is discontinuous at every integer, where the greatest integer function $\lfloor t \rfloor$ is discontinuous. ∎

Derivatives and Motion

Suppose that $\mathbf{r}(t) = f(t)\mathbf{i} + g(t)\mathbf{j} + h(t)\mathbf{k}$ is the position vector of a particle moving along a curve in space and that f, g, and h are differentiable functions of t. Then the difference between the particle's positions at time t and time $t + \Delta t$ is

$$\Delta \mathbf{r} = \mathbf{r}(t + \Delta t) - \mathbf{r}(t)$$

(Figure 11.5a). In terms of components,

$$\begin{aligned}
\Delta \mathbf{r} &= \mathbf{r}(t + \Delta t) - \mathbf{r}(t) \\
&= [f(t + \Delta t)\mathbf{i} + g(t + \Delta t)\mathbf{j} + h(t + \Delta t)\mathbf{k}] \\
&\quad - [f(t)\mathbf{i} + g(t)\mathbf{j} + h(t)\mathbf{k}] \\
&= [f(t + \Delta t) - f(t)]\mathbf{i} + [g(t + \Delta t) - g(t)]\mathbf{j} + [h(t + \Delta t) - h(t)]\mathbf{k}.
\end{aligned}$$

As Δt approaches zero, three things seem to happen simultaneously. First, Q approaches P along the curve. Second, the secant line PQ seems to approach a limiting position tangent to the curve at P. Third, the quotient $\Delta \mathbf{r}/\Delta t$ (Figure 11.5b) approaches the limit

$$\begin{aligned}
\lim_{\Delta t \to 0} \frac{\Delta \mathbf{r}}{\Delta t} &= \left[\lim_{\Delta t \to 0} \frac{f(t + \Delta t) - f(t)}{\Delta t} \right]\mathbf{i} + \left[\lim_{\Delta t \to 0} \frac{g(t + \Delta t) - g(t)}{\Delta t} \right]\mathbf{j} \\
&\quad + \left[\lim_{\Delta t \to 0} \frac{h(t + \Delta t) - h(t)}{\Delta t} \right]\mathbf{k} \\
&= \left[\frac{df}{dt} \right]\mathbf{i} + \left[\frac{dg}{dt} \right]\mathbf{j} + \left[\frac{dh}{dt} \right]\mathbf{k}.
\end{aligned}$$

We are therefore led by past experience to the following definition.

DEFINITION The vector function $\mathbf{r}(t) = f(t)\mathbf{i} + g(t)\mathbf{j} + h(t)\mathbf{k}$ has a **derivative (is differentiable) at t** if f, g, and h have derivatives at t. The derivative is the vector function

$$\mathbf{r}'(t) = \frac{d\mathbf{r}}{dt} = \lim_{\Delta t \to 0} \frac{\mathbf{r}(t + \Delta t) - \mathbf{r}(t)}{\Delta t} = \frac{df}{dt}\mathbf{i} + \frac{dg}{dt}\mathbf{j} + \frac{dh}{dt}\mathbf{k}.$$

A vector function $\mathbf{r}$ is **differentiable** if it is differentiable at every point of its domain. The curve traced by $\mathbf{r}$ is **smooth** if $d\mathbf{r}/dt$ is continuous and never $\mathbf{0}$, that is, if f, g, and h have continuous first derivatives that are not simultaneously 0.

The geometric significance of the definition of derivative is shown in Figure 11.5. The points P and Q have position vectors $\mathbf{r}(t)$ and $\mathbf{r}(t + \Delta t)$, and the vector $\vec{PQ}$ is represented by $\mathbf{r}(t + \Delta t) - \mathbf{r}(t)$. For $\Delta t > 0$, the scalar multiple $(1/\Delta t)(\mathbf{r}(t + \Delta t) - \mathbf{r}(t))$

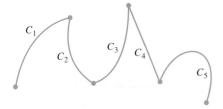

FIGURE 11.6 A piecewise smooth curve made up of five smooth curves connected end to end in continuous fashion. The curve here is not smooth at the points joining the five smooth curves.

points in the same direction as the vector $\overrightarrow{PQ}$. As $\Delta t \to 0$, this vector approaches a vector that is tangent to the curve at P (Figure 11.5b). The vector $\mathbf{r}'(t)$, when different from $\mathbf{0}$, is defined to be the vector **tangent** to the curve at P. The **tangent line** to the curve at a point $(f(t_0), g(t_0), h(t_0))$ is defined to be the line through the point parallel to $\mathbf{r}'(t_0)$. We require $d\mathbf{r}/dt \neq \mathbf{0}$ for a smooth curve to make sure the curve has a continuously turning tangent at each point. On a smooth curve, there are no sharp corners or cusps.

A curve that is made up of a finite number of smooth curves pieced together in a continuous fashion is called **piecewise smooth** (Figure 11.6).

Look once again at Figure 11.5. We drew the figure for Δt positive, so $\Delta\mathbf{r}$ points forward, in the direction of the motion. The vector $\Delta\mathbf{r}/\Delta t$, having the same direction as $\Delta\mathbf{r}$, points forward too. Had Δt been negative, $\Delta\mathbf{r}$ would have pointed backward, against the direction of motion. The quotient $\Delta\mathbf{r}/\Delta t$, however, being a negative scalar multiple of $\Delta\mathbf{r}$, would once again have pointed forward. No matter how $\Delta\mathbf{r}$ points, $\Delta\mathbf{r}/\Delta t$ points forward and we expect the vector $d\mathbf{r}/dt = \lim_{\Delta t \to 0} \Delta\mathbf{r}/\Delta t$, when different from $\mathbf{0}$, to do the same. This means that the derivative $d\mathbf{r}/dt$, which is the rate of change of position with respect to time, always points in the direction of motion. For a smooth curve, $d\mathbf{r}/dt$ is never zero; the particle does not stop or reverse direction.

DEFINITIONS If $\mathbf{r}$ is the position vector of a particle moving along a smooth curve in space, then

$$\mathbf{v}(t) = \frac{d\mathbf{r}}{dt}$$

is the particle's **velocity vector**, tangent to the curve. At any time t, the direction of $\mathbf{v}$ is the **direction of motion**, the magnitude of $\mathbf{v}$ is the particle's **speed**, and the derivative $\mathbf{a} = d\mathbf{v}/dt$, when it exists, is the particle's **acceleration vector**. In summary,

1. Velocity is the derivative of position: $\quad \mathbf{v} = \dfrac{d\mathbf{r}}{dt}$.

2. Speed is the magnitude of velocity: $\quad$ Speed $= |\mathbf{v}|$.

3. Acceleration is the derivative of velocity: $\quad \mathbf{a} = \dfrac{d\mathbf{v}}{dt} = \dfrac{d^2\mathbf{r}}{dt^2}$.

4. The unit vector $\mathbf{v}/|\mathbf{v}|$ is the direction of motion at time t.

We can express the velocity of a moving particle as the product of its speed and direction:

$$\text{Velocity} = |\mathbf{v}|\left(\frac{\mathbf{v}}{|\mathbf{v}|}\right) = (\text{speed})(\text{direction}).$$

Differentiation Rules

Because the derivatives of vector functions may be computed component by component, the rules for differentiating vector functions have the same form as the rules for differentiating scalar functions.

Differentiation Rules for Vector Functions

Let $\mathbf{u}$ and $\mathbf{v}$ be differentiable vector functions of t, $\mathbf{C}$ a constant vector, c any scalar, and f any differentiable scalar function.

1. *Constant Function Rule:* $\dfrac{d}{dt}\mathbf{C} = \mathbf{0}$

2. *Scalar Multiple Rules:* $\dfrac{d}{dt}[c\mathbf{u}(t)] = c\mathbf{u}'(t)$

   $\dfrac{d}{dt}[f(t)\mathbf{u}(t)] = f'(t)\mathbf{u}(t) + f(t)\mathbf{u}'(t)$

3. *Sum Rule:* $\dfrac{d}{dt}[\mathbf{u}(t) + \mathbf{v}(t)] = \mathbf{u}'(t) + \mathbf{v}'(t)$

4. *Difference Rule:* $\dfrac{d}{dt}[\mathbf{u}(t) - \mathbf{v}(t)] = \mathbf{u}'(t) - \mathbf{v}'(t)$

5. *Dot Product Rule:* $\dfrac{d}{dt}[\mathbf{u}(t) \cdot \mathbf{v}(t)] = \mathbf{u}'(t) \cdot \mathbf{v}(t) + \mathbf{u}(t) \cdot \mathbf{v}'(t)$

6. *Cross Product Rule:* $\dfrac{d}{dt}[\mathbf{u}(t) \times \mathbf{v}(t)] = \mathbf{u}'(t) \times \mathbf{v}(t) + \mathbf{u}(t) \times \mathbf{v}'(t)$

7. *Chain Rule:* $\dfrac{d}{dt}[\mathbf{u}(f(t))] = f'(t)\mathbf{u}'(f(t))$

When you use the Cross Product Rule, remember to preserve the order of the factors. If $\mathbf{u}$ comes first on the left side of the equation, it must also come first on the right or the signs will be wrong.

We will prove the product rules and Chain Rule but leave the rules for constants, scalar multiples, sums, and differences as exercises.

Proof of the Dot Product Rule   Suppose that

$$\mathbf{u} = u_1(t)\mathbf{i} + u_2(t)\mathbf{j} + u_3(t)\mathbf{k}$$

and

$$\mathbf{v} = v_1(t)\mathbf{i} + v_2(t)\mathbf{j} + v_3(t)\mathbf{k}.$$

Then

$$\frac{d}{dt}(\mathbf{u} \cdot \mathbf{v}) = \frac{d}{dt}(u_1 v_1 + u_2 v_2 + u_3 v_3)$$

$$= \underbrace{u_1' v_1 + u_2' v_2 + u_3' v_3}_{\mathbf{u}' \cdot \mathbf{v}} + \underbrace{u_1 v_1' + u_2 v_2' + u_3 v_3'}_{\mathbf{u} \cdot \mathbf{v}'}.$$

∎

Proof of the Cross Product Rule   We model the proof after the proof of the Product Rule for scalar functions. According to the definition of derivative,

$$\frac{d}{dt}(\mathbf{u} \times \mathbf{v}) = \lim_{h \to 0} \frac{\mathbf{u}(t+h) \times \mathbf{v}(t+h) - \mathbf{u}(t) \times \mathbf{v}(t)}{h}.$$

To change this fraction into an equivalent one that contains the difference quotients for the derivatives of $\mathbf{u}$ and $\mathbf{v}$, we subtract and add $\mathbf{u}(t) \times \mathbf{v}(t + h)$ in the numerator. Then

$$\frac{d}{dt}(\mathbf{u} \times \mathbf{v})$$

$$= \lim_{h \to 0} \frac{\mathbf{u}(t + h) \times \mathbf{v}(t + h) - \mathbf{u}(t) \times \mathbf{v}(t + h) + \mathbf{u}(t) \times \mathbf{v}(t + h) - \mathbf{u}(t) \times \mathbf{v}(t)}{h}$$

$$= \lim_{h \to 0} \left[\frac{\mathbf{u}(t + h) - \mathbf{u}(t)}{h} \times \mathbf{v}(t + h) + \mathbf{u}(t) \times \frac{\mathbf{v}(t + h) - \mathbf{v}(t)}{h} \right]$$

$$= \lim_{h \to 0} \frac{\mathbf{u}(t + h) - \mathbf{u}(t)}{h} \times \lim_{h \to 0} \mathbf{v}(t + h) + \lim_{h \to 0} \mathbf{u}(t) \times \lim_{h \to 0} \frac{\mathbf{v}(t + h) - \mathbf{v}(t)}{h}.$$

The last of these equalities holds because the limit of the cross product of two vector functions is the cross product of their limits if the latter exist (Exercise 32). As h approaches zero, $\mathbf{v}(t + h)$ approaches $\mathbf{v}(t)$ because $\mathbf{v}$, being differentiable at t, is continuous at t (Exercise 33). The two fractions approach the values of $d\mathbf{u}/dt$ and $d\mathbf{v}/dt$ at t. In short,

$$\frac{d}{dt}(\mathbf{u} \times \mathbf{v}) = \frac{d\mathbf{u}}{dt} \times \mathbf{v} + \mathbf{u} \times \frac{d\mathbf{v}}{dt}. \qquad \blacksquare$$

As an algebraic convenience, we sometimes write the product of a scalar c and a vector $\mathbf{v}$ as $\mathbf{v}c$ instead of $c\mathbf{v}$. This permits us, for instance, to write the Chain Rule in a familiar form:

$$\frac{d\mathbf{u}}{dt} = \frac{d\mathbf{u}}{ds}\frac{ds}{dt},$$

where $s = f(t)$.

Proof of the Chain Rule Suppose that $\mathbf{u}(s) = a(s)\mathbf{i} + b(s)\mathbf{j} + c(s)\mathbf{k}$ is a differentiable vector function of s and that $s = f(t)$ is a differentiable scalar function of t. Then a, b, and c are differentiable functions of t, and the Chain Rule for differentiable real-valued functions gives

$$\frac{d}{dt}[\mathbf{u}(s)] = \frac{da}{dt}\mathbf{i} + \frac{db}{dt}\mathbf{j} + \frac{dc}{dt}\mathbf{k}$$

$$= \frac{da}{ds}\frac{ds}{dt}\mathbf{i} + \frac{db}{ds}\frac{ds}{dt}\mathbf{j} + \frac{dc}{ds}\frac{ds}{dt}\mathbf{k}$$

$$= \frac{ds}{dt}\left(\frac{da}{ds}\mathbf{i} + \frac{db}{ds}\mathbf{j} + \frac{dc}{ds}\mathbf{k} \right)$$

$$= \frac{ds}{dt}\frac{d\mathbf{u}}{ds}$$

$$= f'(t)\mathbf{u}'(f(t)). \qquad \qquad s = f(t) \qquad \blacksquare$$

Vector Functions of Constant Length

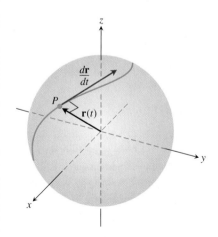

FIGURE 11.7 If a particle moves on a sphere in such a way that its position $\mathbf{r}$ is a differentiable function of time, then $\mathbf{r} \cdot (d\mathbf{r}/dt) = 0$.

When we track a particle moving on a sphere centered at the origin (Figure 11.7), the position vector has a constant length equal to the radius of the sphere. The velocity vector $d\mathbf{r}/dt$, tangent to the path of motion, is tangent to the sphere and hence perpendicular to $\mathbf{r}$. This is always the case for a differentiable vector function of constant length: The vector and its first derivative are orthogonal. By direct calculation,

$$\mathbf{r}(t) \cdot \mathbf{r}(t) = c^2 \qquad |\mathbf{r}(t)| = c \text{ is constant.}$$

$$\frac{d}{dt}[\mathbf{r}(t) \cdot \mathbf{r}(t)] = 0 \qquad \text{Differentiate both sides.}$$

$$\mathbf{r}'(t) \cdot \mathbf{r}(t) + \mathbf{r}(t) \cdot \mathbf{r}'(t) = 0 \qquad \text{Rule 5 with } \mathbf{r}(t) = \mathbf{u}(t) = \mathbf{v}(t)$$

$$2\mathbf{r}'(t) \cdot \mathbf{r}(t) = 0.$$

The vectors $\mathbf{r}'(t)$ and $\mathbf{r}(t)$ are orthogonal because their dot product is 0. In summary,

---

If $\mathbf{r}$ is a differentiable vector function of t of constant length, then

$$\mathbf{r} \cdot \frac{d\mathbf{r}}{dt} = 0. \qquad (4)$$

We will use this observation repeatedly in Section 11.4.

EXERCISES 11.1

In Exercises 1–4, $\mathbf{r}(t)$ is the position of a particle in the xy-plane at time t. Find an equation in x and y whose graph is the path of the particle. Then find the particle's velocity and acceleration vectors at the given value of t.

1. $\mathbf{r}(t) = (t + 1)\mathbf{i} + (t^2 - 1)\mathbf{j}, \quad t = 1$

2. $\mathbf{r}(t) = (t^2 + 1)\mathbf{i} + (2t - 1)\mathbf{j}, \quad t = 1/2$

3. $\mathbf{r}(t) = e^t\mathbf{i} + \frac{2}{9}e^{2t}\mathbf{j}, \quad t = \ln 3$

4. $\mathbf{r}(t) = (\cos 2t)\mathbf{i} + (3 \sin 2t)\mathbf{j}, \quad t = 0$

Exercises 5–8 give the position vectors of particles moving along various curves in the xy-plane. In each case, find the particle's velocity and acceleration vectors at the stated times and sketch them as vectors on the curve.

5. Motion on the circle $x^2 + y^2 = 1$

$$\mathbf{r}(t) = (\sin t)\mathbf{i} + (\cos t)\mathbf{j}; \quad t = \pi/4 \text{ and } \pi/2$$

6. Motion on the circle $x^2 + y^2 = 16$

$$\mathbf{r}(t) = \left(4 \cos \frac{t}{2}\right)\mathbf{i} + \left(4 \sin \frac{t}{2}\right)\mathbf{j}; \quad t = \pi \text{ and } 3\pi/2$$

7. Motion on the cycloid $x = t - \sin t, y = 1 - \cos t$

$$\mathbf{r}(t) = (t - \sin t)\mathbf{i} + (1 - \cos t)\mathbf{j}; \quad t = \pi \text{ and } 3\pi/2$$

8. Motion on the parabola $y = x^2 + 1$

$$\mathbf{r}(t) = t\mathbf{i} + (t^2 + 1)\mathbf{j}; \quad t = -1, 0, \text{ and } 1$$

In Exercises 9–14, $\mathbf{r}(t)$ is the position of a particle in space at time t. Find the particle's velocity and acceleration vectors. Then find the particle's speed and direction of motion at the given value of t. Write the particle's velocity at that time as the product of its speed and direction.

9. $\mathbf{r}(t) = (t + 1)\mathbf{i} + (t^2 - 1)\mathbf{j} + 2t\mathbf{k}, \quad t = 1$

10. $\mathbf{r}(t) = (1 + t)\mathbf{i} + \frac{t^2}{\sqrt{2}}\mathbf{j} + \frac{t^3}{3}\mathbf{k}, \quad t = 1$

11. $\mathbf{r}(t) = (2 \cos t)\mathbf{i} + (3 \sin t)\mathbf{j} + 4t\mathbf{k}, \quad t = \pi/2$

12. $\mathbf{r}(t) = (\sec t)\mathbf{i} + (\tan t)\mathbf{j} + \frac{4}{3}t\mathbf{k}, \quad t = \pi/6$

13. $\mathbf{r}(t) = (2 \ln (t + 1))\mathbf{i} + t^2\mathbf{j} + \frac{t^2}{2}\mathbf{k}, \quad t = 1$

14. $\mathbf{r}(t) = (e^{-t})\mathbf{i} + (2 \cos 3t)\mathbf{j} + (2 \sin 3t)\mathbf{k}, \quad t = 0$

In Exercises 15–18, $\mathbf{r}(t)$ is the position of a particle in space at time t. Find the angle between the velocity and acceleration vectors at time $t = 0$.

15. $\mathbf{r}(t) = (3t + 1)\mathbf{i} + \sqrt{3}t\mathbf{j} + t^2\mathbf{k}$

16. $\mathbf{r}(t) = \left(\frac{\sqrt{2}}{2}t\right)\mathbf{i} + \left(\frac{\sqrt{2}}{2}t - 16t^2\right)\mathbf{j}$

17. $\mathbf{r}(t) = (\ln (t^2 + 1))\mathbf{i} + (\tan^{-1} t)\mathbf{j} + \sqrt{t^2 + 1}\,\mathbf{k}$

18. $\mathbf{r}(t) = \frac{4}{9}(1 + t)^{3/2}\mathbf{i} + \frac{4}{9}(1 - t)^{3/2}\mathbf{j} + \frac{1}{3}t\mathbf{k}$

As mentioned in the text, the **tangent line** to a smooth curve $\mathbf{r}(t) = f(t)\mathbf{i} + g(t)\mathbf{j} + h(t)\mathbf{k}$ at $t = t_0$ is the line that passes through the point $(f(t_0), g(t_0), h(t_0))$ parallel to $\mathbf{v}(t_0)$, the curve's velocity vector at t_0. In Exercises 19–22, find parametric equations for the line that is tangent to the given curve at the given parameter value $t = t_0$.

19. $\mathbf{r}(t) = (\sin t)\mathbf{i} + (t^2 - \cos t)\mathbf{j} + e^t\mathbf{k}, \quad t_0 = 0$

20. $\mathbf{r}(t) = (2 \sin t)\mathbf{i} + (2 \cos t)\mathbf{j} + 5t\mathbf{k}, \quad t_0 = 4\pi$

21. $\mathbf{r}(t) = (a \sin t)\mathbf{i} + (a \cos t)\mathbf{j} + bt\mathbf{k}, \quad t_0 = 2\pi$

22. $\mathbf{r}(t) = (\cos t)\mathbf{i} + (\sin t)\mathbf{j} + (\sin 2t)\mathbf{k}, \quad t_0 = \frac{\pi}{2}$

23. Each of the following equations in parts (a)–(e) describes the motion of a particle having the same path, namely the unit circle $x^2 + y^2 = 1$. Although the path of each particle in parts (a)–(e) is the same, the behavior, or "dynamics," of each particle is different. For each particle, answer the following questions.

   i. Does the particle have constant speed? If so, what is its constant speed?

   ii. Is the particle's acceleration vector always orthogonal to its velocity vector?

   iii. Does the particle move clockwise or counterclockwise around the circle?

   iv. Does the particle begin at the point $(1, 0)$?

  a. $\mathbf{r}(t) = (\cos t)\mathbf{i} + (\sin t)\mathbf{j}, \quad t \geq 0$

  b. $\mathbf{r}(t) = \cos(2t)\mathbf{i} + \sin(2t)\mathbf{j}, \quad t \geq 0$

  c. $\mathbf{r}(t) = \cos(t - \pi/2)\mathbf{i} + \sin(t - \pi/2)\mathbf{j}, \quad t \geq 0$

  d. $\mathbf{r}(t) = (\cos t)\mathbf{i} - (\sin t)\mathbf{j}, \quad t \geq 0$

  e. $\mathbf{r}(t) = \cos(t^2)\mathbf{i} + \sin(t^2)\mathbf{j}, \quad t \geq 0$

24. Show that the vector-valued function

$$\mathbf{r}(t) = (2\mathbf{i} + 2\mathbf{j} + \mathbf{k})$$
$$+ \cos t \left(\frac{1}{\sqrt{2}}\mathbf{i} - \frac{1}{\sqrt{2}}\mathbf{j} \right) + \sin t \left(\frac{1}{\sqrt{3}}\mathbf{i} + \frac{1}{\sqrt{3}}\mathbf{j} + \frac{1}{\sqrt{3}}\mathbf{k} \right)$$

describes the motion of a particle moving in the circle of radius 1 centered at the point $(2, 2, 1)$ and lying in the plane $x + y - 2z = 2$.

25. Motion along a parabola   A particle moves along the top of the parabola $y^2 = 2x$ from left to right at a constant speed of 5 units per second. Find the velocity of the particle as it moves through the point $(2, 2)$.

26. Motion along a cycloid   A particle moves in the xy-plane in such a way that its position at time t is

$$\mathbf{r}(t) = (t - \sin t)\mathbf{i} + (1 - \cos t)\mathbf{j}.$$

[T] **a.** Graph $\mathbf{r}(t)$. The resulting curve is a cycloid.

  b. Find the maximum and minimum values of $|\mathbf{v}|$ and $|\mathbf{a}|$. (*Hint:* Find the extreme values of $|\mathbf{v}|^2$ and $|\mathbf{a}|^2$ first and take square roots later.)

27. Let $\mathbf{v}$ be a differentiable vector function of t. Show that if $\mathbf{v} \cdot (d\mathbf{v}/dt) = 0$ for all t, then $|\mathbf{v}|$ is constant.

28. Derivatives of triple scalar products

  a. Show that if $\mathbf{u}$, $\mathbf{v}$, and $\mathbf{w}$ are differentiable vector functions of t, then

$$\frac{d}{dt}(\mathbf{u} \cdot \mathbf{v} \times \mathbf{w}) = \frac{d\mathbf{u}}{dt} \cdot \mathbf{v} \times \mathbf{w} + \mathbf{u} \cdot \frac{d\mathbf{v}}{dt} \times \mathbf{w} + \mathbf{u} \cdot \mathbf{v} \times \frac{d\mathbf{w}}{dt}.$$

  b. Show that

$$\frac{d}{dt}\left(\mathbf{r} \cdot \frac{d\mathbf{r}}{dt} \times \frac{d^2\mathbf{r}}{dt^2} \right) = \mathbf{r} \cdot \left(\frac{d\mathbf{r}}{dt} \times \frac{d^3\mathbf{r}}{dt^3} \right).$$

(*Hint:* Differentiate on the left and look for vectors whose products are zero.)

29. Scalar Multiple Rules

  a. Prove that if $\mathbf{u}$ is a differentiable function of t and c is any real number, then

$$\frac{d(c\mathbf{u})}{dt} = c\frac{d\mathbf{u}}{dt}.$$

  b. Prove that if $\mathbf{u}$ is a differentiable function of t and f is a differentiable scalar function of t, then

$$\frac{d}{dt}(f\mathbf{u}) = \frac{df}{dt}\mathbf{u} + f\frac{d\mathbf{u}}{dt}.$$

30. Sum and Difference Rules   Prove that if $\mathbf{u}$ and $\mathbf{v}$ are differentiable functions of t, then

$$\frac{d}{dt}(\mathbf{u} + \mathbf{v}) = \frac{d\mathbf{u}}{dt} + \frac{d\mathbf{v}}{dt}$$

and

$$\frac{d}{dt}(\mathbf{u} - \mathbf{v}) = \frac{d\mathbf{u}}{dt} - \frac{d\mathbf{v}}{dt}.$$

31. Component Test for Continuity at a Point   Show that the vector function $\mathbf{r}$ defined by $\mathbf{r}(t) = f(t)\mathbf{i} + g(t)\mathbf{j} + h(t)\mathbf{k}$ is continuous at $t = t_0$ if and only if f, g, and h are continuous at t_0.

32. Limits of cross products of vector functions   Suppose that $\mathbf{r}_1(t) = f_1(t)\mathbf{i} + f_2(t)\mathbf{j} + f_3(t)\mathbf{k}$, $\mathbf{r}_2(t) = g_1(t)\mathbf{i} + g_2(t)\mathbf{j} + g_3(t)\mathbf{k}$, $\lim_{t \to t_0} \mathbf{r}_1(t) = \mathbf{A}$, and $\lim_{t \to t_0} \mathbf{r}_2(t) = \mathbf{B}$. Use the determinant formula for cross products and the Limit Product Rule for scalar functions to show that

$$\lim_{t \to t_0}(\mathbf{r}_1(t) \times \mathbf{r}_2(t)) = \mathbf{A} \times \mathbf{B}$$

33. Differentiable vector functions are continuous   Show that if $\mathbf{r}(t) = f(t)\mathbf{i} + g(t)\mathbf{j} + h(t)\mathbf{k}$ is differentiable at $t = t_0$, then it is continuous at t_0 as well.

34. Constant Function Rule   Prove that if $\mathbf{u}$ is the vector function with the constant value $\mathbf{C}$, then $d\mathbf{u}/dt = \mathbf{0}$.

COMPUTER EXPLORATIONS

Use a CAS to perform the following steps in exercises 35–38.

  a. Plot the space curve traced out by the position vector $\mathbf{r}$.

  b. Find the components of the velocity vector $d\mathbf{r}/dt$.

  c. Evaluate $d\mathbf{r}/dt$ at the given point t_0 and determine the equation of the tangent line to the curve at $\mathbf{r}(t_0)$.

  d. Plot the tangent line together with the curve over the given interval.

35. $\mathbf{r}(t) = (\sin t - t \cos t)\mathbf{i} + (\cos t + t \sin t)\mathbf{j} + t^2\mathbf{k}$, $\quad 0 \leq t \leq 6\pi, \quad t_0 = 3\pi/2$

36. $\mathbf{r}(t) = \sqrt{2}t\mathbf{i} + e^t\mathbf{j} + e^{-t}\mathbf{k}, \quad -2 \leq t \leq 3, \quad t_0 = 1$

37. $\mathbf{r}(t) = (\sin 2t)\mathbf{i} + (\ln(1 + t))\mathbf{j} + t\mathbf{k}, \quad 0 \leq t \leq 4\pi,$ $\quad t_0 = \pi/4$

38. $\mathbf{r}(t) = (\ln(t^2 + 2))\mathbf{i} + (\tan^{-1} 3t)\mathbf{j} + \sqrt{t^2 + 1}\ \mathbf{k},$ $\quad -3 \leq t \leq 5, \quad t_0 = 3$

In Exercises 39 and 40, you will explore graphically the behavior of the helix

$$\mathbf{r}(t) = (\cos at)\mathbf{i} + (\sin at)\mathbf{j} + bt\mathbf{k}$$

as you change the values of the constants a and b. Use a CAS to perform the steps in each exercise.

39. Set $b = 1$. Plot the helix $\mathbf{r}(t)$ together with the tangent line to the curve at $t = 3\pi/2$ for $a = 1, 2, 4$, and 6 over the interval $0 \le t \le 4\pi$. Describe in your own words what happens to the graph of the helix and the position of the tangent line as a increases through these positive values.

40. Set $a = 1$. Plot the helix $\mathbf{r}(t)$ together with the tangent line to the curve at $t = 3\pi/2$ for $b = 1/4, 1/2, 2$, and 4 over the interval $0 \le t \le 4\pi$. Describe in your own words what happens to the graph of the helix and the position of the tangent line as b increases through these positive values.

11.2 Integrals of Vector Functions

In this section we investigate integrals of vector functions and their application to motion along a path in space or in the plane.

Integrals of Vector Functions

A differentiable vector function $\mathbf{R}(t)$ is an **antiderivative** of a vector function $\mathbf{r}(t)$ on an interval I if $d\mathbf{R}/dt = \mathbf{r}$ at each point of I. If $\mathbf{R}$ is an antiderivative of $\mathbf{r}$ on I, it can be shown, working one component at a time, that every antiderivative of $\mathbf{r}$ on I has the form $\mathbf{R} + \mathbf{C}$ for some constant vector $\mathbf{C}$ (Exercise 29). The set of all antiderivatives of $\mathbf{r}$ on I is the **indefinite integral** of $\mathbf{r}$ on I.

> **DEFINITION**   The **indefinite integral** of $\mathbf{r}$ with respect to t is the set of all antiderivatives of $\mathbf{r}$, denoted by $\int \mathbf{r}(t)\, dt$. If $\mathbf{R}$ is any antiderivative of $\mathbf{r}$, then
>
> $$\int \mathbf{r}(t)\, dt = \mathbf{R}(t) + \mathbf{C}.$$

The usual arithmetic rules for indefinite integrals apply.

EXAMPLE 1

$$\int ((\cos t)\mathbf{i} + \mathbf{j} - 2t\mathbf{k})\, dt = \left(\int \cos t\, dt \right)\mathbf{i} + \left(\int dt \right)\mathbf{j} - \left(\int 2t\, dt \right)\mathbf{k} \qquad (1)$$

$$= (\sin t + C_1)\mathbf{i} + (t + C_2)\mathbf{j} - (t^2 + C_3)\mathbf{k} \qquad (2)$$

$$= (\sin t)\mathbf{i} + t\mathbf{j} - t^2\mathbf{k} + \mathbf{C} \qquad \mathbf{C} = C_1\mathbf{i} + C_2\mathbf{j} - C_3\mathbf{k}$$

As in the integration of scalar functions, we recommend that you skip the steps in Equations (1) and (2) and go directly to the final form. Find an antiderivative for each component and add a constant vector at the end. ∎

Definite integrals of vector functions are best defined in terms of components.

> **DEFINITION**   If the components of $\mathbf{r}(t) = f(t)\mathbf{i} + g(t)\mathbf{j} + h(t)\mathbf{k}$ are integrable over $[a, b]$, then so is $\mathbf{r}$, and the **definite integral** of $\mathbf{r}$ from a to b is
>
> $$\int_a^b \mathbf{r}(t)\, dt = \left(\int_a^b f(t)\, dt \right)\mathbf{i} + \left(\int_a^b g(t)\, dt \right)\mathbf{j} + \left(\int_a^b h(t)\, dt \right)\mathbf{k}.$$

EXAMPLE 2

$$\int_0^\pi ((\cos t)\mathbf{i} + \mathbf{j} - 2t\mathbf{k})\,dt = \left(\int_0^\pi \cos t\,dt\right)\mathbf{i} + \left(\int_0^\pi dt\right)\mathbf{j} - \left(\int_0^\pi 2t\,dt\right)\mathbf{k}$$

$$= \left[\sin t\right]_0^\pi \mathbf{i} + \left[t\right]_0^\pi \mathbf{j} - \left[t^2\right]_0^\pi \mathbf{k}$$

$$= [0 - 0]\mathbf{i} + [\pi - 0]\mathbf{j} - [\pi^2 - 0^2]\mathbf{k}$$

$$= \pi\mathbf{j} - \pi^2\mathbf{k} \qquad\blacksquare$$

The Fundamental Theorem of Calculus for continuous vector functions says that

$$\int_a^b \mathbf{r}(t)\,dt = \mathbf{R}(t)\Big]_a^b = \mathbf{R}(b) - \mathbf{R}(a)$$

where $\mathbf{R}$ is any antiderivative of $\mathbf{r}$, so that $\mathbf{R}'(t) = \mathbf{r}(t)$ (Exercise 30).

EXAMPLE 3   Suppose we do not know the path of a hang glider, but only its acceleration vector $\mathbf{a}(t) = -(3\cos t)\mathbf{i} - (3\sin t)\mathbf{j} + 2\mathbf{k}$. We also know that initially (at time $t = 0$), the glider departed from the point (3, 0, 0) with velocity $\mathbf{v}(0) = 3\mathbf{j}$. Find the glider's position as a function of t.

Solution   Our goal is to find $\mathbf{r}(t)$ knowing

The differential equation: $\mathbf{a} = \dfrac{d^2\mathbf{r}}{dt^2} = -(3\cos t)\mathbf{i} - (3\sin t)\mathbf{j} + 2\mathbf{k}$

The initial conditions: $\mathbf{v}(0) = 3\mathbf{j}$ and $\mathbf{r}(0) = 3\mathbf{i} + 0\mathbf{j} + 0\mathbf{k}$.

Integrating both sides of the differential equation with respect to t gives

$$\mathbf{v}(t) = -(3\sin t)\mathbf{i} + (3\cos t)\mathbf{j} + 2t\mathbf{k} + \mathbf{C}_1.$$

We use $\mathbf{v}(0) = 3\mathbf{j}$ to find $\mathbf{C}_1$:

$$3\mathbf{j} = -(3\sin 0)\mathbf{i} + (3\cos 0)\mathbf{j} + (0)\mathbf{k} + \mathbf{C}_1$$
$$3\mathbf{j} = 3\mathbf{j} + \mathbf{C}_1$$
$$\mathbf{C}_1 = \mathbf{0}.$$

The glider's velocity as a function of time is

$$\frac{d\mathbf{r}}{dt} = \mathbf{v}(t) = -(3\sin t)\mathbf{i} + (3\cos t)\mathbf{j} + 2t\mathbf{k}.$$

Integrating both sides of this last differential equation gives

$$\mathbf{r}(t) = (3\cos t)\mathbf{i} + (3\sin t)\mathbf{j} + t^2\mathbf{k} + \mathbf{C}_2.$$

We then use the initial condition $\mathbf{r}(0) = 3\mathbf{i}$ to find $\mathbf{C}_2$:

$$3\mathbf{i} = (3\cos 0)\mathbf{i} + (3\sin 0)\mathbf{j} + (0^2)\mathbf{k} + \mathbf{C}_2$$
$$3\mathbf{i} = 3\mathbf{i} + (0)\mathbf{j} + (0)\mathbf{k} + \mathbf{C}_2$$
$$\mathbf{C}_2 = \mathbf{0}.$$

The glider's position as a function of t is

$$\mathbf{r}(t) = (3\cos t)\mathbf{i} + (3\sin t)\mathbf{j} + t^2\mathbf{k}.$$

This is the path of the glider shown in Figure 11.8.

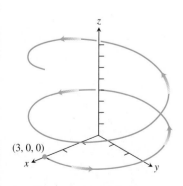

FIGURE 11.8 The path of a hang glider with position vector $\mathbf{r}(t) = (3\cos t)\mathbf{i} + (3\sin t)\mathbf{j} + t^2\mathbf{k}$ (Example 3).

Note: It turned out in this example that both of the constant vectors of integration, $\mathbf{C}_1$ and $\mathbf{C}_2$, are $\mathbf{0}$. Exercises 11 and 12 give examples for which the constant vectors of integration are not $\mathbf{0}$. ∎

The Vector and Parametric Equations for Ideal Projectile Motion

To derive equations for projectile motion, we assume that the projectile behaves like a particle moving in a vertical coordinate plane and that the only force acting on the projectile during its flight is the constant force of gravity, which always points straight down. (In practice, none of these assumptions really holds. The ground moves beneath the projectile as the earth turns, the air creates a frictional force that varies with the projectile's speed and altitude, and the force of gravity changes as the projectile moves along.)

We assume that the projectile is launched from the origin at time $t = 0$ into the first quadrant with an initial velocity $\mathbf{v}_0$ (Figure 11.9). If $\mathbf{v}_0$ makes an angle α with the horizontal, then

$$\mathbf{v}_0 = (|\mathbf{v}_0|\cos\alpha)\mathbf{i} + (|\mathbf{v}_0|\sin\alpha)\mathbf{j}.$$

If we use the simpler notation v_0 for the initial speed $|\mathbf{v}_0|$, then

$$\mathbf{v}_0 = (v_0\cos\alpha)\mathbf{i} + (v_0\sin\alpha)\mathbf{j}. \tag{3}$$

The projectile's initial position is

$$\mathbf{r}_0 = 0\mathbf{i} + 0\mathbf{j} = \mathbf{0}. \tag{4}$$

Newton's second law of motion says that the force acting on the projectile is equal to the projectile's mass m times its acceleration, or $m(d^2\mathbf{r}/dt^2)$ if $\mathbf{r}$ is the projectile's position vector and t is time. If the force is solely the gravitational force $-mg\mathbf{j}$, then

$$m\frac{d^2\mathbf{r}}{dt^2} = -mg\mathbf{j} \quad \text{and} \quad \frac{d^2\mathbf{r}}{dt^2} = -g\mathbf{j}.$$

We find $\mathbf{r}$ as a function of t by solving the following initial value problem.

Differential equation: $\quad \dfrac{d^2\mathbf{r}}{dt^2} = -g\mathbf{j}$

Initial conditions: $\quad \mathbf{r} = \mathbf{r}_0 \quad$ and $\quad \dfrac{d\mathbf{r}}{dt} = \mathbf{v}_0 \quad$ when $t = 0$

The first integration gives

$$\frac{d\mathbf{r}}{dt} = -(gt)\mathbf{j} + \mathbf{v}_0.$$

A second integration gives

$$\mathbf{r} = -\frac{1}{2}gt^2\mathbf{j} + \mathbf{v}_0 t + \mathbf{r}_0.$$

Substituting the values of $\mathbf{v}_0$ and $\mathbf{r}_0$ from Equations (3) and (4) gives

$$\mathbf{r} = -\frac{1}{2}gt^2\mathbf{j} + \underbrace{(v_0\cos\alpha)t\mathbf{i} + (v_0\sin\alpha)t\mathbf{j}}_{\mathbf{v}_0 t} + \mathbf{0}.$$

Collecting terms, we have

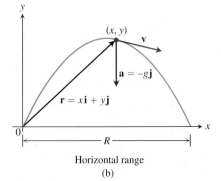

FIGURE 11.9 (a) Position, velocity, acceleration, and launch angle at $t = 0$. (b) Position, velocity, and acceleration at a later time t.

Ideal Projectile Motion Equation

$$\mathbf{r} = (v_0\cos\alpha)t\mathbf{i} + \left((v_0\sin\alpha)t - \frac{1}{2}gt^2\right)\mathbf{j}. \tag{5}$$

Equation (5) is the *vector equation* for ideal projectile motion. The angle α is the projectile's **launch angle (firing angle, angle of elevation)**, and v_0, as we said before, is the projectile's **initial speed**. The components of **r** give the parametric equations

$$x = (v_0 \cos \alpha)t \quad \text{and} \quad y = (v_0 \sin \alpha)t - \frac{1}{2}gt^2, \tag{6}$$

where x is the distance downrange and y is the height of the projectile at time $t \geq 0$.

EXAMPLE 4   A projectile is fired from the origin over horizontal ground at an initial speed of 500 m/sec and a launch angle of 60°. Where will the projectile be 10 sec later?

Solution   We use Equation (5) with $v_0 = 500$, $\alpha = 60°$, $g = 9.8$, and $t = 10$ to find the projectile's components 10 sec after firing.

$$\mathbf{r} = (v_0 \cos \alpha)t\mathbf{i} + \left((v_0 \sin \alpha)t - \frac{1}{2}gt^2 \right)\mathbf{j}$$

$$= (500)\left(\frac{1}{2}\right)(10)\mathbf{i} + \left((500)\left(\frac{\sqrt{3}}{2}\right)10 - \left(\frac{1}{2}\right)(9.8)(100) \right)\mathbf{j}$$

$$\approx 2500\mathbf{i} + 3840\mathbf{j}.$$

Ten seconds after firing, the projectile is about 3840 m in the air and 2500 m downrange. ∎

Ideal projectiles move along parabolas, as we can see from Equations (6). If we substitute $t = x/(v_0 \cos \alpha)$ from the first equation into the second, we obtain the Cartesian-coordinate equation

$$y = -\left(\frac{g}{2v_0^2 \cos^2 \alpha} \right)x^2 + (\tan \alpha)x.$$

This equation has the form $y = ax^2 + bx$, so its graph is a parabola.

A projectile reaches its highest point when its vertical velocity component is zero. When fired over horizontal ground, the projectile lands when its vertical component equals zero in Equation (5), and the **range R** is the distance from the origin to the point of impact. We summarize the results here, which you are asked to verify in Exercise 23.

Height, Flight Time, and Range for Ideal Projectile Motion

For ideal projectile motion when an object is launched from the origin over a horizontal surface with initial speed v_0 and launch angle α:

$$\text{Maximum height:} \quad y_{\text{max}} = \frac{(v_0 \sin \alpha)^2}{2g}$$

$$\text{Flight time:} \quad t = \frac{2v_0 \sin \alpha}{g}$$

$$\text{Range:} \quad R = \frac{v_0^2}{g} \sin 2\alpha.$$

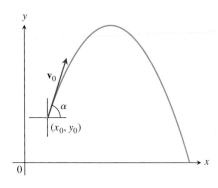

FIGURE 11.10   The path of a projectile fired from (x_0, y_0) with an initial velocity $\mathbf{v}_0$ at an angle of α degrees with the horizontal.

If we fire our ideal projectile from the point (x_0, y_0) instead of the origin (Figure 11.10), the position vector for the path of motion is

$$\mathbf{r} = (x_0 + (v_0 \cos \alpha)t)\mathbf{i} + \left(y_0 + (v_0 \sin \alpha)t - \frac{1}{2}gt^2\right)\mathbf{j}, \qquad (7)$$

as you are asked to show in Exercise 25.

EXERCISES 11.2

Evaluate the integrals in Exercises 1–6.

1. $\displaystyle\int_0^1 [t^3\mathbf{i} + 7\mathbf{j} + (t + 1)\mathbf{k}]\, dt$

2. $\displaystyle\int_1^2 \left[(6 - 6t)\mathbf{i} + 3\sqrt{t}\,\mathbf{j} + \left(\frac{4}{t^2}\right)\mathbf{k}\right] dt$

3. $\displaystyle\int_{-\pi/4}^{\pi/4} [(\sin t)\mathbf{i} + (1 + \cos t)\mathbf{j} + (\sec^2 t)\mathbf{k}]\, dt$

4. $\displaystyle\int_0^{\pi/3} [(\sec t \tan t)\mathbf{i} + (\tan t)\mathbf{j} + (2 \sin t \cos t)\mathbf{k}]\, dt$

5. $\displaystyle\int_1^4 \left[\frac{1}{t}\mathbf{i} + \frac{1}{5 - t}\mathbf{j} + \frac{1}{2t}\mathbf{k}\right] dt$

6. $\displaystyle\int_0^1 \left[\frac{2}{\sqrt{1 - t^2}}\mathbf{i} + \frac{\sqrt{3}}{1 + t^2}\mathbf{k}\right] dt$

Solve the initial value problems in Exercises 7–12 for $\mathbf{r}$ as a vector function of t.

7. Differential equation: $\dfrac{d\mathbf{r}}{dt} = -t\mathbf{i} - t\mathbf{j} - t\mathbf{k}$

   Initial condition: $\mathbf{r}(0) = \mathbf{i} + 2\mathbf{j} + 3\mathbf{k}$

8. Differential equation: $\dfrac{d\mathbf{r}}{dt} = (180t)\mathbf{i} + (180t - 16t^2)\mathbf{j}$

   Initial condition: $\mathbf{r}(0) = 100\mathbf{j}$

9. Differential equation: $\dfrac{d\mathbf{r}}{dt} = \dfrac{3}{2}(t + 1)^{1/2}\mathbf{i} + e^{-t}\mathbf{j} + \dfrac{1}{t + 1}\mathbf{k}$

   Initial condition: $\mathbf{r}(0) = \mathbf{k}$

10. Differential equation: $\dfrac{d\mathbf{r}}{dt} = (t^3 + 4t)\mathbf{i} + t\mathbf{j} + 2t^2\mathbf{k}$

   Initial condition: $\mathbf{r}(0) = \mathbf{i} + \mathbf{j}$

11. Differential equation: $\dfrac{d^2\mathbf{r}}{dt^2} = -32\mathbf{k}$

   Initial conditions: $\mathbf{r}(0) = 100\mathbf{k}$ and

   $\left.\dfrac{d\mathbf{r}}{dt}\right|_{t=0} = 8\mathbf{i} + 8\mathbf{j}$

12. Differential equation: $\dfrac{d^2\mathbf{r}}{dt^2} = -(\mathbf{i} + \mathbf{j} + \mathbf{k})$

   Initial conditions: $\mathbf{r}(0) = 10\mathbf{i} + 10\mathbf{j} + 10\mathbf{k}$ and

   $\left.\dfrac{d\mathbf{r}}{dt}\right|_{t=0} = \mathbf{0}$

13. At time $t = 0$, a particle is located at the point $(1, 2, 3)$. It travels in a straight line to the point $(4, 1, 4)$, has speed 2 at $(1, 2, 3)$ and constant acceleration $3\mathbf{i} - \mathbf{j} + \mathbf{k}$. Find an equation for the position vector $\mathbf{r}(t)$ of the particle at time t.

14. A particle traveling in a straight line is located at the point $(1, -1, 2)$ and has speed 2 at time $t = 0$. The particle moves toward the point $(3, 0, 3)$ with constant acceleration $2\mathbf{i} + \mathbf{j} + \mathbf{k}$. Find its position vector $\mathbf{r}(t)$ at time t.

Projectile flights in the following exercises are to be treated as ideal unless stated otherwise. All launch angles are assumed to be measured from the horizontal. All projectiles are assumed to be launched from the origin over a horizontal surface unless stated otherwise.

15. Travel time   A projectile is fired at a speed of 840 m/sec at an angle of 60°. How long will it take to get 21 km downrange?

16. Finding muzzle speed   Find the muzzle speed of a gun whose maximum range is 24.5 km.

17. Flight time and height   A projectile is fired with an initial speed of 500 m/sec at an angle of elevation of 45°.

   a. When and how far away will the projectile strike?

b. How high overhead will the projectile be when it is 5 km downrange?

c. What is the greatest height reached by the projectile?

18. **Throwing a baseball** A baseball is thrown from the stands 32 ft above the field at an angle of 30° up from the horizontal. When and how far away will the ball strike the ground if its initial speed is 32 ft/sec?

19. **Firing golf balls** A spring gun at ground level fires a golf ball at an angle of 45°. The ball lands 10 m away.

a. What was the ball's initial speed?

b. For the same initial speed, find the two firing angles that make the range 6 m.

20. **Beaming electrons** An electron in a TV tube is beamed horizontally at a speed of 5×10^6 m/sec toward the face of the tube 40 cm away. About how far will the electron drop before it hits?

21. **Equal-range firing angles** What two angles of elevation will enable a projectile to reach a target 16 km downrange on the same level as the gun if the projectile's initial speed is 400 m/sec?

22. **Range and height versus speed**

a. Show that doubling a projectile's initial speed at a given launch angle multiplies its range by 4.

b. By about what percentage should you increase the initial speed to double the height and range?

23. Verify the results given in the text (following Example 4) for the maximum height, flight time, and range for ideal projectile motion.

24. **Colliding marbles** The figure shows an experiment with two marbles. Marble A was launched toward marble B with launch angle α and initial speed v_0. At the same instant, marble B was released to fall from rest at $R \tan \alpha$ units directly above a spot R units downrange from A. The marbles were found to collide regardless of the value of v_0. Was this mere coincidence, or must this happen? Give reasons for your answer.

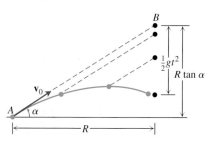

25. **Firing from (x_0, y_0)** Derive the equations

$$x = x_0 + (v_0 \cos \alpha)t,$$

$$y = y_0 + (v_0 \sin \alpha)t - \frac{1}{2}gt^2,$$

(see Equation (7) in the text) by solving the following initial value problem for a vector $\mathbf{r}$ in the plane.

Differential equation: $\quad \dfrac{d^2\mathbf{r}}{dt^2} = -g\mathbf{j}$

Initial conditions: $\quad \mathbf{r}(0) = x_0\mathbf{i} + y_0\mathbf{j}$

$$\frac{d\mathbf{r}}{dt}(0) = (v_0 \cos \alpha)\mathbf{i} + (v_0 \sin \alpha)\mathbf{j}$$

26. **Where trajectories crest** For a projectile fired from the ground at launch angle α with initial speed v_0, consider α as a variable and v_0 as a fixed constant. For each α, $0 < \alpha < \pi/2$, we obtain a parabolic trajectory as shown in the accompanying figure. Show that the points in the plane that give the maximum heights of these parabolic trajectories all lie on the ellipse

$$x^2 + 4\left(y - \frac{v_0^2}{4g}\right)^2 = \frac{v_0^4}{4g^2},$$

where $x \geq 0$.

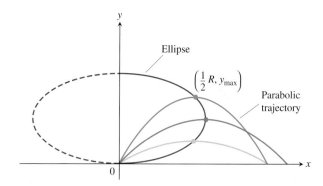

27. Establish the following properties of integrable vector functions.

a. The *Constant Scalar Multiple Rule:*

$$\int_a^b k\mathbf{r}(t)\, dt = k \int_a^b \mathbf{r}(t)\, dt \quad \text{(any scalar } k\text{)}$$

The *Rule for Negatives,*

$$\int_a^b (-\mathbf{r}(t))\, dt = -\int_a^b \mathbf{r}(t)\, dt,$$

is obtained by taking $k = -1$.

b. The *Sum and Difference Rules:*

$$\int_a^b (\mathbf{r}_1(t) \pm \mathbf{r}_2(t))\, dt = \int_a^b \mathbf{r}_1(t)\, dt \pm \int_a^b \mathbf{r}_2(t)\, dt$$

c. The *Constant Vector Multiple Rules:*

$$\int_a^b \mathbf{C} \cdot \mathbf{r}(t)\, dt = \mathbf{C} \cdot \int_a^b \mathbf{r}(t)\, dt \quad \text{(any constant vector } \mathbf{C}\text{)}$$

and

$$\int_a^b \mathbf{C} \times \mathbf{r}(t)\, dt = \mathbf{C} \times \int_a^b \mathbf{r}(t)\, dt \quad \text{(any constant vector } \mathbf{C}\text{)}$$

28. Products of scalar and vector functions Suppose that the scalar function $u(t)$ and the vector function $\mathbf{r}(t)$ are both defined for $a \leq t \leq b$.

a. Show that $u\mathbf{r}$ is continuous on $[a, b]$ if u and $\mathbf{r}$ are continuous on $[a, b]$.

b. If u and $\mathbf{r}$ are both differentiable on $[a, b]$, show that $u\mathbf{r}$ is differentiable on $[a, b]$ and that

$$\frac{d}{dt}(u\mathbf{r}) = u\frac{d\mathbf{r}}{dt} + \mathbf{r}\frac{du}{dt}.$$

29. Antiderivatives of vector functions

a. Use Corollary 2 of the Mean Value Theorem for scalar functions to show that if two vector functions $\mathbf{R}_1(t)$ and $\mathbf{R}_2(t)$ have identical derivatives on an interval I, then the functions differ by a constant vector value throughout I.

b. Use the result in part (a) to show that if $\mathbf{R}(t)$ is any antiderivative of $\mathbf{r}(t)$ on I, then any other antiderivative of $\mathbf{r}$ on I equals $\mathbf{R}(t) + \mathbf{C}$ for some constant vector $\mathbf{C}$.

30. The Fundamental Theorem of Calculus The Fundamental Theorem of Calculus for scalar functions of a real variable holds for vector functions of a real variable as well. Prove this by using the theorem for scalar functions to show first that if a vector function $\mathbf{r}(t)$ is continuous for $a \leq t \leq b$, then

$$\frac{d}{dt}\int_a^t \mathbf{r}(\tau)\, d\tau = \mathbf{r}(t)$$

at every point t of (a, b). Then use the conclusion in part (b) of Exercise 29 to show that if $\mathbf{R}$ is any antiderivative of $\mathbf{r}$ on $[a, b]$ then

$$\int_a^b \mathbf{r}(t)\, dt = \mathbf{R}(b) - \mathbf{R}(a).$$

31. Volleyball A volleyball is hit when it is 4 ft above the ground and 12 ft from a 6-ft-high net. It leaves the point of impact with an initial velocity of 35 ft/sec at an angle of 27° and slips by the opposing team untouched.

a. Find a vector equation for the path of the volleyball.

b. How high does the volleyball go, and when does it reach maximum height?

c. Find its range and flight time.

d. When is the volleyball 7 ft above the ground? How far (ground distance) is the volleyball from where it will land?

e. Suppose that the net is raised to 8 ft. Does this change things? Explain.

32. Shot put In Moscow in 1987, Natalya Lisouskaya set a women's world record by putting an 8 lb 13 oz shot 73 ft 10 in. Assuming that she launched the shot at a 40° angle to the horizontal from 6.5 ft above the ground, what was the shot's initial speed?

11.3 Arc Length in Space

In this and the next two sections, we study the mathematical features of a curve's shape that describe the sharpness of its turning and its twisting.

Arc Length Along a Space Curve

One of the features of smooth space curves is that they have a measurable length. This enables us to locate points along these curves by giving their directed distance s along the curve from some base point, the way we locate points on coordinate axes by giving their directed distance from the origin (Figure 11.11).

To measure distance along a smooth curve in space, we add a z-term to the formula we use for curves in the plane.

FIGURE 11.11 Smooth curves can be scaled like number lines, the coordinate of each point being its directed distance along the curve from a preselected base point.

> **DEFINITION** The **length** of a smooth curve $\mathbf{r}(t) = x(t)\mathbf{i} + y(t)\mathbf{j} + z(t)\mathbf{k}$, $a \leq t \leq b$, that is traced exactly once as t increases from $t = a$ to $t = b$, is
>
> $$L = \int_a^b \sqrt{\left(\frac{dx}{dt}\right)^2 + \left(\frac{dy}{dt}\right)^2 + \left(\frac{dz}{dt}\right)^2}\, dt. \tag{1}$$

Just as for plane curves, we can calculate the length of a curve in space from any convenient parametrization that meets the stated conditions. We omit the proof.

The square root in Equation (1) is $|\mathbf{v}|$, the length of a velocity vector $d\mathbf{r}/dt$. This enables us to write the formula for length a shorter way.

Arc Length Formula

$$L = \int_a^b |\mathbf{v}|\, dt \qquad\qquad (2)$$

EXAMPLE 1 A glider is soaring upward along the helix $\mathbf{r}(t) = (\cos t)\mathbf{i} + (\sin t)\mathbf{j} + t\mathbf{k}$. How long is the glider's path from $t = 0$ to $t = 2\pi$?

Solution The path segment during this time corresponds to one full turn of the helix (Figure 11.12). The length of this portion of the curve is

$$L = \int_a^b |\mathbf{v}|\, dt = \int_0^{2\pi} \sqrt{(-\sin t)^2 + (\cos t)^2 + (1)^2}\, dt$$

$$= \int_0^{2\pi} \sqrt{2}\, dt = 2\pi\sqrt{2} \text{ units of length}.$$

This is $\sqrt{2}$ times the circumference of the circle in the xy-plane over which the helix stands. ∎

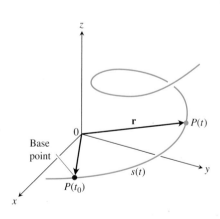

FIGURE 11.12 The helix $\mathbf{r}(t) = (\cos t)\mathbf{i} + (\sin t)\mathbf{j} + t\mathbf{k}$ in Example 1.

If we choose a base point $P(t_0)$ on a smooth curve C parametrized by t, each value of t determines a point $P(t) = (x(t), y(t), z(t))$ on C and a "directed distance"

$$s(t) = \int_{t_0}^t |\mathbf{v}(\tau)|\, d\tau,$$

measured along C from the base point (Figure 11.13). If $t > t_0$, $s(t)$ is the distance along the curve from $P(t_0)$ to $P(t)$. If $t < t_0$, $s(t)$ is the negative of the distance. Each value of s determines a point on C and this parametrizes C with respect to s. We call s an **arc length parameter** for the curve. The parameter's value increases in the direction of increasing t. The arc length parameter is particularly effective for investigating the turning and twisting nature of a space curve.

Arc Length Parameter with Base Point $P(t_0)$

$$s(t) = \int_{t_0}^t \sqrt{[x'(\tau)]^2 + [y'(\tau)]^2 + [z'(\tau)]^2}\, d\tau = \int_{t_0}^t |\mathbf{v}(\tau)|\, d\tau \qquad (3)$$

We use the Greek letter τ ("tau") as the variable of integration in Equation (3) because the letter t is already in use as the upper limit.

If a curve $\mathbf{r}(t)$ is already given in terms of some parameter t and $s(t)$ is the arc length function given by Equation (3), then we may be able to solve for t as a function of s: $t = t(s)$. Then the curve can be reparametrized in terms of s by substituting for t: $\mathbf{r} = \mathbf{r}(t(s))$. The new parametrization identifies a point on the curve with its directed distance along the curve from the base point.

FIGURE 11.13 The directed distance along the curve from $P(t_0)$ to any point $P(t)$ is

$$s(t) = \int_{t_0}^t |\mathbf{v}(\tau)|\, d\tau.$$

EXAMPLE 2   If $t_0 = 0$, the arc length parameter along the helix

$$\mathbf{r}(t) = (\cos t)\mathbf{i} + (\sin t)\mathbf{j} + t\mathbf{k}$$

from t_0 to t is

$$
\begin{aligned}
s(t) &= \int_{t_0}^{t} |\mathbf{v}(\tau)| \, d\tau && \text{Equation (3)} \\
&= \int_{0}^{t} \sqrt{2} \, d\tau && \text{Value from Example 1} \\
&= \sqrt{2}\, t.
\end{aligned}
$$

Solving this equation for t gives $t = s/\sqrt{2}$. Substituting into the position vector $\mathbf{r}$ gives the following arc length parametrization for the helix:

$$\mathbf{r}(t(s)) = \left(\cos \frac{s}{\sqrt{2}}\right)\mathbf{i} + \left(\sin \frac{s}{\sqrt{2}}\right)\mathbf{j} + \frac{s}{\sqrt{2}}\mathbf{k}. \qquad \blacksquare$$

Unlike Example 2, the arc length parametrization is generally difficult to find analytically for a curve already given in terms of some other parameter t. Fortunately, however, we rarely need an exact formula for $s(t)$ or its inverse $t(s)$.

HISTORICAL BIOGRAPHY

Josiah Willard Gibbs
(1839–1903)

Speed on a Smooth Curve

Since the derivatives beneath the radical in Equation (3) are continuous (the curve is smooth), the Fundamental Theorem of Calculus tells us that s is a differentiable function of t with derivative

$$\frac{ds}{dt} = |\mathbf{v}(t)|. \tag{4}$$

As we already knew, the speed with which a particle moves along its path is the magnitude of $\mathbf{v}$.

Although the base point $P(t_0)$ plays a role in defining s in Equation (3), it plays no role in Equation (4). The rate at which a moving particle covers distance along its path is independent of how far away it is from the base point.

Notice that $ds/dt > 0$ since, by definition, $|\mathbf{v}|$ is never zero for a smooth curve. We see once again that s is an increasing function of t.

Unit Tangent Vector T

We already know the velocity vector $\mathbf{v} = d\mathbf{r}/dt$ is tangent to the curve and that the vector

$$\mathbf{T} = \frac{\mathbf{v}}{|\mathbf{v}|}$$

is therefore a unit vector tangent to the (smooth) curve. Since $ds/dt > 0$ for the curves we are considering, s is one-to-one and has an inverse that gives t as a differentiable function of s (Section 3.7). The derivative of the inverse is

$$\frac{dt}{ds} = \frac{1}{ds/dt} = \frac{1}{|\mathbf{v}|}.$$

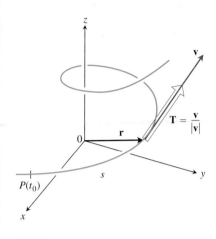

FIGURE 11.14 We find the unit tangent vector **T** by dividing **v** by $|\mathbf{v}|$.

This makes **r** a differentiable function of s whose derivative can be calculated with the Chain Rule to be

$$\frac{d\mathbf{r}}{ds} = \frac{d\mathbf{r}}{dt}\frac{dt}{ds} = \mathbf{v}\frac{1}{|\mathbf{v}|} = \frac{\mathbf{v}}{|\mathbf{v}|} = \mathbf{T}.$$

This equation says that $d\mathbf{r}/ds$ is the unit tangent vector in the direction of the velocity vector **v** (Figure 11.14).

DEFINITION   The **unit tangent vector** of a smooth curve $\mathbf{r}(t)$ is

$$\mathbf{T} = \frac{d\mathbf{r}}{ds} = \frac{d\mathbf{r}/dt}{ds/dt} = \frac{\mathbf{v}}{|\mathbf{v}|}. \tag{5}$$

The unit tangent vector **T** is a differentiable function of t whenever **v** is a differentiable function of t. As we will see in Section 11.5, **T** is one of three unit vectors in a traveling reference frame that is used to describe the motion of objects traveling in three dimensions.

EXAMPLE 3   Find the unit tangent vector of the curve

$$\mathbf{r}(t) = (3\cos t)\mathbf{i} + (3\sin t)\mathbf{j} + t^2\mathbf{k}$$

representing the path of the glider in Example 3, Section 11.2.

Solution   In that example, we found

$$\mathbf{v} = \frac{d\mathbf{r}}{dt} = -(3\sin t)\mathbf{i} + (3\cos t)\mathbf{j} + 2t\mathbf{k}$$

and

$$|\mathbf{v}| = \sqrt{9 + 4t^2}.$$

Thus,

$$\mathbf{T} = \frac{\mathbf{v}}{|\mathbf{v}|} = -\frac{3\sin t}{\sqrt{9 + 4t^2}}\mathbf{i} + \frac{3\cos t}{\sqrt{9 + 4t^2}}\mathbf{j} + \frac{2t}{\sqrt{9 + 4t^2}}\mathbf{k}. \qquad \blacksquare$$

For the counterclockwise motion

$$\mathbf{r}(t) = (\cos t)\mathbf{i} + (\sin t)\mathbf{j}$$

around the unit circle, we see that

$$\mathbf{v} = (-\sin t)\mathbf{i} + (\cos t)\mathbf{j}$$

is already a unit vector, so $\mathbf{T} = \mathbf{v}$ (Figure 11.15).

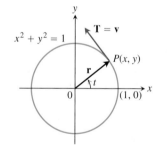

FIGURE 11.15 Counterclockwise motion around the unit circle.

EXERCISES 11.3

In Exercises 1–8, find the curve's unit tangent vector. Also, find the length of the indicated portion of the curve.

1. $\mathbf{r}(t) = (2\cos t)\mathbf{i} + (2\sin t)\mathbf{j} + \sqrt{5}t\mathbf{k}, \quad 0 \le t \le \pi$

2. $\mathbf{r}(t) = (6\sin 2t)\mathbf{i} + (6\cos 2t)\mathbf{j} + 5t\mathbf{k}, \quad 0 \le t \le \pi$

3. $\mathbf{r}(t) = t\mathbf{i} + (2/3)t^{3/2}\mathbf{k}, \quad 0 \le t \le 8$

4. $\mathbf{r}(t) = (2 + t)\mathbf{i} - (t + 1)\mathbf{j} + t\mathbf{k}, \quad 0 \le t \le 3$

5. $\mathbf{r}(t) = (\cos^3 t)\mathbf{j} + (\sin^3 t)\mathbf{k}, \quad 0 \le t \le \pi/2$

6. $\mathbf{r}(t) = 6t^3\mathbf{i} - 2t^3\mathbf{j} - 3t^3\mathbf{k}, \quad 1 \le t \le 2$

7. $\mathbf{r}(t) = (t \cos t)\mathbf{i} + (t \sin t)\mathbf{j} + \left(2\sqrt{2}/3\right)t^{3/2}\mathbf{k}, \quad 0 \le t \le \pi$

8. $\mathbf{r}(t) = (t \sin t + \cos t)\mathbf{i} + (t \cos t - \sin t)\mathbf{j}, \quad \sqrt{2} \le t \le 2$

9. Find the point on the curve

$$\mathbf{r}(t) = (5 \sin t)\mathbf{i} + (5 \cos t)\mathbf{j} + 12t\mathbf{k}$$

at a distance 26π units along the curve from the point $(0,5,0)$ in the direction of increasing arc length.

10. Find the point on the curve

$$\mathbf{r}(t) = (12 \sin t)\mathbf{i} - (12 \cos t)\mathbf{j} + 5t\mathbf{k}$$

at a distance 13π units along the curve from the point $(0,-12,0)$ in the direction opposite to the direction of increasing arc length.

In Exercises 11–14, find the arc length parameter along the curve from the point where $t = 0$ by evaluating the integral

$$s = \int_0^t |\mathbf{v}(\tau)| \, d\tau$$

from Equation (3). Then find the length of the indicated portion of the curve.

11. $\mathbf{r}(t) = (4 \cos t)\mathbf{i} + (4 \sin t)\mathbf{j} + 3t\mathbf{k}, \quad 0 \le t \le \pi/2$

12. $\mathbf{r}(t) = (\cos t + t \sin t)\mathbf{i} + (\sin t - t \cos t)\mathbf{j}, \quad \pi/2 \le t \le \pi$

13. $\mathbf{r}(t) = (e^t \cos t)\mathbf{i} + (e^t \sin t)\mathbf{j} + e^t\mathbf{k}, \quad -\ln 4 \le t \le 0$

14. $\mathbf{r}(t) = (1 + 2t)\mathbf{i} + (1 + 3t)\mathbf{j} + (6 - 6t)\mathbf{k}, \quad -1 \le t \le 0$

15. **Arc length**   Find the length of the curve

$$\mathbf{r}(t) = \left(\sqrt{2}t\right)\mathbf{i} + \left(\sqrt{2}t\right)\mathbf{j} + (1 - t^2)\mathbf{k}$$

from $(0, 0, 1)$ to $\left(\sqrt{2}, \sqrt{2}, 0\right)$.

16. **Length of helix**   The length $2\pi\sqrt{2}$ of the turn of the helix in Example 1 is also the length of the diagonal of a square 2π units on a side. Show how to obtain this square by cutting away and flattening a portion of the cylinder around which the helix winds.

17. **Ellipse**
   a. Show that the curve $\mathbf{r}(t) = (\cos t)\mathbf{i} + (\sin t)\mathbf{j} + (1 - \cos t)\mathbf{k}$, $0 \le t \le 2\pi$, is an ellipse by showing that it is the intersection of a right circular cylinder and a plane. Find equations for the cylinder and plane.
   b. Sketch the ellipse on the cylinder. Add to your sketch the unit tangent vectors at $t = 0, \pi/2, \pi$, and $3\pi/2$.
   c. Show that the acceleration vector always lies parallel to the plane (orthogonal to a vector normal to the plane). Thus, if you draw the acceleration as a vector attached to the ellipse, it will lie in the plane of the ellipse. Add the acceleration vectors for $t = 0, \pi/2, \pi$, and $3\pi/2$ to your sketch.
   d. Write an integral for the length of the ellipse. Do not try to evaluate the integral; it is nonelementary.

T e. **Numerical integrator**   Estimate the length of the ellipse to two decimal places.

18. **Length is independent of parametrization**   To illustrate that the length of a smooth space curve does not depend on the parametrization you use to compute it, calculate the length of one turn of the helix in Example 1 with the following parametrizations.
   a. $\mathbf{r}(t) = (\cos 4t)\mathbf{i} + (\sin 4t)\mathbf{j} + 4t\mathbf{k}, \quad 0 \le t \le \pi/2$
   b. $\mathbf{r}(t) = [\cos (t/2)]\mathbf{i} + [\sin (t/2)]\mathbf{j} + (t/2)\mathbf{k}, \quad 0 \le t \le 4\pi$
   c. $\mathbf{r}(t) = (\cos t)\mathbf{i} - (\sin t)\mathbf{j} - t\mathbf{k}, \quad -2\pi \le t \le 0$

19. **The involute of a circle**   If a string wound around a fixed circle is unwound while held taut in the plane of the circle, its end P traces an *involute* of the circle. In the accompanying figure, the circle in question is the circle $x^2 + y^2 = 1$ and the tracing point starts at $(1, 0)$. The unwound portion of the string is tangent to the circle at Q, and t is the radian measure of the angle from the positive x-axis to segment OQ. Derive the parametric equations

$$x = \cos t + t \sin t, \quad y = \sin t - t \cos t, \quad t > 0$$

of the point $P(x, y)$ for the involute.

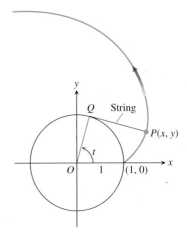

20. (*Continuation of Exercise 19.*) Find the unit tangent vector to the involute of the circle at the point $P(x, y)$.

21. **Distance along a line**   Show that if $\mathbf{u}$ is a unit vector, then the arc length parameter along the line $\mathbf{r}(t) = P_0 + t\mathbf{u}$ from the point $P_0(x_0, y_0, z_0)$ where $t = 0$, is t itself.

22. Use Simpson's Rule with $n = 10$ to approximate the length of arc of $\mathbf{r}(t) = t\mathbf{i} + t^2\mathbf{j} + t^3\mathbf{k}$ from the origin to the point $(2, 4, 8)$.

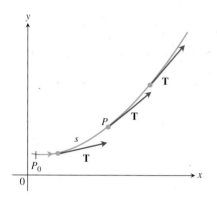

FIGURE 11.16   As P moves along the curve in the direction of increasing arc length, the unit tangent vector turns. The value of $|d\mathbf{T}/ds|$ at P is called the *curvature* of the curve at P.

In this section we study how a curve turns or bends. We look first at curves in the coordinate plane, and then at curves in space.

Curvature of a Plane Curve

As a particle moves along a smooth curve in the plane, $\mathbf{T} = d\mathbf{r}/ds$ turns as the curve bends. Since $\mathbf{T}$ is a unit vector, its length remains constant and only its direction changes as the particle moves along the curve. The rate at which $\mathbf{T}$ turns per unit of length along the curve is called the *curvature* (Figure 11.16). The traditional symbol for the curvature function is the Greek letter κ ("kappa").

> **DEFINITION**   If $\mathbf{T}$ is the unit vector of a smooth curve, the **curvature** function of the curve is
> $$\kappa = \left| \frac{d\mathbf{T}}{ds} \right|.$$

If $|d\mathbf{T}/ds|$ is large, $\mathbf{T}$ turns sharply as the particle passes through P, and the curvature at P is large. If $|d\mathbf{T}/ds|$ is close to zero, $\mathbf{T}$ turns more slowly and the curvature at P is smaller.

If a smooth curve $\mathbf{r}(t)$ is already given in terms of some parameter t other than the arc length parameter s, we can calculate the curvature as

$$\kappa = \left| \frac{d\mathbf{T}}{ds} \right| = \left| \frac{d\mathbf{T}}{dt} \frac{dt}{ds} \right| \qquad \text{Chain Rule}$$

$$= \frac{1}{|ds/dt|} \left| \frac{d\mathbf{T}}{dt} \right|$$

$$= \frac{1}{|\mathbf{v}|} \left| \frac{d\mathbf{T}}{dt} \right|. \qquad \frac{ds}{dt} = |\mathbf{v}|$$

> **Formula for Calculating Curvature**
> If $\mathbf{r}(t)$ is a smooth curve, then the curvature is
> $$\kappa = \frac{1}{|\mathbf{v}|} \left| \frac{d\mathbf{T}}{dt} \right|, \qquad (1)$$
> where $\mathbf{T} = \mathbf{v}/|\mathbf{v}|$ is the unit tangent vector.

Testing the definition, we see in Examples 1 and 2 below that the curvature is constant for straight lines and circles.

FIGURE 11.17   Along a straight line, $\mathbf{T}$ always points in the same direction. The curvature, $|d\mathbf{T}/ds|$, is zero (Example 1).

EXAMPLE 1   A straight line is parametrized by $\mathbf{r}(t) = \mathbf{C} + t\mathbf{v}$ for constant vectors $\mathbf{C}$ and $\mathbf{v}$. Thus, $\mathbf{r}'(t) = \mathbf{v}$, and the unit tangent vector $\mathbf{T} = \mathbf{v}/|\mathbf{v}|$ is a constant vector which always points in the same direction and has derivative $\mathbf{0}$ (Figure 11.17). It follows that, for any value of the parameter t, the curvature of the straight line is

$$\kappa = \frac{1}{|\mathbf{v}|} \left| \frac{d\mathbf{T}}{dt} \right| = \frac{1}{|\mathbf{v}|} |\mathbf{0}| = 0. \qquad \blacksquare$$

EXAMPLE 2 Here we find the curvature of a circle. We begin with the parametrization

$$\mathbf{r}(t) = (a \cos t)\mathbf{i} + (a \sin t)\mathbf{j}$$

of a circle of radius a. Then,

$$\mathbf{v} = \frac{d\mathbf{r}}{dt} = -(a \sin t)\mathbf{i} + (a \cos t)\mathbf{j}$$

$$|\mathbf{v}| = \sqrt{(-a \sin t)^2 + (a \cos t)^2} = \sqrt{a^2} = |a| = a. \qquad \text{Since } a > 0, \\ |a| = a.$$

From this we find

$$\mathbf{T} = \frac{\mathbf{v}}{|\mathbf{v}|} = -(\sin t)\mathbf{i} + (\cos t)\mathbf{j}$$

$$\frac{d\mathbf{T}}{dt} = -(\cos t)\mathbf{i} - (\sin t)\mathbf{j}$$

$$\left|\frac{d\mathbf{T}}{dt}\right| = \sqrt{\cos^2 t + \sin^2 t} = 1.$$

Hence, for any value of the parameter t, the curvature of the circle is

$$\kappa = \frac{1}{|\mathbf{v}|}\left|\frac{d\mathbf{T}}{dt}\right| = \frac{1}{a}(1) = \frac{1}{a} = \frac{1}{\text{radius}}. \qquad \blacksquare$$

Although the formula for calculating κ in Equation (1) is also valid for space curves, in the next section we find a computational formula that is usually more convenient to apply.

Among the vectors orthogonal to the unit tangent vector $\mathbf{T}$ is one of particular significance because it points in the direction in which the curve is turning. Since $\mathbf{T}$ has constant length (namely, 1), the derivative $d\mathbf{T}/ds$ is orthogonal to $\mathbf{T}$ (Equation 4, Section 11.1). Therefore, if we divide $d\mathbf{T}/ds$ by its length κ, we obtain a *unit* vector $\mathbf{N}$ orthogonal to $\mathbf{T}$ (Figure 11.18).

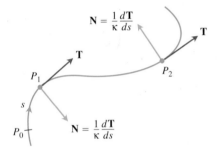

FIGURE 11.18 The vector $d\mathbf{T}/ds$, normal to the curve, always points in the direction in which $\mathbf{T}$ is turning. The unit normal vector $\mathbf{N}$ is the direction of $d\mathbf{T}/ds$.

DEFINITION At a point where $\kappa \neq 0$, the **principal unit normal** vector for a smooth curve in the plane is

$$\mathbf{N} = \frac{1}{\kappa}\frac{d\mathbf{T}}{ds}.$$

The vector $d\mathbf{T}/ds$ points in the direction in which $\mathbf{T}$ turns as the curve bends. Therefore, if we face in the direction of increasing arc length, the vector $d\mathbf{T}/ds$ points toward the right if $\mathbf{T}$ turns clockwise and toward the left if $\mathbf{T}$ turns counterclockwise. In other words, the principal normal vector $\mathbf{N}$ will point toward the concave side of the curve (Figure 11.18).

If a smooth curve $\mathbf{r}(t)$ is already given in terms of some parameter t other than the arc length parameter s, we can use the Chain Rule to calculate $\mathbf{N}$ directly:

$$\mathbf{N} = \frac{d\mathbf{T}/ds}{|d\mathbf{T}/ds|}$$

$$= \frac{(d\mathbf{T}/dt)(dt/ds)}{|d\mathbf{T}/dt||dt/ds|}$$

$$= \frac{d\mathbf{T}/dt}{|d\mathbf{T}/dt|}. \qquad \frac{dt}{ds} = \frac{1}{ds/dt} > 0 \text{ cancels}$$

This formula enables us to find $\mathbf{N}$ without having to find κ and s first.

> **Formula for Calculating N**
> If $\mathbf{r}(t)$ is a smooth curve, then the principal unit normal is
>
> $$\mathbf{N} = \frac{d\mathbf{T}/dt}{|d\mathbf{T}/dt|}, \qquad (2)$$
>
> where $\mathbf{T} = \mathbf{v}/|\mathbf{v}|$ is the unit tangent vector.

EXAMPLE 3   Find $\mathbf{T}$ and $\mathbf{N}$ for the circular motion

$$\mathbf{r}(t) = (\cos 2t)\mathbf{i} + (\sin 2t)\mathbf{j}.$$

Solution   We first find $\mathbf{T}$:

$$\mathbf{v} = -(2\sin 2t)\mathbf{i} + (2\cos 2t)\mathbf{j}$$

$$|\mathbf{v}| = \sqrt{4\sin^2 2t + 4\cos^2 2t} = 2$$

$$\mathbf{T} = \frac{\mathbf{v}}{|\mathbf{v}|} = -(\sin 2t)\mathbf{i} + (\cos 2t)\mathbf{j}.$$

From this we find

$$\frac{d\mathbf{T}}{dt} = -(2\cos 2t)\mathbf{i} - (2\sin 2t)\mathbf{j}$$

$$\left|\frac{d\mathbf{T}}{dt}\right| = \sqrt{4\cos^2 2t + 4\sin^2 2t} = 2$$

and

$$\mathbf{N} = \frac{d\mathbf{T}/dt}{|d\mathbf{T}/dt|}$$

$$= -(\cos 2t)\mathbf{i} - (\sin 2t)\mathbf{j}. \qquad \text{Equation (2)}$$

Notice that $\mathbf{T} \cdot \mathbf{N} = 0$, verifying that $\mathbf{N}$ is orthogonal to $\mathbf{T}$. Notice too, that for the circular motion here, $\mathbf{N}$ points from $\mathbf{r}(t)$ towards the circle's center at the origin.   ■

Circle of Curvature for Plane Curves

The **circle of curvature** or **osculating circle** at a point P on a plane curve where $\kappa \neq 0$ is the circle in the plane of the curve that

1.   is tangent to the curve at P (has the same tangent line the curve has)
2.   has the same curvature the curve has at P
3.   lies toward the concave or inner side of the curve (as in Figure 11.19).

The **radius of curvature** of the curve at P is the radius of the circle of curvature, which, according to Example 2, is

$$\text{Radius of curvature} = \rho = \frac{1}{\kappa}.$$

To find ρ, we find κ and take the reciprocal. The **center of curvature** of the curve at P is the center of the circle of curvature.

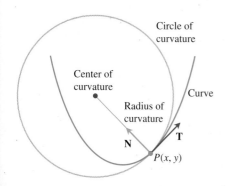

FIGURE 11.19   The osculating circle at $P(x, y)$ lies toward the inner side of the curve.

EXAMPLE 4   Find and graph the osculating circle of the parabola $y = x^2$ at the origin.

Solution   We parametrize the parabola using the parameter $t = x$ (Section 9.6, Example 1)

$$\mathbf{r}(t) = t\mathbf{i} + t^2\mathbf{j}.$$

First we find the curvature of the parabola at the origin, using Equation (1):

$$\mathbf{v} = \frac{d\mathbf{r}}{dt} = \mathbf{i} + 2t\mathbf{j}$$

$$|\mathbf{v}| = \sqrt{1 + 4t^2}$$

so that

$$\mathbf{T} = \frac{\mathbf{v}}{|\mathbf{v}|} = (1 + 4t^2)^{-1/2}\mathbf{i} + 2t(1 + 4t^2)^{-1/2}\mathbf{j}.$$

From this we find

$$\frac{d\mathbf{T}}{dt} = -4t(1 + 4t^2)^{-3/2}\mathbf{i} + [2(1 + 4t^2)^{-1/2} - 8t^2(1 + 4t^2)^{-3/2}]\mathbf{j}.$$

At the origin, $t = 0$, so the curvature is

$$\kappa(0) = \frac{1}{|\mathbf{v}(0)|}\left|\frac{d\mathbf{T}}{dt}(0)\right| \qquad \text{Equation (1)}$$

$$= \frac{1}{\sqrt{1}}|0\mathbf{i} + 2\mathbf{j}|$$

$$= (1)\sqrt{0^2 + 2^2} = 2.$$

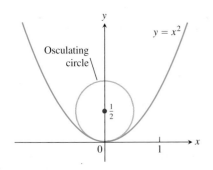

FIGURE 11.20   The osculating circle for the parabola $y = x^2$ at the origin (Example 4).

Therefore, the radius of curvature is $1/\kappa = 1/2$. We see in Figure 11.20 that at the origin, $\mathbf{T} = \mathbf{i}$ so $\mathbf{N} = \mathbf{j}$ and the center of the circle is $(0, 1/2)$. The equation of the osculating circle is

$$(x - 0)^2 + \left(y - \frac{1}{2}\right)^2 = \left(\frac{1}{2}\right)^2.$$

You can see from Figure 11.20 that the osculating circle is a better approximation to the parabola at the origin than is the tangent line approximation $y = 0$.   ∎

Curvature and Normal Vectors for Space Curves

If a smooth curve in space is specified by the position vector $\mathbf{r}(t)$ as a function of some parameter t, and if s is the arc length parameter of the curve, then the unit tangent vector $\mathbf{T}$ is $d\mathbf{r}/ds = \mathbf{v}/|\mathbf{v}|$. The **curvature** in space is then defined to be

$$\kappa = \left|\frac{d\mathbf{T}}{ds}\right| = \frac{1}{|\mathbf{v}|}\left|\frac{d\mathbf{T}}{dt}\right| \tag{3}$$

just as for plane curves. The vector $d\mathbf{T}/ds$ is orthogonal to $\mathbf{T}$, and we define the **principal unit normal** to be

$$\mathbf{N} = \frac{1}{\kappa}\frac{d\mathbf{T}}{ds} = \frac{d\mathbf{T}/dt}{|d\mathbf{T}/dt|}. \tag{4}$$

EXAMPLE 5   Find the curvature for the helix (Figure 11.21)

$$\mathbf{r}(t) = (a\cos t)\mathbf{i} + (a\sin t)\mathbf{j} + bt\mathbf{k}, \qquad a, b \geq 0, \qquad a^2 + b^2 \neq 0.$$

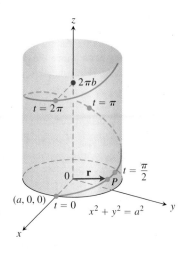

FIGURE 11.21 The helix

$\mathbf{r}(t) = (a \cos t)\mathbf{i} + (a \sin t)\mathbf{j} + bt\mathbf{k}$,

drawn with a and b positive and $t \geq 0$ (Example 5).

Solution We calculate $\mathbf{T}$ from the velocity vector $\mathbf{v}$:

$$\mathbf{v} = -(a \sin t)\mathbf{i} + (a \cos t)\mathbf{j} + b\mathbf{k}$$

$$|\mathbf{v}| = \sqrt{a^2 \sin^2 t + a^2 \cos^2 t + b^2} = \sqrt{a^2 + b^2}$$

$$\mathbf{T} = \frac{\mathbf{v}}{|\mathbf{v}|} = \frac{1}{\sqrt{a^2 + b^2}}[-(a \sin t)\mathbf{i} + (a \cos t)\mathbf{j} + b\mathbf{k}].$$

Then using Equation (3),

$$\kappa = \frac{1}{|\mathbf{v}|}\left|\frac{d\mathbf{T}}{dt}\right|$$

$$= \frac{1}{\sqrt{a^2 + b^2}}\left|\frac{1}{\sqrt{a^2 + b^2}}[-(a \cos t)\mathbf{i} - (a \sin t)\mathbf{j}]\right|$$

$$= \frac{a}{a^2 + b^2}|-(\cos t)\mathbf{i} - (\sin t)\mathbf{j}|$$

$$= \frac{a}{a^2 + b^2}\sqrt{(\cos t)^2 + (\sin t)^2} = \frac{a}{a^2 + b^2}.$$

From this equation, we see that increasing b for a fixed a decreases the curvature. Decreasing a for a fixed b eventually decreases the curvature as well.

If $b = 0$, the helix reduces to a circle of radius a and its curvature reduces to $1/a$, as it should. If $a = 0$, the helix becomes the z-axis, and its curvature reduces to 0, again as it should. ∎

EXAMPLE 6 Find $\mathbf{N}$ for the helix in Example 5 and describe how the vector is pointing.

Solution We have

$$\frac{d\mathbf{T}}{dt} = -\frac{1}{\sqrt{a^2 + b^2}}[(a \cos t)\mathbf{i} + (a \sin t)\mathbf{j}] \qquad \text{Example 5}$$

$$\left|\frac{d\mathbf{T}}{dt}\right| = \frac{1}{\sqrt{a^2 + b^2}}\sqrt{a^2 \cos^2 t + a^2 \sin^2 t} = \frac{a}{\sqrt{a^2 + b^2}}$$

$$\mathbf{N} = \frac{d\mathbf{T}/dt}{|d\mathbf{T}/dt|} \qquad \text{Equation (4)}$$

$$= -\frac{\sqrt{a^2 + b^2}}{a} \cdot \frac{1}{\sqrt{a^2 + b^2}}[(a \cos t)\mathbf{i} + (a \sin t)\mathbf{j}]$$

$$= -(\cos t)\mathbf{i} - (\sin t)\mathbf{j}.$$

Thus, $\mathbf{N}$ is parallel to the xy-plane and always points toward the z-axis. ∎

EXERCISES 11.4

Find $\mathbf{T}$, $\mathbf{N}$, and κ for the plane curves in Exercises 1–4.

1. $\mathbf{r}(t) = t\mathbf{i} + (\ln \cos t)\mathbf{j}, \quad -\pi/2 < t < \pi/2$

2. $\mathbf{r}(t) = (\ln \sec t)\mathbf{i} + t\mathbf{j}, \quad -\pi/2 < t < \pi/2$

3. $\mathbf{r}(t) = (2t + 3)\mathbf{i} + (5 - t^2)\mathbf{j}$

4. $\mathbf{r}(t) = (\cos t + t \sin t)\mathbf{i} + (\sin t - t \cos t)\mathbf{j}, \quad t > 0$

5. A formula for the curvature of the graph of a function in the xy-plane

a. The graph $y = f(x)$ in the xy-plane automatically has the parametrization $x = x, y = f(x)$, and the vector formula $\mathbf{r}(x) = x\mathbf{i} + f(x)\mathbf{j}$. Use this formula to show that if f is a twice-differentiable function of x, then

$$\kappa(x) = \frac{|f''(x)|}{\left[1 + (f'(x))^2\right]^{3/2}}.$$

b. Use the formula for κ in part (a) to find the curvature of $y = \ln(\cos x), -\pi/2 < x < \pi/2$. Compare your answer with the answer in Exercise 1.

c. Show that the curvature is zero at a point of inflection.

6. A formula for the curvature of a parametrized plane curve

a. Show that the curvature of a smooth curve $\mathbf{r}(t) = f(t)\mathbf{i} + g(t)\mathbf{j}$ defined by twice-differentiable functions $x = f(t)$ and $y = g(t)$ is given by the formula

$$\kappa = \frac{|\dot{x}\ddot{y} - \ddot{x}\dot{y}|}{(\dot{x}^2 + \dot{y}^2)^{3/2}}.$$

The dots in the formula denote differentiation with respect to t, one derivative for each dot. Apply the formula to find the curvatures of the following curves.

b. $\mathbf{r}(t) = t\mathbf{i} + (\ln \sin t)\mathbf{j}, \quad 0 < t < \pi$

c. $\mathbf{r}(t) = [\tan^{-1}(\sinh t)]\mathbf{i} + (\ln \cosh t)\mathbf{j}$.

7. Normals to plane curves

a. Show that $\mathbf{n}(t) = -g'(t)\mathbf{i} + f'(t)\mathbf{j}$ and $-\mathbf{n}(t) = g'(t)\mathbf{i} - f'(t)\mathbf{j}$ are both normal to the curve $\mathbf{r}(t) = f(t)\mathbf{i} + g(t)\mathbf{j}$ at the point $(f(t), g(t))$.

To obtain $\mathbf{N}$ for a particular plane curve, we can choose the one of $\mathbf{n}$ or $-\mathbf{n}$ from part (a) that points toward the concave side of the curve, and make it into a unit vector. (See Figure 11.18.) Apply this method to find $\mathbf{N}$ for the following curves.

b. $\mathbf{r}(t) = t\mathbf{i} + e^{2t}\mathbf{j}$

c. $\mathbf{r}(t) = \sqrt{4 - t^2}\,\mathbf{i} + t\mathbf{j}, \quad -2 \le t \le 2$

8. *(Continuation of Exercise 7.)*

a. Use the method of Exercise 7 to find $\mathbf{N}$ for the curve $\mathbf{r}(t) = t\mathbf{i} + (1/3)t^3\mathbf{j}$ when $t < 0$; when $t > 0$.

b. Calculate $\mathbf{N}$ for $t \ne 0$ directly from $\mathbf{T}$ using Eq. (4) for the curve in part (a). Does $\mathbf{N}$ exist at $t = 0$? Graph the curve and explain what is happening to $\mathbf{N}$ as t passes from negative to positive values.

Find $\mathbf{T}, \mathbf{N}$, and κ for the space curves in Exercises 9–16.

9. $\mathbf{r}(t) = (3 \sin t)\mathbf{i} + (3 \cos t)\mathbf{j} + 4t\mathbf{k}$

10. $\mathbf{r}(t) = (\cos t + t \sin t)\mathbf{i} + (\sin t - t \cos t)\mathbf{j} + 3\mathbf{k}$

11. $\mathbf{r}(t) = (e^t \cos t)\mathbf{i} + (e^t \sin t)\mathbf{j} + 2\mathbf{k}$

12. $\mathbf{r}(t) = (6 \sin 2t)\mathbf{i} + (6 \cos 2t)\mathbf{j} + 5t\mathbf{k}$

13. $\mathbf{r}(t) = (t^3/3)\mathbf{i} + (t^2/2)\mathbf{j}, \quad t > 0$

14. $\mathbf{r}(t) = (\cos^3 t)\mathbf{i} + (\sin^3 t)\mathbf{j}, \quad 0 < t < \pi/2$

15. $\mathbf{r}(t) = t\mathbf{i} + (a \cosh(t/a))\mathbf{j}, \quad a > 0$

16. $\mathbf{r}(t) = (\cosh t)\mathbf{i} - (\sinh t)\mathbf{j} + t\mathbf{k}$

17. Show that the parabola $y = ax^2, a \ne 0$, has its largest curvature at its vertex and has no minimum curvature. (*Note:* Since the curvature of a curve remains the same if the curve is translated or rotated, this result is true for any parabola.)

18. Show that the ellipse $x = a \cos t, y = b \sin t, a > b > 0$, has its largest curvature on its major axis and its smallest curvature on its minor axis. (As in Exercise 17, the same is true for any ellipse.)

19. Maximizing the curvature of a helix In Example 5, we found the curvature of the helix $\mathbf{r}(t) = (a \cos t)\mathbf{i} + (a \sin t)\mathbf{j} + bt\mathbf{k}$ $(a, b \ge 0)$ to be $\kappa = a/(a^2 + b^2)$. What is the largest value κ can have for a given value of b? Give reasons for your answer.

20. Total curvature We find the **total curvature** of the portion of a smooth curve that runs from $s = s_0$ to $s = s_1 > s_0$ by integrating κ from s_0 to s_1. If the curve has some other parameter, say t, then the total curvature is

$$K = \int_{s_0}^{s_1} \kappa\, ds = \int_{t_0}^{t_1} \kappa \frac{ds}{dt}\, dt = \int_{t_0}^{t_1} \kappa |\mathbf{v}|\, dt,$$

where t_0 and t_1 correspond to s_0 and s_1. Find the total curvatures of

a. The portion of the helix $\mathbf{r}(t) = (3 \cos t)\mathbf{i} + (3 \sin t)\mathbf{j} + t\mathbf{k}$, $0 \le t \le 4\pi$.

b. The parabola $y = x^2, -\infty < x < \infty$.

21. Find an equation for the circle of curvature of the curve $\mathbf{r}(t) = t\mathbf{i} + (\sin t)\mathbf{j}$ at the point $(\pi/2, 1)$. (The curve parametrizes the graph of $y = \sin x$ in the xy-plane.)

22. Find an equation for the circle of curvature of the curve $\mathbf{r}(t) = (2 \ln t)\mathbf{i} - [t + (1/t)]\mathbf{j}, e^{-2} \le t \le e^2$, at the point $(0, -2)$, where $t = 1$.

T The formula

$$\kappa(x) = \frac{|f''(x)|}{\left[1 + (f'(x))^2\right]^{3/2}},$$

derived in Exercise 5, expresses the curvature $\kappa(x)$ of a twice-differentiable plane curve $y = f(x)$ as a function of x. Find the curvature function of each of the curves in Exercises 23–26. Then graph $f(x)$ together with $\kappa(x)$ over the given interval. You will find some surprises.

23. $y = x^2, \quad -2 \le x \le 2$ **24.** $y = x^4/4, \quad -2 \le x \le 2$

25. $y = \sin x, \quad 0 \le x \le 2\pi$ **26.** $y = e^x, \quad -1 \le x \le 2$

COMPUTER EXPLORATIONS

In Exercises 27–34 you will use a CAS to explore the osculating circle at a point P on a plane curve where $\kappa \ne 0$. Use a CAS to perform the following steps:

a. Plot the plane curve given in parametric or function form over the specified interval to see what it looks like.

b. Calculate the curvature κ of the curve at the given value t_0 using the appropriate formula from Exercise 5 or 6. Use the parametrization $x = t$ and $y = f(t)$ if the curve is given as a function $y = f(x)$.

c. Find the unit normal vector **N** at t_0. Notice that the signs of the components of **N** depend on whether the unit tangent vector **T** is turning clockwise or counterclockwise at $t = t_0$. (See Exercise 7.)

d. If $\mathbf{C} = a\mathbf{i} + b\mathbf{j}$ is the vector from the origin to the center (a, b) of the osculating circle, find the center **C** from the vector equation

$$\mathbf{C} = \mathbf{r}(t_0) + \frac{1}{\kappa(t_0)}\mathbf{N}(t_0).$$

The point $P(x_0, y_0)$ on the curve is given by the position vector $\mathbf{r}(t_0)$.

e. Plot implicitly the equation $(x - a)^2 + (y - b)^2 = 1/\kappa^2$ of the osculating circle. Then plot the curve and osculating circle

together. You may need to experiment with the size of the viewing window, but be sure it is square.

27. $\mathbf{r}(t) = (3 \cos t)\mathbf{i} + (5 \sin t)\mathbf{j}, \quad 0 \le t \le 2\pi, \quad t_0 = \pi/4$

28. $\mathbf{r}(t) = (\cos^3 t)\mathbf{i} + (\sin^3 t)\mathbf{j}, \quad 0 \le t \le 2\pi, \quad t_0 = \pi/4$

29. $\mathbf{r}(t) = t^2\mathbf{i} + (t^3 - 3t)\mathbf{j}, \quad -4 \le t \le 4, \quad t_0 = 3/5$

30. $\mathbf{r}(t) = (t^3 - 2t^2 - t)\mathbf{i} + \dfrac{3t}{\sqrt{1 + t^2}}\mathbf{j}, \quad -2 \le t \le 5, \quad t_0 = 1$

31. $\mathbf{r}(t) = (2t - \sin t)\mathbf{i} + (2 - 2 \cos t)\mathbf{j}, \quad 0 \le t \le 3\pi,$ $t_0 = 3\pi/2$

32. $\mathbf{r}(t) = (e^{-t} \cos t)\mathbf{i} + (e^{-t} \sin t)\mathbf{j}, \quad 0 \le t \le 6\pi, \quad t_0 = \pi/4$

33. $y = x^2 - x, \quad -2 \le x \le 5, \quad x_0 = 1$

34. $y = x(1 - x)^{2/5}, \quad -1 \le x \le 2, \quad x_0 = 1/2$

11.5   Tangential and Normal Components of Acceleration

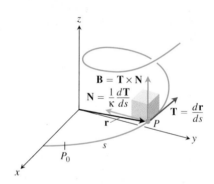

FIGURE 11.22 The **TNB** frame of mutually orthogonal unit vectors traveling along a curve in space.

If you are traveling along a space curve, the Cartesian **i**, **j**, and **k** coordinate system for representing the vectors describing your motion is not truly relevant to you. What is meaningful instead are the vectors representative of your forward direction (the unit tangent vector **T**), the direction in which your path is turning (the unit normal vector **N**), and the tendency of your motion to "twist" out of the plane created by these vectors in the direction perpendicular to this plane (defined by the *unit binormal vector* $\mathbf{B} = \mathbf{T} \times \mathbf{N}$). Expressing the acceleration vector along the curve as a linear combination of this **TNB** frame of mutually orthogonal unit vectors traveling with the motion (Figure 11.22) is particularly revealing of the nature of the path and motion along it.

The TNB Frame

The **binormal vector** of a curve in space is $\mathbf{B} = \mathbf{T} \times \mathbf{N}$, a unit vector orthogonal to both **T** and **N** (Figure 11.23). Together **T**, **N**, and **B** define a moving right-handed vector frame that plays a significant role in calculating the paths of particles moving through space. It is called the **Frenet** ("fre-*nay*") **frame** (after Jean-Frédéric Frenet, 1816–1900), or the **TNB frame**.

Tangential and Normal Components of Acceleration

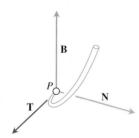

FIGURE 11.23 The vectors **T**, **N**, and **B** (in that order) make a right-handed frame of mutually orthogonal unit vectors in space.

When a body is accelerated by gravity, brakes, or a combination of rocket motors, we usually want to know how much of the acceleration acts in the direction of motion, in the tangential direction **T**. We can calculate this using the Chain Rule to rewrite **v** as

$$\mathbf{v} = \frac{d\mathbf{r}}{dt} = \frac{d\mathbf{r}}{ds}\frac{ds}{dt} = \mathbf{T}\frac{ds}{dt}$$

and differentiating both ends of this string of equalities to get

$$\mathbf{a} = \frac{d\mathbf{v}}{dt} = \frac{d}{dt}\left(\mathbf{T}\frac{ds}{dt}\right) = \frac{d^2s}{dt^2}\mathbf{T} + \frac{ds}{dt}\frac{d\mathbf{T}}{dt}$$

$$= \frac{d^2s}{dt^2}\mathbf{T} + \frac{ds}{dt}\left(\frac{d\mathbf{T}}{ds}\frac{ds}{dt}\right) = \frac{d^2s}{dt^2}\mathbf{T} + \frac{ds}{dt}\left(\kappa\mathbf{N}\frac{ds}{dt}\right) \qquad \frac{d\mathbf{T}}{ds} = \kappa\mathbf{N}$$

$$= \frac{d^2s}{dt^2}\mathbf{T} + \kappa\left(\frac{ds}{dt}\right)^2\mathbf{N}.$$

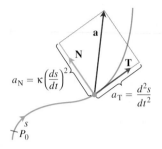

$a_N = \kappa\left(\dfrac{ds}{dt}\right)^2$ $a_T = \dfrac{d^2s}{dt^2}$

s
P_0

FIGURE 11.24 The tangential and normal components of acceleration. The acceleration **a** always lies in the plane of **T** and **N**, orthogonal to **B**.

> **DEFINITION**   If the acceleration vector is written as
>
> $$\mathbf{a} = a_T\mathbf{T} + a_N\mathbf{N}, \tag{1}$$
>
> then
>
> $$a_T = \frac{d^2s}{dt^2} = \frac{d}{dt}\,|\mathbf{v}| \quad \text{and} \quad a_N = \kappa\left(\frac{ds}{dt}\right)^2 = \kappa|\mathbf{v}|^2 \tag{2}$$
>
> are the **tangential** and **normal** scalar components of acceleration.

Notice that the binormal vector **B** does not appear in Equation (1). No matter how the path of the moving body we are watching may appear to twist and turn in space, the acceleration **a** *always lies in the plane of* **T** *and* **N** orthogonal to **B**. The equation also tells us exactly how much of the acceleration takes place tangent to the motion (d^2s/dt^2) and how much takes place normal to the motion $[\kappa(ds/dt)^2]$ (Figure 11.24).

What information can we discover from Equations (2)? By definition, acceleration **a** is the rate of change of velocity **v**, and in general, both the length and direction of **v** change as a body moves along its path. The tangential component of acceleration a_T measures the rate of change of the *length* of **v** (that is, the change in the speed). The normal component of acceleration a_N measures the rate of change of the *direction* of **v**.

Notice that the normal scalar component of the acceleration is the curvature times the *square* of the speed. This explains why you have to hold on when your car makes a sharp (large κ), high-speed (large $|\mathbf{v}|$) turn. If you double the speed of your car, you will experience four times the normal component of acceleration for the same curvature.

If a body moves in a circle at a constant speed, d^2s/dt^2 is zero and all the acceleration points along **N** toward the circle's center. If the body is speeding up or slowing down, **a** has a nonzero tangential component (Figure 11.25).

To calculate a_N, we usually use the formula $a_N = \sqrt{|\mathbf{a}|^2 - a_T{}^2}$, which comes from solving the equation $|\mathbf{a}|^2 = \mathbf{a} \cdot \mathbf{a} = a_T{}^2 + a_N{}^2$ for a_N. With this formula, we can find a_N without having to calculate κ first.

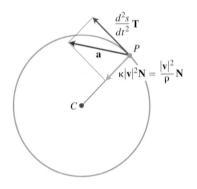

$\dfrac{d^2s}{dt^2}\mathbf{T}$

$\mathbf{a}$

P

$\kappa|\mathbf{v}|^2\mathbf{N} = \dfrac{|\mathbf{v}|^2}{\rho}\mathbf{N}$

C

FIGURE 11.25 The tangential and normal components of the acceleration of a body that is speeding up as it moves counterclockwise around a circle of radius ρ.

> **Formula for Calculating the Normal Component of Acceleration**
>
> $$a_N = \sqrt{|\mathbf{a}|^2 - a_T{}^2} \tag{3}$$

EXAMPLE 1   Without finding **T** and **N**, write the acceleration of the motion

$$\mathbf{r}(t) = (\cos t + t\sin t)\mathbf{i} + (\sin t - t\cos t)\mathbf{j}, \qquad t > 0$$

in the form $\mathbf{a} = a_T\mathbf{T} + a_N\mathbf{N}$. (The path of the motion is the involute of the circle in Figure 11.26.)

Solution   We use the first of Equations (2) to find a_T:

$$\mathbf{v} = \frac{d\mathbf{r}}{dt} = (-\sin t + \sin t + t\cos t)\mathbf{i} + (\cos t - \cos t + t\sin t)\mathbf{j}$$

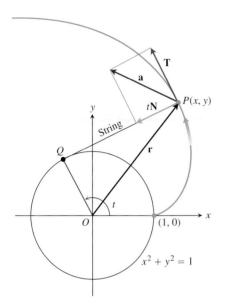

FIGURE 11.26 The tangential and normal components of the acceleration of the motion $\mathbf{r}(t) = (\cos t + t \sin t)\mathbf{i} + (\sin t - t \cos t)\mathbf{j}$, for $t > 0$. If a string wound around a fixed circle is unwound while held taut in the plane of the circle, its end P traces an involute of the circle (Example 1).

$$= (t \cos t)\mathbf{i} + (t \sin t)\mathbf{j}$$

$$|\mathbf{v}| = \sqrt{t^2 \cos^2 t + t^2 \sin^2 t} = \sqrt{t^2} = |t| = t \qquad t > 0$$

$$a_{\mathrm{T}} = \frac{d}{dt}|\mathbf{v}| = \frac{d}{dt}(t) = 1. \qquad \text{Equation (2)}$$

Knowing a_{T}, we use Equation (3) to find a_{N}:

$$\mathbf{a} = (\cos t - t \sin t)\mathbf{i} + (\sin t + t \cos t)\mathbf{j}$$

$$|\mathbf{a}|^2 = t^2 + 1 \qquad \text{After some algebra}$$

$$a_{\mathrm{N}} = \sqrt{|\mathbf{a}|^2 - a_{\mathrm{T}}{}^2}$$

$$= \sqrt{(t^2 + 1) - (1)} = \sqrt{t^2} = t.$$

We then use Equation (1) to find $\mathbf{a}$:

$$\mathbf{a} = a_{\mathrm{T}}\mathbf{T} + a_{\mathrm{N}}\mathbf{N} = (1)\mathbf{T} + (t)\mathbf{N} = \mathbf{T} + t\mathbf{N}. \qquad \blacksquare$$

Torsion

How does $d\mathbf{B}/ds$ behave in relation to $\mathbf{T}$, $\mathbf{N}$, and $\mathbf{B}$? From the rule for differentiating a cross product, we have

$$\frac{d\mathbf{B}}{ds} = \frac{d(\mathbf{T} \times \mathbf{N})}{ds} = \frac{d\mathbf{T}}{ds} \times \mathbf{N} + \mathbf{T} \times \frac{d\mathbf{N}}{ds}.$$

Since $\mathbf{N}$ is the direction of $d\mathbf{T}/ds$, $(d\mathbf{T}/ds) \times \mathbf{N} = \mathbf{0}$ and

$$\frac{d\mathbf{B}}{ds} = \mathbf{0} + \mathbf{T} \times \frac{d\mathbf{N}}{ds} = \mathbf{T} \times \frac{d\mathbf{N}}{ds}.$$

From this we see that $d\mathbf{B}/ds$ is orthogonal to $\mathbf{T}$ since a cross product is orthogonal to its factors.

Since $d\mathbf{B}/ds$ is also orthogonal to $\mathbf{B}$ (the latter has constant length), it follows that $d\mathbf{B}/ds$ is orthogonal to the plane of $\mathbf{B}$ and $\mathbf{T}$. In other words, $d\mathbf{B}/ds$ is parallel to $\mathbf{N}$, so $d\mathbf{B}/ds$ is a scalar multiple of $\mathbf{N}$. In symbols,

$$\frac{d\mathbf{B}}{ds} = -\tau\mathbf{N}.$$

The negative sign in this equation is traditional. The scalar τ is called the *torsion* along the curve. Notice that

$$\frac{d\mathbf{B}}{ds} \cdot \mathbf{N} = -\tau\mathbf{N} \cdot \mathbf{N} = -\tau(1) = -\tau.$$

We use this equation for our next definition.

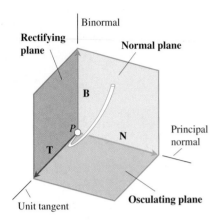

FIGURE 11.27 The names of the three planes determined by **T**, **N**, and **B**.

> **DEFINITION** Let $\mathbf{B} = \mathbf{T} \times \mathbf{N}$. The **torsion** function of a smooth curve is
>
> $$\tau = -\frac{d\mathbf{B}}{ds} \cdot \mathbf{N}. \qquad (4)$$

Unlike the curvature κ, which is never negative, the torsion τ may be positive, negative, or zero.

The three planes determined by **T**, **N**, and **B** are named and shown in Figure 11.27. The curvature $\kappa = |d\mathbf{T}/ds|$ can be thought of as the rate at which the normal plane turns as the point P moves along its path. Similarly, the torsion $\tau = -(d\mathbf{B}/ds) \cdot \mathbf{N}$ is the rate at which the osculating plane turns about **T** as P moves along the curve. Torsion measures how the curve twists.

Look at Figure 11.28. If P is a train climbing up a curved track, the rate at which the headlight turns from side to side per unit distance is the curvature of the track. The rate at which the engine tends to twist out of the plane formed by **T** and **N** is the torsion.

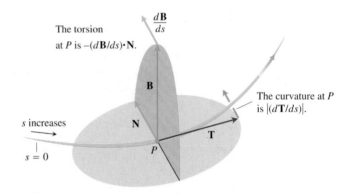

FIGURE 11.28 Every moving body travels with a **TNB** frame that characterizes the geometry of its path of motion.

Computational Formulas

The most widely used formula for torsion, derived in more advanced texts, is

$$\tau = \frac{\begin{vmatrix} \dot{x} & \dot{y} & \dot{z} \\ \ddot{x} & \ddot{y} & \ddot{z} \\ \dddot{x} & \dddot{y} & \dddot{z} \end{vmatrix}}{|\mathbf{v} \times \mathbf{a}|^2} \qquad (\text{if } \mathbf{v} \times \mathbf{a} \neq \mathbf{0}). \qquad (5)$$

The dots in Equation (5) denote differentiation with respect to t, one derivative for each dot. Thus, $\dot{x}$ ("x dot") means dx/dt, $\ddot{x}$ ("x double dot") means d^2x/dt^2, and $\dddot{x}$ ("x triple dot") means d^3x/dt^3. Similarly, $\dot{y} = dy/dt$, and so on.

There is also an easy-to-use formula for curvature, as given in the following summary table (see Exercise 21).

Formulas for Curves in Space

Unit tangent vector:
$$\mathbf{T} = \frac{\mathbf{v}}{|\mathbf{v}|}$$

Principal unit normal vector:
$$\mathbf{N} = \frac{d\mathbf{T}/dt}{|d\mathbf{T}/dt|}$$

Binormal vector:
$$\mathbf{B} = \mathbf{T} \times \mathbf{N}$$

Curvature:
$$\kappa = \left|\frac{d\mathbf{T}}{ds}\right| = \frac{|\mathbf{v} \times \mathbf{a}|}{|\mathbf{v}|^3}$$

Torsion:
$$\tau = -\frac{d\mathbf{B}}{ds} \cdot \mathbf{N} = \frac{\begin{vmatrix} \dot{x} & \dot{y} & \dot{z} \\ \ddot{x} & \ddot{y} & \ddot{z} \\ \dddot{x} & \dddot{y} & \dddot{z} \end{vmatrix}}{|\mathbf{v} \times \mathbf{a}|^2}$$

Tangential and normal scalar components of acceleration:
$$\mathbf{a} = a_T\mathbf{T} + a_N\mathbf{N}$$
$$a_T = \frac{d}{dt}|\mathbf{v}|$$
$$a_N = \kappa|\mathbf{v}|^2 = \sqrt{|\mathbf{a}|^2 - a_T^2}$$

EXERCISES 11.5

In Exercises 1 and 2, write $\mathbf{a}$ in the form $\mathbf{a} = a_T\mathbf{T} + a_N\mathbf{N}$ without finding $\mathbf{T}$ and $\mathbf{N}$.

1. $\mathbf{r}(t) = (a\cos t)\mathbf{i} + (a\sin t)\mathbf{j} + bt\mathbf{k}$

2. $\mathbf{r}(t) = (1 + 3t)\mathbf{i} + (t - 2)\mathbf{j} - 3t\mathbf{k}$

In Exercises 3–6, write $\mathbf{a}$ in the form $\mathbf{a} = a_T\mathbf{T} + a_N\mathbf{N}$ at the given value of t without finding $\mathbf{T}$ and $\mathbf{N}$.

3. $\mathbf{r}(t) = (t + 1)\mathbf{i} + 2t\mathbf{j} + t^2\mathbf{k}, \quad t = 1$

4. $\mathbf{r}(t) = (t\cos t)\mathbf{i} + (t\sin t)\mathbf{j} + t^2\mathbf{k}, \quad t = 0$

5. $\mathbf{r}(t) = t^2\mathbf{i} + (t + (1/3)t^3)\mathbf{j} + (t - (1/3)t^3)\mathbf{k}, \quad t = 0$

6. $\mathbf{r}(t) = (e^t\cos t)\mathbf{i} + (e^t\sin t)\mathbf{j} + \sqrt{2}e^t\mathbf{k}, \quad t = 0$

In Exercises 7 and 8, find $\mathbf{r}$, $\mathbf{T}$, $\mathbf{N}$, and $\mathbf{B}$ at the given value of t. Then find equations for the osculating, normal, and rectifying planes at that value of t.

7. $\mathbf{r}(t) = (\cos t)\mathbf{i} + (\sin t)\mathbf{j} - \mathbf{k}, \quad t = \pi/4$

8. $\mathbf{r}(t) = (\cos t)\mathbf{i} + (\sin t)\mathbf{j} + t\mathbf{k}, \quad t = 0$

In Exercises 9–16 of Section 11.4, you found $\mathbf{T}$, $\mathbf{N}$, and κ. Now, in Exercises 9–16 of Section 11.5, find $\mathbf{B}$ and τ for these space curves.

9. $\mathbf{r}(t) = (3\sin t)\mathbf{i} + (3\cos t)\mathbf{j} + 4t\mathbf{k}$

10. $\mathbf{r}(t) = (\cos t + t\sin t)\mathbf{i} + (\sin t - t\cos t)\mathbf{j} + 3\mathbf{k}$

11. $\mathbf{r}(t) = (e^t\cos t)\mathbf{i} + (e^t\sin t)\mathbf{j} + 2\mathbf{k}$

12. $\mathbf{r}(t) = (6\sin 2t)\mathbf{i} + (6\cos 2t)\mathbf{j} + 5t\mathbf{k}$

13. $\mathbf{r}(t) = (t^3/3)\mathbf{i} + (t^2/2)\mathbf{j}, \quad t > 0$

14. $\mathbf{r}(t) = (\cos^3 t)\mathbf{i} + (\sin^3 t)\mathbf{j}, \quad 0 < t < \pi/2$

15. $\mathbf{r}(t) = t\mathbf{i} + (a\cosh(t/a))\mathbf{j}, \quad a > 0$

16. $\mathbf{r}(t) = (\cosh t)\mathbf{i} - (\sinh t)\mathbf{j} + t\mathbf{k}$

17. The speedometer on your car reads a steady 35 mph. Could you be accelerating? Explain.

18. Can anything be said about the acceleration of a particle that is moving at a constant speed? Give reasons for your answer.

19. Can anything be said about the speed of a particle whose acceleration is always orthogonal to its velocity? Give reasons for your answer.

20. An object of mass m travels along the parabola $y = x^2$ with a constant speed of 10 units/sec. What is the force on the object due to its acceleration at $(0, 0)$? at $(2^{1/2}, 2)$? Write your answers in terms of $\mathbf{i}$ and $\mathbf{j}$. (Remember Newton's law, $\mathbf{F} = m\mathbf{a}$.)

21. Vector formula for curvature For a smooth curve, derive the curvature formula

$$\kappa = \frac{|\mathbf{v} \times \mathbf{a}|}{|\mathbf{v}|^3}.$$

22. Show that a moving particle will move in a straight line if the normal component of its acceleration is zero.

23. A sometime shortcut to curvature If you already know $|a_N|$ and $|\mathbf{v}|$, then the formula $a_N = \kappa|\mathbf{v}|^2$ gives a convenient way to find the curvature. Use it to find the curvature and radius of curvature of the curve

$$\mathbf{r}(t) = (\cos t + t \sin t)\mathbf{i} + (\sin t - t \cos t)\mathbf{j}, \quad t > 0.$$

(Take a_N and $|\mathbf{v}|$ from Example 1.)

24. Show that κ and τ are both zero for the line

$$\mathbf{r}(t) = (x_0 + At)\mathbf{i} + (y_0 + Bt)\mathbf{j} + (z_0 + Ct)\mathbf{k}.$$

25. What can be said about the torsion of a smooth plane curve $\mathbf{r}(t) = f(t)\mathbf{i} + g(t)\mathbf{j}$? Give reasons for your answer.

26. The torsion of a helix Show that the torsion of the helix

$$\mathbf{r}(t) = (a \cos t)\mathbf{i} + (a \sin t)\mathbf{j} + bt\mathbf{k}, \quad a, b \geq 0$$

is $\tau = b/(a^2 + b^2)$. What is the largest value τ can have for a given value of a? Give reasons for your answer.

27. Differentiable curves with zero torsion lie in planes That a sufficiently differentiable curve with zero torsion lies in a plane is a special case of the fact that a particle whose velocity remains perpendicular to a fixed vector $\mathbf{C}$ moves in a plane perpendicular to $\mathbf{C}$. This, in turn, can be viewed as the solution of the following problem in calculus.

Suppose $\mathbf{r}(t) = f(t)\mathbf{i} + g(t)\mathbf{j} + h(t)\mathbf{k}$ is twice differentiable for all t in an interval $[a, b]$, that $\mathbf{r} = 0$ when $t = a$, and that $\mathbf{v} \cdot \mathbf{k} = 0$ for all t in $[a, b]$. Then $h(t) = 0$ for all t in $[a, b]$.

Solve this problem. (*Hint:* Start with $\mathbf{a} = d^2\mathbf{r}/dt^2$ and apply the initial conditions in reverse order.)

28. A formula that calculates τ from B and v If we start with the definition $\tau = -(d\mathbf{B}/ds) \cdot \mathbf{N}$ and apply the Chain Rule to rewrite $d\mathbf{B}/ds$ as

$$\frac{d\mathbf{B}}{ds} = \frac{d\mathbf{B}}{dt}\frac{dt}{ds} = \frac{d\mathbf{B}}{dt}\frac{1}{|\mathbf{v}|},$$

we arrive at the formula

$$\tau = -\frac{1}{|\mathbf{v}|}\left(\frac{d\mathbf{B}}{dt} \cdot \mathbf{N}\right).$$

The advantage of this formula over Equation (5) is that it is easier to derive and state. The disadvantage is that it can take a lot of work to evaluate without a computer. Use the new formula to find the torsion of the helix in Exercise 26.

COMPUTER EXPLORATIONS

Rounding the answers to four decimal places, use a CAS to find $\mathbf{v}$, $\mathbf{a}$, speed, $\mathbf{T}$, $\mathbf{N}$, $\mathbf{B}$, κ, τ, and the tangential and normal components of acceleration for the curves in Exercises 29–32 at the given values of t.

29. $\mathbf{r}(t) = (t \cos t)\mathbf{i} + (t \sin t)\mathbf{j} + t\mathbf{k}, \quad t = \sqrt{3}$

30. $\mathbf{r}(t) = (e^t \cos t)\mathbf{i} + (e^t \sin t)\mathbf{j} + e^t\mathbf{k}, \quad t = \ln 2$

31. $\mathbf{r}(t) = (t - \sin t)\mathbf{i} + (1 - \cos t)\mathbf{j} + \sqrt{-t}\mathbf{k}, \quad t = -3\pi$

32. $\mathbf{r}(t) = (3t - t^2)\mathbf{i} + (3t^2)\mathbf{j} + (3t + t^3)\mathbf{k}, \quad t = 1$

11.6 Velocity and Acceleration in Polar Coordinates

In this section we derive equations for velocity and acceleration in polar coordinates. We then apply them to establish Kepler's second law of planetary motion as a consequence of Newton's law of gravitation.

Motion in Polar and Cylindrical Coordinates

When a particle at $P(r, \theta)$ moves along a curve in the polar coordinate plane, we express its position, velocity, and acceleration in terms of the moving unit vectors

$$\mathbf{u}_r = (\cos \theta)\mathbf{i} + (\sin \theta)\mathbf{j}, \qquad \mathbf{u}_\theta = -(\sin \theta)\mathbf{i} + (\cos \theta)\mathbf{j}, \tag{1}$$

shown in Figure 11.29. The vector $\mathbf{u}_r$ points along the position vector $\overrightarrow{OP}$, so $\mathbf{r} = r\mathbf{u}_r$. The vector $\mathbf{u}_\theta$, orthogonal to $\mathbf{u}_r$, points in the direction of increasing θ.

We find from Equations (1) that

$$\frac{d\mathbf{u}_r}{d\theta} = -(\sin \theta)\mathbf{i} + (\cos \theta)\mathbf{j} = \mathbf{u}_\theta$$

$$\frac{d\mathbf{u}_\theta}{d\theta} = -(\cos \theta)\mathbf{i} - (\sin \theta)\mathbf{j} = -\mathbf{u}_r.$$

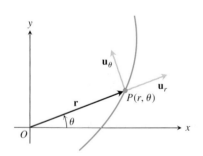

FIGURE 11.29 The length of r is the positive polar coordinate r of the point P. Thus, $\mathbf{u}_r$, which is $\mathbf{r}/|\mathbf{r}|$, is also $\mathbf{r}/r$. Equations (1) express $\mathbf{u}_r$ and $\mathbf{u}_\theta$ in terms of $\mathbf{i}$ and $\mathbf{j}$.

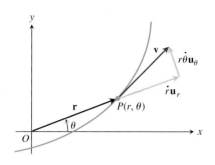

FIGURE 11.30   In polar coordinates, the velocity vector is

$$\mathbf{v} = \dot{r}\,\mathbf{u}_r + r\dot{\theta}\,\mathbf{u}_\theta$$

Notice that $|\mathbf{r}| \neq r$ if $z \neq 0$.

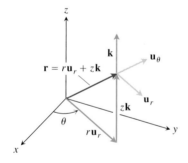

FIGURE 11.31   Position vector and basic unit vectors in cylindrical coordinates.

When we differentiate $\mathbf{u}_r$ and $\mathbf{u}_\theta$ with respect to t to find how they change with time, the Chain Rule gives

$$\dot{\mathbf{u}}_r = \frac{d\mathbf{u}_r}{d\theta}\dot{\theta} = \dot{\theta}\mathbf{u}_\theta, \qquad \dot{\mathbf{u}}_\theta = \frac{d\mathbf{u}_\theta}{d\theta}\dot{\theta} = -\dot{\theta}\mathbf{u}_r. \tag{2}$$

Hence,

$$\mathbf{v} = \dot{\mathbf{r}} = \frac{d}{dt}\left(r\mathbf{u}_r\right) = \dot{r}\mathbf{u}_r + r\dot{\mathbf{u}}_r = \dot{r}\mathbf{u}_r + r\dot{\theta}\,\mathbf{u}_\theta.$$

See Figure 11.30. As in the previous section, we use Newton's dot notation for time derivatives to keep the formulas as simple as we can: $\dot{\mathbf{u}}_r$ means $d\mathbf{u}_r/dt$, $\dot{\theta}$ means $d\theta/dt$, and so on.

The acceleration is

$$\mathbf{a} = \dot{\mathbf{v}} = (\ddot{r}\mathbf{u}_r + \dot{r}\dot{\mathbf{u}}_r) + (\dot{r}\dot{\theta}\mathbf{u}_\theta + r\ddot{\theta}\mathbf{u}_\theta + r\dot{\theta}\dot{\mathbf{u}}_\theta).$$

When Equations (2) are used to evaluate $\dot{\mathbf{u}}_r$ and $\dot{\mathbf{u}}_\theta$ and the components are separated, the equation for acceleration becomes

$$\mathbf{a} = (\ddot{r} - r\dot{\theta}^2)\mathbf{u}_r + (r\ddot{\theta} + 2\dot{r}\dot{\theta})\mathbf{u}_\theta.$$

To extend these equations of motion to space, we add $z\mathbf{k}$ to the right-hand side of the equation $\mathbf{r} = r\mathbf{u}_r$. Then, in these *cylindrical coordinates*, we have

$$\mathbf{r} = r\mathbf{u}_r + z\mathbf{k}$$

$$\mathbf{v} = \dot{r}\mathbf{u}_r + r\dot{\theta}\mathbf{u}_\theta + \dot{z}\mathbf{k} \tag{3}$$

$$\mathbf{a} = (\ddot{r} - r\dot{\theta}^2)\mathbf{u}_r + (r\ddot{\theta} + 2\dot{r}\dot{\theta})\mathbf{u}_\theta + \ddot{z}\mathbf{k}.$$

The vectors $\mathbf{u}_r$, $\mathbf{u}_\theta$, and $\mathbf{k}$ make a right-handed frame (Figure 11.31) in which

$$\mathbf{u}_r \times \mathbf{u}_\theta = \mathbf{k}, \qquad \mathbf{u}_\theta \times \mathbf{k} = \mathbf{u}_r, \qquad \mathbf{k} \times \mathbf{u}_r = \mathbf{u}_\theta.$$

Planets Move in Planes

Newton's law of gravitation says that if $\mathbf{r}$ is the radius vector from the center of a sun of mass M to the center of a planet of mass m, then the force $\mathbf{F}$ of the gravitational attraction between the planet and sun is

$$\mathbf{F} = -\frac{GmM}{|\mathbf{r}|^2}\frac{\mathbf{r}}{|\mathbf{r}|}$$

(Figure 11.32). The number G is the **universal gravitational constant**. If we measure mass in kilograms, force in newtons, and distance in meters, G is about $6.6726 \times 10^{-11}\ \mathrm{Nm^2\ kg^{-2}}$.

Combining the gravitation law with Newton's second law, $\mathbf{F} = m\ddot{\mathbf{r}}$, for the force acting on the planet gives

$$m\ddot{\mathbf{r}} = -\frac{GmM}{|\mathbf{r}|^2}\frac{\mathbf{r}}{|\mathbf{r}|},$$

$$\ddot{\mathbf{r}} = -\frac{GM}{|\mathbf{r}|^2}\frac{\mathbf{r}}{|\mathbf{r}|}.$$

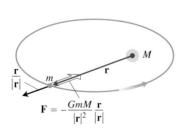

FIGURE 11.32   The force of gravity is directed along the line joining the centers of mass.

The planet is accelerated toward the sun's center of mass at all times.

FIGURE 11.33 A planet that obeys Newton's laws of gravitation and motion travels in the plane through the sun's center of mass perpendicular to $\mathbf{C} = \mathbf{r} \times \dot{\mathbf{r}}$.

HISTORICAL BIOGRAPHY

Johannes Kepler
(1571–1630)

Since $\ddot{\mathbf{r}}$ is a scalar multiple of $\mathbf{r}$, we have

$$\mathbf{r} \times \ddot{\mathbf{r}} = \mathbf{0}.$$

From this last equation,

$$\frac{d}{dt}(\mathbf{r} \times \dot{\mathbf{r}}) = \underbrace{\dot{\mathbf{r}} \times \dot{\mathbf{r}}}_{0} + \mathbf{r} \times \ddot{\mathbf{r}} = \mathbf{r} \times \ddot{\mathbf{r}} = \mathbf{0}.$$

It follows that

$$\mathbf{r} \times \dot{\mathbf{r}} = \mathbf{C} \tag{4}$$

for some constant vector C.

Equation (4) tells us that $\mathbf{r}$ and $\dot{\mathbf{r}}$ always lie in a plane perpendicular to $\mathbf{C}$. Hence, the planet moves in a fixed plane through the center of its sun (Figure 11.33).

Kepler's First Law (Ellipse Law)

Kepler's first law says that a planet's path is an ellipse with the sun at one focus. The eccentricity of the ellipse is

$$e = \frac{r_0 v_0^2}{GM} - 1 \tag{5}$$

and the polar equation is

$$r = \frac{(1 + e)r_0}{1 + e \cos \theta}. \tag{6}$$

Here v_0 is the speed when the planet is positioned at its minimum distance r_0 from the sun. We omit the lengthy proof. The sun's mass M is 1.99×10^{30} kg.

Kepler's Second Law (Equal Area Law)

Kepler's second law says that the radius vector from the sun to a planet (the vector $\mathbf{r}$ in our model) sweeps out equal areas in equal times (Figure 11.34). To derive the law, we use Equation (3) to evaluate the cross product $\mathbf{C} = \mathbf{r} \times \dot{\mathbf{r}}$ from Equation (4):

$$\begin{aligned}
\mathbf{C} = \mathbf{r} \times \dot{\mathbf{r}} &= \mathbf{r} \times \mathbf{v} \\
&= r\mathbf{u}_r \times (\dot{r}\mathbf{u}_r + r\dot{\theta}\mathbf{u}_\theta) \qquad \text{Equation (3), } \dot{z} = 0 \\
&= r\dot{r}\underbrace{(\mathbf{u}_r \times \mathbf{u}_r)}_{0} + r(r\dot{\theta})\underbrace{(\mathbf{u}_r \times \mathbf{u}_\theta)}_{\mathbf{k}} \\
&= r(r\dot{\theta})\mathbf{k}. \tag{7}
\end{aligned}$$

Setting t equal to zero shows that

$$\mathbf{C} = [r(r\dot{\theta})]_{t=0}\, \mathbf{k} = r_0 v_0 \mathbf{k}.$$

Substituting this value for $\mathbf{C}$ in Equation (7) gives

$$r_0 v_0 \mathbf{k} = r^2\dot{\theta}\mathbf{k}, \qquad \text{or} \qquad r^2\dot{\theta} = r_0 v_0.$$

This is where the area comes in. The area differential in polar coordinates is

$$dA = \frac{1}{2}r^2\, d\theta$$

FIGURE 11.34 The line joining a planet to its sun sweeps over equal areas in equal times.

(Section 9.3). Accordingly, dA/dt has the constant value

$$\frac{dA}{dt} = \frac{1}{2} r^2 \dot{\theta} = \frac{1}{2} r_0 v_0. \tag{8}$$

So dA/dt is constant, giving Kepler's second law.

Kepler's Third Law (Time–Distance Law)

The time T it takes a planet to go around its sun once is the planet's **orbital period**. *Kepler's third law* says that T and the orbit's semimajor axis a are related by the equation

$$\frac{T^2}{a^3} = \frac{4\pi^2}{GM}.$$

Since the right-hand side of this equation is constant within a given solar system, the ratio of T^2 to a^3 *is the same for every planet in the system.*

Here is a partial derivation of Kepler's third law. The area enclosed by the planet's elliptical orbit is calculated as follows:

$$\text{Area} = \int_0^T dA$$

$$= \int_0^T \frac{1}{2} r_0 v_0 \, dt \qquad \text{Equation (8)}$$

$$= \frac{1}{2} T r_0 v_0.$$

If b is the semiminor axis, the area of the ellipse is πab, so

$$T = \frac{2\pi ab}{r_0 v_0} = \frac{2\pi a^2}{r_0 v_0} \sqrt{1 - e^2}. \qquad \begin{array}{l} \text{For any ellipse,} \\ b = a\sqrt{1-e^2} \end{array} \tag{9}$$

It remains only to express a and e in terms of r_0, v_0, G, and M. Equation (5) does this for e. For a, we observe that setting θ equal to π in Equation (6) gives

$$r_{max} = r_0 \frac{1 + e}{1 - e}.$$

Hence, from Figure 11.34,

$$2a = r_0 + r_{max} = \frac{2r_0}{1 - e} = \frac{2r_0 GM}{2GM - r_0 v_0^2}. \tag{10}$$

Squaring both sides of Equation (9) and substituting the results of Equations (5) and (10) produces Kepler's third law (Exercise 9).

EXERCISES 11.6

In Problems 1–5, find the velocity and acceleration vectors in terms of $\mathbf{u}_r$ and $\mathbf{u}_\theta$.

1. $r = a(1 - \cos \theta)$ and $\dfrac{d\theta}{dt} = 3$

2. $r = a \sin 2\theta$ and $\dfrac{d\theta}{dt} = 2t$

3. $r = e^{a\theta}$ and $\dfrac{d\theta}{dt} = 2$

4. $r = a(1 + \sin t)$ and $\theta = 1 - e^{-t}$

5. $r = 2 \cos 4t$ and $\theta = 2t$

6. Type of orbit For what values of v_0 in Equation (5) is the orbit in Equation (6) a circle? An ellipse? A parabola? A hyperbola?

7. **Circular orbits**   Show that a planet in a circular orbit moves with a constant speed. (*Hint:* This is a consequence of one of Kepler's laws.)

8. Suppose that **r** is the position vector of a particle moving along a plane curve and dA/dt is the rate at which the vector sweeps out area. Without introducing coordinates, and assuming the necessary derivatives exist, give a geometric argument based on increments and limits for the validity of the equation

$$\frac{dA}{dt} = \frac{1}{2}|\mathbf{r} \times \dot{\mathbf{r}}|.$$

9. **Kepler's third law**   Complete the derivation of Kepler's third law (the part following Equation (10)).

10. Find the length of the major axis of the Earth's orbit using Kepler's third law and the fact that the Earth's orbital period is 365.256 days.

Chapter 11   Questions to Guide Your Review

1. State the rules for differentiating and integrating vector functions. Give examples.

2. How do you define and calculate the velocity, speed, direction of motion, and acceleration of a body moving along a sufficiently differentiable space curve? Give an example.

3. What is special about the derivatives of vector functions of constant length? Give an example.

4. What are the vector and parametric equations for ideal projectile motion? How do you find a projectile's maximum height, flight time, and range? Give examples.

5. How do you define and calculate the length of a segment of a smooth space curve? Give an example. What mathematical assumptions are involved in the definition?

6. How do you measure distance along a smooth curve in space from a preselected base point? Give an example.

7. What is a differentiable curve's unit tangent vector? Give an example.

8. Define curvature, circle of curvature (osculating circle), center of curvature, and radius of curvature for twice-differentiable curves in the plane. Give examples. What curves have zero curvature? Constant curvature?

9. What is a plane curve's principal normal vector? When is it defined? Which way does it point? Give an example.

10. How do you define **N** and κ for curves in space? How are these quantities related? Give examples.

11. What is a curve's binormal vector? Give an example. How is this vector related to the curve's torsion? Give an example.

12. What formulas are available for writing a moving body's acceleration as a sum of its tangential and normal components? Give an example. Why might one want to write the acceleration this way? What if the body moves at a constant speed? At a constant speed around a circle?

13. State Kepler's laws.

Chapter 11   Practice Exercises

In Exercises 1 and 2, graph the curves and sketch their velocity and acceleration vectors at the given values of t. Then write **a** in the form $\mathbf{a} = a_T\mathbf{T} + a_N\mathbf{N}$ without finding **T** and **N**, and find the value of κ at the given values of t.

1. $\mathbf{r}(t) = (4\cos t)\mathbf{i} + \left(\sqrt{2}\sin t\right)\mathbf{j}, \quad t = 0$ and $\pi/4$

2. $\mathbf{r}(t) = \left(\sqrt{3}\sec t\right)\mathbf{i} + \left(\sqrt{3}\tan t\right)\mathbf{j}, \quad t = 0$

3. The position of a particle in the plane at time t is

$$\mathbf{r} = \frac{1}{\sqrt{1 + t^2}}\mathbf{i} + \frac{t}{\sqrt{1 + t^2}}\mathbf{j}.$$

Find the particle's highest speed.

4. Suppose $\mathbf{r}(t) = (e^t\cos t)\mathbf{i} + (e^t\sin t)\mathbf{j}$. Show that the angle between **r** and **a** never changes. What *is* the angle?

5. **Finding curvature**   At point P, the velocity and acceleration of a particle moving in the plane are $\mathbf{v} = 3\mathbf{i} + 4\mathbf{j}$ and $\mathbf{a} = 5\mathbf{i} + 15\mathbf{j}$. Find the curvature of the particle's path at P.

6. Find the point on the curve $y = e^x$ where the curvature is greatest.

7. A particle moves around the unit circle in the xy-plane. Its position at time t is $\mathbf{r} = x\mathbf{i} + y\mathbf{j}$, where x and y are differentiable functions of t. Find dy/dt if $\mathbf{v} \cdot \mathbf{i} = y$. Is the motion clockwise, or counterclockwise?

8. You send a message through a pneumatic tube that follows the curve $9y = x^3$ (distance in meters). At the point $(3, 3)$, $\mathbf{v} \cdot \mathbf{i} = 4$ and $\mathbf{a} \cdot \mathbf{i} = -2$. Find the values of $\mathbf{v} \cdot \mathbf{j}$ and $\mathbf{a} \cdot \mathbf{j}$ at $(3, 3)$.

9. **Characterizing circular motion**   A particle moves in the plane so that its velocity and position vectors are always orthogonal. Show that the particle moves in a circle centered at the origin.

10. Speed along a cycloid A circular wheel with radius 1 ft and center C rolls to the right along the x-axis at a half-turn per second. (See the accompanying figure.) At time t seconds, the position vector of the point P on the wheel's circumference is

$$\mathbf{r} = (\pi t - \sin \pi t)\mathbf{i} + (1 - \cos \pi t)\mathbf{j}.$$

a. Sketch the curve traced by P during the interval $0 \le t \le 3$.

b. Find $\mathbf{v}$ and $\mathbf{a}$ at $t = 0, 1, 2$, and 3 and add these vectors to your sketch.

c. At any given time, what is the forward speed of the topmost point of the wheel? Of C?

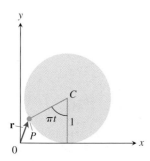

11. Shot put A shot leaves the thrower's hand 6.5 ft above the ground at a 45° angle at 44 ft/sec. Where is it 3 sec later?

12. Javelin A javelin leaves the thrower's hand 7 ft above the ground at a 45° angle at 80 ft/sec. How high does it go?

13. A golf ball is hit with an initial speed v_0 at an angle α to the horizontal from a point that lies at the foot of a straight-sided hill that is inclined at an angle ϕ to the horizontal, where

$$0 < \phi < \alpha < \frac{\pi}{2}.$$

Show that the ball lands at a distance

$$\frac{2v_0^2 \cos \alpha}{g \cos^2 \phi} \sin (\alpha - \phi),$$

measured up the face of the hill. Hence, show that the greatest range that can be achieved for a given v_0 occurs when $\alpha = (\phi/2) + (\pi/4)$, i.e., when the initial velocity vector bisects the angle between the vertical and the hill.

T **14. Javelin** In Potsdam in 1988, Petra Felke of (then) East Germany set a women's world record by throwing a javelin 262 ft 5 in.

a. Assuming that Felke launched the javelin at a 40° angle to the horizontal 6.5 ft above the ground, what was the javelin's initial speed?

b. How high did the javelin go?

Find the lengths of the curves in Exercises 15 and 16.

15. $\mathbf{r}(t) = (2 \cos t)\mathbf{i} + (2 \sin t)\mathbf{j} + t^2\mathbf{k}, \quad 0 \le t \le \pi/4$

16. $\mathbf{r}(t) = (3 \cos t)\mathbf{i} + (3 \sin t)\mathbf{j} + 2t^{3/2}\mathbf{k}, \quad 0 \le t \le 3$

In Exercises 17–20, find $\mathbf{T}, \mathbf{N}, \mathbf{B}, \kappa$, and τ at the given value of t.

17. $\mathbf{r}(t) = \dfrac{4}{9}(1 + t)^{3/2}\mathbf{i} + \dfrac{4}{9}(1 - t)^{3/2}\mathbf{j} + \dfrac{1}{3}t\mathbf{k}, \quad t = 0$

18. $\mathbf{r}(t) = (e^t \sin 2t)\mathbf{i} + (e^t \cos 2t)\mathbf{j} + 2e^t\mathbf{k}, \quad t = 0$

19. $\mathbf{r}(t) = t\mathbf{i} + \dfrac{1}{2}e^{2t}\mathbf{j}, \quad t = \ln 2$

20. $\mathbf{r}(t) = (3 \cosh 2t)\mathbf{i} + (3 \sinh 2t)\mathbf{j} + 6t\mathbf{k}, \quad t = \ln 2$

In Exercises 21 and 22, write $\mathbf{a}$ in the form $\mathbf{a} = a_T\mathbf{T} + a_N\mathbf{N}$ at $t = 0$ without finding $\mathbf{T}$ and $\mathbf{N}$.

21. $\mathbf{r}(t) = (2 + 3t + 3t^2)\mathbf{i} + (4t + 4t^2)\mathbf{j} - (6 \cos t)\mathbf{k}$

22. $\mathbf{r}(t) = (2 + t)\mathbf{i} + (t + 2t^2)\mathbf{j} + (1 + t^2)\mathbf{k}$

23. Find $\mathbf{T}, \mathbf{N}, \mathbf{B}, \kappa$, and τ as functions of t if $\mathbf{r}(t) = (\sin t)\mathbf{i} + \left(\sqrt{2} \cos t\right)\mathbf{j} + (\sin t)\mathbf{k}$.

24. At what times in the interval $0 \le t \le \pi$ are the velocity and acceleration vectors of the motion $\mathbf{r}(t) = \mathbf{i} + (5 \cos t)\mathbf{j} + (3 \sin t)\mathbf{k}$ orthogonal?

25. The position of a particle moving in space at time $t \ge 0$ is

$$\mathbf{r}(t) = 2\mathbf{i} + \left(4 \sin \frac{t}{2}\right)\mathbf{j} + \left(3 - \frac{t}{\pi}\right)\mathbf{k}.$$

Find the first time $\mathbf{r}$ is orthogonal to the vector $\mathbf{i} - \mathbf{j}$.

26. Find equations for the osculating, normal, and rectifying planes of the curve $\mathbf{r}(t) = t\mathbf{i} + t^2\mathbf{j} + t^3\mathbf{k}$ at the point $(1, 1, 1)$.

27. Find parametric equations for the line that is tangent to the curve $\mathbf{r}(t) = e^t\mathbf{i} + (\sin t)\mathbf{j} + \ln (1 - t)\mathbf{k}$ at $t = 0$.

28. Find parametric equations for the line tangent to the helix $\mathbf{r}(t) = \left(\sqrt{2} \cos t\right)\mathbf{i} + \left(\sqrt{2} \sin t\right)\mathbf{j} + t\mathbf{k}$ at the point where $t = \pi/4$.

29. Synchronous curves By eliminating α from the ideal projectile equations

$$x = (v_0 \cos \alpha)t, \quad y = (v_0 \sin \alpha)t - \frac{1}{2}gt^2,$$

show that $x^2 + (y + gt^2/2)^2 = v_0^2 t^2$. This shows that projectiles launched simultaneously from the origin at the same initial speed will, at any given instant, all lie on the circle of radius $v_0 t$ centered at $(0, -gt^2/2)$, regardless of their launch angle. These circles are the *synchronous curves* of the launching.

30. Radius of curvature Show that the radius of curvature of a twice-differentiable plane curve $\mathbf{r}(t) = f(t)\mathbf{i} + g(t)\mathbf{j}$ is given by the formula

$$\rho = \frac{\dot{x}^2 + \dot{y}^2}{\sqrt{\dot{x}^2 + \ddot{y}^2 - \ddot{s}^2}}, \quad \text{where} \quad \ddot{s} = \frac{d}{dt}\sqrt{\dot{x}^2 + \dot{y}^2}.$$

31. An alternative definition of curvature in the plane   An alternative definition gives the curvature of a sufficiently differentiable plane curve to be $|d\phi/ds|$, where ϕ is the angle between $\mathbf{T}$ and $\mathbf{i}$ (Figure 11.35a). Figure 11.35b shows the distance s measured counterclockwise around the circle $x^2 + y^2 = a^2$ from the point $(a, 0)$ to a point P, along with the angle ϕ at P. Calculate the circle's curvature using the alternative definition. (*Hint:* $\phi = \theta + \pi/2$.)

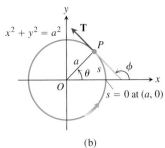

FIGURE 11.35   Figures for Exercise 31.

32. The view from *Skylab 4*   What percentage of Earth's surface area could the astronauts see when *Skylab 4* was at its apogee height, 437 km above the surface? To find out, model the visible surface as the surface generated by revolving the circular arc GT, shown here, about the y-axis. Then carry out these steps:

1. Use similar triangles in the figure to show that $y_0/6380 = 6380/(6380 + 437)$. Solve for y_0.

2. To four significant digits, calculate the visible area as

$$VA = \int_{y_0}^{6380} 2\pi x \sqrt{1 + \left(\frac{dx}{dy}\right)^2}\, dy.$$

3. Express the result as a percentage of Earth's surface area.

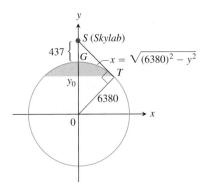

Chapter **11**   Additional and Advanced Exercises

1. A frictionless particle P, starting from rest at time $t = 0$ at the point $(a, 0, 0)$, slides down the helix

$$\mathbf{r}(\theta) = (a \cos \theta)\mathbf{i} + (a \sin \theta)\mathbf{j} + b\theta\mathbf{k} \quad (a, b > 0)$$

under the influence of gravity, as in the accompanying figure. The θ in this equation is the cylindrical coordinate θ and the helix is the curve $r = a, z = b\theta, \theta \geq 0$, in cylindrical coordinates. We assume θ to be a differentiable function of t for the motion. The law of conservation of energy tells us that the particle's speed after it has fallen straight down a distance z is $\sqrt{2gz}$, where g is the constant acceleration of gravity.

a. Find the angular velocity $d\theta/dt$ when $\theta = 2\pi$.

b. Express the particle's θ- and z-coordinates as functions of t.

c. Express the tangential and normal components of the velocity $d\mathbf{r}/dt$ and acceleration $d^2\mathbf{r}/dt^2$ as functions of t. Does the acceleration have any nonzero component in the direction of the binormal vector $\mathbf{B}$?

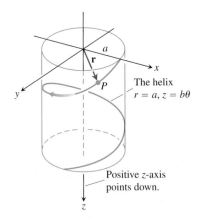

2. Suppose the curve in Exercise 1 is replaced by the conical helix $r = a\theta, z = b\theta$ shown in the next accompanying figure.

a. Express the angular velocity $d\theta/dt$ as a function of θ.

b. Express the distance the particle travels along the helix as a function of θ.

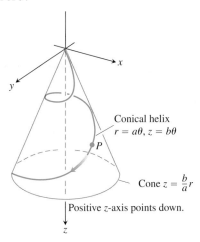

Conical helix
$r = a\theta, z = b\theta$

P

Cone $z = \dfrac{b}{a}r$

Positive z-axis points down.

3. Deduce from the orbit equation

$$r = \frac{(1 + e)r_0}{1 + e \cos \theta}$$

that a planet is closest to its sun when $\theta = 0$ and show that $r = r_0$ at that time.

T **4. A Kepler equation** The problem of locating a planet in its orbit at a given time and date eventually leads to solving "Kepler" equations of the form

$$f(x) = x - 1 - \frac{1}{2} \sin x = 0.$$

a. Show that this particular equation has a solution between $x = 0$ and $x = 2$.

b. With your computer or calculator in radian mode, use Newton's method to find the solution to as many places as you can.

5. In Section 11.6, we found the velocity of a particle moving in the plane to be

$$\mathbf{v} = \dot{x}\,\mathbf{i} + \dot{y}\,\mathbf{j} = \dot{r}\,\mathbf{u}_r + r\dot{\theta}\,\mathbf{u}_\theta.$$

a. Express $\dot{x}$ and $\dot{y}$ in terms of $\dot{r}$ and $r\dot{\theta}$ by evaluating the dot products $\mathbf{v} \cdot \mathbf{i}$ and $\mathbf{v} \cdot \mathbf{j}$.

b. Express $\dot{r}$ and $r\dot{\theta}$ in terms of $\dot{x}$ and $\dot{y}$ by evaluating the dot products $\mathbf{v} \cdot \mathbf{u}_r$ and $\mathbf{v} \cdot \mathbf{u}_\theta$.

6. Express the curvature of a twice-differentiable curve $r = f(\theta)$ in the polar coordinate plane in terms of f and its derivatives.

7. A slender rod through the origin of the polar coordinate plane rotates (in the plane) about the origin at the rate of 3 rad/min. A beetle starting from the point $(2, 0)$ crawls along the rod toward the origin at the rate of 1 in./min.

a. Find the beetle's acceleration and velocity in polar form when it is halfway to (1 in. from) the origin.

T **b.** To the nearest tenth of an inch, what will be the length of the path the beetle has traveled by the time it reaches the origin?

8. Conservation of angular momentum Let $\mathbf{r}(t)$ denote the position in space of a moving object at time t. Suppose the force acting on the object at time t is

$$\mathbf{F}(t) = -\frac{c}{|\mathbf{r}(t)|^3}\,\mathbf{r}(t),$$

where c is a constant. In physics the **angular momentum** of an object at time t is defined to be $\mathbf{L}(t) = \mathbf{r}(t) \times m\mathbf{v}(t)$, where m is the mass of the object and $\mathbf{v}(t)$ is the velocity. Prove that angular momentum is a conserved quantity; i.e., prove that $\mathbf{L}(t)$ is a constant vector, independent of time. Remember Newton's law $\mathbf{F} = m\mathbf{a}$. (This is a calculus problem, not a physics problem.)

9. Unit vectors for position and motion in cylindrical coordinates When the position of a particle moving in space is given in cylindrical coordinates, the unit vectors we use to describe its position and motion are

$$\mathbf{u}_r = (\cos \theta)\mathbf{i} + (\sin \theta)\mathbf{j}, \qquad \mathbf{u}_\theta = -(\sin \theta)\mathbf{i} + (\cos \theta)\mathbf{j},$$

and $\mathbf{k}$ (see accompanying figure). The particle's position vector is then $\mathbf{r} = r\mathbf{u}_r + z\mathbf{k}$, where r is the positive polar distance coordinate of the particle's position.

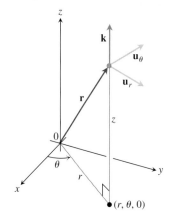

$(r, \theta, 0)$

a. Show that $\mathbf{u}_r$, $\mathbf{u}_\theta$, and $\mathbf{k}$, in this order, form a right-handed frame of unit vectors.

b. Show that

$$\frac{d\mathbf{u}_r}{d\theta} = \mathbf{u}_\theta \quad \text{and} \quad \frac{d\mathbf{u}_\theta}{d\theta} = -\mathbf{u}_r.$$

c. Assuming that the necessary derivatives with respect to t exist, express $\mathbf{v} = \dot{\mathbf{r}}$ and $\mathbf{a} = \ddot{\mathbf{r}}$ in terms of $\mathbf{u}_r$, $\mathbf{u}_\theta$, $\mathbf{k}$, $\dot{r}$, and $\dot{\theta}$.

10. Arc length in cylindrical coordinates

a. Show that when you express $ds^2 = dx^2 + dy^2 + dz^2$ in terms of cylindrical coordinates, you get $ds^2 = dr^2 + r^2\,d\theta^2 + dz^2$.

b. Interpret this result geometrically in terms of the edges and a diagonal of a box. Sketch the box.

c. Use the result in part (a) to find the length of the curve $r = e^\theta, z = e^\theta, 0 \le \theta \le \theta \ln 8$.

12

PARTIAL DERIVATIVES

OVERVIEW Many functions depend on more than one independent variable. For instance, the volume of a right circular cylinder is a function $V = \pi r^2 h$ of its radius and its height, so it is a function $V(r, h)$ of two variables. In this chapter we extend the basic ideas of single variable calculus to functions of several variables. Their derivatives are more varied and interesting because of the different ways the variables can interact. Their integrals (studied in the next chapter) lead to a great variety of applications.

12.1 Functions of Several Variables

In this section we define functions of more than one independent variable and discuss ways to graph them.

Real-valued functions of several independent real variables are defined similarly to functions in the single-variable case. The domains are sets of ordered pairs (triples, quadruples, n-tuples) of real numbers, and the ranges are sets of real numbers of the kind we have worked with all along.

DEFINITIONS Suppose D is a set of n-tuples of real numbers $(x_1, x_2, \ldots, x_n)$. A **real-valued function** f on D is a rule that assigns a unique (single) real number

$$w = f(x_1, x_2, \ldots, x_n)$$

to each element in D. The set D is the function's **domain**. The set of w-values taken on by f is the function's **range**. The symbol w is the **dependent variable** of f, and f is said to be a function of the n **independent variables** x_1 to x_n. We also call the x_j's the function's **input variables** and call w the function's **output variable**.

If f is a function of two independent variables, we usually call the independent variables x and y and picture the domain of f as a region in the xy-plane. If f is a function of three independent variables, we call the variables x, y, and z and picture the domain as a region in space.

In applications, we tend to use letters that remind us of what the variables stand for. To say that the volume of a right circular cylinder is a function of its radius and height, we might write $V = f(r, h)$. To be more specific, we might replace the notation $f(r, h)$ by the formula that calculates the value of V from the values of r and h, and write $V = \pi r^2 h$. In either case, r and h would be the independent variables and V the dependent variable of the function.

As usual, we evaluate functions defined by formulas by substituting the values of the independent variables in the formula and calculating the corresponding value of the dependent variable. For example, the value of $f(x, y, z) = \sqrt{x^2 + y^2 + z^2}$ at the point $(3, 0, 4)$ is

$$f(3, 0, 4) = \sqrt{(3)^2 + (0)^2 + (4)^2} = \sqrt{25} = 5.$$

Domains and Ranges

In defining a function of more than one variable, we follow the usual practice of excluding inputs that lead to complex numbers or division by zero. If $f(x, y) = \sqrt{y - x^2}$, y cannot be less than x^2. If $f(x, y) = 1/(xy)$, xy cannot be zero. The domain of a function is assumed to be the largest set for which the defining rule generates real numbers, unless the domain is otherwise specified explicitly. The range consists of the set of output values for the dependent variable.

EXAMPLE 1(a)   These are functions of two variables.

Function	Domain	Range
$w = \sqrt{y - x^2}$	$y \geq x^2$	$[0, \infty)$
$w = \dfrac{1}{xy}$	$xy \neq 0$	$(-\infty, 0) \cup (0, \infty)$
$w = \sin xy$	Entire plane	$[-1, 1]$

(b)   These are functions of three variables.

Function	Domain	Range
$w = \sqrt{x^2 + y^2 + z^2}$	Entire space	$[0, \infty)$
$w = \dfrac{1}{x^2 + y^2 + z^2}$	$(x, y, z) \neq (0, 0, 0)$	$(0, \infty)$
$w = xy \ln z$	Half-space $z > 0$	$(-\infty, \infty)$

Functions of Two Variables

Regions in the plane can have interior points and boundary points just like intervals on the real line. Closed intervals $[a, b]$ include their boundary points, open intervals (a, b) don't include their boundary points, and intervals such as $[a, b)$ are neither open nor closed.

(a) Interior point

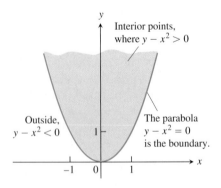

(b) Boundary point

FIGURE 12.1 Interior points and boundary points of a plane region R. An interior point is necessarily a point of R. A boundary point of R need not belong to R.

DEFINITIONS A point (x_0, y_0) in a region (set) R in the xy-plane is an **interior point** of R if it is the center of a disk of positive radius that lies entirely in R (Figure 12.1). A point (x_0, y_0) is a **boundary point** of R if every disk centered at (x_0, y_0) contains points that lie outside of R as well as points that lie in R. (The boundary point itself need not belong to R.)

The interior points of a region, as a set, make up the **interior** of the region. The region's boundary points make up its **boundary**. A region is **open** if it consists entirely of interior points. A region is **closed** if it contains all its boundary points (Figure 12.2).

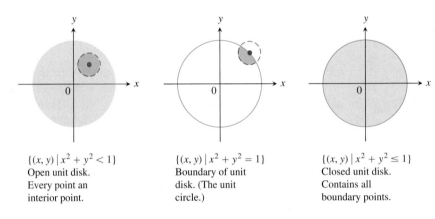

| $\{(x, y) \mid x^2 + y^2 < 1\}$
Open unit disk.
Every point an
interior point. | $\{(x, y) \mid x^2 + y^2 = 1\}$
Boundary of unit
disk. (The unit
circle.) | $\{(x, y) \mid x^2 + y^2 \le 1\}$
Closed unit disk.
Contains all
boundary points. |

FIGURE 12.2 Interior points and boundary points of the unit disk in the plane.

As with a half-open interval of real numbers [a, b), some regions in the plane are neither open nor closed. If you start with the open disk in Figure 12.2 and add to it some of but not all its boundary points, the resulting set is neither open nor closed. The boundary points that *are* there keep the set from being open. The absence of the remaining boundary points keeps the set from being closed.

DEFINITIONS A region in the plane is **bounded** if it lies inside a disk of fixed radius. A region is **unbounded** if it is not bounded.

Examples of *bounded* sets in the plane include line segments, triangles, interiors of triangles, rectangles, circles, and disks. Examples of *unbounded* sets in the plane include lines, coordinate axes, the graphs of functions defined on infinite intervals, quadrants, half-planes, and the plane itself.

EXAMPLE 2 Describe the domain of the function $f(x, y) = \sqrt{y - x^2}$.

FIGURE 12.3 The domain of $f(x, y)$ in Example 2 consists of the shaded region and its bounding parabola.

Solution Since f is defined only where $y - x^2 \ge 0$, the domain is the closed, unbounded region shown in Figure 12.3. The parabola $y = x^2$ is the boundary of the domain. The points above the parabola make up the domain's interior. ∎

Graphs, Level Curves, and Contours of Functions of Two Variables

There are two standard ways to picture the values of a function $f(x, y)$. One is to draw and label curves in the domain on which f has a constant value. The other is to sketch the surface $z = f(x, y)$ in space.

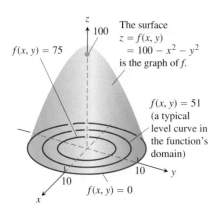

FIGURE 12.4 The graph and selected level curves of the function $f(x, y)$ in Example 3.

> **DEFINITIONS** The set of points in the plane where a function $f(x, y)$ has a constant value $f(x, y) = c$ is called a **level curve** of f. The set of all points $(x, y, f(x, y))$ in space, for (x, y) in the domain of f, is called the **graph** of f. The graph of f is also called the **surface $z = f(x, y)$**.

EXAMPLE 3 Graph $f(x, y) = 100 - x^2 - y^2$ and plot the level curves $f(x, y) = 0$, $f(x, y) = 51$, and $f(x, y) = 75$ in the domain of f in the plane.

Solution The domain of f is the entire xy-plane, and the range of f is the set of real numbers less than or equal to 100. The graph is the paraboloid $z = 100 - x^2 - y^2$, a portion of which is shown in Figure 12.4.

The level curve $f(x, y) = 0$ is the set of points in the xy-plane at which

$$f(x, y) = 100 - x^2 - y^2 = 0, \quad \text{or} \quad x^2 + y^2 = 100,$$

which is the circle of radius 10 centered at the origin. Similarly, the level curves $f(x, y) = 51$ and $f(x, y) = 75$ (Figure 12.4) are the circles

$$f(x, y) = 100 - x^2 - y^2 = 51, \quad \text{or} \quad x^2 + y^2 = 49$$
$$f(x, y) = 100 - x^2 - y^2 = 75, \quad \text{or} \quad x^2 + y^2 = 25.$$

The level curve $f(x, y) = 100$ consists of the origin alone. (It is still a level curve.) ∎

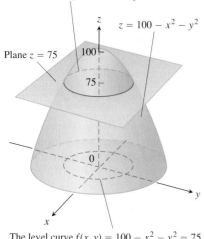

The contour curve $f(x, y) = 100 - x^2 - y^2 = 75$ is the circle $x^2 + y^2 = 25$ in the plane $z = 75$.

The level curve $f(x, y) = 100 - x^2 - y^2 = 75$ is the circle $x^2 + y^2 = 25$ in the xy-plane.

FIGURE 12.5 A plane $z = c$ parallel to the xy-plane intersecting a surface $z = f(x, y)$ produces a contour curve.

The curve in space in which the plane $z = c$ cuts a surface $z = f(x, y)$ is made up of the points that represent the function value $f(x, y) = c$. It is called the **contour curve** $f(x, y) = c$ to distinguish it from the level curve $f(x, y) = c$ in the domain of f. Figure 12.5 shows the contour curve $f(x, y) = 75$ on the surface $z = 100 - x^2 - y^2$ defined by the function $f(x, y) = 100 - x^2 - y^2$. The contour curve lies directly above the circle $x^2 + y^2 = 25$, which is the level curve $f(x, y) = 75$ in the function's domain.

Not everyone makes this distinction, however, and you may wish to call both kinds of curves by a single name and rely on context to convey which one you have in mind. On most maps, for example, the curves that represent constant elevation (height above sea level) are called contours, not level curves (Figure 12.6).

Functions of Three Variables

In the plane, the points where a function of two independent variables has a constant value $f(x, y) = c$ make a curve in the function's domain. In space, the points where a function of three independent variables has a constant value $f(x, y, z) = c$ make a surface in the function's domain.

FIGURE 12.6 Contours on Mt. Washington in New Hampshire. (Reproduced by permission from the Appalachian Mountain Club.)

DEFINITION The set of points (x, y, z) in space where a function of three independent variables has a constant value $f(x, y, z) = c$ is called a **level surface** of f.

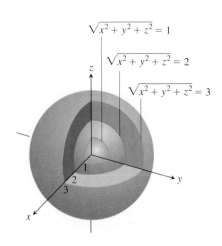

FIGURE 12.7 The level surfaces of $f(x, y, z) = \sqrt{x^2 + y^2 + z^2}$ are concentric spheres (Example 4).

Since the graphs of functions of three variables consist of points $(x, y, z, f(x, y, z))$ lying in a four-dimensional space, we cannot sketch them effectively in our three-dimensional frame of reference. We can see how the function behaves, however, by looking at its three-dimensional level surfaces.

EXAMPLE 4 Describe the level surfaces of the function

$$f(x, y, z) = \sqrt{x^2 + y^2 + z^2}.$$

Solution The value of f is the distance from the origin to the point (x, y, z). Each level surface $\sqrt{x^2 + y^2 + z^2} = c$, $c > 0$, is a sphere of radius c centered at the origin. Figure 12.7 shows a cutaway view of three of these spheres. The level surface $\sqrt{x^2 + y^2 + z^2} = 0$ consists of the origin alone.

We are not graphing the function here; we are looking at level surfaces in the function's domain. The level surfaces show how the function's values change as we move through its domain. If we remain on a sphere of radius c centered at the origin, the function maintains a constant value, namely c. If we move from a point on one sphere to a point on

(a) Interior point

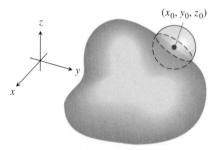

(b) Boundary point

FIGURE 12.8 Interior points and boundary points of a region in space.

another, the function's value changes. It increases if we move away from the origin and decreases if we move toward the origin. The way the values change depends on the direction we take. The dependence of change on direction is important. We return to it in Section 12.5. ∎

The definitions of interior, boundary, open, closed, bounded, and unbounded for regions in space are similar to those for regions in the plane. To accommodate the extra dimension, we use solid balls of positive radius instead of disks.

DEFINITIONS A point (x_0, y_0, z_0) in a region R in space is an **interior point** of R if it is the center of a solid ball that lies entirely in R (Figure 12.8a). A point (x_0, y_0, z_0) is a **boundary point** of R if every solid ball centered at (x_0, y_0, z_0) contain points that lie outside of R as well as points that lie inside R (Figure 12.8b). The **interior** of R is the set of interior points of R. The **boundary** of R is the set of boundary points of R.

A region is **open** if it consists entirely of interior points. A region is **closed** if it contains its entire boundary.

Examples of *open* sets in space include the interior of a sphere, the open half-space $z > 0$, the first octant (where x, y, and z are all positive), and space itself.

Examples of *closed* sets in space include lines, planes, and the closed half-space $z \geq 0$.

A solid sphere with part of its boundary removed or a solid cube with a missing face, edge, or corner point is *neither open nor closed*.

Functions of more than three independent variables are also important. For example, the temperature on a surface in space may depend not only on the location of the point $P(x, y, z)$ on the surface, but also on the time t when it is visited, so we would write $T = f(x, y, z, t)$.

Computer Graphing

Three-dimensional graphing programs for computers and calculators make it possible to graph functions of two variables with only a few keystrokes. We can often get information more quickly from a graph than from a formula.

EXAMPLE 5 The temperature w beneath the Earth's surface is a function of the depth x beneath the surface and the time t of the year. If we measure x in feet and t as the number of days elapsed from the expected date of the yearly highest surface temperature, we can model the variation in temperature with the function

$$w = \cos\left(1.7 \times 10^{-2}t - 0.2x\right)e^{-0.2x}.$$

(The temperature at 0 ft is scaled to vary from $+1$ to -1, so that the variation at x feet can be interpreted as a fraction of the variation at the surface.)

Figure 12.9 shows a computer-generated graph of the function. At a depth of 15 ft, the variation (change in vertical amplitude in the figure) is about 5% of the surface variation. At 25 ft, there is almost no variation during the year.

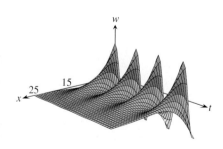

FIGURE 12.9 This computer-generated graph shows the seasonal variation of the temperature below ground as a fraction of surface temperature. (Example 5). (Adapted from art provided by Norton Starr.)

The graph also shows that the temperature 15 ft below the surface is about half a year out of phase with the surface temperature. When the temperature is lowest on the surface (late January, say), it is at its highest 15 ft below. Fifteen feet below the ground, the seasons are reversed. ■

Figure 12.10 shows computer-generated graphs of a number of functions of two variables together with their level curves.

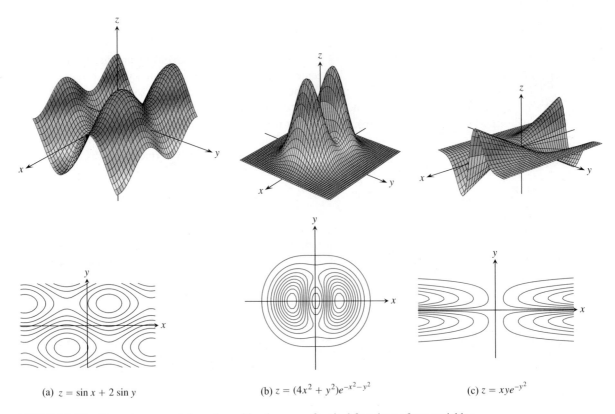

(a) $z = \sin x + 2 \sin y$ (b) $z = (4x^2 + y^2)e^{-x^2 - y^2}$ (c) $z = xye^{-y^2}$

FIGURE 12.10   Computer-generated graphs and level curves of typical functions of two variables.

EXERCISES 12.1

In Exercises 1–12, **(a)** find the function's domain, **(b)** find the function's range, **(c)** describe the function's level curves, **(d)** find the boundary of the function's domain, **(e)** determine if the domain is an open region, a closed region, or neither, and **(f)** decide if the domain is bounded or unbounded.

1. $f(x, y) = y - x$

2. $f(x, y) = \sqrt{y - x}$

3. $f(x, y) = 4x^2 + 9y^2$

4. $f(x, y) = x^2 - y^2$

5. $f(x, y) = xy$

6. $f(x, y) = y/x^2$

7. $f(x, y) = \dfrac{1}{\sqrt{16 - x^2 - y^2}}$

8. $f(x, y) = \sqrt{9 - x^2 - y^2}$

9. $f(x, y) = \ln(x^2 + y^2)$

10. $f(x, y) = e^{-(x^2 + y^2)}$

11. $f(x, y) = \sin^{-1}(y - x)$

12. $f(x, y) = \tan^{-1}\left(\dfrac{y}{x}\right)$

Exercises 13–18 show level curves for the functions graphed in (a)–(f). Match each set of curves with the appropriate function.

13.

14.

c.

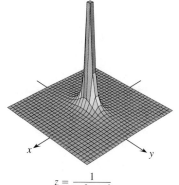

$$z = \frac{1}{4x^2 + y^2}$$

15.

16.

d.

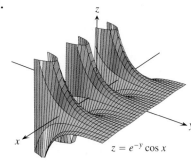

$$z = e^{-y} \cos x$$

17.

18.

e.

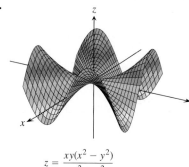

$$z = \frac{xy(x^2 - y^2)}{x^2 + y^2}$$

a.

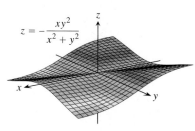

$$z = (\cos x)(\cos y)\, e^{-\sqrt{x^2 + y^2}/4}$$

b.

$$z = -\frac{xy^2}{x^2 + y^2}$$

f.

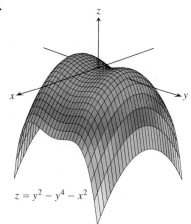

$$z = y^2 - y^4 - x^2$$

Display the values of the functions in Exercises 19–28 in two ways:
(a) by sketching the surface $z = f(x, y)$ and (b) by drawing an assortment of level curves in the function's domain. Label each level curve with its function value.

19. $f(x, y) = y^2$

20. $f(x, y) = 4 - y^2$

21. $f(x, y) = x^2 + y^2$

22. $f(x, y) = \sqrt{x^2 + y^2}$

23. $f(x, y) = -(x^2 + y^2)$

24. $f(x, y) = 4 - x^2 - y^2$

25. $f(x, y) = 4x^2 + y^2$

26. $f(x, y) = 4x^2 + y^2 + 1$

27. $f(x, y) = 1 - |y|$

28. $f(x, y) = 1 - |x| - |y|$

In Exercises 29–32, find an equation for the level curve of the function $f(x, y)$ that passes through the given point.

29. $f(x, y) = 16 - x^2 - y^2$, $\left(2\sqrt{2}, \sqrt{2}\right)$

30. $f(x, y) = \sqrt{x^2 - 1}$, $(1, 0)$

31. $f(x, y) = \int_x^y \dfrac{dt}{1 + t^2}$, $\left(-\sqrt{2}, \sqrt{2}\right)$

32. $f(x, y) = \sum_{n=0}^{\infty} \left(\dfrac{x}{y}\right)^n$, $(1, 2)$

In Exercises 33–40, sketch a typical level surface for the function.

33. $f(x, y, z) = x^2 + y^2 + z^2$ 34. $f(x, y, z) = \ln(x^2 + y^2 + z^2)$

35. $f(x, y, z) = x + z$

36. $f(x, y, z) = z$

37. $f(x, y, z) = x^2 + y^2$

38. $f(x, y, z) = y^2 + z^2$

39. $f(x, y, z) = z - x^2 - y^2$

40. $f(x, y, z) = (x^2/25) + (y^2/16) + (z^2/9)$

In Exercises 41–44, find an equation for the level surface of the function through the given point.

41. $f(x, y, z) = \sqrt{x - y} - \ln z$, $(3, -1, 1)$

42. $f(x, y, z) = \ln(x^2 + y + z^2)$, $(-1, 2, 1)$

43. $g(x, y, z) = \sum_{n=0}^{\infty} \dfrac{(x + y)^n}{n! z^n}$, $(\ln 2, \ln 4, 3)$

44. $g(x, y, z) = \int_x^y \dfrac{d\theta}{\sqrt{1 - \theta^2}} + \int_{\sqrt{2}}^z \dfrac{dt}{t\sqrt{t^2 - 1}}$, $(0, 1/2, 2)$

COMPUTER EXPLORATIONS

Use a CAS to perform the following steps for each of the functions in Exercises 45–48.

a. Plot the surface over the given rectangle.

b. Plot several level curves in the rectangle.

c. Plot the level curve of f through the given point.

45. $f(x, y) = x \sin \dfrac{y}{2} + y \sin 2x$, $0 \le x \le 5\pi$, $0 \le y \le 5\pi$, $P(3\pi, 3\pi)$

46. $f(x, y) = (\sin x)(\cos y)e^{\sqrt{x^2 + y^2}/8}$, $0 \le x \le 5\pi$, $0 \le y \le 5\pi$, $P(4\pi, 4\pi)$

47. $f(x, y) = \sin(x + 2\cos y)$, $-2\pi \le x \le 2\pi$, $-2\pi \le y \le 2\pi$, $P(\pi, \pi)$

48. $f(x, y) = e^{(x^{0.1} - y)} \sin(x^2 + y^2)$, $0 \le x \le 2\pi$, $-2\pi \le y \le \pi$, $P(\pi, -\pi)$

Use a CAS to plot the implicitly defined level surfaces in Exercises 49–52.

49. $4 \ln(x^2 + y^2 + z^2) = 1$ 50. $x^2 + z^2 = 1$

51. $x + y^2 - 3z^2 = 1$

52. $\sin\left(\dfrac{x}{2}\right) - (\cos y)\sqrt{x^2 + z^2} = 2$

Parametrized Surfaces   Just as you describe curves in the plane parametrically with a pair of equations $x = f(t), y = g(t)$ defined on some parameter interval I, you can sometimes describe surfaces in space with a triple of equations $x = f(u, v), y = g(u, v), z = h(u, v)$ defined on some parameter rectangle $a \le u \le b, c \le v \le d$. Many computer algebra systems permit you to plot such surfaces in *parametric mode*. (Parametrized surfaces are discussed in detail in Section 14.5.) Use a CAS to plot the surfaces in Exercises 53–56. Also plot several level curves in the *xy*-plane.

53. $x = u \cos v$, $y = u \sin v$, $z = u$, $0 \le u \le 2$, $0 \le v \le 2\pi$

54. $x = u \cos v$, $y = u \sin v$, $z = v$, $0 \le u \le 2$, $0 \le v \le 2\pi$

55. $x = (2 + \cos u) \cos v$, $y = (2 + \cos u) \sin v$, $z = \sin u$, $0 \le u \le 2\pi$, $0 \le v \le 2\pi$

56. $x = 2 \cos u \cos v$, $y = 2 \cos u \sin v$, $z = 2 \sin u$, $0 \le u \le 2\pi$, $0 \le v \le \pi$

12.2	Limits and Continuity in Higher Dimensions

This section treats limits and continuity for multivariable functions.

Limits

If the values of $f(x, y)$ lie arbitrarily close to a fixed real number L for all points (x, y) sufficiently close to a point (x_0, y_0), we say that f approaches the limit L as (x, y) approaches (x_0, y_0). This is similar to the informal definition for the limit of a function of a single variable. Notice, however, that if (x_0, y_0) lies in the interior of f's domain, (x, y) can approach (x_0, y_0) from any direction. The direction of approach can be an issue, as in some of the examples that follow.

DEFINITION   We say that a function $f(x, y)$ approaches the **limit** L as (x, y) approaches (x_0, y_0), and write

$$\lim_{(x, y) \to (x_0, y_0)} f(x, y) = L$$

if, for every number $\epsilon > 0$, there exists a corresponding number $\delta > 0$ such that for all (x, y) in the domain of f,

$$|f(x, y) - L| < \epsilon \qquad \text{whenever} \qquad 0 < \sqrt{(x - x_0)^2 + (y - y_0)^2} < \delta.$$

The definition of limit says that the distance between $f(x, y)$ and L becomes arbitrarily small whenever the distance from (x, y) to (x_0, y_0) is made sufficiently small (but not 0).

The definition of limit applies to boundary points (x_0, y_0) as well as interior points of the domain of f. The only requirement is that the point (x, y) remain in the domain at all times. It can be shown, as for functions of a single variable, that

$$\lim_{(x, y) \to (x_0, y_0)} x = x_0$$

$$\lim_{(x, y) \to (x_0, y_0)} y = y_0$$

$$\lim_{(x, y) \to (x_0, y_0)} k = k \qquad \text{(any number } k\text{)}.$$

For example, in the first limit statement above, $f(x, y) = x$ and $L = x_0$. Using the definition of limit, suppose that $\epsilon > 0$ is chosen. If we let δ equal this ϵ, we see that

$$0 < \sqrt{(x - x_0)^2 + (y - y_0)^2} < \delta = \epsilon$$

implies

$$\sqrt{(x - x_0)^2} < \epsilon \qquad (x - x_0)^2 \le (x - x_0)^2 + (y - y_0)^2$$

$$|x - x_0| < \epsilon \qquad \sqrt{a^2} = |a|$$

$$|f(x, y) - x_0| < \epsilon \qquad x = f(x, y)$$

That is,

$$|f(x, y) - x_0| < \epsilon \qquad \text{whenever} \qquad 0 < \sqrt{(x - x_0)^2 + (y - y_0)^2} < \delta.$$

So

$$\lim_{(x, y) \to (x_0, y_0)} f(x, y) = \lim_{(x, y) \to (x_0, y_0)} x = x_0.$$

As with single-variable functions, the limit of the sum of two functions is the sum of their limits (when they both exist), with similar results for the limits of the differences, products, constant multiples, quotients, and powers.

THEOREM 1—Properties of Limits of Functions of Two Variables   The following rules hold if L, M, and k are real numbers and

$$\lim_{(x, y) \to (x_0, y_0)} f(x, y) = L \qquad \text{and} \qquad \lim_{(x, y) \to (x_0, y_0)} g(x, y) = M.$$

1.   *Sum Rule:* $\qquad\qquad\qquad \displaystyle\lim_{(x, y) \to (x_0, y_0)} (f(x, y) + g(x, y)) = L + M$

2.   *Difference Rule:* $\qquad\quad \displaystyle\lim_{(x, y) \to (x_0, y_0)} (f(x, y) - g(x, y)) = L - M$

3.   *Product Rule:* $\qquad\qquad \displaystyle\lim_{(x, y) \to (x_0, y_0)} (f(x, y) \cdot g(x, y)) = L \cdot M$

4.   *Constant Multiple Rule:* $\quad \displaystyle\lim_{(x, y) \to (x_0, y_0)} (kf(x, y)) = kL \qquad \text{(any number } k\text{)}$

5.   *Quotient Rule:* $\qquad\qquad \displaystyle\lim_{(x, y) \to (x_0, y_0)} \frac{f(x, y)}{g(x, y)} = \frac{L}{M} \qquad M \neq 0$

6.   *Power Rule:* If r and s are integers with no common factors, and $s \neq 0$, then

$$\lim_{(x, y) \to (x_0, y_0)} (f(x, y))^{r/s} = L^{r/s}$$

provided $L^{r/s}$ is a real number. (If s is even, we assume that $L > 0$.)

While we won't prove Theorem 1 here, we give an informal discussion of why it's true. If (x, y) is sufficiently close to (x_0, y_0), then $f(x, y)$ is close to L and $g(x, y)$ is close to M (from the informal interpretation of limits). It is then reasonable that $f(x, y) + g(x, y)$ is close to $L + M$; $f(x, y) - g(x, y)$ is close to $L - M$; $f(x, y)g(x, y)$ is close to LM; $kf(x, y)$ is close to kL; and that $f(x, y)/g(x, y)$ is close to L/M if $M \neq 0$.

When we apply Theorem 1 to polynomials and rational functions, we obtain the useful result that the limits of these functions as $(x, y) \to (x_0, y_0)$ can be calculated by evaluating the functions at (x_0, y_0). The only requirement is that the rational functions be defined at (x_0, y_0).

EXAMPLE 1

(a) $\displaystyle\lim_{(x, y) \to (0,1)} \frac{x - xy + 3}{x^2 y + 5xy - y^3} = \frac{0 - (0)(1) + 3}{(0)^2(1) + 5(0)(1) - (1)^3} = -3$

(b) $\displaystyle\lim_{(x, y) \to (3, -4)} \sqrt{x^2 + y^2} = \sqrt{(3)^2 + (-4)^2} = \sqrt{25} = 5$ ∎

EXAMPLE 2   Find

$$\lim_{(x,y)\to(0,0)} \frac{x^2 - xy}{\sqrt{x} - \sqrt{y}}.$$

Solution   Since the denominator $\sqrt{x} - \sqrt{y}$ approaches 0 as $(x, y) \to (0, 0)$, we cannot use the Quotient Rule from Theorem 1. If we multiply numerator and denominator by $\sqrt{x} + \sqrt{y}$, however, we produce an equivalent fraction whose limit we *can* find:

$$\lim_{(x,y)\to(0,0)} \frac{x^2 - xy}{\sqrt{x} - \sqrt{y}} = \lim_{(x,y)\to(0,0)} \frac{\left(x^2 - xy\right)\left(\sqrt{x} + \sqrt{y}\right)}{\left(\sqrt{x} - \sqrt{y}\right)\left(\sqrt{x} + \sqrt{y}\right)}$$

$$= \lim_{(x,y)\to(0,0)} \frac{x\left(x - y\right)\left(\sqrt{x} + \sqrt{y}\right)}{x - y} \qquad \text{Algebra}$$

$$= \lim_{(x,y)\to(0,0)} x\left(\sqrt{x} + \sqrt{y}\right) \qquad \begin{array}{l}\text{Cancel the nonzero} \\ \text{factor } (x - y).\end{array}$$

$$= 0\left(\sqrt{0} + \sqrt{0}\right) = 0$$

We can cancel the factor $(x - y)$ because the path $y = x$ (along which $x - y = 0$) is *not* in the domain of the function

$$\frac{x^2 - xy}{\sqrt{x} - \sqrt{y}}. \qquad \blacksquare$$

EXAMPLE 3   Find $\displaystyle\lim_{(x,y)\to(0,0)} \frac{4xy^2}{x^2 + y^2}$ if it exists.

Solution   We first observe that along the line $x = 0$, the function always has value 0 when $y \neq 0$. Likewise, along the line $y = 0$, the function has value 0 provided $x \neq 0$. So if the limit does exist as (x, y) approaches $(0, 0)$, the value of the limit must be 0. To see if this is true, we apply the definition of limit.

Let $\epsilon > 0$ be given, but arbitrary. We want to find a $\delta > 0$ such that

$$\left| \frac{4xy^2}{x^2 + y^2} - 0 \right| < \epsilon \qquad \text{whenever} \qquad 0 < \sqrt{x^2 + y^2} < \delta$$

or

$$\frac{4|x|y^2}{x^2 + y^2} < \epsilon \qquad \text{whenever} \qquad 0 < \sqrt{x^2 + y^2} < \delta.$$

Since $y^2 \leq x^2 + y^2$ we have that

$$\frac{4|x|y^2}{x^2 + y^2} \leq 4|x| = 4\sqrt{x^2} \leq 4\sqrt{x^2 + y^2}. \qquad \frac{y^2}{x^2 + y^2} \leq 1$$

So if we choose $\delta = \epsilon/4$ and let $0 < \sqrt{x^2 + y^2} < \delta$, we get

$$\left| \frac{4xy^2}{x^2 + y^2} - 0 \right| \leq 4\sqrt{x^2 + y^2} < 4\delta = 4\left(\frac{\epsilon}{4}\right) = \epsilon.$$

It follows from the definition that

$$\lim_{(x, y)\to(0,0)} \frac{4xy^2}{x^2 + y^2} = 0. \qquad \blacksquare$$

Continuity

As with functions of a single variable, continuity is defined in terms of limits.

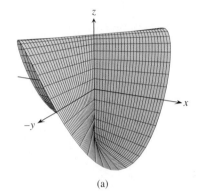

(a)

> **DEFINITION**   A function $f(x, y)$ is **continuous at the point (x_0, y_0)** if
>
> 1. f is defined at (x_0, y_0),
>
> 2. $\displaystyle\lim_{(x, y)\to(x_0, y_0)} f(x, y)$ exists,
>
> 3. $\displaystyle\lim_{(x, y)\to(x_0, y_0)} f(x, y) = f(x_0, y_0)$.
>
> A function is **continuous** if it is continuous at every point of its domain.

As with the definition of limit, the definition of continuity applies at boundary points as well as interior points of the domain of f. The only requirement is that the point (x, y) be in the domain of f.

A consequence of Theorem 1 is that algebraic combinations of continuous functions are continuous at every point at which all the functions involved are defined. This means that sums, differences, products, constant multiples, quotients, and powers of continuous functions are continuous where defined. In particular, polynomials and rational functions of two variables are continuous at every point at which they are defined.

EXAMPLE 4   Show that

$$f(x, y) = \begin{cases} \dfrac{2xy}{x^2 + y^2}, & (x, y) \neq (0, 0) \\ 0, & (x, y) = (0, 0) \end{cases}$$

is continuous at every point except the origin (Figure 12.11).

Solution   The function f is continuous at any point $(x, y) \neq (0, 0)$ because its values are then given by a rational function of x and y.

At $(0, 0)$, the value of f is defined, but f, we claim, has no limit as $(x, y) \to (0, 0)$. The reason is that different paths of approach to the origin can lead to different results, as we now see.

(b)

FIGURE 12.11   (a) The graph of

$$f(x, y) = \begin{cases} \dfrac{2xy}{x^2 + y^2}, & (x, y) \neq (0, 0) \\ 0, & (x, y) = (0, 0). \end{cases}$$

The function is continuous at every point except the origin. (b) The level curves of f (Example 4).

For every value of m, the function f has a constant value on the "punctured" line $y = mx, x \neq 0$, because

$$f(x, y)\Big|_{y=mx} = \frac{2xy}{x^2 + y^2}\Big|_{y=mx} = \frac{2x(mx)}{x^2 + (mx)^2} = \frac{2mx^2}{x^2 + m^2x^2} = \frac{2m}{1 + m^2}.$$

Therefore, f has this number as its limit as (x, y) approaches $(0, 0)$ along the line:

$$\lim_{\substack{(x, y)\to(0,0) \\ \text{along } y=mx}} f(x, y) = \lim_{(x, y)\to(0,0)}\left[f(x, y)\Big|_{y=mx}\right] = \frac{2m}{1 + m^2}.$$

This limit changes with m. There is therefore no single number we may call the limit of f as (x, y) approaches the origin. The limit fails to exist, and the function is not continuous. ∎

Example 4 illustrates an important point about limits of functions of two variables (or even more variables, for that matter). For a limit to exist at a point, the limit must be the same along every approach path. This result is analogous to the single-variable case where both the left- and right-sided limits had to have the same value. For functions of two or more variables, if we ever find paths with different limits, we know the function has no limit at the point they approach.

> **Two-Path Test for Nonexistence of a Limit**
> If a function $f(x, y)$ has different limits along two different paths in the domain of f as (x, y) approaches (x_0, y_0), then $\lim_{(x, y)\to(x_0, y_0)} f(x, y)$ does not exist.

EXAMPLE 5   Show that the function

$$f(x, y) = \frac{2x^2y}{x^4 + y^2}$$

(Figure 12.12) has no limit as (x, y) approaches $(0, 0)$.

Solution   The limit cannot be found by direct substitution, which gives the indeterminate form $0/0$. We examine the values of f along curves that end at $(0, 0)$. Along the curve $y = kx^2, x \neq 0$, the function has the constant value

$$f(x, y)\Big|_{y=kx^2} = \frac{2x^2y}{x^4 + y^2}\Big|_{y=kx^2} = \frac{2x^2(kx^2)}{x^4 + (kx^2)^2} = \frac{2kx^4}{x^4 + k^2x^4} = \frac{2k}{1 + k^2}.$$

Therefore,

$$\lim_{\substack{(x, y)\to(0,0) \\ \text{along } y=kx^2}} f(x, y) = \lim_{(x, y)\to(0,0)}\left[f(x, y)\Big|_{y=kx^2}\right] = \frac{2k}{1 + k^2}.$$

This limit varies with the path of approach. If (x, y) approaches $(0, 0)$ along the parabola $y = x^2$, for instance, $k = 1$ and the limit is 1. If (x, y) approaches $(0, 0)$ along the x-axis, $k = 0$ and the limit is 0. By the two-path test, f has no limit as (x, y) approaches $(0, 0)$. ∎

It can be shown that the function in Example 5 has limit 0 along every path $y = mx$ (Exercise 43). Having the same limit along all straight lines approaching (x_0, y_0) does not imply a limit exists at (x_0, y_0).

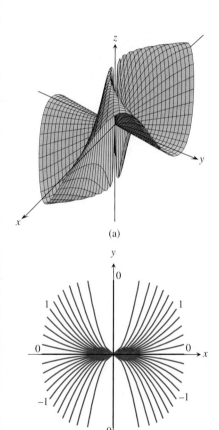

(a)

(b)

FIGURE 12.12   (a) The graph of $f(x, y) = 2x^2y/(x^4 + y^2)$. As the graph suggests and the level-curve values in part (b) confirm, $\lim_{(x, y)\to(0,0)} f(x, y)$ does not exist (Example 5).

Compositions of continuous functions are also continuous. The proof, omitted here, is similar to that for functions of a single variable (Theorem 10 in Section 2.6).

Continuity of Composites

If f is continuous at (x_0, y_0) and g is a single-variable function continuous at $f(x_0, y_0)$, then the composite function $h = g \circ f$ defined by $h(x, y) = g(f(x, y))$ is continuous at (x_0, y_0).

For example, the composite functions

$$e^{x-y}, \qquad \cos \frac{xy}{x^2 + 1}, \qquad \ln\left(1 + x^2 y^2\right)$$

are continuous at every point (x, y).

As with functions of a single variable, the general rule is that composites of continuous functions are continuous. The only requirement is that each function be continuous where it is applied.

Functions of More Than Two Variables

The definitions of limit and continuity for functions of two variables and the conclusions about limits and continuity for sums, products, quotients, powers, and composites all extend to functions of three or more variables. Functions like

$$\ln\left(x + y + z\right) \qquad \text{and} \qquad \frac{y \sin z}{x - 1}$$

are continuous throughout their domains, and limits like

$$\lim_{P \to (1,0,-1)} \frac{e^{x+z}}{z^2 + \cos \sqrt{xy}} = \frac{e^{1-1}}{(-1)^2 + \cos 0} = \frac{1}{2},$$

where P denotes the point (x, y, z), may be found by direct substitution.

Extreme Values of Continuous Functions on Closed, Bounded Sets

The Extreme Value Theorem (Theorem 1, Section 4.1) states that a function of a single variable that is continuous throughout a closed, bounded interval $[a, b]$ takes on an absolute maximum value and an absolute minimum value at least once in $[a, b]$. The same is true of a function $z = f(x, y)$ that is continuous on a closed, bounded set R in the plane (like a line segment, a disk, or a filled-in triangle). The function takes on an absolute maximum value at some point in R and an absolute minimum value at some point in R.

Similar results hold for functions of three or more variables. A continuous function $w = f(x, y, z)$, for example, must take on absolute maximum and minimum values on any closed, bounded set (solid ball or cube, spherical shell, rectangular solid) on which it is defined. We will learn how to find these extreme values in Section 12.7.

EXERCISES 12.2

Find the limits in Exercises 1–12.

1. $\displaystyle\lim_{(x,y)\to(0,0)} \frac{3x^2 - y^2 + 5}{x^2 + y^2 + 2}$

2. $\displaystyle\lim_{(x,y)\to(0,4)} \frac{x}{\sqrt{y}}$

3. $\displaystyle\lim_{(x,y)\to(3,4)} \sqrt{x^2 + y^2 - 1}$

4. $\displaystyle\lim_{(x,y)\to(2,-3)} \left(\frac{1}{x} + \frac{1}{y}\right)^2$

5. $\displaystyle\lim_{(x,y)\to(0,\pi/4)} \sec x \tan y$

6. $\displaystyle\lim_{(x,y)\to(0,0)} \cos\frac{x^2 + y^3}{x + y + 1}$

7. $\displaystyle\lim_{(x,y)\to(0,\ln 2)} e^{x-y}$

8. $\displaystyle\lim_{(x,y)\to(1,1)} \ln|1 + x^2 y^2|$

9. $\displaystyle\lim_{(x,y)\to(0,0)} \frac{e^y \sin x}{x}$

10. $\displaystyle\lim_{(x,y)\to(1,1)} \cos\sqrt[3]{|xy| - 1}$

11. $\displaystyle\lim_{(x,y)\to(1,0)} \frac{x \sin y}{x^2 + 1}$

12. $\displaystyle\lim_{(x,y)\to(\pi/2,0)} \frac{\cos y + 1}{y - \sin x}$

Find the limits in Exercises 13–20 by rewriting the fractions first.

13. $\displaystyle\lim_{\substack{(x,y)\to(1,1)\\x\neq y}} \frac{x^2 - 2xy + y^2}{x - y}$

14. $\displaystyle\lim_{\substack{(x,y)\to(1,1)\\x\neq y}} \frac{x^2 - y^2}{x - y}$

15. $\displaystyle\lim_{\substack{(x,y)\to(1,1)\\x\neq 1}} \frac{xy - y - 2x + 2}{x - 1}$

16. $\displaystyle\lim_{\substack{(x,y)\to(2,-4)\\y\neq-4,\,x\neq x^2}} \frac{y + 4}{x^2 y - xy + 4x^2 - 4x}$

17. $\displaystyle\lim_{\substack{(x,y)\to(0,0)\\x\neq y}} \frac{x - y + 2\sqrt{x} - 2\sqrt{y}}{\sqrt{x} - \sqrt{y}}$

18. $\displaystyle\lim_{\substack{(x,y)\to(2,2)\\x+y\neq4}} \frac{x + y - 4}{\sqrt{x + y} - 2}$

19. $\displaystyle\lim_{\substack{(x,y)\to(2,0)\\2x-y\neq4}} \frac{\sqrt{2x - y} - 2}{2x - y - 4}$

20. $\displaystyle\lim_{\substack{(x,y)\to(4,3)\\x\neq y+1}} \frac{\sqrt{x} - \sqrt{y + 1}}{x - y - 1}$

Find the limits in Exercises 21–26.

21. $\displaystyle\lim_{P\to(1,3,4)} \left(\frac{1}{x} + \frac{1}{y} + \frac{1}{z}\right)$

22. $\displaystyle\lim_{P\to(1,-1,-1)} \frac{2xy + yz}{x^2 + z^2}$

23. $\displaystyle\lim_{P\to(3,3,0)} (\sin^2 x + \cos^2 y + \sec^2 z)$

24. $\displaystyle\lim_{P\to(-1/4,\pi/2,2)} \tan^{-1} xyz$

25. $\displaystyle\lim_{P\to(\pi,0,3)} ze^{-2y}\cos 2x$

26. $\displaystyle\lim_{P\to(0,-2,0)} \ln\sqrt{x^2 + y^2 + z^2}$

At what points (x, y) in the plane are the functions in Exercises 27–30 continuous?

27. a. $f(x, y) = \sin(x + y)$ **b.** $f(x, y) = \ln(x^2 + y^2)$

28. a. $f(x, y) = \dfrac{x + y}{x - y}$ **b.** $f(x, y) = \dfrac{y}{x^2 + 1}$

29. a. $g(x, y) = \sin\dfrac{1}{xy}$ **b.** $g(x, y) = \dfrac{x + y}{2 + \cos x}$

30. a. $g(x, y) = \dfrac{x^2 + y^2}{x^2 - 3x + 2}$ **b.** $g(x, y) = \dfrac{1}{x^2 - y}$

At what points (x, y, z) in space are the functions in Exercises 31–34 continuous?

31. a. $f(x, y, z) = x^2 + y^2 - 2z^2$

 b. $f(x, y, z) = \sqrt{x^2 + y^2 - 1}$

32. a. $f(x, y, z) = \ln xyz$ **b.** $f(x, y, z) = e^{x+y}\cos z$

33. a. $h(x, y, z) = xy\sin\dfrac{1}{z}$ **b.** $h(x, y, z) = \dfrac{1}{x^2 + z^2 - 1}$

34. a. $h(x, y, z) = \dfrac{1}{|y| + |z|}$ **b.** $h(x, y, z) = \dfrac{1}{|xy| + |z|}$

By considering different paths of approach, show that the functions in Exercises 35–42 have no limit as $(x, y) \to (0, 0)$.

35. $f(x, y) = -\dfrac{x}{\sqrt{x^2 + y^2}}$ **36.** $f(x, y) = \dfrac{x^4}{x^4 + y^2}$

37. $f(x, y) = \dfrac{x^4 - y^2}{x^4 + y^2}$ **38.** $f(x, y) = \dfrac{xy}{|xy|}$

39. $g(x, y) = \dfrac{x - y}{x + y}$ **40.** $g(x, y) = \dfrac{x + y}{x - y}$

41. $h(x, y) = \dfrac{x^2 + y}{y}$ **42.** $h(x, y) = \dfrac{x^2}{x^2 - y}$

43. Show that the function in Example 5 has limit 0 along every straight line approaching $(0, 0)$.

44. If $f(x_0, y_0) = 3$, what can you say about

$$\lim_{(x,y)\to(x_0,y_0)} f(x, y)$$

if f is continuous at (x_0, y_0)? If f is not continuous at (x_0, y_0)? Give reasons for your answers.

The Sandwich Theorem   for functions of two variables states that if $g(x, y) \le f(x, y) \le h(x, y)$ for all $(x, y) \ne (x_0, y_0)$ in a disk centered at (x_0, y_0) and if g and h have the same finite limit L as $(x, y) \to (x_0, y_0)$, then

$$\lim_{(x, y) \to (x_0, y_0)} f(x, y) = L.$$

Use this result to support your answers to the questions in Exercises 45–48.

45. Does knowing that

$$1 - \frac{x^2 y^2}{3} < \frac{\tan^{-1} xy}{xy} < 1$$

tell you anything about

$$\lim_{(x, y) \to (0,0)} \frac{\tan^{-1} xy}{xy}?$$

Give reasons for your answer.

46. Does knowing that

$$2|xy| - \frac{x^2 y^2}{6} < 4 - 4 \cos \sqrt{|xy|} < 2|xy|$$

tell you anything about

$$\lim_{(x, y) \to (0,0)} \frac{4 - 4 \cos \sqrt{|xy|}}{|xy|}?$$

Give reasons for your answer.

47. Does knowing that $|\sin (1/x)| \le 1$ tell you anything about

$$\lim_{(x, y) \to (0,0)} y \sin \frac{1}{x}?$$

Give reasons for your answer.

48. Does knowing that $|\cos (1/y)| \le 1$ tell you anything about

$$\lim_{(x, y) \to (0,0)} x \cos \frac{1}{y}?$$

Give reasons for your answer.

49. (*Continuation of Example 4.*)

a. Reread Example 4. Then substitute $m = \tan \theta$ into the formula

$$f(x, y)\Big|_{y=mx} = \frac{2m}{1 + m^2}$$

and simplify the result to show how the value of f varies with the line's angle of inclination.

b. Use the formula you obtained in part (a) to show that the limit of f as $(x, y) \to (0, 0)$ along the line $y = mx$ varies from -1 to 1 depending on the angle of approach.

50. Continuous extension   Define $f(0, 0)$ in a way that extends

$$f(x, y) = xy \frac{x^2 - y^2}{x^2 + y^2}$$

to be continuous at the origin.

Changing to Polar Coordinates   If you cannot make any headway with $\lim_{(x, y) \to (0,0)} f(x, y)$ in rectangular coordinates, try changing to polar coordinates. Substitute $x = r \cos \theta$, $y = r \sin \theta$, and investigate the limit of the resulting expression as $r \to 0$. In other words, try to decide whether there exists a number L satisfying the following criterion:

Given $\epsilon > 0$, there exists a $\delta > 0$ such that for all r and θ,

$$|r| < \delta \quad \Rightarrow \quad |f(r, \theta) - L| < \epsilon. \tag{1}$$

If such an L exists, then

$$\lim_{(x, y) \to (0,0)} f(x, y) = \lim_{r \to 0} f(r, \theta) = L.$$

For instance,

$$\lim_{(x, y) \to (0,0)} \frac{x^3}{x^2 + y^2} = \lim_{r \to 0} \frac{r^3 \cos^3 \theta}{r^2} = \lim_{r \to 0} r \cos^3 \theta = 0.$$

To verify the last of these equalities, we need to show that Equation (1) is satisfied with $f(r, \theta) = r \cos^3 \theta$ and $L = 0$. That is, we need to show that given any $\epsilon > 0$ there exists a $\delta > 0$ such that for all r and θ,

$$|r| < \delta \quad \Rightarrow \quad |r \cos^3 \theta - 0| < \epsilon.$$

Since

$$|r \cos^3 \theta| = |r||\cos^3 \theta| \le |r| \cdot 1 = |r|,$$

the implication holds for all r and θ if we take $\delta = \epsilon$.

In contrast,

$$\frac{x^2}{x^2 + y^2} = \frac{r^2 \cos^2 \theta}{r^2} = \cos^2 \theta$$

takes on all values from 0 to 1 regardless of how small $|r|$ is, so that $\lim_{(x, y) \to (0,0)} x^2/(x^2 + y^2)$ does not exist.

In each of these instances, the existence or nonexistence of the limit as $r \to 0$ is fairly clear. Shifting to polar coordinates does not always help, however, and may even tempt us to false conclusions. For example, the limit may exist along every straight line (or ray) $\theta = \text{constant}$ and yet fail to exist in the broader sense. Example 4 illustrates this point. In polar coordinates, $f(x, y) = (2x^2 y)/(x^4 + y^2)$ becomes

$$f(r \cos \theta, r \sin \theta) = \frac{r \cos \theta \sin 2\theta}{r^2 \cos^4 \theta + \sin^2 \theta}$$

for $r \ne 0$. If we hold θ constant and let $r \to 0$, the limit is 0. On the path $y = x^2$, however, we have $r \sin \theta = r^2 \cos^2 \theta$ and

$$f(r \cos \theta, r \sin \theta) = \frac{r \cos \theta \sin 2\theta}{r^2 \cos^4 \theta + (r \cos^2 \theta)^2}$$

$$= \frac{2r \cos^2 \theta \sin \theta}{2r^2 \cos^4 \theta} = \frac{r \sin \theta}{r^2 \cos^2 \theta} = 1.$$

In Exercises 51–56, find the limit of f as $(x, y) \to (0, 0)$ or show that the limit does not exist.

51. $f(x, y) = \dfrac{x^3 - xy^2}{x^2 + y^2}$ **52.** $f(x, y) = \cos \left(\dfrac{x^3 - y^3}{x^2 + y^2} \right)$

53. $f(x, y) = \dfrac{y^2}{x^2 + y^2}$

54. $f(x, y) = \dfrac{2x}{x^2 + x + y^2}$

55. $f(x, y) = \tan^{-1}\left(\dfrac{|x| + |y|}{x^2 + y^2}\right)$

56. $f(x, y) = \dfrac{x^2 - y^2}{x^2 + y^2}$

In Exercises 57 and 58, define $f(0, 0)$ in a way that extends f to be continuous at the origin.

57. $f(x, y) = \ln\left(\dfrac{3x^2 - x^2y^2 + 3y^2}{x^2 + y^2}\right)$

58. $f(x, y) = \dfrac{3x^2y}{x^2 + y^2}$

Each of Exercises 59–62 gives a function $f(x, y)$ and a positive number ϵ. In each exercise, show that there exists a $\delta > 0$ such that for all (x, y),

$$\sqrt{x^2 + y^2} < \delta \quad \Rightarrow \quad |f(x, y) - f(0, 0)| < \epsilon.$$

59. $f(x, y) = x^2 + y^2, \quad \epsilon = 0.01$

60. $f(x, y) = y/(x^2 + 1), \quad \epsilon = 0.05$

61. $f(x, y) = (x + y)/(x^2 + 1), \quad \epsilon = 0.01$

62. $f(x, y) = (x + y)/(2 + \cos x), \quad \epsilon = 0.02$

Each of Exercises 63–66 gives a function $f(x, y, z)$ and a positive number ϵ. In each exercise, show that there exists a $\delta > 0$ such that for all (x, y, z),

$$\sqrt{x^2 + y^2 + z^2} < \delta \quad \Rightarrow \quad |f(x, y, z) - f(0, 0, 0)| < \epsilon.$$

63. $f(x, y, z) = x^2 + y^2 + z^2, \quad \epsilon = 0.015$

64. $f(x, y, z) = xyz, \quad \epsilon = 0.008$

65. $f(x, y, z) = \dfrac{x + y + z}{x^2 + y^2 + z^2 + 1}, \quad \epsilon = 0.015$

66. $f(x, y, z) = \tan^2 x + \tan^2 y + \tan^2 z, \quad \epsilon = 0.03$

67. Show that $f(x, y, z) = x + y - z$ is continuous at every point (x_0, y_0, z_0).

68. Show that $f(x, y, z) = x^2 + y^2 + z^2$ is continuous at the origin.

12.3 Partial Derivatives

The calculus of several variables is basically single-variable calculus applied to several variables one at a time. When we hold all but one of the independent variables of a function constant and differentiate with respect to that one variable, we get a "partial" derivative. This section shows how partial derivatives are defined and interpreted geometrically, and how to calculate them by applying the rules for differentiating functions of a single variable.

Partial Derivatives of a Function of Two Variables

If (x_0, y_0) is a point in the domain of a function $f(x, y)$, the vertical plane $y = y_0$ will cut the surface $z = f(x, y)$ in the curve $z = f(x, y_0)$ (Figure 12.13). This curve is the graph of the function $z = f(x, y_0)$ in the plane $y = y_0$. The horizontal coordinate in this plane is x; the vertical coordinate is z. The y-value is held constant at y_0, so y is not a variable.

We define the partial derivative of f with respect to x at the point (x_0, y_0) as the ordinary derivative of $f(x, y_0)$ with respect to x at the point $x = x_0$. To distinguish partial derivatives from ordinary derivatives we use the symbol ∂ rather than the d previously used.

DEFINITION The **partial derivative of $f(x, y)$ with respect to x** at the point (x_0, y_0) is

$$\left.\dfrac{\partial f}{\partial x}\right|_{(x_0, y_0)} = \lim_{h \to 0} \dfrac{f(x_0 + h, y_0) - f(x_0, y_0)}{h},$$

provided the limit exists.

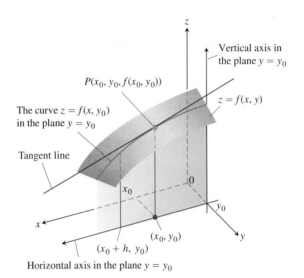

FIGURE 12.13 The intersection of the plane $y = y_0$ with the surface $z = f(x, y)$, viewed from above the first quadrant of the xy-plane.

An equivalent expression for the partial derivative is

$$\left. \frac{d}{dx} f(x, y_0) \right|_{x=x_0}.$$

The slope of the curve $z = f(x, y_0)$ at the point $P(x_0, y_0, f(x_0, y_0))$ in the plane $y = y_0$ is the value of the partial derivative of f with respect to x at (x_0, y_0). (In Figure 12.13 this slope is negative.) The tangent line to the curve at P is the line in the plane $y = y_0$ that passes through P with this slope. The partial derivative $\partial f / \partial x$ at (x_0, y_0) gives the rate of change of f with respect to x when y is held fixed at the value y_0.

The notation for a partial derivative depends on what we want to emphasize:

$\dfrac{\partial f}{\partial x}(x_0, y_0)$ or $f_x(x_0, y_0)$	"Partial derivative of f with respect to x at (x_0, y_0)" or "f sub x at (x_0, y_0)." Convenient for stressing the point (x_0, y_0).
$\left. \dfrac{\partial z}{\partial x} \right|_{(x_0, y_0)}$	"Partial derivative of z with respect to x at (x_0, y_0)." Common in science and engineering when you are dealing with variables and do not mention the function explicitly.
$f_x, \dfrac{\partial f}{\partial x}, z_x,$ or $\dfrac{\partial z}{\partial x}$	"Partial derivative of f (or z) with respect to x." Convenient when you regard the partial derivative as a function in its own right.

The definition of the partial derivative of $f(x, y)$ with respect to y at a point (x_0, y_0) is similar to the definition of the partial derivative of f with respect to x. We hold x fixed at the value x_0 and take the ordinary derivative of $f(x_0, y)$ with respect to y at y_0.

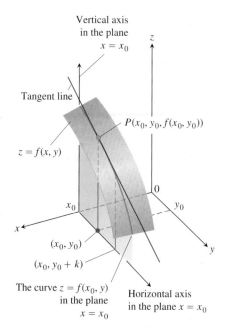

FIGURE 12.14 The intersection of the plane $x = x_0$ with the surface $z = f(x, y)$, viewed from above the first quadrant of the xy-plane.

> **DEFINITION** The **partial derivative of $f(x, y)$ with respect to y** at the point (x_0, y_0) is
>
> $$\left.\frac{\partial f}{\partial y}\right|_{(x_0, y_0)} = \left.\frac{d}{dy} f(x_0, y)\right|_{y=y_0} = \lim_{h \to 0} \frac{f(x_0, y_0 + h) - f(x_0, y_0)}{h},$$
>
> provided the limit exists.

The slope of the curve $z = f(x_0, y)$ at the point $P(x_0, y_0, f(x_0, y_0))$ in the vertical plane $x = x_0$ (Figure 12.14) is the partial derivative of f with respect to y at (x_0, y_0). The tangent line to the curve at P is the line in the plane $x = x_0$ that passes through P with this slope. The partial derivative gives the rate of change of f with respect to y at (x_0, y_0) when x is held fixed at the value x_0.

The partial derivative with respect to y is denoted the same way as the partial derivative with respect to x:

$$\frac{\partial f}{\partial y}(x_0, y_0), \qquad f_y(x_0, y_0), \qquad \frac{\partial f}{\partial y}, \qquad f_y.$$

Notice that we now have two tangent lines associated with the surface $z = f(x, y)$ at the point $P(x_0, y_0, f(x_0, y_0))$ (Figure 12.15). Is the plane they determine tangent to the surface at P? We will see that it is, but we have to learn more about partial derivatives before we can find out why.

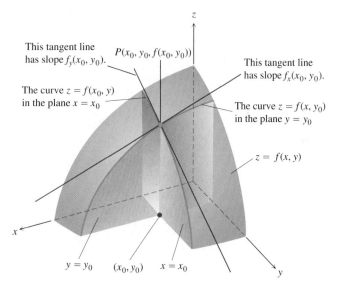

FIGURE 12.15 Figures 12.13 and 12.14 combined. The tangent lines at the point $(x_0, y_0, f(x_0, y_0))$ determine a plane that, in this picture at least, appears to be tangent to the surface.

Calculations

The definitions of $\partial f/\partial x$ and $\partial f/\partial y$ give us two different ways of differentiating f at a point: with respect to x in the usual way while treating y as a constant and with respect to y

in the usual way while treating x as constant. As the following examples show, the values of these partial derivatives are usually different at a given point (x_0, y_0).

EXAMPLE 1   Find the values of $\partial f/\partial x$ and $\partial f/\partial y$ at the point $(4, -5)$ if

$$f(x, y) = x^2 + 3xy + y - 1.$$

Solution   To find $\partial f/\partial x$, we treat y as a constant and differentiate with respect to x:

$$\frac{\partial f}{\partial x} = \frac{\partial}{\partial x}(x^2 + 3xy + y - 1) = 2x + 3 \cdot 1 \cdot y + 0 - 0 = 2x + 3y.$$

The value of $\partial f/\partial x$ at $(4, -5)$ is $2(4) + 3(-5) = -7$.
   To find $\partial f/\partial y$, we treat x as a constant and differentiate with respect to y:

$$\frac{\partial f}{\partial y} = \frac{\partial}{\partial y}(x^2 + 3xy + y - 1) = 0 + 3 \cdot x \cdot 1 + 1 - 0 = 3x + 1.$$

The value of $\partial f/\partial y$ at $(4, -5)$ is $3(4) + 1 = 13$.   ∎

EXAMPLE 2   Find $\partial f/\partial y$ as a function if $f(x, y) = y \sin xy$.

Solution   We treat x as a constant and f as a product of y and $\sin xy$:

$$\frac{\partial f}{\partial y} = \frac{\partial}{\partial y}(y \sin xy) = y \frac{\partial}{\partial y} \sin xy + (\sin xy) \frac{\partial}{\partial y}(y)$$

$$= (y \cos xy) \frac{\partial}{\partial y}(xy) + \sin xy = xy \cos xy + \sin xy.$$ ∎

EXAMPLE 3   Find f_x and f_y as functions if

$$f(x, y) = \frac{2y}{y + \cos x}.$$

Solution   We treat f as a quotient. With y held constant, we get

$$f_x = \frac{\partial}{\partial x}\left(\frac{2y}{y + \cos x}\right) = \frac{(y + \cos x)\frac{\partial}{\partial x}(2y) - 2y\frac{\partial}{\partial x}(y + \cos x)}{(y + \cos x)^2}$$

$$= \frac{(y + \cos x)(0) - 2y(-\sin x)}{(y + \cos x)^2} = \frac{2y \sin x}{(y + \cos x)^2}.$$

With x held constant, we get

$$f_y = \frac{\partial}{\partial y}\left(\frac{2y}{y + \cos x}\right) = \frac{(y + \cos x)\frac{\partial}{\partial y}(2y) - 2y\frac{\partial}{dy}(y + \cos x)}{(y + \cos x)^2}$$

$$= \frac{(y + \cos x)(2) - 2y(1)}{(y + \cos x)^2} = \frac{2 \cos x}{(y + \cos x)^2}.$$ ∎

Implicit differentiation works for partial derivatives the way it works for ordinary derivatives, as the next example illustrates.

EXAMPLE 4   Find $\partial z / \partial x$ if the equation

$$yz - \ln z = x + y$$

defines z as a function of the two independent variables x and y and the partial derivative exists.

Solution   We differentiate both sides of the equation with respect to x, holding y constant and treating z as a differentiable function of x:

$$\frac{\partial}{\partial x}(yz) - \frac{\partial}{\partial x}\ln z = \frac{\partial x}{\partial x} + \frac{\partial y}{\partial x}$$

$$y\frac{\partial z}{\partial x} - \frac{1}{z}\frac{\partial z}{\partial x} = 1 + 0 \qquad \text{With } y \text{ constant,} \quad \frac{\partial}{\partial x}(yz) = y\frac{\partial z}{\partial x}.$$

$$\left(y - \frac{1}{z}\right)\frac{\partial z}{\partial x} = 1$$

$$\frac{\partial z}{\partial x} = \frac{z}{yz - 1}. \qquad \blacksquare$$

EXAMPLE 5   The plane $x = 1$ intersects the paraboloid $z = x^2 + y^2$ in a parabola. Find the slope of the tangent to the parabola at $(1, 2, 5)$ (Figure 12.16).

Solution   The slope is the value of the partial derivative $\partial z / \partial y$ at $(1, 2)$:

$$\left.\frac{\partial z}{\partial y}\right|_{(1,2)} = \left.\frac{\partial}{\partial y}(x^2 + y^2)\right|_{(1,2)} = \left.2y\right|_{(1,2)} = 2(2) = 4.$$

As a check, we can treat the parabola as the graph of the single-variable function $z = (1)^2 + y^2 = 1 + y^2$ in the plane $x = 1$ and ask for the slope at $y = 2$. The slope, calculated now as an ordinary derivative, is

$$\left.\frac{dz}{dy}\right|_{y=2} = \left.\frac{d}{dy}(1 + y^2)\right|_{y=2} = \left.2y\right|_{y=2} = 4. \qquad \blacksquare$$

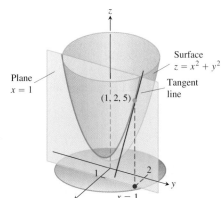

FIGURE 12.16   The tangent to the curve of intersection of the plane $x = 1$ and surface $z = x^2 + y^2$ at the point $(1, 2, 5)$ (Example 5).

Functions of More Than Two Variables

The definitions of the partial derivatives of functions of more than two independent variables are like the definitions for functions of two variables. They are ordinary derivatives with respect to one variable, taken while the other independent variables are held constant.

EXAMPLE 6   If $x, y,$ and z are independent variables and

$$f(x, y, z) = x \sin(y + 3z),$$

then

$$\frac{\partial f}{\partial z} = \frac{\partial}{\partial z}[x \sin(y + 3z)] = x\frac{\partial}{\partial z}\sin(y + 3z)$$

$$= x \cos(y + 3z)\frac{\partial}{\partial z}(y + 3z) = 3x\cos(y + 3z). \qquad \blacksquare$$

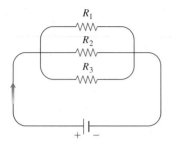

FIGURE 12.17   Resistors arranged this way are said to be connected in parallel (Example 7). Each resistor lets a portion of the current through. Their equivalent resistance R is calculated with the formula

$$\frac{1}{R} = \frac{1}{R_1} + \frac{1}{R_2} + \frac{1}{R_3}.$$

EXAMPLE 7   If resistors of R_1, R_2, and R_3 ohms are connected in parallel to make an R-ohm resistor, the value of R can be found from the equation

$$\frac{1}{R} = \frac{1}{R_1} + \frac{1}{R_2} + \frac{1}{R_3}$$

(Figure 12.17). Find the value of $\partial R/\partial R_2$ when $R_1 = 30$, $R_2 = 45$, and $R_3 = 90$ ohms.

Solution   To find $\partial R/\partial R_2$, we treat R_1 and R_3 as constants and, using implicit differentiation, differentiate both sides of the equation with respect to R_2:

$$\frac{\partial}{\partial R_2}\left(\frac{1}{R}\right) = \frac{\partial}{\partial R_2}\left(\frac{1}{R_1} + \frac{1}{R_2} + \frac{1}{R_3}\right)$$

$$-\frac{1}{R^2}\frac{\partial R}{\partial R_2} = 0 - \frac{1}{R_2^{\,2}} + 0$$

$$\frac{\partial R}{\partial R_2} = \frac{R^2}{R_2^{\,2}} = \left(\frac{R}{R_2}\right)^2.$$

When $R_1 = 30$, $R_2 = 45$, and $R_3 = 90$,

$$\frac{1}{R} = \frac{1}{30} + \frac{1}{45} + \frac{1}{90} = \frac{3 + 2 + 1}{90} = \frac{6}{90} = \frac{1}{15},$$

so $R = 15$ and

$$\frac{\partial R}{\partial R_2} = \left(\frac{15}{45}\right)^2 = \left(\frac{1}{3}\right)^2 = \frac{1}{9}.$$

Thus at the given values, a small change in the resistance R_2 leads to a change in R about 1/9 as large. ∎

Partial Derivatives and Continuity

A function $f(x, y)$ can have partial derivatives with respect to both x and y at a point without the function being continuous there. This is different from functions of a single variable, where the existence of a derivative implies continuity. If the partial derivatives of $f(x, y)$ exist and are continuous throughout a disk centered at (x_0, y_0), however, then f is continuous at (x_0, y_0), as we see at the end of this section.

EXAMPLE 8   Let

$$f(x, y) = \begin{cases} 0, & xy \neq 0 \\ 1, & xy = 0 \end{cases}$$

(Figure 12.18).

(a) Find the limit of f as (x, y) approaches $(0, 0)$ along the line $y = x$.

(b) Prove that f is not continuous at the origin.

(c) Show that both partial derivatives $\partial f/\partial x$ and $\partial f/\partial y$ exist at the origin.

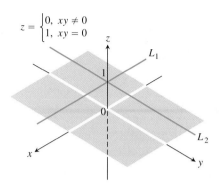

$$z = \begin{cases} 0, & xy \neq 0 \\ 1, & xy = 0 \end{cases}$$

FIGURE 12.18 The graph of

$$f(x, y) = \begin{cases} 0, & xy \neq 0 \\ 1, & xy = 0 \end{cases}$$

consists of the lines L_1 and L_2 and the four open quadrants of the xy-plane. The function has partial derivatives at the origin but is not continuous there (Example 8).

Solution

(a) Since $f(x, y)$ is constantly zero along the line $y = x$ (except at the origin), we have

$$\lim_{(x, y) \to (0,0)} f(x, y)\Big|_{y=x} = \lim_{(x, y) \to (0,0)} 0 = 0.$$

(b) Since $f(0, 0) = 1$, the limit in part (a) proves that f is not continuous at $(0, 0)$.

(c) To find $\partial f/\partial x$ at $(0, 0)$, we hold y fixed at $y = 0$. Then $f(x, y) = 1$ for all x, and the graph of f is the line L_1 in Figure 12.18. The slope of this line at any x is $\partial f/\partial x = 0$. In particular, $\partial f/\partial x = 0$ at $(0, 0)$. Similarly, $\partial f/\partial y$ is the slope of line L_2 at any y, so $\partial f/\partial y = 0$ at $(0, 0)$. ∎

Example 8 notwithstanding, it is still true in higher dimensions that *differentiability* at a point implies continuity. What Example 8 suggests is that we need a stronger requirement for differentiability in higher dimensions than the mere existence of the partial derivatives. We define differentiability for functions of two variables at the end of this section and revisit the connection to continuity.

Second-Order Partial Derivatives

When we differentiate a function $f(x, y)$ twice, we produce its second-order derivatives. These derivatives are usually denoted by

$$\frac{\partial^2 f}{\partial x^2} \qquad \text{"}d \text{ squared } f dx \text{ squared"} \qquad \text{or} \qquad f_{xx} \qquad \text{"}f \text{ sub } xx\text{"}$$

$$\frac{\partial^2 f}{\partial y^2} \qquad \text{"}d \text{ squared } f dy \text{ squared"} \qquad \text{or} \qquad f_{yy} \qquad \text{"}f \text{ sub } yy\text{"}$$

$$\frac{\partial^2 f}{\partial x \partial y} \qquad \text{"}d \text{ squared } f dx \, dy\text{"} \qquad \text{or} \qquad f_{yx} \qquad \text{"}f \text{ sub } yx\text{"}$$

$$\frac{\partial^2 f}{\partial y \partial x} \qquad \text{"}d \text{ squared } f dy \, dx\text{"} \qquad \text{or} \qquad f_{xy} \qquad \text{"}f \text{ sub } xy\text{"}$$

The defining equations are

$$\frac{\partial^2 f}{\partial x^2} = \frac{\partial}{\partial x}\left(\frac{\partial f}{\partial x}\right), \qquad \frac{\partial^2 f}{\partial x \partial y} = \frac{\partial}{\partial x}\left(\frac{\partial f}{\partial y}\right),$$

and so on. Notice the order in which the derivatives are taken:

$$\frac{\partial^2 f}{\partial x \partial y} \qquad \qquad \text{Differentiate first with respect to } y, \text{ then with respect to } x.$$

$$f_{yx} = (f_y)_x \qquad \text{Means the same thing.}$$

HISTORICAL BIOGRAPHY

Pierre-Simon Laplace
(1749–1827)

EXAMPLE 9 If $f(x, y) = x \cos y + y e^x$, find the second-order derivatives

$$\frac{\partial^2 f}{\partial x^2}, \qquad \frac{\partial^2 f}{\partial y \partial x}, \qquad \frac{\partial^2 f}{\partial y^2}, \qquad \text{and} \qquad \frac{\partial^2 f}{\partial x \partial y}.$$

Solution

$$\frac{\partial f}{\partial x} = \frac{\partial}{\partial x}(x \cos y + ye^x) \qquad \frac{\partial f}{\partial y} = \frac{\partial}{\partial y}(x \cos y + ye^x)$$

$$= \cos y + ye^x \qquad\qquad = -x \sin y + e^x$$

So

$$\frac{\partial^2 f}{\partial y \partial x} = \frac{\partial}{\partial y}\left(\frac{\partial f}{\partial x}\right) = -\sin y + e^x \qquad \frac{\partial^2 f}{\partial x \partial y} = \frac{\partial}{\partial x}\left(\frac{\partial f}{\partial y}\right) = -\sin y + e^x$$

$$\frac{\partial^2 f}{\partial x^2} = \frac{\partial}{\partial x}\left(\frac{\partial f}{\partial x}\right) = ye^x. \qquad \frac{\partial^2 f}{\partial y^2} = \frac{\partial}{\partial y}\left(\frac{\partial f}{\partial y}\right) = -x \cos y. \qquad ∎$$

The Mixed Derivative Theorem

You may have noticed that the "mixed" second-order partial derivatives

$$\frac{\partial^2 f}{\partial y \partial x} \qquad \text{and} \qquad \frac{\partial^2 f}{\partial x \partial y}$$

in Example 9 were equal. This was not a coincidence. They must be equal whenever $f, f_x, f_y, f_{xy},$ and f_{yx} are continuous, as stated in the following theorem.

THEOREM 2—The Mixed Derivative Theorem If $f(x, y)$ and its partial derivatives $f_x, f_y, f_{xy},$ and f_{yx} are defined throughout an open region containing a point (a, b) and are all continuous at (a, b), then

$$f_{xy}(a, b) = f_{yx}(a, b).$$

HISTORICAL BIOGRAPHY

Alexis Clairaut
(1713–1765)

Theorem 2 is also known as Clairaut's Theorem, named after the French mathematician Alexis Clairaut who discovered it. A proof is given in Appendix 9. Theorem 2 says that to calculate a mixed second-order derivative, we may differentiate in either order, provided the continuity conditions are satisfied. This can work to our advantage.

EXAMPLE 10 Find $\partial^2 w/\partial x \partial y$ if

$$w = xy + \frac{e^y}{y^2 + 1}.$$

Solution The symbol $\partial^2 w/\partial x \partial y$ tells us to differentiate first with respect to y and then with respect to x. If we postpone the differentiation with respect to y and differentiate first with respect to x, however, we get the answer more quickly. In two steps,

$$\frac{\partial w}{\partial x} = y \qquad \text{and} \qquad \frac{\partial^2 w}{\partial y \partial x} = 1.$$

If we differentiate first with respect to y, we obtain $\partial^2 w/\partial x \partial y = 1$ as well. We can differentiate in either order because the conditions of Theorem 2 hold for w at all points (x_0, y_0). ∎

Partial Derivatives of Still Higher Order

Although we will deal mostly with first- and second-order partial derivatives, because these appear the most frequently in applications, there is no theoretical limit to how many times we can differentiate a function as long as the derivatives involved exist. Thus, we get third- and fourth-order derivatives denoted by symbols like

$$\frac{\partial^3 f}{\partial x \partial y^2} = f_{yyx}$$

$$\frac{\partial^4 f}{\partial x^2 \partial y^2} = f_{yyxx},$$

and so on. As with second-order derivatives, the order of differentiation is immaterial as long as all the derivatives through the order in question are continuous.

EXAMPLE 11 Find f_{yxyz} if $f(x, y, z) = 1 - 2xy^2z + x^2y$.

Solution We first differentiate with respect to the variable y, then x, then y again, and finally with respect to z:

$$f_y = -4xyz + x^2$$
$$f_{yx} = -4yz + 2x$$
$$f_{yxy} = -4z$$
$$f_{yxyz} = -4$$

∎

Differentiability

The starting point for differentiability is not the difference quotient we saw in studying single-variable functions, but rather the idea of increment. Recall from our work with functions of a single variable in Section 3.10 that if $y = f(x)$ is differentiable at $x = x_0$, then the change in the value of f that results from changing x from x_0 to $x_0 + \Delta x$ is given by an equation of the form

$$\Delta y = f'(x_0)\Delta x + \epsilon \Delta x$$

in which $\epsilon \to 0$ as $\Delta x \to 0$. For functions of two variables, the analogous property becomes the definition of differentiability. The Increment Theorem (from advanced calculus) tells us when to expect the property to hold.

THEOREM 3—The Increment Theorem for Functions of Two Variables Suppose that the first partial derivatives of $f(x, y)$ are defined throughout an open region R containing the point (x_0, y_0) and that f_x and f_y are continuous at (x_0, y_0). Then the change

$$\Delta z = f(x_0 + \Delta x, y_0 + \Delta y) - f(x_0, y_0)$$

in the value of f that results from moving from (x_0, y_0) to another point $(x_0 + \Delta x, y_0 + \Delta y)$ in R satisfies an equation of the form

$$\Delta z = f_x(x_0, y_0)\Delta x + f_y(x_0, y_0)\Delta y + \epsilon_1 \Delta x + \epsilon_2 \Delta y,$$

in which each of $\epsilon_1, \epsilon_2 \to 0$ as both $\Delta x, \Delta y \to 0$.

You can see where the epsilons come from in the proof in Appendix 9. You will also see that similar results hold for functions of more than two independent variables.

DEFINITION A function $z = f(x, y)$ is **differentiable at** (x_0, y_0) if $f_x(x_0, y_0)$ and $f_y(x_0, y_0)$ exist and Δz satisfies an equation of the form

$$\Delta z = f_x(x_0, y_0)\Delta x + f_y(x_0, y_0)\Delta y + \epsilon_1 \Delta x + \epsilon_2 \Delta y,$$

in which each of $\epsilon_1, \epsilon_2 \rightarrow 0$ as both $\Delta x, \Delta y \rightarrow 0$. We call f **differentiable** if it is differentiable at every point in its domain.

In light of this definition, we have the immediate corollary of Theorem 3 that a function is differentiable if its first partial derivatives are *continuous*.

COROLLARY OF THEOREM 3 If the partial derivatives f_x and f_y of a function $f(x, y)$ are continuous throughout an open region R, then f is differentiable at every point of R.

If $z = f(x, y)$ is differentiable, then the definition of differentiability assures that $\Delta z = f(x_0 + \Delta x, y_0 + \Delta y) - f(x_0, y_0)$ approaches 0 as Δx and Δy approach 0. This tells us that a function of two variables is continuous at every point where it is differentiable.

THEOREM 4—Differentiability Implies Continuity If a function $f(x, y)$ is differentiable at (x_0, y_0), then f is continuous at (x_0, y_0).

As we can see from Theorems 3 and 4, a function $f(x, y)$ must be continuous at a point (x_0, y_0) if f_x and f_y are continuous throughout an open region containing (x_0, y_0). Remember, however, that it is still possible for a function of two variables to be discontinuous at a point where its first partial derivatives exist, as we saw in Example 8. Existence alone of the partial derivatives at that point is not enough.

EXERCISES 12.3

In Exercises 1–22, find $\partial f/\partial x$ and $\partial f/\partial y$.

1. $f(x, y) = 2x^2 - 3y - 4$ **2.** $f(x, y) = x^2 - xy + y^2$

3. $f(x, y) = (x^2 - 1)(y + 2)$

4. $f(x, y) = 5xy - 7x^2 - y^2 + 3x - 6y + 2$

5. $f(x, y) = (xy - 1)^2$ **6.** $f(x, y) = (2x - 3y)^3$

7. $f(x, y) = \sqrt{x^2 + y^2}$ **8.** $f(x, y) = (x^3 + (y/2))^{2/3}$

9. $f(x, y) = 1/(x + y)$ **10.** $f(x, y) = x/(x^2 + y^2)$

11. $f(x, y) = (x + y)/(xy - 1)$ **12.** $f(x, y) = \tan^{-1}(y/x)$

13. $f(x, y) = e^{(x+y+1)}$ **14.** $f(x, y) = e^{-x} \sin(x + y)$

15. $f(x, y) = \ln(x + y)$ **16.** $f(x, y) = e^{xy} \ln y$

17. $f(x, y) = \sin^2(x - 3y)$ **18.** $f(x, y) = \cos^2(3x - y^2)$

19. $f(x, y) = x^y$ **20.** $f(x, y) = \log_y x$

21. $f(x, y) = \displaystyle\int_x^y g(t)\, dt$ (g continuous for all t)

22. $f(x, y) = \displaystyle\sum_{n=0}^{\infty} (xy)^n$ ($|xy| < 1$)

In Exercises 23–34, find f_x, f_y, and f_z.

23. $f(x, y, z) = 1 + xy^2 - 2z^2$ **24.** $f(x, y, z) = xy + yz + xz$

25. $f(x, y, z) = x - \sqrt{y^2 + z^2}$

26. $f(x, y, z) = (x^2 + y^2 + z^2)^{-1/2}$

27. $f(x, y, z) = \sin^{-1}(xyz)$ **28.** $f(x, y, z) = \sec^{-1}(x + yz)$

29. $f(x, y, z) = \ln(x + 2y + 3z)$

30. $f(x, y, z) = yz \ln(xy)$ **31.** $f(x, y, z) = e^{-(x^2+y^2+z^2)}$

32. $f(x, y, z) = e^{-xyz}$

33. $f(x, y, z) = \tanh(x + 2y + 3z)$

34. $f(x, y, z) = \sinh(xy - z^2)$

In Exercises 35–40, find the partial derivative of the function with respect to each variable.

35. $f(t, \alpha) = \cos(2\pi t - \alpha)$ **36.** $g(u, v) = v^2 e^{(2u/v)}$

37. $h(\rho, \phi, \theta) = \rho \sin \phi \cos \theta$ **38.** $g(r, \theta, z) = r(1 - \cos \theta) - z$

39. Work done by the heart (Section 3.10, Exercise 59)

$$W(P, V, \delta, v, g) = PV + \frac{V\delta v^2}{2g}$$

40. Wilson lot size formula (Section 4.5, Exercise 45)

$$A(c, h, k, m, q) = \frac{km}{q} + cm + \frac{hq}{2}$$

Find all the second-order partial derivatives of the functions in Exercises 41–46.

41. $f(x, y) = x + y + xy$ **42.** $f(x, y) = \sin xy$

43. $g(x, y) = x^2 y + \cos y + y \sin x$

44. $h(x, y) = xe^y + y + 1$ **45.** $r(x, y) = \ln(x + y)$

46. $s(x, y) = \tan^{-1}(y/x)$

In Exercises 47–50, verify that $w_{xy} = w_{yx}$.

47. $w = \ln(2x + 3y)$ **48.** $w = e^x + x \ln y + y \ln x$

49. $w = xy^2 + x^2 y^3 + x^3 y^4$ **50.** $w = x \sin y + y \sin x + xy$

51. Which order of differentiation will calculate f_{xy} faster: x first or y first? Try to answer without writing anything down.

  a. $f(x, y) = x \sin y + e^y$

  b. $f(x, y) = 1/x$

  c. $f(x, y) = y + (x/y)$

  d. $f(x, y) = y + x^2 y + 4y^3 - \ln(y^2 + 1)$

  e. $f(x, y) = x^2 + 5xy + \sin x + 7e^x$

  f. $f(x, y) = x \ln xy$

52. The fifth-order partial derivative $\partial^5 f/\partial x^2 \partial y^3$ is zero for each of the following functions. To show this as quickly as possible, which variable would you differentiate with respect to first: x or y? Try to answer without writing anything down.

  a. $f(x, y) = y^2 x^4 e^x + 2$

  b. $f(x, y) = y^2 + y(\sin x - x^4)$

  c. $f(x, y) = x^2 + 5xy + \sin x + 7e^x$

  d. $f(x, y) = xe^{y^2/2}$

In Exercises 53 and 54, use the limit definition of partial derivative to compute the partial derivatives of the functions at the specified points.

53. $f(x, y) = 1 - x + y - 3x^2 y$, $\dfrac{\partial f}{\partial x}$ and $\dfrac{\partial f}{\partial y}$ at $(1, 2)$

54. $f(x, y) = 4 + 2x - 3y - xy^2$, $\dfrac{\partial f}{\partial x}$ and $\dfrac{\partial f}{\partial y}$ at $(-2, 1)$

55. Three variables Let $w = f(x, y, z)$ be a function of three independent variables and write the formal definition of the partial derivative $\partial f/\partial z$ at (x_0, y_0, z_0). Use this definition to find $\partial f/\partial z$ at $(1, 2, 3)$ for $f(x, y, z) = x^2 y z^2$.

56. Three variables Let $w = f(x, y, z)$ be a function of three independent variables and write the formal definition of the partial derivative $\partial f/\partial y$ at (x_0, y_0, z_0). Use this definition to find $\partial f/\partial y$ at $(-1, 0, 3)$ for $f(x, y, z) = -2xy^2 + yz^2$.

57. Find the value of $\partial z/\partial x$ at the point $(1, 1, 1)$ if the equation

$$xy + z^3 x - 2yz = 0$$

defines z as a function of the two independent variables x and y and the partial derivative exists.

58. Find the value of $\partial x/\partial z$ at the point $(1, -1, -3)$ if the equation

$$xz + y \ln x - x^2 + 4 = 0$$

defines x as a function of the two independent variables y and z and the partial derivative exists.

Exercises 59 and 60 are about the triangle shown here.

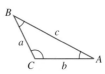

59. Express A implicitly as a function of a, b, and c and calculate $\partial A/\partial a$ and $\partial A/\partial b$.

60. Express a implicitly as a function of A, b, and B and calculate $\partial a/\partial A$ and $\partial a/\partial B$.

61. Two dependent variables Express v_x in terms of u and y if the equations $x = v \ln u$ and $y = u \ln v$ define u and v as functions of the independent variables x and y, and if v_x exists. (*Hint:* Differentiate both equations with respect to x and solve for v_x by eliminating u_x.)

62. **Two dependent variables** Find $\partial x/\partial u$ and $\partial y/\partial u$ if the equations $u = x^2 - y^2$ and $v = x^2 - y$ define x and y as functions of the independent variables u and v, and the partial derivatives exist. (See the hint in Exercise 61.) Then let $s = x^2 + y^2$ and find $\partial s/\partial u$.

The **three-dimensional Laplace equation**

$$\frac{\partial^2 f}{\partial x^2} + \frac{\partial^2 f}{\partial y^2} + \frac{\partial^2 f}{\partial z^2} = 0$$

is satisfied by steady-state temperature distributions $T = f(x, y, z)$ in space, by gravitational potentials, and by electrostatic potentials. The **two-dimensional Laplace equation**

$$\frac{\partial^2 f}{\partial x^2} + \frac{\partial^2 f}{\partial y^2} = 0,$$

obtained by dropping the $\partial^2 f/\partial z^2$ term from the previous equation, describes potentials and steady-state temperature distributions in a plane (see the accompanying figure). The plane (a) may be treated as a thin slice of the solid (b) perpendicular to the z-axis.

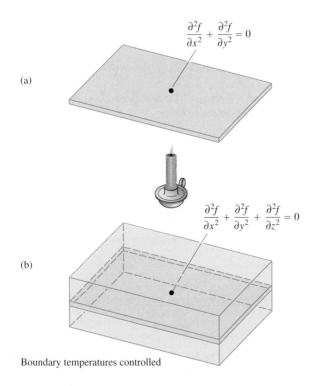

Boundary temperatures controlled

Show that each function in Exercises 63–68 satisfies a Laplace equation.

63. $f(x, y, z) = x^2 + y^2 - 2z^2$
64. $f(x, y, z) = 2z^3 - 3(x^2 + y^2)z$

65. $f(x, y) = e^{-2y} \cos 2x$
66. $f(x, y) = \ln\sqrt{x^2 + y^2}$
67. $f(x, y, z) = (x^2 + y^2 + z^2)^{-1/2}$
68. $f(x, y, z) = e^{3x+4y} \cos 5z$

The Wave Equation If we stand on an ocean shore and take a snapshot of the waves, the picture shows a regular pattern of peaks and valleys in an instant of time. We see periodic vertical motion in space, with respect to distance. If we stand in the water, we can feel the rise and fall of the water as the waves go by. We see periodic vertical motion in time. In physics, this beautiful symmetry is expressed by the **one-dimensional wave equation**

$$\frac{\partial^2 w}{\partial t^2} = c^2 \frac{\partial^2 w}{\partial x^2},$$

where w is the wave height, x is the distance variable, t is the time variable, and c is the velocity with which the waves are propagated.

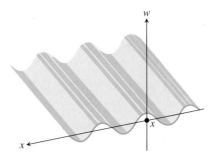

In our example, x is the distance across the ocean's surface, but in other applications, x might be the distance along a vibrating string, distance through air (sound waves), or distance through space (light waves). The number c varies with the medium and type of wave.

Show that the functions in Exercises 69–75 are all solutions of the wave equation.

69. $w = \sin(x + ct)$
70. $w = \cos(2x + 2ct)$
71. $w = \sin(x + ct) + \cos(2x + 2ct)$
72. $w = \ln(2x + 2ct)$ 73. $w = \tan(2x - 2ct)$
74. $w = 5\cos(3x + 3ct) + e^{x+ct}$
75. $w = f(u)$, where f is a differentiable function of u, and $u = a(x + ct)$, where a is a constant

76. Does a function $f(x, y)$ with continuous first partial derivatives throughout an open region R have to be continuous on R? Give reasons for your answer.

77. If a function $f(x, y)$ has continuous second partial derivatives throughout an open region R, must the first-order partial derivatives of f be continuous on R? Give reasons for your answer.

12.4 The Chain Rule

The Chain Rule for functions of a single variable studied in Section 3.5 said that when $w = f(x)$ was a differentiable function of x and $x = g(t)$ was a differentiable function of t, w became a differentiable function of t and dw/dt could be calculated with the formula

$$\frac{dw}{dt} = \frac{dw}{dx}\frac{dx}{dt}.$$

For functions of two or more variables the Chain Rule has several forms. The form depends on how many variables are involved but works like the Chain Rule in Section 3.5 once we account for the presence of additional variables.

Functions of Two Variables

The Chain Rule formula for a function $w = f(x, y)$ when $x = x(t)$ and $y = y(t)$ are both differentiable functions of t is given in the following theorem.

THEOREM 5—Chain Rule for Functions of Two Independent Variables   If $w = f(x, y)$ has continuous partial derivatives f_x and f_y and if $x = x(t), y = y(t)$ are differentiable functions of t, then the composite $w = f(x(t), y(t))$ is a differentiable function of t and

$$\frac{dw}{dt} = f_x(x(t), y(t)) \cdot x'(t) + f_y(x(t), y(t)) \cdot y'(t),$$

or

$$\frac{dw}{dt} = \frac{\partial f}{\partial x}\frac{dx}{dt} + \frac{\partial f}{\partial y}\frac{dy}{dt}.$$

Each of $\dfrac{\partial f}{\partial x}, \dfrac{\partial w}{\partial x}, f_x$ indicates the partial derivative of f with respect to x.

Proof   The proof consists of showing that if x and y are differentiable at $t = t_0$, then w is differentiable at t_0 and

$$\left(\frac{dw}{dt}\right)_{t_0} = \left(\frac{\partial w}{\partial x}\right)_{P_0}\left(\frac{dx}{dt}\right)_{t_0} + \left(\frac{\partial w}{\partial y}\right)_{P_0}\left(\frac{dy}{dt}\right)_{t_0},$$

where $P_0 = (x(t_0), y(t_0))$. The subscripts indicate where each of the derivatives are to be evaluated.

Let $\Delta x, \Delta y,$ and Δw be the increments that result from changing t from t_0 to $t_0 + \Delta t$. Since f is differentiable (see the definition in Section 12.3),

$$\Delta w = \left(\frac{\partial w}{\partial x}\right)_{P_0}\Delta x + \left(\frac{\partial w}{\partial y}\right)_{P_0}\Delta y + \epsilon_1\Delta x + \epsilon_2\Delta y,$$

where $\epsilon_1, \epsilon_2 \rightarrow 0$ as $\Delta x, \Delta y \rightarrow 0$. To find dw/dt, we divide this equation through by Δt and let Δt approach zero. The division gives

$$\frac{\Delta w}{\Delta t} = \left(\frac{\partial w}{\partial x}\right)_{P_0}\frac{\Delta x}{\Delta t} + \left(\frac{\partial w}{\partial y}\right)_{P_0}\frac{\Delta y}{\Delta t} + \epsilon_1\frac{\Delta x}{\Delta t} + \epsilon_2\frac{\Delta y}{\Delta t}.$$

To remember the Chain Rule picture the diagram below. To find dw/dt, start at w and read down each route to t, multiplying derivatives along the way. Then add the products.

Chain Rule

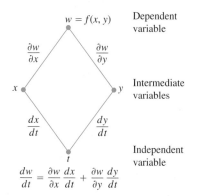

$$w = f(x, y) \quad \text{Dependent variable}$$

$$\frac{\partial w}{\partial x} \qquad \frac{\partial w}{\partial y}$$

$$x \qquad y \quad \text{Intermediate variables}$$

$$\frac{dx}{dt} \qquad \frac{dy}{dt}$$

$$t \quad \text{Independent variable}$$

$$\frac{dw}{dt} = \frac{\partial w}{\partial x}\frac{dx}{dt} + \frac{\partial w}{\partial y}\frac{dy}{dt}$$

Letting Δt approach zero gives

$$\left(\frac{dw}{dt}\right)_{t_0} = \lim_{\Delta t \to 0} \frac{\Delta w}{\Delta t}$$

$$= \left(\frac{\partial w}{\partial x}\right)_{P_0}\left(\frac{dx}{dt}\right)_{t_0} + \left(\frac{\partial w}{\partial y}\right)_{P_0}\left(\frac{dy}{dt}\right)_{t_0} + 0 \cdot \left(\frac{dx}{dt}\right)_{t_0} + 0 \cdot \left(\frac{dy}{dt}\right)_{t_0}. \quad \blacksquare$$

The **tree diagram** in the margin provides a convenient way to remember the Chain Rule. Notice that w is both a function of x and y and a function of t. In the diagram, when $t = t_0$, the derivatives dx/dt and dy/dt are evaluated at t_0. The value of t_0 then determines the value x_0 for the differentiable function x and the value y_0 for the differentiable function y. The partial derivatives $\partial w/\partial x$ and $\partial w/\partial y$ (which are themselves functions of x and y) are evaluated at the point $P_0(x_0, y_0)$ corresponding to t_0. The "true" independent variable is t, whereas x and y are *intermediate variables* (controlled by t) and w is the dependent variable.

A more precise notation for the Chain Rule shows where the various derivatives in Theorem 5 are evaluated:

$$\frac{dw}{dt}(t_0) = \frac{\partial f}{\partial x}(x_0, y_0) \cdot \frac{dx}{dt}(t_0) + \frac{\partial f}{\partial y}(x_0, y_0) \cdot \frac{dy}{dt}(t_0).$$

EXAMPLE 1 Use the Chain Rule to find the derivative of

$$w = xy$$

with respect to t along the path $x = \cos t, y = \sin t$. What is the derivative's value at $t = \pi/2$?

Solution We apply the Chain Rule to find dw/dt as follows:

$$\frac{dw}{dt} = \frac{\partial w}{\partial x}\frac{dx}{dt} + \frac{\partial w}{\partial y}\frac{dy}{dt}$$

$$= \frac{\partial(xy)}{\partial x} \cdot \frac{d}{dt}(\cos t) + \frac{\partial(xy)}{\partial y} \cdot \frac{d}{dt}(\sin t)$$

$$= (y)(-\sin t) + (x)(\cos t)$$

$$= (\sin t)(-\sin t) + (\cos t)(\cos t)$$

$$= -\sin^2 t + \cos^2 t$$

$$= \cos 2t.$$

In this example, we can check the result with a more direct calculation. As a function of t,

$$w = xy = \cos t \sin t = \frac{1}{2}\sin 2t,$$

so

$$\frac{dw}{dt} = \frac{d}{dt}\left(\frac{1}{2}\sin 2t\right) = \frac{1}{2} \cdot 2\cos 2t = \cos 2t.$$

In either case, at the given value of t,

$$\left(\frac{dw}{dt}\right)_{t=\pi/2} = \cos\left(2 \cdot \frac{\pi}{2}\right) = \cos \pi = -1.$$ ∎

Functions of Three Variables

You can probably predict the Chain Rule for functions of three variables, as it only involves adding the expected third term to the two-variable formula.

> **THEOREM 6—Chain Rule for Functions of Three Independent Variables** If $w = f(x, y, z)$ is differentiable and x, y, and z are differentiable functions of t, then w is a differentiable function of t and
>
> $$\frac{dw}{dt} = \frac{\partial f}{\partial x}\frac{dx}{dt} + \frac{\partial f}{\partial y}\frac{dy}{dt} + \frac{\partial f}{\partial z}\frac{dz}{dt}.$$

The proof is identical with the proof of Theorem 5 except that there are now three intermediate variables instead of two. The diagram we use for remembering the new equation is similar as well, with three routes from w to t.

Here we have three routes from w to t instead of two, but finding dw/dt is still the same. Read down each route, multiplying derivatives along the way; then add.

Chain Rule

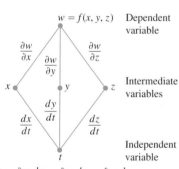

$w = f(x, y, z)$ Dependent variable

Intermediate variables

Independent variable

$$\frac{dw}{dt} = \frac{\partial w}{\partial x}\frac{dx}{dt} + \frac{\partial w}{\partial y}\frac{dy}{dt} + \frac{\partial w}{\partial z}\frac{dz}{dt}$$

EXAMPLE 2 Find dw/dt if

$$w = xy + z, \qquad x = \cos t, \qquad y = \sin t, \qquad z = t.$$

In this example the values of $w(t)$ are changing along the path of a helix (Section 11.1) as t changes. What is the derivative's value at $t = 0$?

Solution

$$\frac{dw}{dt} = \frac{\partial w}{\partial x}\frac{dx}{dt} + \frac{\partial w}{\partial y}\frac{dy}{dt} + \frac{\partial w}{\partial z}\frac{dz}{dt}$$

$$= (y)(-\sin t) + (x)(\cos t) + (1)(1)$$

$$= (\sin t)(-\sin t) + (\cos t)(\cos t) + 1 \qquad \text{Substitute for the intermediate variables.}$$

$$= -\sin^2 t + \cos^2 t + 1 = 1 + \cos 2t.$$

$$\left(\frac{dw}{dt}\right)_{t=0} = 1 + \cos(0) = 2.$$ ∎

For a physical interpretation of change along a curve think of an object whose position is changing with time t. If $w = T(x, y, z)$ is the temperature at each point (x, y, z) along a curve C with parametric equations $x = x(t)$, $y = y(t)$, and $z = z(t)$, then the composite function $w = T(x(t), y(t), z(t))$ represents the temperature relative to t along the curve. The derivative dw/dt is then the instantaneous rate of change of temperature due to the motion along the curve, as calculated in Theorem 6.

Functions Defined on Surfaces

If we are interested in the temperature $w = f(x, y, z)$ at points (x, y, z) on a globe in space, we might prefer to think of x, y, and z as functions of the variables r and s that give the points' longitudes and latitudes. If $x = g(r, s)$, $y = h(r, s)$, and $z = k(r, s)$, we could then express the temperature as a function of r and s with the composite function

$$w = f(g(r, s), h(r, s), k(r, s)).$$

Under the conditions stated below, w has partial derivatives with respect to both r and s that can be calculated in the following way.

THEOREM 7—Chain Rule for Two Independent Variables and Three Intermediate Variables   Suppose that $w = f(x, y, z)$, $x = g(r, s)$, $y = h(r, s)$, and $z = k(r, s)$. If all four functions are differentiable, then w has partial derivatives with respect to r and s, given by the formulas

$$\frac{\partial w}{\partial r} = \frac{\partial w}{\partial x}\frac{\partial x}{\partial r} + \frac{\partial w}{\partial y}\frac{\partial y}{\partial r} + \frac{\partial w}{\partial z}\frac{\partial z}{\partial r}$$

$$\frac{\partial w}{\partial s} = \frac{\partial w}{\partial x}\frac{\partial x}{\partial s} + \frac{\partial w}{\partial y}\frac{\partial y}{\partial s} + \frac{\partial w}{\partial z}\frac{\partial z}{\partial s}.$$

The first of these equations can be derived from the Chain Rule in Theorem 6 by holding s fixed and treating r as t. The second can be derived in the same way, holding r fixed and treating s as t. The tree diagrams for both equations are shown in Figure 12.19.

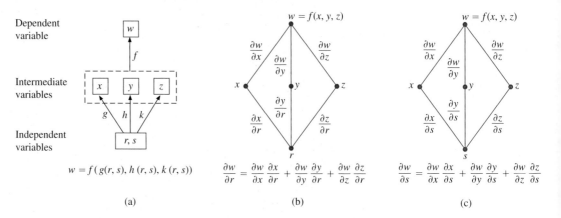

FIGURE 12.19 Composite function and tree diagrams for Theorem 7.

EXAMPLE 3   Express $\partial w/\partial r$ and $\partial w/\partial s$ in terms of r and s if

$$w = x + 2y + z^2, \qquad x = \frac{r}{s}, \qquad y = r^2 + \ln s, \qquad z = 2r.$$

Solution

$$\frac{\partial w}{\partial r} = \frac{\partial w}{\partial x}\frac{\partial x}{\partial r} + \frac{\partial w}{\partial y}\frac{\partial y}{\partial r} + \frac{\partial w}{\partial z}\frac{\partial z}{\partial r}$$

$$= (1)\left(\frac{1}{s}\right) + (2)(2r) + (2z)(2)$$

$$= \frac{1}{s} + 4r + (4r)(2) = \frac{1}{s} + 12r \qquad \text{Substitute for intermediate variable } z.$$

$$\frac{\partial w}{\partial s} = \frac{\partial w}{\partial x}\frac{\partial x}{\partial s} + \frac{\partial w}{\partial y}\frac{\partial y}{\partial s} + \frac{\partial w}{\partial z}\frac{\partial z}{\partial s}$$

$$= (1)\left(-\frac{r}{s^2}\right) + (2)\left(\frac{1}{s}\right) + (2z)(0) = \frac{2}{s} - \frac{r}{s^2} \qquad \blacksquare$$

If f is a function of two variables instead of three, each equation in Theorem 7 becomes correspondingly one term shorter.

Chain Rule

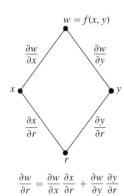

$$\frac{\partial w}{\partial r} = \frac{\partial w}{\partial x}\frac{\partial x}{\partial r} + \frac{\partial w}{\partial y}\frac{\partial y}{\partial r}.$$

FIGURE 12.20 Tree diagram for the equation

$$\frac{\partial w}{\partial r} = \frac{\partial w}{\partial x}\frac{\partial x}{\partial r} + \frac{\partial w}{\partial y}\frac{\partial y}{\partial r}.$$

If $w = f(x, y)$, $x = g(r, s)$, and $y = h(r, s)$, then

$$\frac{\partial w}{\partial r} = \frac{\partial w}{\partial x}\frac{\partial x}{\partial r} + \frac{\partial w}{\partial y}\frac{\partial y}{\partial r} \quad \text{and} \quad \frac{\partial w}{\partial s} = \frac{\partial w}{\partial x}\frac{\partial x}{\partial s} + \frac{\partial w}{\partial y}\frac{\partial y}{\partial s}.$$

Figure 12.20 shows the tree diagram for the first of these equations. The diagram for the second equation is similar; just replace r with s.

EXAMPLE 4   Express $\partial w/\partial r$ and $\partial w/\partial s$ in terms of r and s if

$$w = x^2 + y^2, \qquad x = r - s, \qquad y = r + s.$$

Solution

$$\frac{\partial w}{\partial r} = \frac{\partial w}{\partial x}\frac{\partial x}{\partial r} + \frac{\partial w}{\partial y}\frac{\partial y}{\partial r} \qquad\qquad \frac{\partial w}{\partial s} = \frac{\partial w}{\partial x}\frac{\partial x}{\partial s} + \frac{\partial w}{\partial y}\frac{\partial y}{\partial s}$$

$$= (2x)(1) + (2y)(1) \qquad\qquad = (2x)(-1) + (2y)(1) \qquad \text{Substitute for the intermediate variables.}$$

$$= 2(r - s) + 2(r + s) \qquad\qquad = -2(r - s) + 2(r + s)$$

$$= 4r \qquad\qquad\qquad\qquad\quad = 4s \qquad\qquad\qquad \blacksquare$$

If f is a function of x alone, our equations are even simpler.

Chain Rule

$w = f(x)$

$\dfrac{dw}{dx}$

x

$\dfrac{\partial x}{\partial r}$ $\dfrac{\partial x}{\partial s}$

r s

$$\frac{\partial w}{\partial r} = \frac{dw}{dx}\frac{\partial x}{\partial r}$$

$$\frac{\partial w}{\partial s} = \frac{dw}{dx}\frac{\partial x}{\partial s}$$

FIGURE 12.21 Tree diagram for differentiating f as a composite function of r and s with one intermediate variable.

If $w = f(x)$ and $x = g(r, s)$, then

$$\frac{\partial w}{\partial r} = \frac{dw}{dx}\frac{\partial x}{\partial r} \quad \text{and} \quad \frac{\partial w}{\partial s} = \frac{dw}{dx}\frac{\partial x}{\partial s}.$$

In this case, we use the ordinary (single-variable) derivative, dw/dx. The tree diagram is shown in Figure 12.21.

Implicit Differentiation Revisited

The two-variable Chain Rule in Theorem 5 leads to a formula that takes some of the algebra out of implicit differentiation. Suppose that

1. The function $F(x, y)$ is differentiable and
2. The equation $F(x, y) = 0$ defines y implicitly as a differentiable function of x, say $y = h(x)$.

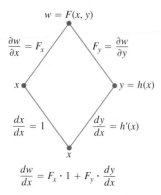

$$\frac{dw}{dx} = F_x \cdot 1 + F_y \cdot \frac{dy}{dx}$$

FIGURE 12.22 Tree diagram for differentiating $w = F(x, y)$ with respect to x. Setting $dw/dx = 0$ leads to a simple computational formula for implicit differentiation (Theorem 8).

Since $w = F(x, y) = 0$, the derivative dw/dx must be zero. Computing the derivative from the Chain Rule (tree diagram in Figure 12.22), we find

$$0 = \frac{dw}{dx} = F_x \frac{dx}{dx} + F_y \frac{dy}{dx} \qquad \text{Theorem 5 with } t = x \text{ and } f = F$$

$$= F_x \cdot 1 + F_y \cdot \frac{dy}{dx}.$$

If $F_y = \partial w / \partial y \neq 0$, we can solve this equation for dy/dx to get

$$\frac{dy}{dx} = -\frac{F_x}{F_y}.$$

We state this result formally.

THEOREM 8—A Formula for Implicit Differentiation   Suppose that $F(x, y)$ is differentiable and that the equation $F(x, y) = 0$ defines y as a differentiable function of x. Then at any point where $F_y \neq 0$,

$$\frac{dy}{dx} = -\frac{F_x}{F_y}.$$

EXAMPLE 5   Use Theorem 8 to find dy/dx if $y^2 - x^2 - \sin xy = 0$.

Solution   Take $F(x, y) = y^2 - x^2 - \sin xy$. Then

$$\frac{dy}{dx} = -\frac{F_x}{F_y} = -\frac{-2x - y \cos xy}{2y - x \cos xy}$$

$$= \frac{2x + y \cos xy}{2y - x \cos xy}.$$

This calculation is significantly shorter than the single-variable calculation with which we found dy/dx in Section 3.6, Example 3. ■

Functions of Many Variables

We have seen several different forms of the Chain Rule in this section, but you do not have to memorize them all if you can see them as special cases of the same general formula. When solving particular problems, it may help to draw the appropriate tree diagram by placing the dependent variable on top, the intermediate variables in the middle, and the selected independent variable at the bottom. To find the derivative of the dependent variable with respect to the selected independent variable, start at the dependent variable and read down each route of the tree to the independent variable, calculating and multiplying the derivatives along each route. Then add the products you found for the different routes.

In general, suppose that $w = f(x, y, \dots, v)$ is a differentiable function of the variables $x, y, \dots, v$ (a finite set) and the $x, y, \dots, v$ are differentiable functions of $p, q, \dots, t$ (another finite set). Then w is a differentiable function of the variables p through t and the partial derivatives of w with respect to these variables are given by equations of the form

$$\frac{\partial w}{\partial p} = \frac{\partial w}{\partial x}\frac{\partial x}{\partial p} + \frac{\partial w}{\partial y}\frac{\partial y}{\partial p} + \cdots + \frac{\partial w}{\partial v}\frac{\partial v}{\partial p}.$$

The other equations are obtained by replacing p by $q, \dots, t$, one at a time.

One way to remember this equation is to think of the right-hand side as the dot product of two vectors with components

$$\underbrace{\left(\frac{\partial w}{\partial x}, \frac{\partial w}{\partial y}, \dots, \frac{\partial w}{\partial v}\right)}_{\substack{\text{Derivatives of } w \text{ with} \\ \text{respect to the} \\ \text{intermediate variables}}} \quad \text{and} \quad \underbrace{\left(\frac{\partial x}{\partial p}, \frac{\partial y}{\partial p}, \dots, \frac{\partial v}{\partial p}\right)}_{\substack{\text{Derivatives of the intermediate} \\ \text{variables with respect to the} \\ \text{selected independent variable}}}.$$

EXERCISES 12.4

In Exercises 1–6, **(a)** express dw/dt as a function of t, both by using the Chain Rule and by expressing w in terms of t and differentiating directly with respect to t. Then **(b)** evaluate dw/dt at the given value of t.

1. $w = x^2 + y^2$, $\quad x = \cos t$, $\quad y = \sin t$; $\quad t = \pi$

2. $w = x^2 + y^2$, $\quad x = \cos t + \sin t$, $\quad y = \cos t - \sin t$; $\quad t = 0$

3. $w = \frac{x}{z} + \frac{y}{z}$, $\quad x = \cos^2 t$, $\quad y = \sin^2 t$, $\quad z = 1/t$; $\quad t = 3$

4. $w = \ln(x^2 + y^2 + z^2)$, $\quad x = \cos t$, $\quad y = \sin t$, $\quad z = 4\sqrt{t}$; $\quad t = 3$

5. $w = 2ye^x - \ln z$, $\quad x = \ln(t^2 + 1)$, $\quad y = \tan^{-1} t$, $\quad z = e^t$; $\quad t = 1$

6. $w = z - \sin xy$, $\quad x = t$, $\quad y = \ln t$, $\quad z = e^{t-1}$; $\quad t = 1$

In Exercises 7 and 8, **(a)** express $\partial z/\partial u$ and $\partial z/\partial v$ as functions of u and v both by using the Chain Rule and by expressing z directly in terms of u and v before differentiating. Then **(b)** evaluate $\partial z/\partial u$ and $\partial z/\partial v$ at the given point (u, v).

7. $z = 4e^x \ln y$, $\quad x = \ln(u \cos v)$, $\quad y = u \sin v$; $\quad (u, v) = (2, \pi/4)$

8. $z = \tan^{-1}(x/y)$, $\quad x = u \cos v$, $\quad y = u \sin v$; $\quad (u, v) = (1.3, \pi/6)$

In Exercises 9 and 10, **(a)** express $\partial w/\partial u$ and $\partial w/\partial v$ as functions of u and v both by using the Chain Rule and by expressing w directly in terms of u and v before differentiating. Then **(b)** evaluate $\partial w/\partial u$ and $\partial w/\partial v$ at the given point (u, v).

9. $w = xy + yz + xz$, $\quad x = u + v$, $\quad y = u - v$, $\quad z = uv$; $\quad (u, v) = (1/2, 1)$

10. $w = \ln(x^2 + y^2 + z^2)$, $\quad x = ue^v \sin u$, $\quad y = ue^v \cos u$, $\quad z = ue^v$; $\quad (u, v) = (-2, 0)$

In Exercises 11 and 12, **(a)** express $\partial u/\partial x$, $\partial u/\partial y$, and $\partial u/\partial z$ as functions of x, y, and z both by using the Chain Rule and by expressing u directly in terms of x, y, and z before differentiating. Then **(b)** evaluate $\partial u/\partial x$, $\partial u/\partial y$, and $\partial u/\partial z$ at the given point (x, y, z).

11. $u = \frac{p - q}{q - r}$, $\quad p = x + y + z$, $\quad q = x - y + z$, $\quad r = x + y - z$; $\quad (x, y, z) = \left(\sqrt{3}, 2, 1\right)$

12. $u = e^{qr} \sin^{-1} p$, $\quad p = \sin x$, $\quad q = z^2 \ln y$, $\quad r = 1/z$; $\quad (x, y, z) = (\pi/4, 1/2, -1/2)$

In Exercises 13–24, draw a tree diagram and write a Chain Rule formula for each derivative.

13. $\dfrac{dz}{dt}$ for $z = f(x, y)$, $\quad x = g(t)$, $\quad y = h(t)$

14. $\dfrac{dz}{dt}$ for $z = f(u, v, w)$, $\quad u = g(t)$, $\quad v = h(t)$, $\quad w = k(t)$

15. $\dfrac{\partial w}{\partial u}$ and $\dfrac{\partial w}{\partial v}$ for $w = h(x, y, z)$, $\quad x = f(u, v)$, $\quad y = g(u, v)$, $\quad z = k(u, v)$

16. $\dfrac{\partial w}{\partial x}$ and $\dfrac{\partial w}{\partial y}$ for $w = f(r, s, t)$, $\quad r = g(x, y)$, $\quad s = h(x, y)$, $\quad t = k(x, y)$

17. $\dfrac{\partial w}{\partial u}$ and $\dfrac{\partial w}{\partial v}$ for $w = g(x, y)$, $\quad x = h(u, v)$, $\quad y = k(u, v)$

18. $\dfrac{\partial w}{\partial x}$ and $\dfrac{\partial w}{\partial y}$ for $w = g(u, v)$, $\quad u = h(x, y)$, $\quad v = k(x, y)$

19. $\frac{\partial z}{\partial t}$ and $\frac{\partial z}{\partial s}$ for $z = f(x, y)$, $x = g(t, s)$, $y = h(t, s)$

20. $\frac{\partial y}{\partial r}$ for $y = f(u)$, $u = g(r, s)$

21. $\frac{\partial w}{\partial s}$ and $\frac{\partial w}{\partial t}$ for $w = g(u)$, $u = h(s, t)$

22. $\frac{\partial w}{\partial p}$ for $w = f(x, y, z, v)$, $x = g(p, q)$, $y = h(p, q)$,
    $z = j(p, q)$, $v = k(p, q)$

23. $\frac{\partial w}{\partial r}$ and $\frac{\partial w}{\partial s}$ for $w = f(x, y)$, $x = g(r)$, $y = h(s)$

24. $\frac{\partial w}{\partial s}$ for $w = g(x, y)$, $x = h(r, s, t)$, $y = k(r, s, t)$

Assuming that the equations in Exercises 25–28 define y as a differentiable function of x, use Theorem 8 to find the value of dy/dx at the given point.

25. $x^3 - 2y^2 + xy = 0$, $(1, 1)$
26. $xy + y^2 - 3x - 3 = 0$, $(-1, 1)$
27. $x^2 + xy + y^2 - 7 = 0$, $(1, 2)$
28. $xe^y + \sin xy + y - \ln 2 = 0$, $(0, \ln 2)$

Three-Variable Implicit Differentiation   Theorem 8 can be generalized to functions of three variables and even more. The three-variable version goes like this: If the equation $F(x, y, z) = 0$ determines z as a differentiable function of x and y, then, at points where $F_z \neq 0$,

$$\frac{\partial z}{\partial x} = -\frac{F_x}{F_z} \quad \text{and} \quad \frac{\partial z}{\partial y} = -\frac{F_y}{F_z}.$$

Use these equations to find the values of $\partial z/\partial x$ and $\partial z/\partial y$ at the points in Exercises 29–32.

29. $z^3 - xy + yz + y^3 - 2 = 0$, $(1, 1, 1)$
30. $\frac{1}{x} + \frac{1}{y} + \frac{1}{z} - 1 = 0$, $(2, 3, 6)$
31. $\sin(x + y) + \sin(y + z) + \sin(x + z) = 0$, (π, π, π)
32. $xe^y + ye^z + 2\ln x - 2 - 3\ln 2 = 0$, $(1, \ln 2, \ln 3)$
33. Find $\partial w/\partial r$ when $r = 1$, $s = -1$ if $w = (x + y + z)^2$, $x = r - s$, $y = \cos(r + s)$, $z = \sin(r + s)$.
34. Find $\partial w/\partial v$ when $u = -1$, $v = 2$ if $w = xy + \ln z$, $x = v^2/u$, $y = u + v$, $z = \cos u$.
35. Find $\partial w/\partial v$ when $u = 0$, $v = 0$ if $w = x^2 + (y/x)$, $x = u - 2v + 1$, $y = 2u + v - 2$.
36. Find $\partial z/\partial u$ when $u = 0$, $v = 1$ if $z = \sin xy + x \sin y$, $x = u^2 + v^2$, $y = uv$.
37. Find $\partial z/\partial u$ and $\partial z/\partial v$ when $u = \ln 2$, $v = 1$ if $z = 5 \tan^{-1} x$ and $x = e^u + \ln v$.
38. Find $\partial z/\partial u$ and $\partial z/\partial v$ when $u = 1$ and $v = -2$ if $z = \ln q$ and $q = \sqrt{v} + 3 \tan^{-1} u$.

39. **Changing voltage in a circuit**   The voltage V in a circuit that satisfies the law $V = IR$ is slowly dropping as the battery wears out. At the same time, the resistance R is increasing as the resistor heats up. Use the equation

$$\frac{dV}{dt} = \frac{\partial V}{\partial I}\frac{dI}{dt} + \frac{\partial V}{\partial R}\frac{dR}{dt}$$

to find how the current is changing at the instant when $R = 600$ ohms, $I = 0.04$ amp, $dR/dt = 0.5$ ohm/sec, and $dV/dt = -0.01$ volt/sec.

40. **Changing dimensions in a box**   The lengths a, b, and c of the edges of a rectangular box are changing with time. At the instant in question, $a = 1$ m, $b = 2$ m, $c = 3$ m, $da/dt = db/dt = 1$ m/sec, and $dc/dt = -3$ m/sec. At what rates are the box's volume V and surface area S changing at that instant? Are the box's interior diagonals increasing in length or decreasing?

41. If $f(u, v, w)$ is differentiable and $u = x - y$, $v = y - z$, and $w = z - x$, show that

$$\frac{\partial f}{\partial x} + \frac{\partial f}{\partial y} + \frac{\partial f}{\partial z} = 0.$$

42. **Polar coordinates**   Suppose that we substitute polar coordinates $x = r \cos \theta$ and $y = r \sin \theta$ in a differentiable function $w = f(x, y)$.

   a. Show that

   $$\frac{\partial w}{\partial r} = f_x \cos \theta + f_y \sin \theta$$

   and

   $$\frac{1}{r}\frac{\partial w}{\partial \theta} = -f_x \sin \theta + f_y \cos \theta.$$

   b. Solve the equations in part (a) to express f_x and f_y in terms of $\partial w/\partial r$ and $\partial w/\partial \theta$.

   c. Show that

   $$(f_x)^2 + (f_y)^2 = \left(\frac{\partial w}{\partial r}\right)^2 + \frac{1}{r^2}\left(\frac{\partial w}{\partial \theta}\right)^2.$$

43. **Laplace equations**   Show that if $w = f(u, v)$ satisfies the Laplace equation $f_{uu} + f_{vv} = 0$ and if $u = (x^2 - y^2)/2$ and $v = xy$, then w satisfies the Laplace equation $w_{xx} + w_{yy} = 0$.

44. **Laplace equations**   Let $w = f(u) + g(v)$, where $u = x + iy$ and $v = x - iy$ and $i = \sqrt{-1}$. Show that w satisfies the Laplace equation $w_{xx} + w_{yy} = 0$ if all the necessary functions are differentiable.

45. Extreme values on a helix Suppose that the partial derivatives of a function $f(x, y, z)$ at points on the helix $x = \cos t$, $y = \sin t$, $z = t$ are

$$f_x = \cos t, \qquad f_y = \sin t, \qquad f_z = t^2 + t - 2.$$

At what points on the curve, if any, can f take on extreme values?

46. A space curve Let $w = x^2 e^{2y} \cos 3z$. Find the value of dw/dt at the point $(1, \ln 2, 0)$ on the curve $x = \cos t$, $y = \ln(t + 2)$, $z = t$.

47. Temperature on a circle Let $T = f(x, y)$ be the temperature at the point (x, y) on the circle $x = \cos t$, $y = \sin t$, $0 \le t \le 2\pi$ and suppose that

$$\frac{\partial T}{\partial x} = 8x - 4y, \qquad \frac{\partial T}{\partial y} = 8y - 4x.$$

a. Find where the maximum and minimum temperatures on the circle occur by examining the derivatives dT/dt and d^2T/dt^2.

b. Suppose that $T = 4x^2 - 4xy + 4y^2$. Find the maximum and minimum values of T on the circle.

48. Temperature on an ellipse Let $T = g(x, y)$ be the temperature at the point (x, y) on the ellipse

$$x = 2\sqrt{2}\cos t, \qquad y = \sqrt{2}\sin t, \qquad 0 \le t \le 2\pi,$$

and suppose that

$$\frac{\partial T}{\partial x} = y, \qquad \frac{\partial T}{\partial y} = x.$$

a. Locate the maximum and minimum temperatures on the ellipse by examining dT/dt and d^2T/dt^2.

b. Suppose that $T = xy - 2$. Find the maximum and minimum values of T on the ellipse.

Differentiating Integrals Under mild continuity restrictions, it is true that if

$$F(x) = \int_a^b g(t, x)\, dt,$$

then $F'(x) = \int_a^b g_x(t, x)\, dt$. Using this fact and the Chain Rule, we can find the derivative of

$$F(x) = \int_a^{f(x)} g(t, x)\, dt$$

by letting

$$G(u, x) = \int_a^u g(t, x)\, dt,$$

where $u = f(x)$. Find the derivatives of the functions in Exercises 49 and 50.

49. $F(x) = \int_0^{x^2} \sqrt{t^4 + x^3}\, dt$

50. $F(x) = \int_{x^2}^1 \sqrt{t^3 + x^2}\, dt$

12.5 Directional Derivatives and Gradient Vectors

If you look at the map (Figure 12.23) showing contours on the West Point Area along the Hudson River in New York, you will notice that the tributary streams flow perpendicular to the contours. The streams are following paths of steepest descent so the waters reach the Hudson as quickly as possible. Therefore, the instantaneous rate of change in a stream's altitude above sea level has a particular direction. In this section, you see why this direction, called the "downhill" direction, is perpendicular to the contours.

Directional Derivatives in the Plane

We know from Section 12.4 that if $f(x, y)$ is differentiable, then the rate at which f changes with respect to t along a differentiable curve $x = g(t)$, $y = h(t)$ is

$$\frac{df}{dt} = \frac{\partial f}{\partial x}\frac{dx}{dt} + \frac{\partial f}{\partial y}\frac{dy}{dt}.$$

At any point $P_0(x_0, y_0) = P_0(g(t_0), h(t_0))$, this equation gives the rate of change of f with respect to increasing t and therefore depends, among other things, on the direction of motion along the curve. If the curve is a straight line and t is the arc length parameter along

FIGURE 12.23 Contours of the West Point Area in New York show streams, which follow paths of steepest descent, running perpendicular to the contours.

the line measured from P_0 in the direction of a given unit vector $\mathbf{u}$, then df/dt is the rate of change of f with respect to distance in its domain in the direction of $\mathbf{u}$. By varying $\mathbf{u}$, we find the rates at which f changes with respect to distance as we move through P_0 in different directions. We now define this idea more precisely.

Suppose that the function $f(x, y)$ is defined throughout a region R in the xy-plane, that $P_0(x_0, y_0)$ is a point in R, and that $\mathbf{u} = u_1\mathbf{i} + u_2\mathbf{j}$ is a unit vector. Then the equations

$$x = x_0 + su_1, \qquad y = y_0 + su_2$$

parametrize the line through P_0 parallel to $\mathbf{u}$. If the parameter s measures arc length from P_0 in the direction of $\mathbf{u}$, we find the rate of change of f at P_0 in the direction of $\mathbf{u}$ by calculating df/ds at P_0 (Figure 12.24).

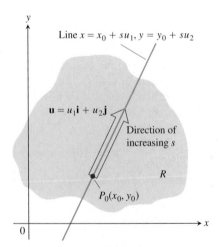

FIGURE 12.24 The rate of change of f in the direction of $\mathbf{u}$ at a point P_0 is the rate at which f changes along this line at P_0.

DEFINITION The **derivative of f at $P_0(x_0, y_0)$ in the direction of the unit vector $\mathbf{u} = u_1\mathbf{i} + u_2\mathbf{j}$** is the number

$$\left(\frac{df}{ds}\right)_{\mathbf{u}, P_0} = \lim_{s \to 0} \frac{f(x_0 + su_1, y_0 + su_2) - f(x_0, y_0)}{s}, \tag{1}$$

provided the limit exists.

The directional derivative is also denoted by

$$(D_{\mathbf{u}} f)_{P_0}.$$

"The derivative of f at P_0 in the direction of $\mathbf{u}$"

EXAMPLE 1   Using the definition, find the derivative of

$$f(x, y) = x^2 + xy$$

at $P_0(1, 2)$ in the direction of the unit vector $\mathbf{u} = \left(1/\sqrt{2}\right)\mathbf{i} + \left(1/\sqrt{2}\right)\mathbf{j}$.

Solution

$$\left(\frac{df}{ds}\right)_{\mathbf{u},P_0} = \lim_{s \to 0} \frac{f(x_0 + su_1, y_0 + su_2) - f(x_0, y_0)}{s} \qquad \text{Equation (1)}$$

$$= \lim_{s \to 0} \frac{f\left(1 + s \cdot \dfrac{1}{\sqrt{2}}, 2 + s \cdot \dfrac{1}{\sqrt{2}}\right) - f(1, 2)}{s}$$

$$= \lim_{s \to 0} \frac{\left(1 + \dfrac{s}{\sqrt{2}}\right)^2 + \left(1 + \dfrac{s}{\sqrt{2}}\right)\left(2 + \dfrac{s}{\sqrt{2}}\right) - (1^2 + 1 \cdot 2)}{s}$$

$$= \lim_{s \to 0} \frac{\left(1 + \dfrac{2s}{\sqrt{2}} + \dfrac{s^2}{2}\right) + \left(2 + \dfrac{3s}{\sqrt{2}} + \dfrac{s^2}{2}\right) - 3}{s}$$

$$= \lim_{s \to 0} \frac{\dfrac{5s}{\sqrt{2}} + s^2}{s} = \lim_{s \to 0} \left(\frac{5}{\sqrt{2}} + s\right) = \frac{5}{\sqrt{2}}.$$

The rate of change of $f(x, y) = x^2 + xy$ at $P_0(1, 2)$ in the direction $\mathbf{u}$ is $5/\sqrt{2}$. ∎

Interpretation of the Directional Derivative

The equation $z = f(x, y)$ represents a surface S in space. If $z_0 = f(x_0, y_0)$, then the point $P(x_0, y_0, z_0)$ lies on S. The vertical plane that passes through P and $P_0(x_0, y_0)$ parallel to $\mathbf{u}$ intersects S in a curve C (Figure 12.25). The rate of change of f in the direction of $\mathbf{u}$ is the slope of the tangent to C at P in the right-handed system formed by the vectors $\mathbf{u}$ and $\mathbf{k}$.

When $\mathbf{u} = \mathbf{i}$, the directional derivative at P_0 is $\partial f/\partial x$ evaluated at (x_0, y_0). When $\mathbf{u} = \mathbf{j}$, the directional derivative at P_0 is $\partial f/\partial y$ evaluated at (x_0, y_0). The directional derivative generalizes the two partial derivatives. We can now ask for the rate of change of f in any direction $\mathbf{u}$, not just the directions $\mathbf{i}$ and $\mathbf{j}$.

Here's a physical interpretation of the directional derivative. Suppose that $T = f(x, y)$ is the temperature at each point (x, y) over a region in the plane. Then $f(x_0, y_0)$ is the temperature at the point $P_0(x_0, y_0)$ and $(D_{\mathbf{u}}f)_{P_0}$ is the instantaneous rate of change of the temperature at P_0 stepping off in the direction $\mathbf{u}$.

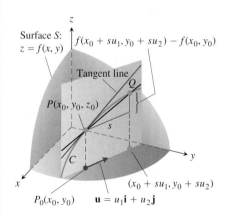

FIGURE 12.25 The slope of curve C at P_0 is $\displaystyle\lim_{Q \to P}$ slope (PQ); this is the directional derivative

$$\left(\frac{df}{ds}\right)_{\mathbf{u},P_0} = (D_{\mathbf{u}}f)_{P_0}.$$

Calculation and Gradients

We now develop an efficient formula to calculate the directional derivative for a differentiable function f. We begin with the line

$$x = x_0 + su_1, \qquad y = y_0 + su_2, \qquad (2)$$

through $P_0(x_0, y_0)$, parametrized with the arc length parameter s increasing in the direction of the unit vector $\mathbf{u} = u_1\mathbf{i} + u_2\mathbf{j}$. Then

$$\left(\frac{df}{ds}\right)_{\mathbf{u},P_0} = \left(\frac{\partial f}{\partial x}\right)_{P_0}\frac{dx}{ds} + \left(\frac{\partial f}{\partial y}\right)_{P_0}\frac{dy}{ds} \qquad \text{Chain Rule for differentiable } f$$

$$= \left(\frac{\partial f}{\partial x}\right)_{P_0}\cdot u_1 + \left(\frac{\partial f}{\partial y}\right)_{P_0}\cdot u_2 \qquad \begin{array}{l}\text{From Equations (2),}\\ dx/ds = u_1 \text{ and } dy/ds = u_2\end{array}$$

$$= \underbrace{\left[\left(\frac{\partial f}{\partial x}\right)_{P_0}\mathbf{i} + \left(\frac{\partial f}{\partial y}\right)_{P_0}\mathbf{j}\right]}_{\text{Gradient of } f \text{ at } P_0} \cdot \underbrace{\left[u_1\mathbf{i} + u_2\mathbf{j}\right]}_{\text{Direction } \mathbf{u}}. \qquad (3)$$

DEFINITION   The **gradient vector (gradient)** of $f(x, y)$ at a point $P_0(x_0, y_0)$ is the vector

$$\nabla f = \frac{\partial f}{\partial x}\mathbf{i} + \frac{\partial f}{\partial y}\mathbf{j}$$

obtained by evaluating the partial derivatives of f at P_0.

The notation ∇f is read "grad f" as well as "gradient of f" and "del f." The symbol ∇ by itself is read "del." Another notation for the gradient is grad f, read the way it is written.

Equation (3) says that the derivative of a differentiable function f in the direction of $\mathbf{u}$ at P_0 is the dot product of $\mathbf{u}$ with the gradient of f at P_0.

THEOREM 9—The Directional Derivative Is a Dot Product   If $f(x, y)$ is differentiable in an open region containing $P_0(x_0, y_0)$, then

$$\left(\frac{df}{ds}\right)_{\mathbf{u},P_0} = (\nabla f)_{P_0}\cdot\mathbf{u}, \qquad (4)$$

the dot product of the gradient f at P_0 and $\mathbf{u}$.

EXAMPLE 2   Find the derivative of $f(x, y) = xe^y + \cos(xy)$ at the point $(2, 0)$ in the direction of $\mathbf{v} = 3\mathbf{i} - 4\mathbf{j}$.

Solution   The direction of $\mathbf{v}$ is the unit vector obtained by dividing $\mathbf{v}$ by its length:

$$\mathbf{u} = \frac{\mathbf{v}}{|\mathbf{v}|} = \frac{\mathbf{v}}{5} = \frac{3}{5}\mathbf{i} - \frac{4}{5}\mathbf{j}.$$

The partial derivatives of f are everywhere continuous and at $(2, 0)$ are given by

$$f_x(2, 0) = (e^y - y\sin(xy))_{(2,0)} = e^0 - 0 = 1$$

$$f_y(2, 0) = (xe^y - x\sin(xy))_{(2,0)} = 2e^0 - 2\cdot 0 = 2.$$

The gradient of f at $(2, 0)$ is

$$\nabla f|_{(2,0)} = f_x(2, 0)\mathbf{i} + f_y(2, 0)\mathbf{j} = \mathbf{i} + 2\mathbf{j}$$

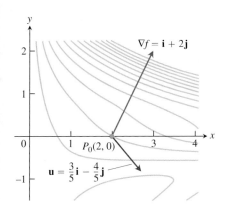

FIGURE 12.26 Picture ∇f as a vector in the domain of f. In the case of $f(x, y) = xe^{y} + \cos(xy)$, the domain is the entire plane. The rate at which f changes at $(2, 0)$ in the direction $\mathbf{u} = (3/5)\mathbf{i} - (4/5)\mathbf{j}$ is $\nabla f \cdot \mathbf{u} = -1$ (Example 2).

(Figure 12.26). The derivative of f at $(2, 0)$ in the direction of $\mathbf{v}$ is therefore

$$(D_{\mathbf{u}}f)|_{(2,0)} = \nabla f|_{(2,0)} \cdot \mathbf{u} \qquad \text{Equation (4)}$$

$$= (\mathbf{i} + 2\mathbf{j}) \cdot \left(\frac{3}{5}\mathbf{i} - \frac{4}{5}\mathbf{j}\right) = \frac{3}{5} - \frac{8}{5} = -1 \,. \qquad ■$$

Evaluating the dot product in the formula

$$D_{\mathbf{u}}f = \nabla f \cdot \mathbf{u} = |\nabla f||\mathbf{u}|\cos\theta = |\nabla f|\cos\theta,$$

where θ is the angle between the vectors $\mathbf{u}$ and ∇f, reveals the following properties.

Properties of the Directional Derivative $D_{\mathbf{u}}f = \nabla f \cdot \mathbf{u} = |\nabla f|\cos\theta$

1. The function f increases most rapidly when $\cos\theta = 1$ or when $\mathbf{u}$ is the direction of ∇f. That is, at each point P in its domain, f increases most rapidly in the direction of the gradient vector ∇f at P. The derivative in this direction is

$$D_{\mathbf{u}}f = |\nabla f|\cos(0) = |\nabla f|.$$

2. Similarly, f decreases most rapidly in the direction of $-\nabla f$. The derivative in this direction is $D_{\mathbf{u}}f = |\nabla f|\cos(\pi) = -|\nabla f|$.

3. Any direction $\mathbf{u}$ orthogonal to a gradient $\nabla f \neq 0$ is a direction of zero change in f because θ then equals $\pi/2$ and

$$D_{\mathbf{u}}f = |\nabla f|\cos(\pi/2) = |\nabla f|\cdot 0 = 0.$$

As we discuss later, these properties hold in three dimensions as well as two.

EXAMPLE 3 Find the directions in which $f(x, y) = (x^2/2) + (y^2/2)$

(a) Increases most rapidly at the point $(1, 1)$

(b) Decreases most rapidly at $(1, 1)$.

(c) What are the directions of zero change in f at $(1, 1)$?

Solution

(a) The function increases most rapidly in the direction of ∇f at $(1, 1)$. The gradient there is

$$(\nabla f)_{(1,1)} = (x\mathbf{i} + y\mathbf{j})_{(1,1)} = \mathbf{i} + \mathbf{j}.$$

Its direction is

$$\mathbf{u} = \frac{\mathbf{i} + \mathbf{j}}{|\mathbf{i} + \mathbf{j}|} = \frac{\mathbf{i} + \mathbf{j}}{\sqrt{(1)^2 + (1)^2}} = \frac{1}{\sqrt{2}}\mathbf{i} + \frac{1}{\sqrt{2}}\mathbf{j}.$$

(b) The function decreases most rapidly in the direction of $-\nabla f$ at $(1, 1)$, which is

$$-\mathbf{u} = -\frac{1}{\sqrt{2}}\mathbf{i} - \frac{1}{\sqrt{2}}\mathbf{j}.$$

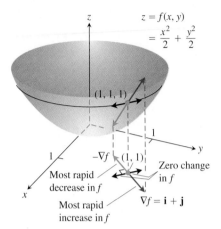

FIGURE 12.27 The direction in which $f(x, y)$ increases most rapidly at $(1, 1)$ is the direction of $\nabla f|_{(1,1)} = \mathbf{i} + \mathbf{j}$. It corresponds to the direction of steepest ascent on the surface at $(1, 1, 1)$ (Example 3).

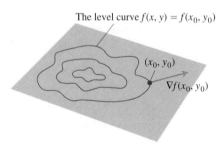

FIGURE 12.28 The gradient of a differentiable function of two variables at a point is always normal to the function's level curve through that point.

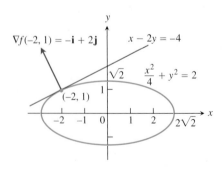

FIGURE 12.29 We can find the tangent to the ellipse $(x^2/4) + y^2 = 2$ by treating the ellipse as a level curve of the function $f(x, y) = (x^2/4) + y^2$ (Example 4).

(c) The directions of zero change at $(1, 1)$ are the directions orthogonal to ∇f:

$$\mathbf{n} = -\frac{1}{\sqrt{2}}\mathbf{i} + \frac{1}{\sqrt{2}}\mathbf{j} \quad \text{and} \quad -\mathbf{n} = \frac{1}{\sqrt{2}}\mathbf{i} - \frac{1}{\sqrt{2}}\mathbf{j}.$$

See Figure 12.27.

Gradients and Tangents to Level Curves

If a differentiable function $f(x, y)$ has a constant value c along a smooth curve $\mathbf{r} = g(t)\mathbf{i} + h(t)\mathbf{j}$ (making the curve a level curve of f), then $f(g(t), h(t)) = c$. Differentiating both sides of this equation with respect to t leads to the equations

$$\frac{d}{dt}f(g(t), h(t)) = \frac{d}{dt}(c)$$

$$\frac{\partial f}{\partial x}\frac{dg}{dt} + \frac{\partial f}{\partial y}\frac{dh}{dt} = 0 \qquad \text{Chain Rule}$$

$$\underbrace{\left(\frac{\partial f}{\partial x}\mathbf{i} + \frac{\partial f}{\partial y}\mathbf{j}\right)}_{\nabla f} \cdot \underbrace{\left(\frac{dg}{dt}\mathbf{i} + \frac{dh}{dt}\mathbf{j}\right)}_{\frac{d\mathbf{r}}{dt}} = 0. \tag{5}$$

Equation (5) says that ∇f is normal to the tangent vector $d\mathbf{r}/dt$, so it is normal to the curve.

> At every point (x_0, y_0) in the domain of a differentiable function $f(x, y)$, the gradient of f is normal to the level curve through (x_0, y_0) (Figure 12.28).

Equation (5) validates our observation that streams flow perpendicular to the contours in topographical maps (see Figure 12.23). Since the downflowing stream will reach its destination in the fastest way, it must flow in the direction of the negative gradient vectors from Property 2 for the directional derivative. Equation (5) tells us these directions are perpendicular to the level curves.

This observation also enables us to find equations for tangent lines to level curves. They are the lines normal to the gradients. The line through a point $P_0(x_0, y_0)$ normal to a vector $\mathbf{N} = A\mathbf{i} + B\mathbf{j}$ has the equation

$$A(x - x_0) + B(y - y_0) = 0$$

(Exercise 35). If $\mathbf{N}$ is the gradient $(\nabla f)_{(x_0, y_0)} = f_x(x_0, y_0)\mathbf{i} + f_y(x_0, y_0)\mathbf{j}$, the equation is the tangent line given by

$$f_x(x_0, y_0)(x - x_0) + f_y(x_0, y_0)(y - y_0) = 0. \tag{6}$$

EXAMPLE 4 Find an equation for the tangent to the ellipse

$$\frac{x^2}{4} + y^2 = 2$$

(Figure 12.29) at the point $(-2, 1)$.

Solution The ellipse is a level curve of the function

$$f(x, y) = \frac{x^2}{4} + y^2.$$

The gradient of f at $(-2, 1)$ is

$$\nabla f\big|_{(-2,1)} = \left(\frac{x}{2}\mathbf{i} + 2y\mathbf{j}\right)_{(-2,1)} = -\mathbf{i} + 2\mathbf{j}.$$

The tangent is the line

$$(-1)(x + 2) + (2)(y - 1) = 0 \qquad \text{Equation (6)}$$

$$x - 2y = -4. \qquad\qquad\qquad\qquad ■$$

If we know the gradients of two functions f and g, we automatically know the gradients of their constant multiples, sum, difference, product, and quotient. You are asked to establish the following rules in Exercise 36. Notice that these rules have the same form as the corresponding rules for derivatives of single-variable functions.

Algebra Rules for Gradients

1. *Constant Multiple Rule:* $\nabla(kf) = k\nabla f$ (any number k)

2. *Sum Rule:* $\nabla(f + g) = \nabla f + \nabla g$

3. *Difference Rule:* $\nabla(f - g) = \nabla f - \nabla g$

4. *Product Rule:* $\nabla(fg) = f\nabla g + g\nabla f$

5. *Quotient Rule:* $\nabla\left(\dfrac{f}{g}\right) = \dfrac{g\nabla f - f\nabla g}{g^2}$

EXAMPLE 5   We illustrate two of the rules with

$$f(x, y) = x - y \qquad g(x, y) = 3y$$
$$\nabla f = \mathbf{i} - \mathbf{j} \qquad\qquad \nabla g = 3\mathbf{j}.$$

We have

1. $\nabla(f - g) = \nabla(x - 4y) = \mathbf{i} - 4\mathbf{j} = \nabla f - \nabla g$ Rule 3

2. $\nabla(fg) = \nabla(3xy - 3y^2) = 3y\mathbf{i} + (3x - 6y)\mathbf{j}$

   $\quad\quad = 3y(\mathbf{i} - \mathbf{j}) + 3y\mathbf{j} + (3x - 6y)\mathbf{j}$

   $\quad\quad = 3y(\mathbf{i} - \mathbf{j}) + (3x - 3y)\mathbf{j}$

   $\quad\quad = 3y(\mathbf{i} - \mathbf{j}) + (x - y)3\mathbf{j} = g\nabla f + f\nabla g$ Rule 4   ■

Functions of Three Variables

For a differentiable function $f(x, y, z)$ and a unit vector $\mathbf{u} = u_1\mathbf{i} + u_2\mathbf{j} + u_3\mathbf{k}$ in space, we have

$$\nabla f = \frac{\partial f}{\partial x}\mathbf{i} + \frac{\partial f}{\partial y}\mathbf{j} + \frac{\partial f}{\partial z}\mathbf{k}$$

and

$$D_{\mathbf{u}}f = \nabla f \cdot \mathbf{u} = \frac{\partial f}{\partial x}u_1 + \frac{\partial f}{\partial y}u_2 + \frac{\partial f}{\partial z}u_3.$$

The directional derivative can once again be written in the form

$$D_{\mathbf{u}}f = \nabla f \cdot \mathbf{u} = |\nabla f||u|\cos\theta = |\nabla f|\cos\theta,$$

so the properties listed earlier for functions of two variables continue to hold. At any given point, f increases most rapidly in the direction of ∇f and decreases most rapidly in the direction of $-\nabla f$. In any direction orthogonal to ∇f, the derivative is zero.

EXAMPLE 6

(a) Find the derivative of $f(x, y, z) = x^3 - xy^2 - z$ at $P_0(1, 1, 0)$ in the direction of $\mathbf{v} = 2\mathbf{i} - 3\mathbf{j} + 6\mathbf{k}$.

(b) In what directions does f change most rapidly at P_0, and what are the rates of change in these directions?

Solution

(a) The direction of $\mathbf{v}$ is obtained by dividing $\mathbf{v}$ by its length:

$$|\mathbf{v}| = \sqrt{(2)^2 + (-3)^2 + (6)^2} = \sqrt{49} = 7$$

$$\mathbf{u} = \frac{\mathbf{v}}{|\mathbf{v}|} = \frac{2}{7}\mathbf{i} - \frac{3}{7}\mathbf{j} + \frac{6}{7}\mathbf{k}.$$

The partial derivatives of f at P_0 are

$$f_x = (3x^2 - y^2)_{(1,1,0)} = 2, \qquad f_y = -2xy|_{(1,1,0)} = -2, \qquad f_z = -1|_{(1,1,0)} = -1.$$

The gradient of f at P_0 is

$$\nabla f|_{(1,1,0)} = 2\mathbf{i} - 2\mathbf{j} - \mathbf{k}.$$

The derivative of f at P_0 in the direction of $\mathbf{v}$ is therefore

$$(D_{\mathbf{u}}f)_{(1,1,0)} = \nabla f|_{(1,1,0)} \cdot \mathbf{u} = (2\mathbf{i} - 2\mathbf{j} - \mathbf{k}) \cdot \left(\frac{2}{7}\mathbf{i} - \frac{3}{7}\mathbf{j} + \frac{6}{7}\mathbf{k} \right)$$

$$= \frac{4}{7} + \frac{6}{7} - \frac{6}{7} = \frac{4}{7}.$$

(b) The function increases most rapidly in the direction of $\nabla f = 2\mathbf{i} - 2\mathbf{j} - \mathbf{k}$ and decreases most rapidly in the direction of $-\nabla f$. The rates of change in the directions are, respectively,

$$|\nabla f| = \sqrt{(2)^2 + (-2)^2 + (-1)^2} = \sqrt{9} = 3 \qquad \text{and} \qquad -|\nabla f| = -3. \qquad ■$$

EXERCISES 12.5

In Exercises 1–4, find the gradient of the function at the given point. Then sketch the gradient together with the level curve that passes through the point.

1. $f(x, y) = y - x$, $(2, 1)$ **2.** $f(x, y) = \ln(x^2 + y^2)$, $(1, 1)$

3. $g(x, y) = y - x^2$, $(-1, 0)$ **4.** $g(x, y) = \dfrac{x^2}{2} - \dfrac{y^2}{2}$, $(\sqrt{2}, 1)$

In Exercises 5–8, find ∇f at the given point.

5. $f(x, y, z) = x^2 + y^2 - 2z^2 + z \ln x$, $(1, 1, 1)$

6. $f(x, y, z) = 2z^3 - 3(x^2 + y^2)z + \tan^{-1} xz$, $(1, 1, 1)$

7. $f(x, y, z) = (x^2 + y^2 + z^2)^{-1/2} + \ln(xyz)$, $(-1, 2, -2)$

8. $f(x, y, z) = e^{x+y} \cos z + (y + 1) \sin^{-1} x$, $(0, 0, \pi/6)$

In Exercises 9–16, find the derivative of the function at P_0 in the direction of $\mathbf{A}$.

9. $f(x, y) = 2xy - 3y^2$, $P_0(5, 5)$, $\mathbf{A} = 4\mathbf{i} + 3\mathbf{j}$

10. $f(x, y) = 2x^2 + y^2$, $P_0(-1, 1)$, $\mathbf{A} = 3\mathbf{i} - 4\mathbf{j}$

11. $g(x, y) = x - (y^2/x) + \sqrt{3} \sec^{-1}(2xy)$, $P_0(1, 1)$, $\mathbf{A} = 12\mathbf{i} + 5\mathbf{j}$

12. $h(x, y) = \tan^{-1}(y/x) + \sqrt{3}\sin^{-1}(xy/2)$, $P_0(1, 1)$,
$\mathbf{A} = 3\mathbf{i} - 2\mathbf{j}$

13. $f(x, y, z) = xy + yz + zx$, $P_0(1, -1, 2)$, $\mathbf{A} = 3\mathbf{i} + 6\mathbf{j} - 2\mathbf{k}$

14. $f(x, y, z) = x^2 + 2y^2 - 3z^2$, $P_0(1, 1, 1)$, $\mathbf{A} = \mathbf{i} + \mathbf{j} + \mathbf{k}$

15. $g(x, y, z) = 3e^x \cos yz$, $P_0(0, 0, 0)$, $\mathbf{A} = 2\mathbf{i} + \mathbf{j} - 2\mathbf{k}$

16. $h(x, y, z) = \cos xy + e^{yz} + \ln zx$, $P_0(1, 0, 1/2)$,
$\mathbf{A} = \mathbf{i} + 2\mathbf{j} + 2\mathbf{k}$

In Exercises 17–22, find the directions in which the functions increase and decrease most rapidly at P_0. Then find the derivatives of the functions in these directions.

17. $f(x, y) = x^2 + xy + y^2$, $P_0(-1, 1)$

18. $f(x, y) = x^2 y + e^{xy} \sin y$, $P_0(1, 0)$

19. $f(x, y, z) = (x/y) - yz$, $P_0(4, 1, 1)$

20. $g(x, y, z) = xe^y + z^2$, $P_0(1, \ln 2, 1/2)$

21. $f(x, y, z) = \ln xy + \ln yz + \ln xz$, $P_0(1, 1, 1)$

22. $h(x, y, z) = \ln(x^2 + y^2 - 1) + y + 6z$, $P_0(1, 1, 0)$

In Exercises 23–26, sketch the curve $f(x, y) = c$ together with ∇f and the tangent line at the given point. Then write an equation for the tangent line.

23. $x^2 + y^2 = 4$, $\left(\sqrt{2}, \sqrt{2}\right)$

24. $x^2 - y = 1$, $\left(\sqrt{2}, 1\right)$

25. $xy = -4$, $(2, -2)$

26. $x^2 - xy + y^2 = 7$, $(-1, 2)$

27. Zero directional derivative In what direction is the derivative of $f(x, y) = xy + y^2$ at $P(3, 2)$ equal to zero?

28. Zero directional derivative In what directions is the derivative of $f(x, y) = (x^2 - y^2)/(x^2 + y^2)$ at $P(1, 1)$ equal to zero?

29. Is there a direction $\mathbf{u}$ in which the rate of change of $f(x, y) = x^2 - 3xy + 4y^2$ at $P(1, 2)$ equals 14? Give reasons for your answer.

30. Changing temperature along a circle Is there a direction $\mathbf{u}$ in which the rate of change of the temperature function $T(x, y, z) = 2xy - yz$ (temperature in degrees Celsius, distance in feet) at $P(1, -1, 1)$ is $-3°C/\text{ft}$? Give reasons for your answer.

31. The derivative of $f(x, y)$ at $P_0(1, 2)$ in the direction of $\mathbf{i} + \mathbf{j}$ is $2\sqrt{2}$ and in the direction of $-2\mathbf{j}$ is -3. What is the derivative of f in the direction of $-\mathbf{i} - 2\mathbf{j}$? Give reasons for your answer.

32. The derivative of $f(x, y, z)$ at a point P is greatest in the direction of $\mathbf{v} = \mathbf{i} + \mathbf{j} - \mathbf{k}$. In this direction, the value of the derivative is $2\sqrt{3}$.

a. What is ∇f at P? Give reasons for your answer.

b. What is the derivative of f at P in the direction of $\mathbf{i} + \mathbf{j}$?

33. Directional derivatives and scalar components How is the derivative of a differentiable function $f(x, y, z)$ at a point P_0 in the direction of a unit vector $\mathbf{u}$ related to the scalar component of $(\nabla f)_{P_0}$ in the direction of $\mathbf{u}$? Give reasons for your answer.

34. Directional derivatives and partial derivatives Assuming that the necessary derivatives of $f(x, y, z)$ are defined, how are $D_\mathbf{i} f$, $D_\mathbf{j} f$, and $D_\mathbf{k} f$ related to f_x, f_y, and f_z? Give reasons for your answer.

35. Lines in the xy-plane Show that $A(x - x_0) + B(y - y_0) = 0$ is an equation for the line in the xy-plane through the point (x_0, y_0) normal to the vector $\mathbf{N} = A\mathbf{i} + B\mathbf{j}$.

36. The algebra rules for gradients Given a constant k and the gradients

$$\nabla f = \frac{\partial f}{\partial x}\mathbf{i} + \frac{\partial f}{\partial y}\mathbf{j} + \frac{\partial f}{\partial z}\mathbf{k}, \qquad \nabla g = \frac{\partial g}{\partial x}\mathbf{i} + \frac{\partial g}{\partial y}\mathbf{j} + \frac{\partial g}{\partial z}\mathbf{k},$$

establish the algebra rules for gradients.

12.6 Tangent Planes and Differentials

In this section we define the tangent plane at a point on a smooth surface in space. Then we show how to calculate an equation of the tangent plane from the partial derivatives of the function defining the surface. This idea is similar to the definition of the tangent line at a point on a curve in the coordinate plane for single-variable functions (Section 2.7). We then study the total differential and linearization of functions of several variables.

Tangent Planes and Normal Lines

If $\mathbf{r} = g(t)\mathbf{i} + h(t)\mathbf{j} + k(t)\mathbf{k}$ is a smooth curve on the level surface $f(x, y, z) = c$ of a differentiable function f, then $f(g(t), h(t), k(t)) = c$. Differentiating both sides of this

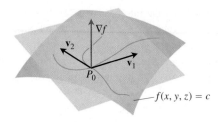

FIGURE 12.30 The gradient ∇f is orthogonal to the velocity vector of every smooth curve in the surface through P_0. The velocity vectors at P_0 therefore lie in a common plane, which we call the tangent plane at P_0.

equation with respect to t leads to

$$\frac{d}{dt} f(g(t), h(t), k(t)) = \frac{d}{dt}(c)$$

$$\frac{\partial f}{\partial x}\frac{dg}{dt} + \frac{\partial f}{\partial y}\frac{dh}{dt} + \frac{\partial f}{\partial z}\frac{dk}{dt} = 0 \qquad \text{Chain Rule}$$

$$\underbrace{\left(\frac{\partial f}{\partial x}\mathbf{i} + \frac{\partial f}{\partial y}\mathbf{j} + \frac{\partial f}{\partial z}\mathbf{k}\right)}_{\nabla f} \cdot \underbrace{\left(\frac{dg}{dt}\mathbf{i} + \frac{dh}{dt}\mathbf{j} + \frac{dk}{dt}\mathbf{k}\right)}_{d\mathbf{r}/dt} = 0. \qquad (1)$$

At every point along the curve, ∇f is orthogonal to the curve's velocity vector.

Now let us restrict our attention to the curves that pass through P_0 (Figure 12.30). All the velocity vectors at P_0 are orthogonal to ∇f at P_0, so the curves' tangent lines all lie in the plane through P_0 normal to ∇f. We call this plane the tangent plane of the surface at P_0. The line through P_0 perpendicular to the plane is the surface's normal line at P_0.

> **DEFINITIONS** The **tangent plane** at the point $P_0(x_0, y_0, z_0)$ on the level surface $f(x, y, z) = c$ of a differentiable function f is the plane through P_0 normal to $\nabla f|_{P_0}$.
>
> The **normal line** of the surface at P_0 is the line through P_0 parallel to $\nabla f|_{P_0}$.

Thus, from Section 10.5, the tangent plane and normal line have the following equations:

> **Tangent Plane to $f(x, y, z) = c$ at $P_0(x_0, y_0, z_0)$**
>
> $$f_x(P_0)(x - x_0) + f_y(P_0)(y - y_0) + f_z(P_0)(z - z_0) = 0 \qquad (2)$$
>
> **Normal Line to $f(x, y, z) = c$ at $P_0(x_0, y_0, z_0)$**
>
> $$x = x_0 + f_x(P_0)t, \qquad y = y_0 + f_y(P_0)t, \qquad z = z_0 + f_z(P_0)t \qquad (3)$$

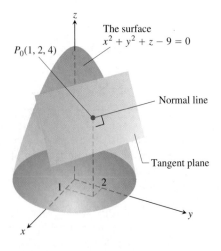

FIGURE 12.31 The tangent plane and normal line to this surface at P_0 (Example 1).

EXAMPLE 1 Find the tangent plane and normal line of the surface

$$f(x, y, z) = x^2 + y^2 + z - 9 = 0 \qquad \text{A circular paraboloid}$$

at the point $P_0(1, 2, 4)$.

Solution The surface is shown in Figure 12.31.

The tangent plane is the plane through P_0 perpendicular to the gradient of f at P_0. The gradient is

$$\nabla f|_{P_0} = (2x\mathbf{i} + 2y\mathbf{j} + \mathbf{k})_{(1,2,4)} = 2\mathbf{i} + 4\mathbf{j} + \mathbf{k}.$$

The tangent plane is therefore the plane

$$2(x - 1) + 4(y - 2) + (z - 4) = 0, \qquad \text{or} \qquad 2x + 4y + z = 14.$$

The line normal to the surface at P_0 is

$$x = 1 + 2t, \qquad y = 2 + 4t, \qquad z = 4 + t. \qquad \blacksquare$$

To find an equation for the plane tangent to a smooth surface $z = f(x, y)$ at a point $P_0(x_0, y_0, z_0)$ where $z_0 = f(x_0, y_0)$, we first observe that the equation $z = f(x, y)$ is equivalent to $f(x, y) - z = 0$. The surface $z = f(x, y)$ is therefore the zero level surface of the function $F(x, y, z) = f(x, y) - z$. The partial derivatives of F are

$$F_x = \frac{\partial}{\partial x}(f(x, y) - z) = f_x - 0 = f_x$$

$$F_y = \frac{\partial}{\partial y}(f(x, y) - z) = f_y - 0 = f_y$$

$$F_z = \frac{\partial}{\partial z}(f(x, y) - z) = 0 - 1 = -1.$$

The formula

$$F_x(P_0)(x - x_0) + F_y(P_0)(y - y_0) + F_z(P_0)(z - z_0) = 0$$

for the plane tangent to the level surface at P_0 therefore reduces to

$$f_x(x_0, y_0)(x - x_0) + f_y(x_0, y_0)(y - y_0) - (z - z_0) = 0.$$

Plane Tangent to a Surface $z = f(x, y)$ at $(x_0, y_0, f(x_0, y_0))$

The plane tangent to the surface $z = f(x, y)$ of a differentiable function f at the point $P_0(x_0, y_0, z_0) = (x_0, y_0, f(x_0, y_0))$ is

$$f_x(x_0, y_0)(x - x_0) + f_y(x_0, y_0)(y - y_0) - (z - z_0) = 0. \qquad (4)$$

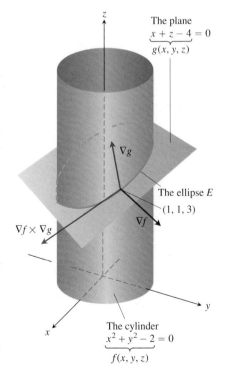

FIGURE 12.32 This cylinder and plane intersect in an ellipse E (Example 3).

EXAMPLE 2 Find the plane tangent to the surface $z = x \cos y - ye^x$ at $(0, 0, 0)$.

Solution We calculate the partial derivatives of $f(x, y) = x \cos y - ye^x$ and use Equation (4):

$$f_x(0, 0) = (\cos y - ye^x)_{(0,0)} = 1 - 0 \cdot 1 = 1$$

$$f_y(0, 0) = (-x \sin y - e^x)_{(0,0)} = 0 - 1 = -1.$$

The tangent plane is therefore

$$1 \cdot (x - 0) - 1 \cdot (y - 0) - (z - 0) = 0, \qquad \text{Equation (4)}$$

or

$$x - y - z = 0. \qquad \blacksquare$$

EXAMPLE 3 The surfaces

$$f(x, y, z) = x^2 + y^2 - 2 = 0 \qquad \text{A cylinder}$$

and

$$g(x, y, z) = x + z - 4 = 0 \qquad \text{A plane}$$

meet in an ellipse E (Figure 12.32). Find parametric equations for the line tangent to E at the point $P_0(1, 1, 3)$.

Solution The tangent line is orthogonal to both ∇f and ∇g at P_0, and therefore parallel to $\mathbf{v} = \nabla f \times \nabla g$. The components of $\mathbf{v}$ and the coordinates of P_0 give us equations for the line. We have

$$\nabla f|_{(1,1,3)} = (2x\mathbf{i} + 2y\mathbf{j})_{(1,1,3)} = 2\mathbf{i} + 2\mathbf{j}$$

$$\nabla g|_{(1,1,3)} = (\mathbf{i} + \mathbf{k})_{(1,1,3)} = \mathbf{i} + \mathbf{k}$$

$$\mathbf{v} = (2\mathbf{i} + 2\mathbf{j}) \times (\mathbf{i} + \mathbf{k}) = \begin{vmatrix} \mathbf{i} & \mathbf{j} & \mathbf{k} \\ 2 & 2 & 0 \\ 1 & 0 & 1 \end{vmatrix} = 2\mathbf{i} - 2\mathbf{j} - 2\mathbf{k}.$$

The tangent line is

$$x = 1 + 2t, \qquad y = 1 - 2t, \qquad z = 3 - 2t. \qquad \blacksquare$$

Estimating Change in a Specific Direction

The directional derivative plays the role of an ordinary derivative when we want to estimate how much the value of a function f changes if we move a small distance ds from a point P_0 to another point nearby. If f were a function of a single variable, we would have

$$df = f'(P_0)\, ds. \qquad \text{Ordinary derivative} \times \text{increment}$$

For a function of two or more variables, we use the formula

$$df = (\nabla f|_{P_0} \cdot \mathbf{u})\, ds, \qquad \text{Directional derivative} \times \text{increment}$$

where $\mathbf{u}$ is the direction of the motion away from P_0.

Estimating the Change in f in a Direction u

To estimate the change in the value of a differentiable function f when we move a small distance ds from a point P_0 in a particular direction $\mathbf{u}$, use the formula

$$df = \underbrace{(\nabla f|_{P_0} \cdot \mathbf{u})}_{\substack{\text{Directional} \\ \text{derivative}}} \cdot \underbrace{ds}_{\substack{\text{Distance} \\ \text{increment}}}$$

EXAMPLE 4   Estimate how much the value of

$$f(x, y, z) = y \sin x + 2yz$$

will change if the point $P(x, y, z)$ moves 0.1 unit from $P_0(0, 1, 0)$ straight toward $P_1(2, 2, -2)$.

Solution   We first find the derivative of f at P_0 in the direction of the vector $\overrightarrow{P_0P_1} = 2\mathbf{i} + \mathbf{j} - 2\mathbf{k}$. The direction of this vector is

$$\mathbf{u} = \frac{\overrightarrow{P_0P_1}}{|\overrightarrow{P_0P_1}|} = \frac{\overrightarrow{P_0P_1}}{3} = \frac{2}{3}\mathbf{i} + \frac{1}{3}\mathbf{j} - \frac{2}{3}\mathbf{k}.$$

The gradient of f at P_0 is

$$\nabla f|_{(0,1,0)} = ((y \cos x)\mathbf{i} + (\sin x + 2z)\mathbf{j} + 2y\mathbf{k})_{(0,1,0)} = \mathbf{i} + 2\mathbf{k}.$$

Therefore,

$$\nabla f|_{P_0} \cdot \mathbf{u} = (\mathbf{i} + 2\mathbf{k}) \cdot \left(\frac{2}{3}\mathbf{i} + \frac{1}{3}\mathbf{j} - \frac{2}{3}\mathbf{k} \right) = \frac{2}{3} - \frac{4}{3} = -\frac{2}{3}.$$

The change df in f that results from moving $ds = 0.1$ unit away from P_0 in the direction of $\mathbf{u}$ is approximately

$$df = (\nabla f|_{P_0} \cdot \mathbf{u})(ds) = \left(-\frac{2}{3} \right)(0.1) \approx -0.067 \text{ unit.} \qquad \blacksquare$$

How to Linearize a Function of Two Variables

Functions of two variables can be complicated, and we sometimes need to replace them with simpler ones that give the accuracy required for specific applications without being so difficult to work with. We do this in a way that is similar to the way we find linear replacements for functions of a single variable (Section 3.10).

Suppose the function we wish to replace is $z = f(x, y)$ and that we want the replacement to be effective near a point (x_0, y_0) at which we know the values of f, f_x, and f_y and at which f is differentiable. If we move from (x_0, y_0) to any point (x, y) by increments $\Delta x = x - x_0$ and $\Delta y = y - y_0$, then the definition of differentiability from Section 12.3 gives the change

$$f(x, y) - f(x_0, y_0) = f_x(x_0, y_0)\Delta x + f_y(x_0, y_0)\Delta y + \epsilon_1 \Delta x + \epsilon_2 \Delta y,$$

where $\epsilon_1, \epsilon_2 \to 0$ as $\Delta x, \Delta y \to 0$. If the increments Δx and Δy are small, the products $\epsilon_1 \Delta x$ and $\epsilon_2 \Delta y$ will eventually be smaller still and we will have

$$f(x, y) \approx \underbrace{f(x_0, y_0) + f_x(x_0, y_0)(x - x_0) + f_y(x_0, y_0)(y - y_0)}_{L(x, y)}.$$

In other words, as long as Δx and Δy are small, f will have approximately the same value as the linear function L. If f is hard to use, and our work can tolerate the error involved, we may approximate f by L (Figure 12.33).

DEFINITIONS   The **linearization** of a function $f(x, y)$ at a point (x_0, y_0) where f is differentiable is the function

$$L(x, y) = f(x_0, y_0) + f_x(x_0, y_0)(x - x_0) + f_y(x_0, y_0)(y - y_0). \qquad (5)$$

The approximation

$$f(x, y) \approx L(x, y)$$

is the **standard linear approximation** of f at (x_0, y_0).

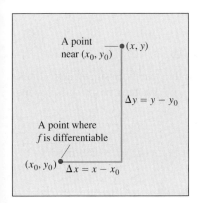

FIGURE 12.33   If f is differentiable at (x_0, y_0), then the value of f at any point (x, y) nearby is approximately $f(x_0, y_0) + f_x(x_0, y_0)\Delta x + f_y(x_0, y_0)\Delta y$.

From Equation (4), we see that the plane $z = L(x, y)$ is tangent to the surface $z = f(x, y)$ at the point (x_0, y_0). Thus, the linearization of a function of two variables is a tangent-*plane* approximation in the same way that the linearization of a function of a single variable is a tangent-*line* approximation.

When approximating a differentiable function $f(x, y)$ by its linearization $L(x, y)$ at (x_0, y_0), an important question is how accurate the approximation might be.

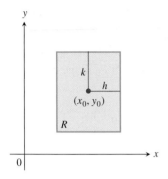

FIGURE 12.34 The rectangular region R: $|x - x_0| \leq h$, $|y - y_0| \leq k$ in the xy-plane.

If we can find a common upper bound M for $|f_{xx}|, |f_{yy}|$, and $|f_{xy}|$ on a rectangle R centered at (x_0, y_0) (Figure 12.34), then we can bound the error E throughout R by using a simple formula (derived in Section 12.9). The **error** is defined by $E(x, y) = f(x, y) - L(x, y)$.

The Error in the Standard Linear Approximation

If f has continuous first and second partial derivatives throughout an open set containing a rectangle R centered at (x_0, y_0) and if M is any upper bound for the values of $|f_{xx}|, |f_{yy}|$, and $|f_{xy}|$ on R, then the error $E(x, y)$ incurred in replacing $f(x, y)$ on R by its linearization

$$L(x, y) = f(x_0, y_0) + f_x(x_0, y_0)(x - x_0) + f_y(x_0, y_0)(y - y_0)$$

satisfies the inequality

$$|E(x, y)| \leq \frac{1}{2} M(|x - x_0| + |y - y_0|)^2.$$

To make $|E(x, y)|$ small for a given M, we just make $|x - x_0|$ and $|y - y_0|$ small.

Differentials

Recall from Section 3.10 that for a function of a single variable, $y = f(x)$, we defined the change in f as x changes from a to $a + \Delta x$ by

$$\Delta f = f(a + \Delta x) - f(a)$$

and the differential of f as

$$df = f'(a)\Delta x.$$

We now consider a function of two variables.

Suppose a differentiable function $f(x, y)$ and its partial derivatives exist at a point (x_0, y_0). If we move to a nearby point $(x_0 + \Delta x, y_0 + \Delta y)$, the change in f is

$$\Delta f = f(x_0 + \Delta x, y_0 + \Delta y) - f(x_0, y_0).$$

A straightforward calculation from the definition of $L(x, y)$, using the notation $x - x_0 = \Delta x$ and $y - y_0 = \Delta y$, shows that the corresponding change in L is

$$\Delta L = L(x_0 + \Delta x, y_0 + \Delta y) - L(x_0, y_0)$$

$$= f_x(x_0, y_0)\Delta x + f_y(x_0, y_0)\Delta y.$$

The **differentials** dx and dy are independent variables, so they can be assigned any values. Often we take $dx = \Delta x = x - x_0$, and $dy = \Delta y = y - y_0$. We then have the following definition of the differential or *total* differential of f.

DEFINITION If we move from (x_0, y_0) to a point $(x_0 + dx, y_0 + dy)$ nearby, the resulting change

$$df = f_x(x_0, y_0)\, dx + f_y(x_0, y_0)\, dy$$

in the linearization of f is called the **total differential of f**.

EXAMPLE 5 Your company manufactures right circular cylindrical molasses storage tanks that are 25 ft high with a radius of 5 ft. How sensitive are the tanks' volumes to small variations in height and radius?

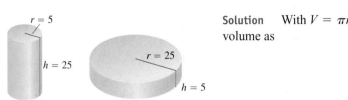

$r = 5$

$h = 25$

$r = 25$

$h = 5$

(a)   (b)

FIGURE 12.35 The volume of cylinder
(a) is more sensitive to a small change in r
than it is to an equally small change in h.
The volume of cylinder (b) is more
sensitive to small changes in h than it
is to small changes in r (Example 5).

Solution With $V = \pi r^2 h$, the total differential gives the approximation for the change in
volume as

$$dV = V_r(5, 25)\, dr + V_h(5, 25)\, dh$$
$$= (2\pi rh)_{(5,25)}\, dr + (\pi r^2)_{(5,25)}\, dh$$
$$= 250\pi\, dr + 25\pi\, dh.$$

Thus, a 1-unit change in r will change V by about 250π units. A 1-unit change in h will
change V by about 25π units. The tank's volume is 10 times more sensitive to a small
change in r than it is to a small change of equal size in h. As a quality control engineer
concerned with being sure the tanks have the correct volume, you would want to pay special attention to their radii.

In contrast, if the values of r and h are reversed to make $r = 25$ and $h = 5$, then the
total differential in V becomes

$$dV = (2\pi rh)_{(25,5)}\, dr + (\pi r^2)_{(25,5)}\, dh = 250\pi\, dr + 625\pi\, dh.$$

Now the volume is more sensitive to changes in h than to changes in r (Figure 12.35).

The general rule is that functions are most sensitive to small changes in the variables
that generate the largest partial derivatives.   ∎

Functions of More Than Two Variables

Analogous results hold for differentiable functions of more than two variables.

1. The **linearization** of $f(x, y, z)$ at a point $P_0(x_0, y_0, z_0)$ is

    $$L(x, y, z) = f(P_0) + f_x(P_0)(x - x_0) + f_y(P_0)(y - y_0) + f_z(P_0)(z - z_0).$$

2. Suppose that R is a closed rectangular solid centered at P_0 and lying in an open region
    on which the second partial derivatives of f are continuous. Suppose also that
    $|f_{xx}|, |f_{yy}|, |f_{zz}|, |f_{xy}|, |f_{xz}|$, and $|f_{yz}|$ are all less than or equal to M throughout R. Then
    the **error** $E(x, y, z) = f(x, y, z) - L(x, y, z)$ in the approximation of f by L is bounded
    throughout R by the inequality

    $$|E| \le \frac{1}{2} M(|x - x_0| + |y - y_0| + |z - z_0|)^2.$$

3. If the second partial derivatives of f are continuous and if x, y, and z change from
    $x_0, y_0,$ and z_0 by small amounts $dx, dy,$ and dz, the **total differential**

    $$df = f_x(P_0)\, dx + f_y(P_0)\, dy + f_z(P_0)\, dz$$

    gives a good approximation of the resulting change in f.

EXAMPLE 6 Find the linearization $L(x, y, z)$ of

$$f(x, y, z) = x^2 - xy + 3 \sin z$$

at the point $(x_0, y_0, z_0) = (2, 1, 0)$. Find an upper bound for the error incurred in replacing
f by L on the rectangle

$$R: \quad |x - 2| \le 0.01, \qquad |y - 1| \le 0.02, \qquad |z| \le 0.01.$$

Solution Routine calculations give

$$f(2, 1, 0) = 2, \qquad f_x(2, 1, 0) = 3, \qquad f_y(2, 1, 0) = -2, \qquad f_z(2, 1, 0) = 3.$$

Thus,

$$L(x, y, z) = 2 + 3(x - 2) + (-2)(y - 1) + 3(z - 0) = 3x - 2y + 3z - 2.$$

Since

$$f_{xx} = 2, \qquad f_{yy} = 0, \qquad f_{zz} = -3 \sin z, \qquad f_{xy} = -1, \qquad f_{xz} = 0, \qquad f_{yz} = 0,$$

we may safely take M to be max $|-3 \sin z| = 3$. Hence, the error incurred by replacing f by L on R satisfies

$$|E| \le \frac{1}{2}(3)(0.01 + 0.02 + 0.01)^2 = 0.0024 .$$ ∎

EXERCISES 12.6

In Exercises 1–8, find equations for the

(a) tangent plane and   (b) normal line at the point P_0 on the given surface.

1. $x^2 + y^2 + z^2 = 3$, $P_0(1, 1, 1)$

2. $x^2 + y^2 - z^2 = 18$, $P_0(3, 5, -4)$

3. $2z - x^2 = 0$, $P_0(2, 0, 2)$

4. $x^2 + 2xy - y^2 + z^2 = 7$, $P_0(1, -1, 3)$

5. $\cos \pi x - x^2 y + e^{xz} + yz = 4$, $P_0(0, 1, 2)$

6. $x^2 - xy - y^2 - z = 0$, $P_0(1, 1, -1)$

7. $x + y + z = 1$, $P_0(0, 1, 0)$

8. $x^2 + y^2 - 2xy - x + 3y - z = -4$, $P_0(2, -3, 18)$

In Exercises 9–12, find an equation for the plane that is tangent to the given surface at the given point.

9. $z = \ln(x^2 + y^2)$, $(1, 0, 0)$ 10. $z = e^{-(x^2+y^2)}$, $(0, 0, 1)$

11. $z = \sqrt{y - x}$, $(1, 2, 1)$ 12. $z = 4x^2 + y^2$, $(1, 1, 5)$

In Exercises 13–18, find parametric equations for the line tangent to the curve of intersection of the surfaces at the given point.

13. Surfaces: $x + y^2 + 2z = 4$, $x = 1$

   Point: $(1, 1, 1)$

14. Surfaces: $xyz = 1$, $x^2 + 2y^2 + 3z^2 = 6$

   Point: $(1, 1, 1)$

15. Surfaces: $x^2 + 2y + 2z = 4$, $y = 1$

   Point: $(1, 1, 1/2)$

16. Surfaces: $x + y^2 + z = 2$, $y = 1$

   Point: $(1/2, 1, 1/2)$

17. Surfaces: $x^3 + 3x^2 y^2 + y^3 + 4xy - z^2 = 0$, $x^2 + y^2 + z^2 = 11$

   Point: $(1, 1, 3)$

18. Surfaces: $x^2 + y^2 = 4$, $x^2 + y^2 - z = 0$

   Point: $(\sqrt{2}, \sqrt{2}, 4)$

19. By about how much will

$$f(x, y, z) = \ln\sqrt{x^2 + y^2 + z^2}$$

change if the point $P(x, y, z)$ moves from $P_0(3, 4, 12)$ a distance of $ds = 0.1$ unit in the direction of $3\mathbf{i} + 6\mathbf{j} - 2\mathbf{k}$?

20. By about how much will

$$f(x, y, z) = e^x \cos yz$$

change as the point $P(x, y, z)$ moves from the origin a distance of $ds = 0.1$ unit in the direction of $2\mathbf{i} + 2\mathbf{j} - 2\mathbf{k}$?

21. By about how much will

$$g(x, y, z) = x + x \cos z - y \sin z + y$$

change if the point $P(x, y, z)$ moves from $P_0(2, -1, 0)$ a distance of $ds = 0.2$ unit toward the point $P_1(0, 1, 2)$?

22. By about how much will

$$h(x, y, z) = \cos(\pi xy) + xz^2$$

change if the point $P(x, y, z)$ moves from $P_0(-1, -1, -1)$ a distance of $ds = 0.1$ unit toward the origin?

23. **Temperature change along a circle**   Suppose that the Celsius temperature at the point (x, y) in the xy-plane is $T(x, y) = x \sin 2y$ and that distance in the xy-plane is measured in meters. A particle is moving *clockwise* around the circle of radius 1 m centered at the origin at the constant rate of 2 m/sec.

   a. How fast is the temperature experienced by the particle changing in degrees Celsius per meter at the point $P\left(1/2, \sqrt{3}/2\right)$?

   b. How fast is the temperature experienced by the particle changing in degrees Celsius per second at P?

24. **Changing temperature along a space curve**   The Celsius temperature in a region in space is given by $T(x, y, z) = 2x^2 - xyz$. A particle is moving in this region and its position at time t is given by $x = 2t^2, y = 3t, z = -t^2$, where time is measured in seconds and distance in meters.

   a. How fast is the temperature experienced by the particle changing in degrees Celsius per meter when the particle is at the point $P(8, 6, -4)$?

   b. How fast is the temperature experienced by the particle changing in degrees Celsius per second at P?

In Exercises 25–30, find the linearization $L(x, y)$ of the function at each point.

25. $f(x, y) = x^2 + y^2 + 1$ at   **a.** $(0, 0)$,   **b.** $(1, 1)$

26. $f(x, y) = (x + y + 2)^2$ at   **a.** $(0, 0)$,   **b.** $(1, 2)$

27. $f(x, y) = 3x - 4y + 5$ at   **a.** $(0, 0)$,   **b.** $(1, 1)$
28. $f(x, y) = x^3 y^4$ at   **a.** $(1, 1)$,   **b.** $(0, 0)$
29. $f(x, y) = e^x \cos y$ at   **a.** $(0, 0)$,   **b.** $(0, \pi/2)$
30. $f(x, y) = e^{2y - x}$ at   **a.** $(0, 0)$,   **b.** $(1, 2)$

In Exercises 31–36, find the linearization $L(x, y)$ of the function $f(x, y)$ at P_0. Then find an upper bound for the magnitude $|E|$ of the error in the approximation $f(x, y) \approx L(x, y)$ over the rectangle R.

31. $f(x, y) = x^2 - 3xy + 5$ at $P_0(2, 1)$,
   R: $|x - 2| \le 0.1$, $|y - 1| \le 0.1$
32. $f(x, y) = (1/2)x^2 + xy + (1/4)y^2 + 3x - 3y + 4$ at $P_0(2, 2)$,
   R: $|x - 2| \le 0.1$, $|y - 2| \le 0.1$
33. $f(x, y) = 1 + y + x \cos y$ at $P_0(0, 0)$,
   R: $|x| \le 0.2$, $|y| \le 0.2$
   (Use $|\cos y| \le 1$ and $|\sin y| \le 1$ in estimating E.)
34. $f(x, y) = xy^2 + y \cos (x - 1)$ at $P_0(1, 2)$,
   R: $|x - 1| \le 0.1$, $|y - 2| \le 0.1$
35. $f(x, y) = e^x \cos y$ at $P_0(0, 0)$,
   R: $|x| \le 0.1$, $|y| \le 0.1$
   (Use $e^x \le 1.11$ and $|\cos y| \le 1$ in estimating E.)
36. $f(x, y) = \ln x + \ln y$ at $P_0(1, 1)$,
   R: $|x - 1| \le 0.2$, $|y - 1| \le 0.2$

Find the linearizations $L(x, y, z)$ of the functions in Exercises 37–42 at the given points.

37. $f(x, y, z) = xy + yz + xz$ at
   a. $(1, 1, 1)$ **b.** $(1, 0, 0)$ **c.** $(0, 0, 0)$
38. $f(x, y, z) = x^2 + y^2 + z^2$ at
   a. $(1, 1, 1)$ **b.** $(0, 1, 0)$ **c.** $(1, 0, 0)$
39. $f(x, y, z) = \sqrt{x^2 + y^2 + z^2}$ at
   a. $(1, 0, 0)$ **b.** $(1, 1, 0)$ **c.** $(1, 2, 2)$
40. $f(x, y, z) = (\sin xy)/z$ at
   a. $(\pi/2, 1, 1)$ **b.** $(2, 0, 1)$
41. $f(x, y, z) = e^x + \cos (y + z)$ at
   a. $(0, 0, 0)$ **b.** $\left(0, \dfrac{\pi}{2}, 0\right)$ **c.** $\left(0, \dfrac{\pi}{4}, \dfrac{\pi}{4}\right)$
42. $f(x, y, z) = \tan^{-1}(xyz)$ at
   a. $(1, 0, 0)$ **b.** $(1, 1, 0)$ **c.** $(1, 1, 1)$

In Exercises 43–46, find the linearization $L(x, y, z)$ of the function $f(x, y, z)$ at P_0. Then find an upper bound for the magnitude of the error E in the approximation $f(x, y, z) \approx L(x, y, z)$ over the region R.

43. $f(x, y, z) = xz - 3yz + 2$ at $P_0(1, 1, 2)$
   R: $|x - 1| \le 0.01$, $|y - 1| \le 0.01$, $|z - 2| \le 0.02$
44. $f(x, y, z) = x^2 + xy + yz + (1/4)z^2$ at $P_0(1, 1, 2)$
   R: $|x - 1| \le 0.01$, $|y - 1| \le 0.01$, $|z - 2| \le 0.08$

45. $f(x, y, z) = xy + 2yz - 3xz$ at $P_0(1, 1, 0)$
   R: $|x - 1| \le 0.01$, $|y - 1| \le 0.01$, $|z| \le 0.01$
46. $f(x, y, z) = \sqrt{2} \cos x \sin (y + z)$ at $P_0(0, 0, \pi/4)$
   R: $|x| \le 0.01$, $|y| \le 0.01$, $|z - \pi/4| \le 0.01$

47. **Estimating maximum error** Suppose that T is to be found from the formula $T = x(e^y + e^{-y})$, where x and y are found to be 2 and ln 2 with maximum possible errors of $|dx| = 0.1$ and $|dy| = 0.02$. Estimate the maximum possible error in the computed value of T.

48. **Variation in electrical resistance** The resistance R produced by wiring resistors of R_1 and R_2 ohms in parallel (see accompanying figure) can be calculated from the formula

$$\frac{1}{R} = \frac{1}{R_1} + \frac{1}{R_2}.$$

   a. Show that

$$dR = \left(\frac{R}{R_1}\right)^2 dR_1 + \left(\frac{R}{R_2}\right)^2 dR_2.$$

   b. You have designed a two-resistor circuit like the one shown here to have resistances of $R_1 = 100$ ohms and $R_2 = 400$ ohms, but there is always some variation in manufacturing and the resistors received by your firm will probably not have these exact values. Will the value of R be more sensitive to variation in R_1 or to variation in R_2? Give reasons for your answer.

49. You plan to calculate the area of a long, thin rectangle from measurements of its length and width. Which dimension should you measure more carefully? Give reasons for your answer.

50. **a.** Around the point $(1, 0)$, is $f(x, y) = x^2(y + 1)$ more sensitive to changes in x or to changes in y? Give reasons for your answer.
   b. What ratio of dx to dy will make df equal zero at $(1, 0)$?

51. **Value of a 2 × 2 determinant** If $|a|$ is much greater than $|b|, |c|$, and $|d|$, to which of a, b, c, and d is the value of the determinant

$$f(a, b, c, d) = \begin{vmatrix} a & b \\ c & d \end{vmatrix}$$

most sensitive? Give reasons for your answer.

52. **The Wilson lot size formula** The Wilson lot size formula in economics says that the most economical quantity Q of goods (radios, shoes, brooms, whatever) for a store to order is given by the formula $Q = \sqrt{2KM/h}$, where K is the cost of placing the order,

M is the number of items sold per week, and h is the weekly holding cost for each item (cost of space, utilities, security, and so on). To which of the variables K, M, and h is Q most sensitive near the point $(K_0, M_0, h_0) = (2, 20, 0.05)$? Give reasons for your answer.

53. The linearization of $f(x, y)$ is a tangent-plane approximation Show that the tangent plane at the point $P_0(x_0, y_0, f(x_0, y_0))$ on the surface $z = f(x, y)$ defined by a differentiable function f is the plane

$$f_x(x_0, y_0)(x - x_0) + f_y(x_0, y_0)(y - y_0) - (z - f(x_0, y_0)) = 0$$

or

$$z = f(x_0, y_0) + f_x(x_0, y_0)(x - x_0) + f_y(x_0, y_0)(y - y_0).$$

Thus, the tangent plane at P_0 is the graph of the linearization of f at P_0 (see accompanying figure).

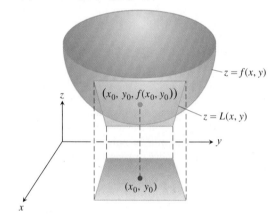

54. Change along the involute of a circle Find the derivative of $f(x, y) = x^2 + y^2$ in the direction of the unit tangent vector of the curve

$$\mathbf{r}(t) = (\cos t + t \sin t)\mathbf{i} + (\sin t - t \cos t)\mathbf{j}, \qquad t > 0.$$

55. Change along a helix Find the derivative of $f(x, y, z) = x^2 + y^2 + z^2$ in the direction of the unit tangent vector of the helix

$$\mathbf{r}(t) = (\cos t)\mathbf{i} + (\sin t)\mathbf{j} + t\mathbf{k}$$

at the points where $t = -\pi/4$, 0, and $\pi/4$. The function f gives the square of the distance from a point $P(x, y, z)$ on the helix to the origin. The derivatives calculated here give the rates at which the square of the distance is changing with respect to t as P moves through the points where $t = -\pi/4$, 0, and $\pi/4$.

56. Normal curves A smooth curve is *normal* to a surface $f(x, y, z) = c$ at a point of intersection if the curve's velocity vector is a nonzero scalar multiple of ∇f at the point.

Show that the curve

$$\mathbf{r}(t) = \sqrt{t}\mathbf{i} + \sqrt{t}\mathbf{j} - \frac{1}{4}(t + 3)\mathbf{k}$$

is normal to the surface $x^2 + y^2 - z = 3$ when $t = 1$.

57. Tangent curves A smooth curve is *tangent* to the surface at a point of intersection if its velocity vector is orthogonal to ∇f there.

Show that the curve

$$\mathbf{r}(t) = \sqrt{t}\mathbf{i} + \sqrt{t}\mathbf{j} + (2t - 1)\mathbf{k}$$

is tangent to the surface $x^2 + y^2 - z = 1$ when $t = 1$.

12.7 Extreme Values and Saddle Points

FIGURE 12.36 The function

$$z = (\cos x)(\cos y)e^{-\sqrt{x^2+y^2}}$$

has a maximum value of 1 and a minimum value of about -0.067 on the square region $|x| \le 3\pi/2, |y| \le 3\pi/2$.

Continuous functions of two variables assume extreme values on closed, bounded domains (see Figures 12.36 and 12.37). We see in this section that we can narrow the search for these extreme values by examining the functions' first partial derivatives. A function of two variables can assume extreme values only at domain boundary points or at interior domain points where both first partial derivatives are zero or where one or both of the first partial derivatives fails to exist. However, the vanishing of derivatives at an interior point (a, b) does not always signal the presence of an extreme value. The surface that is the graph of the function might be shaped like a saddle right above (a, b) and cross its tangent plane there.

Derivative Tests for Local Extreme Values

To find the local extreme values of a function of a single variable, we look for points where the graph has a horizontal tangent line. At such points, we then look for local maxima, local minima, and points of inflection. For a function $f(x, y)$ of two variables, we look for points where the surface $z = f(x, y)$ has a horizontal tangent *plane*. At such points, we then look for local maxima, local minima, and saddle points. We begin by defining maxima and minima.

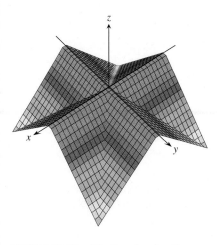

FIGURE 12.37 The "roof surface"

$$z = \frac{1}{2}\left(\big||x| - |y|\big| - |x| - |y|\right)$$

viewed from the point (10, 15, 20). The defining function has a maximum value of 0 and a minimum value of $-a$ on the square region $|x| \le a, |y| \le a$.

DEFINITIONS   Let $f(x, y)$ be defined on a region R containing the point (a, b). Then

1. $f(a, b)$ is a **local maximum** value of f if $f(a, b) \ge f(x, y)$ for all domain points (x, y) in an open disk centered at (a, b).
2. $f(a, b)$ is a **local minimum** value of f if $f(a, b) \le f(x, y)$ for all domain points (x, y) in an open disk centered at (a, b).

Local maxima correspond to mountain peaks on the surface $z = f(x, y)$ and local minima correspond to valley bottoms (Figure 12.38). At such points the tangent planes, when they exist, are horizontal. Local extrema are also called **relative extrema**.

As with functions of a single variable, the key to identifying the local extrema is a first derivative test.

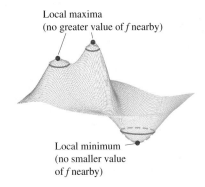

Local maxima
(no greater value of f nearby)

Local minimum
(no smaller value
of f nearby)

FIGURE 12.38   A local maximum is a mountain peak and a local minimum is a valley low.

HISTORICAL BIOGRAPHY

Siméon-Denis Poisson
(1781–1840)

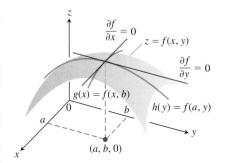

FIGURE 12.39   If a local maximum of f occurs at $x = a, y = b$, then the first partial derivatives $f_x(a, b)$ and $f_y(a, b)$ are both zero.

THEOREM 10—First Derivative Test for Local Extreme Values   If $f(x, y)$ has a local maximum or minimum value at an interior point (a, b) of its domain and if the first partial derivatives exist there, then $f_x(a, b) = 0$ and $f_y(a, b) = 0$.

Proof   If f has a local extremum at (a, b), then the function $g(x) = f(x, b)$ has a local extremum at $x = a$ (Figure 12.39). Therefore, $g'(a) = 0$ (Chapter 4, Theorem 2). Now $g'(a) = f_x(a, b)$, so $f_x(a, b) = 0$. A similar argument with the function $h(y) = f(a, y)$ shows that $f_y(a, b) = 0$.   ∎

If we substitute the values $f_x(a, b) = 0$ and $f_y(a, b) = 0$ into the equation

$$f_x(a, b)(x - a) + f_y(a, b)(y - b) - (z - f(a, b)) = 0$$

for the tangent plane to the surface $z = f(x, y)$ at (a, b), the equation reduces to

$$0 \cdot (x - a) + 0 \cdot (y - b) - z + f(a, b) = 0$$

or

$$z = f(a, b).$$

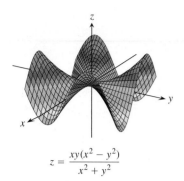

$$z = \frac{xy(x^2 - y^2)}{x^2 + y^2}$$

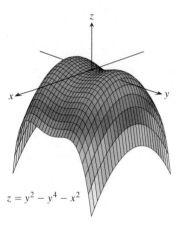

$$z = y^2 - y^4 - x^2$$

FIGURE 12.40 Saddle points at the origin.

Thus, Theorem 10 says that the surface does indeed have a horizontal tangent plane at a local extremum, provided there is a tangent plane there.

DEFINITION   An interior point of the domain of a function $f(x, y)$ where both f_x and f_y are zero or where one or both of f_x and f_y do not exist is a **critical point** of f.

Theorem 10 says that the only points where a function $f(x, y)$ can assume extreme values are critical points and boundary points. As with differentiable functions of a single variable, not every critical point gives rise to a local extremum. A differentiable function of a single variable might have a point of inflection. A differentiable function of two variables might have a *saddle point*.

DEFINITION   A differentiable function $f(x, y)$ has a **saddle point** at a critical point (a, b) if in every open disk centered at (a, b) there are domain points (x, y) where $f(x, y) > f(a, b)$ and domain points (x, y) where $f(x, y) < f(a, b)$. The corresponding point $(a, b, f(a, b))$ on the surface $z = f(x, y)$ is called a saddle point of the surface (Figure 12.40).

EXAMPLE 1   Find the local extreme values of $f(x, y) = x^2 + y^2$.

Solution   The domain of f is the entire plane (so there are no boundary points) and the partial derivatives $f_x = 2x$ and $f_y = 2y$ exist everywhere. Therefore, local extreme values can occur only where

$$f_x = 2x = 0 \qquad \text{and} \qquad f_y = 2y = 0.$$

The only possibility is the origin, where the value of f is zero. Since f is never negative, we see that the origin gives a local minimum (Figure 12.41).   ∎

EXAMPLE 2   Find the local extreme values (if any) of $f(x, y) = y^2 - x^2$.

Solution   The domain of f is the entire plane (so there are no boundary points) and the partial derivatives $f_x = -2x$ and $f_y = 2y$ exist everywhere. Therefore, local extrema can occur only at the origin $(0, 0)$. Along the positive x-axis, however, f has the value $f(x, 0) = -x^2 < 0$; along the positive y-axis, f has the value $f(0, y) = y^2 > 0$. Therefore, every open disk in the xy-plane centered at $(0, 0)$ contains points where the function is positive and points where it is negative. The function has a saddle point at the origin (Figure 12.42) instead of a local extreme value. We conclude that the function has no local extreme values.   ∎

That $f_x = f_y = 0$ at an interior point (a, b) of R does not guarantee f has a local extreme value there. If f and its first and second partial derivatives are continuous on R, however, we may be able to learn more from the following theorem, proved in Section 12.9.

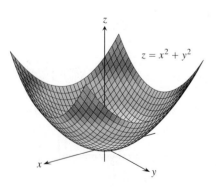

$$z = x^2 + y^2$$

FIGURE 12.41   The graph of the function $f(x, y) = x^2 + y^2$ is the paraboloid $z = x^2 + y^2$. The function has a local minimum value of 0 at the origin (Example 1).

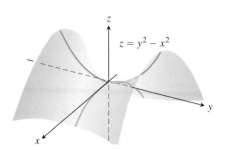

FIGURE 12.42 The origin is a saddle point of the function $f(x, y) = y^2 - x^2$. There are no local extreme values (Example 2).

THEOREM 11—Second Derivative Test for Local Extreme Values   Suppose that $f(x, y)$ and its first and second partial derivatives are continuous throughout a disk centered at (a, b) and that $f_x(a, b) = f_y(a, b) = 0$. Then

  i. f has a **local maximum** at (a, b) if $f_{xx} < 0$ and $f_{xx}f_{yy} - f_{xy}{}^2 > 0$ at (a, b).

  ii. f has a **local minimum** at (a, b) if $f_{xx} > 0$ and $f_{xx}f_{yy} - f_{xy}{}^2 > 0$ at (a, b).

  iii. f has a **saddle point** at (a, b) if $f_{xx}f_{yy} - f_{xy}{}^2 < 0$ at (a, b).

  iv.   **The test is inconclusive** at (a, b) if $f_{xx}f_{yy} - f_{xy}{}^2 = 0$ at (a, b). In this case, we must find some other way to determine the behavior of f at (a, b).

The expression $f_{xx}f_{yy} - f_{xy}{}^2$ is called the **discriminant** or **Hessian** of f. It is sometimes easier to remember it in determinant form,

$$f_{xx}f_{yy} - f_{xy}{}^2 = \begin{vmatrix} f_{xx} & f_{xy} \\ f_{xy} & f_{yy} \end{vmatrix}.$$

Theorem 11 says that if the discriminant is positive at the point (a, b), then the surface curves the same way in all directions: downward if $f_{xx} < 0$, giving rise to a local maximum, and upward if $f_{xx} > 0$, giving a local minimum. On the other hand, if the discriminant is negative at (a, b), then the surface curves up in some directions and down in others, so we have a saddle point.

EXAMPLE 3   Find the local extreme values of the function

$$f(x, y) = xy - x^2 - y^2 - 2x - 2y + 4.$$

Solution   The function is defined and differentiable for all x and y and its domain has no boundary points. The function therefore has extreme values only at the points where f_x and f_y are simultaneously zero. This leads to

$$f_x = y - 2x - 2 = 0, \qquad f_y = x - 2y - 2 = 0,$$

or

$$x = y = -2.$$

Therefore, the point $(-2, -2)$ is the only point where f may take on an extreme value. To see if it does so, we calculate

$$f_{xx} = -2, \qquad f_{yy} = -2, \qquad f_{xy} = 1.$$

The discriminant of f at $(a, b) = (-2, -2)$ is

$$f_{xx}f_{yy} - f_{xy}{}^2 = (-2)(-2) - (1)^2 = 4 - 1 = 3.$$

The combination

$$f_{xx} < 0 \qquad \text{and} \qquad f_{xx}f_{yy} - f_{xy}{}^2 > 0$$

tells us that f has a local maximum at $(-2, -2)$. The value of f at this point is $f(-2, -2) = 8$. ∎

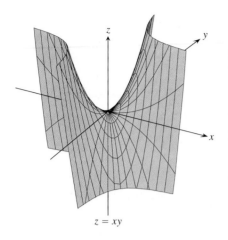

FIGURE 12.43 The surface $z = xy$ has a saddle point at the origin (Example 4).

EXAMPLE 4 Find the local extreme values of $f(x, y) = xy$.

Solution Since f is differentiable everywhere (Figure 12.43), it can assume extreme values only where

$$f_x = y = 0 \quad \text{and} \quad f_y = x = 0.$$

Thus, the origin is the only point where f might have an extreme value. To see what happens there, we calculate

$$f_{xx} = 0, \quad f_{yy} = 0, \quad f_{xy} = 1.$$

The discriminant,

$$f_{xx}f_{yy} - f_{xy}{}^2 = -1,$$

is negative. Therefore, the function has a saddle point at $(0, 0)$. We conclude that $f(x, y) = xy$ has no local extreme values. ∎

Absolute Maxima and Minima on Closed Bounded Regions

The definitions of **absolute maximum** and **absolute minimum** values for a function $f(x, y)$ on a region R are exactly the same as those for a single variable function (see Section 4.1). We organize the search for the absolute extrema of a continuous function $f(x, y)$ on a closed and bounded region R into three steps.

1. *List the interior points of R* where f may have local maxima and minima and evaluate f at these points. These are the critical points of f.

2. *List the boundary points of R* where f has local maxima and minima and evaluate f at these points. We show how to do this shortly.

3. *Look through the lists* for the maximum and minimum values of f. These will be the absolute maximum and minimum values of f on R. Since absolute maxima and minima are also local maxima and minima, the absolute maximum and minimum values of f appear somewhere in the lists made in Steps 1 and 2.

EXAMPLE 5 Find the absolute maximum and minimum values of

$$f(x, y) = 2 + 2x + 2y - x^2 - y^2$$

on the triangular region in the first quadrant bounded by the lines $x = 0, y = 0, y = 9 - x$.

Solution Since f is differentiable, the only places where f can assume these values are points inside the triangle (Figure 12.44) where $f_x = f_y = 0$ and points on the boundary.

(a) Interior points. For these we have

$$f_x = 2 - 2x = 0, \quad f_y = 2 - 2y = 0,$$

yielding the single point $(x, y) = (1, 1)$. The value of f there is

$$f(1, 1) = 4.$$

(b) Boundary points. We take the triangle one side at a time:

(i) On the segment OA, $y = 0$. The function

$$f(x, y) = f(x, 0) = 2 + 2x - x^2$$

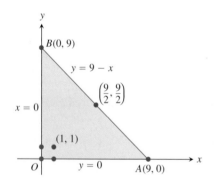

FIGURE 12.44 This triangular region is the domain of the function in Example 5.

may now be regarded as a function of x defined on the closed interval $0 \le x \le 9$. Its extreme values (we know from Chapter 4) may occur at the endpoints

$$x = 0 \qquad \text{where} \qquad f(0, 0) = 2$$
$$x = 9 \qquad \text{where} \qquad f(9, 0) = 2 + 18 - 81 = -61$$

and at the interior points where $f'(x, 0) = 2 - 2x = 0$. The only interior point where $f'(x, 0) = 0$ is $x = 1$, where

$$f(x, 0) = f(1, 0) = 3.$$

(ii) On the segment OB, $x = 0$ and

$$f(x, y) = f(0, y) = 2 + 2y - y^2.$$

We know from the symmetry of f in x and y and from the analysis we just carried out that the candidates on this segment are

$$f(0, 0) = 2, \qquad f(0, 9) = -61, \qquad f(0, 1) = 3.$$

(iii) We have already accounted for the values of f at the endpoints of AB, so we need only look at the interior points of AB. With $y = 9 - x$, we have

$$f(x, y) = 2 + 2x + 2(9 - x) - x^2 - (9 - x)^2 = -61 + 18x - 2x^2.$$

Setting $f'(x, 9 - x) = 18 - 4x = 0$ gives

$$x = \frac{18}{4} = \frac{9}{2}.$$

At this value of x,

$$y = 9 - \frac{9}{2} = \frac{9}{2} \qquad \text{and} \qquad f(x, y) = f\left(\frac{9}{2}, \frac{9}{2}\right) = -\frac{41}{2}.$$

Summary We list all the candidates: 4, 2, -61, 3, $-(41/2)$. The maximum is 4, which f assumes at $(1, 1)$. The minimum is -61, which f assumes at $(0, 9)$ and $(9, 0)$. ∎

Solving extreme value problems with algebraic constraints on the variables usually requires the method of Lagrange multipliers introduced in the next section. But sometimes we can solve such problems directly, as in the next example.

EXAMPLE 6 A delivery company accepts only rectangular boxes the sum of whose length and girth (perimeter of a cross-section) does not exceed 108 in. Find the dimensions of an acceptable box of largest volume.

Solution Let x, y, and z represent the length, width, and height of the rectangular box, respectively. Then the girth is $2y + 2z$. We want to maximize the volume $V = xyz$ of the box (Figure 12.45) satisfying $x + 2y + 2z = 108$ (the largest box accepted by the delivery company). Thus, we can write the volume of the box as a function of two variables:

$$V(y, z) = (108 - 2y - 2z)yz \qquad {\scriptstyle V = xyz \text{ and}} \atop {\scriptstyle x = 108 - 2y - 2z}$$
$$= 108yz - 2y^2z - 2yz^2.$$

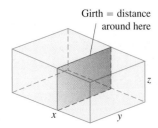

Girth = distance around here

FIGURE 12.45 The box in Example 6.

Setting the first partial derivatives equal to zero,

$$V_y(y, z) = 108z - 4yz - 2z^2 = (108 - 4y - 2z)z = 0$$

$$V_z(y, z) = 108y - 2y^2 - 4yz = (108 - 2y - 4z)y = 0,$$

gives the critical points $(0, 0)$, $(0, 54)$, $(54, 0)$, and $(18, 18)$. The volume is zero at $(0, 0)$, $(0, 54)$, $(54, 0)$, which are not maximum values. At the point $(18, 18)$, we apply the Second Derivative Test (Theorem 11):

$$V_{yy} = -4z, \qquad V_{zz} = -4y, \qquad V_{yz} = 108 - 4y - 4z.$$

Then

$$V_{yy}V_{zz} - V_{yz}^2 = 16yz - 16(27 - y - z)^2.$$

Thus,

$$V_{yy}(18, 18) = -4(18) < 0$$

and

$$\left[V_{yy}V_{zz} - V_{yz}^2\right]_{(18,18)} = 16(18)(18) - 16(-9)^2 > 0$$

imply that $(18, 18)$ gives a maximum volume. The dimensions of the package are $x = 108 - 2(18) - 2(18) = 36$ in., $y = 18$ in., and $z = 18$ in. The maximum volume is $V = (36)(18)(18) = 11{,}664$ in.^{3}, or 6.75 ft^3. ∎

Despite the power of Theorem 10, we urge you to remember its limitations. It does not apply to boundary points of a function's domain, where it is possible for a function to have extreme values along with nonzero derivatives. Also, it does not apply to points where either f_x or f_y fails to exist.

Summary of Max-Min Tests

The extreme values of $f(x, y)$ can occur only at

  i.   **boundary points** of the domain of f

  ii. **critical points** (interior points where $f_x = f_y = 0$ or points where f_x or f_y fail to exist).

If the first- and second-order partial derivatives of f are continuous throughout a disk centered at a point (a, b) and $f_x(a, b) = f_y(a, b) = 0$, the nature of $f(a, b)$ can be tested with the **Second Derivative Test**:

  i. $f_{xx} < 0$ and $f_{xx}f_{yy} - f_{xy}^2 > 0$ at (a, b) $\Rightarrow$ **local maximum**

  ii. $f_{xx} > 0$ and $f_{xx}f_{yy} - f_{xy}^2 > 0$ at (a, b) $\Rightarrow$ **local minimum**

  iii. $f_{xx}f_{yy} - f_{xy}^2 < 0$ at (a, b) $\Rightarrow$ **saddle point**

  iv. $f_{xx}f_{yy} - f_{xy}^2 = 0$ at (a, b) $\Rightarrow$ **test is inconclusive**.

EXERCISES 12.7

Find all the local maxima, local minima, and saddle points of the functions in Exercises 1–30.

1. $f(x, y) = x^2 + xy + y^2 + 3x - 3y + 4$
2. $f(x, y) = x^2 + 3xy + 3y^2 - 6x + 3y - 6$
3. $f(x, y) = 2xy - 5x^2 - 2y^2 + 4x + 4y - 4$
4. $f(x, y) = 2xy - 5x^2 - 2y^2 + 4x - 4$
5. $f(x, y) = x^2 + xy + 3x + 2y + 5$
6. $f(x, y) = y^2 + xy - 2x - 2y + 2$
7. $f(x, y) = 5xy - 7x^2 + 3x - 6y + 2$
8. $f(x, y) = 2xy - x^2 - 2y^2 + 3x + 4$
9. $f(x, y) = x^2 - 4xy + y^2 + 6y + 2$
10. $f(x, y) = 3x^2 + 6xy + 7y^2 - 2x + 4y$
11. $f(x, y) = 2x^2 + 3xy + 4y^2 - 5x + 2y$
12. $f(x, y) = 4x^2 - 6xy + 5y^2 - 20x + 26y$
13. $f(x, y) = x^2 - y^2 - 2x + 4y + 6$
14. $f(x, y) = x^2 - 2xy + 2y^2 - 2x + 2y + 1$
15. $f(x, y) = x^2 + 2xy$
16. $f(x, y) = 3 + 2x + 2y - 2x^2 - 2xy - y^2$
17. $f(x, y) = x^3 - y^3 - 2xy + 6$
18. $f(x, y) = x^3 + 3xy + y^3$
19. $f(x, y) = 6x^2 - 2x^3 + 3y^2 + 6xy$
20. $f(x, y) = 3y^2 - 2y^3 - 3x^2 + 6xy$
21. $f(x, y) = 9x^3 + y^3/3 - 4xy$
22. $f(x, y) = 8x^3 + y^3 + 6xy$
23. $f(x, y) = x^3 + y^3 + 3x^2 - 3y^2 - 8$
24. $f(x, y) = 2x^3 + 2y^3 - 9x^2 + 3y^2 - 12y$
25. $f(x, y) = 4xy - x^4 - y^4$
26. $f(x, y) = x^4 + y^4 + 4xy$
27. $f(x, y) = \dfrac{1}{x^2 + y^2 - 1}$
28. $f(x, y) = \dfrac{1}{x} + xy + \dfrac{1}{y}$
29. $f(x, y) = y \sin x$
30. $f(x, y) = e^{2x} \cos y$

In Exercises 31–38, find the absolute maxima and minima of the functions on the given domains.

31. $f(x, y) = 2x^2 - 4x + y^2 - 4y + 1$ on the closed triangular plate bounded by the lines $x = 0$, $y = 2$, $y = 2x$ in the first quadrant
32. $D(x, y) = x^2 - xy + y^2 + 1$ on the closed triangular plate in the first quadrant bounded by the lines $x = 0, y = 4, y = x$
33. $f(x, y) = x^2 + y^2$ on the closed triangular plate bounded by the lines $x = 0, y = 0, y + 2x = 2$ in the first quadrant
34. $T(x, y) = x^2 + xy + y^2 - 6x$ on the rectangular plate $0 \le x \le 5, -3 \le y \le 3$

35. $T(x, y) = x^2 + xy + y^2 - 6x + 2$ on the rectangular plate $0 \le x \le 5, -3 \le y \le 0$
36. $f(x, y) = 48xy - 32x^3 - 24y^2$ on the rectangular plate $0 \le x \le 1, 0 \le y \le 1$
37. $f(x, y) = (4x - x^2) \cos y$ on the rectangular plate $1 \le x \le 3$, $-\pi/4 \le y \le \pi/4$ (see accompanying figure).

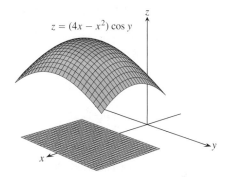

$z = (4x - x^2) \cos y$

38. $f(x, y) = 4x - 8xy + 2y + 1$ on the triangular plate bounded by the lines $x = 0, y = 0, x + y = 1$ in the first quadrant
39. Find two numbers a and b with $a \le b$ such that

$$\int_a^b (6 - x - x^2) \, dx$$

has its largest value.

40. Find two numbers a and b with $a \le b$ such that

$$\int_a^b (24 - 2x - x^2)^{1/3} \, dx$$

has its largest value.

41. Find the maxima, minima, and saddle points of $f(x, y)$, if any, given that

   a. $f_x = 2x - 4y$ and $f_y = 2y - 4x$
   b. $f_x = 2x - 2$ and $f_y = 2y - 4$
   c. $f_x = 9x^2 - 9$ and $f_y = 2y + 4$

   Describe your reasoning in each case.

42. The discriminant $f_{xx}f_{yy} - f_{xy}^2$ is zero at the origin for each of the following functions, so the Second Derivative Test fails there. Determine whether the function has a maximum, a minimum, or neither at the origin by imagining what the surface $z = f(x, y)$ looks like. Describe your reasoning in each case.

   a. $f(x, y) = x^2 y^2$ b. $f(x, y) = 1 - x^2 y^2$
   c. $f(x, y) = xy^2$ d. $f(x, y) = x^3 y^2$
   e. $f(x, y) = x^3 y^3$ f. $f(x, y) = x^4 y^4$

43. Show that $(0, 0)$ is a critical point of $f(x, y) = x^2 + kxy + y^2$ no matter what value the constant k has. (*Hint:* Consider two cases: $k = 0$ and $k \neq 0$.)

44. For what values of the constant k does the Second Derivative Test guarantee that $f(x, y) = x^2 + kxy + y^2$ will have a saddle point at $(0, 0)$? A local minimum at $(0, 0)$? For what values of k is the Second Derivative Test inconclusive? Give reasons for your answers.

45. If $f_x(a, b) = f_y(a, b) = 0$, must f have a local maximum or minimum value at (a, b)? Give reasons for your answer.

46. Can you conclude anything about $f(a, b)$ if f and its first and second partial derivatives are continuous throughout a disk centered at the critical point (a, b) and $f_{xx}(a, b)$ and $f_{yy}(a, b)$ differ in sign? Give reasons for your answer.

47. Among all the points on the graph of $z = 10 - x^2 - y^2$ that lie above the plane $x + 2y + 3z = 0$, find the point farthest from the plane.

48. Find the point on the graph of $z = x^2 + y^2 + 10$ nearest the plane $x + 2y - z = 0$.

49. The function $f(x, y) = x + y$ fails to have an absolute maximum value in the closed first quadrant $x \geq 0$ and $y \geq 0$. Does this contradict the discussion on finding absolute extrema given in the text? Give reasons for your answer.

50. Consider the function $f(x, y) = x^2 + y^2 + 2xy - x - y + 1$ over the square $0 \leq x \leq 1$ and $0 \leq y \leq 1$.

  a. Show that f has an absolute minimum along the line segment $2x + 2y = 1$ in this square. What *is* the absolute minimum value?

  b. Find the absolute maximum value of f over the square.

Extreme Values on Parametrized Curves To find the extreme values of a function $f(x, y)$ on a curve $x = x(t), y = y(t)$, we treat f as a function of the single variable t and use the Chain Rule to find where df/dt is zero. As in any other single-variable case, the extreme values of f are then found among the values at the

  a. critical points (points where df/dt is zero or fails to exist), and

  b. endpoints of the parameter domain.

Find the absolute maximum and minimum values of the following functions on the given curves.

51. Functions:

  a. $f(x, y) = x + y$ **b.** $g(x, y) = xy$

  c. $h(x, y) = 2x^2 + y^2$

  Curves:

  i. The semicircle $x^2 + y^2 = 4, \quad y \geq 0$

  ii. The quarter circle $x^2 + y^2 = 4, \quad x \geq 0, \quad y \geq 0$

  Use the parametric equations $x = 2 \cos t, y = 2 \sin t$.

52. Functions:

  a. $f(x, y) = 2x + 3y$ **b.** $g(x, y) = xy$

  c. $h(x, y) = x^2 + 3y^2$

  Curves:

  i. The semi-ellipse $(x^2/9) + (y^2/4) = 1, \quad y \geq 0$

  ii. The quarter ellipse $(x^2/9) + (y^2/4) = 1, \quad x \geq 0, \quad y \geq 0$

  Use the parametric equations $x = 3 \cos t, y = 2 \sin t$.

53. Function: $f(x, y) = xy$

  Curves:

  i. The line $x = 2t, \quad y = t + 1$

  ii. The line segment $x = 2t, \quad y = t + 1, \quad -1 \leq t \leq 0$

  iii. The line segment $x = 2t, \quad y = t + 1, \quad 0 \leq t \leq 1$

54. Functions:

  a. $f(x, y) = x^2 + y^2$

  b. $g(x, y) = 1/(x^2 + y^2)$

  Curves:

  i. The line $x = t, \quad y = 2 - 2t$

  ii. The line segment $x = t, \quad y = 2 - 2t, \quad 0 \leq t \leq 1$

55. Least Squares and Regression Lines When we try to fit a line $y = mx + b$ to a set of numerical data points $(x_1, y_1),$ $(x_2, y_2), \ldots, (x_n, y_n)$ (Figure 12.46), we usually choose the line that minimizes the sum of the squares of the vertical distances from the points to the line. In theory, this means finding the values of m and b that minimize the value of the function

$$w = (mx_1 + b - y_1)^2 + \cdots + (mx_n + b - y_n)^2. \quad (1)$$

Show that the values of m and b that do this are

$$m = \frac{\left(\sum x_k\right)\left(\sum y_k\right) - n \sum x_k y_k}{\left(\sum x_k\right)^2 - n \sum x_k^2}, \quad (2)$$

$$b = \frac{1}{n}\left(\sum y_k - m \sum x_k\right), \quad (3)$$

with all sums running from $k = 1$ to $k = n$. Many scientific calculators have these formulas built in, enabling you to find m and b with only a few key strokes after you have entered the data.

The line $y = mx + b$ determined by these values of m and b is called the **least squares line, regression line,** or **trend line** for the data under study. Finding a least squares line lets you

1. summarize data with a simple expression,

2. predict values of y for other, experimentally untried values of x,

3. handle data analytically.

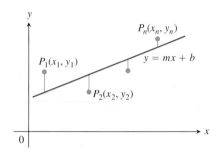

FIGURE 12.46 To fit a line to noncollinear points, we choose the line that minimizes the sum of the squares of the deviations.

In Exercises 56–58, use Equations (2) and (3) to find the least squares line for each set of data points. Then use the linear equation you obtain to predict the value of y that would correspond to $x = 4$.

56. $(-2, 0)$, $(0, 2)$, $(2, 3)$ **57.** $(-1, 2)$, $(0, 1)$, $(3, -4)$

58. $(0, 0)$, $(1, 2)$, $(2, 3)$

COMPUTER EXPLORATIONS

In Exercises 59–64, you will explore functions to identify their local extrema. Use a CAS to perform the following steps:

a. Plot the function over the given rectangle.

b. Plot some level curves in the rectangle.

c. Calculate the function's first partial derivatives and use the CAS equation solver to find the critical points. How do the critical points relate to the level curves plotted in part (b)? Which critical points, if any, appear to give a saddle point? Give reasons for your answer.

d. Calculate the function's second partial derivatives and find the discriminant $f_{xx}f_{yy} - f_{xy}^2$.

e. Using the max-min tests, classify the critical points found in part (c). Are your findings consistent with your discussion in part (c)?

59. $f(x, y) = x^2 + y^3 - 3xy$, $-5 \le x \le 5$, $-5 \le y \le 5$

60. $f(x, y) = x^3 - 3xy^2 + y^2$, $-2 \le x \le 2$, $-2 \le y \le 2$

61. $f(x, y) = x^4 + y^2 - 8x^2 - 6y + 16$, $-3 \le x \le 3$, $-6 \le y \le 6$

62. $f(x, y) = 2x^4 + y^4 - 2x^2 - 2y^2 + 3$, $-3/2 \le x \le 3/2$, $-3/2 \le y \le 3/2$

63. $f(x, y) = 5x^6 + 18x^5 - 30x^4 + 30xy^2 - 120x^3$, $-4 \le x \le 3$, $-2 \le y \le 2$

64. $f(x, y) = \begin{cases} x^5 \ln (x^2 + y^2), & (x, y) \ne (0, 0) \\ 0, & (x, y) = (0, 0) \end{cases}$, $-2 \le x \le 2$, $-2 \le y \le 2$

12.8 Lagrange Multipliers

HISTORICAL BIOGRAPHY

Joseph Louis Lagrange
(1736–1813)

Sometimes we need to find the extreme values of a function whose domain is constrained to lie within some particular subset of the plane—a disk, for example, a closed triangular region, or along a curve. In this section, we explore a powerful method for finding extreme values of constrained functions: the method of *Lagrange multipliers*.

Constrained Maxima and Minima

We first consider a problem where a constrained minimum can be found by eliminating a variable.

EXAMPLE 1   Find the point $P(x, y, z)$ closest to the origin on the plane $2x + y - z - 5 = 0$.

Solution   The problem asks us to find the minimum value of the function

$$|\overrightarrow{OP}| = \sqrt{(x - 0)^2 + (y - 0)^2 + (z - 0)^2}$$
$$= \sqrt{x^2 + y^2 + z^2}$$

subject to the constraint that

$$2x + y - z - 5 = 0.$$

Since $|\overrightarrow{OP}|$ has a minimum value wherever the function

$$f(x, y, z) = x^2 + y^2 + z^2$$

has a minimum value, we may solve the problem by finding the minimum value of $f(x, y, z)$ subject to the constraint $2x + y - z - 5 = 0$ (thus avoiding square roots). If we regard x and y as the independent variables in this equation and write z as

$$z = 2x + y - 5,$$

our problem reduces to one of finding the points (x, y) at which the function

$$h(x, y) = f(x, y, 2x + y - 5) = x^2 + y^2 + (2x + y - 5)^2$$

has its minimum value or values. Since the domain of h is the entire xy-plane, the First Derivative Test of Section 12.7 tells us that any minima that h might have must occur at points where

$$h_x = 2x + 2(2x + y - 5)(2) = 0, \qquad h_y = 2y + 2(2x + y - 5) = 0.$$

This leads to

$$10x + 4y = 20, \qquad 4x + 4y = 10,$$

and the solution

$$x = \frac{5}{3}, \qquad y = \frac{5}{6}.$$

We may apply a geometric argument together with the Second Derivative Test to show that these values minimize h. The z-coordinate of the corresponding point on the plane $z = 2x + y - 5$ is

$$z = 2\left(\frac{5}{3}\right) + \frac{5}{6} - 5 = -\frac{5}{6}.$$

Therefore, the point we seek is

Closest point: $\qquad P\left(\frac{5}{3}, \frac{5}{6}, -\frac{5}{6}\right).$

The distance from P to the origin is $5/\sqrt{6} \approx 2.04$.   ∎

Attempts to solve a constrained maximum or minimum problem by substitution, as we might call the method of Example 1, do not always go smoothly. This is one of the reasons for learning the new method of this section.

EXAMPLE 2   Find the points closest to the origin on the hyperbolic cylinder $x^2 - z^2 - 1 = 0$.

Solution 1   The cylinder is shown in Figure 12.47. We seek the points on the cylinder closest to the origin. These are the points whose coordinates minimize the value of the function

$$f(x, y, z) = x^2 + y^2 + z^2 \qquad \text{Square of the distance}$$

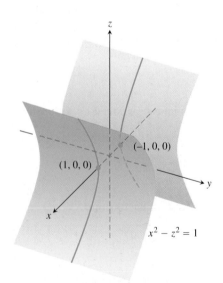

($-1, 0, 0$)

($1, 0, 0$)

$x^2 - z^2 = 1$

FIGURE 12.47   The hyperbolic cylinder $x^2 - z^2 - 1 = 0$ in Example 2.

The hyperbolic cylinder $x^2 - z^2 = 1$

On this part,
$x = \sqrt{z^2 + 1}$

On this part,
$x = -\sqrt{z^2 + 1}$

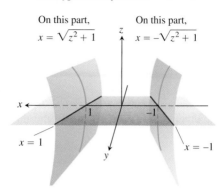

$x = 1$ $x = -1$

FIGURE 12.48 The region in the xy-plane from which the first two coordinates of the points (x, y, z) on the hyperbolic cylinder $x^2 - z^2 = 1$ are selected excludes the band $-1 < x < 1$ in the xy-plane (Example 2).

subject to the constraint that $x^2 - z^2 - 1 = 0$. If we regard x and y as independent variables in the constraint equation, then

$$z^2 = x^2 - 1$$

and the values of $f(x, y, z) = x^2 + y^2 + z^2$ on the cylinder are given by the function

$$h(x, y) = x^2 + y^2 + (x^2 - 1) = 2x^2 + y^2 - 1.$$

To find the points on the cylinder whose coordinates minimize f, we look for the points in the xy-plane whose coordinates minimize h. The only extreme value of h occurs where

$$h_x = 4x = 0 \qquad \text{and} \qquad h_y = 2y = 0,$$

that is, at the point $(0, 0)$. But there are no points on the cylinder where both x and y are zero. What went wrong?

What happened was that the First Derivative Test found (as it should have) the point *in the domain of h* where h has a minimum value. We, on the other hand, want the points *on the cylinder* where h has a minimum value. Although the domain of h is the entire xy-plane, the domain from which we can select the first two coordinates of the points (x, y, z) on the cylinder is restricted to the "shadow" of the cylinder on the xy-plane; it does not include the band between the lines $x = -1$ and $x = 1$ (Figure 12.48).

We can avoid this problem if we treat y and z as independent variables (instead of x and y) and express x in terms of y and z as

$$x^2 = z^2 + 1.$$

With this substitution, $f(x, y, z) = x^2 + y^2 + z^2$ becomes

$$k(y, z) = (z^2 + 1) + y^2 + z^2 = 1 + y^2 + 2z^2$$

and we look for the points where k takes on its smallest value. The domain of k in the yz-plane now matches the domain from which we select the y- and z-coordinates of the points (x, y, z) on the cylinder. Hence, the points that minimize k in the plane will have corresponding points on the cylinder. The smallest values of k occur where

$$k_y = 2y = 0 \qquad \text{and} \qquad k_z = 4z = 0,$$

or where $y = z = 0$. This leads to

$$x^2 = z^2 + 1 = 1, \qquad x = \pm 1.$$

The corresponding points on the cylinder are $(\pm 1, 0, 0)$. We can see from the inequality

$$k(y, z) = 1 + y^2 + 2z^2 \geq 1$$

that the points $(\pm 1, 0, 0)$ give a minimum value for k. We can also see that the minimum distance from the origin to a point on the cylinder is 1 unit.

Solution 2 Another way to find the points on the cylinder closest to the origin is to imagine a small sphere centered at the origin expanding like a soap bubble until it just touches the cylinder (Figure 12.49). At each point of contact, the cylinder and sphere have the same tangent plane and normal line. Therefore, if the sphere and cylinder are represented as the level surfaces obtained by setting

$$f(x, y, z) = x^2 + y^2 + z^2 - a^2 \qquad \text{and} \qquad g(x, y, z) = x^2 - z^2 - 1$$

$x^2 - z^2 - 1 = 0$

$x^2 + y^2 + z^2 - a^2 = 0$

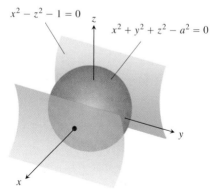

FIGURE 12.49 A sphere expanding like a soap bubble centered at the origin until it just touches the hyperbolic cylinder $x^2 - z^2 - 1 = 0$ (Example 2).

equal to 0, then the gradients ∇f and ∇g will be parallel where the surfaces touch. At any point of contact, we should therefore be able to find a scalar λ ("lambda") such that

$$\nabla f = \lambda \nabla g,$$

or

$$2x\mathbf{i} + 2y\mathbf{j} + 2z\mathbf{k} = \lambda(2x\mathbf{i} - 2z\mathbf{k}).$$

Thus, the coordinates x, y, and z of any point of tangency will have to satisfy the three scalar equations

$$2x = 2\lambda x, \qquad 2y = 0, \qquad 2z = -2\lambda z.$$

For what values of λ will a point (x, y, z) whose coordinates satisfy these scalar equations also lie on the surface $x^2 - z^2 - 1 = 0$? To answer this question, we use our knowledge that no point on the surface has a zero x-coordinate to conclude that $x \neq 0$. Hence, $2x = 2\lambda x$ only if

$$2 = 2\lambda, \qquad \text{or} \qquad \lambda = 1.$$

For $\lambda = 1$, the equation $2z = -2\lambda z$ becomes $2z = -2z$. If this equation is to be satisfied as well, z must be zero. Since $y = 0$ also (from the equation $2y = 0$), we conclude that the points we seek all have coordinates of the form

$$(x, 0, 0).$$

What points on the surface $x^2 - z^2 = 1$ have coordinates of this form? The answer is the points $(x, 0, 0)$ for which

$$x^2 - (0)^2 = 1, \qquad x^2 = 1, \qquad \text{or} \qquad x = \pm 1.$$

The points on the cylinder closest to the origin are the points $(\pm 1, 0, 0)$. ∎

The Method of Lagrange Multipliers

In Solution 2 of Example 2, we used the **method of Lagrange multipliers**. The method says that the extreme values of a function $f(x, y, z)$ whose variables are subject to a constraint $g(x, y, z) = 0$ are to be found on the surface $g = 0$ among the points where

$$\nabla f = \lambda \nabla g$$

for some scalar λ (called a **Lagrange multiplier**).

To explore the method further and see why it works, we first make the following observation, which we state as a theorem.

THEOREM 12—The Orthogonal Gradient Theorem   Suppose that $f(x, y, z)$ is differentiable in a region whose interior contains a smooth curve

$$C: \quad \mathbf{r}(t) = g(t)\mathbf{i} + h(t)\mathbf{j} + k(t)\mathbf{k}.$$

If P_0 is a point on C where f has a local maximum or minimum relative to its values on C, then ∇f is orthogonal to C at P_0.

Proof We show that ∇f is orthogonal to the curve's velocity vector at P_0. The values of f on C are given by the composite $f(g(t), h(t), k(t))$, whose derivative with respect to t is

$$\frac{df}{dt} = \frac{\partial f}{\partial x}\frac{dg}{dt} + \frac{\partial f}{\partial y}\frac{dh}{dt} + \frac{\partial f}{\partial z}\frac{dk}{dt} = \nabla f \cdot \mathbf{v}.$$

At any point P_0 where f has a local maximum or minimum relative to its values on the curve, $df/dt = 0$, so

$$\nabla f \cdot \mathbf{v} = 0. \qquad\blacksquare$$

By dropping the z-terms in Theorem 12, we obtain a similar result for functions of two variables.

COROLLARY OF THEOREM 12 At the points on a smooth curve $\mathbf{r}(t) = g(t)\mathbf{i} + h(t)\mathbf{j}$ where a differentiable function $f(x, y)$ takes on its local maxima and minima relative to its values on the curve, $\nabla f \cdot \mathbf{v} = 0$, where $\mathbf{v} = d\mathbf{r}/dt$.

Theorem 12 is the key to the method of Lagrange multipliers. Suppose that $f(x, y, z)$ and $g(x, y, z)$ are differentiable and that P_0 is a point on the surface $g(x, y, z) = 0$ where f has a local maximum or minimum value relative to its other values on the surface. We assume also that $\nabla g \neq \mathbf{0}$ at points on the surface $g(x, y, z) = 0$. Then f takes on a local maximum or minimum at P_0 relative to its values on every differentiable curve through P_0 on the surface $g(x, y, z) = 0$. Therefore, ∇f is orthogonal to the velocity vector of every such differentiable curve through P_0. So is ∇g, moreover (because ∇g is orthogonal to the level surface $g = 0$, as we saw in Section 12.5). Therefore, at P_0, ∇f is some scalar multiple λ of ∇g.

The Method of Lagrange Multipliers

Suppose that $f(x, y, z)$ and $g(x, y, z)$ are differentiable and $\nabla g \neq \mathbf{0}$ when $g(x, y, z) = 0$. To find the local maximum and minimum values of f subject to the constraint $g(x, y, z) = 0$ (if these exist), find the values of x, y, z, and λ that simultaneously satisfy the equations

$$\nabla f = \lambda \nabla g \qquad \text{and} \qquad g(x, y, z) = 0. \qquad (1)$$

For functions of two independent variables, the condition is similar, but without the variable z.

Some care must be used in applying this method. An extreme value may not actually exist (Exercise 41).

EXAMPLE 3 Find the greatest and smallest values that the function

$$f(x, y) = xy$$

takes on the ellipse (Figure 12.50)

$$\frac{x^2}{8} + \frac{y^2}{2} = 1.$$

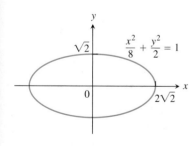

FIGURE 12.50 Example 3 shows how to find the largest and smallest values of the product xy on this ellipse.

Solution   We want the extreme values of $f(x, y) = xy$ subject to the constraint

$$g(x, y) = \frac{x^2}{8} + \frac{y^2}{2} - 1 = 0.$$

To do so, we first find the values of x, y, and λ for which

$$\nabla f = \lambda \nabla g \quad \text{and} \quad g(x, y) = 0.$$

The gradient equation in Equations (1) gives

$$y\mathbf{i} + x\mathbf{j} = \frac{\lambda}{4} x\mathbf{i} + \lambda y\mathbf{j},$$

from which we find

$$y = \frac{\lambda}{4} x, \quad x = \lambda y, \quad \text{and} \quad y = \frac{\lambda}{4} (\lambda y) = \frac{\lambda^2}{4} y,$$

so that $y = 0$ or $\lambda = \pm 2$. We now consider these two cases.

Case 1: If $y = 0$, then $x = y = 0$. But $(0, 0)$ is not on the ellipse. Hence, $y \neq 0$.
Case 2: If $y \neq 0$, then $\lambda = \pm 2$ and $x = \pm 2y$. Substituting this in the equation $g(x, y) = 0$ gives

$$\frac{(\pm 2y)^2}{8} + \frac{y^2}{2} = 1, \quad 4y^2 + 4y^2 = 8 \quad \text{and} \quad y = \pm 1.$$

The function $f(x, y) = xy$ therefore takes on its extreme values on the ellipse at the four points $(\pm 2, 1)$, $(\pm 2, -1)$. The extreme values are $xy = 2$ and $xy = -2$.

The Geometry of the Solution   The level curves of the function $f(x, y) = xy$ are the hyperbolas $xy = c$ (Figure 12.51). The farther the hyperbolas lie from the origin, the larger the absolute value of f. We want to find the extreme values of $f(x, y)$, given that the point (x, y) also lies on the ellipse $x^2 + 4y^2 = 8$. Which hyperbolas intersecting the ellipse lie farthest from the origin? The hyperbolas that just graze the ellipse, the ones that are tangent to it, are farthest. At these points, any vector normal to the hyperbola is normal to the ellipse, so $\nabla f = y\mathbf{i} + x\mathbf{j}$ is a multiple ($\lambda = \pm 2$) of $\nabla g = (x/4)\mathbf{i} + y\mathbf{j}$. At the point $(2, 1)$, for example,

$$\nabla f = \mathbf{i} + 2\mathbf{j}, \quad \nabla g = \frac{1}{2}\mathbf{i} + \mathbf{j}, \quad \text{and} \quad \nabla f = 2\nabla g.$$

At the point $(-2, 1)$,

$$\nabla f = \mathbf{i} - 2\mathbf{j}, \quad \nabla g = -\frac{1}{2}\mathbf{i} + \mathbf{j}, \quad \text{and} \quad \nabla f = -2\nabla g. \qquad \blacksquare$$

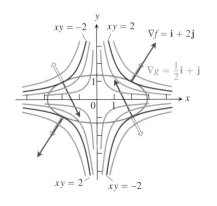

FIGURE 12.51   When subjected to the constraint $g(x, y) = x^2/8 + y^2/2 - 1 = 0$, the function $f(x, y) = xy$ takes on extreme values at the four points $(\pm 2, \pm 1)$. These are the points on the ellipse when ∇f (red) is a scalar multiple of ∇g (blue) (Example 3).

EXAMPLE 4   Find the maximum and minimum values of the function $f(x, y) = 3x + 4y$ on the circle $x^2 + y^2 = 1$.

Solution   We model this as a Lagrange multiplier problem with

$$f(x, y) = 3x + 4y, \quad g(x, y) = x^2 + y^2 - 1$$

and look for the values of x, y, and λ that satisfy the equations

$$\nabla f = \lambda \nabla g: \quad 3\mathbf{i} + 4\mathbf{j} = 2x\lambda\mathbf{i} + 2y\lambda\mathbf{j}$$
$$g(x, y) = 0: \quad x^2 + y^2 - 1 = 0.$$

The gradient equation in Equations (1) implies that $\lambda \neq 0$ and gives

$$x = \frac{3}{2\lambda}, \quad y = \frac{2}{\lambda}.$$

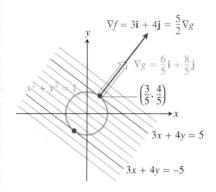

FIGURE 12.52 The function $f(x, y) = 3x + 4y$ takes on its largest value on the unit circle $g(x, y) = x^2 + y^2 - 1 = 0$ at the point $(3/5, 4/5)$ and its smallest value at the point $(-3/5, -4/5)$ (Example 4). At each of these points, ∇f is a scalar multiple of ∇g. The figure shows the gradients at the first point but not the second.

These equations tell us, among other things, that x and y have the same sign. With these values for x and y, the equation $g(x, y) = 0$ gives

$$\left(\frac{3}{2\lambda}\right)^2 + \left(\frac{2}{\lambda}\right)^2 - 1 = 0,$$

so

$$\frac{9}{4\lambda^2} + \frac{4}{\lambda^2} = 1, \qquad 9 + 16 = 4\lambda^2, \qquad 4\lambda^2 = 25, \qquad \text{and} \qquad \lambda = \pm\frac{5}{2}.$$

Thus,

$$x = \frac{3}{2\lambda} = \pm\frac{3}{5}, \qquad y = \frac{2}{\lambda} = \pm\frac{4}{5},$$

and $f(x, y) = 3x + 4y$ has extreme values at $(x, y) = \pm(3/5, 4/5)$.

By calculating the value of $3x + 4y$ at the points $\pm(3/5, 4/5)$, we see that its maximum and minimum values on the circle $x^2 + y^2 = 1$ are

$$3\left(\frac{3}{5}\right) + 4\left(\frac{4}{5}\right) = \frac{25}{5} = 5 \qquad \text{and} \qquad 3\left(-\frac{3}{5}\right) + 4\left(-\frac{4}{5}\right) = -\frac{25}{5} = -5.$$

The Geometry of the Solution The level curves of $f(x, y) = 3x + 4y$ are the lines $3x + 4y = c$ (Figure 12.52). The farther the lines lie from the origin, the larger the absolute value of f. We want to find the extreme values of $f(x, y)$ given that the point (x, y) also lies on the circle $x^2 + y^2 = 1$. Which lines intersecting the circle lie farthest from the origin? The lines tangent to the circle are farthest. At the points of tangency, any vector normal to the line is normal to the circle, so the gradient $\nabla f = 3\mathbf{i} + 4\mathbf{j}$ is a multiple ($\lambda = \pm5/2$) of the gradient $\nabla g = 2x\mathbf{i} + 2y\mathbf{j}$. At the point $(3/5, 4/5)$, for example,

$$\nabla f = 3\mathbf{i} + 4\mathbf{j}, \qquad \nabla g = \frac{6}{5}\mathbf{i} + \frac{8}{5}\mathbf{j}, \qquad \text{and} \qquad \nabla f = \frac{5}{2}\nabla g. \qquad \blacksquare$$

Lagrange Multipliers with Two Constraints

Many problems require us to find the extreme values of a differentiable function $f(x, y, z)$ whose variables are subject to two constraints. If the constraints are

$$g_1(x, y, z) = 0 \qquad \text{and} \qquad g_2(x, y, z) = 0$$

and g_1 and g_2 are differentiable, with ∇g_1 not parallel to ∇g_2, we find the constrained local maxima and minima of f by introducing two Lagrange multipliers λ and μ (mu, pronounced "mew"). That is, we locate the points $P(x, y, z)$ where f takes on its constrained extreme values by finding the values of $x, y, z, \lambda,$ and μ that simultaneously satisfy the equations

$$\nabla f = \lambda \nabla g_1 + \mu \nabla g_2, \qquad g_1(x, y, z) = 0, \qquad g_2(x, y, z) = 0 \qquad (2)$$

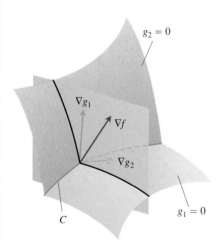

FIGURE 12.53 The vectors ∇g_1 and ∇g_2 lie in a plane perpendicular to the curve C because ∇g_1 is normal to the surface $g_1 = 0$ and ∇g_2 is normal to the surface $g_2 = 0$.

Equations (2) have a nice geometric interpretation. The surfaces $g_1 = 0$ and $g_2 = 0$ (usually) intersect in a smooth curve, say C (Figure 12.53). Along this curve we seek the points where f has local maximum and minimum values relative to its other values on the curve. These are the points where ∇f is normal to C, as we saw in Theorem 12. But ∇g_1 and ∇g_2

are also normal to C at these points because C lies in the surfaces $g_1 = 0$ and $g_2 = 0$. Therefore, ∇f lies in the plane determined by ∇g_1 and ∇g_2, which means that $\nabla f = \lambda \nabla g_1 + \mu \nabla g_2$ for some λ and μ. Since the points we seek also lie in both surfaces, their coordinates must satisfy the equations $g_1(x, y, z) = 0$ and $g_2(x, y, z) = 0$, which are the remaining requirements in Equations (2).

EXAMPLE 5   The plane $x + y + z = 1$ cuts the cylinder $x^2 + y^2 = 1$ in an ellipse (Figure 12.54). Find the points on the ellipse that lie closest to and farthest from the origin.

Solution   We find the extreme values of

$$f(x, y, z) = x^2 + y^2 + z^2$$

(the square of the distance from (x, y, z) to the origin) subject to the constraints

$$g_1(x, y, z) = x^2 + y^2 - 1 = 0 \tag{3}$$

$$g_2(x, y, z) = x + y + z - 1 = 0. \tag{4}$$

The gradient equation in Equations (2) then gives

$$\nabla f = \lambda \nabla g_1 + \mu \nabla g_2$$
$$2x\mathbf{i} + 2y\mathbf{j} + 2z\mathbf{k} = \lambda(2x\mathbf{i} + 2y\mathbf{j}) + \mu(\mathbf{i} + \mathbf{j} + \mathbf{k})$$
$$2x\mathbf{i} + 2y\mathbf{j} + 2z\mathbf{k} = (2\lambda x + \mu)\mathbf{i} + (2\lambda y + \mu)\mathbf{j} + \mu\mathbf{k}$$

or

$$2x = 2\lambda x + \mu, \qquad 2y = 2\lambda y + \mu, \qquad 2z = \mu. \tag{5}$$

The scalar equations in Equations (5) yield

$$2x = 2\lambda x + 2z \Rightarrow (1 - \lambda)x = z, \tag{6}$$
$$2y = 2\lambda y + 2z \Rightarrow (1 - \lambda)y = z.$$

Equations (6) are satisfied simultaneously if either $\lambda = 1$ and $z = 0$ or $\lambda \neq 1$ and $x = y = z/(1 - \lambda)$.

If $z = 0$, then solving Equations (3) and (4) simultaneously to find the corresponding points on the ellipse gives the two points $(1, 0, 0)$ and $(0, 1, 0)$. This makes sense when you look at Figure 12.54.

If $x = y$, then Equations (3) and (4) give

$$x^2 + x^2 - 1 = 0 \qquad\qquad x + x + z - 1 = 0$$
$$2x^2 = 1 \qquad\qquad\qquad z = 1 - 2x$$
$$x = \pm\frac{\sqrt{2}}{2} \qquad\qquad\qquad z = 1 \mp \sqrt{2}.$$

The corresponding points on the ellipse are

$$P_1 = \left(\frac{\sqrt{2}}{2}, \frac{\sqrt{2}}{2}, 1 - \sqrt{2}\right) \qquad \text{and} \qquad P_2 = \left(-\frac{\sqrt{2}}{2}, -\frac{\sqrt{2}}{2}, 1 + \sqrt{2}\right).$$

Here we need to be careful, however. Although P_1 and P_2 both give local maxima of f on the ellipse, P_2 is farther from the origin than P_1.

The points on the ellipse closest to the origin are $(1, 0, 0)$ and $(0, 1, 0)$. The point on the ellipse farthest from the origin is P_2.   ∎

FIGURE 12.54   On the ellipse where the plane and cylinder meet, we find the points closest to and farthest from the origin. (Example 5).

EXERCISES 12.8

1. **Extrema on an ellipse** Find the points on the ellipse $x^2 + 2y^2 = 1$ where $f(x, y) = xy$ has its extreme values.

2. **Extrema on a circle** Find the extreme values of $f(x, y) = xy$ subject to the constraint $g(x, y) = x^2 + y^2 - 10 = 0$.

3. **Maximum on a line** Find the maximum value of $f(x, y) = 49 - x^2 - y^2$ on the line $x + 3y = 10$.

4. **Extrema on a line** Find the local extreme values of $f(x, y) = x^2 y$ on the line $x + y = 3$.

5. **Constrained minimum** Find the points on the curve $xy^2 = 54$ nearest the origin.

6. **Constrained minimum** Find the points on the curve $x^2 y = 2$ nearest the origin.

7. Use the method of Lagrange multipliers to find

   a. **Minimum on a hyperbola** The minimum value of $x + y$, subject to the constraints $xy = 16$, $x > 0$, $y > 0$

   b. **Maximum on a line** The maximum value of xy, subject to the constraint $x + y = 16$.

   Comment on the geometry of each solution.

8. **Extrema on a curve** Find the points on the curve $x^2 + xy + y^2 = 1$ in the xy-plane that are nearest to and farthest from the origin.

9. **Minimum surface area with fixed volume** Find the dimensions of the closed right circular cylindrical can of smallest surface area whose volume is 16π cm^3.

10. **Cylinder in a sphere** Find the radius and height of the open right circular cylinder of largest surface area that can be inscribed in a sphere of radius a. What *is* the largest surface area?

11. **Rectangle of greatest area in an ellipse** Use the method of Lagrange multipliers to find the dimensions of the rectangle of greatest area that can be inscribed in the ellipse $x^2/16 + y^2/9 = 1$ with sides parallel to the coordinate axes.

12. **Rectangle of longest perimeter in an ellipse** Find the dimensions of the rectangle of largest perimeter that can be inscribed in the ellipse $x^2/a^2 + y^2/b^2 = 1$ with sides parallel to the coordinate axes. What *is* the largest perimeter?

13. **Extrema on a circle** Find the maximum and minimum values of $x^2 + y^2$ subject to the constraint $x^2 - 2x + y^2 - 4y = 0$.

14. **Extrema on a circle** Find the maximum and minimum values of $3x - y + 6$ subject to the constraint $x^2 + y^2 = 4$.

15. **Ant on a metal plate** The temperature at a point (x, y) on a metal plate is $T(x, y) = 4x^2 - 4xy + y^2$. An ant on the plate walks around the circle of radius 5 centered at the origin. What are the highest and lowest temperatures encountered by the ant?

16. **Cheapest storage tank** Your firm has been asked to design a storage tank for liquid petroleum gas. The customer's specifications call for a cylindrical tank with hemispherical ends, and the

tank is to hold 8000 m^3 of gas. The customer also wants to use the smallest amount of material possible in building the tank. What radius and height do you recommend for the cylindrical portion of the tank?

17. **Minimum distance to a point** Find the point on the plane $x + 2y + 3z = 13$ closest to the point $(1, 1, 1)$.

18. **Maximum distance to a point** Find the point on the sphere $x^2 + y^2 + z^2 = 4$ farthest from the point $(1, -1, 1)$.

19. **Minimum distance to the origin** Find the minimum distance from the surface $x^2 - y^2 - z^2 = 1$ to the origin.

20. **Minimum distance to the origin** Find the point on the surface $z = xy + 1$ nearest the origin.

21. **Minimum distance to the origin** Find the points on the surface $z^2 = xy + 4$ closest to the origin.

22. **Minimum distance to the origin** Find the point(s) on the surface $xyz = 1$ closest to the origin.

23. **Extrema on a sphere** Find the maximum and minimum values of

$$f(x, y, z) = x - 2y + 5z$$

on the sphere $x^2 + y^2 + z^2 = 30$.

24. **Extrema on a sphere** Find the points on the sphere $x^2 + y^2 + z^2 = 25$ where $f(x, y, z) = x + 2y + 3z$ has its maximum and minimum values.

25. **Minimizing a sum of squares** Find three real numbers whose sum is 9 and the sum of whose squares is as small as possible.

26. **Maximizing a product** Find the largest product the positive numbers x, y, and z can have if $x + y + z^2 = 16$.

27. **Rectangular box of longest volume in a sphere** Find the dimensions of the closed rectangular box with maximum volume that can be inscribed in the unit sphere.

28. **Box with vertex on a plane** Find the volume of the largest closed rectangular box in the first octant having three faces in the coordinate planes and a vertex on the plane $x/a + y/b + z/c = 1$, where $a > 0$, $b > 0$, and $c > 0$.

29. **Hottest point on a space probe** A space probe in the shape of the ellipsoid

$$4x^2 + y^2 + 4z^2 = 16$$

enters Earth's atmosphere and its surface begins to heat. After 1 hour, the temperature at the point (x, y, z) on the probe's surface is

$$T(x, y, z) = 8x^2 + 4yz - 16z + 600.$$

Find the hottest point on the probe's surface.

30. **Extreme temperatures on a sphere** Suppose that the Celsius temperature at the point (x, y, z) on the sphere $x^2 + y^2 + z^2 = 1$ is $T = 400xyz^2$. Locate the highest and lowest temperatures on the sphere.

31. **Maximizing a utility function: an example from economics**
In economics, the usefulness or *utility* of amounts x and y of two capital goods G_1 and G_2 is sometimes measured by a function $U(x, y)$. For example, G_1 and G_2 might be two chemicals a pharmaceutical company needs to have on hand and $U(x, y)$ the gain from manufacturing a product whose synthesis requires different amounts of the chemicals depending on the process used. If G_1 costs a dollars per kilogram, G_2 costs b dollars per kilogram, and the total amount allocated for the purchase of G_1 and G_2 together is c dollars, then the company's managers want to maximize $U(x, y)$ given that $ax + by = c$. Thus, they need to solve a typical Lagrange multiplier problem.

Suppose that

$$U(x, y) = xy + 2x$$

and that the equation $ax + by = c$ simplifies to

$$2x + y = 30.$$

Find the maximum value of U and the corresponding values of x and y subject to this latter constraint.

32. **Locating a radio telescope** You are in charge of erecting a radio telescope on a newly discovered planet. To minimize interference, you want to place it where the magnetic field of the planet is weakest. The planet is spherical, with a radius of 6 units. Based on a coordinate system whose origin is at the center of the planet, the strength of the magnetic field is given by $M(x, y, z) = 6x - y^2 + xz + 60$. Where should you locate the radio telescope?

33. Maximize the function $f(x, y, z) = x^2 + 2y - z^2$ subject to the constraints $2x - y = 0$ and $y + z = 0$.

34. Minimize the function $f(x, y, z) = x^2 + y^2 + z^2$ subject to the constraints $x + 2y + 3z = 6$ and $x + 3y + 9z = 9$.

35. **Minimum distance to the origin** Find the point closest to the origin on the line of intersection of the planes $y + 2z = 12$ and $x + y = 6$.

36. **Maximum value on line of intersection** Find the maximum value that $f(x, y, z) = x^2 + 2y - z^2$ can have on the line of intersection of the planes $2x - y = 0$ and $y + z = 0$.

37. **Extrema on a curve of intersection** Find the extreme values of $f(x, y, z) = x^2yz + 1$ on the intersection of the plane $z = 1$ with the sphere $x^2 + y^2 + z^2 = 10$.

38. a. **Maximum on line of intersection** Find the maximum value of $w = xyz$ on the line of intersection of the two planes $x + y + z = 40$ and $x + y - z = 0$.

    b. Give a geometric argument to support your claim that you have found a maximum, and not a minimum, value of w.

39. **Extrema on a circle of intersection** Find the extreme values of the function $f(x, y, z) = xy + z^2$ on the circle in which the plane $y - x = 0$ intersects the sphere $x^2 + y^2 + z^2 = 4$.

40. **Minimum distance to the origin** Find the point closest to the origin on the curve of intersection of the plane $2y + 4z = 5$ and the cone $z^2 = 4x^2 + 4y^2$.

41. **The condition $\nabla f = \lambda \nabla g$ is not sufficient** Although $\nabla f = \lambda \nabla g$ is a necessary condition for the occurrence of an extreme value of $f(x, y)$ subject to the conditions $g(x, y) = 0$ and $\nabla g \neq \mathbf{0}$, it does not in itself guarantee that one exists. As a case in point, try using the method of Lagrange multipliers to find a maximum value of $f(x, y) = x + y$ subject to the constraint that $xy = 16$. The method will identify the two points $(4, 4)$ and $(-4, -4)$ as candidates for the location of extreme values. Yet the sum $(x + y)$ has no maximum value on the hyperbola $xy = 16$. The farther you go from the origin on this hyperbola in the first quadrant, the larger the sum $f(x, y) = x + y$ becomes.

42. **A least squares plane** The plane $z = Ax + By + C$ is to be "fitted" to the following points (x_k, y_k, z_k):

    $$(0, 0, 0), \quad (0, 1, 1), \quad (1, 1, 1), \quad (1, 0, -1).$$

    Find the values of A, B, and C that minimize

    $$\sum_{k=1}^{4} (Ax_k + By_k + C - z_k)^2,$$

    the sum of the squares of the deviations.

43. a. **Maximum on a sphere** Show that the maximum value of $a^2b^2c^2$ on a sphere of radius r centered at the origin of a Cartesian abc-coordinate system is $(r^2/3)^3$.

    b. **Geometric and arithmetic means** Using part (a), show that for nonnegative numbers a, b, and c,

    $$(abc)^{1/3} \leq \frac{a + b + c}{3};$$

    that is, the *geometric mean* of three nonnegative numbers is less than or equal to their *arithmetic mean*.

44. **Sum of products** Let $a_1, a_2, \ldots, a_n$ be n positive numbers. Find the maximum of $\sum_{i=1}^{n} a_i x_i$ subject to the constraint $\sum_{i=1}^{n} x_i^2 = 1$.

COMPUTER EXPLORATIONS

In Exercises 45–50, use a CAS to perform the following steps implementing the method of Lagrange multipliers for finding constrained extrema:

a. Form the function $h = f - \lambda_1 g_1 - \lambda_2 g_2$, where f is the function to optimize subject to the constraints $g_1 = 0$ and $g_2 = 0$.

b. Determine all the first partial derivatives of h, including the partials with respect to λ_1 and λ_2, and set them equal to 0.

c. Solve the system of equations found in part (b) for all the unknowns, including λ_1 and λ_2.

d. Evaluate f at each of the solution points found in part (c) and select the extreme value subject to the constraints asked for in the exercise.

45. Minimize $f(x, y, z) = xy + yz$ subject to the constraints $x^2 + y^2 - 2 = 0$ and $x^2 + z^2 - 2 = 0$.

46. Minimize $f(x, y, z) = xyz$ subject to the constraints $x^2 + y^2 - 1 = 0$ and $x - z = 0$.

47. Maximize $f(x, y, z) = x^2 + y^2 + z^2$ subject to the constraints $2y + 4z - 5 = 0$ and $4x^2 + 4y^2 - z^2 = 0$.

48. Minimize $f(x, y, z) = x^2 + y^2 + z^2$ subject to the constraints $x^2 - xy + y^2 - z^2 - 1 = 0$ and $x^2 + y^2 - 1 = 0$.

49. Minimize $f(x, y, z, w) = x^2 + y^2 + z^2 + w^2$ subject to the constraints $2x - y + z - w - 1 = 0$ and $x + y - z + w - 1 = 0$.

50. Determine the distance from the line $y = x + 1$ to the parabola $y^2 = x$. (*Hint:* Let (x, y) be a point on the line and (w, z) a point on the parabola. You want to minimize $(x - w)^2 + (y - z)^2$.)

12.9 Taylor's Formula for Two Variables

This section uses Taylor's formula to derive the Second Derivative Test for local extreme values (Section 12.7) and the error formula for linearizations of functions of two independent variables (Section 12.6). The use of Taylor's formula in these derivations leads to an extension of the formula that provides polynomial approximations of all orders for functions of two independent variables.

Derivation of the Second Derivative Test

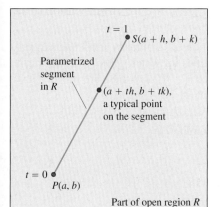

FIGURE 12.55 We begin the derivation of the second derivative test at $P(a, b)$ by parametrizing a typical line segment from P to a point S nearby.

Let $f(x, y)$ have continuous partial derivatives in an open region R containing a point $P(a, b)$ where $f_x = f_y = 0$ (Figure 12.55). Let h and k be increments small enough to put the point $S(a + h, b + k)$ and the line segment joining it to P inside R. We parametrize the segment PS as

$$x = a + th, \qquad y = b + tk, \qquad 0 \le t \le 1.$$

If $F(t) = f(a + th, b + tk)$, the Chain Rule gives

$$F'(t) = f_x \frac{dx}{dt} + f_y \frac{dy}{dt} = hf_x + kf_y.$$

Since f_x and f_y are differentiable (they have continuous partial derivatives), F' is a differentiable function of t and

$$F'' = \frac{\partial F'}{\partial x} \frac{dx}{dt} + \frac{\partial F'}{\partial y} \frac{dy}{dt} = \frac{\partial}{\partial x}(hf_x + kf_y) \cdot h + \frac{\partial}{\partial y}(hf_x + kf_y) \cdot k$$

$$= h^2 f_{xx} + 2hk f_{xy} + k^2 f_{yy}. \qquad f_{xy} = f_{yx}$$

Since F and F' are continuous on $[0, 1]$ and F' is differentiable on $(0, 1)$, we can apply Taylor's formula with $n = 2$ and $a = 0$ to obtain

$$F(1) = F(0) + F'(0)(1 - 0) + F''(c)\frac{(1 - 0)^2}{2}$$

$$F(1) = F(0) + F'(0) + \frac{1}{2}F''(c) \tag{1}$$

for some c between 0 and 1. Writing Equation (1) in terms of f gives

$$f(a + h, b + k) = f(a, b) + hf_x(a, b) + kf_y(a, b)$$

$$+ \frac{1}{2}\left(h^2 f_{xx} + 2hk f_{xy} + k^2 f_{yy}\right)\Big|_{(a+ch,\, b+ck)}. \tag{2}$$

Since $f_x(a, b) = f_y(a, b) = 0$, this reduces to

$$f(a + h, b + k) - f(a, b) = \frac{1}{2}\left(h^2 f_{xx} + 2hk f_{xy} + k^2 f_{yy}\right)\Big|_{(a+ch, b+ck)}. \tag{3}$$

The presence of an extremum of f at (a, b) is determined by the sign of $f(a + h, b + k) - f(a, b)$. By Equation (3), this is the same as the sign of

$$Q(c) = (h^2 f_{xx} + 2hk f_{xy} + k^2 f_{yy})|_{(a+ch, b+ck)}.$$

Now, if $Q(0) \neq 0$, the sign of $Q(c)$ will be the same as the sign of $Q(0)$ for sufficiently small values of h and k. We can predict the sign of

$$Q(0) = h^2 f_{xx}(a, b) + 2hk f_{xy}(a, b) + k^2 f_{yy}(a, b) \tag{4}$$

from the signs of f_{xx} and $f_{xx} f_{yy} - f_{xy}^2$ at (a, b). Multiply both sides of Equation (4) by f_{xx} and rearrange the right-hand side to get

$$f_{xx} Q(0) = (hf_{xx} + kf_{xy})^2 + (f_{xx} f_{yy} - f_{xy}^2)k^2. \tag{5}$$

From Equation (5) we see that

1. If $f_{xx} < 0$ and $f_{xx} f_{yy} - f_{xy}^2 > 0$ at (a, b), then $Q(0) < 0$ for all sufficiently small nonzero values of h and k, and f has a *local maximum* value at (a, b).

2. If $f_{xx} > 0$ and $f_{xx} f_{yy} - f_{xy}^2 > 0$ at (a, b), then $Q(0) > 0$ for all sufficiently small nonzero values of h and k and f has a *local minimum* value at (a, b).

3. If $f_{xx} f_{yy} - f_{xy}^2 < 0$ at (a, b), there are combinations of arbitrarily small nonzero values of h and k for which $Q(0) > 0$, and other values for which $Q(0) < 0$. Arbitrarily close to the point $P_0(a, b, f(a, b))$ on the surface $z = f(x, y)$ there are points above P_0 and points below P_0, so f has a *saddle point* at (a, b).

4. If $f_{xx} f_{yy} - f_{xy}^2 = 0$ at (a, b), another test is needed. The possibility that $Q(0)$ equals zero prevents us from drawing conclusions about the sign of $Q(c)$.

The Error Formula for Linear Approximations

We want to show that the difference $E(x, y)$, between the values of a function $f(x, y)$, and its linearization $L(x, y)$ at (x_0, y_0) satisfies the inequality

$$|E(x, y)| \leq \frac{1}{2} M(|x - x_0| + |y - y_0|)^2.$$

The function f is assumed to have continuous second partial derivatives throughout an open set containing a closed rectangular region R centered at (x_0, y_0). The number M is an upper bound for $|f_{xx}|, |f_{yy}|$, and $|f_{xy}|$ on R.

The inequality we want comes from Equation (2). We substitute x_0 and y_0 for a and b, and $x - x_0$ and $y - y_0$ for h and k, respectively, and rearrange the result as

$$f(x, y) = \underbrace{f(x_0, y_0) + f_x(x_0, y_0)(x - x_0) + f_y(x_0, y_0)(y - y_0)}_{\text{linearization } L(x, y)}$$

$$+ \underbrace{\frac{1}{2}\left((x - x_0)^2 f_{xx} + 2(x - x_0)(y - y_0)f_{xy} + (y - y_0)^2 f_{yy}\right)\Big|_{(x_0+c(x-x_0),\, y_0+c(y-y_0))}}_{\text{error } E(x, y)}.$$

This equation reveals that

$$|E| \leq \frac{1}{2}\left(|x - x_0|^2 |f_{xx}| + 2|x - x_0||y - y_0||f_{xy}| + |y - y_0|^2 |f_{yy}|\right).$$

Hence, if M is an upper bound for the values of $|f_{xx}|$, $|f_{xy}|$, and $|f_{yy}|$ on R,

$$|E| \le \frac{1}{2}\left(|x - x_0|^2 M + 2|x - x_0||y - y_0|M + |y - y_0|^2 M\right)$$

$$= \frac{1}{2}M(|x - x_0| + |y - y_0|)^2.$$

Taylor's Formula for Functions of Two Variables

The formulas derived earlier for F' and F'' can be obtained by applying to $f(x, y)$ the operators

$$\left(h\frac{\partial}{\partial x} + k\frac{\partial}{\partial y}\right) \quad \text{and} \quad \left(h\frac{\partial}{\partial x} + k\frac{\partial}{\partial y}\right)^2 = h^2\frac{\partial^2}{\partial x^2} + 2hk\frac{\partial^2}{\partial x\,\partial y} + k^2\frac{\partial^2}{\partial y^2}.$$

These are the first two instances of a more general formula,

$$F^{(n)}(t) = \frac{d^n}{dt^n}F(t) = \left(h\frac{\partial}{\partial x} + k\frac{\partial}{\partial y}\right)^n f(x, y), \tag{6}$$

which says that applying d^n/dt^n to $F(t)$ gives the same result as applying the operator

$$\left(h\frac{\partial}{\partial x} + k\frac{\partial}{\partial y}\right)^n$$

to $f(x, y)$ after expanding it by the Binomial Theorem.

If partial derivatives of f through order $n + 1$ are continuous throughout a rectangular region centered at (a, b), we may extend the Taylor formula for $F(t)$ to

$$F(t) = F(0) + F'(0)t + \frac{F''(0)}{2!}t^2 + \cdots + \frac{F^{(n)}(0)}{n!}t^{(n)} + \text{ remainder,}$$

and take $t = 1$ to obtain

$$F(1) = F(0) + F'(0) + \frac{F''(0)}{2!} + \cdots + \frac{F^{(n)}(0)}{n!} + \text{ remainder.}$$

When we replace the first n derivatives on the right of this last series by their equivalent expressions from Equation (6) evaluated at $t = 0$ and add the appropriate remainder term, we arrive at the following formula.

Taylor's Formula for $f(x, y)$ at the Point (a, b)

Suppose $f(x, y)$ and its partial derivatives through order $n + 1$ are continuous throughout an open rectangular region R centered at a point (a, b). Then, throughout R,

$$f(a + h, b + k) = f(a, b) + (hf_x + kf_y)|_{(a,b)} + \frac{1}{2!}(h^2 f_{xx} + 2hk f_{xy} + k^2 f_{yy})|_{(a,b)}$$

$$+ \frac{1}{3!}(h^3 f_{xxx} + 3h^2 k f_{xxy} + 3hk^2 f_{xyy} + k^3 f_{yyy})|_{(a,b)} + \cdots + \frac{1}{n!}\left(h\frac{\partial}{\partial x} + k\frac{\partial}{\partial y}\right)^n f\bigg|_{(a,b)}$$

$$+ \frac{1}{(n+1)!}\left(h\frac{\partial}{\partial x} + k\frac{\partial}{\partial y}\right)^{n+1} f\bigg|_{(a+ch, b+ck)}. \tag{7}$$

The first n derivative terms are evaluated at (a, b). The last term is evaluated at some point $(a + ch, b + ck)$ on the line segment joining (a, b) and $(a + h, b + k)$.

If $(a, b) = (0, 0)$ and we treat h and k as independent variables (denoting them now by x and y), then Equation (7) assumes the following simpler form.

Taylor's Formula for $f(x, y)$ at the Origin

$$f(x, y) = f(0, 0) + xf_x + yf_y + \frac{1}{2!}(x^2 f_{xx} + 2xy f_{xy} + y^2 f_{yy})$$

$$+ \frac{1}{3!}(x^3 f_{xxx} + 3x^2 y f_{xxy} + 3xy^2 f_{xyy} + y^3 f_{yyy}) + \cdots + \frac{1}{n!}\left(x\frac{\partial}{\partial x} + y\frac{\partial}{\partial y}\right)^n f$$

$$+ \frac{1}{(n+1)!}\left(x\frac{\partial}{\partial x} + y\frac{\partial}{\partial y}\right)^{n+1} f \Bigg|_{(cx, cy)} \tag{8}$$

The first n derivative terms are evaluated at $(0, 0)$. The last term is evaluated at a point on the line segment joining the origin and (x, y).

Taylor's formula provides polynomial approximations of two-variable functions. The first n derivative terms give the polynomial; the last term gives the approximation error. The first three terms of Taylor's formula give the function's linearization. To improve on the linearization, we add higher power terms.

EXAMPLE 1   Find a quadratic approximation to $f(x, y) = \sin x \sin y$ near the origin. How accurate is the approximation if $|x| \le 0.1$ and $|y| \le 0.1$?

Solution   We take $n = 2$ in Equation (8):

$$f(x, y) = f(0, 0) + (xf_x + yf_y) + \frac{1}{2}(x^2 f_{xx} + 2xy f_{xy} + y^2 f_{yy})$$

$$+ \frac{1}{6}(x^3 f_{xxx} + 3x^2 y f_{xxy} + 3xy^2 f_{xyy} + y^3 f_{yyy})_{(cx, cy)}$$

with

$$f(0, 0) = \sin x \sin y|_{(0,0)} = 0, \qquad f_{xx}(0, 0) = -\sin x \sin y|_{(0,0)} = 0,$$

$$f_x(0, 0) = \cos x \sin y|_{(0,0)} = 0, \qquad f_{xy}(0, 0) = \cos x \cos y|_{(0,0)} = 1,$$

$$f_y(0, 0) = \sin x \cos y|_{(0,0)} = 0, \qquad f_{yy}(0, 0) = -\sin x \sin y|_{(0,0)} = 0,$$

we have

$$\sin x \sin y \approx 0 + 0 + 0 + \frac{1}{2}(x^2(0) + 2xy(1) + y^2(0)),$$

$$\sin x \sin y \approx xy.$$

The error in the approximation is

$$E(x, y) = \frac{1}{6}(x^3 f_{xxx} + 3x^2 y f_{xxy} + 3xy^2 f_{xyy} + y^3 f_{yyy})|_{(cx, cy)}.$$

The third derivatives never exceed 1 in absolute value because they are products of sines and cosines. Also, $|x| \leq 0.1$ and $|y| \leq 0.1$. Hence

$$|E(x, y)| \leq \frac{1}{6}((0.1)^3 + 3(0.1)^3 + 3(0.1)^3 + (0.1)^3) = \frac{8}{6}(0.1)^3 \leq 0.00134$$

(rounded up). The error will not exceed 0.00134 if $|x| \leq 0.1$ and $|y| \leq 0.1$. ∎

EXERCISES 12.9

In Exercises 1–10, use Taylor's formula for $f(x, y)$ at the origin to find quadratic and cubic approximations of f near the origin.

1. $f(x, y) = xe^y$

2. $f(x, y) = e^x \cos y$

3. $f(x, y) = y \sin x$

4. $f(x, y) = \sin x \cos y$

5. $f(x, y) = e^x \ln(1 + y)$

6. $f(x, y) = \ln(2x + y + 1)$

7. $f(x, y) = \sin(x^2 + y^2)$

8. $f(x, y) = \cos(x^2 + y^2)$

9. $f(x, y) = \dfrac{1}{1 - x - y}$

10. $f(x, y) = \dfrac{1}{1 - x - y + xy}$

11. Use Taylor's formula to find a quadratic approximation of $f(x, y) = \cos x \cos y$ at the origin. Estimate the error in the approximation if $|x| \leq 0.1$ and $|y| \leq 0.1$.

12. Use Taylor's formula to find a quadratic approximation of $e^x \sin y$ at the origin. Estimate the error in the approximation if $|x| \leq 0.1$ and $|y| \leq 0.1$.

Chapter 12   Questions to Guide Your Review

1. What is a real-valued function of two independent variables? Three independent variables? Give examples.

2. What does it mean for sets in the plane or in space to be open? Closed? Give examples. Give examples of sets that are neither open nor closed.

3. How can you display the values of a function $f(x, y)$ of two independent variables graphically? How do you do the same for a function $f(x, y, z)$ of three independent variables?

4. What does it mean for a function $f(x, y)$ to have limit L as $(x, y) \to (x_0, y_0)$? What are the basic properties of limits of functions of two independent variables?

5. When is a function of two (three) independent variables continuous at a point in its domain? Give examples of functions that are continuous at some points but not others.

6. What can be said about algebraic combinations and composites of continuous functions?

7. Explain the two-path test for nonexistence of limits.

8. How are the partial derivatives $\partial f/\partial x$ and $\partial f/\partial y$ of a function $f(x, y)$ defined? How are they interpreted and calculated?

9. How does the relation between first partial derivatives and continuity of functions of two independent variables differ from the relation between first derivatives and continuity for real-valued functions of a single independent variable? Give an example.

10. What is the Mixed Derivative Theorem for mixed second-order partial derivatives? How can it help in calculating partial derivatives of second and higher orders? Give examples.

11. What does it mean for a function $f(x, y)$ to be differentiable? What does the Increment Theorem say about differentiability?

12. How can you sometimes decide from examining f_x and f_y that a function $f(x, y)$ is differentiable? What is the relation between the differentiability of f and the continuity of f at a point?

13. What is the Chain Rule? What form does it take for functions of two independent variables? Three independent variables? Functions defined on surfaces? How do you diagram these different forms? Give examples. What pattern enables one to remember all the different forms?

14. What is the derivative of a function $f(x, y)$ at a point P_0 in the direction of a unit vector $\mathbf{u}$? What rate does it describe? What geometric interpretation does it have? Give examples.

15. What is the gradient vector of a differentiable function $f(x, y)$? How is it related to the function's directional derivatives? State the analogous results for functions of three independent variables.

16. How do you find the tangent line at a point on a level curve of a differentiable function $f(x, y)$? How do you find the tangent plane and normal line at a point on a level surface of a differentiable function $f(x, y, z)$? Give examples.

17. How can you use directional derivatives to estimate change?

18. How do you linearize a function $f(x, y)$ of two independent variables at a point (x_0, y_0)? Why might you want to do this? How do you linearize a function of three independent variables?

19. What can you say about the accuracy of linear approximations of functions of two (three) independent variables?

20. If (x, y) moves from (x_0, y_0) to a point $(x_0 + dx, y_0 + dy)$ nearby, how can you estimate the resulting change in the value of a differentiable function $f(x, y)$? Give an example.

21. How do you define local maxima, local minima, and saddle points for a differentiable function $f(x, y)$? Give examples.

22. What derivative tests are available for determining the local extreme values of a function $f(x, y)$? How do they enable you to narrow your search for these values? Give examples.

23. How do you find the extrema of a continuous function $f(x, y)$ on a closed bounded region of the xy-plane? Give an example.

24. Describe the method of Lagrange multipliers and give examples.

25. How does Taylor's formula for a function $f(x, y)$ generate polynomial approximations and error estimates?

Chapter 12   Practice Exercises

In Exercises 1–4, find the domain and range of the given function and identify its level curves. Sketch a typical level curve.

1. $f(x, y) = 9x^2 + y^2$ **2.** $f(x, y) = e^{x+y}$

3. $g(x, y) = 1/xy$ **4.** $g(x, y) = \sqrt{x^2 - y}$

In Exercises 5–8, find the domain and range of the given function and identify its level surfaces. Sketch a typical level surface.

5. $f(x, y, z) = x^2 + y^2 - z$ **6.** $g(x, y, z) = x^2 + 4y^2 + 9z^2$

7. $h(x, y, z) = \dfrac{1}{x^2 + y^2 + z^2}$

8. $k(x, y, z) = \dfrac{1}{x^2 + y^2 + z^2 + 1}$

Find the limits in Exercises 9–14.

9. $\displaystyle\lim_{(x,y)\to(\pi,\,\ln 2)} e^y \cos x$ **10.** $\displaystyle\lim_{(x,y)\to(0,0)} \dfrac{2 + y}{x + \cos y}$

11. $\displaystyle\lim_{(x,y)\to(1,1)} \dfrac{x - y}{x^2 - y^2}$ **12.** $\displaystyle\lim_{(x,y)\to(1,1)} \dfrac{x^3 y^3 - 1}{xy - 1}$

13. $\displaystyle\lim_{P\to(1,-1,e)} \ln|x + y + z|$ **14.** $\displaystyle\lim_{P\to(1,-1,-1)} \tan^{-1}(x + y + z)$

By considering different paths of approach, show that the limits in Exercises 15 and 16 do not exist.

15. $\displaystyle\lim_{\substack{(x,y)\to(0,0) \\ y\neq x^2}} \dfrac{y}{x^2 - y}$ **16.** $\displaystyle\lim_{\substack{(x,y)\to(0,0) \\ xy\neq 0}} \dfrac{x^2 + y^2}{xy}$

17. Continuous extension Let $f(x, y) = (x^2 - y^2)/(x^2 + y^2)$ for $(x, y) \neq (0, 0)$. Is it possible to define $f(0, 0)$ in a way that makes f continuous at the origin? Why?

18. Continuous extension Let

$$f(x, y) = \begin{cases} \dfrac{\sin(x - y)}{|x| + |y|}, & |x| + |y| \neq 0 \\ 0, & (x, y) = (0, 0). \end{cases}$$

Is f continuous at the origin? Why?

In Exercises 19–24, find the partial derivative of the function with respect to each variable.

19. $g(r, \theta) = r \cos \theta + r \sin \theta$

20. $f(x, y) = \dfrac{1}{2} \ln(x^2 + y^2) + \tan^{-1} \dfrac{y}{x}$

21. $f(R_1, R_2, R_3) = \dfrac{1}{R_1} + \dfrac{1}{R_2} + \dfrac{1}{R_3}$

22. $h(x, y, z) = \sin(2\pi x + y - 3z)$

23. $P(n, R, T, V) = \dfrac{nRT}{V}$ (the ideal gas law)

24. $f(r, l, T, w) = \dfrac{1}{2rl} \sqrt{\dfrac{T}{\pi w}}$

Find the second-order partial derivatives of the functions in Exercises 25–28.

25. $g(x, y) = y + \dfrac{x}{y}$

26. $g(x, y) = e^x + y \sin x$

27. $f(x, y) = x + xy - 5x^3 + \ln(x^2 + 1)$

28. $f(x, y) = y^2 - 3xy + \cos y + 7e^y$

29. Find dw/dt at $t = 0$ if $w = \sin(xy + \pi)$, $x = e^t$, and $y = \ln(t + 1)$.

30. Find dw/dt at $t = 1$ if $w = xe^y + y \sin z - \cos z$, $x = 2\sqrt{t}$, $y = t - 1 + \ln t$, and $z = \pi t$.

31. Find $\partial w/\partial r$ and $\partial w/\partial s$ when $r = \pi$ and $s = 0$ if $w = \sin(2x - y)$, $x = r + \sin s$, $y = rs$.

32. Find $\partial w/\partial u$ and $\partial w/\partial v$ when $u = v = 0$ if $w = \ln\sqrt{1 + x^2} - \tan^{-1} x$ and $x = 2e^u \cos v$.

33. Find the value of the derivative of $f(x, y, z) = xy + yz + xz$ with respect to t on the curve $x = \cos t$, $y = \sin t$, $z = \cos 2t$ at $t = 1$.

34. Show that if $w = f(s)$ is any differentiable function of s and if $s = y + 5x$, then

$$\frac{\partial w}{\partial x} - 5\frac{\partial w}{\partial y} = 0.$$

Assuming that the equations in Exercises 35 and 36 define y as a differentiable function of x, find the value of dy/dx at point P.

35. $1 - x - y^2 - \sin xy = 0$, $P(0, 1)$

36. $2xy + e^{x+y} - 2 = 0$, $P(0, \ln 2)$

In Exercises 37–40, find the directions in which f increases and decreases most rapidly at P_0 and find the derivative of f in each direction. Also, find the derivative of f at P_0 in the direction of the vector $\mathbf{v}$.

37. $f(x, y) = \cos x \cos y$, $P_0(\pi/4, \pi/4)$, $\mathbf{v} = 3\mathbf{i} + 4\mathbf{j}$

38. $f(x, y) = x^2 e^{-2y}$, $P_0(1, 0)$, $\mathbf{v} = \mathbf{i} + \mathbf{j}$

39. $f(x, y, z) = \ln(2x + 3y + 6z)$, $P_0(-1, -1, 1)$,
$\mathbf{v} = 2\mathbf{i} + 3\mathbf{j} + 6\mathbf{k}$

40. $f(x, y, z) = x^2 + 3xy - z^2 + 2y + z + 4$, $P_0(0, 0, 0)$,
$\mathbf{v} = \mathbf{i} + \mathbf{j} + \mathbf{k}$

41. Derivative in velocity direction   Find the derivative of $f(x, y, z) = xyz$ in the direction of the velocity vector of the helix

$$\mathbf{r}(t) = (\cos 3t)\mathbf{i} + (\sin 3t)\mathbf{j} + 3t\mathbf{k}$$

at $t = \pi/3$.

42. Maximum directional derivative   What is the largest value that the directional derivative of $f(x, y, z) = xyz$ can have at the point $(1, 1, 1)$?

43. Directional derivatives with given values   At the point $(1, 2)$, the function $f(x, y)$ has a derivative of 2 in the direction toward $(2, 2)$ and a derivative of -2 in the direction toward $(1, 1)$.

  a. Find $f_x(1, 2)$ and $f_y(1, 2)$.

  b. Find the derivative of f at $(1, 2)$ in the direction toward the point $(4, 6)$.

44. Which of the following statements are true if $f(x, y)$ is differentiable at (x_0, y_0)? Give reasons for your answers.

  a. If $\mathbf{u}$ is a unit vector, the derivative of f at (x_0, y_0) in the direction of $\mathbf{u}$ is $(f_x(x_0, y_0)\mathbf{i} + f_y(x_0, y_0)\mathbf{j}) \cdot \mathbf{u}$.

  b. The derivative of f at (x_0, y_0) in the direction of $\mathbf{u}$ is a vector.

  c. The directional derivative of f at (x_0, y_0) has its greatest value in the direction of ∇f.

  d. At (x_0, y_0), vector ∇f is normal to the curve $f(x, y) = f(x_0, y_0)$.

In Exercises 45 and 46, sketch the surface $f(x, y, z) = c$ together with ∇f at the given points.

45. $x^2 + y + z^2 = 0$; $(0, -1, \pm 1)$, $(0, 0, 0)$

46. $y^2 + z^2 = 4$; $(2, \pm 2, 0)$, $(2, 0, \pm 2)$

In Exercises 47 and 48, find an equation for the plane tangent to the level surface $f(x, y, z) = c$ at the point P_0. Also, find parametric equations for the line that is normal to the surface at P_0.

47. $x^2 - y - 5z = 0$, $P_0(2, -1, 1)$

48. $x^2 + y^2 + z = 4$, $P_0(1, 1, 2)$

In Exercises 49 and 50, find an equation for the plane tangent to the surface $z = f(x, y)$ at the given point.

49. $z = \ln(x^2 + y^2)$, $(0, 1, 0)$

50. $z = 1/(x^2 + y^2)$, $(1, 1, 1/2)$

In Exercises 51 and 52, find equations for the lines that are tangent and normal to the level curve $f(x, y) = c$ at the point P_0. Then sketch the lines and level curve together with ∇f at P_0.

51. $y - \sin x = 1$, $P_0(\pi, 1)$ **52.** $\dfrac{y^2}{2} - \dfrac{x^2}{2} = \dfrac{3}{2}$, $P_0(1, 2)$

In Exercises 53 and 54, find parametric equations for the line that is tangent to the curve of intersection of the surfaces at the given point.

53. Surfaces: $x^2 + 2y + 2z = 4$, $y = 1$

  Point: $(1, 1, 1/2)$

54. Surfaces: $x + y^2 + z = 2$, $y = 1$

  Point: $(1/2, 1, 1/2)$

In Exercises 55 and 56, find the linearization $L(x, y)$ of the function $f(x, y)$ at the point P_0. Then find an upper bound for the magnitude of the error E in the approximation $f(x, y) \approx L(x, y)$ over the rectangle R.

55. $f(x, y) = \sin x \cos y$, $P_0(\pi/4, \pi/4)$

  R: $\left|x - \dfrac{\pi}{4}\right| \le 0.1$, $\left|y - \dfrac{\pi}{4}\right| \le 0.1$

56. $f(x, y) = xy - 3y^2 + 2$, $P_0(1, 1)$

  R: $|x - 1| \le 0.1$, $|y - 1| \le 0.2$

Find the linearizations of the functions in Exercises 57 and 58 at the given points.

57. $f(x, y, z) = xy + 2yz - 3xz$ at $(1, 0, 0)$ and $(1, 1, 0)$

58. $f(x, y, z) = \sqrt{2}\cos x \sin(y + z)$ at $(0, 0, \pi/4)$ and $(\pi/4, \pi/4, 0)$

Test the functions in Exercises 59–64 for local maxima and minima and saddle points. Find each function's value at these points.

59. $f(x, y) = x^2 - xy + y^2 + 2x + 2y - 4$

60. $f(x, y) = 5x^2 + 4xy - 2y^2 + 4x - 4y$

61. $f(x, y) = 2x^3 + 3xy + 2y^3$

62. $f(x, y) = x^3 + y^3 - 3xy + 15$

63. $f(x, y) = x^3 + y^3 + 3x^2 - 3y^2$

64. $f(x, y) = x^4 - 8x^2 + 3y^2 - 6y$

In Exercises 65–72, find the absolute maximum and minimum values of f on the region R.

65. $f(x, y) = x^2 + xy + y^2 - 3x + 3y$

  R: The triangular region cut from the first quadrant by the line $x + y = 4$

66. $f(x, y) = x^2 - y^2 - 2x + 4y + 1$

  R: The rectangular region in the first quadrant bounded by the co-ordinate axes and the lines $x = 4$ and $y = 2$

67. $f(x, y) = y^2 - xy - 3y + 2x$

  R: The square region enclosed by the lines $x = \pm 2$ and $y = \pm 2$

68. $f(x, y) = 2x + 2y - x^2 - y^2$

  R: The square region bounded by the coordinate axes and the lines $x = 2, y = 2$ in the first quadrant

69. $f(x, y) = x^2 - y^2 - 2x + 4y$

  R: The triangular region bounded below by the x-axis, above by the line $y = x + 2$, and on the right by the line $x = 2$

70. $f(x, y) = 4xy - x^4 - y^4 + 16$

  R: The triangular region bounded below by the line $y = -2$, above by the line $y = x$, and on the right by the line $x = 2$

71. $f(x, y) = x^3 + y^3 + 3x^2 - 3y^2$

  R: The square region enclosed by the lines $x = \pm 1$ and $y = \pm 1$

72. $f(x, y) = x^3 + 3xy + y^3 + 1$

  R: The square region enclosed by the lines $x = \pm 1$ and $y = \pm 1$

73. Extrema on a circle Find the extreme values of $f(x, y) = x^3 + y^2$ on the circle $x^2 + y^2 = 1$.

74. Extrema on a circle Find the extreme values of $f(x, y) = xy$ on the circle $x^2 + y^2 = 1$.

75. Extrema in a disk Find the extreme values of $f(x, y) = x^2 + 3y^2 + 2y$ on the unit disk $x^2 + y^2 \le 1$.

76. Extrema in a disk Find the extreme values of $f(x, y) = x^2 + y^2 - 3x - xy$ on the disk $x^2 + y^2 \le 9$.

77. Extrema on a sphere Find the extreme values of $f(x, y, z) = x - y + z$ on the unit sphere $x^2 + y^2 + z^2 = 1$.

78. Minimum distance to origin Find the points on the surface $x^2 - zy = 4$ closest to the origin.

79. Minimizing cost of a box A closed rectangular box is to have volume $V \text{ cm}^3$. The cost of the material used in the box is $a \text{ cents/cm}^2$ for top and bottom, $b \text{ cents/cm}^2$ for front and back, and $c \text{ cents/cm}^2$ for the remaining sides. What dimensions minimize the total cost of materials?

80. Least volume Find the plane $x/a + y/b + z/c = 1$ that passes through the point $(2, 1, 2)$ and cuts off the least volume from the first octant.

81. Extrema on curve of intersecting surfaces Find the extreme values of $f(x, y, z) = x(y + z)$ on the curve of intersection of the right circular cylinder $x^2 + y^2 = 1$ and the hyperbolic cylinder $xz = 1$.

82. Minimum distance to origin on curve of intersecting plane and cone Find the point closest to the origin on the curve of intersection of the plane $x + y + z = 1$ and the cone $z^2 = 2x^2 + 2y^2$.

83. Let $w = f(r, \theta)$, $r = \sqrt{x^2 + y^2}$, and $\theta = \tan^{-1}(y/x)$. Find $\partial w/\partial x$ and $\partial w/\partial y$ and express your answers in terms of r and θ.

84. Let $z = f(u, v)$, $u = ax + by$, and $v = ax - by$. Express z_x and z_y in terms of f_u, f_v, and the constants a and b.

85. If a and b are constants, $w = u^3 + \tanh u + \cos u$, and $u = ax + by$, show that

$$a \frac{\partial w}{\partial y} = b \frac{\partial w}{\partial x}.$$

86. Using the Chain Rule If $w = \ln(x^2 + y^2 + 2z)$, $x = r + s$, $y = r - s$, and $z = 2rs$, find w_r and w_s by the Chain Rule. Then check your answer another way.

87. Angle between vectors The equations $e^u \cos v - x = 0$ and $e^u \sin v - y = 0$ define u and v as differentiable functions of x and y. Show that the angle between the vectors

$$\frac{\partial u}{\partial x} \mathbf{i} + \frac{\partial u}{\partial y} \mathbf{j} \quad \text{and} \quad \frac{\partial v}{\partial x} \mathbf{i} + \frac{\partial v}{\partial y} \mathbf{j}$$

is constant.

88. Polar coordinates and second derivatives Introducing polar coordinates $x = r \cos \theta$ and $y = r \sin \theta$ changes $f(x, y)$ to $g(r, \theta)$. Find the value of $\partial^2 g/\partial \theta^2$ at the point $(r, \theta) = (2, \pi/2)$, given that

$$\frac{\partial f}{\partial x} = \frac{\partial f}{\partial y} = \frac{\partial^2 f}{\partial x^2} = \frac{\partial^2 f}{\partial y^2} = 1$$

at that point.

89. Normal line parallel to a plane Find the points on the surface

$$(y + z)^2 + (z - x)^2 = 16$$

where the normal line is parallel to the yz-plane.

90. Tangent plane parallel to xy-plane Find the points on the surface

$$xy + yz + zx - x - z^2 = 0$$

where the tangent plane is parallel to the xy-plane.

91. When gradient is parallel to position vector Suppose that $\nabla f(x, y, z)$ is always parallel to the position vector $x\mathbf{i} + y\mathbf{j} + z\mathbf{k}$. Show that $f(0, 0, a) = f(0, 0, -a)$ for any a.

92. One-sided directional derivative in all directions, but no gradient The one-sided directional derivative of f at P (x_0, y_0, z_0) in the direction $\mathbf{u} = u_1\mathbf{i}, u_2\mathbf{j}, u_3\mathbf{k}$ is the number

$$\lim_{s \to 0^+} \frac{f(x_0 + su_1, y_0 + su_2, z_0 + su_3) - f(x_0, y_0, z_0)}{s}.$$

Show that the one-sided directional derivative of

$$f(x, y, z) = \sqrt{x^2 + y^2 + z^2}$$

at the origin equals 1 in any direction but that f has no gradient vector at the origin.

93. Normal line through origin Show that the line normal to the surface $xy + z = 2$ at the point $(1, 1, 1)$ passes through the origin.

94. Tangent plane and normal line

   a. Sketch the surface $x^2 - y^2 + z^2 = 4$.

   b. Find a vector normal to the surface at $(2, -3, 3)$. Add the vector to your sketch.

   c. Find equations for the tangent plane and normal line at $(2, -3, 3)$.

95. Measuring the volume of a pipeline You plan to calculate the volume inside a stretch of pipeline that is about 36 in. in diameter and 1 mile long. With which measurement should you be more careful, the length or the diameter? Why?

96. Sensitivity to change Near the point $(1, 2)$, is $f(x, y) = x^2 - xy + y^2 - 3$ more sensitive to changes in x or to changes in y? How do you know?

Chapter 12   Additional and Advanced Exercises

1. Function with saddle at the origin If you did Exercise 50 in Section 12.2, you know that the function

$$f(x, y) = \begin{cases} xy\dfrac{x^2 - y^2}{x^2 + y^2}, & (x, y) \neq (0, 0) \\ 0, & (x, y) = (0, 0) \end{cases}$$

(see the accompanying figure) is continuous at $(0, 0)$. Find $f_{xy}(0, 0)$ and $f_{yx}(0, 0)$.

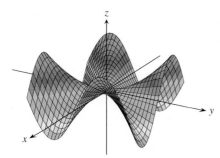

2. Finding a function from second partials Find a function $w = f(x, y)$ whose first partial derivatives are $\partial w/\partial x = 1 + e^x \cos y$ and $\partial w/\partial y = 2y - e^x \sin y$ and whose value at the point $(\ln 2, 0)$ is $\ln 2$.

3. A proof of Leibniz's Rule Leibniz's Rule says that if f is continuous on $[a, b]$ and if $u(x)$ and $v(x)$ are differentiable functions of x whose values lie in $[a, b]$, then

$$\frac{d}{dx}\int_{u(x)}^{v(x)} f(t)\, dt = f(v(x))\frac{dv}{dx} - f(u(x))\frac{du}{dx}.$$

Prove the rule by setting

$$g(u, v) = \int_u^v f(t)\, dt, \quad u = u(x), \quad v = v(x)$$

and calculating dg/dx with the Chain Rule.

4. Finding a function with constrained second partials Suppose that f is a twice-differentiable function of r, that $r = \sqrt{x^2 + y^2 + z^2}$, and that

$$f_{xx} + f_{yy} + f_{zz} = 0.$$

Show that for some constants a and b,

$$f(r) = \frac{a}{r} + b.$$

5. Homogeneous functions A function $f(x, y)$ is *homogeneous of degree n* (n a nonnegative integer) if $f(tx, ty) = t^n f(x, y)$ for all t, x, and y. For such a function (sufficiently differentiable), prove that

   a. $x\dfrac{\partial f}{\partial x} + y\dfrac{\partial f}{\partial y} = nf(x, y)$

   b. $x^2\left(\dfrac{\partial^2 f}{\partial x^2}\right) + 2xy\left(\dfrac{\partial^2 f}{\partial x \partial y}\right) + y^2\left(\dfrac{\partial^2 f}{\partial y^2}\right) = n(n-1)f.$

6. Surface in polar coordinates Let

$$f(r, \theta) = \begin{cases} \dfrac{\sin 6r}{6r}, & r \neq 0 \\ 1, & r = 0, \end{cases}$$

where r and θ are polar coordinates. Find

   a. $\lim\limits_{r \to 0} f(r, \theta)$ **b.** $f_r(0, 0)$ **c.** $f_\theta(r, \theta), \quad r \neq 0.$

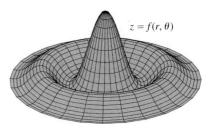

$z = f(r, \theta)$

7. Properties of position vectors Let $\mathbf{r} = x\mathbf{i} + y\mathbf{j} + z\mathbf{k}$ and let $r = |\mathbf{r}|$.

   a. Show that $\nabla r = \mathbf{r}/r$.

   b. Show that $\nabla(r^n) = nr^{n-2}\mathbf{r}$.

   c. Find a function whose gradient equals $\mathbf{r}$.

   d. Show that $\mathbf{r} \cdot d\mathbf{r} = r\, dr$.

   e. Show that $\nabla(\mathbf{A} \cdot \mathbf{r}) = \mathbf{A}$ for any constant vector $\mathbf{A}$.

8. **Gradient orthogonal to tangent**   Suppose that a differentiable function $f(x, y)$ has the constant value c along the differentiable curve $x = g(t), y = h(t)$; that is

$$f(g(t), h(t)) = c$$

for all values of t. Differentiate both sides of this equation with respect to t to show that ∇f is orthogonal to the curve's tangent vector at every point on the curve.

9. **Curve tangent to a surface**   Show that the curve

$$\mathbf{r}(t) = (\ln t)\mathbf{i} + (t \ln t)\mathbf{j} + t\mathbf{k}$$

is tangent to the surface

$$xz^2 - yz + \cos xy = 1$$

at $(0, 0, 1)$.

10. **Curve tangent to a surface**   Show that the curve

$$\mathbf{r}(t) = \left(\frac{t^3}{4} - 2\right)\mathbf{i} + \left(\frac{4}{t} - 3\right)\mathbf{j} + \cos(t - 2)\mathbf{k}$$

is tangent to the surface

$$x^3 + y^3 + z^3 - xyz = 0$$

at $(0, -1, 1)$.

11. **Extrema on a surface**   Show that the only possible maxima and minima of z on the surface $z = x^3 + y^3 - 9xy + 27$ occur at $(0, 0)$ and $(3, 3)$. Show that neither a maximum nor a minimum occurs at $(0, 0)$. Determine whether z has a maximum or a minimum at $(3, 3)$.

12. **Maximum in closed first quadrant**   Find the maximum value of $f(x, y) = 6xye^{-(2x+3y)}$ in the closed first quadrant (includes the nonnegative axes).

13. **Minimum volume cut from first octant**   Find the minimum volume for a region bounded by the planes $x = 0, y = 0, z = 0$ and a plane tangent to the ellipsoid

$$\frac{x^2}{a^2} + \frac{y^2}{b^2} + \frac{z^2}{c^2} = 1$$

at a point in the first octant.

14. **Minimum distance from line to parabola in xy-plane**   By minimizing the function $f(x, y, u, v) = (x - u)^2 + (y - v)^2$ subject to the constraints $y = x + 1$ and $u = v^2$, find the minimum distance in the xy-plane from the line $y = x + 1$ to the parabola $y^2 = x$.

15. **Boundedness of first partials implies continuity**   Prove the following theorem: If $f(x, y)$ is defined in an open region R of the xy-plane and if f_x and f_y are bounded on R, then $f(x, y)$ is continuous on R. (The assumption of boundedness is essential.)

16. Suppose that $\mathbf{r}(t) = g(t)\mathbf{i} + h(t)\mathbf{j} + k(t)\mathbf{k}$ is a smooth curve in the domain of a differentiable function $f(x, y, z)$. Describe the relation between df/dt, ∇f, and $\mathbf{v} = d\mathbf{r}/dt$. What can be said about ∇f and $\mathbf{v}$ at interior points of the curve where f has extreme values relative to its other values on the curve? Give reasons for your answer.

17. **Finding functions from partial derivatives**   Suppose that f and g are functions of x and y such that

$$\frac{\partial f}{\partial y} = \frac{\partial g}{\partial x} \quad \text{and} \quad \frac{\partial f}{\partial x} = \frac{\partial g}{\partial y},$$

and suppose that

$$\frac{\partial f}{\partial x} = 0, \quad f(1, 2) = g(1, 2) = 5 \quad \text{and} \quad f(0, 0) = 4.$$

Find $f(x, y)$ and $g(x, y)$.

18. **Rate of change of the rate of change**   We know that if $f(x, y)$ is a function of two variables and if $\mathbf{u} = a\mathbf{i} + b\mathbf{j}$ is a unit vector, then $D_\mathbf{u} f(x, y) = f_x(x, y)a + f_y(x, y)b$ is the rate of change of $f(x, y)$ at (x, y) in the direction of $\mathbf{u}$. Give a similar formula for the rate of change *of the rate of change* of $f(x, y)$ at (x, y) in the direction $\mathbf{u}$.

19. **Path of a heat-seeking particle**   A heat-seeking particle has the property that at any point (x, y) in the plane it moves in the direction of maximum temperature increase. If the temperature at (x, y) is $T(x, y) = -e^{-2y} \cos x$, find an equation $y = f(x)$ for the path of a heat-seeking particle at the point $(\pi/4, 0)$.

20. **Velocity after a ricochet**   A particle traveling in a straight line with constant velocity $\mathbf{i} + \mathbf{j} - 5\mathbf{k}$ passes through the point $(0, 0, 30)$ and hits the surface $z = 2x^2 + 3y^2$. The particle ricochets off the surface, the angle of reflection being equal to the angle of incidence. Assuming no loss of speed, what is the velocity of the particle after the ricochet? Simplify your answer.

21. **Directional derivatives tangent to a surface**   Let S be the surface that is the graph of $f(x, y) = 10 - x^2 - y^2$. Suppose that the temperature in space at each point (x, y, z) is $T(x, y, z) = x^2y + y^2z + 4x + 14y + z$.

   a. Among all the possible directions tangential to the surface S at the point $(0, 0, 10)$, which direction will make the rate of change of temperature at $(0, 0, 10)$ a maximum?

   b. Which direction tangential to S at the point $(1, 1, 8)$ will make the rate of change of temperature a maximum?

22. **Drilling another borehole**   On a flat surface of land, geologists drilled a borehole straight down and hit a mineral deposit at 1000 ft. They drilled a second borehole 100 ft to the north of the first and hit the mineral deposit at 950 ft. A third borehole 100 ft east of the first borehole struck the mineral deposit at 1025 ft. The geologists have reasons to believe that the mineral deposit is in the shape of a dome, and for the sake of economy, they would like to find where the deposit is closest to the surface. Assuming the surface to be the xy-plane, in what direction from the first borehole would you suggest the geologists drill their fourth borehole?

Chapter

13

MULTIPLE INTEGRALS

OVERVIEW In this chapter we consider the integral of a function of two variables $f(x, y)$ over a region in the plane and the integral of a function of three variables $f(x, y, z)$ over a region in space. These *multiple integrals* are defined as the limit of approximating Riemann sums, much like the single-variable integrals presented in Chapter 5.

13.1 Double and Iterated Integrals over Rectangles

In Chapter 5 we defined the definite integral of a continuous function $f(x)$ over an interval $[a, b]$ as a limit of Riemann sums. In this section we extend this idea to define the *double integral* of a continuous function of two variables $f(x, y)$ over a bounded rectangle R in the plane. In both cases the integrals are limits of approximating Riemann sums. The Riemann sums for the integral of a single-variable function $f(x)$ are obtained by partitioning a finite interval into thin subintervals, multiplying the width of each subinterval by the value of f at a point c_k inside that subinterval, and then adding together all the products. A similar method of partitioning, multiplying, and summing is used to construct double integrals.

Double Integrals

We begin our investigation of double integrals by considering the simplest type of planar region, a rectangle. We consider a function $f(x, y)$ defined on a rectangular region R,

$$R: \quad a \le x \le b, \quad c \le y \le d.$$

We subdivide R into small rectangles using a network of lines parallel to the x- and y-axes (Figure 13.1). The lines divide R into n rectangular pieces, where the number of such pieces n gets large as the width and height of each piece gets small. These rectangles form a **partition** of R. A small rectangular piece of width Δx and height Δy has area $\Delta A = \Delta x \Delta y$. If we number the small pieces partitioning R in some order, then their areas are given by numbers $\Delta A_1, \Delta A_2, \ldots, \Delta A_n$, where ΔA_k is the area of the kth small rectangle.

To form a Riemann sum over R, we choose a point (x_k, y_k) in the kth small rectangle, multiply the value of f at that point by the area ΔA_k, and add together the products:

$$S_n = \sum_{k=1}^{n} f(x_k, y_k) \, \Delta A_k.$$

Depending on how we pick (x_k, y_k) in the kth small rectangle, we may get different values for S_n.

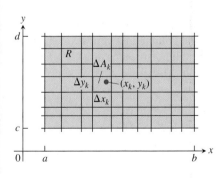

FIGURE 13.1 Rectangular grid partitioning the region R into small rectangles of area $\Delta A_k = \Delta x_k \, \Delta y_k$.

We are interested in what happens to these Riemann sums as the widths and heights of all the small rectangles in the partition of R approach zero. The **norm** of a partition P, written $\|P\|$, is the largest width or height of any rectangle in the partition. If $\|P\| = 0.1$ then all the rectangles in the partition of R have width at most 0.1 and height at most 0.1. Sometimes the Riemann sums converge as the norm of P goes to zero, written $\|P\| \to 0$. The resulting limit is then written as

$$\lim_{\|P\| \to 0} \sum_{k=1}^{n} f(x_k, y_k)\, \Delta A_k.$$

As $\|P\| \to 0$ and the rectangles get narrow and short, their number n increases, so we can also write this limit as

$$\lim_{n \to \infty} \sum_{k=1}^{n} f(x_k, y_k)\, \Delta A_k.$$

with the understanding that $\Delta A_k \to 0$ as $n \to \infty$ and $\|P\| \to 0$.

There are many choices involved in a limit of this kind. The collection of small rectangles is determined by the grid of vertical and horizontal lines that determine a rectangular partition of R. In each of the resulting small rectangles there is a choice of an arbitrary point (x_k, y_k) at which f is evaluated. These choices together determine a single Riemann sum. To form a limit, we repeat the whole process again and again, choosing partitions whose rectangle widths and heights both go to zero and whose number goes to infinity.

When a limit of the sums S_n exists, giving the same limiting value no matter what choices are made, then the function f is said to be **integrable** and the limit is called the **double integral** of f over R, written as

$$\iint\limits_{R} f(x, y)\, dA \qquad \text{or} \qquad \iint\limits_{R} f(x, y)\, dx\, dy.$$

It can be shown that if $f(x, y)$ is a continuous function throughout R, then f is integrable, as in the single-variable case discussed in Chapter 5. Many discontinuous functions are also integrable, including functions which are discontinuous only on a finite number of points or smooth curves. We leave the proof of these facts to a more advanced text.

Double Integrals as Volumes

When $f(x, y)$ is a positive function over a rectangular region R in the xy-plane, we may interpret the double integral of f over R as the volume of the 3-dimensional solid region over the xy-plane bounded below by R and above by the surface $z = f(x, y)$ (Figure 13.2). Each term $f(x_k, y_k)\Delta A_k$ in the sum $S_n = \sum f(x_k, y_k)\Delta A_k$ is the volume of a vertical rectangular box that approximates the volume of the portion of the solid that stands directly above the base ΔA_k. The sum S_n thus approximates what we want to call the total volume of the solid. We *define* this volume to be

$$\text{Volume} = \lim_{n \to \infty} S_n = \iint\limits_{R} f(x, y)\, dA,$$

where $\Delta A_k \to 0$ as $n \to \infty$.

As you might expect, this more general method of calculating volume agrees with the methods in Chapter 6, but we do not prove this here. Figure 13.3 shows Riemann sum approximations to the volume becoming more accurate as the number n of boxes increases.

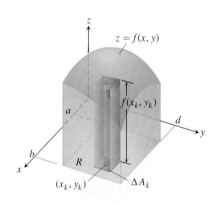

FIGURE 13.2 Approximating solids with rectangular boxes leads us to define the volumes of more general solids as double integrals. The volume of the solid shown here is the double integral of $f(x, y)$ over the base region R.

(a) $n = 16$

(b) $n = 64$

(c) $n = 256$

FIGURE 13.3 As n increases, the Riemann sum approximations approach the total volume of the solid shown in Figure 13.2.

Fubini's Theorem for Calculating Double Integrals

Suppose that we wish to calculate the volume under the plane $z = 4 - x - y$ over the rectangular region R: $0 \le x \le 2$, $0 \le y \le 1$ in the xy-plane. If we apply the method of slicing from Section 6.1, with slices perpendicular to the x-axis (Figure 13.4), then the volume is

$$\int_{x=0}^{x=2} A(x)\,dx, \tag{1}$$

where $A(x)$ is the cross-sectional area at x. For each value of x, we may calculate $A(x)$ as the integral

$$A(x) = \int_{y=0}^{y=1} (4 - x - y)\,dy, \tag{2}$$

which is the area under the curve $z = 4 - x - y$ in the plane of the cross-section at x. In calculating $A(x)$, x is held fixed and the integration takes place with respect to y. Combining Equations (1) and (2), we see that the volume of the entire solid is

$$\text{Volume} = \int_{x=0}^{x=2} A(x)\,dx = \int_{x=0}^{x=2} \left(\int_{y=0}^{y=1} (4 - x - y)\,dy \right) dx$$

$$= \int_{x=0}^{x=2} \left[4y - xy - \frac{y^2}{2} \right]_{y=0}^{y=1} dx = \int_{x=0}^{x=2} \left(\frac{7}{2} - x \right) dx$$

$$= \left[\frac{7}{2}x - \frac{x^2}{2} \right]_{0}^{2} = 5. \tag{3}$$

If we just wanted to write a formula for the volume, without carrying out any of the integrations, we could write

$$\text{Volume} = \int_{0}^{2} \int_{0}^{1} (4 - x - y)\,dy\,dx.$$

The expression on the right, called an **iterated** or **repeated integral**, says that the volume is obtained by integrating $4 - x - y$ with respect to y from $y = 0$ to $y = 1$, holding x fixed, and then integrating the resulting expression in x with respect to x from $x = 0$ to $x = 2$. The limits of integration 0 and 1 are associated with y, so they are placed on the integral closest to dy. The other limits of integration, 0 and 2, are associated with the variable x, so they are placed on the outside integral symbol that is paired with dx.

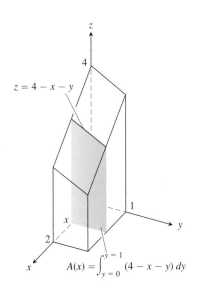

FIGURE 13.4 To obtain the cross-sectional area $A(x)$, we hold x fixed and integrate with respect to y.

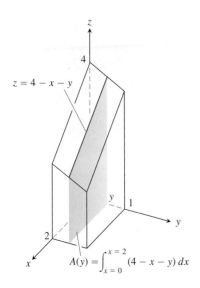

FIGURE 13.5 To obtain the cross-sectional area $A(y)$, we hold y fixed and integrate with respect to x.

HISTORICAL BIOGRAPHY

Guido Fubini
(1879–1943)

What would have happened if we had calculated the volume by slicing with planes perpendicular to the y-axis (Figure 13.5)? As a function of y, the typical cross-sectional area is

$$A(y) = \int_{x=0}^{x=2} (4 - x - y)\, dx = \left[4x - \frac{x^2}{2} - xy \right]_{x=0}^{x=2} = 6 - 2y. \tag{4}$$

The volume of the entire solid is therefore

$$\text{Volume} = \int_{y=0}^{y=1} A(y)\, dy = \int_{y=0}^{y=1} (6 - 2y)\, dy = \left[6y - y^2 \right]_0^1 = 5,$$

in agreement with our earlier calculation.

Again, we may give a formula for the volume as an iterated integral by writing

$$\text{Volume} = \int_0^1 \int_0^2 (4 - x - y)\, dx\, dy.$$

The expression on the right says we can find the volume by integrating $4 - x - y$ with respect to x from $x = 0$ to $x = 2$ as in Equation (4) and integrating the result with respect to y from $y = 0$ to $y = 1$. In this iterated integral, the order of integration is first x and then y, the reverse of the order in Equation (3).

What do these two volume calculations with iterated integrals have to do with the double integral

$$\iint_R (4 - x - y)\, dA$$

over the rectangle $R: 0 \le x \le 2, 0 \le y \le 1$? The answer is that both iterated integrals give the value of the double integral. This is what we would reasonably expect, since the double integral measures the volume of the same region as the two iterated integrals. A theorem published in 1907 by Guido Fubini says that the double integral of any continuous function over a rectangle can be calculated as an iterated integral in either order of integration. (Fubini proved his theorem in greater generality, but this is what it says in our setting.)

THEOREM 1—Fubini's Theorem (First Form)   If $f(x, y)$ is continuous throughout the rectangular region $R: a \le x \le b, c \le y \le d$, then

$$\iint_R f(x, y)\, dA = \int_c^d \int_a^b f(x, y)\, dx\, dy = \int_a^b \int_c^d f(x, y)\, dy\, dx.$$

Fubini's Theorem says that double integrals over rectangles can be calculated as iterated integrals. Thus, we can evaluate a double integral by integrating with respect to one variable at a time.

Fubini's Theorem also says that we may calculate the double integral by integrating in *either* order, a genuine convenience. When we calculate a volume by slicing, we may use either planes perpendicular to the x-axis or planes perpendicular to the y-axis.

EXAMPLE 1   Calculate $\iint_R f(x, y)\, dA$ for

$$f(x, y) = 1 - 6x^2 y \quad \text{and} \quad R: \ 0 \le x \le 2, \ -1 \le y \le 1.$$

Solution   By Fubini's Theorem,

$$\iint\limits_R f(x, y)\, dA = \int_{-1}^{1}\int_0^2 (1 - 6x^2 y)\, dx\, dy = \int_{-1}^{1}\left[x - 2x^3 y\right]_{x=0}^{x=2} dy$$

$$= \int_{-1}^{1} (2 - 16y)\, dy = \left[2y - 8y^2\right]_{-1}^{1} = 4.$$

Reversing the order of integration gives the same answer:

$$\int_0^2\int_{-1}^{1} (1 - 6x^2 y)\, dy\, dx = \int_0^2 \left[y - 3x^2 y^2\right]_{y=-1}^{y=1} dx$$

$$= \int_0^2 [(1 - 3x^2) - (-1 - 3x^2)]\, dx$$

$$= \int_0^2 2\, dx = 4. \qquad \blacksquare$$

EXAMPLE 2   Find the volume of the region bounded above by the ellipitical paraboloid $z = x^2 + 3y^2$ and below by the rectangle $R: 0 \le x \le 1, 0 \le y \le 2$.

Solution   The volume is given by the double integral

$$V = \iint\limits_R (x^2 + 3y^2)\, dA = \int_0^1\int_0^2 (x^2 + 3y^2)\, dy\, dx$$

$$= \int_0^1 \left[x^2 y + y^3\right]_{y=0}^{y=2} dx$$

$$= \int_0^1 (2x^2 + 8)\, dx = \left[\frac{2}{3}x^3 + 8x\right]_0^1 = \frac{26}{3}. \qquad \blacksquare$$

EXERCISES 13.1

In Exercises 1–12, evaluate the iterated integral.

1. $\displaystyle\int_1^2\int_0^4 2xy\, dy\, dx$

2. $\displaystyle\int_0^2\int_{-1}^{1} (x - y)\, dy\, dx$

3. $\displaystyle\int_{-1}^{0}\int_0^1 (x + y + 1)\, dx\, dy$

4. $\displaystyle\int_0^1\int_0^1 \left(1 - \frac{x^2 + y^2}{2}\right) dx\, dy$

5. $\displaystyle\int_0^3\int_0^2 (4 - y^2)\, dy\, dx$

6. $\displaystyle\int_0^3\int_{-2}^{0} (x^2 y - 2xy)\, dy\, dx$

7. $\displaystyle\int_0^1\int_0^1 \frac{y}{1 + xy}\, dx\, dy$

8. $\displaystyle\int_1^4\int_0^4 \left(\frac{x}{2} + \sqrt{y}\right) dx\, dy$

9. $\displaystyle\int_0^{\ln 2}\int_1^{\ln 5} e^{2x + y}\, dy\, dx$

10. $\displaystyle\int_0^1\int_1^2 xye^x\, dy\, dx$

11. $\displaystyle\int_{-1}^{2}\int_0^{\pi/2} y \sin x\, dx\, dy$

12. $\displaystyle\int_{\pi}^{2\pi}\int_0^{\pi} (\sin x + \cos y)\, dx\, dy$

In Exercises 13–20, evaluate the double integral over the given region R.

13. $\displaystyle\iint\limits_R (6y^2 - 2x)\, dA, \qquad R: \ 0 \le x \le 1, \ \ 0 \le y \le 2$

14. $\displaystyle\iint\limits_R \left(\frac{\sqrt{x}}{y^2}\right) dA, \qquad R: \ 0 \le x \le 4, \ \ 1 \le y \le 2$

15. $\displaystyle\iint\limits_R xy \cos y\, dA, \qquad R: \ -1 \le x \le 1, \ \ 0 \le y \le \pi$

16. $\iint\limits_R y \sin(x+y)\, dA,$ $R: \; -\pi \le x \le 0, \;\; 0 \le y \le \pi$

17. $\iint\limits_R e^{x-y}\, dA,$ $R: \; 0 \le x \le \ln 2, \;\; 0 \le y \le \ln 2$

18. $\iint\limits_R xye^{xy^2}\, dA,$ $R: \; 0 \le x \le 2, \;\; 0 \le y \le 1$

19. $\iint\limits_R \dfrac{xy^3}{x^2+1}\, dA,$ $R: \; 0 \le x \le 1, \;\; 0 \le y \le 2$

20. $\iint\limits_R \dfrac{y}{x^2y^2+1}\, dA,$ $R: \; 0 \le x \le 1, \;\; 0 \le y \le 1$

In Exercises 21 and 22, integrate f over the given region.

21. Square $f(x,y) = 1/(xy)$ over the square $1 \le x \le 2, 1 \le y \le 2$

22. Rectangle $f(x,y) = y \cos xy$ over the rectangle $0 \le x \le \pi$, $0 \le y \le 1$

23. Find the volume of the region bounded above by the paraboloid $z = x^2 + y^2$ and below by the square $R: -1 \le x \le 1$, $-1 \le y \le 1$.

24. Find the volume of the region bounded above by the elliptical paraboloid $z = 16 - x^2 - y^2$ and below by the square $R: 0 \le x \le 2, 0 \le y \le 2$.

25. Find the volume of the region bounded above by the plane $z = 2 - x - y$ and below by the square $R: 0 \le x \le 1$, $0 \le y \le 1$.

26. Find the volume of the region bounded above by the plane $z = y/2$ and below by the rectangle $R: 0 \le x \le 4, 0 \le y \le 2$.

27. Find the volume of the region bounded above by the surface $z = 2 \sin x \cos y$ and below by the rectangle $R: 0 \le x \le \pi/2$, $0 \le y \le \pi/4$.

28. Find the volume of the region bounded above by the surface $z = 4 - y^2$ and below by the rectangle $R: 0 \le x \le 1$, $0 \le y \le 2$.

13.2   Double Integrals over General Regions

In this section we define and evaluate double integrals over bounded regions in the plane which are more general than rectangles. These double integrals are also evaluated as iterated integrals, with the main practical problem being that of determining the limits of integration. Since the region of integration may have boundaries other than line segments parallel to the coordinate axes, the limits of integration often involve variables, not just constants.

Double Integrals over Bounded Nonrectangular Regions

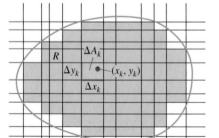

FIGURE 13.6   A rectangular grid partitioning a bounded nonrectangular region into rectangular cells.

To define the double integral of a function $f(x, y)$ over a bounded, nonrectangular region R, such as the one in Figure 13.6, we again begin by covering R with a grid of small rectangular cells whose union contains all points of R. This time, however, we cannot exactly fill R with a finite number of rectangles lying inside R, since its boundary is curved, and some of the small rectangles in the grid lie partly outside R. A partition of R is formed by taking the rectangles that lie completely inside it, not using any that are either partly or completely outside. For commonly arising regions, more and more of R is included as the norm of a partition (the largest width or height of any rectangle used) approaches zero.

Once we have a partition of R, we number the rectangles in some order from 1 to n and let ΔA_k be the area of the kth rectangle. We then choose a point (x_k, y_k) in the kth rectangle and form the Riemann sum

$$S_n = \sum_{k=1}^{n} f(x_k, y_k)\, \Delta A_k.$$

As the norm of the partition forming S_n goes to zero, $\|P\| \to 0$, the width and height of each enclosed rectangle goes to zero and their number goes to infinity. If $f(x, y)$ is a continuous function, then these Riemann sums converge to a limiting value, not dependent on any of the choices we made. This limit is called the **double integral** of $f(x, y)$ over R:

$$\lim_{\|P\| \to 0} \sum_{k=1}^{n} f(x_k, y_k)\, \Delta A_k = \iint\limits_R f(x, y)\, dA.$$

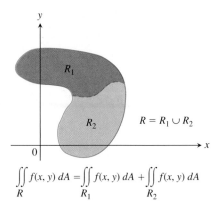

$$\iint\limits_{R} f(x, y)\, dA = \iint\limits_{R_1} f(x, y)\, dA + \iint\limits_{R_2} f(x, y)\, dA$$

FIGURE 13.7 The Additivity Property for rectangular regions holds for regions bounded by continuous curves.

The nature of the boundary of R introduces issues not found in integrals over an interval. When R has a curved boundary, the n rectangles of a partition lie inside R but do not cover all of R. In order for a partition to approximate R well, the parts of R covered by small rectangles lying partly outside R must become negligible as the norm of the partition approaches zero. This property of being nearly filled in by a partition of small norm is satisfied by all the regions that we will encounter. There is no problem with boundaries made from polygons, circles, ellipses, and from continuous graphs over an interval, joined end to end. A curve with a "fractal" type of shape would be problematic, but such curves are not relevant for most applications. A careful discussion of which type of regions R can be used for computing double integrals is left to a more advanced text.

Double integrals of continuous functions over nonrectangular regions have algebraic properties (summarized further on) similar to those for single integrals. The Domain Additivity Property says that if R is decomposed into nonoverlapping regions R_1 and R_2 with boundaries that are made of a finite number of line segments or smooth curves (see Figure 13.7 for an example), then

$$\iint\limits_{R} f(x, y)\, dA = \iint\limits_{R_1} f(x, y)\, dA + \iint\limits_{R_2} f(x, y)\, dA.$$

Volumes

If $f(x, y)$ is positive and continuous over R we define the volume of the solid region between R and the surface $z = f(x, y)$ to be $\iint_{R} f(x, y)\, dA$, as before (Figure 13.8).

If R is a region like the one shown in the xy-plane in Figure 13.9, bounded "above" and "below" by the curves $y = g_2(x)$ and $y = g_1(x)$ and on the sides by the lines $x = a, x = b$, we may again calculate the volume by the method of slicing. We first calculate the cross-sectional area

$$A(x) = \int_{y=g_1(x)}^{y=g_2(x)} f(x, y)\, dy$$

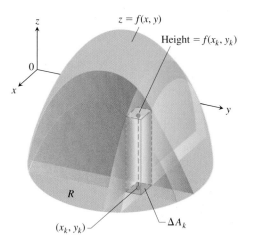

$$\text{Volume} = \lim \sum f(x_k, y_k)\, \Delta A_k = \iint\limits_{R} f(x, y)\, dA$$

FIGURE 13.8 We define the volumes of solids with curved bases the same way we define the volumes of solids with rectangular bases.

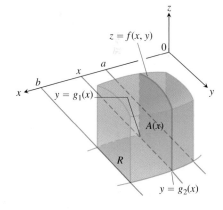

FIGURE 13.9 The area of the vertical slice shown here is $A(x)$. To calculate the volume of the solid, we integrate this area from $x = a$ to $x = b$:

$$\int_{a}^{b} A(x)\, dx = \int_{a}^{b} \int_{g_1(x)}^{g_2(x)} f(x, y)\, dy\, dx.$$

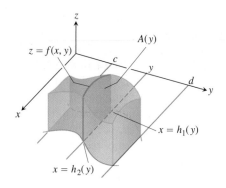

FIGURE 13.10 The volume of the solid shown here is

$$\int_c^d A(y) \, dy = \int_c^d \int_{h_1(y)}^{h_2(y)} f(x, y) \, dx \, dy.$$

It is the same volume as calculated in Figure 13.9.

and then integrate $A(x)$ from $x = a$ to $x = b$ to get the volume as an iterated integral:

$$V = \int_a^b A(x) \, dx = \int_a^b \int_{g_1(x)}^{g_2(x)} f(x, y) \, dy \, dx. \tag{1}$$

Similarly, if R is a region like the one shown in Figure 13.10, bounded by the curves $x = h_2(y)$ and $x = h_1(y)$ and the lines $y = c$ and $y = d$, then the volume calculated by slicing is given by the iterated integral

$$\text{Volume} = \int_c^d \int_{h_1(y)}^{h_2(y)} f(x, y) \, dx \, dy. \tag{2}$$

That the iterated integrals in Equations (1) and (2) both give the volume that we defined to be the double integral of f over R is a consequence of the following stronger form of Fubini's Theorem.

THEOREM 2—Fubini's Theorem (Stronger Form) Let $f(x, y)$ be continuous on a region R.

1. If R is defined by $a \le x \le b$, $g_1(x) \le y \le g_2(x)$, with g_1 and g_2 continuous on $[a, b]$, then

$$\iint\limits_R f(x, y) \, dA = \int_a^b \int_{g_1(x)}^{g_2(x)} f(x, y) \, dy \, dx.$$

2. If R is defined by $c \le y \le d$, $h_1(y) \le x \le h_2(y)$, with h_1 and h_2 continuous on $[c, d]$, then

$$\iint\limits_R f(x, y) \, dA = \int_c^d \int_{h_1(y)}^{h_2(y)} f(x, y) \, dx \, dy.$$

EXAMPLE 1 Find the volume of the prism whose base is the triangle in the xy-plane bounded by the x-axis and the lines $y = x$ and $x = 1$ and whose top lies in the plane

$$z = f(x, y) = 3 - x - y.$$

Solution See Figure 13.11 on the next page. For any x between 0 and 1, y may vary from $y = 0$ to $y = x$ (Figure 13.11b). Hence,

$$V = \int_0^1 \int_0^x (3 - x - y) \, dy \, dx = \int_0^1 \left[3y - xy - \frac{y^2}{2} \right]_{y=0}^{y=x} dx$$

$$= \int_0^1 \left(3x - \frac{3x^2}{2} \right) dx = \left[\frac{3x^2}{2} - \frac{x^3}{2} \right]_{x=0}^{x=1} = 1.$$

When the order of integration is reversed (Figure 13.11c), the integral for the volume is

$$V = \int_0^1 \int_y^1 (3 - x - y) \, dx \, dy = \int_0^1 \left[3x - \frac{x^2}{2} - xy \right]_{x=y}^{x=1} dy$$

$$= \int_0^1 \left(3 - \frac{1}{2} - y - 3y + \frac{y^2}{2} + y^2 \right) dy$$

$$= \int_0^1 \left(\frac{5}{2} - 4y + \frac{3}{2}y^2 \right) dy = \left[\frac{5}{2}y - 2y^2 + \frac{y^3}{2} \right]_{y=0}^{y=1} = 1.$$

The two integrals are equal, as they should be. ∎

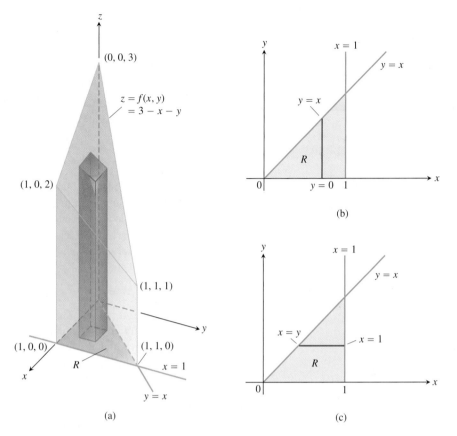

FIGURE 13.11 (a) Prism with a triangular base in the xy-plane. The volume of this prism is defined as a double integral over R. To evaluate it as an iterated integral, we may integrate first with respect to y and then with respect to x, or the other way around (Example 1). (b) Integration limits of

$$\int_{x=0}^{x=1} \int_{y=0}^{y=x} f(x, y) \, dy \, dx.$$

If we integrate first with respect to y, we integrate along a vertical line through R and then integrate from left to right to include all the vertical lines in R. (c) Integration limits of

$$\int_{y=0}^{y=1} \int_{x=y}^{x=1} f(x, y) \, dx \, dy.$$

If we integrate first with respect to x, we integrate along a horizontal line through R and then integrate from bottom to top to include all the horizontal lines in R.

Although Fubini's Theorem assures us that a double integral may be calculated as an iterated integral in either order of integration, the value of one integral may be easier to find than the value of the other. The next example shows how this can happen.

EXAMPLE 2   Calculate

$$\iint\limits_R \frac{\sin x}{x} \, dA,$$

where R is the triangle in the xy-plane bounded by the x-axis, the line $y = x$, and the line $x = 1$.

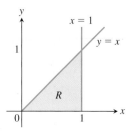

FIGURE 13.12 The region of integration in Example 2.

(a)

(b)

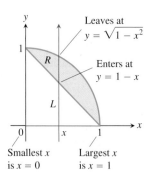

(c)

FIGURE 13.13 Finding the limits of integration when integrating first with respect to y and then with respect to x.

Solution The region of integration is shown in Figure 13.12. If we integrate first with respect to y and then with respect to x, we find

$$\int_0^1 \left(\int_0^x \frac{\sin x}{x} \, dy \right) dx = \int_0^1 \left(y \frac{\sin x}{x} \right]_{y=0}^{y=x} \right) dx = \int_0^1 \sin x \, dx$$
$$= -\cos(1) + 1 \approx 0.46.$$

If we reverse the order of integration and attempt to calculate

$$\int_0^1 \int_y^1 \frac{\sin x}{x} \, dx \, dy,$$

we run into a problem, because $\int ((\sin x)/x) \, dx$ cannot be expressed in terms of elementary functions (there is no simple antiderivative).

There is no general rule for predicting which order of integration will be the good one in circumstances like these. If the order you first choose doesn't work, try the other. Sometimes neither order will work, and then we need to use numerical approximations. ∎

Finding Limits of Integration

We now give a procedure for finding limits of integration that applies for many regions in the plane. Regions that are more complicated, and for which this procedure fails, can often be split up into pieces on which the procedure works.

When faced with evaluating $\iint_R f(x, y) \, dA$, integrating first with respect to y and then with respect to x, do the following:

1. *Sketch.* Sketch the region of integration and label the bounding curves (Figure 13.13a).

2. *Find the y-limits of integration.* Imagine a vertical line L cutting through R in the direction of increasing y. Mark the y-values where L enters and leaves. These are the y-limits of integration and are usually functions of x (instead of constants) (Figure 13.13b).

3. *Find the x-limits of integration.* Choose x-limits that include all the vertical lines through R. The integral shown here (see Figure 13.13c) is

$$\iint_R f(x, y) \, dA = \int_{x=0}^{x=1} \int_{y=1-x}^{y=\sqrt{1-x^2}} f(x, y) \, dy \, dx.$$

To evaluate the same double integral as an iterated integral with the order of integration reversed, use horizontal lines instead of vertical lines in Steps 2 and 3. The integral is

$$\iint_R f(x, y) \, dA = \int_0^1 \int_{1-y}^{\sqrt{1-y^2}} f(x, y) \, dx \, dy.$$

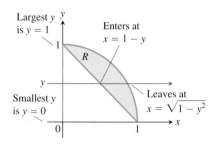

EXAMPLE 3   Sketch the region of integration for the integral

$$\int_0^2 \int_{x^2}^{2x} (4x + 2) \, dy \, dx$$

and write an equivalent integral with the order of integration reversed.

Solution   The region of integration is given by the inequalities $x^2 \leq y \leq 2x$ and $0 \leq x \leq 2$. It is therefore the region bounded by the curves $y = x^2$ and $y = 2x$ between $x = 0$ and $x = 2$ (Figure 13.14a).

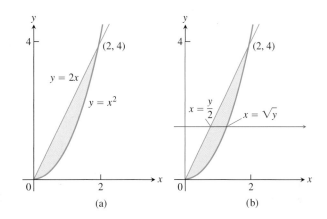

FIGURE 13.14   Region of integration for Example 3.

To find limits for integrating in the reverse order, we imagine a horizontal line passing from left to right through the region. It enters at $x = y/2$ and leaves at $x = \sqrt{y}$. To include all such lines, we let y run from $y = 0$ to $y = 4$ (Figure 13.14b). The integral is

$$\int_0^4 \int_{y/2}^{\sqrt{y}} (4x + 2) \, dx \, dy.$$

The common value of these integrals is 8.   ∎

USING TECHNOLOGY   Multiple Integration

Most CAS can calculate both multiple and iterated integrals. The typical procedure is to apply the CAS integrate command in nested iterations according to the order of integration you specify.

Integral	Typical CAS Formulation
$\iint x^2 y \, dx \, dy$	int (int $(x \wedge 2 * y, x), y$) ;
$\int_{-\pi/3}^{\pi/4} \int_0^1 x \cos y \, dx \, dy$	int (int $(x * \cos (y), x = 0 .. 1), y = -\text{Pi}/3 .. \text{Pi}/4$);

If a CAS cannot produce an exact value for a definite integral, it can usually find an approximate value numerically. Setting up a multiple integral for a CAS to solve can be a difficult task, and requires an understanding of how to describe the boundaries of the region and set up an appropriate integral.

Properties of Double Integrals

Like single integrals, double integrals of continuous functions have algebraic properties that are useful in computations and applications.

If $f(x, y)$ and $g(x, y)$ are continuous, then

1. *Constant Multiple:* $\displaystyle\iint\limits_{R} cf(x, y)\, dA = c \iint\limits_{R} f(x, y)\, dA$ (any number c)

2. *Sum and Difference:*

$$\iint\limits_{R} (f(x, y) \pm g(x, y))\, dA = \iint\limits_{R} f(x, y)\, dA \pm \iint\limits_{R} g(x, y)\, dA$$

3. *Domination:*

   (a) $\displaystyle\iint\limits_{R} f(x, y)\, dA \geq 0$ if $f(x, y) \geq 0$ on R

   (b) $\displaystyle\iint\limits_{R} f(x, y)\, dA \geq \iint\limits_{R} g(x, y)\, dA$ if $f(x, y) \geq g(x, y)$ on R

4. *Additivity:* $\displaystyle\iint\limits_{R} f(x, y)\, dA = \iint\limits_{R_1} f(x, y)\, dA + \iint\limits_{R_2} f(x, y)\, dA$

if R is the union of two nonoverlapping regions R_1 and R_2 (Figure 13.7).

The idea behind these properties is that integrals behave like sums. If the function $f(x, y)$ is replaced by its constant multiple $cf(x, y)$, then a Riemann sum for f

$$S_n = \sum_{k=1}^{n} f(x_k, y_k)\, \Delta A_k$$

is replaced by a Riemann sum for cf

$$\sum_{k=1}^{n} cf(x_k, y_k)\, \Delta A_k = c \sum_{k=1}^{n} f(x_k, y_k)\, \Delta A_k = cS_n.$$

Taking limits as $n \to \infty$ shows that $c \lim_{n\to\infty} S_n = c \iint_{R} f\, dA$ and $\lim_{n\to\infty} cS_n = \iint_{R} cf\, dA$ are equal. It follows that the constant multiple property carries over from sums to double integrals.

The other properties are also easy to verify for Riemann sums, and carry over to double integrals for the same reason. While this discussion gives the idea, an actual proof that these properties hold requires a more careful analysis of how Riemann sums converge.

EXERCISES 13.2

In Exercises 1–6, sketch the region of integration and evaluate the integral.

1. $\int_0^\pi \int_0^x x \sin y \, dy \, dx$

2. $\int_0^\pi \int_0^{\sin x} y \, dy \, dx$

3. $\int_1^{\ln 8} \int_0^{\ln y} e^{x+y} \, dx \, dy$

4. $\int_1^2 \int_y^{y^2} dx \, dy$

5. $\int_0^1 \int_0^{y^2} 3y^3 e^{xy} \, dx \, dy$

6. $\int_1^4 \int_0^{\sqrt{x}} \frac{3}{2} e^{y/\sqrt{x}} \, dy \, dx$

In Exercises 7–10, integrate f over the given region.

7. Quadrilateral $f(x, y) = x/y$ over the region in the first quadrant bounded by the lines $y = x$, $y = 2x$, $x = 1$, $x = 2$

8. Triangle $f(x, y) = x^2 + y^2$ over the triangular region with vertices $(0, 0)$, $(1, 0)$, and $(0, 1)$

9. Triangle $f(u, v) = v - \sqrt{u}$ over the triangular region cut from the first quadrant of the uv-plane by the line $u + v = 1$

10. Curved region $f(s, t) = e^s \ln t$ over the region in the first quadrant of the st-plane that lies above the curve $s = \ln t$ from $t = 1$ to $t = 2$

Each of Exercises 11–14 gives an integral over a region in a Cartesian coordinate plane. Sketch the region and evaluate the integral.

11. $\int_{-2}^0 \int_v^{-v} 2 \, dp \, dv$ (the pv-plane)

12. $\int_0^1 \int_0^{\sqrt{1-s^2}} 8t \, dt \, ds$ (the st-plane)

13. $\int_{-\pi/3}^{\pi/3} \int_0^{\sec t} 3 \cos t \, du \, dt$ (the tu-plane)

14. $\int_0^3 \int_1^{4-2u} \frac{4 - 2u}{v^2} \, dv \, du$ (the uv-plane)

In Exercises 15–24, sketch the region of integration and write an equivalent double integral with the order of integration reversed.

15. $\int_0^1 \int_2^{4-2x} dy \, dx$

16. $\int_0^2 \int_{y-2}^0 dx \, dy$

17. $\int_0^1 \int_y^{\sqrt{y}} dx \, dy$

18. $\int_0^1 \int_{1-x}^{1-x^2} dy \, dx$

19. $\int_0^1 \int_1^{e^x} dy \, dx$

20. $\int_0^{\ln 2} \int_{e^y}^2 dx \, dy$

21. $\int_0^{3/2} \int_0^{9-4x^2} 16x \, dy \, dx$

22. $\int_0^2 \int_0^{4-y^2} y \, dx \, dy$

23. $\int_0^1 \int_{-\sqrt{1-y^2}}^{\sqrt{1-y^2}} 3y \, dx \, dy$

24. $\int_0^2 \int_{-\sqrt{4-x^2}}^{\sqrt{4-x^2}} 6x \, dy \, dx$

In Exercises 25–34, sketch the region of integration, reverse the order of integration, and evaluate the integral.

25. $\int_0^\pi \int_x^\pi \frac{\sin y}{y} \, dy \, dx$

26. $\int_0^2 \int_x^2 2y^2 \sin xy \, dy \, dx$

27. $\int_0^1 \int_y^1 x^2 e^{xy} \, dx \, dy$

28. $\int_0^2 \int_0^{4-x^2} \frac{xe^{2y}}{4 - y} \, dy \, dx$

29. $\int_0^{2\sqrt{\ln 3}} \int_{y/2}^{\sqrt{\ln 3}} e^{x^2} \, dx \, dy$

30. $\int_0^3 \int_{\sqrt{x/3}}^1 e^{y^3} \, dy \, dx$

31. $\int_0^{1/16} \int_{y^{1/4}}^{1/2} \cos(16\pi x^5) \, dx \, dy$

32. $\int_0^8 \int_{\sqrt[3]{x}}^2 \frac{dy \, dx}{y^4 + 1}$

33. Square region $\iint_R (y - 2x^2) \, dA$ where R is the region bounded by the square $|x| + |y| = 1$

34. Triangular region $\iint_R xy \, dA$ where R is the region bounded by the lines $y = x$, $y = 2x$, and $x + y = 2$

35. Find the volume of the region bounded by the paraboloid $z = x^2 + y^2$ and below by the triangle enclosed by the lines $y = x$, $x = 0$, and $x + y = 2$ in the xy-plane.

36. Find the volume of the solid that is bounded above by the cylinder $z = x^2$ and below by the region enclosed by the parabola $y = 2 - x^2$ and the line $y = x$ in the xy-plane.

37. Find the volume of the solid whose base is the region in the xy-plane that is bounded by the parabola $y = 4 - x^2$ and the line $y = 3x$, while the top of the solid is bounded by the plane $z = x + 4$.

38. Find the volume of the solid in the first octant bounded by the coordinate planes, the cylinder $x^2 + y^2 = 4$, and the plane $z + y = 3$.

39. Find the volume of the solid in the first octant bounded by the coordinate planes, the plane $x = 3$, and the parabolic cylinder $z = 4 - y^2$.

40. Find the volume of the solid cut from the first octant by the surface $z = 4 - x^2 - y$.

41. Find the volume of the wedge cut from the first octant by the cylinder $z = 12 - 3y^2$ and the plane $x + y = 2$.

42. Find the volume of the solid cut from the square column $|x| + |y| \le 1$ by the planes $z = 0$ and $3x + z = 3$.

43. Find the volume of the solid that is bounded on the front and back by the planes $x = 2$ and $x = 1$, on the sides by the cylinders $y = \pm 1/x$, and above and below by the planes $z = x + 1$ and $z = 0$.

44. Find the volume of the solid bounded on the front and back by the planes $x = \pm\pi/3$, on the sides by the cylinders $y = \pm\sec x$, above by the cylinder $z = 1 + y^2$, and below by the xy-plane.

Improper double integrals can often be computed similarly to improper integrals of one variable. The first iteration of the following improper integrals is conducted just as if they were proper integrals. One then evaluates an improper integral of a single variable by taking appropriate limits, as in Section 7.7. Evaluate the improper integrals in Exercises 45–48 as iterated integrals.

45. $\displaystyle\int_{1}^{\infty}\int_{e^{-x}}^{1}\frac{1}{x^3 y}\,dy\,dx$

46. $\displaystyle\int_{-1}^{1}\int_{-1/\sqrt{1-x^2}}^{1/\sqrt{1-x^2}}(2y+1)\,dy\,dx$

47. $\displaystyle\int_{-\infty}^{\infty}\int_{-\infty}^{\infty}\frac{1}{(x^2+1)(y^2+1)}\,dx\,dy$

48. $\displaystyle\int_{0}^{\infty}\int_{0}^{\infty}xe^{-(x+2y)}\,dx\,dy$

In Exercises 49 and 50, approximate the double integral of $f(x, y)$ over the region R partitioned by the given vertical lines $x = a$ and horizontal lines $y = c$. In each subrectangle, use (x_k, y_k) as indicated for your approximation.

$$\iint_{R} f(x, y)\,dA \approx \sum_{k=1}^{n} f(x_k, y_k)\,\Delta A_k$$

49. $f(x, y) = x + y$ over the region R bounded above by the semicircle $y = \sqrt{1 - x^2}$ and below by the x-axis, using the partition $x = -1, -1/2, 0, 1/4, 1/2, 1$ and $y = 0, 1/2, 1$ with (x_k, y_k) the lower left corner in the kth subrectangle (provided the subrectangle lies within R)

50. $f(x, y) = x + 2y$ over the region R inside the circle $(x - 2)^2 + (y - 3)^2 = 1$ using the partition $x = 1, 3/2, 2, 5/2, 3$ and $y = 2, 5/2, 3, 7/2, 4$ with (x_k, y_k) the center (centroid) in the kth subrectangle (provided the subrectangle lies within R)

51. Circular sector Integrate $f(x, y) = \sqrt{4 - x^2}$ over the smaller sector cut from the disk $x^2 + y^2 \le 4$ by the rays $\theta = \pi/6$ and $\theta = \pi/2$.

52. Unbounded region Integrate $f(x, y) = 1/[(x^2 - x)(y - 1)^{2/3}]$ over the infinite rectangle $2 \le x < \infty, 0 \le y \le 2$.

53. Noncircular cylinder A solid right (noncircular) cylinder has its base R in the xy-plane and is bounded above by the paraboloid $z = x^2 + y^2$. The cylinder's volume is

$$V = \int_{0}^{1}\int_{0}^{y}(x^2 + y^2)\,dx\,dy + \int_{1}^{2}\int_{0}^{2-y}(x^2 + y^2)\,dx\,dy.$$

Sketch the base region R and express the cylinder's volume as a single iterated integral with the order of integration reversed. Then evaluate the integral to find the volume.

54. Converting to a double integral Evaluate the integral

$$\int_{0}^{2}(\tan^{-1}\pi x - \tan^{-1}x)\,dx.$$

(*Hint:* Write the integrand as an integral.)

55. Maximizing a double integral What region R in the xy-plane maximizes the value of

$$\iint_{R}(4 - x^2 - 2y^2)\,dA?$$

Give reasons for your answer.

56. Minimizing a double integral What region R in the xy-plane minimizes the value of

$$\iint_{R}(x^2 + y^2 - 9)\,dA?$$

Give reasons for your answer.

57. Is it possible to evaluate the integral of a continuous function $f(x, y)$ over a rectangular region in the xy-plane and get different answers depending on the order of integration? Give reasons for your answer.

58. How would you evaluate the double integral of a continuous function $f(x, y)$ over the region R in the xy-plane enclosed by the triangle with vertices $(0, 1), (2, 0)$, and $(1, 2)$? Give reasons for your answer.

59. Unbounded region Prove that

$$\int_{-\infty}^{\infty}\int_{-\infty}^{\infty} e^{-x^2 - y^2}\,dx\,dy = \lim_{b\to\infty}\int_{-b}^{b}\int_{-b}^{b} e^{-x^2 - y^2}\,dx\,dy$$

$$= 4\left(\int_{0}^{\infty} e^{-x^2}\,dx\right)^2.$$

60. Improper double integral Evaluate the improper integral

$$\int_{0}^{1}\int_{0}^{3}\frac{x^2}{(y - 1)^{2/3}}\,dy\,dx.$$

COMPUTER EXPLORATIONS

Use a CAS double-integral evaluator to estimate the values of the integrals in Exercises 61–64.

61. $\displaystyle\int_{1}^{3}\int_{1}^{x}\frac{1}{xy}\,dy\,dx$ **62.** $\displaystyle\int_{0}^{1}\int_{0}^{1} e^{-(x^2 + y^2)}\,dy\,dx$

63. $\displaystyle\int_{0}^{1}\int_{0}^{1}\tan^{-1}xy\,dy\,dx$ **64.** $\displaystyle\int_{-1}^{1}\int_{0}^{\sqrt{1-x^2}}3\sqrt{1 - x^2 - y^2}\,dy\,dx$

Use a CAS double-integral evaluator to find the integrals in Exercises 65–70. Then reverse the order of integration and evaluate, again with a CAS.

65. $\displaystyle\int_{0}^{1}\int_{2y}^{4} e^{x^2}\,dx\,dy$ **66.** $\displaystyle\int_{0}^{3}\int_{x^2}^{9} x\cos(y^2)\,dy\,dx$

67. $\displaystyle\int_{0}^{2}\int_{y^3}^{4\sqrt{2y}}(x^2 y - xy^2)\,dx\,dy$ **68.** $\displaystyle\int_{0}^{2}\int_{0}^{4-y^2} e^{xy}\,dx\,dy$

69. $\displaystyle\int_{1}^{2}\int_{0}^{x^2}\frac{1}{x + y}\,dy\,dx$ **70.** $\displaystyle\int_{1}^{2}\int_{y^3}^{8}\frac{1}{\sqrt{x^2 + y^2}}\,dx\,dy$

13.3 Area by Double Integration

In this section we show how to use double integrals to calculate the areas of bounded regions in the plane, and to find the average value of a function of two variables.

Areas of Bounded Regions in the Plane

If we take $f(x, y) = 1$ in the definition of the double integral over a region R in the preceding section, the Riemann sums reduce to

$$S_n = \sum_{k=1}^{n} f(x_k, y_k)\,\Delta A_k = \sum_{k=1}^{n} \Delta A_k. \tag{1}$$

This is simply the sum of the areas of the small rectangles in the partition of R, and approximates what we would like to call the area of R. As the norm of a partition of R approaches zero, the height and width of all rectangles in the partition approach zero, and the coverage of R becomes increasingly complete (Figure 13.6). We define the area of R to be the limit

$$\lim_{\|P\| \to 0} \sum_{k=1}^{n} \Delta A_k = \iint\limits_{R} dA \tag{2}$$

DEFINITION   The **area** of a closed, bounded plane region R is

$$A = \iint\limits_{R} dA.$$

As with the other definitions in this chapter, the definition here applies to a greater variety of regions than does the earlier single-variable definition of area, but it agrees with the earlier definition on regions to which they both apply. To evaluate the integral in the definition of area, we integrate the constant function $f(x, y) = 1$ over R.

EXAMPLE 1   Find the area of the region R bounded by $y = x$ and $y = x^2$ in the first quadrant.

Solution   We sketch the region (Figure 13.15), noting where the two curves intersect, and calculate the area as

$$A = \int_{0}^{1} \int_{x^2}^{x} dy\,dx = \int_{0}^{1} \left[y \right]_{x^2}^{x} dx$$

$$= \int_{0}^{1} (x - x^2)\,dx = \left[\frac{x^2}{2} - \frac{x^3}{3} \right]_{0}^{1} = \frac{1}{6}.$$

Notice that the single integral $\int_{0}^{1} (x - x^2)\,dx$, obtained from evaluating the inside iterated integral, is the integral for the area between these two curves using the method of Section 5.6. ∎

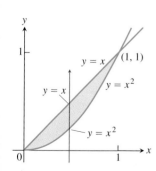
FIGURE 13.15   The region in Example 1.

EXAMPLE 2   Find the area of the region R enclosed by the parabola $y = x^2$ and the line $y = x + 2$.

Solution   If we divide R into the regions R_1 and R_2 shown in Figure 13.16a, we may calculate the area as

$$A = \iint\limits_{R_1} dA + \iint\limits_{R_2} dA = \int_0^1 \int_{-\sqrt{y}}^{\sqrt{y}} dx\, dy + \int_1^4 \int_{y-2}^{\sqrt{y}} dx\, dy.$$

On the other hand, reversing the order of integration (Figure 13.16b) gives

$$A = \int_{-1}^2 \int_{x^2}^{x+2} dy\, dx.$$

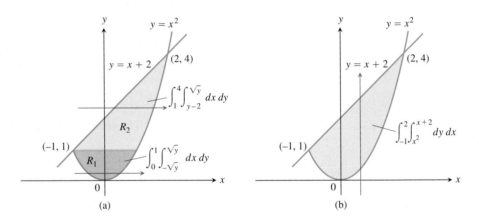

(a)   (b)

FIGURE 13.16   Calculating this area takes (a) two double integrals if the first integration is with respect to x, but (b) only one if the first integration is with respect to y (Example 2).

This second result, which requires only one integral, is simpler and is the only one we would bother to write down in practice. The area is

$$A = \int_{-1}^2 \left[y \right]_{x^2}^{x+2} dx = \int_{-1}^2 (x + 2 - x^2)\, dx = \left[\frac{x^2}{2} + 2x - \frac{x^3}{3} \right]_{-1}^2 = \frac{9}{2}. \qquad \blacksquare$$

Average Value

The average value of an integrable function of one variable on a closed interval is the integral of the function over the interval divided by the length of the interval. For an integrable function of two variables defined on a bounded region in the plane, the average value is the integral over the region divided by the area of the region. This can be visualized by thinking of the function as giving the height at one instant of some water sloshing around in a tank whose vertical walls lie over the boundary of the region. The average height of the water in the tank can be found by letting the water settle down to a constant height. The height is then equal to the volume of water in the tank divided by the area of R. We are led to define the average value of an integrable function f over a region R as follows:

$$\text{\textbf{Average value} of } f \text{ over } R = \frac{1}{\text{area of } R} \iint\limits_R f \, dA. \qquad (3)$$

If f is the temperature of a thin plate covering R, then the double integral of f over R divided by the area of R is the plate's average temperature. If $f(x, y)$ is the distance from the point (x, y) to a fixed point P, then the average value of f over R is the average distance of points in R from P.

EXAMPLE 3 Find the average value of $f(x, y) = x \cos xy$ over the rectangle $R: 0 \le x \le \pi, \ 0 \le y \le 1$.

Solution The value of the integral of f over R is

$$\int_0^\pi \int_0^1 x \cos xy \, dy \, dx = \int_0^\pi \left[\sin xy \right]_{y=0}^{y=1} dx \qquad \int x \cos xy \, dy = \sin xy + C$$

$$= \int_0^\pi (\sin x - 0) \, dx = -\cos x \Big]_0^\pi = 1 + 1 = 2.$$

The area of R is π. The average value of f over R is $2/\pi$. ∎

EXERCISES 13.3

In Exercises 1–8, sketch the region bounded by the given lines and curves. Then express the region's area as an iterated double integral and evaluate the integral.

1. The coordinate axes and the line $x + y = 2$
2. The lines $x = 0, y = 2x$, and $y = 4$
3. The parabola $x = -y^2$ and the line $y = x + 2$
4. The parabola $x = y - y^2$ and the line $y = -x$
5. The curve $y = e^x$ and the lines $y = 0, x = 0$, and $x = \ln 2$
6. The curves $y = \ln x$ and $y = 2 \ln x$ and the line $x = e$, in the first quadrant
7. The parabolas $x = y^2$ and $x = 2y - y^2$
8. The parabolas $x = y^2 - 1$ and $x = 2y^2 - 2$

The integrals and sums of integrals in Exercises 9–14 give the areas of regions in the xy-plane. Sketch each region, label each bounding curve with its equation, and give the coordinates of the points where the curves intersect. Then find the area of the region.

9. $\int_0^6 \int_{y^2/3}^{2y} dx \, dy$

10. $\int_0^3 \int_{-x}^{x(2-x)} dy \, dx$

11. $\int_0^{\pi/4} \int_{\sin x}^{\cos x} dy \, dx$

12. $\int_{-1}^2 \int_{y^2}^{y+2} dx \, dy$

13. $\int_{-1}^0 \int_{-2x}^{1-x} dy \, dx + \int_0^2 \int_{-x/2}^{1-x} dy \, dx$

14. $\int_0^2 \int_{x^2-4}^0 dy \, dx + \int_0^4 \int_0^{\sqrt{x}} dy \, dx$

15. Find the average value of $f(x, y) = \sin (x + y)$ over
    a. the rectangle $0 \le x \le \pi, \ 0 \le y \le \pi$
    b. the rectangle $0 \le x \le \pi, \ 0 \le y \le \pi/2$

16. Which do you think will be larger, the average value of $f(x, y) = xy$ over the square $0 \le x \le 1, 0 \le y \le 1$, or the average value of f over the quarter circle $x^2 + y^2 \le 1$ in the first quadrant? Calculate them to find out.

17. Find the average height of the paraboloid $z = x^2 + y^2$ over the square $0 \le x \le 2, 0 \le y \le 2$.

18. Find the average value of $f(x, y) = 1/(xy)$ over the square $\ln 2 \le x \le 2 \ln 2, \ln 2 \le y \le 2 \ln 2$.

19. **Bacterium population** If $f(x, y) = (10{,}000e^y)/(1 + |x|/2)$ represents the "population density" of a certain bacterium on the xy-plane, where x and y are measured in centimeters, find the total population of bacteria within the rectangle $-5 \le x \le 5$ and $-2 \le y \le 0$.

20. **Regional population** If $f(x, y) = 100\,(y + 1)$ represents the population density of a planar region on Earth, where x and y are measured in miles, find the number of people in the region bounded by the curves $x = y^2$ and $x = 2y - y^2$.

21. **Average temperature in Texas** According to the *Texas Almanac*, Texas has 254 counties and a National Weather Service station in each county. Assume that at time t_0, each of the 254 weather stations recorded the local temperature. Find a formula that would give a reasonable approximation to the average temperature in Texas at time t_0. Your answer should involve information that you would expect to be readily available in the *Texas Almanac*.

22. If $y = f(x)$ is a nonnegative continuous function over the closed interval $a \le x \le b$, show that the double integral definition of area for the closed plane region bounded by the graph of f, the vertical lines $x = a$ and $x = b$, and the x-axis, agrees with the definition for area beneath the curve in Section 5.3.

13.4 Double Integrals in Polar Form

Integrals are sometimes easier to evaluate if we change to polar coordinates. This section shows how to accomplish the change and how to evaluate integrals over regions whose boundaries are given by polar equations.

Integrals in Polar Coordinates

When we defined the double integral of a function over a region R in the xy-plane, we began by cutting R into rectangles whose sides were parallel to the coordinate axes. These were the natural shapes to use because their sides have either constant x-values or constant y-values. In polar coordinates, the natural shape is a "polar rectangle" whose sides have constant r- and θ-values.

Suppose that a function $f(r, \theta)$ is defined over a region R that is bounded by the rays $\theta = \alpha$ and $\theta = \beta$ and by the continuous curves $r = g_1(\theta)$ and $r = g_2(\theta)$. Suppose also that $0 \le g_1(\theta) \le g_2(\theta) \le a$ for every value of θ between α and β. Then R lies in a fan-shaped region Q defined by the inequalities $0 \le r \le a$ and $\alpha \le \theta \le \beta$. See Figure 13.17.

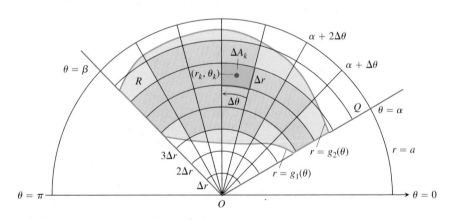

FIGURE 13.17 The region R: $g_1(\theta) \le r \le g_2(\theta)$, $\alpha \le \theta \le \beta$, is contained in the fan-shaped region Q: $0 \le r \le a$, $\alpha \le \theta \le \beta$. The partition of Q by circular arcs and rays induces a partition of R.

We cover Q by a grid of circular arcs and rays. The arcs are cut from circles centered at the origin, with radii $\Delta r, 2\Delta r, \ldots, m\Delta r$, where $\Delta r = a/m$. The rays are given by

$$\theta = \alpha, \qquad \theta = \alpha + \Delta\theta, \qquad \theta = \alpha + 2\Delta\theta, \qquad \ldots, \qquad \theta = \alpha + m'\Delta\theta = \beta,$$

where $\Delta\theta = (\beta - \alpha)/m'$. The arcs and rays partition Q into small patches called "polar rectangles."

We number the polar rectangles that lie inside R (the order does not matter), calling their areas $\Delta A_1, \Delta A_2, \ldots, \Delta A_n$. We let (r_k, θ_k) be any point in the polar rectangle whose area is ΔA_k. We then form the sum

$$S_n = \sum_{k=1}^{n} f(r_k, \theta_k)\, \Delta A_k.$$

If f is continuous throughout R, this sum will approach a limit as we refine the grid to make Δr and $\Delta\theta$ go to zero. The limit is called the double integral of f over R. In symbols,

$$\lim_{n \to \infty} S_n = \iint_R f(r, \theta)\, dA.$$

To evaluate this limit, we first have to write the sum S_n in a way that expresses ΔA_k in terms of Δr and $\Delta\theta$. For convenience we choose r_k to be the average of the radii of the inner and outer arcs bounding the kth polar rectangle ΔA_k. The radius of the inner arc bounding ΔA_k is then $r_k - (\Delta r/2)$ (Figure 13.18). The radius of the outer arc is $r_k + (\Delta r/2)$.

The area of a wedge-shaped sector of a circle having radius r and angle θ is

$$A = \frac{1}{2}\theta \cdot r^2,$$

as can be seen by multiplying πr^2, the area of the circle, by $\theta/2\pi$, the fraction of the circle's area contained in the wedge. So the areas of the circular sectors subtended by these arcs at the origin are

Inner radius: $\quad \dfrac{1}{2}\left(r_k - \dfrac{\Delta r}{2}\right)^2 \Delta\theta$

Outer radius: $\quad \dfrac{1}{2}\left(r_k + \dfrac{\Delta r}{2}\right)^2 \Delta\theta.$

Therefore,

$$\Delta A_k = \text{area of large sector} - \text{area of small sector}$$

$$= \frac{\Delta\theta}{2}\left[\left(r_k + \frac{\Delta r}{2}\right)^2 - \left(r_k - \frac{\Delta r}{2}\right)^2\right] = \frac{\Delta\theta}{2}(2r_k\,\Delta r) = r_k\,\Delta r\,\Delta\theta.$$

Combining this result with the sum defining S_n gives

$$S_n = \sum_{k=1}^{n} f(r_k, \theta_k) r_k\, \Delta r\, \Delta\theta.$$

As $n \to \infty$ and the values of Δr and $\Delta\theta$ approach zero, these sums converge to the double integral

$$\lim_{n \to \infty} S_n = \iint_R f(r, \theta)\, r\, dr\, d\theta.$$

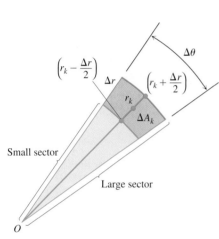

FIGURE 13.18 The observation that

$$\Delta A_k = \begin{pmatrix} \text{area of} \\ \text{large sector} \end{pmatrix} - \begin{pmatrix} \text{area of} \\ \text{small sector} \end{pmatrix}$$

leads to the formula $\Delta A_k = r_k\,\Delta r\,\Delta\theta$.

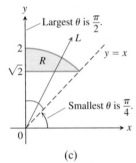

FIGURE 13.19 Finding the limits of integration in polar coordinates.

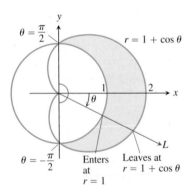

FIGURE 13.20 Finding the limits of integration in polar coordinates for the region in Example 1.

A version of Fubini's Theorem says that the limit approached by these sums can be evaluated by repeated single integrations with respect to r and θ as

$$\iint_R f(r, \theta) \, dA = \int_{\theta=\alpha}^{\theta=\beta} \int_{r=g_1(\theta)}^{r=g_2(\theta)} f(r, \theta) \, r \, dr \, d\theta.$$

Finding Limits of Integration

The procedure for finding limits of integration in rectangular coordinates also works for polar coordinates. To evaluate $\iint_R f(r, \theta) \, dA$ over a region R in polar coordinates, integrating first with respect to r and then with respect to θ, take the following steps.

1. *Sketch*: Sketch the region and label the bounding curves (Figure 13.19a).

2. *Find the r-limits of integration*: Imagine a ray L from the origin cutting through R in the direction of increasing r. Mark the r-values where L enters and leaves R. These are the r-limits of integration. They usually depend on the angle θ that L makes with the positive x-axis (Figure 13.19b).

3. *Find the θ-limits of integration*: Find the smallest and largest θ-values that bound R. These are the θ-limits of integration (Figure 13.19c). The integral is

$$\iint_R f(r, \theta) \, dA = \int_{\theta=\pi/4}^{\theta=\pi/2} \int_{r=\sqrt{2}\csc\theta}^{r=2} f(r, \theta) \, r \, dr \, d\theta.$$

EXAMPLE 1   Find the limits of integration for integrating $f(r, \theta)$ over the region R that lies inside the cardioid $r = 1 + \cos\theta$ and outside the circle $r = 1$.

Solution

1. We first sketch the region and label the bounding curves (Figure 13.20).

2. Next we find the *r-limits of integration*. A typical ray from the origin enters R where $r = 1$ and leaves where $r = 1 + \cos\theta$.

3. Finally we find the *θ-limits of integration*. The rays from the origin that intersect R run from $\theta = -\pi/2$ to $\theta = \pi/2$. The integral is

$$\int_{-\pi/2}^{\pi/2} \int_1^{1+\cos\theta} f(r, \theta) \, r \, dr \, d\theta. \qquad \blacksquare$$

If $f(r, \theta)$ is the constant function whose value is 1, then the integral of f over R is the area of R.

Area in Polar Coordinates

The area of a closed and bounded region R in the polar coordinate plane is

$$A = \iint_R r \, dr \, d\theta.$$

This formula for area is consistent with all earlier formulas, although we do not prove this fact.

EXAMPLE 2   Find the area enclosed by the lemniscate $r^2 = 4 \cos 2\theta$.

Solution   We graph the lemniscate to determine the limits of integration (Figure 13.21) and see from the symmetry of the region that the total area is 4 times the first-quadrant portion.

$$A = 4 \int_0^{\pi/4} \int_0^{\sqrt{4 \cos 2\theta}} r \, dr \, d\theta = 4 \int_0^{\pi/4} \left[\frac{r^2}{2} \right]_{r=0}^{r=\sqrt{4 \cos 2\theta}} d\theta$$

$$= 4 \int_0^{\pi/4} 2 \cos 2\theta \, d\theta = 4 \sin 2\theta \Big]_0^{\pi/4} = 4.$$

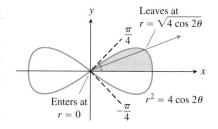

FIGURE 13.21   To integrate over the shaded region, we run r from 0 to $\sqrt{4 \cos 2\theta}$ and θ from 0 to $\pi/4$ (Example 2).

Changing Cartesian Integrals into Polar Integrals

The procedure for changing a Cartesian integral $\iint_R f(x, y) \, dx \, dy$ into a polar integral has two steps. First substitute $x = r \cos \theta$ and $y = r \sin \theta$, and replace $dx \, dy$ by $r \, dr \, d\theta$ in the Cartesian integral. Then supply polar limits of integration for the boundary of R. The Cartesian integral then becomes

$$\iint_R f(x, y) \, dx \, dy = \iint_G f(r \cos \theta, r \sin \theta) \, r \, dr \, d\theta,$$

where G denotes the region of integration in polar coordinates. This is like the substitution method in Chapter 5 except that there are now two variables to substitute for instead of one. Notice that $dx \, dy$ is not replaced by $dr \, d\theta$ but by $r \, dr \, d\theta$. A more general discussion of changes of variables (substitutions) in multiple integrals is given in Section 13.8.

EXAMPLE 3   Evaluate

$$\iint_R e^{x^2+y^2} \, dy \, dx,$$

where R is the semicircular region bounded by the x-axis and the curve $y = \sqrt{1 - x^2}$ (Figure 13.22).

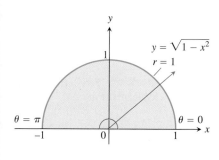

FIGURE 13.22   The semicircular region in Example 3 is the region

$$0 \le r \le 1, \qquad 0 \le \theta \le \pi.$$

Solution   In Cartesian coordinates, the integral in question is a nonelementary integral and there is no direct way to integrate $e^{x^2+y^2}$ with respect to either x or y. Yet this integral and others like it are important in mathematics—in statistics, for example—and we need to find a way to evaluate it. Polar coordinates save the day. Substituting $x = r \cos \theta, y = r \sin \theta$ and replacing $dy \, dx$ by $r \, dr \, d\theta$ enables us to evaluate the integral as

$$\iint_R e^{x^2+y^2} \, dy \, dx = \int_0^{\pi} \int_0^1 e^{r^2} r \, dr \, d\theta = \int_0^{\pi} \left[\frac{1}{2} e^{r^2} \right]_0^1 d\theta$$

$$= \int_0^{\pi} \frac{1}{2} (e - 1) \, d\theta = \frac{\pi}{2} (e - 1).$$

The r in the $r \, dr \, d\theta$ was just what we needed to integrate e^{r^2}. Without it, we would have been unable to proceed.

EXERCISES 13.4

In Exercises 1–16, change the Cartesian integral into an equivalent polar integral. Then evaluate the polar integral.

1. $\displaystyle\int_{-1}^{1}\int_{0}^{\sqrt{1-x^2}} dy\, dx$

2. $\displaystyle\int_{-1}^{1}\int_{-\sqrt{1-x^2}}^{\sqrt{1-x^2}} dy\, dx$

3. $\displaystyle\int_{0}^{1}\int_{0}^{\sqrt{1-y^2}} (x^2+y^2)\, dx\, dy$

4. $\displaystyle\int_{-1}^{1}\int_{-\sqrt{1-y^2}}^{\sqrt{1-y^2}} (x^2+y^2)\, dy\, dx$

5. $\displaystyle\int_{-a}^{a}\int_{-\sqrt{a^2-x^2}}^{\sqrt{a^2-x^2}} dy\, dx$

6. $\displaystyle\int_{0}^{2}\int_{0}^{\sqrt{4-y^2}} (x^2+y^2)\, dx\, dy$

7. $\displaystyle\int_{0}^{6}\int_{0}^{y} x\, dx\, dy$

8. $\displaystyle\int_{0}^{2}\int_{0}^{x} y\, dy\, dx$

9. $\displaystyle\int_{-1}^{0}\int_{-\sqrt{1-x^2}}^{0} \frac{2}{1+\sqrt{x^2+y^2}}\, dy\, dx$

10. $\displaystyle\int_{-1}^{1}\int_{-\sqrt{1-y^2}}^{0} \frac{4\sqrt{x^2+y^2}}{1+x^2+y^2}\, dx\, dy$

11. $\displaystyle\int_{0}^{\ln 2}\int_{0}^{\sqrt{(\ln 2)^2-y^2}} e^{\sqrt{x^2+y^2}}\, dx\, dy$

12. $\displaystyle\int_{0}^{1}\int_{0}^{\sqrt{1-x^2}} e^{-(x^2+y^2)}\, dy\, dx$

13. $\displaystyle\int_{0}^{2}\int_{0}^{\sqrt{1-(x-1)^2}} \frac{x+y}{x^2+y^2}\, dy\, dx$

14. $\displaystyle\int_{0}^{2}\int_{-\sqrt{1-(y-1)^2}}^{0} xy^2\, dx\, dy$

15. $\displaystyle\int_{-1}^{1}\int_{-\sqrt{1-y^2}}^{\sqrt{1-y^2}} \ln(x^2+y^2+1)\, dx\, dy$

16. $\displaystyle\int_{-1}^{1}\int_{-\sqrt{1-x^2}}^{\sqrt{1-x^2}} \frac{2}{(1+x^2+y^2)^2}\, dy\, dx$

17. Find the area of the region cut from the first quadrant by the curve $r = 2(2 - \sin 2\theta)^{1/2}$.

18. Cardioid overlapping a circle   Find the area of the region that lies inside the cardioid $r = 1 + \cos\theta$ and outside the circle $r = 1$.

19. One leaf of a rose   Find the area enclosed by one leaf of the rose $r = 12 \cos 3\theta$.

20. Snail shell   Find the area of the region enclosed by the positive x-axis and spiral $r = 4\theta/3,\, 0 \le \theta \le 2\pi$. The region looks like a snail shell.

21. Cardioid in the first quadrant   Find the area of the region cut from the first quadrant by the cardioid $r = 1 + \sin\theta$.

22. Overlapping cardioids   Find the area of the region common to the interiors of the cardioids $r = 1 + \cos\theta$ and $r = 1 - \cos\theta$.

In polar coordinates, the **average value** of a function over a region R (Section 13.3) is given by

$$\frac{1}{\text{Area}(R)}\iint_{R} f(r,\theta)\, r\, dr\, d\theta.$$

23. Average height of a hemisphere   Find the average height of the hemispherical surface $z = \sqrt{a^2 - x^2 - y^2}$ above the disk $x^2 + y^2 \le a^2$ in the xy-plane.

24. Average height of a cone   Find the average height of the (single) cone $z = \sqrt{x^2 + y^2}$ above the disk $x^2 + y^2 \le a^2$ in the xy-plane.

25. Average distance from interior of disk to center   Find the average distance from a point $P(x, y)$ in the disk $x^2 + y^2 \le a^2$ to the origin.

26. Average distance squared from a point in a disk to a point in its boundary   Find the average value of the *square* of the distance from the point $P(x, y)$ in the disk $x^2 + y^2 \le 1$ to the boundary point $A(1, 0)$.

27. Converting to a polar integral   Integrate $f(x, y) = [\ln(x^2 + y^2)]/\sqrt{x^2 + y^2}$ over the region $1 \le x^2 + y^2 \le e$.

28. Converting to a polar integral   Integrate $f(x, y) = [\ln(x^2 + y^2)]/(x^2 + y^2)$ over the region $1 \le x^2 + y^2 \le e^2$.

29. Volume of noncircular right cylinder   The region that lies inside the cardioid $r = 1 + \cos\theta$ and outside the circle $r = 1$ is the base of a solid right cylinder. The top of the cylinder lies in the plane $z = x$. Find the cylinder's volume.

30. Volume of noncircular right cylinder   The region enclosed by the lemniscate $r^2 = 2 \cos 2\theta$ is the base of a solid right cylinder whose top is bounded by the sphere $z = \sqrt{2 - r^2}$. Find the cylinder's volume.

31. Converting to polar integrals

a. The usual way to evaluate the improper integral $I = \int_{0}^{\infty} e^{-x^2}\, dx$ is first to calculate its square:
$$I^2 = \left(\int_{0}^{\infty} e^{-x^2}\, dx\right)\left(\int_{0}^{\infty} e^{-y^2}\, dy\right) = \int_{0}^{\infty}\int_{0}^{\infty} e^{-(x^2+y^2)}\, dx\, dy.$$
Evaluate the last integral using polar coordinates and solve the resulting equation for I.

b. Evaluate
$$\lim_{x\to\infty}\mathrm{erf}(x) = \lim_{x\to\infty}\int_{0}^{x}\frac{2e^{-t^2}}{\sqrt{\pi}}\, dt.$$

32. Converting to a polar integral   Evaluate the integral
$$\int_{0}^{\infty}\int_{0}^{\infty}\frac{1}{(1+x^2+y^2)^2}\, dx\, dy.$$

33. Existence   Integrate the function $f(x, y) = 1/(1 - x^2 - y^2)$ over the disk $x^2 + y^2 \le 3/4$. Does the integral of $f(x, y)$ over the disk $x^2 + y^2 \le 1$ exist? Give reasons for your answer.

34. Area formula in polar coordinates   Use the double integral in polar coordinates to derive the formula
$$A = \int_{\alpha}^{\beta}\frac{1}{2}r^2\, d\theta$$
for the area of the fan-shaped region between the origin and polar curve $r = f(\theta),\, \alpha \le \theta \le \beta$.

35. Average distance to a given point inside a disk   Let P_0 be a point inside a circle of radius a and let h denote the distance from

P_0 to the center of the circle. Let d denote the distance from an arbitrary point P to P_0. Find the average value of d^2 over the region enclosed by the circle. (*Hint:* Simplify your work by placing the center of the circle at the origin and P_0 on the x-axis.)

36. Area Suppose that the area of a region in the polar coordinate plane is

$$A = \int_{\pi/4}^{3\pi/4} \int_{\csc\theta}^{2\sin\theta} r \, dr \, d\theta.$$

Sketch the region and find its area.

COMPUTER EXPLORATIONS

In Exercises 37–40, use a CAS to change the Cartesian integrals into an equivalent polar integral and evaluate the polar integral. Perform the following steps in each exercise.

a. Plot the Cartesian region of integration in the xy-plane.

b. Change each boundary curve of the Cartesian region in part (a) to its polar representation by solving its Cartesian equation for r and θ.

c. Using the results in part (b), plot the polar region of integration in the $r\theta$-plane.

d. Change the integrand from Cartesian to polar coordinates. Determine the limits of integration from your plot in part (c) and evaluate the polar integral using the CAS integration utility.

37. $\displaystyle \int_0^1 \int_x^1 \frac{y}{x^2 + y^2} \, dy \, dx$ **38.** $\displaystyle \int_0^1 \int_0^{x/2} \frac{x}{x^2 + y^2} \, dy \, dx$

39. $\displaystyle \int_0^1 \int_{-y/3}^{y/3} \frac{y}{\sqrt{x^2 + y^2}} \, dx \, dy$ **40.** $\displaystyle \int_0^1 \int_y^{2-y} \sqrt{x + y} \, dx \, dy$

13.5   Triple Integrals in Rectangular Coordinates

Just as double integrals allow us to deal with more general situations than could be handled by single integrals, triple integrals enable us to solve still more general problems. We use triple integrals to calculate the volumes of three-dimensional shapes and the average value of a function over a three-dimensional region. Triple integrals also arise in the study of vector fields and fluid flow in three dimensions, as we will see in Chapter 14.

Triple Integrals

If $F(x, y, z)$ is a function defined on a closed bounded region D in space, such as the region occupied by a solid ball or a lump of clay, then the integral of F over D may be defined in the following way. We partition a rectangular boxlike region containing D into rectangular cells by planes parallel to the coordinate axes (Figure 13.23). We number the cells that lie inside D from 1 to n in some order, the kth cell having dimensions Δx_k by Δy_k by Δz_k and volume $\Delta V_k = \Delta x_k \Delta y_k \Delta z_k$. We choose a point (x_k, y_k, z_k) in each cell and form the sum

$$S_n = \sum_{k=1}^{n} F(x_k, y_k, z_k) \, \Delta V_k. \tag{1}$$

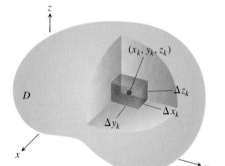

FIGURE 13.23 Partitioning a solid with rectangular cells of volume ΔV_k.

We are interested in what happens as D is partitioned by smaller and smaller cells, so that Δx_k, Δy_k, Δz_k and the norm of the partition $\|P\|$, the largest value among Δx_k, Δy_k, Δz_k, all approach zero. When a single limiting value is attained, no matter how the partitions and points (x_k, y_k, z_k) are chosen, we say that F is **integrable** over D. As before, it can be shown that when F is continuous and the bounding surface of D is formed from finitely many smooth surfaces joined together along finitely many smooth curves, then F is integrable. As $\|P\| \to 0$ and the number of cells n goes to ∞, the sums S_n approach a limit. We call this limit the **triple integral of F over D** and write

$$\lim_{n \to \infty} S_n = \iiint_D F(x, y, z) \, dV \quad \text{or} \quad \lim_{\|P\| \to 0} S_n = \iiint_D F(x, y, z) \, dx \, dy \, dz.$$

The regions D over which continuous functions are integrable are those having "reasonably smooth" boundaries.

Volume of a Region in Space

If F is the constant function whose value is 1, then the sums in Equation (1) reduce to

$$S_n = \sum F(x_k, y_k, z_k) \,\Delta V_k = \sum 1 \cdot \Delta V_k = \sum \Delta V_k.$$

As Δx_k, Δy_k, and Δz_k approach zero, the cells ΔV_k become smaller and more numerous and fill up more and more of D. We therefore define the volume of D to be the triple integral

$$\lim_{n \to \infty} \sum_{k=1}^{n} \Delta V_k = \iiint\limits_{D} dV.$$

> **DEFINITION**   The **volume** of a closed, bounded region D in space is
>
> $$V = \iiint\limits_{D} dV.$$

This definition is in agreement with our previous definitions of volume, although we omit the verification of this fact. As we see in a moment, this integral enables us to calculate the volumes of solids enclosed by curved surfaces.

Finding Limits of Integration

We evaluate a triple integral by applying a three-dimensional version of Fubini's Theorem (Section 13.2) to evaluate it by three repeated single integrations. As with double integrals, there is a geometric procedure for finding the limits of integration for these single integrals.

To evaluate

$$\iiint\limits_{D} F(x, y, z) \, dV$$

over a region D, integrate first with respect to z, then with respect to y, finally with x.

1. *Sketch:* Sketch the region D along with its "shadow" R (vertical projection) in the xy-plane. Label the upper and lower bounding surfaces of D and the upper and lower bounding curves of R.

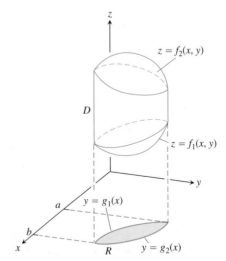

2. *Find the z-limits of integration:* Draw a line M passing through a typical point (x, y) in R parallel to the z-axis. As z increases, M enters D at $z = f_1(x, y)$ and leaves at $z = f_2(x, y)$. These are the z-limits of integration.

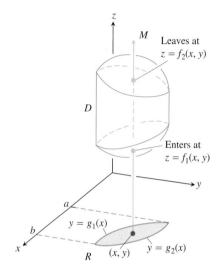

3. *Find the y-limits of integration:* Draw a line L through (x, y) parallel to the y-axis. As y increases, L enters R at $y = g_1(x)$ and leaves at $y = g_2(x)$. These are the y-limits of integration.

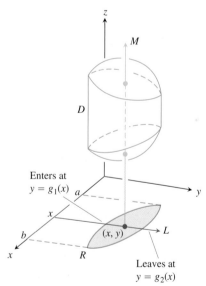

4. *Find the x-limits of integration:* Choose x-limits that include all lines through R parallel to the y-axis ($x = a$ and $x = b$ in the preceding figure). These are the x-limits of integration. The integral is

$$\int_{x=a}^{x=b} \int_{y=g_1(x)}^{y=g_2(x)} \int_{z=f_1(x,y)}^{z=f_2(x,y)} F(x, y, z) \, dz \, dy \, dx.$$

Follow similar procedures if you change the order of integration. The "shadow" of region D lies in the plane of the last two variables with respect to which the iterated integration takes place.

The preceding procedure applies whenever a solid region D is bounded above and below by a surface, and when the "shadow" region R is bounded by a lower and upper curve. It does not apply to regions with complicated holes through them, although sometimes such regions can be subdivided into simpler regions for which the procedure does apply.

EXAMPLE 1   Find the volume of the region D enclosed by the surfaces $z = x^2 + 3y^2$ and $z = 8 - x^2 - y^2$.

Solution   The volume is

$$V = \iiint\limits_{D} dz\,dy\,dx,$$

the integral of $F(x, y, z) = 1$ over D. To find the limits of integration for evaluating the integral, we first sketch the region. The surfaces (Figure 13.24) intersect on the elliptical cylinder $x^2 + 3y^2 = 8 - x^2 - y^2$ or $x^2 + 2y^2 = 4$, $z > 0$. The boundary of the region R, the projection of D onto the xy-plane, is an ellipse with the same equation: $x^2 + 2y^2 = 4$. The "upper" boundary of R is the curve $y = \sqrt{(4 - x^2)/2}$. The lower boundary is the curve $y = -\sqrt{(4 - x^2)/2}$.

Now we find the z-limits of integration. The line M passing through a typical point (x, y) in R parallel to the z-axis enters D at $z = x^2 + 3y^2$ and leaves at $z = 8 - x^2 - y^2$.

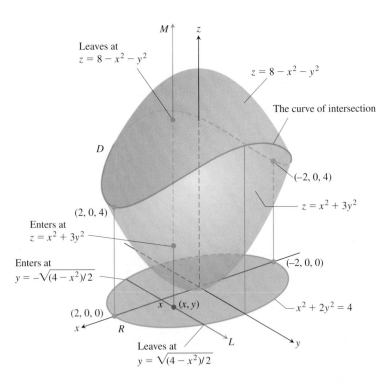

FIGURE 13.24   The volume of the region enclosed by two paraboloids, calculated in Example 1.

Next we find the y-limits of integration. The line L through (x, y) parallel to the y-axis enters R at $y = -\sqrt{(4 - x^2)/2}$ and leaves at $y = \sqrt{(4 - x^2)/2}$.

Finally we find the x-limits of integration. As L sweeps across R, the value of x varies from $x = -2$ at $(-2, 0, 0)$ to $x = 2$ at $(2, 0, 0)$. The volume of D is

$$
\begin{aligned}
V &= \iiint\limits_{D} dz\, dy\, dx \\
&= \int_{-2}^{2} \int_{-\sqrt{(4-x^2)/2}}^{\sqrt{(4-x^2)/2}} \int_{x^2+3y^2}^{8-x^2-y^2} dz\, dy\, dx \\
&= \int_{-2}^{2} \int_{-\sqrt{(4-x^2)/2}}^{\sqrt{(4-x^2)/2}} (8 - 2x^2 - 4y^2)\, dy\, dx \\
&= \int_{-2}^{2} \left[(8 - 2x^2)y - \frac{4}{3}y^3 \right]_{y=-\sqrt{(4-x^2)/2}}^{y=\sqrt{(4-x^2)/2}} dx \\
&= \int_{-2}^{2} \left(2(8 - 2x^2)\sqrt{\frac{4 - x^2}{2}} - \frac{8}{3}\left(\frac{4 - x^2}{2}\right)^{3/2} \right) dx \\
&= \int_{-2}^{2} \left[8\left(\frac{4 - x^2}{2}\right)^{3/2} - \frac{8}{3}\left(\frac{4 - x^2}{2}\right)^{3/2} \right] dx = \frac{4\sqrt{2}}{3} \int_{-2}^{2} (4 - x^2)^{3/2}\, dx \\
&= 8\pi\sqrt{2}. \qquad \text{\small After integration with the substitution } x = 2 \sin u.
\end{aligned}
$$
∎

In the next example, we project D onto the xz-plane instead of the xy-plane, to show how to use a different order of integration.

EXAMPLE 2   Set up the limits of integration for evaluating the triple integral of a function $F(x, y, z)$ over the tetrahedron D with vertices $(0, 0, 0)$, $(1, 1, 0)$, $(0, 1, 0)$, and $(0, 1, 1)$. Use the order of integration $dy\, dz\, dx$.

Solution   We sketch D along with its "shadow" R in the xz-plane (Figure 13.25). The upper (right-hand) bounding surface of D lies in the plane $y = 1$. The lower (left-hand) bounding surface lies in the plane $y = x + z$. The upper boundary of R is the line $z = 1 - x$. The lower boundary is the line $z = 0$.

First we find the y-limits of integration. The line through a typical point (x, z) in R parallel to the y-axis enters D at $y = x + z$ and leaves at $y = 1$.

Next we find the z-limits of integration. The line L through (x, z) parallel to the z-axis enters R at $z = 0$ and leaves at $z = 1 - x$.

Finally we find the x-limits of integration. As L sweeps across R, the value of x varies from $x = 0$ to $x = 1$. The integral is

$$
\int_{0}^{1} \int_{0}^{1-x} \int_{x+z}^{1} F(x, y, z)\, dy\, dz\, dx.
$$
∎

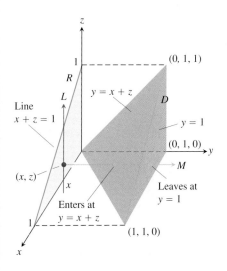

FIGURE 13.25   Finding the limits of integration for evaluating the triple integral of a function defined over the tetrahedron D (Examples 2 and 3).

EXAMPLE 3   Integrate $F(x, y, z)$ over the tetrahedron D in Example 2 in the order $dz\, dy\, dx$.

Solution   First we find the z-limits of integration. A line parallel to the z-axis through a typical point (x, y) in the xy-plane "shadow" enters the tetrahedron at $z = 0$ and exits through the upper plane where $z = y - x$ (Figure 13.25).

Next we find the y-limits of integration. On the xy-plane, where $z = 0$, the sloped side of the tetrahedron crosses the plane along the line $y = x$. A line through (x, y) parallel to the y-axis enters the shadow in the xy-plane at $y = x$ and exits at $y = 1$.

Finally we find the x-limits of integration. As the line parallel to the y-axis in the previous step sweeps out the shadow, the value of x varies from $x = 0$ to $x = 1$ at the point $(1, 1, 0)$. The integral is

$$\int_0^1 \int_x^1 \int_0^{y-x} F(x, y, z)\, dz\, dy\, dx.$$

For example, if $F(x, y, z) = 1$, we would find the volume of the tetrahedron to be

$$V = \int_0^1 \int_x^1 \int_0^{y-x} dz\, dy\, dx$$

$$= \int_0^1 \int_x^1 (y - x)\, dy\, dx$$

$$= \int_0^1 \left[\frac{1}{2} y^2 - xy \right]_{y=x}^{y=1} dx$$

$$= \int_0^1 \left(\frac{1}{2} - x + \frac{1}{2} x^2 \right) dx$$

$$= \left[\frac{1}{2} x - \frac{1}{2} x^2 + \frac{1}{6} x^3 \right]_0^1$$

$$= \frac{1}{6}.$$

We get the same result by integrating with the order $dy\, dz\, dx$. From Example 2,

$$V = \int_0^1 \int_0^{1-x} \int_{x+z}^1 dy\, dz\, dx = \frac{1}{6}. \qquad \blacksquare$$

Average Value of a Function in Space

The average value of a function F over a region D in space is defined by the formula

$$\text{\textbf{Average value} of } F \text{ over } D = \frac{1}{\text{volume of } D} \iiint_D F\, dV. \qquad (2)$$

For example, if $F(x, y, z) = \sqrt{x^2 + y^2 + z^2}$, then the average value of F over D is the average distance of points in D from the origin. If $F(x, y, z)$ is the temperature at (x, y, z) on a solid that occupies a region D in space, then the average value of F over D is the average temperature of the solid.

EXAMPLE 4   Find the average value of $F(x, y, z) = xyz$ over the cube bounded by the coordinate planes and the planes $x = 2$, $y = 2$, and $z = 2$ in the first octant.

Solution   We sketch the cube with enough detail to show the limits of integration (Figure 13.26). We then use Equation (2) to calculate the average value of F over the cube.

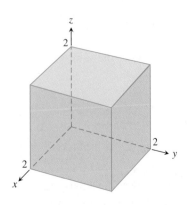

FIGURE 13.26   The region of integration in Example 4.

The volume of the cube is $(2)(2)(2) = 8$. The value of the integral of F over the cube is

$$\int_0^2 \int_0^2 \int_0^2 xyz \, dx \, dy \, dz = \int_0^2 \int_0^2 \left[\frac{x^2}{2} yz \right]_{x=0}^{x=2} dy \, dz = \int_0^2 \int_0^2 2yz \, dy \, dz$$

$$= \int_0^2 \left[y^2 z \right]_{y=0}^{y=2} dz = \int_0^2 4z \, dz = \left[2z^2 \right]_0^2 = 8.$$

With these values, Equation (2) gives

$$\text{Average value of} \atop xyz \text{ over the cube} = \frac{1}{\text{volume}} \iiint_{\text{cube}} xyz \, dV = \left(\frac{1}{8} \right)(8) = 1.$$

In evaluating the integral, we chose the order $dx \, dy \, dz$, but any of the other five possible orders would have done as well. ∎

Properties of Triple Integrals

Triple integrals have the same algebraic properties as double and single integrals.

If $F = F(x, y, z)$ and $G = G(x, y, z)$ are continuous, then

1. *Constant Multiple:* $\iiint_D kF \, dV = k \iiint_D F \, dV$ (any number k)

2. *Sum and Difference:* $\iiint_D (F \pm G) \, dV = \iiint_D F \, dV \pm \iiint_D G \, dV$

3. *Domination:*

    (a) $\iiint_D F \, dV \geq 0$ if $F \geq 0$ on D

    (b) $\iiint_D F \, dV \geq \iiint_D G \, dV$ if $F \geq G$ on D

4. *Additivity:* $\iiint_D F \, dV = \iiint_{D_1} E \, dV + \iiint_{D_2} F \, dV$

    if D is the union of two nonoverlapping regions D_1 and D_2.

EXERCISES 13.5

1. Evaluate the integral in Example 2 taking $F(x, y, z) = 1$ to find the volume of the tetrahedron.

2. **Volume of rectangular solid** Write six different iterated triple integrals for the volume of the rectangular solid in the first octant bounded by the coordinate planes and the planes $x = 1$, $y = 2$, and $z = 3$. Evaluate one of the integrals.

3. **Volume of tetrahedron** Write six different iterated triple integrals for the volume of the tetrahedron cut from the first octant by the plane $6x + 3y + 2z = 6$. Evaluate one of the integrals.

4. Volume of solid Write six different iterated triple integrals for the volume of the region in the first octant enclosed by the cylinder $x^2 + z^2 = 4$ and the plane $y = 3$. Evaluate one of the integrals.

5. Volume enclosed by paraboloids Let D be the region bounded by the paraboloids $z = 8 - x^2 - y^2$ and $z = x^2 + y^2$. Write six different triple iterated integrals for the volume of D. Evaluate one of the integrals.

6. Volume inside paraboloid beneath a plane Let D be the region bounded by the paraboloid $z = x^2 + y^2$ and the plane $z = 2y$. Write triple iterated integrals in the order $dz\,dx\,dy$ and $dz\,dy\,dx$ that give the volume of D. Do not evaluate either integral.

Evaluate the integrals in Exercises 7–20.

7. $\displaystyle\int_0^1\int_0^1\int_0^1 (x^2 + y^2 + z^2)\,dz\,dy\,dx$

8. $\displaystyle\int_0^{\sqrt{2}}\int_0^{3y}\int_{x^2+3y^2}^{8-x^2-y^2} dz\,dx\,dy$ **9.** $\displaystyle\int_1^e\int_1^e\int_1^e \frac{1}{xyz}\,dx\,dy\,dz$

10. $\displaystyle\int_0^1\int_0^{3-3x}\int_0^{3-3x-y} dz\,dy\,dx$ **11.** $\displaystyle\int_0^1\int_0^\pi\int_0^\pi y\sin z\,dx\,dy\,dz$

12. $\displaystyle\int_{-1}^1\int_{-1}^1\int_{-1}^1 (x + y + z)\,dy\,dx\,dz$

13. $\displaystyle\int_0^3\int_0^{\sqrt{9-x^2}}\int_0^{\sqrt{9-x^2}} dz\,dy\,dx$ **14.** $\displaystyle\int_0^2\int_{-\sqrt{4-y^2}}^{\sqrt{4-y^2}}\int_0^{2x+y} dz\,dx\,dy$

15. $\displaystyle\int_0^1\int_0^{2-x}\int_0^{2-x-y} dz\,dy\,dx$ **16.** $\displaystyle\int_0^1\int_0^{1-x^2}\int_3^{4-x^2-y} x\,dz\,dy\,dx$

17. $\displaystyle\int_0^\pi\int_0^\pi\int_0^\pi \cos(u + v + w)\,du\,dv\,dw$ (*uvw*-space)

18. $\displaystyle\int_1^e\int_1^e\int_1^e \ln r \ln s \ln t\,dt\,dr\,ds$ (*rst*-space)

19. $\displaystyle\int_0^{\pi/4}\int_0^{\ln\sec v}\int_{-\infty}^{2t} e^x\,dx\,dt\,dv$ (*tvx*-space)

20. $\displaystyle\int_0^7\int_0^2\int_0^{\sqrt{4-q^2}} \frac{q}{r+1}\,dp\,dq\,dr$ (*pqr*-space)

21. Here is the region of integration of the integral

$$\int_{-1}^1\int_{x^2}^1\int_0^{1-y} dz\,dy\,dx.$$

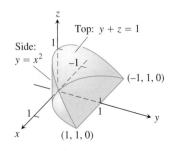

Top: $y + z = 1$
Side: $y = x^2$
$(-1, 1, 0)$
$(1, 1, 0)$

Rewrite the integral as an equivalent iterated integral in the order

a. $dy\,dz\,dx$ **b.** $dy\,dx\,dz$

c. $dx\,dy\,dz$ **d.** $dx\,dz\,dy$

e. $dz\,dx\,dy$.

22. Here is the region of integration of the integral

$$\int_0^1\int_{-1}^0\int_0^{y^2} dz\,dy\,dx.$$

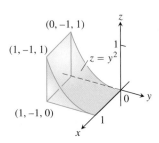

$(0, -1, 1)$
$(1, -1, 1)$
$z = y^2$
$(1, -1, 0)$

Rewrite the integral as an equivalent iterated integral in the order

a. $dy\,dz\,dx$ **b.** $dy\,dx\,dz$

c. $dx\,dy\,dz$ **d.** $dx\,dz\,dy$

e. $dz\,dx\,dy$.

Find the volumes of the regions in Exercises 23–36.

23. The region between the cylinder $z = y^2$ and the xy-plane that is bounded by the planes $x = 0, x = 1, y = -1, y = 1$

24. The region in the first octant bounded by the coordinate planes and the planes $x + z = 1, y + 2z = 2$

25. The region in the first octant bounded by the coordinate planes, the plane $y + z = 2$, and the cylinder $x = 4 - y^2$

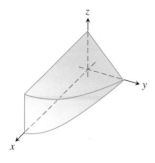

26. The wedge cut from the cylinder $x^2 + y^2 = 1$ by the planes $z = -y$ and $z = 0$

27. The tetrahedron in the first octant bounded by the coordinate planes and the plane passing through $(1, 0, 0)$, $(0, 2, 0)$, and $(0, 0, 3)$.

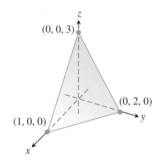

(0, 0, 3)

(0, 2, 0)

(1, 0, 0)

28. The region in the first octant bounded by the coordinate planes, the plane $y = 1 - x$, and the surface $z = \cos(\pi x/2)$, $0 \le x \le 1$

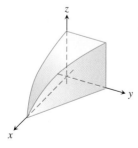

29. The region common to the interiors of the cylinders $x^2 + y^2 = 1$ and $x^2 + z^2 = 1$, one-eighth of which is shown in the accompanying figure.

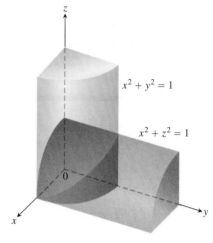

$x^2 + y^2 = 1$

$x^2 + z^2 = 1$

0

30. The region in the first octant bounded by the coordinate planes and the surface $z = 4 - x^2 - y$

31. The region in the first octant bounded by the coordinate planes, the plane $x + y = 4$, and the cylinder $y^2 + 4z^2 = 16$

32. The region cut from the cylinder $x^2 + y^2 = 4$ by the plane $z = 0$ and the plane $x + z = 3$

33. The region between the planes $x + y + 2z = 2$ and $2x + 2y + z = 4$ in the first octant

34. The finite region bounded by the planes $z = x$, $x + z = 8$, $z = y$, $y = 8$, and $z = 0$.

35. The region cut from the solid elliptical cylinder $x^2 + 4y^2 \leq 4$ by the xy-plane and the plane $z = x + 2$

36. The region bounded in back by the plane $x = 0$, on the front and sides by the parabolic cylinder $x = 1 - y^2$, on the top by the paraboloid $z = x^2 + y^2$, and on the bottom by the xy-plane

In Exercises 37–40, find the average value of $F(x, y, z)$ over the given region.

37. $F(x, y, z) = x^2 + 9$ over the cube in the first octant bounded by the coordinate planes and the planes $x = 2$, $y = 2$, and $z = 2$

38. $F(x, y, z) = x + y - z$ over the rectangular solid in the first octant bounded by the coordinate planes and the planes $x = 1$, $y = 1$, and $z = 2$

39. $F(x, y, z) = x^2 + y^2 + z^2$ over the cube in the first octant bounded by the coordinate planes and the planes $x = 1$, $y = 1$, and $z = 1$

40. $F(x, y, z) = xyz$ over the cube in the first octant bounded by the coordinate planes and the planes $x = 2$, $y = 2$, and $z = 2$

Evaluate the integrals in Exercises 41–44 by changing the order of integration in an appropriate way.

41. $\displaystyle\int_0^4 \int_0^1 \int_{2y}^2 \frac{4\cos(x^2)}{2\sqrt{z}}\, dx\, dy\, dz$

42. $\displaystyle\int_0^1 \int_0^1 \int_{x^2}^1 12xze^{zy^2}\, dy\, dx\, dz$

43. $\displaystyle\int_0^1 \int_{\sqrt[3]{z}}^1 \int_0^{\ln 3} \frac{\pi e^{2x}\sin \pi y^2}{y^2}\, dx\, dy\, dz$

44. $\displaystyle\int_0^2 \int_0^{4-x^2} \int_0^x \frac{\sin 2z}{4 - z}\, dy\, dz\, dx$

45. Finding an upper limit of an iterated integral Solve for a:

$$\int_0^1 \int_0^{4-a-x^2} \int_a^{4-x^2-y} dz\, dy\, dx = \frac{4}{15}.$$

46. Ellipsoid For what value of c is the volume of the ellipsoid $x^2 + (y/2)^2 + (z/c)^2 = 1$ equal to 8π?

47. Minimizing a triple integral What domain D in space minimizes the value of the integral

$$\iiint_D (4x^2 + 4y^2 + z^2 - 4)\, dV\,?$$

Give reasons for your answer.

48. Maximizing a triple integral What domain D in space maximizes the value of the integral

$$\iiint_D (1 - x^2 - y^2 - z^2)\, dV\,?$$

Give reasons for your answer.

COMPUTER EXPLORATIONS

In Exercises 49–52, use a CAS integration utility to evaluate the triple integral of the given function over the specified solid region.

49. $F(x, y, z) = x^2 y^2 z$ over the solid cylinder bounded by $x^2 + y^2 = 1$ and the planes $z = 0$ and $z = 1$

50. $F(x, y, z) = |xyz|$ over the solid bounded below by the paraboloid $z = x^2 + y^2$ and above by the plane $z = 1$

51. $F(x, y, z) = \dfrac{z}{(x^2 + y^2 + z^2)^{3/2}}$ over the solid bounded below by the cone $z = \sqrt{x^2 + y^2}$ and above by the plane $z = 1$

52. $F(x, y, z) = x^4 + y^2 + z^2$ over the solid sphere $x^2 + y^2 + z^2 \leq 1$

13.6 Moments and Centers of Mass

This section shows how to calculate the masses and moments of two- and three-dimensional objects in Cartesian coordinates. Section 13.7 gives the calculations for cylindrical and spherical coordinates.

Masses and First Moments

If $\delta(x, y, z)$ is the density (mass per unit volume) of an object occupying a region D in space, the integral of δ over D gives the **mass** of the object. To see why, imagine partitioning the object into n mass elements like the one in Figure 13.27. The object's mass is the limit

$$M = \lim_{n \to \infty} \sum_{k=1}^n \Delta m_k = \lim_{n \to \infty} \sum_{k=1}^n \delta(x_k, y_k, z_k)\, \Delta V_k = \iiint_D \delta(x, y, z)\, dV.$$

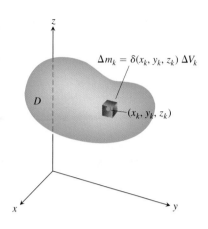

FIGURE 13.27 To define an object's mass, we first imagine it to be partitioned into a finite number of mass elements Δm_k.

The *first moment* of a solid region D about a coordinate plane is defined as the triple integral over D of the distance from a point (x, y, z) in D to the plane multiplied by the density of the solid at that point. For instance, the first moment about the yz-plane is the integral

$$M_{yz} = \iiint_D x\delta(x, y, z)\,dV.$$

The *center of mass* is found from the first moments. For instance, the x-coordinate of the center of mass is $\bar{x} = M_{yz}/M$.

For a two-dimensional object, such as a thin flat plate, we calculate first moments about the coordinate axes by simply dropping the z-coordinate. So the first moment about the y-axis is the double integral over the region R forming the plate of the distance from the axis multiplied by the density, or

$$M_y = \iint_R x\delta(x, y)\,dA.$$

Table 13.1 summarizes the formulas.

TABLE 13.1 Mass and first moment formulas

THREE-DIMENSIONAL SOLID

Mass: $M = \iiint_D \delta\,dV$ $\delta = \delta(x, y, z)$ is the density at (x, y, z).

First moments about the coordinate planes:

$$M_{yz} = \iiint_D x\,\delta\,dV, \qquad M_{xz} = \iiint_D y\,\delta\,dV, \qquad M_{xy} = \iiint_D z\,\delta\,dV$$

Center of mass:

$$\bar{x} = \frac{M_{yz}}{M}, \qquad \bar{y} = \frac{M_{xz}}{M}, \qquad \bar{z} = \frac{M_{xy}}{M}$$

TWO-DIMENSIONAL PLATE

Mass: $M = \iint_R \delta(x, y)\,dA$ $\delta(x, y)$ is the density at (x, y).

First moments: $M_y = \iint_R x\,\delta(x, y)\,dA, \qquad M_x = \iint_R y\,\delta(x, y)\,dA$

Center of mass: $\bar{x} = \dfrac{M_y}{M}, \qquad \bar{y} = \dfrac{M_x}{M}$

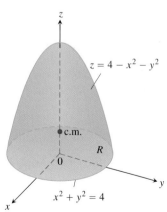

FIGURE 13.28 Finding the center of mass of a solid (Example 1).

EXAMPLE 1 Find the center of mass of a solid of constant density δ bounded below by the disk $R: x^2 + y^2 \leq 4$ in the plane $z = 0$ and above by the paraboloid $z = 4 - x^2 - y^2$ (Figure 13.28).

Solution   By symmetry $\bar{x} = \bar{y} = 0$. To find $\bar{z}$, we first calculate

$$M_{xy} = \iiint\limits_R \int_{z=0}^{z=4-x^2-y^2} z\,\delta\,dz\,dy\,dx = \iint\limits_R \left[\frac{z^2}{2}\right]_{z=0}^{z=4-x^2-y^2} \delta\,dy\,dx$$

$$= \frac{\delta}{2}\iint\limits_R (4 - x^2 - y^2)^2\,dy\,dx$$

$$= \frac{\delta}{2}\int_0^{2\pi}\int_0^2 (4 - r^2)^2\,r\,dr\,d\theta \qquad \text{\small Polar coordinates simplify the integration.}$$

$$= \frac{\delta}{2}\int_0^{2\pi}\left[-\frac{1}{6}(4 - r^2)^3\right]_{r=0}^{r=2}\,d\theta = \frac{16\delta}{3}\int_0^{2\pi}d\theta = \frac{32\pi\delta}{3}.$$

A similar calculation gives the mass

$$M = \iiint\limits_R \int_0^{4-x^2-y^2} \delta\,dz\,dy\,dx = 8\pi\delta.$$

Therefore $\bar{z} = (M_{xy}/M) = 4/3$ and the center of mass is $(\bar{x}, \bar{y}, \bar{z}) = (0, 0, 4/3)$. ∎

When the density of a solid object or plate is constant (as in Example 1), the center of mass is called the **centroid** of the object. To find a centroid, we set δ equal to 1 and proceed to find $\bar{x}$, $\bar{y}$, and $\bar{z}$ as before, by dividing first moments by masses.

EXAMPLE 2   Find the centroid of the region in the first quadrant that is bounded above by the line $y = x$ and below by the parabola $y = x^2$. These calculations are also valid for two-dimensional objects.

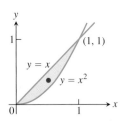

FIGURE 13.29   The centroid of this region is found in Example 2.

Solution   We sketch the region and include enough detail to determine the limits of integration (Figure 13.29). We then set δ equal to 1 and evaluate the appropriate formulas from Table 13.1:

$$M = \int_0^1\int_{x^2}^x 1\,dy\,dx = \int_0^1\left[y\right]_{y=x^2}^{y=x}\,dx = \int_0^1 (x - x^2)\,dx = \left[\frac{x^2}{2} - \frac{x^3}{3}\right]_0^1 = \frac{1}{6}$$

$$M_x = \int_0^1\int_{x^2}^x y\,dy\,dx = \int_0^1\left[\frac{y^2}{2}\right]_{y=x^2}^{y=x}\,dx$$

$$= \int_0^1\left(\frac{x^2}{2} - \frac{x^4}{2}\right)dx = \left[\frac{x^3}{6} - \frac{x^5}{10}\right]_0^1 = \frac{1}{15}$$

$$M_y = \int_0^1\int_{x^2}^x x\,dy\,dx = \int_0^1\left[xy\right]_{y=x^2}^{y=x}\,dx = \int_0^1 (x^2 - x^3)\,dx = \left[\frac{x^3}{3} - \frac{x^4}{4}\right]_0^1 = \frac{1}{12}.$$

From these values of M, M_x, and M_y, we find

$$\bar{x} = \frac{M_y}{M} = \frac{1/12}{1/6} = \frac{1}{2} \qquad \text{and} \qquad \bar{y} = \frac{M_x}{M} = \frac{1/15}{1/6} = \frac{2}{5}.$$

The centroid is the point $(1/2, 2/5)$. ∎

Moments of Inertia

A body's first moments (Table 13.1) tell us about balance and about the torque the body exerts about different axes in a gravitational field. If the body is a rotating shaft, however, we are more likely to be interested in how much energy is stored in the shaft or about how much energy it will take to accelerate the shaft to a particular angular velocity. This is where the second moment or moment of inertia comes in.

Think of partitioning the shaft into small blocks of mass Δm_k and let r_k denote the distance from the kth block's center of mass to the axis of rotation (Figure 13.30). If the shaft rotates at a constant angular velocity of $\omega = d\theta/dt$ radians per second, the block's center of mass will trace its orbit at a linear speed of

$$v_k = \frac{d}{dt}(r_k\theta) = r_k\frac{d\theta}{dt} = r_k\omega.$$

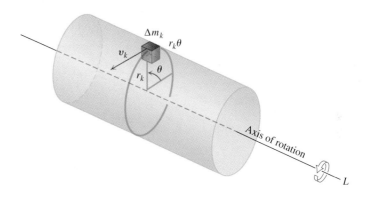

FIGURE 13.30 To find an integral for the amount of energy stored in a rotating shaft, we first imagine the shaft to be partitioned into small blocks. Each block has its own kinetic energy. We add the contributions of the individual blocks to find the kinetic energy of the shaft.

The block's kinetic energy will be approximately

$$\frac{1}{2}\Delta m_k v_k^2 = \frac{1}{2}\Delta m_k(r_k\omega)^2 = \frac{1}{2}\omega^2 r_k^2\,\Delta m_k.$$

The kinetic energy of the shaft will be approximately

$$\sum\frac{1}{2}\omega^2 r_k^2\,\Delta m_k.$$

The integral approached by these sums as the shaft is partitioned into smaller and smaller blocks gives the shaft's kinetic energy:

$$\text{KE}_{\text{shaft}} = \int\frac{1}{2}\omega^2 r^2\,dm = \frac{1}{2}\omega^2\int r^2\,dm. \qquad (1)$$

The factor

$$I = \int r^2\,dm$$

is the *moment of inertia* of the shaft about its axis of rotation, and we see from Equation (1) that the shaft's kinetic energy is

$$\text{KE}_{\text{shaft}} = \frac{1}{2} I\omega^2.$$

The moment of inertia of a shaft resembles in some ways the inertial mass of a locomotive. To start a locomotive with mass m moving at a linear velocity v, we need to provide a kinetic energy of $\text{KE} = (1/2)mv^2$. To stop the locomotive we have to remove this amount of energy. To start a shaft with moment of inertia I rotating at an angular velocity ω, we need to provide a kinetic energy of $\text{KE} = (1/2)I\omega^2$. To stop the shaft we have to take this amount of energy back out. The shaft's moment of inertia is analogous to the locomotive's mass. What makes the locomotive hard to start or stop is its mass. What makes the shaft hard to start or stop is its moment of inertia. The moment of inertia depends not only on the mass of the shaft, but also its distribution.

We now derive a formula for the moment of inertia for a solid in space. If $r(x, y, z)$ is the distance from the point (x, y, z) in D to a line L, then the moment of inertia of the mass $\Delta m_k = \delta(x_k, y_k, z_k)\Delta V_k$ about the line L (as in Figure 13.30) is approximately $\Delta I_k = r^2(x_k, y_k, z_k)\Delta m_k$. **The moment of inertia about L** of the entire object is

$$I_L = \lim_{n\to\infty} \sum_{k=1}^{n} \Delta I_k = \lim_{n\to\infty} \sum_{k=1}^{n} r^2(x_k, y_k, z_k)\,\delta(x_k, y_k, z_k)\,\Delta V_k = \iiint_D r^2\delta\,dV.$$

If L is the x-axis, then $r^2 = y^2 + z^2$ (Figure 13.31) and

$$I_x = \iiint_D (y^2 + z^2)\,\delta\,dV.$$

Similarly, if L is the y-axis or z-axis we have

$$I_y = \iiint_D (x^2 + z^2)\,\delta\,dV \qquad \text{and} \qquad I_z = \iiint_D (x^2 + y^2)\,\delta\,dV.$$

Table 13.2 summarizes the formulas for these moments of inertia (second moments because they invoke the *squares* of the distances). It shows the definition of the *polar moment* about the origin as well.

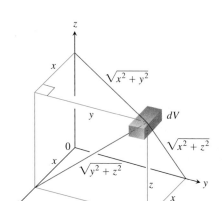

FIGURE 13.31 Distances from dV to the coordinate planes and axes.

TABLE 13.2 Moments of inertia (second moments) formulas

THREE-DIMENSIONAL SOLID

About the x-axis:	$I_x = \iiint (y^2 + z^2)\,\delta\,dV$	$\delta = \delta(x, y, z)$
About the y-axis:	$I_y = \iiint (x^2 + z^2)\,\delta\,dV$	
About the z-axis:	$I_z = \iiint (x^2 + y^2)\,\delta\,dV$	
About a line L:	$I_L = \iiint r^2\,\delta\,dV$	$r(x, y, z) =$ distance from the point (x, y, z) to line L

TWO-DIMENSIONAL PLATE

About the x-axis: $\qquad I_x = \iint y^2 \, \delta \, dA \qquad\qquad \delta = \delta(x, y)$

About the y-axis: $\qquad I_y = \iint x^2 \, \delta \, dA$

About a line L: $\qquad I_L = \iint r^2(x, y) \, \delta \, dA, \qquad r(x, y) = \text{distance from}$
$\qquad\qquad\qquad\qquad\qquad\qquad\qquad\qquad\qquad\qquad (x, y) \text{ to } L$

About the origin
(polar moment): $\qquad I_0 = \iint (x^2 + y^2) \, \delta \, dA = I_x + I_y$

EXAMPLE 3   Find I_x, I_y, I_z for the rectangular solid of constant density δ shown in Figure 13.32.

Solution   The formula for I_x gives

$$I_x = \int_{-c/2}^{c/2} \int_{-b/2}^{b/2} \int_{-a/2}^{a/2} (y^2 + z^2) \, \delta \, dx \, dy \, dz.$$

We can avoid some of the work of integration by observing that $(y^2 + z^2)\delta$ is an even function of x, y, and z. The rectangular solid consists of eight symmetric pieces, one in each octant. We can evaluate the integral on one of these pieces and then multiply by 8 to get the total value.

$$I_x = 8 \int_0^{c/2} \int_0^{b/2} \int_0^{a/2} (y^2 + z^2) \, \delta \, dx \, dy \, dz = 4a\delta \int_0^{c/2} \int_0^{b/2} (y^2 + z^2) \, dy \, dz$$

$$= 4a\delta \int_0^{c/2} \left[\frac{y^3}{3} + z^2 y \right]_{y=0}^{y=b/2} dz$$

$$= 4a\delta \int_0^{c/2} \left(\frac{b^3}{24} + \frac{z^2 b}{2} \right) dz$$

$$= 4a\delta \left(\frac{b^3 c}{48} + \frac{c^3 b}{48} \right) = \frac{abc\delta}{12} (b^2 + c^2) = \frac{M}{12} (b^2 + c^2).$$

Similarly,

$$I_y = \frac{M}{12} (a^2 + c^2) \qquad \text{and} \qquad I_z = \frac{M}{12} (a^2 + b^2). \qquad \blacksquare$$

EXAMPLE 4   A thin plate covers the triangular region bounded by the x-axis and the lines $x = 1$ and $y = 2x$ in the first quadrant. The plate's density at the point (x, y) is $\delta(x, y) = 6x + 6y + 6$. Find the plate's moments of inertia about the coordinate axes and the origin.

Solution   We sketch the plate and put in enough detail to determine the limits of integration for the integrals we have to evaluate (Figure 13.33). The moment of inertia about the x-axis is

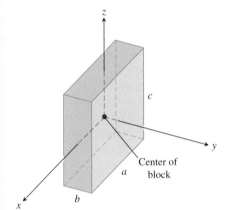

FIGURE 13.32   Finding I_x, I_y, and I_z for the block shown here. The origin lies at the center of the block (Example 3).

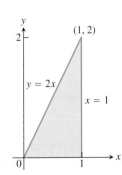

FIGURE 13.33   The triangular region covered by the plate in Example 4.

Beam A

Axis

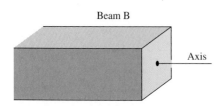

Beam B

Axis

FIGURE 13.34 The greater the polar moment of inertia of the cross-section of a beam about the beam's longitudinal axis, the stiffer the beam. Beams A and B have the same cross-sectional area, but A is stiffer.

$$I_x = \int_0^1 \int_0^{2x} y^2 \delta(x, y)\, dy\, dx = \int_0^1 \int_0^{2x} (6xy^2 + 6y^3 + 6y^2)\, dy\, dx$$

$$= \int_0^1 \left[2xy^3 + \frac{3}{2} y^4 + 2y^3 \right]_{y=0}^{y=2x} dx = \int_0^1 (40x^4 + 16x^3)\, dx$$

$$= \left[8x^5 + 4x^4 \right]_0^1 = 12.$$

Similarly, the moment of inertia about the y-axis is

$$I_y = \int_0^1 \int_0^{2x} x^2 \delta(x, y)\, dy\, dx = \frac{39}{5}.$$

Notice that we integrate y^2 times density in calculating I_x and x^2 times density to find I_y.

Since we know I_x and I_y, we do not need to evaluate an integral to find I_0; we can use the equation $I_0 = I_x + I_y$ from Table 13.2 instead:

$$I_0 = 12 + \frac{39}{5} = \frac{60 + 39}{5} = \frac{99}{5}. \qquad \blacksquare$$

The moment of inertia also plays a role in determining how much a horizontal metal beam will bend under a load. The stiffness of the beam is a constant times I, the moment of inertia of a typical cross-section of the beam about the beam's longitudinal axis. The greater the value of I, the stiffer the beam and the less it will bend under a given load. That is why we use I-beams instead of beams whose cross-sections are square. The flanges at the top and bottom of the beam hold most of the beam's mass away from the longitudinal axis to maximize the value of I (Figure 13.34).

EXERCISES 13.6

1. **Finding center of mass** Find the center of mass of a thin plate of density $\delta = 3$ bounded by the lines $x = 0, y = x$, and the parabola $y = 2 - x^2$ in the first quadrant.

2. **Finding moments of inertia** Find the moments of inertia about the coordinate axes of a thin rectangular plate of constant density δ bounded by the lines $x = 3$ and $y = 3$ in the first quadrant.

3. **Finding a centroid** Find the centroid of the region in the first quadrant bounded by the x-axis, the parabola $y^2 = 2x$, and the line $x + y = 4$.

4. **Finding a centroid** Find the centroid of the triangular region cut from the first quadrant by the line $x + y = 3$.

5. **Finding a centroid** Find the centroid of the region cut from the first quadrant by the circle $x^2 + y^2 = a^2$.

6. **Finding a centroid** Find the centroid of the region between the x-axis and the arch $y = \sin x, 0 \le x \le \pi$.

7. **Finding moments of inertia** Find the moment of inertia about the x-axis of a thin plate of density $\delta = 1$ bounded by the circle $x^2 + y^2 = 4$. Then use your result to find I_y and I_0 for the plate.

8. **Finding a moment of inertia** Find the moment of inertia with respect to the y-axis of a thin sheet of constant density $\delta = 1$ bounded by the curve $y = (\sin^2 x)/x^2$ and the interval $\pi \le x \le 2\pi$ of the x-axis.

9. **The centroid of an infinite region** Find the centroid of the infinite region in the second quadrant enclosed by the coordinate axes and the curve $y = e^x$. (Use improper integrals in the mass-moment formulas.)

10. **The first moment of an infinite plate** Find the first moment about the y-axis of a thin plate of density $\delta(x, y) = 1$ covering the infinite region under the curve $y = e^{-x^2/2}$ in the first quadrant.

11. **Finding a moment of inertia** Find the moment of inertia about the x-axis of a thin plate bounded by the parabola $x = y - y^2$ and the line $x + y = 0$ if $\delta(x, y) = x + y$.

12. **Finding mass** Find the mass of a thin plate occupying the smaller region cut from the ellipse $x^2 + 4y^2 = 12$ by the parabola $x = 4y^2$ if $\delta(x, y) = 5x$.

13. **Finding a center of mass** Find the center of mass of a thin triangular plate bounded by the y-axis and the lines $y = x$ and $y = 2 - x$ if $\delta(x, y) = 6x + 3y + 3$.

14. **Finding a center of mass and moment of inertia** Find the center of mass and moment of inertia about the x-axis of a thin

plate bounded by the curves $x = y^2$ and $x = 2y - y^2$ if the density at the point (x, y) is $\delta(x, y) = y + 1$.

15. **Center of mass, moment of inertia** Find the center of mass and the moment of inertia about the y-axis of a thin rectangular plate cut from the first quadrant by the lines $x = 6$ and $y = 1$ if $\delta(x, y) = x + y + 1$.

16. **Center of mass, moment of inertia** Find the center of mass and the moment of inertia about the y-axis of a thin plate bounded by the line $y = 1$ and the parabola $y = x^2$ if the density is $\delta(x, y) = y + 1$.

17. **Center of mass, moment of inertia** Find the center of mass and the moment of inertia about the y-axis of a thin plate bounded by the x-axis, the lines $x = \pm 1$, and the parabola $y = x^2$ if $\delta(x, y) = 7y + 1$.

18. **Center of mass, moment of inertia** Find the center of mass and the moment of inertia about the x-axis of a thin rectangular plate bounded by the lines $x = 0, x = 20, y = -1$, and $y = 1$ if $\delta(x, y) = 1 + (x/20)$.

19. **Center of mass, moments of inertia** Find the center of mass, the moment of inertia about the coordinate axes, and the polar moment of inertia of a thin triangular plate bounded by the lines $y = x, y = -x$, and $y = 1$ if $\delta(x, y) = y + 1$.

20. **Center of mass, moments of inertia** Repeat Exercise 19 for $\delta(x, y) = 3x^2 + 1$.

21. **Moments of inertia** Find the moments of inertia of the rectangular solid shown here with respect to its edges by calculating I_x, I_y, and I_z.

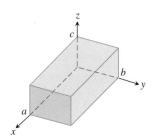

22. **Moments of inertia** The coordinate axes in the figure run through the centroid of a solid wedge parallel to the labeled edges. Find I_x, I_y, and I_z if $a = b = 6$ and $c = 4$.

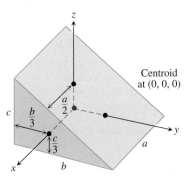

Centroid at $(0, 0, 0)$

23. **Center of mass and moments of inertia** A solid "trough" of constant density is bounded below by the surface $z = 4y^2$, above by the plane $z = 4$, and on the ends by the planes $x = 1$ and $x = -1$. Find the center of mass and the moments of inertia with respect to the three axes.

24. **Center of mass** A solid of constant density is bounded below by the plane $z = 0$, on the sides by the elliptical cylinder $x^2 + 4y^2 = 4$, and above by the plane $z = 2 - x$ (see the accompanying figure).

   a. Find $\bar{x}$ and $\bar{y}$.

   b. Evaluate the integral

   $$M_{xy} = \int_{-2}^{2} \int_{-(1/2)\sqrt{4-x^2}}^{(1/2)\sqrt{4-x^2}} \int_{0}^{2-x} z \, dz \, dy \, dx$$

   using integral tables to carry out the final integration with respect to x. Then divide M_{xy} by M to verify that $\bar{z} = 5/4$.

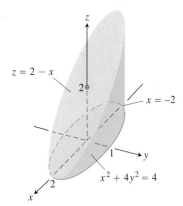

25. **a. Center of mass** Find the center of mass of a solid of constant density bounded below by the paraboloid $z = x^2 + y^2$ and above by the plane $z = 4$.

   b. Find the plane $z = c$ that divides the solid into two parts of equal volume. This plane does not pass through the center of mass.

26. **Moments** A solid cube, 2 units on a side, is bounded by the planes $x = \pm 1, z = \pm 1, y = 3$, and $y = 5$. Find the center of mass and the moments of inertia about the coordinate axes.

27. **Moment of inertia about a line** A wedge like the one in Exercise 22 has $a = 4, b = 6$, and $c = 3$. Make a quick sketch to check for yourself that the square of the distance from a typical point (x, y, z) of the wedge to the line $L: z = 0, y = 6$ is $r^2 = (y - 6)^2 + z^2$. Then calculate the moment of inertia of the wedge about L.

28. **Moment of inertia about a line** A wedge like the one in Exercise 22 has $a = 4, b = 6$, and $c = 3$. Make a quick sketch to check for yourself that the square of the distance from a typical point (x, y, z) of the wedge to the line $L: x = 4, y = 0$ is $r^2 = (x - 4)^2 + y^2$. Then calculate the moment of inertia of the wedge about L.

In Exercises 29 and 30, find

    a. the mass of the solid.   **b.** the center of mass.

29. A solid region in the first octant is bounded by the coordinate planes and the plane $x + y + z = 2$. The density of the solid is $\delta(x, y, z) = 2x$.

30. A solid in the first octant is bounded by the planes $y = 0$ and $z = 0$ and by the surfaces $z = 4 - x^2$ and $x = y^2$ (see the accompanying figure). Its density function is $\delta(x, y, z) = kxy$, k a constant.

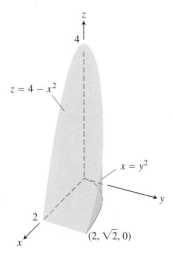

In Exercises 31 and 32, find

    a. the mass of the solid.

    b. the center of mass.

    c. the moments of inertia about the coordinate axes.

31. A solid cube in the first octant is bounded by the coordinate planes and by the planes $x = 1$, $y = 1$, and $z = 1$. The density of the cube is $\delta(x, y, z) = x + y + z + 1$.

32. A wedge like the one in Exercise 22 has dimensions $a = 2, b = 6$, and $c = 3$. The density is $\delta(x, y, z) = x + 1$. Notice that if the density is constant, the center of mass will be $(0, 0, 0)$.

33. Mass Find the mass of the solid bounded by the planes $x + z = 1, x - z = -1, y = 0$ and the surface $y = \sqrt{z}$. The density of the solid is $\delta(x, y, z) = 2y + 5$.

34. Mass Find the mass of the solid region bounded by the parabolic surfaces $z = 16 - 2x^2 - 2y^2$ and $z = 2x^2 + 2y^2$ if the density of the solid is $\delta(x, y, z) = \sqrt{x^2 + y^2}$.

The Parallel Axis Theorem Let $L_{c.m.}$ be a line through the center of mass of a body of mass m and let L be a parallel line h units away from $L_{c.m.}$. The *Parallel Axis Theorem* says that the moments of inertia $I_{c.m.}$ and I_L of the body about $L_{c.m.}$ and L satisfy the equation

$$I_L = I_{c.m.} + mh^2. \tag{2}$$

As in the two-dimensional case, the theorem gives a quick way to calculate one moment when the other moment and the mass are known.

35. Proof of the Parallel Axis Theorem

    a. Show that the first moment of a body in space about any plane through the body's center of mass is zero. (*Hint:* Place the body's center of mass at the origin and let the plane be the yz-plane. What does the formula $\bar{x} = M_{yz}/M$ then tell you?)

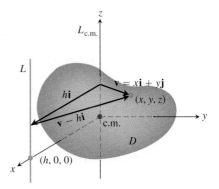

    b. To prove the Parallel Axis Theorem, place the body with its center of mass at the origin, with the line $L_{c.m.}$ along the z-axis and the line L perpendicular to the xy-plane at the point $(h, 0, 0)$. Let D be the region of space occupied by the body. Then, in the notation of the figure,

$$I_L = \iiint\limits_{D} |\mathbf{v} - h\mathbf{i}|^2 \, dm.$$

    Expand the integrand in this integral and complete the proof.

36. The moment of inertia about a diameter of a solid sphere of constant density and radius a is $(2/5)ma^2$, where m is the mass of the sphere. Find the moment of inertia about a line tangent to the sphere.

37. The moment of inertia of the solid in Exercise 21 about the z-axis is $I_z = abc(a^2 + b^2)/3$.

    a. Use Equation (2) to find the moment of inertia of the solid about the line parallel to the z-axis through the solid's center of mass.

    b. Use Equation (2) and the result in part (a) to find the moment of inertia of the solid about the line $x = 0, y = 2b$.

38. If $a = b = 6$ and $c = 4$, the moment of inertia of the solid wedge in Exercise 22 about the x-axis is $I_x = 208$. Find the moment of inertia of the wedge about the line $y = 4, z = -4/3$ (the edge of the wedge's narrow end).

13.7 Triple Integrals in Cylindrical and Spherical Coordinates

When a calculation in physics, engineering, or geometry involves a cylinder, cone, or sphere, we can often simplify our work by using cylindrical or spherical coordinates, which are introduced in this section. The procedure for transforming to these coordinates and evaluating the resulting triple integrals is similar to the transformation to polar coordinates in the plane studied in Section 13.4.

Integration in Cylindrical Coordinates

We obtain cylindrical coordinates for space by combining polar coordinates in the xy-plane with the usual z-axis. This assigns to every point in space one or more coordinate triples of the form (r, θ, z), as shown in Figure 13.35.

> **DEFINITION**   **Cylindrical coordinates** represent a point P in space by ordered triples (r, θ, z) in which
>
> 1. r and θ are polar coordinates for the vertical projection of P on the xy-plane
> 2. z is the rectangular vertical coordinate.

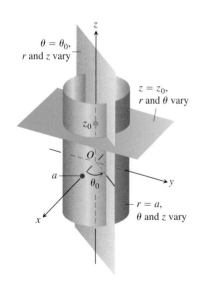

FIGURE 13.35 The cylindrical coordinates of a point in space are r, θ, and z.

The values of x, y, r, and θ in rectangular and cylindrical coordinates are related by the usual equations.

> **Equations Relating Rectangular (x, y, z) and Cylindrical (r, θ, z) Coordinates**
>
> $$x = r \cos \theta, \qquad y = r \sin \theta, \qquad z = z,$$
> $$r^2 = x^2 + y^2, \qquad \tan \theta = y/x$$

In cylindrical coordinates, the equation $r = a$ describes not just a circle in the xy-plane but an entire cylinder about the z-axis (Figure 13.36). The z-axis is given by $r = 0$. The equation $\theta = \theta_0$ describes the plane that contains the z-axis and makes an angle θ_0 with the positive x-axis. And, just as in rectangular coordinates, the equation $z = z_0$ describes a plane perpendicular to the z-axis.

Cylindrical coordinates are good for describing cylinders whose axes run along the z-axis and planes that either contain the z-axis or lie perpendicular to the z-axis. Surfaces like these have equations of constant coordinate value:

$$r = 4. \qquad \text{Cylinder, radius 4, axis the } z\text{-axis}$$
$$\theta = \frac{\pi}{3}. \qquad \text{Plane containing the } z\text{-axis}$$
$$z = 2. \qquad \text{Plane perpendicular to the } z\text{-axis}$$

FIGURE 13.36 Constant-coordinate equations in cylindrical coordinates yield cylinders and planes.

When computing triple integrals over a region D in cylindrical coordinates, we partition the region into n small cylindrical wedges, rather than into rectangular boxes. In the kth cylindrical wedge, r, θ and z change by Δr_k, $\Delta \theta_k$, and Δz_k, and the largest of these numbers among all the cylindrical wedges is called the **norm** of the partition. We define the triple integral as a limit of Riemann sums using these wedges. The volume of such a

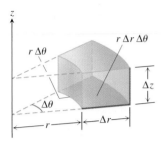

FIGURE 13.37 In cylindrical coordinates the volume of the wedge is approximated by the product $\Delta V = \Delta z \, r \, \Delta r \, \Delta \theta$.

cylindrical wedge ΔV_k is obtained by taking the area ΔA_k of its base in the $r\theta$-plane and multiplying by the height Δz (Figure 13.37).

For a point (r_k, θ_k, z_k) in the center of the kth wedge, we calculated in polar coordinates that $\Delta A_k = r_k \, \Delta r_k \, \Delta \theta_k$. So $\Delta V_k = \Delta z_k \, r_k \, \Delta r_k \, \Delta \theta_k$ and a Riemann sum for f over D has the form

$$S_n = \sum_{k=1}^{n} f(r_k, \theta_k, z_k) \, \Delta z_k \, r_k \, \Delta r_k \, \Delta \theta_k.$$

The triple integral of a function f over D is obtained by taking a limit of such Riemann sums with partitions whose norms approach zero

$$\lim_{n\to\infty} S_n = \iiint_D f \, dV = \iiint_D f \, dz \, r \, dr \, d\theta.$$

Triple integrals in cylindrical coordinates are then evaluated as iterated integrals, as in the following example.

EXAMPLE 1   Find the limits of integration in cylindrical coordinates for integrating a function $f(r, \theta, z)$ over the region D bounded below by the plane $z = 0$, laterally by the circular cylinder $x^2 + (y - 1)^2 = 1$, and above by the paraboloid $z = x^2 + y^2$.

Solution   The base of D is also the region's projection R on the xy-plane. The boundary of R is the circle $x^2 + (y - 1)^2 = 1$. Its polar coordinate equation is

$$x^2 + (y - 1)^2 = 1$$
$$x^2 + y^2 - 2y + 1 = 1$$
$$r^2 - 2r \sin \theta = 0$$
$$r = 2 \sin \theta.$$

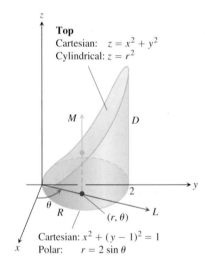

Top
Cartesian: $z = x^2 + y^2$
Cylindrical: $z = r^2$

M D

Cartesian: $x^2 + (y - 1)^2 = 1$
Polar: $r = 2 \sin \theta$

FIGURE 13.38 Finding the limits of integration for evaluating an integral in cylindrical coordinates (Example 1).

The region is sketched in Figure 13.38.

We find the limits of integration, starting with the z-limits. A line M through a typical point (r, θ) in R parallel to the z-axis enters D at $z = 0$ and leaves at $z = x^2 + y^2 = r^2$.

Next we find the r-limits of integration. A ray L through (r, θ) from the origin enters R at $r = 0$ and leaves at $r = 2 \sin \theta$.

Finally we find the θ-limits of integration. As L sweeps across R, the angle θ it makes with the positive x-axis runs from $\theta = 0$ to $\theta = \pi$. The integral is

$$\iiint_D f(r, \theta, z) \, dV = \int_0^\pi \int_0^{2 \sin \theta} \int_0^{r^2} f(r, \theta, z) \, dz \, r \, dr \, d\theta. \qquad \blacksquare$$

Example 1 illustrates a good procedure for finding limits of integration in cylindrical coordinates. The procedure is summarized as follows.

How to Integrate in Cylindrical Coordinates

To evaluate

$$\iiint_D f(r, \theta, z) \, dV$$

over a region D in space in cylindrical coordinates, integrating first with respect to z, then with respect to r, and finally with respect to θ, take the following steps.

1. *Sketch*. Sketch the region D along with its projection R on the xy-plane. Label the surfaces and curves that bound D and R.

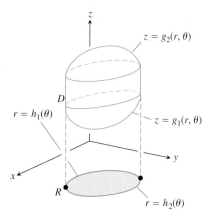

2. *Find the z-limits of integration*. Draw a line M through a typical point (r, θ) of R parallel to the z-axis. As z increases, M enters D at $z = g_1(r, \theta)$ and leaves at $z = g_2(r, \theta)$. These are the z-limits of integration.

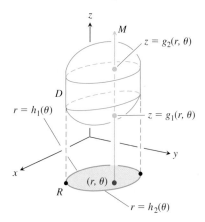

3. *Find the r-limits of integration*. Draw a ray L through (r, θ) from the origin. The ray enters R at $r = h_1(\theta)$ and leaves at $r = h_2(\theta)$. These are the r-limits of integration.

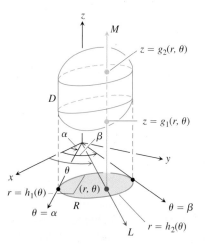

4. *Find the θ-limits of integration.* As L sweeps across R, the angle θ it makes with the positive x-axis runs from $\theta = \alpha$ to $\theta = \beta$. These are the θ-limits of integration. The integral is

$$\iiint\limits_D f(r, \theta, z)\, dV = \int_{\theta=\alpha}^{\theta=\beta} \int_{r=h_1(\theta)}^{r=h_2(\theta)} \int_{z=g_1(r, \theta)}^{z=g_2(r, \theta)} f(r, \theta, z)\, dz\, r\, dr\, d\theta.$$

EXAMPLE 2 Find the centroid ($\delta = 1$) of the solid enclosed by the cylinder $x^2 + y^2 = 4$, bounded above by the paraboloid $z = x^2 + y^2$, and bounded below by the xy-plane.

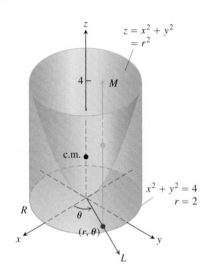

FIGURE 13.39 Example 2 shows how to find the centroid of this solid.

Solution We sketch the solid, bounded above by the paraboloid $z = r^2$ and below by the plane $z = 0$ (Figure 13.39). Its base R is the disk $0 \le r \le 2$ in the xy-plane.

The solid's centroid $(\bar{x}, \bar{y}, \bar{z})$ lies on its axis of symmetry, here the z-axis. This makes $\bar{x} = \bar{y} = 0$. To find $\bar{z}$, we divide the first moment M_{xy} by the mass M.

To find the limits of integration for the mass and moment integrals, we continue with the four basic steps. We completed our initial sketch. The remaining steps give the limits of integration.

The z-limits. A line M through a typical point (r, θ) in the base parallel to the z-axis enters the solid at $z = 0$ and leaves at $z = r^2$.

The r-limits. A ray L through (r, θ) from the origin enters R at $r = 0$ and leaves at $r = 2$.

The θ-limits. As L sweeps over the base like a clock hand, the angle θ it makes with the positive x-axis runs from $\theta = 0$ to $\theta = 2\pi$. The value of M_{xy} is

$$M_{xy} = \int_0^{2\pi}\int_0^2\int_0^{r^2} z\, dz\, r\, dr\, d\theta = \int_0^{2\pi}\int_0^2 \left[\frac{z^2}{2}\right]_0^{r^2} r\, dr\, d\theta$$

$$= \int_0^{2\pi}\int_0^2 \frac{r^5}{2}\, dr\, d\theta = \int_0^{2\pi}\left[\frac{r^6}{12}\right]_0^2 d\theta = \int_0^{2\pi} \frac{16}{3}\, d\theta = \frac{32\pi}{3}.$$

The value of M is

$$M = \int_0^{2\pi}\int_0^2\int_0^{r^2} dz\, r\, dr\, d\theta = \int_0^{2\pi}\int_0^2 \left[z\right]_0^{r^2} r\, dr\, d\theta$$

$$= \int_0^{2\pi}\int_0^2 r^3\, dr\, d\theta = \int_0^{2\pi}\left[\frac{r^4}{4}\right]_0^2 d\theta = \int_0^{2\pi} 4\, d\theta = 8\pi.$$

Therefore,

$$\bar{z} = \frac{M_{xy}}{M} = \frac{32\pi}{3}\frac{1}{8\pi} = \frac{4}{3},$$

and the centroid is $(0, 0, 4/3)$. Notice that the centroid lies outside the solid. ∎

Spherical Coordinates and Integration

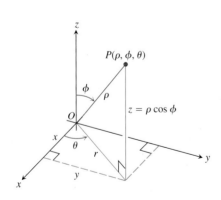

FIGURE 13.40 The spherical coordinates ρ, ϕ, and θ and their relation to x, y, z, and r.

Spherical coordinates locate points in space with two angles and one distance, as shown in Figure 13.40. The first coordinate, $\rho = |\overrightarrow{OP}|$, is the point's distance from the origin. Unlike r, *the variable ρ is never negative.* The second coordinate, ϕ, is the angle $\overrightarrow{OP}$ makes with the positive z-axis. It is required to lie in the interval $[0, \pi]$. The third coordinate is the angle θ as measured in cylindrical coordinates.

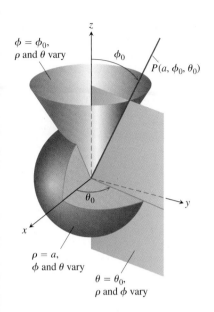

$\phi = \phi_0$, ρ and θ vary

$\rho = a$, ϕ and θ vary

$\theta = \theta_0$, ρ and ϕ vary

FIGURE 13.41 Constant-coordinate equations in spherical coordinates yield spheres, single cones, and half-planes.

DEFINITION **Spherical coordinates** represent a point P in space by ordered triples (ρ, ϕ, θ) in which

1. ρ is the distance from P to the origin.
2. ϕ is the angle $\overrightarrow{OP}$ makes with the positive z-axis ($0 \le \phi \le \pi$).
3. θ is the angle from cylindrical coordinates.

On maps of the Earth, θ is related to the meridian of a point on the Earth and ϕ to its latitude, while ρ is related to elevation above the Earth's surface.

The equation $\rho = a$ describes the sphere of radius a centered at the origin (Figure 13.41). The equation $\phi = \phi_0$ describes a single cone whose vertex lies at the origin and whose axis lies along the z-axis. (We broaden our interpretation to include the xy-plane as the cone $\phi = \pi/2$.) If ϕ_0 is greater than $\pi/2$, the cone $\phi = \phi_0$ opens downward. The equation $\theta = \theta_0$ describes the half-plane that contains the z-axis and makes an angle θ_0 with the positive x-axis.

Equations Relating Spherical Coordinates to Cartesian and Cylindrical Coordinates

$$r = \rho \sin \phi, \qquad x = r \cos \theta = \rho \sin \phi \cos \theta,$$
$$z = \rho \cos \phi, \qquad y = r \sin \theta = \rho \sin \phi \sin \theta, \qquad (1)$$
$$\rho = \sqrt{x^2 + y^2 + z^2} = \sqrt{r^2 + z^2}.$$

EXAMPLE 3 Find a spherical coordinate equation for the sphere $x^2 + y^2 + (z - 1)^2 = 1$.

Solution We use Equations (1) to substitute for x, y, and z:

$$x^2 + y^2 + (z - 1)^2 = 1$$
$$\rho^2 \sin^2 \phi \cos^2 \theta + \rho^2 \sin^2 \phi \sin^2 \theta + (\rho \cos \phi - 1)^2 = 1 \qquad \text{Equations (1)}$$
$$\rho^2 \sin^2 \phi \underbrace{(\cos^2 \theta + \sin^2 \theta)}_{1} + \rho^2 \cos^2 \phi - 2\rho \cos \phi + 1 = 1$$
$$\rho^2 \underbrace{(\sin^2 \phi + \cos^2 \phi)}_{1} = 2\rho \cos \phi$$
$$\rho^2 = 2\rho \cos \phi$$
$$\rho = 2 \cos \phi.$$

$x^2 + y^2 + (z - 1)^2 = 1$
$\rho = 2 \cos \phi$

FIGURE 13.42 The sphere in Example 3. (See Figure 13.42.)

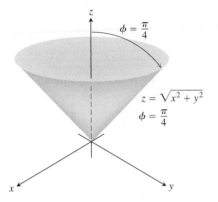

FIGURE 13.43 The cone in Example 4.

EXAMPLE 4 Find a spherical coordinate equation for the cone $z = \sqrt{x^2 + y^2}$.

Solution 1 *Use geometry.* The cone is symmetric with respect to the z-axis and cuts the first quadrant of the yz-plane along the line $z = y$. The angle between the cone and the positive z-axis is therefore $\pi/4$ radians. The cone consists of the points whose spherical coordinates have ϕ equal to $\pi/4$, so its equation is $\phi = \pi/4$. (See Figure 13.43.)

Solution 2 *Use algebra.* If we use Equations (1) to substitute for x, y, and z we obtain the same result:

$$z = \sqrt{x^2 + y^2}$$
$$\rho \cos \phi = \sqrt{\rho^2 \sin^2 \phi} \qquad \text{Example 3}$$
$$\rho \cos \phi = \rho \sin \phi \qquad \rho \geq 0,\ \sin \phi \geq 0$$
$$\cos \phi = \sin \phi$$
$$\phi = \frac{\pi}{4}. \qquad 0 \leq \phi \leq \pi \qquad \blacksquare$$

Spherical coordinates are good for describing spheres centered at the origin, half-planes hinged along the z-axis, and cones whose vertices lie at the origin and whose axes lie along the z-axis. Surfaces like these have equations of constant coordinate value:

$$\rho = 4 \qquad \text{Sphere, radius 4, center at origin}$$

$$\phi = \frac{\pi}{3} \qquad \begin{array}{l}\text{Cone opening up from the origin, making an}\\\text{angle of }\pi/3\text{ radians with the positive }z\text{-axis}\end{array}$$

$$\theta = \frac{\pi}{3}. \qquad \begin{array}{l}\text{Half-plane, hinged along the }z\text{-axis, making an}\\\text{angle of }\pi/3\text{ radians with the positive }x\text{-axis}\end{array}$$

When computing triple integrals over a region D in spherical coordinates, we partition the region into n spherical wedges. The size of the kth spherical wedge, which contains a point $(\rho_k, \phi_k, \theta_k)$, is given by the changes $\Delta\rho_k$, $\Delta\theta_k$, and $\Delta\phi_k$ in ρ, θ, and ϕ. Such a spherical wedge has one edge a circular arc of length $\rho_k \Delta\phi_k$, another edge a circular arc of length $\rho_k \sin \phi_k \Delta\theta_k$, and thickness $\Delta\rho_k$. The spherical wedge closely approximates a cube of these dimensions when $\Delta\rho_k$, $\Delta\theta_k$, and $\Delta\phi_k$ are all small (Figure 13.44). It can be shown that the volume of this spherical wedge ΔV_k is $\Delta V_k = \rho_k^2 \sin \phi_k \Delta\rho_k \Delta\phi_k \Delta\theta_k$ for $(\rho_k, \phi_k, \theta_k)$ a point chosen inside the wedge.

The corresponding Riemann sum for a function $f(\rho, \phi, \theta)$ is

$$S_n = \sum_{k=1}^{n} f(\rho_k, \phi_k, \theta_k)\, \rho_k^2 \sin \phi_k \Delta\rho_k \Delta\phi_k \Delta\theta_k.$$

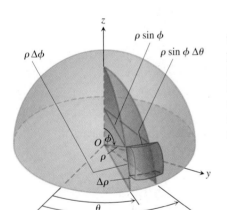

FIGURE 13.44 In spherical coordinates

$$dV = d\rho \cdot \rho\, d\phi \cdot \rho \sin \phi\, d\theta$$
$$= \rho^2 \sin \phi\, d\rho\, d\phi\, d\theta.$$

As the norm of a partition approaches zero, and the spherical wedges get smaller, the Riemann sums have a limit when f is continuous:

$$\lim_{n \to \infty} S_n = \iiint\limits_{D} f(\rho, \phi, \theta)\, dV = \iiint\limits_{D} f(\rho, \phi, \theta)\, \rho^2 \sin \phi\, d\rho\, d\phi\, d\theta.$$

In spherical coordinates, we have

$$dV = \rho^2 \sin \phi\, d\rho\, d\phi\, d\theta.$$

To evaluate integrals in spherical coordinates, we usually integrate first with respect to ρ. The procedure for finding the limits of integration is shown below. We restrict our attention to integrating over domains that are solids of revolution about the z-axis (or portions thereof) and for which the limits for θ and ϕ are constant.

How to Integrate in Spherical Coordinates

To evaluate

$$\iiint\limits_{D} f(\rho, \phi, \theta) \, dV$$

over a region D in space in spherical coordinates, integrating first with respect to ρ, then with respect to ϕ, and finally with respect to θ, take the following steps.

1. *Sketch*. Sketch the region D along with its projection R on the xy-plane. Label the surfaces that bound D.

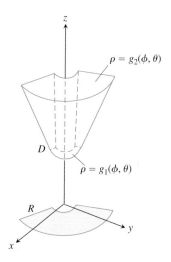

2. *Find the ρ-limits of integration*. Draw a ray M from the origin through D making an angle ϕ with the positive z-axis. Also draw the projection of M on the xy-plane (call the projection L). The ray L makes an angle θ with the positive x-axis. As ρ increases, M enters D at $\rho = g_1(\phi, \theta)$ and leaves at $\rho = g_2(\phi, \theta)$. These are the ρ-limits of integration.

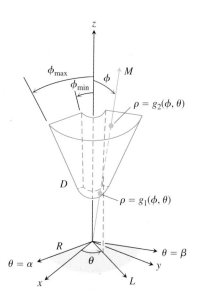

3. *Find the φ-limits of integration.* For any given θ, the angle ϕ that M makes with the z-axis runs from $\phi = \phi_{min}$ to $\phi = \phi_{max}$. These are the ϕ-limits of integration.

4. *Find the θ-limits of integration.* The ray L sweeps over R as θ runs from α to β. These are the θ-limits of integration. The integral is

$$\iiint\limits_{D} f(\rho, \phi, \theta) \, dV = \int_{\theta=\alpha}^{\theta=\beta} \int_{\phi=\phi_{min}}^{\phi=\phi_{max}} \int_{\rho=g_1(\phi, \theta)}^{\rho=g_2(\phi, \theta)} f(\rho, \phi, \theta) \, \rho^2 \sin \phi \, d\rho \, d\phi \, d\theta.$$

EXAMPLE 5 Find the volume of the "ice cream cone" D cut from the solid sphere $\rho \leq 1$ by the cone $\phi = \pi/3$.

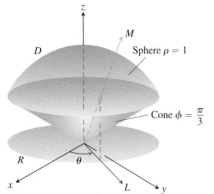

FIGURE 13.45 The ice cream cone in Example 5.

Solution The volume is $V = \iiint_D \rho^2 \sin \phi \, d\rho \, d\phi \, d\theta$, the integral of $f(\rho, \phi, \theta) = 1$ over D.

To find the limits of integration for evaluating the integral, we begin by sketching D and its projection R on the xy-plane (Figure 13.45).

The ρ-limits of integration. We draw a ray M from the origin through D making an angle ϕ with the positive z-axis. We also draw L, the projection of M on the xy-plane, along with the angle θ that L makes with the positive x-axis. Ray M enters D at $\rho = 0$ and leaves at $\rho = 1$.

The φ-limits of integration. The cone $\phi = \pi/3$ makes an angle of $\pi/3$ with the positive z-axis. For any given θ, the angle ϕ can run from $\phi = 0$ to $\phi = \pi/3$.

The θ-limits of integration. The ray L sweeps over R as θ runs from 0 to 2π. The volume is

$$V = \iiint\limits_{D} \rho^2 \sin \phi \, d\rho \, d\phi \, d\theta = \int_0^{2\pi} \int_0^{\pi/3} \int_0^1 \rho^2 \sin \phi \, d\rho \, d\phi \, d\theta$$

$$= \int_0^{2\pi} \int_0^{\pi/3} \left[\frac{\rho^3}{3} \right]_0^1 \sin \phi \, d\phi \, d\theta = \int_0^{2\pi} \int_0^{\pi/3} \frac{1}{3} \sin \phi \, d\phi \, d\theta$$

$$= \int_0^{2\pi} \left[-\frac{1}{3} \cos \phi \right]_0^{\pi/3} d\theta = \int_0^{2\pi} \left(-\frac{1}{6} + \frac{1}{3} \right) d\theta = \frac{1}{6} (2\pi) = \frac{\pi}{3} . \qquad \blacksquare$$

EXAMPLE 6 A solid of constant density $\delta = 1$ occupies the region D in Example 5. Find the solid's moment of inertia about the z-axis.

Solution In rectangular coordinates, the moment is

$$I_z = \iiint (x^2 + y^2) \, dV.$$

In spherical coordinates, $x^2 + y^2 = (\rho \sin \phi \cos \theta)^2 + (\rho \sin \phi \sin \theta)^2 = \rho^2 \sin^2 \phi$. Hence,

$$I_z = \iiint (\rho^2 \sin^2 \phi) \, \rho^2 \sin \phi \, d\rho \, d\phi \, d\theta = \iiint \rho^4 \sin^3 \phi \, d\rho \, d\phi \, d\theta.$$

For the region in Example 5, this becomes

$$I_z = \int_0^{2\pi}\int_0^{\pi/3}\int_0^1 \rho^4 \sin^3\phi\, d\rho\, d\phi\, d\theta = \int_0^{2\pi}\int_0^{\pi/3}\left[\frac{\rho^5}{5}\right]_0^1 \sin^3\phi\, d\phi\, d\theta$$

$$= \frac{1}{5}\int_0^{2\pi}\int_0^{\pi/3}(1 - \cos^2\phi)\sin\phi\, d\phi\, d\theta = \frac{1}{5}\int_0^{2\pi}\left[-\cos\phi + \frac{\cos^3\phi}{3}\right]_0^{\pi/3} d\theta$$

$$= \frac{1}{5}\int_0^{2\pi}\left(-\frac{1}{2} + 1 + \frac{1}{24} - \frac{1}{3}\right) d\theta = \frac{1}{5}\int_0^{2\pi}\frac{5}{24}\, d\theta = \frac{1}{24}(2\pi) = \frac{\pi}{12}. \quad \blacksquare$$

Coordinate Conversion Formulas

CYLINDRICAL TO RECTANGULAR	**SPHERICAL TO RECTANGULAR**	**SPHERICAL TO CYLINDRICAL**
$x = r\cos\theta$	$x = \rho\sin\phi\cos\theta$	$r = \rho\sin\phi$
$y = r\sin\theta$	$y = \rho\sin\phi\sin\theta$	$z = \rho\cos\phi$
$z = z$	$z = \rho\cos\phi$	$\theta = \theta$

Corresponding formulas for dV in triple integrals:

$$dV = dx\, dy\, dz$$
$$= dz\, r\, dr\, d\theta$$
$$= \rho^2 \sin\phi\, d\rho\, d\phi\, d\theta$$

In the next section we offer a more general procedure for determining dV in cylindrical and spherical coordinates. The results, of course, will be the same.

EXERCISES 13.7

Evaluate the cylindrical coordinate integrals in Exercises 1–6.

1. $\displaystyle\int_0^{2\pi}\int_0^1\int_r^{\sqrt{2-r^2}} dz\, r\, dr\, d\theta$ **2.** $\displaystyle\int_0^{2\pi}\int_0^3\int_{r^2/3}^{\sqrt{18-r^2}} dz\, r\, dr\, d\theta$

3. $\displaystyle\int_0^{2\pi}\int_0^{\theta/2\pi}\int_0^{3+24r^2} dz\, r\, dr\, d\theta$ **4.** $\displaystyle\int_0^\pi\int_0^{\theta/\pi}\int_{-\sqrt{4-r^2}}^{3\sqrt{4-r^2}} z\, dz\, r\, dr\, d\theta$

5. $\displaystyle\int_0^{2\pi}\int_0^1\int_r^{1/\sqrt{2-r^2}} 3\, dz\, r\, dr\, d\theta$

6. $\displaystyle\int_0^{2\pi}\int_0^1\int_{-1/2}^{1/2} (r^2\sin^2\theta + z^2)\, dz\, r\, dr\, d\theta$

The integrals we have seen so far suggest that there are preferred orders of integration for cylindrical coordinates, but other orders usually work well and are occasionally easier to evaluate. Evaluate the integrals in Exercises 7–10.

7. $\displaystyle\int_0^{2\pi}\int_0^3\int_0^{z/3} r^3\, dr\, dz\, d\theta$ **8.** $\displaystyle\int_{-1}^1\int_0^{2\pi}\int_0^{1+\cos\theta} 4r\, dr\, d\theta\, dz$

9. $\displaystyle\int_0^1\int_0^{\sqrt{z}}\int_0^{2\pi} (r^2\cos^2\theta + z^2)\, r\, d\theta\, dr\, dz$

10. $\displaystyle\int_0^2\int_{r-2}^{\sqrt{4-r^2}}\int_0^{2\pi} (r\sin\theta + 1)\, r\, d\theta\, dz\, dr$

11. Let D be the region bounded below by the plane $z = 0$, above by the sphere $x^2 + y^2 + z^2 = 4$, and on the sides by the cylinder $x^2 + y^2 = 1$. Set up the triple integrals in cylindrical coordinates that give the volume of D using the following orders of integration.

   a. $dz\, dr\, d\theta$ **b.** $dr\, dz\, d\theta$ **c.** $d\theta\, dz\, dr$

12. Let D be the region bounded below by the cone $z = \sqrt{x^2 + y^2}$ and above by the paraboloid $z = 2 - x^2 - y^2$. Set up the triple integrals in cylindrical coordinates that give the volume of D using the following orders of integration.

   a. $dz\, dr\, d\theta$ **b.** $dr\, dz\, d\theta$ **c.** $d\theta\, dz\, dr$

13. Give the limits of integration for evaluating the integral

$$\iiint f(r, \theta, z) \, dz \, r \, dr \, d\theta$$

as an iterated integral over the region that is bounded below by the plane $z = 0$, on the side by the cylinder $r = \cos \theta$, and on top by the paraboloid $z = 3r^2$.

14. Convert the integral

$$\int_{-1}^{1} \int_{0}^{\sqrt{1-y^2}} \int_{0}^{x} (x^2 + y^2) \, dz \, dx \, dy$$

to an equivalent integral in cylindrical coordinates and evaluate the result.

In Exercises 15–20, set up the iterated integral for evaluating $\iiint_D f(r, \theta, z) \, dz \, r \, dr \, d\theta$ over the given region D.

15. D is the right circular cylinder whose base is the circle $r = 2 \sin \theta$ in the xy-plane and whose top lies in the plane $z = 4 - y$.

16. D is the right circular cylinder whose base is the circle $r = 3 \cos \theta$ and whose top lies in the plane $z = 5 - x$.

17. D is the solid right cylinder whose base is the region in the xy-plane that lies inside the cardioid $r = 1 + \cos \theta$ and outside the circle $r = 1$ and whose top lies in the plane $z = 4$.

18. D is the solid right cylinder whose base is the region between the circles $r = \cos \theta$ and $r = 2 \cos \theta$ and whose top lies in the plane $z = 3 - y$.

19. D is the prism whose base is the triangle in the xy-plane bounded by the x-axis and the lines $y = x$ and $x = 1$ and whose top lies in the plane $z = 2 - y$.

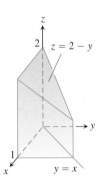

20. D is the prism whose base is the triangle in the xy-plane bounded by the y-axis and the lines $y = x$ and $y = 1$ and whose top lies in the plane $z = 2 - x$.

Evaluate the spherical coordinate integrals in Exercises 21–26.

21. $\displaystyle \int_{0}^{\pi} \int_{0}^{\pi} \int_{0}^{2 \sin \phi} \rho^2 \sin \phi \, d\rho \, d\phi \, d\theta$

22. $\displaystyle \int_{0}^{2\pi} \int_{0}^{\pi/4} \int_{0}^{2} (\rho \cos \phi) \rho^2 \sin \phi \, d\rho \, d\phi \, d\theta$

23. $\int_0^{2\pi} \int_0^{\pi} \int_0^{(1-\cos\phi)/2} \rho^2 \sin\phi \, d\rho \, d\phi \, d\theta$

24. $\int_0^{3\pi/2} \int_0^{\pi} \int_0^{1} 5\rho^3 \sin^3\phi \, d\rho \, d\phi \, d\theta$

25. $\int_0^{2\pi} \int_0^{\pi/3} \int_{\sec\phi}^{2} 3\rho^2 \sin\phi \, d\rho \, d\phi \, d\theta$

26. $\int_0^{2\pi} \int_0^{\pi/4} \int_0^{\sec\phi} (\rho \cos\phi) \, \rho^2 \sin\phi \, d\rho \, d\phi \, d\theta$

The previous integrals suggest there are preferred orders of integration for spherical coordinates, but other orders give the same value and are occasionally easier to evaluate. Evaluate the integrals in Exercises 27–30.

27. $\int_0^2 \int_{-\pi}^0 \int_{\pi/4}^{\pi/2} \rho^3 \sin 2\phi \, d\phi \, d\theta \, d\rho$

28. $\int_{\pi/6}^{\pi/3} \int_{\csc\phi}^{2\csc\phi} \int_0^{2\pi} \rho^2 \sin\phi \, d\theta \, d\rho \, d\phi$

29. $\int_0^1 \int_0^{\pi} \int_0^{\pi/4} 12\rho \sin^3\phi \, d\phi \, d\theta \, d\rho$

30. $\int_{\pi/6}^{\pi/2} \int_{-\pi/2}^{\pi/2} \int_{\csc\phi}^2 5\rho^4 \sin^3\phi \, d\rho \, d\theta \, d\phi$

31. Let D be the region in Exercise 11. Set up the triple integrals in spherical coordinates that give the volume of D using the following orders of integration.

    a. $d\rho \, d\phi \, d\theta$ **b.** $d\phi \, d\rho \, d\theta$

32. Let D be the region bounded below by the cone $z = \sqrt{x^2 + y^2}$ and above by the plane $z = 1$. Set up the triple integrals in spherical coordinates that give the volume of D using the following orders of integration.

    a. $d\rho \, d\phi \, d\theta$ **b.** $d\phi \, d\rho \, d\theta$

In Exercises 33–38, **(a)** find the spherical coordinate limits for the integral that calculates the volume of the given solid and then **(b)** evaluate the integral.

33. The solid between the sphere $\rho = \cos\phi$ and the hemisphere $\rho = 2, z \ge 0$

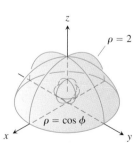

34. The solid bounded below by the hemisphere $\rho = 1, z \ge 0$, and above by the cardioid of revolution $\rho = 1 + \cos\phi$

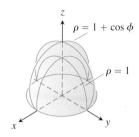

35. The solid enclosed by the cardioid of revolution $\rho = 1 - \cos\phi$

36. The upper portion cut from the solid in Exercise 35 by the xy-plane

37. The solid bounded below by the sphere $\rho = 2\cos\phi$ and above by the cone $z = \sqrt{x^2 + y^2}$

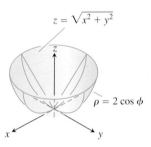

38. The solid bounded below by the xy-plane, on the sides by the sphere $\rho = 2$, and above by the cone $\phi = \pi/3$

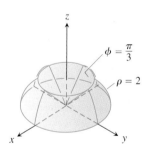

39. Set up triple integrals for the volume of the sphere $\rho = 2$ in **(a)** spherical, **(b)** cylindrical, and **(c)** rectangular coordinates.

40. Let D be the region in the first octant that is bounded below by the cone $\phi = \pi/4$ and above by the sphere $\rho = 3$. Express the volume of D as an iterated triple integral in **(a)** cylindrical and **(b)** spherical coordinates. Then **(c)** find V.

41. Let D be the smaller cap cut from a solid ball of radius 2 units by a plane 1 unit from the center of the sphere. Express the volume of D as an iterated triple integral in **(a)** spherical, **(b)** cylindrical, and **(c)** rectangular coordinates. Then **(d)** find the volume by evaluating one of the three triple integrals.

42. Express the moment of inertia I_z of the solid hemisphere $x^2 + y^2 + z^2 \le 1, z \ge 0$, as an iterated integral in **(a)** cylindrical and **(b)** spherical coordinates. Then **(c)** find I_z.

Find the volumes of the solids in Exercises 43–48.

43.

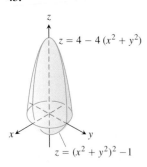

$z = 4 - 4(x^2 + y^2)$

$z = (x^2 + y^2)^2 - 1$

44.

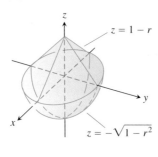

$z = 1 - r$

$z = -\sqrt{1 - r^2}$

45.

$z = -y$

$r = 3 \cos \theta$

46.

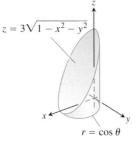

$z = \sqrt{x^2 + y^2}$

$r = -3 \cos \theta$

47.

$z = \sqrt{1 - x^2 - y^2}$

$r = \sin \theta$

48.

$z = 3\sqrt{1 - x^2 - y^2}$

$r = \cos \theta$

49. Sphere and cones Find the volume of the portion of the solid sphere $\rho \le a$ that lies between the cones $\phi = \pi/3$ and $\phi = 2\pi/3$.

50. Sphere and half-planes Find the volume of the region cut from the solid sphere $\rho \le a$ by the half-planes $\theta = 0$ and $\theta = \pi/6$ in the first octant.

51. Sphere and plane Find the volume of the smaller region cut from the solid sphere $\rho \le 2$ by the plane $z = 1$.

52. Cone and planes Find the volume of the solid enclosed by the cone $z = \sqrt{x^2 + y^2}$ between the planes $z = 1$ and $z = 2$.

53. Cylinder and paraboloid Find the volume of the region bounded below by the plane $z = 0$, laterally by the cylinder $x^2 + y^2 = 1$, and above by the paraboloid $z = x^2 + y^2$.

54. Cylinder and paraboloids Find the volume of the region bounded below by the paraboloid $z = x^2 + y^2$, laterally by the cylinder $x^2 + y^2 = 1$, and above by the paraboloid $z = x^2 + y^2 + 1$.

55. Cylinder and cones Find the volume of the solid cut from the thick-walled cylinder $1 \le x^2 + y^2 \le 2$ by the cones $z = \pm\sqrt{x^2 + y^2}$.

56. Sphere and cylinder Find the volume of the region that lies inside the sphere $x^2 + y^2 + z^2 = 2$ and outside the cylinder $x^2 + y^2 = 1$.

57. Cylinder and planes Find the volume of the region enclosed by the cylinder $x^2 + y^2 = 4$ and the planes $z = 0$ and $y + z = 4$.

58. Cylinder and planes Find the volume of the region enclosed by the cylinder $x^2 + y^2 = 4$ and the planes $z = 0$ and $x + y + z = 4$.

59. Region trapped by paraboloids Find the volume of the region bounded above by the paraboloid $z = 5 - x^2 - y^2$ and below by the paraboloid $z = 4x^2 + 4y^2$.

60. Paraboloid and cylinder Find the volume of the region bounded above by the paraboloid $z = 9 - x^2 - y^2$, below by the xy-plane, and lying *outside* the cylinder $x^2 + y^2 = 1$.

61. Cylinder and sphere Find the volume of the region cut from the solid cylinder $x^2 + y^2 \le 1$ by the sphere $x^2 + y^2 + z^2 = 4$.

62. Sphere and paraboloid Find the volume of the region bounded above by the sphere $x^2 + y^2 + z^2 = 2$ and below by the paraboloid $z = x^2 + y^2$.

63. Find the average value of the function $f(r, \theta, z) = r$ over the region bounded by the cylinder $r = 1$ between the planes $z = -1$ and $z = 1$.

64. Find the average value of the function $f(r, \theta, z) = r$ over the solid ball bounded by the sphere $r^2 + z^2 = 1$. (This is the sphere $x^2 + y^2 + z^2 = 1$.)

65. Find the average value of the function $f(\rho, \phi, \theta) = \rho$ over the solid ball $\rho \le 1$.

66. Find the average value of the function $f(\rho, \phi, \theta) = \rho \cos \phi$ over the solid upper ball $\rho \le 1, 0 \le \phi \le \pi/2$.

67. Center of mass A solid of constant density is bounded below by the plane $z = 0$, above by the cone $z = r, r \ge 0$, and on the sides by the cylinder $r = 1$. Find the center of mass.

68. Centroid Find the centroid of the region in the first octant that is bounded above by the cone $z = \sqrt{x^2 + y^2}$, below by the plane $z = 0$, and on the sides by the cylinder $x^2 + y^2 = 4$ and the planes $x = 0$ and $y = 0$.

69. Centroid Find the centroid of the solid in Exercise 38.

70. Centroid Find the centroid of the solid bounded above by the sphere $\rho = a$ and below by the cone $\phi = \pi/4$.

71. Centroid Find the centroid of the region that is bounded above by the surface $z = \sqrt{r}$, on the sides by the cylinder $r = 4$, and below by the xy-plane.

72. Centroid Find the centroid of the region cut from the solid ball $r^2 + z^2 \le 1$ by the half-planes $\theta = -\pi/3, r \ge 0$, and $\theta = \pi/3$, $r \ge 0$.

73. Moment of inertia of solid cone   Find the moment of inertia of a right circular cone of base radius 1 and height 1 about an axis through the vertex parallel to the base. (Take $\delta = 1$.)

74. Moment of inertia of solid sphere   Find the moment of inertia of a solid sphere of radius a about a diameter. (Take $\delta = 1$.)

75. Moment of inertia of solid cone   Find the moment of inertia of a right circular cone of base radius a and height h about its axis. (*Hint:* Place the cone with its vertex at the origin and its axis along the z-axis.)

76. Variable density   A solid is bounded on the top by the paraboloid $z = r^2$, on the bottom by the plane $z = 0$, and on the sides by the cylinder $r = 1$. Find the center of mass and the moment of inertia about the z-axis if the density is

  a. $\delta(r, \theta, z) = z$ **b.** $\delta(r, \theta, z) = r$.

77. Variable density   A solid is bounded below by the cone $z = \sqrt{x^2 + y^2}$ and above by the plane $z = 1$. Find the center of mass and the moment of inertia about the z-axis if the density is

  a. $\delta(r, \theta, z) = z$ **b.** $\delta(r, \theta, z) = z^2$.

78. Variable density   A solid ball is bounded by the sphere $\rho = a$. Find the moment of inertia about the z-axis if the density is

  a. $\delta(\rho, \phi, \theta) = \rho^2$ **b.** $\delta(\rho, \phi, \theta) = r = \rho \sin \phi$.

79. Centroid of solid semiellipsoid   Show that the centroid of the solid semiellipsoid of revolution $(r^2/a^2) + (z^2/h^2) \leq 1, z \geq 0$, lies on the z-axis three-eighths of the way from the base to the top. The special case $h = a$ gives a solid hemisphere. Thus, the centroid of a solid hemisphere lies on the axis of symmetry three-eighths of the way from the base to the top.

80. Centroid of solid cone   Show that the centroid of a solid right circular cone is one-fourth of the way from the base to the vertex. (In general, the centroid of a solid cone or pyramid is one-fourth of the way from the centroid of the base to the vertex.)

81. Density of center of a planet   A planet is in the shape of a sphere of radius R and total mass M with spherically symmetric density distribution that increases linearly as one approaches its center. What is the density at the center of this planet if the density at its edge (surface) is taken to be zero?

82. Mass of planet's atmosphere   A spherical planet of radius R has an atmosphere whose density is $\mu = \mu_0 e^{-ch}$, where h is the altitude above the surface of the planet, μ_0 is the density at sea level, and c is a positive constant. Find the mass of the planet's atmosphere.

83. Vertical planes in cylindrical coordinates

  a. Show that planes perpendicular to the x-axis have equations of the form $r = a \sec \theta$ in cylindrical coordinates.

  b. Show that planes perpendicular to the y-axis have equations of the form $r = b \csc \theta$.

84. (*Continuation of Exercise 83.*) Find an equation of the form $r = f(\theta)$ in cylindrical coordinates for the plane $ax + by = c$, $c \neq 0$.

85. Symmetry   What symmetry will you find in a surface that has an equation of the form $r = f(z)$ in cylindrical coordinates? Give reasons for your answer.

86. Symmetry   What symmetry will you find in a surface that has an equation of the form $\rho = f(\phi)$ in spherical coordinates? Give reasons for your answer.

13.8   Substitutions in Multiple Integrals

This section shows how to evaluate multiple integrals by substitution. As in single integration, the goal of substitution is to replace complicated integrals by ones that are easier to evaluate. Substitutions accomplish this by simplifying the integrand, the limits of integration, or both.

Substitutions in Double Integrals

The polar coordinate substitution of Section 13.4 is a special case of a more general substitution method for double integrals, a method that pictures changes in variables as transformations of regions.

Suppose that a region G in the uv-plane is transformed one-to-one into the region R in the xy-plane by equations of the form

$$x = g(u, v), \qquad y = h(u, v),$$

as suggested in Figure 13.46 on the next page. We call R the **image** of G under the transformation, and G the **preimage** of R. Any function $f(x, y)$ defined on R can be thought of as a function $f(g(u, v), h(u, v))$ defined on G as well. How is the integral of $f(x, y)$ over R related to the integral of $f(g(u, v), h(u, v))$ over G?

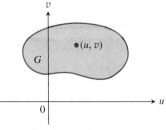

Cartesian uv-plane

$$x = g(u, v)$$
$$y = h(u, v)$$

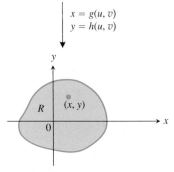

Cartesian xy-plane

FIGURE 13.46 The equations $x = g(u, v)$ and $y = h(u, v)$ allow us to change an integral over a region R in the xy-plane into an integral over a region G in the uv-plane.

HISTORICAL BIOGRAPHY

Carl Gustav Jacob Jacobi
(1804–1851)

The answer is: If g, h, and f have continuous partial derivatives and $J(u, v)$ (to be discussed in a moment) is zero only at isolated points, if at all, then

$$\iint\limits_{R} f(x, y)\, dx\, dy = \iint\limits_{G} f(g(u, v), h(u, v))\,|J(u, v)|\, du\, dv. \qquad (1)$$

The factor $J(u, v)$, whose absolute value appears in Equation (1), is the *Jacobian* of the coordinate transformation, named after German mathematician Carl Jacobi. It measures how much the transformation is expanding or contracting the area around a point in G as G is transformed into R.

DEFINITION   The **Jacobian determinant** or **Jacobian** of the coordinate transformation $x = g(u, v), y = h(u, v)$ is

$$J(u, v) = \begin{vmatrix} \dfrac{\partial x}{\partial u} & \dfrac{\partial x}{\partial v} \\[2mm] \dfrac{\partial y}{\partial u} & \dfrac{\partial y}{\partial v} \end{vmatrix} = \frac{\partial x}{\partial u}\frac{\partial y}{\partial v} - \frac{\partial y}{\partial u}\frac{\partial x}{\partial v}. \qquad (2)$$

The Jacobian is also denoted by

$$J(u, v) = \frac{\partial(x, y)}{\partial(u, v)}$$

to help remember how the determinant in Equation (2) is constructed from the partial derivatives of x and y. The derivation of Equation (1) is intricate and properly belongs to a course in advanced calculus. We do not give the derivation here.

For polar coordinates, we have r and θ in place of u and v. With $x = r\cos\theta$ and $y = r\sin\theta$, the Jacobian is

$$J(r, \theta) = \begin{vmatrix} \dfrac{\partial x}{\partial r} & \dfrac{\partial x}{\partial \theta} \\[2mm] \dfrac{\partial y}{\partial r} & \dfrac{\partial y}{\partial \theta} \end{vmatrix} = \begin{vmatrix} \cos\theta & -r\sin\theta \\ \sin\theta & r\cos\theta \end{vmatrix} = r(\cos^2\theta + \sin^2\theta) = r.$$

Hence, Equation (1) becomes

$$\iint\limits_{R} f(x, y)\, dx\, dy = \iint\limits_{G} f(r\cos\theta, r\sin\theta)\,|r|\, dr\, d\theta$$

$$= \iint\limits_{G} f(r\cos\theta, r\sin\theta)\, r\, dr\, d\theta, \qquad \text{If } r \ge 0 \qquad (3)$$

which is the equation found in Section 13.4.

Figure 13.47 shows how the equations $x = r\cos\theta, y = r\sin\theta$ transform the rectangle $G: 0 \le r \le 1, 0 \le \theta \le \pi/2$ into the quarter circle R bounded by $x^2 + y^2 = 1$ in the first quadrant of the xy-plane.

Notice that the integral on the right-hand side of Equation (3) is not the integral of $f(r\cos\theta, r\sin\theta)$ over a region in the polar coordinate plane. It is the integral of the product of $f(r\cos\theta, r\sin\theta)$ and r over a region G in the *Cartesian $r\theta$-plane*.

Here is an example of another substitution.

Cartesian $r\theta$-plane

$$x = r \cos \theta$$
$$y = r \sin \theta$$

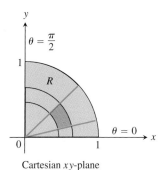

Cartesian xy-plane

FIGURE 13.47 The equations $x = r \cos \theta$, $y = r \sin \theta$ transform G into R.

EXAMPLE 1 Evaluate

$$\int_0^4 \int_{x=y/2}^{x=(y/2)+1} \frac{2x - y}{2} \, dx \, dy$$

by applying the transformation

$$u = \frac{2x - y}{2}, \qquad v = \frac{y}{2} \tag{4}$$

and integrating over an appropriate region in the uv-plane.

Solution We sketch the region R of integration in the xy-plane and identify its boundaries (Figure 13.48).

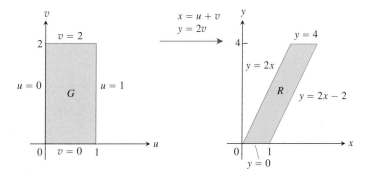

FIGURE 13.48 The equations $x = u + v$ and $y = 2v$ transform G into R. Reversing the transformation by the equations $u = (2x - y)/2$ and $v = y/2$ transforms R into G (Example 1).

To apply Equation (1), we need to find the corresponding uv-region G and the Jacobian of the transformation. To find them, we first solve Equations (4) for x and y in terms of u and v. Routine algebra gives

$$x = u + v \qquad y = 2v. \tag{5}$$

We then find the boundaries of G by substituting these expressions into the equations for the boundaries of R (Figure 13.48).

xy-equations for the boundary of R	Corresponding uv-equations for the boundary of G	Simplified uv-equations
$x = y/2$	$u + v = 2v/2 = v$	$u = 0$
$x = (y/2) + 1$	$u + v = (2v/2) + 1 = v + 1$	$u = 1$
$y = 0$	$2v = 0$	$v = 0$
$y = 4$	$2v = 4$	$v = 2$

The Jacobian of the transformation (again from Equations (5)) is

$$J(u, v) = \begin{vmatrix} \dfrac{\partial x}{\partial u} & \dfrac{\partial x}{\partial v} \\[2mm] \dfrac{\partial y}{\partial u} & \dfrac{\partial y}{\partial v} \end{vmatrix} = \begin{vmatrix} \dfrac{\partial}{\partial u}(u + v) & \dfrac{\partial}{\partial v}(u + v) \\[2mm] \dfrac{\partial}{\partial u}(2v) & \dfrac{\partial}{\partial v}(2v) \end{vmatrix} = \begin{vmatrix} 1 & 1 \\ 0 & 2 \end{vmatrix} = 2.$$

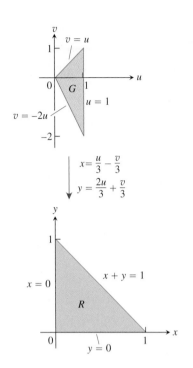

$$x = \frac{u}{3} - \frac{v}{3}$$

$$y = \frac{2u}{3} + \frac{v}{3}$$

FIGURE 13.49 The equations $x = (u/3) - (v/3)$ and $y = (2u/3) + (v/3)$ transform G into R. Reversing the transformation by the equations $u = x + y$ and $v = y - 2x$ transforms R into G (Example 2).

We now have everything we need to apply Equation (1):

$$\int_0^4 \int_{x=y/2}^{x=(y/2)+1} \frac{2x - y}{2} \, dx \, dy = \int_{v=0}^{v=2} \int_{u=0}^{u=1} u \, |J(u, v)| \, du \, dv$$

$$= \int_0^2 \int_0^1 (u)(2) \, du \, dv = \int_0^2 \left[u^2\right]_0^1 dv = \int_0^2 dv = 2.$$

EXAMPLE 2   Evaluate

$$\int_0^1 \int_0^{1-x} \sqrt{x + y} \, (y - 2x)^2 \, dy \, dx.$$

Solution   We sketch the region R of integration in the xy-plane and identify its boundaries (Figure 13.49). The integrand suggests the transformation $u = x + y$ and $v = y - 2x$. Routine algebra produces x and y as functions of u and v:

$$x = \frac{u}{3} - \frac{v}{3}, \qquad y = \frac{2u}{3} + \frac{v}{3}. \tag{6}$$

From Equations (6), we can find the boundaries of the uv-region G (Figure 13.49).

xy-equations for the boundary of R	Corresponding uv-equations for the boundary of G	Simplified uv-equations
$x + y = 1$	$\left(\dfrac{u}{3} - \dfrac{v}{3}\right) + \left(\dfrac{2u}{3} + \dfrac{v}{3}\right) = 1$	$u = 1$
$x = 0$	$\dfrac{u}{3} - \dfrac{v}{3} = 0$	$v = u$
$y = 0$	$\dfrac{2u}{3} + \dfrac{v}{3} = 0$	$v = -2u$

The Jacobian of the transformation in Equations (6) is

$$J(u, v) = \begin{vmatrix} \dfrac{\partial x}{\partial u} & \dfrac{\partial x}{\partial v} \\[2mm] \dfrac{\partial y}{\partial u} & \dfrac{\partial y}{\partial v} \end{vmatrix} = \begin{vmatrix} \dfrac{1}{3} & -\dfrac{1}{3} \\[2mm] \dfrac{2}{3} & \dfrac{1}{3} \end{vmatrix} = \frac{1}{3}.$$

Applying Equation (1), we evaluate the integral:

$$\int_0^1 \int_0^{1-x} \sqrt{x + y} \, (y - 2x)^2 \, dy \, dx = \int_{u=0}^{u=1} \int_{v=-2u}^{v=u} u^{1/2} v^2 \, |J(u, v)| \, dv \, du$$

$$= \int_0^1 \int_{-2u}^u u^{1/2} v^2 \left(\frac{1}{3}\right) dv \, du = \frac{1}{3} \int_0^1 u^{1/2} \left[\frac{1}{3} v^3\right]_{v=-2u}^{v=u} du$$

$$= \frac{1}{9} \int_0^1 u^{1/2}(u^3 + 8u^3) \, du = \int_0^1 u^{7/2} \, du = \frac{2}{9} u^{9/2} \Big]_0^1 = \frac{2}{9}.$$

Substitutions in Triple Integrals

The cylindrical and spherical coordinate substitutions in Section 13.7 are special cases of a substitution method that pictures changes of variables in triple integrals as transformations of three-dimensional regions. The method is like the method for double integrals except that now we work in three dimensions instead of two.

Suppose that a region G in uvw-space is transformed one-to-one into the region D in xyz-space by differentiable equations of the form

$$x = g(u, v, w), \qquad y = h(u, v, w), \qquad z = k(u, v, w),$$

as suggested in Figure 13.50. Then any function $F(x, y, z)$ defined on D can be thought of as a function

$$F(g(u, v, w), h(u, v, w), k(u, v, w)) = H(u, v, w)$$

defined on G. If g, h, and k have continuous first partial derivatives, then the integral of $F(x, y, z)$ over D is related to the integral of $H(u, v, w)$ over G by the equation

$$\iiint\limits_{D} F(x, y, z)\, dx\, dy\, dz = \iiint\limits_{G} H(u, v, w) |J(u, v, w)|\, du\, dv\, dw. \tag{7}$$

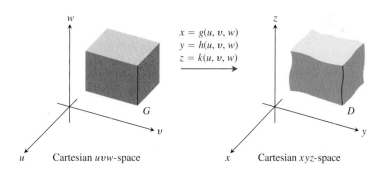

FIGURE 13.50 The equations $x = g(u, v, w)$, $y = h(u, v, w)$, and $z = k(u, v, w)$ allow us to change an integral over a region D in Cartesian xyz-space into an integral over a region G in Cartesian uvw-space.

The factor $J(u, v, w)$, whose absolute value appears in this equation, is the **Jacobian determinant**

$$J(u, v, w) = \begin{vmatrix} \dfrac{\partial x}{\partial u} & \dfrac{\partial x}{\partial v} & \dfrac{\partial x}{\partial w} \\[2mm] \dfrac{\partial y}{\partial u} & \dfrac{\partial y}{\partial v} & \dfrac{\partial y}{\partial w} \\[2mm] \dfrac{\partial z}{\partial u} & \dfrac{\partial z}{\partial v} & \dfrac{\partial z}{\partial w} \end{vmatrix} = \frac{\partial(x, y, z)}{\partial(u, v, w)}.$$

This determinant measures how much the volume near a point in G is being expanded or contracted by the transformation from (u, v, w) to (x, y, z) coordinates. As in the two-dimensional case, the derivation of the change-of-variable formula in Equation (7) is complicated and we do not go into it here.

Cube with sides
parallel to the
coordinate axes

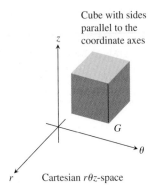

Cartesian $r\theta z$-space

$x = r \cos \theta$
$y = r \sin \theta$
$z = z$

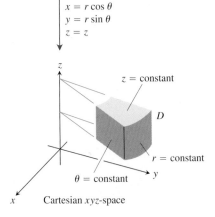

$z = \text{constant}$

D

$r = \text{constant}$

$\theta = \text{constant}$

Cartesian xyz-space

FIGURE 13.51 The equations
$x = r \cos \theta$, $y = r \sin \theta$, and $z = z$
transform the cube G into a cylindrical
wedge D.

For cylindrical coordinates, r, θ, and z take the place of u, v, and w. The transformation from Cartesian $r\theta z$-space to Cartesian xyz-space is given by the equations

$$x = r \cos \theta, \qquad y = r \sin \theta, \qquad z = z$$

(Figure 13.51). The Jacobian of the transformation is

$$J(r, \theta, z) = \begin{vmatrix} \dfrac{\partial x}{\partial r} & \dfrac{\partial x}{\partial \theta} & \dfrac{\partial x}{\partial z} \\[2mm] \dfrac{\partial y}{\partial r} & \dfrac{\partial y}{\partial \theta} & \dfrac{\partial y}{\partial z} \\[2mm] \dfrac{\partial z}{\partial r} & \dfrac{\partial z}{\partial \theta} & \dfrac{\partial z}{\partial z} \end{vmatrix} = \begin{vmatrix} \cos \theta & -r \sin \theta & 0 \\ \sin \theta & r \cos \theta & 0 \\ 0 & 0 & 1 \end{vmatrix}$$

$$= r \cos^2 \theta + r \sin^2 \theta = r.$$

The corresponding version of Equation (7) is

$$\iiint_D F(x, y, z) \, dx \, dy \, dz = \iiint_G H(r, \theta, z) |r| \, dr \, d\theta \, dz.$$

We can drop the absolute value signs whenever $r \geq 0$.

For spherical coordinates, ρ, ϕ, and θ take the place of u, v, and w. The transformation from Cartesian $\rho\phi\theta$-space to Cartesian xyz-space is given by

$$x = \rho \sin \phi \cos \theta, \qquad y = \rho \sin \phi \sin \theta, \qquad z = \rho \cos \phi$$

(Figure 13.52). The Jacobian of the transformation (see Exercise 17) is

$$J(\rho, \phi, \theta) = \begin{vmatrix} \dfrac{\partial x}{\partial \rho} & \dfrac{\partial x}{\partial \phi} & \dfrac{\partial x}{\partial \theta} \\[2mm] \dfrac{\partial y}{\partial \rho} & \dfrac{\partial y}{\partial \phi} & \dfrac{\partial y}{\partial \theta} \\[2mm] \dfrac{\partial z}{\partial \rho} & \dfrac{\partial z}{\partial \phi} & \dfrac{\partial z}{\partial \theta} \end{vmatrix} = \rho^2 \sin \phi.$$

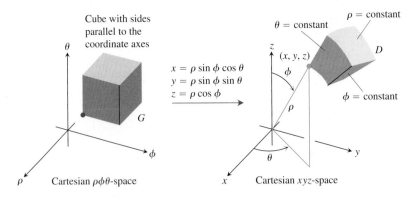

FIGURE 13.52 The equations $x = \rho \sin \phi \cos \theta$, $y = \rho \sin \phi \sin \theta$, and $z = \rho \cos \phi$ transform the cube G into the spherical wedge D.

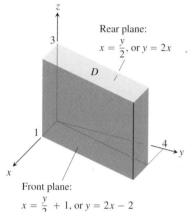

FIGURE 13.53 The equations $x = u + v$, $y = 2v$, and $z = 3w$ transform G into D. Reversing the transformation by the equations $u = (2x - y)/2$, $v = y/2$, and $w = z/3$ transforms D into G (Example 3).

The corresponding version of Equation (7) is

$$\iiint\limits_D F(x, y, z)\, dx\, dy\, dz = \iiint\limits_G H(\rho, \phi, \theta)\, |\rho^2 \sin \phi|\, d\rho\, d\phi\, d\theta.$$

We can drop the absolute value signs because $\sin \phi$ is never negative for $0 \le \phi \le \pi$. Note that this is the same result we obtained in Section 13.7.

Here is an example of another substitution. Although we could evaluate the integral in this example directly, we have chosen it to illustrate the substitution method in a simple (and fairly intuitive) setting.

EXAMPLE 3   Evaluate

$$\int_0^3 \int_0^4 \int_{x=y/2}^{x=(y/2)+1} \left(\frac{2x - y}{2} + \frac{z}{3} \right) dx\, dy\, dz$$

by applying the transformation

$$u = (2x - y)/2, \qquad v = y/2, \qquad w = z/3 \tag{8}$$

and integrating over an appropriate region in uvw-space.

Solution   We sketch the region D of integration in xyz-space and identify its boundaries (Figure 13.53). In this case, the bounding surfaces are planes.

To apply Equation (7), we need to find the corresponding uvw-region G and the Jacobian of the transformation. To find them, we first solve Equations (8) for x, y, and z in terms of u, v, and w. Routine algebra gives

$$x = u + v, \qquad y = 2v, \qquad z = 3w. \tag{9}$$

We then find the boundaries of G by substituting these expressions into the equations for the boundaries of D:

xyz-equations for the boundary of D	Corresponding uvw-equations for the boundary of G	Simplified uvw-equations
$x = y/2$	$u + v = 2v/2 = v$	$u = 0$
$x = (y/2) + 1$	$u + v = (2v/2) + 1 = v + 1$	$u = 1$
$y = 0$	$2v = 0$	$v = 0$
$y = 4$	$2v = 4$	$v = 2$
$z = 0$	$3w = 0$	$w = 0$
$z = 3$	$3w = 3$	$w = 1$

The Jacobian of the transformation, again from Equations (9), is

$$J(u, v, w) = \begin{vmatrix} \dfrac{\partial x}{\partial u} & \dfrac{\partial x}{\partial v} & \dfrac{\partial x}{\partial w} \\[2mm] \dfrac{\partial y}{\partial u} & \dfrac{\partial y}{\partial v} & \dfrac{\partial y}{\partial w} \\[2mm] \dfrac{\partial z}{\partial u} & \dfrac{\partial z}{\partial v} & \dfrac{\partial z}{\partial w} \end{vmatrix} = \begin{vmatrix} 1 & 1 & 0 \\ 0 & 2 & 0 \\ 0 & 0 & 3 \end{vmatrix} = 6.$$

We now have everything we need to apply Equation (7):

$$\int_0^3 \int_0^4 \int_{x=y/2}^{x=(y/2)+1} \left(\frac{2x - y}{2} + \frac{z}{3} \right) dx \, dy \, dz$$

$$= \int_0^1 \int_0^2 \int_0^1 (u + w) |J(u, v, w)| \, du \, dv \, dw$$

$$= \int_0^1 \int_0^2 \int_0^1 (u + w)(6) \, du \, dv \, dw = 6 \int_0^1 \int_0^2 \left[\frac{u^2}{2} + uw \right]_0^1 dv \, dw$$

$$= 6 \int_0^1 \int_0^2 \left(\frac{1}{2} + w \right) dv \, dw = 6 \int_0^1 \left[\frac{v}{2} + vw \right]_0^2 dw = 6 \int_0^1 (1 + 2w) \, dw$$

$$= 6 \left[w + w^2 \right]_0^1 = 6(2) = 12. \qquad ■$$

The goal of this section was to introduce you to the ideas involved in coordinate transformations. A thorough discussion of transformations, the Jacobian, and multivariable substitution is best given in an advanced calculus course after a study of linear algebra.

EXERCISES 13.8

1. a. Solve the system

$$u = x - y, \qquad v = 2x + y$$

for x and y in terms of u and y. Then find the value of the Jacobian $\partial(x, y)/\partial(u, v)$.

b. Find the image under the transformation $u = x - y$, $v = 2x + y$ of the triangular region with vertices $(0, 0)$, $(1, 1)$, and $(1, -2)$ in the xy-plane. Sketch the transformed region in the uv-plane.

2. a. Solve the system

$$u = x + 2y, \qquad v = x - y$$

for x and y in terms of u and v. Then find the value of the Jacobian $\partial(x, y)/\partial(u, v)$.

b. Find the image under the transformation $u = x + 2y$, $v = x - y$ of the triangular region in the xy-plane bounded by the lines $y = 0$, $y = x$, and $x + 2y = 2$. Sketch the transformed region in the uv-plane.

3. a. Solve the system

$$u = 3x + 2y, \qquad v = x + 4y$$

for x and y in terms of u and v. Then find the value of the Jacobian $\partial(x, y)/\partial(u, v)$.

b. Find the image under the transformation $u = 3x + 2y$, $v = x + 4y$ of the triangular region in the xy-plane bounded by the x-axis, the y-axis, and the line $x + y = 1$. Sketch the transformed region in the uv-plane.

4. a. Solve the system

$$u = 2x - 3y, \qquad v = -x + y$$

for x and y in terms of u and v. Then find the value of the Jacobian $\partial(x, y)/\partial(u, v)$.

b. Find the image under the transformation $u = 2x - 3y$, $v = -x + y$ of the parallelogram R in the xy-plane with boundaries $x = -3$, $x = 0$, $y = x$, and $y = x + 1$. Sketch the transformed region in the uv-plane.

5. Evaluate the integral

$$\int_0^4 \int_{x=y/2}^{x=(y/2)+1} \frac{2x - y}{2} \, dx \, dy$$

from Example 1 directly by integration with respect to x and y to confirm that its value is 2.

6. Use the transformation in Exercise 1 to evaluate the integral

$$\iint_R (2x^2 - xy - y^2) \, dx \, dy$$

for the region R in the first quadrant bounded by the lines $y = -2x + 4$, $y = -2x + 7$, $y = x - 2$, and $y = x + 1$.

7. Use the transformation in Exercise 3 to evaluate the integral

$$\iint_R (3x^2 + 14xy + 8y^2) \, dx \, dy$$

for the region R in the first quadrant bounded by the lines $y = -(3/2)x + 1$, $y = -(3/2)x + 3$, $y = -(1/4)x$, and $y = -(1/4)x + 1$.

8. Use the transformation and parallelogram R in Exercise 4 to evaluate the integral

$$\iint_R 2(x - y) \, dx \, dy.$$

9. Let R be the region in the first quadrant of the xy-plane bounded by the hyperbolas $xy = 1, xy = 9$ and the lines $y = x, y = 4x$. Use the transformation $x = u/v, y = uv$ with $u > 0$ and $v > 0$ to rewrite

$$\iint_R \left(\sqrt{\frac{y}{x}} + \sqrt{xy} \right) dx \, dy$$

as an integral over an appropriate region G in the uv-plane. Then evaluate the uv-integral over G.

10. a. Find the Jacobian of the transformation $x = u, y = uv$, and sketch the region $G: 1 \le u \le 2, 1 \le uv \le 2$ in the uv-plane.

   b. Then use Equation (1) to transform the integral

$$\int_1^2 \int_1^2 \frac{y}{x} \, dy \, dx$$

   into an integral over G, and evaluate both integrals.

11. Polar moment of inertia of an elliptical plate   A thin plate of constant density covers the region bounded by the ellipse $x^2/a^2 + y^2/b^2 = 1$, $a > 0$, $b > 0$, in the xy-plane. Find the first moment of the plate about the origin. (*Hint:* Use the transformation $x = ar \cos \theta, y = br \sin \theta$.)

12. The area of an ellipse   The area πab of the ellipse $x^2/a^2 + y^2/b^2 = 1$ can be found by integrating the function $f(x, y) = 1$ over the region bounded by the ellipse in the xy-plane. Evaluating the integral directly requires a trigonometric substitution. An easier way to evaluate the integral is to use the transformation $x = au, y = bv$ and evaluate the transformed integral over the disk $G: u^2 + v^2 \le 1$ in the uv-plane. Find the area this way.

13. Use the transformation in Exercise 2 to evaluate the integral

$$\int_0^{2/3} \int_y^{2-2y} (x + 2y) e^{(y-x)} \, dx \, dy$$

by first writing it as an integral over a region G in the uv-plane.

14. Use the transformation $x = u + (1/2)v, y = v$ to evaluate the integral

$$\int_0^2 \int_{y/2}^{(y+4)/2} y^3 (2x - y) e^{(2x-y)^2} \, dx \, dy$$

by first writing it as an integral over a region G in the uv-plane.

15. Find the Jacobian $\partial(x, y)/\partial(u, v)$ for the transformation

   a. $x = u \cos v, \quad y = u \sin v$

   b. $x = u \sin v, \quad y = u \cos v$.

16. Find the Jacobian $\partial(x, y, z)/\partial(u, v, w)$ of the transformation

   a. $x = u \cos v, \quad y = u \sin v, \quad z = w$

   b. $x = 2u - 1, \quad y = 3v - 4, \quad z = (1/2)(w - 4)$.

17. Evaluate the appropriate determinant to show that the Jacobian of the transformation from Cartesian $\rho\phi\theta$-space to Cartesian xyz-space is $\rho^2 \sin \phi$.

18. Substitutions in single integrals   How can substitutions in single definite integrals be viewed as transformations of regions? What is the Jacobian in such a case? Illustrate with an example.

19. Evaluate the integral in Example 3 by integrating with respect to x, y, and z.

20. Volume of an ellipsoid   Find the volume of the ellipsoid

$$\frac{x^2}{a^2} + \frac{y^2}{b^2} + \frac{z^2}{c^2} = 1.$$

(*Hint:* Let $x = au, y = bv$, and $z = cw$. Then find the volume of an appropriate region in uvw-space.)

21. Evaluate

$$\iiint |xyz| \, dx \, dy \, dz$$

over the solid ellipsoid

$$\frac{x^2}{a^2} + \frac{y^2}{b^2} + \frac{z^2}{c^2} \le 1.$$

(*Hint:* Let $x = au, y = bv$, and $z = cw$. Then integrate over an appropriate region in uvw-space.)

22. Let D be the region in xyz-space defined by the inequalities

$$1 \le x \le 2, \quad 0 \le xy \le 2, \quad 0 \le z \le 1.$$

Evaluate

$$\iiint_D (x^2 y + 3xyz) \, dx \, dy \, dz$$

by applying the transformation

$$u = x, \quad v = xy, \quad w = 3z$$

and integrating over an appropriate region G in uvw-space.

23. Centroid of a solid semiellipsoid   Assuming the result that the centroid of a solid hemisphere lies on the axis of symmetry three-eighths of the way from the base toward the top, show, by transforming the appropriate integrals, that the center of mass of a solid semiellipsoid $(x^2/a^2) + (y^2/b^2) + (z^2/c^2) \le 1, z \ge 0$, lies on the z-axis three-eighths of the way from the base toward the top. (You can do this without evaluating any of the integrals.)

24. Cylindrical shells   In Section 6.2, we learned how to find the volume of a solid of revolution using the shell method; namely, if the region between the curve $y = f(x)$ and the x-axis from a to b ($0 < a < b$) is revolved about the y-axis, the volume of the resulting solid is $\int_a^b 2\pi x f(x) \, dx$. Prove that finding volumes by using triple integrals gives the same result. (*Hint:* Use cylindrical coordinates with the roles of y and z changed.)

1. Define the double integral of a function of two variables over a bounded region in the coordinate plane.

2. How are double integrals evaluated as iterated integrals? Does the order of integration matter? How are the limits of integration determined? Give examples.

3. How are double integrals used to calculate areas and average values. Give examples.

4. How can you change a double integral in rectangular coordinates into a double integral in polar coordinates? Why might it be worthwhile to do so? Give an example.

5. Define the triple integral of a function $f(x, y, z)$ over a bounded region in space.

6. How are triple integrals in rectangular coordinates evaluated? How are the limits of integration determined? Give an example.

7. How are double and triple integrals in rectangular coordinates used to calculate volumes, average values, masses, moments, and centers of mass. Give examples.

8. How are triple integrals defined in cylindrical and spherical coordinates? Why might one prefer working in one of these coordinate systems to working in rectangular coordinates?

9. How are triple integrals in cylindrical and spherical coordinates evaluated? How are the limits of integration found? Give examples.

10. How are substitutions in double integrals pictured as transformations of two-dimensional regions? Give a sample calculation.

11. How are substitutions in triple integrals pictured as transformations of three-dimensional regions? Give a sample calculation.

In Exercises 1–4, sketch the region of integration and evaluate the double integral.

1. $\int_{1}^{10} \int_{0}^{1/y} y e^{xy} \, dx \, dy$

2. $\int_{0}^{1} \int_{0}^{x^3} e^{y/x} \, dy \, dx$

3. $\int_{0}^{3/2} \int_{-\sqrt{9-4t^2}}^{\sqrt{9-4t^2}} t \, ds \, dt$

4. $\int_{0}^{1} \int_{\sqrt{y}}^{2-\sqrt{y}} xy \, dx \, dy$

In Exercises 5–8, sketch the region of integration and write an equivalent integral with the order of integration reversed. Then evaluate both integrals.

5. $\int_{0}^{4} \int_{-\sqrt{4-y}}^{(y-4)/2} dx \, dy$

6. $\int_{0}^{1} \int_{x^2}^{x} \sqrt{x} \, dy \, dx$

7. $\int_{0}^{3/2} \int_{-\sqrt{9-4y^2}}^{\sqrt{9-4y^2}} y \, dx \, dy$

8. $\int_{0}^{2} \int_{0}^{4-x^2} 2x \, dy \, dx$

Evaluate the integrals in Exercises 9–12.

9. $\int_{0}^{1} \int_{2y}^{2} 4 \cos (x^2) \, dx \, dy$

10. $\int_{0}^{2} \int_{y/2}^{1} e^{x^2} \, dx \, dy$

11. $\int_{0}^{8} \int_{\sqrt[3]{x}}^{2} \frac{dy \, dx}{y^4 + 1}$

12. $\int_{0}^{1} \int_{\sqrt[3]{y}}^{1} \frac{2\pi \sin \pi x^2}{x^2} \, dx \, dy$

13. **Area between line and parabola**   Find the area of the region enclosed by the line $y = 2x + 4$ and the parabola $y = 4 - x^2$ in the xy-plane.

14. **Area bounded by lines and parabola**   Find the area of the "triangular" region in the xy-plane that is bounded on the right by the parabola $y = x^2$, on the left by the line $x + y = 2$, and above by the line $y = 4$.

15. **Volume of the region under a paraboloid**   Find the volume under the paraboloid $z = x^2 + y^2$ above the triangle enclosed by the lines $y = x, x = 0$, and $x + y = 2$ in the xy-plane.

16. **Volume of the region under parabolic cylinder**   Find the volume under the parabolic cylinder $z = x^2$ above the region enclosed by the parabola $y = 6 - x^2$ and the line $y = x$ in the xy-plane.

Find the average value of $f(x, y) = xy$ over the regions in Exercises 17 and 18.

17. The square bounded by the lines $x = 1, y = 1$ in the first quadrant

18. The quarter circle $x^2 + y^2 \leq 1$ in the first quadrant

Evaluate the integrals in Exercises 19 and 20 by changing to polar coordinates.

19. $\int_{-1}^{1} \int_{-\sqrt{1-x^2}}^{\sqrt{1-x^2}} \frac{2 \, dy \, dx}{(1 + x^2 + y^2)^2}$

20. $\int_{-1}^{1} \int_{-\sqrt{1-y^2}}^{\sqrt{1-y^2}} \ln (x^2 + y^2 + 1) \, dx \, dy$

21. **Integrating over lemniscate**   Integrate the function $f(x, y) = 1/(1 + x^2 + y^2)^2$ over the region enclosed by one loop of the lemniscate $(x^2 + y^2)^2 - (x^2 - y^2) = 0$.

22. Integrate $f(x, y) = 1/(1 + x^2 + y^2)^2$ over

a. **Triangular region**   The triangle with vertices $(0, 0)$, $(1, 0)$, $\left(1, \sqrt{3}\right)$.

b. **First quadrant**   The first quadrant of the xy-plane.

Evaluate the integrals in Exercises 23–26.

23. $\int_0^\pi \int_0^\pi \int_0^\pi \cos(x + y + z) \, dx \, dy \, dz$

24. $\int_{\ln 6}^{\ln 7} \int_0^{\ln 2} \int_{\ln 4}^{\ln 5} e^{(x+y+z)} \, dz \, dy \, dx$

25. $\int_0^1 \int_0^{x^2} \int_0^{x+y} (2x - y - z) \, dz \, dy \, dx$

26. $\int_1^e \int_1^x \int_0^z \frac{2y}{z^3} \, dy \, dz \, dx$

27. Volume Find the volume of the wedge-shaped region enclosed on the side by the cylinder $x = -\cos y$, $-\pi/2 \le y \le \pi/2$, on the top by the plane $z = -2x$, and below by the xy-plane.

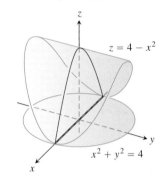

28. Volume Find the volume of the solid that is bounded above by the cylinder $z = 4 - x^2$, on the sides by the cylinder $x^2 + y^2 = 4$, and below by the xy-plane.

29. Average value Find the average value of $f(x, y, z) = 30xz\sqrt{x^2 + y}$ over the rectangular solid in the first octant bounded by the coordinate planes and the planes $x = 1$, $y = 3$, $z = 1$.

30. Average value Find the average value of ρ over the solid sphere $\rho \le a$ (spherical coordinates).

31. Cylindrical to rectangular coordinates Convert

$$\int_0^{2\pi} \int_0^{\sqrt{2}} \int_r^{\sqrt{4-r^2}} 3 \, dz \, r \, dr \, d\theta, \qquad r \ge 0$$

to **(a)** rectangular coordinates with the order of integration $dz \, dx \, dy$ and **(b)** spherical coordinates. Then **(c)** evaluate one of the integrals.

32. Rectangular to cylindrical coordinates **(a)** Convert to cylindrical coordinates. Then **(b)** evaluate the new integral.

$$\int_0^1 \int_{-\sqrt{1-x^2}}^{\sqrt{1-x^2}} \int_{-(x^2+y^2)}^{(x^2+y^2)} 21xy^2 \, dz \, dy \, dx$$

33. Rectangular to spherical coordinates **(a)** Convert to spherical coordinates. Then **(b)** evaluate the new integral.

$$\int_{-1}^1 \int_{-\sqrt{1-x^2}}^{\sqrt{1-x^2}} \int_{\sqrt{x^2+y^2}}^1 dz \, dy \, dx$$

34. Rectangular, cylindrical, and spherical coordinates Write an iterated triple integral for the integral of $f(x, y, z) = 6 + 4y$ over the region in the first octant bounded by the cone $z = \sqrt{x^2 + y^2}$, the cylinder $x^2 + y^2 = 1$, and the coordinate planes in **(a)** rectangular coordinates, **(b)** cylindrical coordinates, and **(c)** spherical coordinates. Then **(d)** find the integral of f by evaluating one of the triple integrals.

35. Cylindrical to rectangular coordinates Set up an integral in rectangular coordinates equivalent to the integral

$$\int_0^{\pi/2} \int_1^{\sqrt{3}} \int_1^{\sqrt{4-r^2}} r^3 (\sin\theta \cos\theta) z^2 \, dz \, dr \, d\theta.$$

Arrange the order of integration to be z first, then y, then x.

36. Rectangular to cylindrical coordinates The volume of a solid is

$$\int_0^2 \int_0^{\sqrt{2x-x^2}} \int_{-\sqrt{4-x^2-y^2}}^{\sqrt{4-x^2-y^2}} dz \, dy \, dx.$$

a. Describe the solid by giving equations for the surfaces that form its boundary.

b. Convert the integral to cylindrical coordinates but do not evaluate the integral.

37. Spherical versus cylindrical coordinates Triple integrals involving spherical shapes do not always require spherical coordinates for convenient evaluation. Some calculations may be accomplished more easily with cylindrical coordinates. As a case in point, find the volume of the region bounded above by the sphere $x^2 + y^2 + z^2 = 8$ and below by the plane $z = 2$ by using **(a)** cylindrical coordinates and **(b)** spherical coordinates.

38. Finding I_z in spherical coordinates Find the moment of inertia about the z-axis of a solid of constant density $\delta = 1$ that is bounded above by the sphere $\rho = 2$ and below by the cone $\phi = \pi/3$ (spherical coordinates).

39. Moment of inertia of a "thick" sphere Find the moment of inertia of a solid of constant density δ bounded by two concentric spheres of radii a and b $(a < b)$ about a diameter.

40. Moment of inertia of an apple Find the moment of inertia about the z-axis of a solid of density $\delta = 1$ enclosed by the spherical coordinate surface $\rho = 1 - \cos\phi$. The solid is the red curve rotated about the z-axis in the accompanying figure.

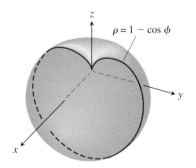

41. Centroid Find the centroid of the "triangular" region bounded by the lines $x = 2, y = 2$ and the hyperbola $xy = 2$ in the xy-plane.

42. Centroid Find the centroid of the region between the parabola $x + y^2 - 2y = 0$ and the line $x + 2y = 0$ in the xy-plane.

43. Polar moment Find the polar moment of inertia about the origin of a thin triangular plate of constant density $\delta = 3$ bounded by the y-axis and the lines $y = 2x$ and $y = 4$ in the xy-plane.

44. Polar moment Find the polar moment of inertia about the center of a thin rectangular sheet of constant density $\delta = 1$ bounded by the lines

a. $x = \pm 2, \quad y = \pm 1$ in the xy-plane

b. $x = \pm a, \quad y = \pm b$ in the xy-plane.

(*Hint:* Find I_x. Then use the formula for I_x to find I_y and add the two to find I_0).

45. Inertial moment Find the moment of inertia about the x-axis of a thin plate of constant density δ covering the triangle with vertices $(0, 0)$, $(3, 0)$, and $(3, 2)$ in the xy-plane.

46. Plate with variable density Find the center of mass and the moments of inertia about the coordinate axes of a thin plate bounded by the line $y = x$ and the parabola $y = x^2$ in the xy-plane if the density is $\delta(x, y) = x + 1$.

47. Plate with variable density Find the mass and first moments about the coordinate axes of a thin square plate bounded by the lines $x = \pm 1, y = \pm 1$ in the xy-plane if the density is $\delta(x, y) = x^2 + y^2 + 1/3$.

48. Triangles with same inertial moment Find the moment of inertia about the x-axis of a thin triangular plate of constant density δ whose base lies along the interval $[0, b]$ on the x-axis and whose vertex lies on the line $y = h$ above the x-axis. As you will see, it does not matter where on the line this vertex lies. All such triangles have the same moment of inertia about the x-axis.

49. Centroid Find the centroid of the region in the polar coordinate plane defined by the inequalities $0 \leq r \leq 3, -\pi/3 \leq \theta \leq \pi/3$.

50. Centroid Find the centroid of the region in the first quadrant bounded by the rays $\theta = 0$ and $\theta = \pi/2$ and the circles $r = 1$ and $r = 3$.

51. a. Centroid Find the centroid of the region in the polar coordinate plane that lies inside the cardioid $r = 1 + \cos \theta$ and outside the circle $r = 1$.

b. Sketch the region and show the centroid in your sketch.

52. a. Centroid Find the centroid of the plane region defined by the polar coordinate inequalities $0 \leq r \leq a, -\alpha \leq \theta \leq \alpha$ $(0 < \alpha \leq \pi)$. How does the centroid move as $\alpha \to \pi^-$?

b. Sketch the region for $\alpha = 5\pi/6$ and show the centroid in your sketch.

53. Show that if $u = x - y$ and $v = y$, then

$$\int_0^\infty \int_0^x e^{-sx} f(x - y, y) \, dy \, dx = \int_0^\infty \int_0^\infty e^{-s(u+v)} f(u, v) \, du \, dv.$$

54. What relationship must hold between the constants a, b, and c to make

$$\int_{-\infty}^\infty \int_{-\infty}^\infty e^{-(ax^2 + 2bxy + cy^2)} \, dx \, dy = 1?$$

(*Hint:* Let $s = \alpha x + \beta y$ and $t = \gamma x + \delta y$, where $(\alpha\delta - \beta\gamma)^2 = ac - b^2$. Then $ax^2 + 2bxy + cy^2 = s^2 + t^2$.)

Chapter 13 Additional and Advanced Exercises

1. Sand pile: double and triple integrals The base of a sand pile covers the region in the xy-plane that is bounded by the parabola $x^2 + y = 6$ and the line $y = x$. The height of the sand above the point (x, y) is x^2. Express the volume of sand as **(a)** a double integral, **(b)** a triple integral. Then **(c)** find the volume.

2. Water in a hemispherical bowl A hemispherical bowl of radius 5 cm is filled with water to within 3 cm of the top. Find the volume of water in the bowl.

3. Solid cylindrical region between two planes Find the volume of the portion of the solid cylinder $x^2 + y^2 \leq 1$ that lies between the planes $z = 0$ and $x + y + z = 2$.

4. Sphere and paraboloid Find the volume of the region bounded above by the sphere $x^2 + y^2 + z^2 = 2$ and below by the paraboloid $z = x^2 + y^2$.

5. Two paraboloids Find the volume of the region bounded above by the paraboloid $z = 3 - x^2 - y^2$ and below by the paraboloid $z = 2x^2 + 2y^2$.

6. Spherical coordinates Find the volume of the region enclosed by the spherical coordinate surface $\rho = 2 \sin \phi$ (see accompanying figure).

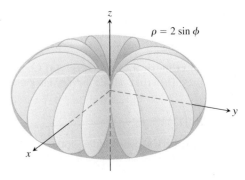

7. Hole in sphere A circular cylindrical hole is bored through a solid sphere, the axis of the hole being a diameter of the sphere.

The volume of the remaining solid is

$$V = 2 \int_0^{2\pi} \int_0^{\sqrt{3}} \int_1^{\sqrt{4-z^2}} r \, dr \, dz \, d\theta.$$

a. Find the radius of the hole and the radius of the sphere.

b. Evaluate the integral.

8. **Sphere and cylinder**   Find the volume of material cut from the solid sphere $r^2 + z^2 \leq 9$ by the cylinder $r = 3 \sin \theta$.

9. **Two paraboloids**   Find the volume of the region enclosed by the surfaces $z = x^2 + y^2$ and $z = (x^2 + y^2 + 1)/2$.

10. **Cylinder and surface $z = xy$**   Find the volume of the region in the first octant that lies between the cylinders $r = 1$ and $r = 2$ and that is bounded below by the xy-plane and above by the surface $z = xy$.

11. Evaluate the integral

$$\int_0^\infty \frac{e^{-ax} - e^{-bx}}{x} \, dx.$$

(*Hint:* Use the relation

$$\frac{e^{-ax} - e^{-bx}}{x} = \int_a^b e^{-xy} \, dy$$

to form a double integral and evaluate the integral by changing the order of integration.)

12. a. **Polar coordinates**   Show, by changing to polar coordinates, that

$$\int_0^{a \sin \beta} \int_{y \cot \beta}^{\sqrt{a^2 - y^2}} \ln (x^2 + y^2) \, dx \, dy = a^2 \beta \left(\ln a - \frac{1}{2} \right),$$

where $a > 0$ and $0 < \beta < \pi/2$.

b. Rewrite the Cartesian integral with the order of integration reversed.

13. **Reducing a double to a single integral**   By changing the order of integration, show that the following double integral can be reduced to a single integral:

$$\int_0^x \int_0^u e^{m(x-t)} f(t) \, dt \, du = \int_0^x (x - t) e^{m(x-t)} f(t) \, dt.$$

Similarly, it can be shown that

$$\int_0^x \int_0^v \int_0^u e^{m(x-t)} f(t) \, dt \, du \, dv = \int_0^x \frac{(x - t)^2}{2} e^{m(x-t)} f(t) \, dt.$$

14. **Transforming a double integral to obtain constant limits**   Sometimes a multiple integral with variable limits can be changed into one with constant limits. By changing the order of integration, show that

$$\int_0^1 f(x) \left(\int_0^x g(x - y) f(y) \, dy \right) dx$$

$$= \int_0^1 f(y) \left(\int_y^1 g(x - y) f(x) \, dx \right) dy$$

$$= \frac{1}{2} \int_0^1 \int_0^1 g(|x - y|) f(x) f(y) \, dx \, dy.$$

15. **Minimizing polar inertia**   A thin plate of constant density is to occupy the triangular region in the first quadrant of the xy-plane having vertices $(0, 0)$, $(a, 0)$, and $(a, 1/a)$. What value of a will minimize the plate's polar moment of inertia about the origin?

16. **Polar inertia of triangular plate**   Find the polar moment of inertia about the origin of a thin triangular plate of constant density $\delta = 3$ bounded by the y-axis and the lines $y = 2x$ and $y = 4$ in the xy-plane.

17. **Mass and polar inertia of a counterweight**   The counterweight of a flywheel of constant density 1 has the form of the smaller segment cut from a circle of radius a by a chord at a distance b from the center $(b < a)$. Find the mass of the counterweight and its polar moment of inertia about the center of the wheel.

18. **Centroid of boomerang**   Find the centroid of the boomerang-shaped region between the parabolas $y^2 = -4(x - 1)$ and $y^2 = -2(x - 2)$ in the xy-plane.

19. Evaluate

$$\int_0^a \int_0^b e^{\max (b^2 x^2, a^2 y^2)} \, dy \, dx,$$

where a and b are positive numbers and

$$\max (b^2 x^2, a^2 y^2) = \begin{cases} b^2 x^2 & \text{if } b^2 x^2 \geq a^2 y^2 \\ a^2 y^2 & \text{if } b^2 x^2 < a^2 y^2. \end{cases}$$

20. Show that

$$\iint \frac{\partial^2 F(x, y)}{\partial x \, \partial y} \, dx \, dy$$

over the rectangle $x_0 \leq x \leq x_1, y_0 \leq y \leq y_1$, is

$$F(x_1, y_1) - F(x_0, y_1) - F(x_1, y_0) + F(x_0, y_0).$$

21. Suppose that $f(x, y)$ can be written as a product $f(x, y) = F(x) G(y)$ of a function of x and a function of y. Then the integral of f over the rectangle $R: a \leq x \leq b, c \leq y \leq d$ can be evaluated as a product as well, by the formula

$$\iint_R f(x, y) \, dA = \left(\int_a^b F(x) \, dx \right) \left(\int_c^d G(y) \, dy \right). \tag{1}$$

The argument is that

$$\iint_R f(x, y) \, dA = \int_c^d \left(\int_a^b F(x) G(y) \, dx \right) dy \tag{i}$$

$$= \int_c^d \left(G(y) \int_a^b F(x) \, dx \right) dy \tag{ii}$$

$$= \int_c^d \left(\int_a^b F(x) \, dx \right) G(y) \, dy \tag{iii}$$

$$= \left(\int_a^b F(x) \, dx \right) \int_c^d G(y) \, dy. \tag{iv}$$

a. Give reasons for steps (i) through (v).

When it applies, Equation (1) can be a time saver. Use it to evaluate the following integrals.

b. $\displaystyle\int_0^{\ln 2}\int_0^{\pi/2} e^x \cos y \, dy \, dx$ **c.** $\displaystyle\int_1^2\int_{-1}^1 \frac{x}{y^2} \, dx \, dy$

22. Let $D_{\mathbf{u}} f$ denote the derivative of $f(x, y) = (x^2 + y^2)/2$ in the direction of the unit vector $\mathbf{u} = u_1\mathbf{i} + u_2\mathbf{j}$.

  a. Finding average value Find the average value of $D_{\mathbf{u}} f$ over the triangular region cut from the first quadrant by the line $x + y = 1$.

  b. Average value and centroid Show in general that the average value of $D_{\mathbf{u}} f$ over a region in the xy-plane is the value of $D_{\mathbf{u}} f$ at the centroid of the region.

23. The value of $\Gamma(1/2)$ The gamma function,

$$\Gamma(x) = \int_0^\infty t^{x-1} e^{-t} \, dt,$$

extends the factorial function from the nonnegative integers to other real values. Of particular interest in the theory of differential equations is the number

$$\Gamma\left(\frac{1}{2}\right) = \int_0^\infty t^{(1/2)-1} e^{-t} \, dt = \int_0^\infty \frac{e^{-t}}{\sqrt{t}} \, dt. \qquad (2)$$

  a. If you have not yet done Exercise 31 in Section 13.4, do it now to show that

$$I = \int_0^\infty e^{-y^2} \, dy = \frac{\sqrt{\pi}}{2}.$$

  b. Substitute $y = \sqrt{t}$ in Equation (2) to show that $\Gamma(1/2) = 2I = \sqrt{\pi}$.

24. Total electrical charge over circular plate The electrical charge distribution on a circular plate of radius R meters is

$\sigma(r, \theta) = kr(1 - \sin\theta)$ coulomb/m^2 (k a constant). Integrate σ over the plate to find the total charge Q.

25. A parabolic rain gauge A bowl is in the shape of the graph of $z = x^2 + y^2$ from $z = 0$ to $z = 10$ in. You plan to calibrate the bowl to make it into a rain gauge. What height in the bowl would correspond to 1 in. of rain? 3 in. of rain?

26. Water in a satellite dish A parabolic satellite dish is 2 m wide and 1/2 m deep. Its axis of symmetry is tilted 30 degrees from the vertical.

  a. Set up, but do not evaluate, a triple integral in rectangular coordinates that gives the amount of water the satellite dish will hold. (*Hint:* Put your coordinate system so that the satellite dish is in "standard position" and the plane of the water level is slanted.) (*Caution:* The limits of integration are not "nice.")

  b. What would be the smallest tilt of the satellite dish so that it holds no water?

27. An infinite half-cylinder Let D be the interior of the infinite right circular half-cylinder of radius 1 with its single-end face suspended 1 unit above the origin and its axis the ray from $(0, 0, 1)$ to ∞. Use cylindrical coordinates to evaluate

$$\iiint_D z(r^2 + z^2)^{-5/2} \, dV.$$

28. Hypervolume We have learned that $\int_a^b 1 \, dx$ is the length of the interval $[a, b]$ on the number line (one-dimensional space), $\iint_R 1 \, dA$ is the area of region R in the xy-plane (two-dimensional space), and $\iiint_D 1 \, dV$ is the volume of the region D in three-dimensional space (xyz-space). We could continue: If Q is a region in 4-space ($xyzw$-space), then $\iiiint_Q 1 \, dV$ is the "hypervolume" of Q. Use your generalizing abilities and a Cartesian coordinate system of 4-space to find the hypervolume inside the unit 3-dimensional sphere $x^2 + y^2 + z^2 + w^2 = 1$.

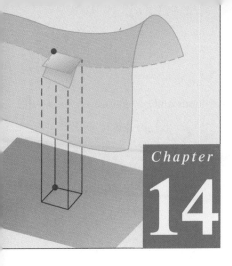

14

INTEGRATION IN VECTOR FIELDS

OVERVIEW In this chapter we combine ideas from the preceding four chapters: space curves, vector functions, partial derivatives, and multiple integrals. The result is the development of line integrals, vector fields, and surface integrals giving powerful mathematical tools for science and engineering. Line integrals are used to find the work done by a force in moving an object along a path, and to find the mass of a wire having a varying density. Surface integrals are used to find the rate at which a fluid flows across a surface. We also present several important theorems that connect these new tools and offer insight into their mathematical calculations and physical applications.

14.1 Line Integrals

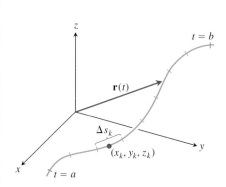

FIGURE 14.1 The curve $\mathbf{r}(t)$ partitioned into small arcs from $t = a$ to $t = b$. The length of a typical subarc is Δs_k.

To calculate the masses of thin rods or wires lying along a *curve* in space, or to find the work done by a variable force acting along such a curve, we need a more general notion of integral than was defined in Chapter 5. We need to integrate over a curve C rather than over an interval $[a, b]$. These more general integrals are called *line integrals* (although "curve" integrals might be more descriptive). We make our definitions for space curves, remembering that curves in the xy-plane are just a special case with z-coordinate identically zero.

Suppose that $f(x, y, z)$ is a real-valued function we wish to integrate over the curve $\mathbf{r}(t) = g(t)\mathbf{i} + h(t)\mathbf{j} + k(t)\mathbf{k}$, $a \leq t \leq b$, lying within the domain of f. The values of f along the curve are given by the composite function $f(g(t), h(t), k(t))$. We are going to integrate this composite with respect to arc length from $t = a$ to $t = b$. To begin, we first partition the curve into a finite number n of subarcs (Figure 14.1). The typical subarc has length Δs_k. In each subarc we choose a point (x_k, y_k, z_k) and form the sum

$$S_n = \sum_{k=1}^{n} f(x_k, y_k, z_k)\, \Delta s_k.$$

If f is continuous and the functions g, h, and k have continuous first derivatives, then these sums approach a limit as n increases and the lengths Δs_k approach zero. We call this limit the **line integral of f over the curve from a to b.** If the curve is denoted by a single letter, C for example, the notation for the integral is

$$\int_C f(x, y, z)\, ds \qquad \text{"The integral of } f \text{ over } C.\text{"} \tag{1}$$

If $\mathbf{r}(t)$ is smooth for $a \leq t \leq b$ ($\mathbf{v} = d\mathbf{r}/dt$ is continuous and never $\mathbf{0}$), we can use the equation

$$s(t) = \int_a^t |\mathbf{v}(\tau)| \, d\tau \qquad \text{Equation (3) of Section} \atop \text{11.3 with } t_0 = a$$

to express ds in Equation (1) as $ds = |\mathbf{v}(t)| \, dt$. We can then evaluate the integral of f over C as

$$\int_C f(x, y, z) \, ds = \int_a^b f(g(t), h(t), k(t)) |\mathbf{v}(t)| \, dt.$$

Notice that the integral on the right side of this last equation is just an ordinary (single) definite integral, as defined in Chapter 5, where we are integrating with respect to the parameter t. The formula evaluates the line integral on the left side correctly no matter what parametrization is used, as long as the parametrization is smooth.

How to Evaluate a Line Integral

To integrate a continuous function $f(x, y, z)$ over a curve C:

1. Find a smooth parametrization of C,
$$\mathbf{r}(t) = g(t)\mathbf{i} + h(t)\mathbf{j} + k(t)\mathbf{k}, \qquad a \leq t \leq b$$

2. Evaluate the integral as
$$\int_C f(x, y, z) \, ds = \int_a^b f(g(t), h(t), k(t)) |\mathbf{v}(t)| \, dt. \qquad (2)$$

If f has the constant value 1, then the integral of f over C gives the length of C.

EXAMPLE 1   Integrate $f(x, y, z) = x - 3y^2 + z$ over the line segment C joining the origin to the point $(1, 1, 1)$ (Figure 14.2).

Solution   We choose the simplest parametrization we can think of:
$$\mathbf{r}(t) = t\mathbf{i} + t\mathbf{j} + t\mathbf{k}, \qquad 0 \leq t \leq 1.$$

The components have continuous first derivatives and $|\mathbf{v}(t)| = |\mathbf{i} + \mathbf{j} + \mathbf{k}| = \sqrt{1^2 + 1^2 + 1^2} = \sqrt{3}$ is never 0, so the parametrization is smooth. The integral of f over C is

$$\int_C f(x, y, z) \, ds = \int_0^1 f(t, t, t)\left(\sqrt{3}\right) dt \qquad \text{Equation (2)}$$

$$= \int_0^1 (t - 3t^2 + t)\sqrt{3} \, dt$$

$$= \sqrt{3} \int_0^1 (2t - 3t^2) \, dt = \sqrt{3}\left[t^2 - t^3\right]_0^1 = 0. \qquad \blacksquare$$

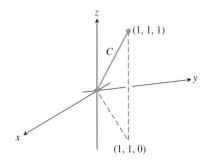

FIGURE 14.2   The integration path in Example 1.

Additivity

Line integrals have the useful property that if a curve C is made by joining a finite number of curves $C_1, C_2, \ldots, C_n$ end to end, then the integral of a function over C is the sum of the integrals over the curves that make it up:

$$\int_C f \, ds = \int_{C_1} f \, ds + \int_{C_2} f \, ds + \cdots + \int_{C_n} f \, ds. \tag{3}$$

EXAMPLE 2 Figure 14.3 shows another path from the origin to $(1, 1, 1)$, the union of line segments C_1 and C_2. Integrate $f(x, y, z) = x - 3y^2 + z$ over $C_1 \cup C_2$.

Solution We choose the simplest parametrizations for C_1 and C_2 we can think of, checking the lengths of the velocity vectors as we go along:

$$C_1: \quad \mathbf{r}(t) = t\mathbf{i} + t\mathbf{j}, \quad 0 \le t \le 1; \quad |\mathbf{v}| = \sqrt{1^2 + 1^2} = \sqrt{2}$$

$$C_2: \quad \mathbf{r}(t) = \mathbf{i} + \mathbf{j} + t\mathbf{k}, \quad 0 \le t \le 1; \quad |\mathbf{v}| = \sqrt{0^2 + 0^2 + 1^2} = 1.$$

With these parametrizations we find that

$$\int_{C_1 \cup C_2} f(x, y, z) \, ds = \int_{C_1} f(x, y, z) \, ds + \int_{C_2} f(x, y, z) \, ds \qquad \text{Equation (3)}$$

$$= \int_0^1 f(t, t, 0)\sqrt{2} \, dt + \int_0^1 f(1, 1, t)(1) \, dt \qquad \text{Equation (2)}$$

$$= \int_0^1 (t - 3t^2 + 0)\sqrt{2} \, dt + \int_0^1 (1 - 3 + t)(1) \, dt$$

$$= \sqrt{2}\left[\frac{t^2}{2} - t^3\right]_0^1 + \left[\frac{t^2}{2} - 2t\right]_0^1 = -\frac{\sqrt{2}}{2} - \frac{3}{2}. \qquad \blacksquare$$

Notice three things about the integrations in Examples 1 and 2. First, as soon as the components of the appropriate curve were substituted into the formula for f, the integration became a standard integration with respect to t. Second, the integral of f over $C_1 \cup C_2$ was obtained by integrating f over each section of the path and adding the results. Third, the integrals of f over C and $C_1 \cup C_2$ had different values. The value of the integral along a path joining two points can change if you change the path between them.

Mass and Moment Calculations

We treat coil springs and wires like masses distributed along smooth curves in space. The distribution is described by a continuous density function $\delta(x, y, z)$ (mass per unit length). The spring's or wire's mass, center of mass, and moments are then calculated with the formulas in Table 14.1. The formulas also apply to thin rods. The derivations of these formulas are similar to those in Section 6.7.

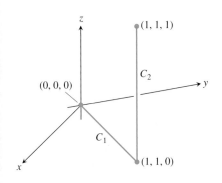

FIGURE 14.3 The path of integration in Example 2.

TABLE 14.1 Mass and moment formulas for coil springs, thin rods, and wires lying along a smooth curve C in space

Mass:
$$M = \int_C \delta(x, y, z)\, ds \qquad \delta = \delta(x, y, z) \text{ is the density at } (x, y, z)$$

First moments about the coordinate planes:
$$M_{yz} = \int_C x\, \delta\, ds, \qquad M_{xz} = \int_C y\, \delta\, ds, \qquad M_{xy} = \int_C z\, \delta\, ds$$

Coordinates of the center of mass:
$$\bar{x} = M_{yz}/M, \qquad \bar{y} = M_{xz}/M, \qquad \bar{z} = M_{xy}/M$$

Moments of inertia about axes and other lines:
$$I_x = \int_C (y^2 + z^2)\, \delta\, ds, \qquad I_y = \int_C (x^2 + z^2)\, \delta\, ds, \qquad I_z = \int_C (x^2 + y^2)\, \delta\, ds,$$

$$I_L = \int_C r^2\, \delta\, ds \qquad r(x, y, z) = \text{distance from the point } (x, y, z) \text{ to line } L$$

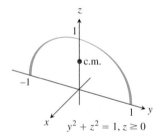

FIGURE 14.4 Example 3 shows how to find the center of mass of a circular arch of variable density.

Notice that the element of mass dm in the table is equal to $\delta\, ds$ rather than $\delta\, dV$ as in Table 13.1, and that the integrals are taken over the curve C.

EXAMPLE 3   A slender metal arch, denser at the bottom than top, lies along the semicircle $y^2 + z^2 = 1$, $z \geq 0$, in the yz-plane (Figure 14.4). Find the center of the arch's mass if the density at the point (x, y, z) on the arch is $\delta(x, y, z) = 2 - z$.

Solution   We know that $\bar{x} = 0$ and $\bar{y} = 0$ because the arch lies in the yz-plane with its mass distributed symmetrically about the z-axis. To find $\bar{z}$, we parametrize the circle as

$$\mathbf{r}(t) = (\cos t)\mathbf{j} + (\sin t)\mathbf{k}, \qquad 0 \leq t \leq \pi.$$

For this parametrization,

$$|\mathbf{v}(t)| = \sqrt{\left(\frac{dx}{dt}\right)^2 + \left(\frac{dy}{dt}\right)^2 + \left(\frac{dz}{dt}\right)^2} = \sqrt{(0)^2 + (-\sin t)^2 + (\cos t)^2} = 1.$$

The formulas in Table 14.1 then give

$$M = \int_C \delta\, ds = \int_C (2 - z)\, ds = \int_0^\pi (2 - \sin t)(1)\, dt = 2\pi - 2$$

$$M_{xy} = \int_C z\delta\, ds = \int_C z(2 - z)\, ds = \int_0^\pi (\sin t)(2 - \sin t)\, dt$$

$$= \int_0^\pi (2\sin t - \sin^2 t)\, dt = \frac{8 - \pi}{2}$$

$$\bar{z} = \frac{M_{xy}}{M} = \frac{8 - \pi}{2} \cdot \frac{1}{2\pi - 2} = \frac{8 - \pi}{4\pi - 4} \approx 0.57.$$

With $\bar{z}$ to the nearest hundredth, the center of mass is $(0, 0, 0.57)$.   ∎

EXERCISES 14.1

Match the vector equations in Exercises 1–8 with the graphs (a)–(h) given here.

a.

b.

c.

d.

e.

f.

g.

h.

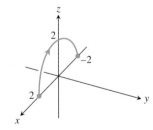

1. $\mathbf{r}(t) = t\mathbf{i} + (1 - t)\mathbf{j}, \quad 0 \le t \le 1$

2. $\mathbf{r}(t) = \mathbf{i} + \mathbf{j} + t\mathbf{k}, \quad -1 \le t \le 1$

3. $\mathbf{r}(t) = (2\cos t)\mathbf{i} + (2\sin t)\mathbf{j}, \quad 0 \le t \le 2\pi$

4. $\mathbf{r}(t) = t\mathbf{i}, \quad -1 \le t \le 1$

5. $\mathbf{r}(t) = t\mathbf{i} + t\mathbf{j} + t\mathbf{k}, \quad 0 \le t \le 2$

6. $\mathbf{r}(t) = t\mathbf{j} + (2 - 2t)\mathbf{k}, \quad 0 \le t \le 1$

7. $\mathbf{r}(t) = (t^2 - 1)\mathbf{j} + 2t\mathbf{k}, \quad -1 \le t \le 1$

8. $\mathbf{r}(t) = (2\cos t)\mathbf{i} + (2\sin t)\mathbf{k}, \quad 0 \le t \le \pi$

9. Evaluate $\int_C (x + y)\, ds$ where C is the straight-line segment $x = t, y = (1 - t), z = 0$, from $(0, 1, 0)$ to $(1, 0, 0)$.

10. Evaluate $\int_C (x - y + z - 2)\, ds$ where C is the straight-line segment $x = t, y = (1 - t), z = 1$, from $(0, 1, 1)$ to $(1, 0, 1)$.

11. Evaluate $\int_C (xy + y + z)\, ds$ along the curve $\mathbf{r}(t) = 2t\mathbf{i} + t\mathbf{j} + (2 - 2t)\mathbf{k}, 0 \le t \le 1$.

12. Evaluate $\int_C \sqrt{x^2 + y^2}\, ds$ along the curve $\mathbf{r}(t) = (4\cos t)\mathbf{i} + (4\sin t)\mathbf{j} + 3t\mathbf{k}, -2\pi \le t \le 2\pi$.

13. Find the line integral of $f(x, y, z) = x + y + z$ over the straight-line segment from $(1, 2, 3)$ to $(0, -1, 1)$.

14. Find the line integral of $f(x, y, z) = \sqrt{3}/(x^2 + y^2 + z^2)$ over the curve $\mathbf{r}(t) = t\mathbf{i} + t\mathbf{j} + t\mathbf{k}, 1 \le t \le \infty$.

15. Integrate $f(x, y, z) = x + \sqrt{y} - z^2$ over the path from $(0, 0, 0)$ to $(1, 1, 1)$ (Figure 14.5a) given by

$$C_1: \quad \mathbf{r}(t) = t\mathbf{i} + t^2\mathbf{j}, \quad 0 \le t \le 1$$
$$C_2: \quad \mathbf{r}(t) = \mathbf{i} + \mathbf{j} + t\mathbf{k}, \quad 0 \le t \le 1$$

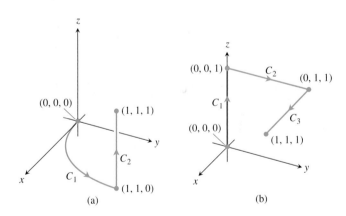

FIGURE 14.5 The paths of integration for Exercises 15 and 16.

16. Integrate $f(x, y, z) = x + \sqrt{y} - z^2$ over the path from $(0, 0, 0)$ to $(1, 1, 1)$ (Figure 14.5b) given by

$$C_1: \quad \mathbf{r}(t) = t\mathbf{k}, \quad 0 \le t \le 1$$
$$C_2: \quad \mathbf{r}(t) = t\mathbf{j} + \mathbf{k}, \quad 0 \le t \le 1$$
$$C_3: \quad \mathbf{r}(t) = t\mathbf{i} + \mathbf{j} + \mathbf{k}, \quad 0 \le t \le 1$$

17. Integrate $f(x, y, z) = (x + y + z)/(x^2 + y^2 + z^2)$ over the path $\mathbf{r}(t) = t\mathbf{i} + t\mathbf{j} + t\mathbf{k}, 0 < a \le t \le b$.

18. Integrate $f(x, y, z) = -\sqrt{x^2 + z^2}$ over the circle

$$\mathbf{r}(t) = (a \cos t)\mathbf{j} + (a \sin t)\mathbf{k}, \qquad 0 \le t \le 2\pi.$$

In Exercises 19–22, integrate f over the given curve.

19. $f(x, y) = x^3/y, \quad C: \quad y = x^2/2, \quad 0 \le x \le 2$

20. $f(x, y) = (x + y^2)/\sqrt{1 + x^2}, \quad C: \quad y = x^2/2$ from $(1, 1/2)$ to $(0, 0)$

21. $f(x, y) = x + y, \quad C: \quad x^2 + y^2 = 4$ in the first quadrant from $(2, 0)$ to $(0, 2)$

22. $f(x, y) = x^2 - y, \quad C: \quad x^2 + y^2 = 4$ in the first quadrant from $(0, 2)$ to $(\sqrt{2}, \sqrt{2})$

23. Mass of a wire   Find the mass of a wire that lies along the curve $\mathbf{r}(t) = (t^2 - 1)\mathbf{j} + 2t\mathbf{k}, 0 \le t \le 1$, if the density is $\delta = (3/2)t$.

24. Center of mass of a curved wire   A wire of density $\delta(x, y, z) = 15\sqrt{y + 2}$ lies along the curve $\mathbf{r}(t) = (t^2 - 1)\mathbf{j} + 2t\mathbf{k}, -1 \le t \le 1$. Find its center of mass. Then sketch the curve and center of mass together.

25. Mass of wire with variable density   Find the mass of a thin wire lying along the curve $\mathbf{r}(t) = \sqrt{2}t\mathbf{i} + \sqrt{2}t\mathbf{j} + (4 - t^2)\mathbf{k}$, $0 \le t \le 1$, if the density is **(a)** $\delta = 3t$ and **(b)** $\delta = 1$.

26. Center of mass of wire with variable density   Find the center of mass of a thin wire lying along the curve $\mathbf{r}(t) = t\mathbf{i} + 2t\mathbf{j} + (2/3)t^{3/2}\mathbf{k}, 0 \le t \le 2$, if the density is $\delta = 3\sqrt{5 + t}$.

27. Moment of inertia of wire hoop   A circular wire hoop of constant density δ lies along the circle $x^2 + y^2 = a^2$ in the xy-plane. Find the hoop's moment of inertia about the z-axis.

28. Inertia of a slender rod   A slender rod of constant density lies along the line segment $\mathbf{r}(t) = t\mathbf{j} + (2 - 2t)\mathbf{k}, 0 \le t \le 1$, in the yz-plane. Find the moments of inertia of the rod about the three coordinate axes.

29. Two springs of constant density   A spring of constant density δ lies along the helix

$$\mathbf{r}(t) = (\cos t)\mathbf{i} + (\sin t)\mathbf{j} + t\mathbf{k}, \qquad 0 \le t \le 2\pi.$$

a. Find I_z.

b. Suppose that you have another spring of constant density δ that is twice as long as the spring in part (a) and lies along the helix for $0 \le t \le 4\pi$. Do you expect I_z for the longer spring to be the same as that for the shorter one, or should it be different? Check your prediction by calculating I_z for the longer spring.

30. Wire of constant density   A wire of constant density $\delta = 1$ lies along the curve

$$\mathbf{r}(t) = (t \cos t)\mathbf{i} + (t \sin t)\mathbf{j} + \left(2\sqrt{2}/3\right)t^{3/2}\mathbf{k}, \qquad 0 \le t \le 1.$$

Find $\bar{z}$ and I_z.

31. The arch in Example 3   Find I_x for the arch in Example 3.

32. Center of mass and moments of inertia for wire with variable density   Find the center of mass and the moments of inertia about the coordinate axes of a thin wire lying along the curve

$$\mathbf{r}(t) = t\mathbf{i} + \frac{2\sqrt{2}}{3}t^{3/2}\mathbf{j} + \frac{t^2}{2}\mathbf{k}, \qquad 0 \le t \le 2,$$

if the density is $\delta = 1/(t + 1)$.

COMPUTER EXPLORATIONS

In Exercises 33–36, use a CAS to perform the following steps to evaluate the line integrals.

a. Find $ds = |\mathbf{v}(t)| \, dt$ for the path $\mathbf{r}(t) = g(t)\mathbf{i} + h(t)\mathbf{j} + k(t)\mathbf{k}$.

b. Express the integrand $f(g(t), h(t), k(t))|\mathbf{v}(t)|$ as a function of the parameter t.

c. Evaluate $\int_C f \, ds$ using Equation (2) in the text.

33. $f(x, y, z) = \sqrt{1 + 30x^2 + 10y}; \quad \mathbf{r}(t) = t\mathbf{i} + t^2\mathbf{j} + 3t^2\mathbf{k},$ $0 \le t \le 2$

34. $f(x, y, z) = \sqrt{1 + x^3 + 5y^3}; \quad \mathbf{r}(t) = t\mathbf{i} + \frac{1}{3}t^2\mathbf{j} + \sqrt{t}\mathbf{k},$ $0 \le t \le 2$

35. $f(x, y, z) = x\sqrt{y} - 3z^2; \quad \mathbf{r}(t) = (\cos 2t)\mathbf{i} + (\sin 2t)\mathbf{j} + 5t\mathbf{k},$ $0 \le t \le 2\pi$

36. $f(x, y, z) = \left(1 + \frac{9}{4}z^{1/3}\right)^{1/4}; \quad \mathbf{r}(t) = (\cos 2t)\mathbf{i} + (\sin 2t)\mathbf{j} + t^{5/2}\mathbf{k}, \quad 0 \le t \le 2\pi$

14.2   Vector Fields, Work, Circulation, and Flux

Gravitational and electric forces have both direction and magnitude. They are represented by a vector at each point in their domain, producing a *vector field.* In this section we show how to compute the work done in moving an object through such a field by using a line integral over the vector field. We also discuss velocity fields, such as the vector field representing the velocity of a flowing fluid in its domain. A line integral can be used to find the rate at which the fluid flows along or across a curve within the domain.

Vector Fields

FIGURE 14.6 Velocity vectors of a flow around an airfoil in a wind tunnel.

FIGURE 14.7 Streamlines in a contracting channel. The water speeds up as the channel narrows and the velocity vectors increase in length.

FIGURE 14.8 Vectors in a gravitational field point toward the center of mass that gives the source of the field.

Suppose a region in the plane or in space is occupied by a moving fluid, such as air or water. The fluid is made up of a large number of particles, and at any instant of time, a particle has a velocity $\mathbf{v}$. At different points of position at a given (same) time, these velocities can vary. We can think of a velocity vector being attached to each point of the fluid representing the velocity of a particle at that point. Such a fluid flow is an example of a *vector field*. Figure 14.6 shows a velocity vector field obtained from air flowing around an airfoil in a wind tunnel. Figure 14.7 shows a vector field of velocity vectors along the streamlines of water moving through a contracting channel. Vector fields are also associated to forces such as gravitational attraction (Figure 14.8), and to magnetic force fields, electric fields, and also purely mathematical fields.

Generally, a **vector field** is a function that assigns a vector to each point in its domain. A vector field on a three-dimensional domain in space might have a formula like

$$\mathbf{F}(x, y, z) = M(x, y, z)\mathbf{i} + N(x, y, z)\mathbf{j} + P(x, y, z)\mathbf{k}.$$

The field is **continuous** if the **component functions** M, N, and P are continuous; it is **differentiable** if each of the component functions is differentiable. The formula for a field of two-dimensional vectors could look like

$$\mathbf{F}(x, y) = M(x, y)\mathbf{i} + N(x, y)\mathbf{j}.$$

We encountered another type of vector field in Chapter 11. The tangent vectors $\mathbf{T}$ and normal vectors $\mathbf{N}$ for a curve in space both form vector fields along the curve. Along a curve $\mathbf{r}(t)$ they might have a formula such as

$$\mathbf{v}(t) = f(t)\mathbf{i} + g(t)\mathbf{j} + h(t)\mathbf{k}.$$

If we attach the gradient vector ∇f of a scalar function $f(x, y, z)$ to each point of a level surface of the function, we obtain a three-dimensional field on the surface. If we attach the velocity vector to each point of a flowing fluid, we have a three-dimensional field defined on a region in space. These and other fields are illustrated in Figures 14.9–14.14. To sketch the fields, we picked a representative selection of domain points and drew the vectors attached to them. The arrows are drawn with their tails, not their heads, attached to the points where the vector functions are evaluated.

FIGURE 14.9 The velocity vectors $\mathbf{v}(t)$ of a projectile's motion make a vector field along the trajectory.

FIGURE 14.10 The field of gradient vectors ∇f on a surface $f(x, y, z) = c$.

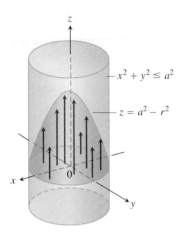

FIGURE 14.11 The flow of fluid in a long cylindrical pipe. The vectors $\mathbf{v} = (a^2 - r^2)\mathbf{k}$ inside the cylinder that have their bases in the xy-plane have their tips on the paraboloid $z = a^2 - r^2$.

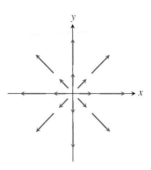

FIGURE 14.12 The radial field $\mathbf{F} = x\mathbf{i} + y\mathbf{j}$ of position vectors of points in the plane. Notice the convention that an arrow is drawn with its tail, not its head, at the point where $\mathbf{F}$ is evaluated.

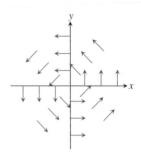

FIGURE 14.13 The circumferential or "spin" field of unit vectors

$$\mathbf{F} = (-y\mathbf{i} + x\mathbf{j})/(x^2 + y^2)^{1/2}$$

in the plane. The field is not defined at the origin.

FIGURE 14.14 NASA's *Seasat* used radar to take 350,000 wind measurements over the world's oceans. The arrows show wind direction; their length and the color contouring indicate speed. Notice the heavy storm south of Greenland.

Gradient Fields

The gradient vector of a differentiable scalar-valued function at a point gives the direction of greatest increase of the function. An important type of vector field is formed by all the gradient vectors of the function.

DEFINITION The **gradient field** of a differentiable function $f(x, y, z)$ is the field of gradient vectors

$$\nabla f = \frac{\partial f}{\partial x}\mathbf{i} + \frac{\partial f}{\partial y}\mathbf{j} + \frac{\partial f}{\partial z}\mathbf{k}.$$

EXAMPLE 1 Suppose that the temperature T at each point (x, y, z) in a region of space is given by

$$T = 100 - x^2 - y^2 - z^2,$$

and that $\mathbf{F}(x, y, z)$ is defined to be the gradient of T. Find the vector field $\mathbf{F}$.

Solution The gradient field $\mathbf{F}$ is the field $\mathbf{F} = \nabla T = -2x\mathbf{i} - 2y\mathbf{j} - 2z\mathbf{k}.$ ∎

At each point in space, the vector field $\mathbf{F}$ in Example 1 gives the direction for which the temperature is increasing most rapidly. It is not a force field and it is not a velocity field.

Work Done by a Force over a Curve in Space

Suppose that the vector field $\mathbf{F} = M(x, y, z)\mathbf{i} + N(x, y, z)\mathbf{j} + P(x, y, z)\mathbf{k}$ represents a force throughout a region in space (it might be the force of gravity or an electromagnetic force of some kind) and that

$$\mathbf{r}(t) = g(t)\mathbf{i} + h(t)\mathbf{j} + k(t)\mathbf{k}, \qquad a \le t \le b,$$

is a smooth curve in the region. The formula for the work done by the force in moving an object along the curve is motivated by the same kind of reasoning we used in Chapter 6 to derive the formula $W = \int_a^b F(x)\, dx$ for the work done by a continuous force of magnitude $F(x)$ directed along an interval of the x-axis. We divide the curve into short segments, apply the (constant-force) × (distance) formula for work to approximate the work over each curved segment, add the results to approximate the work over the entire curve, and calculate the work as the limit of the approximating sums as the segments become shorter and more numerous. To find exactly what the limiting integral should be, we partition the parameter interval $[a, b]$ in the usual way and choose a point c_k in each subinterval $[t_k, t_{k+1}]$. The partition of $[a, b]$ determines ("induces," we say) a partition of the curve, with the point P_k being the tip of the position vector $\mathbf{r}(t_k)$ and Δs_k being the length of the curve segment $P_k P_{k+1}$ (Figure 14.15).

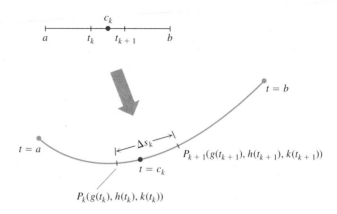

FIGURE 14.15 Each partition of $[a, b]$ induces a partition of the curve $\mathbf{r}(t) = g(t)\mathbf{i} + h(t)\mathbf{j} + k(t)\mathbf{k}$.

FIGURE 14.16 An enlarged view of the curve segment P_kP_{k+1} in Figure 14.15, showing the force and unit tangent vectors at the point on the curve where $t = c_k$.

If $\mathbf{F}_k$ denotes the value of the force field $\mathbf{F}$ at the point on the curve corresponding to $t = c_k$ and $\mathbf{T}_k$ denotes the curve's unit tangent vector at this point, then $\mathbf{F}_k \cdot \mathbf{T}_k$ is the scalar component of $\mathbf{F}$ in the direction of $\mathbf{T}$ at $t = c_k$ (Figure 14.16). The work done by $\mathbf{F}$ along the curve segment P_kP_{k+1} is approximately

$$\begin{pmatrix} \text{Force component in} \\ \text{direction of motion} \end{pmatrix} \times \begin{pmatrix} \text{distance} \\ \text{applied} \end{pmatrix} = \mathbf{F}_k \cdot \mathbf{T}_k \, \Delta s_k.$$

The work done by $\mathbf{F}$ in moving an object along the curve from $t = a$ to $t = b$ is approximately

$$\sum_{k=1}^{n} \mathbf{F}_k \cdot \mathbf{T}_k \, \Delta s_k.$$

As the norm of the partition of $[a, b]$ approaches zero, the norm of the induced partition of the curve approaches zero and these sums approach the line integral

$$\int_{t=a}^{t=b} \mathbf{F} \cdot \mathbf{T} \, ds.$$

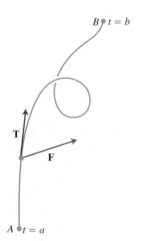

FIGURE 14.17 The work done by a force $\mathbf{F}$ is the line integral of the scalar component $\mathbf{F} \cdot \mathbf{T}$ over the smooth curve from A to B.

DEFINITION The **work** done by a force $\mathbf{F} = M\mathbf{i} + N\mathbf{j} + P\mathbf{k}$ in moving an object over a smooth curve $\mathbf{r}(t)$ from $t = a$ to $t = b$ is

$$W = \int_{t=a}^{t=b} \mathbf{F} \cdot \mathbf{T} \, ds. \tag{1}$$

The sign of the number we calculate with this integral depends on the direction in which the curve is traversed as t increases. If we reverse the direction of motion, we reverse the direction of $\mathbf{T}$ and change the sign of $\mathbf{F} \cdot \mathbf{T}$ and its integral (Figure 14.17).

Table 14.2 shows six ways to write the work integral in Equation (1). Despite their variety, the formulas in Table 14.2 are all evaluated the same way. In the table, $\mathbf{r}(t) = g(t)\mathbf{i} + h(t)\mathbf{j} + k(t)\mathbf{k} = x\mathbf{i} + y\mathbf{j} + z\mathbf{k}$ is a smooth curve, and

$$d\mathbf{r} = \frac{d\mathbf{r}}{dt}\,dt = dg\mathbf{i} + dh\mathbf{j} + dk\mathbf{k}$$

is its differential.

TABLE 14.2 Six different ways to write the work integral

$$\mathbf{W} = \int_{t=a}^{t=b} \mathbf{F} \cdot \mathbf{T}\,ds \qquad \text{The definition}$$

$$= \int_{t=a}^{t=b} \mathbf{F} \cdot d\mathbf{r} \qquad \text{Compact differential form}$$

$$= \int_{a}^{b} \mathbf{F} \cdot \frac{d\mathbf{r}}{dt}\,dt \qquad \begin{array}{l}\text{Expanded to include } dt;\text{ emphasizes}\\ \text{the parameter } t \text{ and velocity vector } d\mathbf{r}/dt\end{array}$$

$$= \int_{a}^{b} \left(M\frac{dg}{dt} + N\frac{dh}{dt} + P\frac{dk}{dt} \right) dt \qquad \text{Emphasizes the component functions of } \mathbf{F}$$

$$= \int_{a}^{b} \left(M\frac{dx}{dt} + N\frac{dy}{dt} + P\frac{dz}{dt} \right) dt \qquad \text{Abbreviates the components of } \mathbf{r}$$

$$= \int_{a}^{b} M\,dx + N\,dy + P\,dz \qquad dt\text{'s canceled; the most common form}$$

To evaluate the work integral along a smooth curve $\mathbf{r}(t)$, take these steps:

1. Evaluate $\mathbf{F}$ on the curve as a function of the parameter t.
2. Find $d\mathbf{r}/dt$.
3. Integrate $\mathbf{F} \cdot d\mathbf{r}/dt$ from $t = a$ to $t = b$.

EXAMPLE 2 Find the work done by $\mathbf{F} = (y - x^2)\mathbf{i} + (z - y^2)\mathbf{j} + (x - z^2)\mathbf{k}$ over the curve $\mathbf{r}(t) = t\mathbf{i} + t^2\mathbf{j} + t^3\mathbf{k}$, $0 \leq t \leq 1$, from $(0, 0, 0)$ to $(1, 1, 1)$ (Figure 14.18).

Solution First we evaluate $\mathbf{F}$ on the curve:

$$\mathbf{F} = (y - x^2)\mathbf{i} + (z - y^2)\mathbf{j} + (x - z^2)\mathbf{k}$$
$$= \underbrace{(t^2 - t^2)}_{0}\mathbf{i} + (t^3 - t^4)\mathbf{j} + (t - t^6)\mathbf{k}$$

Then we find $d\mathbf{r}/dt$,

$$\frac{d\mathbf{r}}{dt} = \frac{d}{dt}(t\mathbf{i} + t^2\mathbf{j} + t^3\mathbf{k}) = \mathbf{i} + 2t\mathbf{j} + 3t^2\mathbf{k}$$

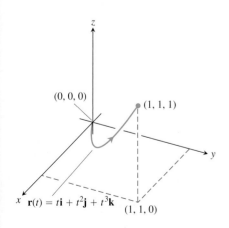

FIGURE 14.18 The curve in Example 2.

Finally, we find $\mathbf{F} \cdot d\mathbf{r}/dt$ and integrate from $t = 0$ to $t = 1$:

$$\mathbf{F} \cdot \frac{d\mathbf{r}}{dt} = [(t^3 - t^4)\mathbf{j} + (t - t^6)\mathbf{k}] \cdot (\mathbf{i} + 2t\mathbf{j} + 3t^2\mathbf{k})$$

$$= (t^3 - t^4)(2t) + (t - t^6)(3t^2) = 2t^4 - 2t^5 + 3t^3 - 3t^8,$$

so

$$\text{Work} = \int_0^1 (2t^4 - 2t^5 + 3t^3 - 3t^8)\, dt$$

$$= \left[\frac{2}{5}t^5 - \frac{2}{6}t^6 + \frac{3}{4}t^4 - \frac{3}{9}t^9\right]_0^1 = \frac{29}{60}. \qquad \blacksquare$$

Flow Integrals and Circulation for Velocity Fields

Suppose that $\mathbf{F}$ represents the velocity field of a fluid flowing through a region in space (a tidal basin or the turbine chamber of a hydroelectric generator, for example). Under these circumstances, the integral of $\mathbf{F} \cdot \mathbf{T}$ along a curve in the region gives the fluid's flow along the curve.

DEFINITIONS If $\mathbf{r}(t)$ is a smooth curve in the domain of a continuous velocity field $\mathbf{F}$, the **flow** along the curve from $t = a$ to $t = b$ is

$$\text{Flow} = \int_a^b \mathbf{F} \cdot \mathbf{T} \, ds. \tag{2}$$

The integral in this case is called a **flow integral**. If the curve is a closed loop, the flow is called the **circulation** around the curve.

We evaluate flow integrals the same way we evaluate work integrals.

EXAMPLE 3 A fluid's velocity field is $\mathbf{F} = x\mathbf{i} + z\mathbf{j} + y\mathbf{k}$. Find the flow along the helix $\mathbf{r}(t) = (\cos t)\mathbf{i} + (\sin t)\mathbf{j} + t\mathbf{k}, 0 \leq t \leq \pi/2$.

Solution We evaluate $\mathbf{F}$ on the curve,

$$\mathbf{F} = x\mathbf{i} + z\mathbf{j} + y\mathbf{k} = (\cos t)\mathbf{i} + t\mathbf{j} + (\sin t)\mathbf{k} \qquad x = \cos t, z = t, \text{ and } y = \sin t \text{ on } \mathbf{r}(t)$$

and then find $d\mathbf{r}/dt$:

$$\frac{d\mathbf{r}}{dt} = (-\sin t)\mathbf{i} + (\cos t)\mathbf{j} + \mathbf{k}.$$

Then we integrate $\mathbf{F} \cdot (d\mathbf{r}/dt)$ from $t = 0$ to $t = \frac{\pi}{2}$:

$$\mathbf{F} \cdot \frac{d\mathbf{r}}{dt} = (\cos t)(-\sin t) + (t)(\cos t) + (\sin t)(1)$$

$$= -\sin t \cos t + t \cos t + \sin t$$

so,

$$\text{Flow} = \int_{t=a}^{t=b} \mathbf{F} \cdot \frac{d\mathbf{r}}{dt} \, dt = \int_0^{\pi/2} (-\sin t \cos t + t \cos t + \sin t) \, dt$$

$$= \left[\frac{\cos^2 t}{2} + t \sin t \right]_0^{\pi/2} = \left(0 + \frac{\pi}{2} \right) - \left(\frac{1}{2} + 0 \right) = \frac{\pi}{2} - \frac{1}{2}. \qquad ■$$

EXAMPLE 4   Find the circulation of the field $\mathbf{F} = (x - y)\mathbf{i} + x\mathbf{j}$ around the circle $\mathbf{r}(t) = (\cos t)\mathbf{i} + (\sin t)\mathbf{j}, 0 \le t \le 2\pi$.

Solution   On the circle, $\mathbf{F} = (x - y)\mathbf{i} + x\mathbf{j} = (\cos t - \sin t)\mathbf{i} + (\cos t)\mathbf{j}$, and

$$\frac{d\mathbf{r}}{dt} = (-\sin t)\mathbf{i} + (\cos t)\mathbf{j}.$$

Then

$$\mathbf{F} \cdot \frac{d\mathbf{r}}{dt} = -\sin t \cos t + \underbrace{\sin^2 t + \cos^2 t}_{1}$$

gives

$$\text{Circulation} = \int_0^{2\pi} \mathbf{F} \cdot \frac{d\mathbf{r}}{dt} \, dt = \int_0^{2\pi} (1 - \sin t \cos t) \, dt$$

$$= \left[t - \frac{\sin^2 t}{2} \right]_0^{2\pi} = 2\pi. \qquad ■$$

Flux Across a Plane Curve

To find the rate at which a fluid is entering or leaving a region enclosed by a smooth curve C in the xy-plane, we calculate the line integral over C of $\mathbf{F} \cdot \mathbf{n}$, the scalar component of the fluid's velocity field in the direction of the curve's outward-pointing normal vector. The value of this integral is the *flux* of $\mathbf{F}$ across C. *Flux* is Latin for *flow*, but many flux calculations involve no motion at all. If $\mathbf{F}$ were an electric field or a magnetic field, for instance, the integral of $\mathbf{F} \cdot \mathbf{n}$ would still be called the flux of the field across C.

DEFINITION   If C is a smooth closed curve in the domain of a continuous vector field $\mathbf{F} = M(x, y)\mathbf{i} + N(x, y)\mathbf{j}$ in the plane and if $\mathbf{n}$ is the outward-pointing unit normal vector on C, the **flux** of $\mathbf{F}$ across C is

$$\text{Flux of } \mathbf{F} \text{ across } C = \int_C \mathbf{F} \cdot \mathbf{n} \, ds. \qquad (3)$$

Notice the difference between flux and circulation. The flux of $\mathbf{F}$ across C is the line integral with respect to arc length of $\mathbf{F} \cdot \mathbf{n}$, the scalar component of $\mathbf{F}$ in the direction of the

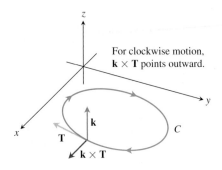

For clockwise motion, $\mathbf{k} \times \mathbf{T}$ points outward.

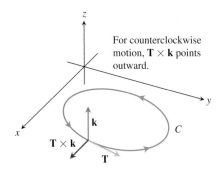

For counterclockwise motion, $\mathbf{T} \times \mathbf{k}$ points outward.

FIGURE 14.19 To find an outward unit normal vector for a smooth curve C in the xy-plane that is traversed counterclockwise as t increases, we take $\mathbf{n} = \mathbf{T} \times \mathbf{k}$. For clockwise motion, we take $\mathbf{n} = \mathbf{k} \times \mathbf{T}$.

outward normal. The circulation of $\mathbf{F}$ around C is the line integral with respect to arc length of $\mathbf{F} \cdot \mathbf{T}$, the scalar component of $\mathbf{F}$ in the direction of the unit tangent vector. Flux is the integral of the normal component of $\mathbf{F}$; circulation is the integral of the tangential component of $\mathbf{F}$.

To evaluate the integral in Equation (3), we begin with a smooth parametrization

$$x = g(t), \qquad y = h(t), \qquad a \le t \le b,$$

that traces the curve C exactly once as t increases from a to b. We can find the outward unit normal vector $\mathbf{n}$ by crossing the curve's unit tangent vector $\mathbf{T}$ with the vector $\mathbf{k}$. But which order do we choose, $\mathbf{T} \times \mathbf{k}$ or $\mathbf{k} \times \mathbf{T}$? Which one points outward? It depends on which way C is traversed as t increases. If the motion is clockwise, $\mathbf{k} \times \mathbf{T}$ points outward; if the motion is counterclockwise, $\mathbf{T} \times \mathbf{k}$ points outward (Figure 14.19). The usual choice is $\mathbf{n} = \mathbf{T} \times \mathbf{k}$, the choice that assumes counterclockwise motion. Thus, although the value of the arc length integral in the definition of flux in Equation (3) does not depend on which way C is traversed, the formulas we are about to derive for evaluating the integral in Equation (3) will assume counterclockwise motion.

In terms of components,

$$\mathbf{n} = \mathbf{T} \times \mathbf{k} = \left(\frac{dx}{ds}\mathbf{i} + \frac{dy}{ds}\mathbf{j} \right) \times \mathbf{k} = \frac{dy}{ds}\mathbf{i} - \frac{dx}{ds}\mathbf{j}.$$

If $\mathbf{F} = M(x, y)\mathbf{i} + N(x, y)\mathbf{j}$, then

$$\mathbf{F} \cdot \mathbf{n} = M(x, y)\frac{dy}{ds} - N(x, y)\frac{dx}{ds}.$$

Hence,

$$\int_C \mathbf{F} \cdot \mathbf{n}\, ds = \int_C \left(M\frac{dy}{ds} - N\frac{dx}{ds} \right) ds = \oint_C M\, dy - N\, dx.$$

We put a directed circle $\circlearrowleft$ on the last integral as a reminder that the integration around the closed curve C is to be in the counterclockwise direction. To evaluate this integral, we express M, dy, N, and dx in terms of t and integrate from $t = a$ to $t = b$. We do not need to know either $\mathbf{n}$ or ds to find the flux.

Calculating Flux Across a Smooth Closed Plane Curve

$$(\text{Flux of } \mathbf{F} = M\mathbf{i} + N\mathbf{j} \text{ across } C) = \oint_C M\, dy - N\, dx \qquad (4)$$

The integral can be evaluated from any smooth parametrization $x = g(t), y = h(t)$, $a \le t \le b$, that traces C counterclockwise exactly once.

EXAMPLE 5 Find the flux of $\mathbf{F} = (x - y)\mathbf{i} + x\mathbf{j}$ across the circle $x^2 + y^2 = 1$ in the xy-plane.

Solution The parametrization $\mathbf{r}(t) = (\cos t)\mathbf{i} + (\sin t)\mathbf{j}, 0 \le t \le 2\pi$, traces the circle counterclockwise exactly once. We can therefore use this parametrization in Equation (4). With

$$M = x - y = \cos t - \sin t, \qquad dy = d(\sin t) = \cos t\, dt$$
$$N = x = \cos t, \qquad dx = d(\cos t) = -\sin t\, dt,$$

we find

$$\text{Flux} = \int_C M\, dy - N\, dx = \int_0^{2\pi} (\cos^2 t - \sin t \cos t + \cos t \sin t)\, dt \quad \text{Equation (4)}$$
$$= \int_0^{2\pi} \cos^2 t\, dt = \int_0^{2\pi} \frac{1 + \cos 2t}{2}\, dt = \left[\frac{t}{2} + \frac{\sin 2t}{4}\right]_0^{2\pi} = \pi.$$

The flux of $\mathbf{F}$ across the circle is π. Since the answer is positive, the net flow across the curve is outward. A net inward flow would have given a negative flux. ∎

EXERCISES 14.2

Find the gradient fields of the functions in Exercises 1–4.

1. $f(x, y, z) = (x^2 + y^2 + z^2)^{-1/2}$

2. $f(x, y, z) = \ln\sqrt{x^2 + y^2 + z^2}$

3. $g(x, y, z) = e^z - \ln(x^2 + y^2)$

4. $g(x, y, z) = xy + yz + xz$

5. Give a formula $\mathbf{F} = M(x, y)\mathbf{i} + N(x, y)\mathbf{j}$ for the vector field in the plane that has the property that $\mathbf{F}$ points toward the origin with magnitude inversely proportional to the square of the distance from (x, y) to the origin. (The field is not defined at $(0, 0)$.)

6. Give a formula $\mathbf{F} = M(x, y)\mathbf{i} + N(x, y)\mathbf{j}$ for the vector field in the plane that has the properties that $\mathbf{F} = \mathbf{0}$ at $(0, 0)$ and that at any other point (a, b), $\mathbf{F}$ is tangent to the circle $x^2 + y^2 = a^2 + b^2$ and points in the clockwise direction with magnitude $|\mathbf{F}| = \sqrt{a^2 + b^2}$.

In Exercises 7–12, find the work done by force $\mathbf{F}$ from $(0, 0, 0)$ to $(1, 1, 1)$ over each of the following paths (Figure 14.20):

 a. The straight-line path C_1: $\mathbf{r}(t) = t\mathbf{i} + t\mathbf{j} + t\mathbf{k}$, $0 \le t \le 1$

 b. The curved path C_2: $\mathbf{r}(t) = t\mathbf{i} + t^2\mathbf{j} + t^4\mathbf{k}$, $0 \le t \le 1$

 c. The path $C_3 \cup C_4$ consisting of the line segment from $(0, 0, 0)$ to $(1, 1, 0)$ followed by the segment from $(1, 1, 0)$ to $(1, 1, 1)$

7. $\mathbf{F} = 3y\mathbf{i} + 2x\mathbf{j} + 4z\mathbf{k}$

8. $\mathbf{F} = [1/(x^2 + 1)]\mathbf{j}$

9. $\mathbf{F} = \sqrt{z}\mathbf{i} - 2x\mathbf{j} + \sqrt{y}\mathbf{k}$

10. $\mathbf{F} = xy\mathbf{i} + yz\mathbf{j} + xz\mathbf{k}$

11. $\mathbf{F} = (3x^2 - 3x)\mathbf{i} + 3z\mathbf{j} + \mathbf{k}$

12. $\mathbf{F} = (y + z)\mathbf{i} + (z + x)\mathbf{j} + (x + y)\mathbf{k}$

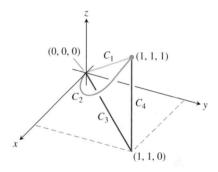

FIGURE 14.20 The paths from $(0, 0, 0)$ to $(1, 1, 1)$.

In Exercises 13–16, find the work done by $\mathbf{F}$ over the curve in the direction of increasing t.

13. $\mathbf{F} = xy\mathbf{i} + y\mathbf{j} - yz\mathbf{k}$
$\mathbf{r}(t) = t\mathbf{i} + t^2\mathbf{j} + t\mathbf{k}$, $0 \le t \le 1$

14. $\mathbf{F} = 2y\mathbf{i} + 3x\mathbf{j} + (x + y)\mathbf{k}$
$\mathbf{r}(t) = (\cos t)\mathbf{i} + (\sin t)\mathbf{j} + (t/6)\mathbf{k}$, $0 \le t \le 2\pi$

15. $\mathbf{F} = z\mathbf{i} + x\mathbf{j} + y\mathbf{k}$
$\mathbf{r}(t) = (\sin t)\mathbf{i} + (\cos t)\mathbf{j} + t\mathbf{k}$, $0 \le t \le 2\pi$

16. $\mathbf{F} = 6z\mathbf{i} + y^2\mathbf{j} + 12x\mathbf{k}$
$\mathbf{r}(t) = (\sin t)\mathbf{i} + (\cos t)\mathbf{j} + (t/6)\mathbf{k}$, $0 \le t \le 2\pi$

17. Evaluate $\int_C xy\,dx + (x + y)\,dy$ along the curve $y = x^2$ from $(-1, 1)$ to $(2, 4)$.

18. Evaluate $\int_C (x - y)\,dx + (x + y)\,dy$ counterclockwise around the triangle with vertices $(0, 0)$, $(1, 0)$, and $(0, 1)$.

19. Evaluate $\int_C \mathbf{F} \cdot \mathbf{T}\,ds$ for the vector field $\mathbf{F} = x^2\mathbf{i} - y\mathbf{j}$ along the curve $x = y^2$ from $(4, 2)$ to $(1, -1)$.

20. Evaluate $\int_C \mathbf{F} \cdot d\mathbf{r}$ for the vector field $\mathbf{F} = y\mathbf{i} - x\mathbf{j}$ counterclockwise along the unit circle $x^2 + y^2 = 1$ from $(1, 0)$ to $(0, 1)$.

21. **Work**   Find the work done by the force $\mathbf{F} = xy\mathbf{i} + (y - x)\mathbf{j}$ over the straight line from $(1, 1)$ to $(2, 3)$.

22. **Work**   Find the work done by the gradient of $f(x, y) = (x + y)^2$ counterclockwise around the circle $x^2 + y^2 = 4$ from $(2, 0)$ to itself.

23. **Circulation and flux**   Find the circulation and flux of the fields

$$\mathbf{F}_1 = x\mathbf{i} + y\mathbf{j} \qquad \text{and} \qquad \mathbf{F}_2 = -y\mathbf{i} + x\mathbf{j}$$

around and across each of the following curves.

   a. The circle $\mathbf{r}(t) = (\cos t)\mathbf{i} + (\sin t)\mathbf{j}, \quad 0 \le t \le 2\pi$

   b. The ellipse $\mathbf{r}(t) = (\cos t)\mathbf{i} + (4 \sin t)\mathbf{j}, \quad 0 \le t \le 2\pi$

12. **Flux across a circle**   Find the flux of the fields

$$\mathbf{F}_1 = 2x\mathbf{i} - 3y\mathbf{j} \qquad \text{and} \qquad \mathbf{F}_2 = 2x\mathbf{i} + (x - y)\mathbf{j}$$

across the circle

$$\mathbf{r}(t) = (a \cos t)\mathbf{i} + (a \sin t)\mathbf{j}, \qquad 0 \le t \le 2\pi.$$

In Exercises 25–28, find the circulation and flux of the field $\mathbf{F}$ around and across the closed semicircular path that consists of the semicircular arch $\mathbf{r}_1(t) = (a \cos t)\mathbf{i} + (a \sin t)\mathbf{j}, 0 \le t \le \pi$, followed by the line segment $\mathbf{r}_2(t) = t\mathbf{i}, -a \le t \le a$.

25. $\mathbf{F} = x\mathbf{i} + y\mathbf{j}$ 26. $\mathbf{F} = x^2\mathbf{i} + y^2\mathbf{j}$

27. $\mathbf{F} = -y\mathbf{i} + x\mathbf{j}$ 28. $\mathbf{F} = -y^2\mathbf{i} + x^2\mathbf{j}$

29. **Flow integrals**   Find the flow of the velocity field $\mathbf{F} = (x + y)\mathbf{i} - (x^2 + y^2)\mathbf{j}$ along each of the following paths from $(1, 0)$ to $(-1, 0)$ in the xy-plane.

   a. The upper half of the circle $x^2 + y^2 = 1$

   b. The line segment from $(1, 0)$ to $(-1, 0)$

   c. The line segment from $(1, 0)$ to $(0, -1)$ followed by the line segment from $(0, -1)$ to $(-1, 0)$.

30. **Flux across a triangle**   Find the flux of the field $\mathbf{F}$ in Exercise 29 outward across the triangle with vertices $(1, 0)$, $(0, 1)$, $(-1, 0)$.

31. **Spin field**   Draw the spin field

$$\mathbf{F} = -\frac{y}{\sqrt{x^2 + y^2}}\mathbf{i} + \frac{x}{\sqrt{x^2 + y^2}}\mathbf{j}$$

(see Figure 14.13) along with its horizontal and vertical components at a representative assortment of points on the circle $x^2 + y^2 = 4$.

32. **Radial field**   Draw the radial field

$$\mathbf{F} = x\mathbf{i} + y\mathbf{j}$$

(see Figure 14.12) along with its horizontal and vertical components at a representative assortment of points on the circle $x^2 + y^2 = 1$.

33. **A field of tangent vectors**

   a. Find a field $\mathbf{G} = P(x, y)\mathbf{i} + Q(x, y)\mathbf{j}$ in the xy-plane with the property that at any point $(a, b) \ne (0, 0)$, $\mathbf{G}$ is a vector of magnitude $\sqrt{a^2 + b^2}$ tangent to the circle $x^2 + y^2 = a^2 + b^2$ and pointing in the counterclockwise direction. (The field is undefined at $(0, 0)$.)

   b. How is $\mathbf{G}$ related to the spin field $\mathbf{F}$ in Figure 14.13?

34. **A field of tangent vectors**

   a. Find a field $\mathbf{G} = P(x, y)\mathbf{i} + Q(x, y)\mathbf{j}$ in the xy-plane with the property that at any point $(a, b) \ne (0, 0)$, $\mathbf{G}$ is a unit vector tangent to the circle $x^2 + y^2 = a^2 + b^2$ and pointing in the clockwise direction.

   b. How is $\mathbf{G}$ related to the spin field $\mathbf{F}$ in Figure 14.13?

35. **Unit vectors pointing toward the origin**   Find a field $\mathbf{F} = M(x, y)\mathbf{i} + N(x, y)\mathbf{j}$ in the xy-plane with the property that at each point $(x, y) \ne (0, 0)$, $\mathbf{F}$ is a unit vector pointing toward the origin. (The field is undefined at $(0, 0)$.)

36. **Two "central" fields**   Find a field $\mathbf{F} = M(x, y)\mathbf{i} + N(x, y)\mathbf{j}$ in the xy-plane with the property that at each point $(x, y) \ne (0, 0)$, $\mathbf{F}$ points toward the origin and $|\mathbf{F}|$ is **(a)** the distance from (x, y) to the origin, **(b)** inversely proportional to the distance from (x, y) to the origin. (The field is undefined at $(0, 0)$.)

In Exercises 37–40, $\mathbf{F}$ is the velocity field of a fluid flowing through a region in space. Find the flow along the given curve in the direction of increasing t.

37. $\mathbf{F} = -4xy\mathbf{i} + 8y\mathbf{j} + 2\mathbf{k}$
    $\mathbf{r}(t) = t\mathbf{i} + t^2\mathbf{j} + \mathbf{k}, \quad 0 \le t \le 2$

38. $\mathbf{F} = x^2\mathbf{i} + yz\mathbf{j} + y^2\mathbf{k}$
    $r(t) = 3t\mathbf{j} + 4t\mathbf{k}, \quad 0 \le t \le 1$

39. $\mathbf{F} = (x - z)\mathbf{i} + x\mathbf{k}$
    $\mathbf{r}(t) = (\cos t)\mathbf{i} + (\sin t)\mathbf{k}, \quad 0 \le t \le \pi$

40. $\mathbf{F} = -y\mathbf{i} + x\mathbf{j} + 2\mathbf{k}$
    $\mathbf{r}(t) = (-2 \cos t)\mathbf{i} + (2 \sin t)\mathbf{j} + 2t\mathbf{k}, \quad 0 \le t \le 2\pi$

41. **Circulation**   Find the circulation of $\mathbf{F} = 2x\mathbf{i} + 2z\mathbf{j} + 2y\mathbf{k}$ around the closed path consisting of the following three curves traversed in the direction of increasing t:

$$C_1: \quad \mathbf{r}(t) = (\cos t)\mathbf{i} + (\sin t)\mathbf{j} + t\mathbf{k}, \quad 0 \le t \le \pi/2$$

$$C_2: \quad \mathbf{r}(t) = \mathbf{j} + (\pi/2)(1 - t)\mathbf{k}, \quad 0 \le t \le 1$$

$$C_3: \quad \mathbf{r}(t) = t\mathbf{i} + (1 - t)\mathbf{j}, \quad 0 \le t \le 1$$

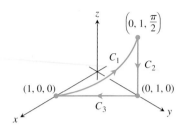

42. Zero circulation Let C be the ellipse in which the plane $2x + 3y - z = 0$ meets the cylinder $x^2 + y^2 = 12$. Show, without evaluating either line integral directly, that the circulation of the field $\mathbf{F} = x\mathbf{i} + y\mathbf{j} + z\mathbf{k}$ around C in either direction is zero.

43. Flow along a curve The field $\mathbf{F} = xy\mathbf{i} + y\mathbf{j} - yz\mathbf{k}$ is the velocity field of a flow in space. Find the flow from $(0, 0, 0)$ to $(1, 1, 1)$ along the curve of intersection of the cylinder $y = x^2$ and the plane $z = x$. (*Hint:* Use $t = x$ as the parameter.)

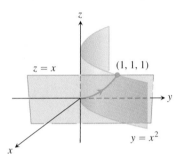

44. Flow of a gradient field Find the flow of the field $\mathbf{F} = \nabla(xy^2z^3)$:

a. Once around the curve C in Exercise 42, clockwise as viewed from above

b. Along the line segment from $(1, 1, 1)$ to $(2, 1, -1)$.

45. Work and area Suppose that $f(t)$ is differentiable and positive for $a \le t \le b$. Let C be the path $\mathbf{r}(t) = t\mathbf{i} + f(t)\mathbf{j}$, $a \le t \le b$, and $\mathbf{F} = y\mathbf{i}$. Is there any relation between the value of the work integral

$$\int_C \mathbf{F} \cdot d\mathbf{r}$$

and the area of the region bounded by the t-axis, the graph of f, and the lines $t = a$ and $t = b$? Give reasons for your answer.

46. Work done by a radial force with constant magnitude A particle moves along the smooth curve $y = f(x)$ from $(a, f(a))$ to $(b, f(b))$. The force moving the particle has constant magnitude k and always points away from the origin. Show that the work done by the force is

$$\int_C \mathbf{F} \cdot \mathbf{T} \, ds = k\big[(b^2 + (f(b))^2)^{1/2} - (a^2 + (f(a))^2)^{1/2}\big].$$

COMPUTER EXPLORATIONS

In Exercises 47–52, use a CAS to perform the following steps for finding the work done by force $\mathbf{F}$ over the given path:

a. Find $d\mathbf{r}$ for the path $\mathbf{r}(t) = g(t)\mathbf{i} + h(t)\mathbf{j} + k(t)\mathbf{k}$.

b. Evaluate the force $\mathbf{F}$ along the path.

c. Evaluate $\displaystyle\int_C \mathbf{F} \cdot d\mathbf{r}$.

47. $\mathbf{F} = xy^6\mathbf{i} + 3x(xy^5 + 2)\mathbf{j}$; $\quad \mathbf{r}(t) = (2\cos t)\mathbf{i} + (\sin t)\mathbf{j}$, $0 \le t \le 2\pi$

48. $\mathbf{F} = \dfrac{3}{1 + x^2}\mathbf{i} + \dfrac{2}{1 + y^2}\mathbf{j}$; $\quad \mathbf{r}(t) = (\cos t)\mathbf{i} + (\sin t)\mathbf{j}$, $0 \le t \le \pi$

49. $\mathbf{F} = (y + yz\cos xyz)\mathbf{i} + (x^2 + xz\cos xyz)\mathbf{j} +$ $(z + xy\cos xyz)\mathbf{k}$; $\quad \mathbf{r}(t) = (2\cos t)\mathbf{i} + (3\sin t)\mathbf{j} + \mathbf{k}$, $0 \le t \le 2\pi$

50. $\mathbf{F} = 2xy\mathbf{i} - y^2\mathbf{j} + ze^x\mathbf{k}$; $\quad \mathbf{r}(t) = -t\mathbf{i} + \sqrt{t}\mathbf{j} + 3t\mathbf{k}$, $1 \le t \le 4$

51. $\mathbf{F} = (2y + \sin x)\mathbf{i} + (z^2 + (1/3)\cos y)\mathbf{j} + x^4\mathbf{k}$; $\mathbf{r}(t) = (\sin t)\mathbf{i} + (\cos t)\mathbf{j} + (\sin 2t)\mathbf{k}$, $\quad -\pi/2 \le t \le \pi/2$

52. $\mathbf{F} = (x^2y)\mathbf{i} + \dfrac{1}{3}x^3\mathbf{j} + xy\mathbf{k}$; $\quad \mathbf{r}(t) = (\cos t)\mathbf{i} + (\sin t)\mathbf{j} +$ $(2\sin^2 t - 1)\mathbf{k}$, $\quad 0 \le t \le 2\pi$

14.3 Path Independence, Potential Functions, and Conservative Fields

A **gravitational field g** is a vector field that represents the effect of gravity on a body placed within it. The gravitational force on a body of mass m placed in the field is given by $\mathbf{F} = m\mathbf{g}$. Similarly, an **electric field E** is a vector field in space that represents the effect of electric forces on a charged particle placed within it. The force on a body of charge q placed in the field is given by $\mathbf{F} = q\mathbf{E}$. In gravitational and electric fields, the amount of work it takes to move a mass or charge from one point to another depends on the initial and final positions of the object—not on which path is taken between these positions. In this section we study fields with this property and the calculation of work integrals associated with them.

Path Independence

If A and B are two points in an open region D in space, the work $\int \mathbf{F} \cdot d\mathbf{r}$ done in moving a particle from A to B by a field $\mathbf{F}$ defined on D usually depends on the path taken. For some special fields, however, the integral's value is the same for all paths from A to B.

DEFINITIONS   Let $\mathbf{F}$ be a field defined on an open region D in space, and suppose that for any two points A and B in D the work $\int_A^B \mathbf{F} \cdot d\mathbf{r}$ done in moving from A to B in D is the same over all paths from A to B. Then the integral $\int \mathbf{F} \cdot d\mathbf{r}$ is **path independent in D** and the field $\mathbf{F}$ is **conservative on D**.

The word *conservative* comes from physics, where it refers to fields in which the principle of conservation of energy holds.

Under differentiability conditions normally met in practice, we will show that a field $\mathbf{F}$ is conservative if and only if it is the gradient field of a scalar function f; that is, if and only if $\mathbf{F} = \nabla f$ for some f. The function f then has a special name.

DEFINITION   If $\mathbf{F}$ is a field defined on D and $\mathbf{F} = \nabla f$ for some scalar function f on D, then f is called a **potential function for $\mathbf{F}$.**

An electric potential is a scalar function whose gradient field is an electric field. A gravitational potential is a scalar function whose gradient field is a gravitational field, and so on. As we will see, once we have found a potential function f for a field $\mathbf{F}$, we can evaluate all the work integrals in the domain of $\mathbf{F}$ over any path between A and B by

$$\int_A^B \mathbf{F} \cdot d\mathbf{r} = \int_A^B \nabla f \cdot d\mathbf{r} = f(B) - f(A). \tag{1}$$

If you think of ∇f for functions of several variables as being something like the derivative f' for functions of a single variable, then you see that Equation (1) is the vector calculus analogue of the Fundamental Theorem of Calculus formula

$$\int_a^b f'(x)\, dx = f(b) - f(a).$$

Conservative fields have other remarkable properties. For example, saying that $\mathbf{F}$ is conservative on D is equivalent to saying that the integral of $\mathbf{F}$ around every closed path in D is zero. Certain conditions on the curves, fields, and domains must be satisfied for Equation (1) to be valid. We discuss these conditions next.

Assumptions on Curves, Vector Fields, and Domains

In order for the computations and results we derive below to be valid, we must assume certain properties for the curves, surfaces, domains, and vector fields we consider. We give these assumptions in the statements of theorems, and they also apply to the examples and exercises unless otherwise stated.

The curves we consider are **piecewise smooth**. Such curves are made up of finitely many smooth pieces connected end to end, as discussed in Section 11.1. We will treat vector fields **F** whose components have continuous first partial derivatives.

The solid domains D we consider are open regions in space, so every point in D is the center of an open ball that lies entirely in D (see Section 12.1). We also assume D to be **connected**. For an open region, this means that any two points in D can be joined by a smooth curve that lies in the region. Finally, we assume D is **simply connected**, which means that every loop in D can be contracted to a point in D without ever leaving D. The plane with a disk removed is a two-dimensional region that is *not* simply connected, for example. A loop in the plane that goes around the disk cannot be contracted to a point without going into the "hole" left by the removed disk. Similarly, if we remove an infinitely long solid cylinder from space, the remaining region D is *not* simply connected. A curve encircling the cylinder cannot be shrunk to a point while remaining inside D.

Connectivity and simple connectivity are not the same, and neither property implies the other. Think of connected regions as being in "one piece" and simply connected regions as not having any "loop-catching holes." All of space itself is both connected and simply connected. Figure 14.21 illustrates some of these properties.

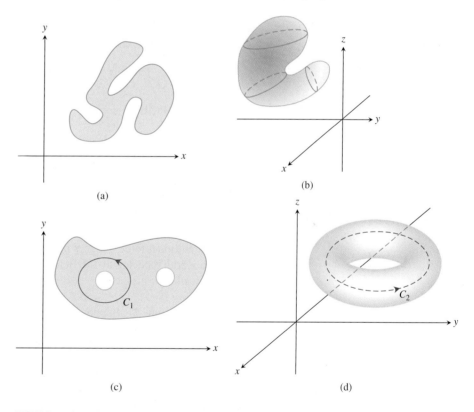

(a)

(b)

(c)

(d)

FIGURE 14.21   Four connected regions. In (a) and (b), the regions are simply connected. In (c) and (d), the regions are not simply connected because the curves C_1 and C_2 cannot be contracted to a point inside the regions containing them.

Caution   Some of the results in this chapter can fail to hold if applied to situations where the conditions we've imposed do not hold. In particular, the component test for conservative fields, given later in this section, is not valid on domains that are not simply connected.

Line Integrals in Conservative Fields

Gradient fields **F** are obtained by differentiating a scalar function f. A theorem analogous to the Fundamental Theorem of Calculus gives a way to evaluate the line integrals of gradient fields.

THEOREM 1—Fundamental Theorem of Line Integrals   Let C be a smooth curve joining the points A and B in the plane or in space and given by $\mathbf{r}(t)$. Let f be a differentiable function with a continuous gradient vector $\mathbf{F} = \nabla f$ on a domain containing C. Then

$$\int_C \mathbf{F} \cdot d\mathbf{r} = f(B) - f(A).$$

Like the Fundamental Theorem, Theorem 1 gives a way to evaluate line integrals without having to take limits of Riemann sums or finding the line integral directly using the method of Section 14.1. Before proving Theorem 1, we give an example.

EXAMPLE 1   Suppose the vector field $\mathbf{F} = \nabla f$ is the gradient of the function

$$f(x, y, z) = -\frac{1}{x^2 + y^2 + z^2}.$$

Find the work done by **F** in moving a particle along a smooth curve C joining $(1, 0, 0)$ to $(0, 0, 2)$.

Solution   There is no need to compute a line integral directly. By applying Theorem 1, we see that the work done by **F** along any smooth curve C joining the two points is

$$\int_C \mathbf{F} \cdot d\mathbf{r} = f(0, 0, 2) - f(1, 0, 0) = -\frac{1}{4} - (-1) = -\frac{3}{4}. \qquad \blacksquare$$

The gravitional force due to a planet, and the electric force associated with a charged particle, are both modeled by the field **F** given in Example 1 up to a constant that depends on the units of measurement.

Proof of Theorem 1   Suppose that A and B are two points in region D and that $C: \mathbf{r}(t) = g(t)\mathbf{i} + h(t)\mathbf{j} + k(t)\mathbf{k}$, $a \le t \le b$, is a smooth curve in D joining A and B. Along the curve, f is a differentiable function of t and

$$\frac{df}{dt} = \frac{\partial f}{\partial x}\frac{dx}{dt} + \frac{\partial f}{\partial y}\frac{dy}{dt} + \frac{\partial f}{\partial z}\frac{dz}{dt} \qquad \text{Chain Rule with } x = g(t),$$
$$y = h(t), z = k(t)$$

$$= \nabla f \cdot \left(\frac{dx}{dt}\mathbf{i} + \frac{dy}{dt}\mathbf{j} + \frac{dz}{dt}\mathbf{k}\right) = \nabla f \cdot \frac{d\mathbf{r}}{dt} = \mathbf{F} \cdot \frac{d\mathbf{r}}{dt}. \qquad \text{Because } \mathbf{F} = \nabla f$$

Therefore,

$$\int_C \mathbf{F} \cdot d\mathbf{r} = \int_{t=a}^{t=b} \mathbf{F} \cdot \frac{d\mathbf{r}}{dt}\, dt = \int_a^b \frac{df}{dt}\, dt$$

$$= f(g(t), h(t), k(t)) \Big]_a^b = f(B) - f(A). \qquad \blacksquare$$

So we see from Theorem 1 that the line integral of a gradient field $\mathbf{F} = \nabla f$ is straight-forward to compute once we know the function f. Many important vector fields arising in nature are indeed gradient fields. The next result, which follows from Theorem 1, shows that any conservative field is of this type. Therefore the value of a line integral of a conservative field along a smooth curve depends only on the endpoints of the curve.

THEOREM 2—Conservative Fields are Gradient Fields   Let $\mathbf{F} = M\mathbf{i} + N\mathbf{j} + P\mathbf{k}$ be a vector field whose components are continuous throughout an open connected region D in space. Then $\mathbf{F}$ is conservative if and only if $\mathbf{F}$ is a gradient field ∇f for a differentiable function f.

Theorem 2 says that $\mathbf{F} = \nabla f$ if and only if for any two points A and B in the region D, the value of line integral $\int_C \mathbf{F} \cdot d\mathbf{r}$ is independent of the path joining A to B in D.

Partial Proof of Theorem 2   If $\mathbf{F}$ is a gradient field, then $\mathbf{F} = \nabla f$ for a differentiable function f, and Theorem 1 shows that $\int_C \mathbf{F} \cdot d\mathbf{r} = f(B) - f(A)$. The value of the line integral does not depend on C, but only on its endpoints A and B. So the line integral is path independent and $\mathbf{F}$ satisfies the definition of a conservative field.

On the other hand, if $\mathbf{F}$ is conservative, we can define a scalar function f on D as follows. Pick a point A in D and set $f(A)$ equal to a constant k. Then define $f(B)$ to equal $f(A) + \int_C \mathbf{F} \cdot d\mathbf{r}$, where C is any smooth path from A to B. Notice that $f(B)$ does not depend on the choice of C since $\mathbf{F}$ is conservative. It can then be shown that $\mathbf{F} = \nabla f$, but we omit the more advanced proof here.   ∎

EXAMPLE 2   Find the work done by the conservative field

$$\mathbf{F} = yz\mathbf{i} + xz\mathbf{j} + xy\mathbf{k} = \nabla(xyz)$$

along any smooth curve C joining the point $A(-1, 3, 9)$ to $B(1, 6, -4)$.

Solution   With $f(x, y, z) = xyz$, we have

$$
\begin{aligned}
\int_A^B \mathbf{F} \cdot d\mathbf{r} &= \int_A^B \nabla f \cdot d\mathbf{r} \qquad && \mathbf{F} = \nabla f \\
&= f(B) - f(A) && \text{Theorem 1} \\
&= xyz\big|_{(1,6,-4)} - xyz\big|_{(-1,3,9)} \\
&= (1)(6)(-4) - (-1)(3)(9) \\
&= -24 + 27 = 3.
\end{aligned}
$$

∎

THEOREM 3—Closed-Loop Property of Conservative Fields   The following statements are equivalent.

1. $\int \mathbf{F} \cdot d\mathbf{r} = 0$ around every closed loop in D.
2. The field $\mathbf{F}$ is conservative on D.

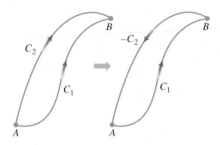

FIGURE 14.22 If we have two paths from A to B, one of them can be reversed to make a loop.

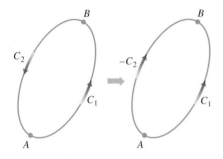

FIGURE 14.23 If A and B lie on a loop, we can reverse part of the loop to make two paths from A to B.

Proof that Part 1 $\Rightarrow$ Part 2 We want to show that for any two points A and B in D, the integral of $\mathbf{F} \cdot d\mathbf{r}$ has the same value over any two paths C_1 and C_2 from A to B. We reverse the direction on C_2 to make a path $-C_2$ from B to A (Figure 14.22). Together, C_1 and $-C_2$ make a closed loop C, and by assumption,

$$\int_{C_1} \mathbf{F} \cdot d\mathbf{r} - \int_{C_2} \mathbf{F} \cdot d\mathbf{r} = \int_{C_1} \mathbf{F} \cdot d\mathbf{r} + \int_{-C_2} \mathbf{F} \cdot d\mathbf{r} = \int_{C} \mathbf{F} \cdot d\mathbf{r} = 0.$$

Thus, the integrals over C_1 and C_2 give the same value. Note that the definition of line integral shows that changing the direction along a curve reverses the sign of the line integral.

Proof that Part 2 $\Rightarrow$ Part 1 We want to show that the integral of $\mathbf{F} \cdot d\mathbf{r}$ is zero over any closed loop C. We pick two points A and B on C and use them to break C into two pieces: C_1 from A to B followed by C_2 from B back to A (Figure 14.23). Then

$$\oint_{C} \mathbf{F} \cdot d\mathbf{r} = \int_{C_1} \mathbf{F} \cdot d\mathbf{r} + \int_{C_2} \mathbf{F} \cdot d\mathbf{r} = \int_{A}^{B} \mathbf{F} \cdot d\mathbf{r} - \int_{A}^{B} \mathbf{F} \cdot d\mathbf{r} = 0. \qquad \blacksquare$$

The following diagram summarizes the results of Theorems 2 and 3.

$$\underset{\mathbf{F} = \nabla f \text{ on } D}{\overset{\text{Theorem 2}}{}} \quad \Leftrightarrow \quad \underset{\substack{\mathbf{F} \text{ conservative} \\ \text{on } D}}{} \quad \overset{\text{Theorem 3}}{\Leftrightarrow} \quad \underset{\substack{\text{over any closed} \\ \text{path in } D}}{\oint_{C} \mathbf{F} \cdot d\mathbf{r} = 0}$$

Two questions arise:

1. How do we know whether a given field $\mathbf{F}$ is conservative?

2. If $\mathbf{F}$ is in fact conservative, how do we find a potential function f (so that $\mathbf{F} = \nabla f$)?

Finding Potentials for Conservative Fields

The test for being conservative is the following.

Component Test for Conservative Fields
Let $\mathbf{F} = M(x, y, z)\mathbf{i} + N(x, y, z)\mathbf{j} + P(x, y, z)\mathbf{k}$ be a field on a connected and simply connected domain whose component functions have continuous first partial derivatives. Then, $\mathbf{F}$ is conservative if and only if

$$\frac{\partial P}{\partial y} = \frac{\partial N}{\partial z}, \qquad \frac{\partial M}{\partial z} = \frac{\partial P}{\partial x}, \qquad \text{and} \qquad \frac{\partial N}{\partial x} = \frac{\partial M}{\partial y}. \qquad (2)$$

Proof that Equations (2) hold if F is conservative There is a potential function f such that

$$\mathbf{F} = M\mathbf{i} + N\mathbf{j} + P\mathbf{k} = \frac{\partial f}{\partial x}\mathbf{i} + \frac{\partial f}{\partial y}\mathbf{j} + \frac{\partial f}{\partial z}\mathbf{k}.$$

Hence,

$$\frac{\partial P}{\partial y} = \frac{\partial}{\partial y}\left(\frac{\partial f}{\partial z}\right) = \frac{\partial^2 f}{\partial y\,\partial z}$$

$$= \frac{\partial^2 f}{\partial z\,\partial y} \qquad \text{Continuity implies that the mixed partial derivatives are equal.}$$

$$= \frac{\partial}{\partial z}\left(\frac{\partial f}{\partial y}\right) = \frac{\partial N}{\partial z}.$$

The others in Equations (2) are proved similarly.   ∎

The second half of the proof, that Equations (2) imply that **F** is conservative, is a consequence of Stokes' Theorem, taken up in Section 14.7, and requires our assumption that the domain of **F** be simply connected.

Once we know that **F** is conservative, we usually want to find a potential function for **F**. This requires solving the equation $\nabla f = \mathbf{F}$ or

$$\frac{\partial f}{\partial x}\mathbf{i} + \frac{\partial f}{\partial y}\mathbf{j} + \frac{\partial f}{\partial z}\mathbf{k} = M\mathbf{i} + N\mathbf{j} + P\mathbf{k}$$

for f. We accomplish this by integrating the three equations

$$\frac{\partial f}{\partial x} = M, \qquad \frac{\partial f}{\partial y} = N, \qquad \frac{\partial f}{\partial z} = P,$$

as illustrated in the next example.

EXAMPLE 3   Show that $\mathbf{F} = (e^x \cos y + yz)\mathbf{i} + (xz - e^x \sin y)\mathbf{j} + (xy + z)\mathbf{k}$ is conservative and find a potential function for it.

Solution   We apply the test in Equations (2) to

$$M = e^x \cos y + yz, \qquad N = xz - e^x \sin y, \qquad P = xy + z$$

and calculate

$$\frac{\partial P}{\partial y} = x = \frac{\partial N}{\partial z}, \qquad \frac{\partial M}{\partial z} = y = \frac{\partial P}{\partial x}, \qquad \frac{\partial N}{\partial x} = -e^x \sin y + z = \frac{\partial M}{\partial y}.$$

Together, these equalities tell us that there is a function f with $\nabla f = \mathbf{F}$.

We find f by integrating the equations

$$\frac{\partial f}{\partial x} = e^x \cos y + yz, \qquad \frac{\partial f}{\partial y} = xz - e^x \sin y, \qquad \frac{\partial f}{\partial z} = xy + z. \tag{3}$$

We integrate the first equation with respect to x, holding y and z fixed, to get

$$f(x, y, z) = e^x \cos y + xyz + g(y, z).$$

We write the constant of integration as a function of y and z because its value may change if y and z change. We then calculate $\partial f/\partial y$ from this equation and match it with the expression for $\partial f/\partial y$ in Equations (3). This gives

$$-e^x \sin y + xz + \frac{\partial g}{\partial y} = xz - e^x \sin y,$$

so $\partial g/\partial y = 0$. Therefore, g is a function of z alone, and

$$f(x, y, z) = e^x \cos y + xyz + h(z).$$

We now calculate $\partial f/\partial z$ from this equation and match it to the formula for $\partial f/\partial z$ in Equations (3). This gives

$$xy + \frac{dh}{dz} = xy + z, \qquad \text{or} \qquad \frac{dh}{dz} = z,$$

so

$$h(z) = \frac{z^2}{2} + C.$$

Hence,

$$f(x, y, z) = e^x \cos y + xyz + \frac{z^2}{2} + C.$$

We have infinitely many potential functions of $\mathbf{F}$, one for each value of C. ∎

EXAMPLE 4   Show that $\mathbf{F} = (2x - 3)\mathbf{i} - z\mathbf{j} + (\cos z)\mathbf{k}$ is not conservative.

Solution   We apply the component test in Equations (2) and find immediately that

$$\frac{\partial P}{\partial y} = \frac{\partial}{\partial y}(\cos z) = 0, \qquad \frac{\partial N}{\partial z} = \frac{\partial}{\partial z}(-z) = -1.$$

The two are unequal, so $\mathbf{F}$ is not conservative. No further testing is required. ∎

Exact Differential Forms

As we see in the next section and again later on, it is often convenient to express work and circulation integrals in the "differential" form

$$\int_A^B M\,dx + N\,dy + P\,dz$$

mentioned in Section 14.2. Such integrals are relatively easy to evaluate if $M\,dx + N\,dy + P\,dz$ is the total differential of a function f. For then

$$\int_A^B M\,dx + N\,dy + P\,dz = \int_A^B \frac{\partial f}{\partial x}\,dx + \frac{\partial f}{\partial y}\,dy + \frac{\partial f}{\partial z}\,dz$$

$$= \int_A^B \nabla f \cdot d\mathbf{r}$$

$$= f(B) - f(A). \qquad \text{Theorem 1}$$

Thus,

$$\int_A^B df = f(B) - f(A),$$

just as with differentiable functions of a single variable.

> **DEFINITIONS**   Any expression $M(x, y, z)\, dx + N(x, y, z)\, dy + P(x, y, z)\, dz$ is a **differential form**. A differential form is **exact** on a domain D in space if
>
> $$M\, dx + N\, dy + P\, dz = \frac{\partial f}{\partial x}\, dx + \frac{\partial f}{\partial y}\, dy + \frac{\partial f}{\partial z}\, dz = df$$
>
> for some scalar function f throughout D.

Notice that if $M\, dx + N\, dy + P\, dz = df$ on D, then $\mathbf{F} = M\mathbf{i} + N\mathbf{j} + P\mathbf{k}$ is the gradient field of f on D. Conversely, if $\mathbf{F} = \nabla f$, then the form $M\, dx + N\, dy + P\, dz$ is exact. The test for the form's being exact is therefore the same as the test for $\mathbf{F}$ being conservative.

> **Component Test for Exactness of $M\, dx + N\, dy + P\, dz$**
> The differential form $M\, dx + N\, dy + P\, dz$ is exact on a connected and simply connected domain if and only if
>
> $$\frac{\partial P}{\partial y} = \frac{\partial N}{\partial z}, \qquad \frac{\partial M}{\partial z} = \frac{\partial P}{\partial x}, \qquad \text{and} \qquad \frac{\partial N}{\partial x} = \frac{\partial M}{\partial y}.$$
>
> This is equivalent to saying that the field $\mathbf{F} = M\mathbf{i} + N\mathbf{j} + P\mathbf{k}$ is conservative.

EXAMPLE 5   Show that $y\, dx + x\, dy + 4\, dz$ is exact and evaluate the integral

$$\int_{(1,1,1)}^{(2,3,-1)} y\, dx + x\, dy + 4\, dz$$

over the line segment from $(1, 1, 1)$ to $(2, 3, -1)$.

Solution   We let $M = y$, $N = x$, $P = 4$ and apply the Test for Exactness:

$$\frac{\partial P}{\partial y} = 0 = \frac{\partial N}{\partial z}, \qquad \frac{\partial M}{\partial z} = 0 = \frac{\partial P}{\partial x}, \qquad \frac{\partial N}{\partial x} = 1 = \frac{\partial M}{\partial y}.$$

These equalities tell us that $y\, dx + x\, dy + 4\, dz$ is exact, so

$$y\, dx + x\, dy + 4\, dz = df$$

for some function f, and the integral's value is $f(2, 3, -1) - f(1, 1, 1)$.

We find f up to a constant by integrating the equations

$$\frac{\partial f}{\partial x} = y, \qquad \frac{\partial f}{\partial y} = x, \qquad \frac{\partial f}{\partial z} = 4. \tag{4}$$

From the first equation we get

$$f(x, y, z) = xy + g(y, z).$$

The second equation tells us that

$$\frac{\partial f}{\partial y} = x + \frac{\partial g}{\partial y} = x, \qquad \text{or} \qquad \frac{\partial g}{\partial y} = 0.$$

Hence, g is a function of z alone, and

$$f(x, y, z) = xy + h(z).$$

The third of Equations (4) tells us that

$$\frac{\partial f}{\partial z} = 0 + \frac{dh}{dz} = 4, \qquad \text{or} \qquad h(z) = 4z + C.$$

Therefore,

$$f(x, y, z) = xy + 4z + C.$$

The value of the integral is

$$f(2, 3, -1) - f(1, 1, 1) = 2 + C - (5 + C) = -3. \qquad \blacksquare$$

EXERCISES 14.3

Which fields in Exercises 1–6 are conservative, and which are not?

1. $\mathbf{F} = yz\mathbf{i} + xz\mathbf{j} + xy\mathbf{k}$

2. $\mathbf{F} = (y \sin z)\mathbf{i} + (x \sin z)\mathbf{j} + (xy \cos z)\mathbf{k}$

3. $\mathbf{F} = y\mathbf{i} + (x + z)\mathbf{j} - y\mathbf{k}$

4. $\mathbf{F} = -y\mathbf{i} + x\mathbf{j}$

5. $\mathbf{F} = (z + y)\mathbf{i} + z\mathbf{j} + (y + x)\mathbf{k}$

6. $\mathbf{F} = (e^x \cos y)\mathbf{i} - (e^x \sin y)\mathbf{j} + z\mathbf{k}$

In Exercises 7–12, find a potential function f for the field $\mathbf{F}$.

7. $\mathbf{F} = 2x\mathbf{i} + 3y\mathbf{j} + 4z\mathbf{k}$

8. $\mathbf{F} = (y + z)\mathbf{i} + (x + z)\mathbf{j} + (x + y)\mathbf{k}$

9. $\mathbf{F} = e^{y+2z}(\mathbf{i} + x\mathbf{j} + 2x\mathbf{k})$

10. $\mathbf{F} = (y \sin z)\mathbf{i} + (x \sin z)\mathbf{j} + (xy \cos z)\mathbf{k}$

11. $\mathbf{F} = (\ln x + \sec^2(x + y))\mathbf{i} +$

$$\left(\sec^2(x + y) + \frac{y}{y^2 + z^2}\right)\mathbf{j} + \frac{z}{y^2 + z^2}\mathbf{k}$$

12. $\mathbf{F} = \dfrac{y}{1 + x^2 y^2}\mathbf{i} + \left(\dfrac{x}{1 + x^2 y^2} + \dfrac{z}{\sqrt{1 - y^2 z^2}}\right)\mathbf{j} +$

$$\left(\frac{y}{\sqrt{1 - y^2 z^2}} + \frac{1}{z}\right)\mathbf{k}$$

In Exercises 13–17, show that the differential forms in the integrals are exact. Then evaluate the integrals.

13. $\displaystyle\int_{(0,0,0)}^{(2,3,-6)} 2x\,dx + 2y\,dy + 2z\,dz$

14. $\displaystyle\int_{(1,1,2)}^{(3,5,0)} yz\,dx + xz\,dy + xy\,dz$

15. $\displaystyle\int_{(0,0,0)}^{(1,2,3)} 2xy\,dx + (x^2 - z^2)\,dy - 2yz\,dz$

16. $\displaystyle\int_{(0,0,0)}^{(3,3,1)} 2x\,dx - y^2\,dy - \frac{4}{1 + z^2}\,dz$

17. $\displaystyle\int_{(1,0,0)}^{(0,1,1)} \sin y \cos x\,dx + \cos y \sin x\,dy + dz$

Although they are not defined on all of space R^3, the fields associated with Exercises 18–22 are simply connected and the Component Test can be used to show they are conservative. Find a potential function for each field and evaluate the integrals as in Example 5.

18. $\displaystyle\int_{(0,2,1)}^{(1,\pi/2,2)} 2 \cos y\,dx + \left(\frac{1}{y} - 2x \sin y\right) dy + \frac{1}{z}\,dz$

19. $\displaystyle\int_{(1,1,1)}^{(1,2,3)} 3x^2\,dx + \frac{z^2}{y}\,dy + 2z \ln y\,dz$

20. $\displaystyle\int_{(1,2,1)}^{(2,1,1)} (2x \ln y - yz)\,dx + \left(\frac{x^2}{y} - xz\right) dy - xy\,dz$

21. $\displaystyle\int_{(1,1,1)}^{(2,2,2)} \frac{1}{y}\,dx + \left(\frac{1}{z} - \frac{x}{y^2}\right) dy - \frac{y}{z^2}\,dz$

22. $\displaystyle\int_{(-1,-1,-1)}^{(2,2,2)} \frac{2x\,dx + 2y\,dy + 2z\,dz}{x^2 + y^2 + z^2}$

23. Revisiting Example 5   Evaluate the integral

$$\int_{(1,1,1)}^{(2,3,-1)} y\,dx + x\,dy + 4\,dz$$

from Example 5 by finding parametric equations for the line segment from $(1, 1, 1)$ to $(2, 3, -1)$ and evaluating the line integral of $\mathbf{F} = y\mathbf{i} + x\mathbf{j} + 4\mathbf{k}$ along the segment. Since $\mathbf{F}$ is conservative, the integral is independent of the path.

24. Evaluate

$$\int_C x^2\,dx + yz\,dy + (y^2/2)\,dz$$

along the line segment C joining $(0, 0, 0)$ to $(0, 3, 4)$.

Independence of path   Show that the values of the integrals in Exercises 25 and 26 do not depend on the path taken from A to B.

25. $\displaystyle\int_A^B z^2\,dx + 2y\,dy + 2xz\,dz$

26. $\int_A^B \dfrac{x\,dx + y\,dy + z\,dz}{\sqrt{x^2 + y^2 + z^2}}$

In Exercises 27 and 28, find a potential function for **F**.

27. $\mathbf{F} = \dfrac{2x}{y}\mathbf{i} + \left(\dfrac{1 - x^2}{y^2}\right)\mathbf{j}, \qquad \{(x, y): y > 0\}$

28. $\mathbf{F} = (e^x \ln y)\mathbf{i} + \left(\dfrac{e^x}{y} + \sin z\right)\mathbf{j} + (y \cos z)\mathbf{k}$

29. Work along different paths Find the work done by $\mathbf{F} = (x^2 + y)\mathbf{i} + (y^2 + x)\mathbf{j} + ze^z\mathbf{k}$ over the following paths from $(1, 0, 0)$ to $(1, 0, 1)$.

a. The line segment $x = 1, y = 0, 0 \le z \le 1$

b. The helix $\mathbf{r}(t) = (\cos t)\mathbf{i} + (\sin t)\mathbf{j} + (t/2\pi)\mathbf{k}, 0 \le t \le 2\pi$

c. The x-axis from $(1, 0, 0)$ to $(0, 0, 0)$ followed by the parabola $z = x^2, y = 0$ from $(0, 0, 0)$ to $(1, 0, 1)$

30. Work along different paths Find the work done by $\mathbf{F} = e^{yz}\mathbf{i} + (xze^{yz} + z \cos y)\mathbf{j} + (xye^{yz} + \sin y)\mathbf{k}$ over the following paths from $(1, 0, 1)$ to $(1, \pi/2, 0)$.

a. The line segment $x = 1, y = \pi t/2, z = 1 - t, 0 \le t \le 1$

b. The line segment from $(1, 0, 1)$ to the origin followed by the line segment from the origin to $(1, \pi/2, 0)$

c. The line segment from $(1, 0, 1)$ to $(1, 0, 0)$, followed by the x-axis from $(1, 0, 0)$ to the origin, followed by the parabola $y = \pi x^2/2, z = 0$ from there to $(1, \pi/2, 0)$

31. Evaluating a work integral two ways Let $\mathbf{F} = \nabla(x^3 y^2)$ and let C be the path in the xy-plane from $(-1, 1)$ to $(1, 1)$ that consists of the line segment from $(-1, 1)$ to $(0, 0)$ followed by the line segment from $(0, 0)$ to $(1, 1)$. Evaluate $\int_C \mathbf{F} \cdot d\mathbf{r}$ in two ways.

a. Find parametrizations for the segments that make up C and evaluate the integral.

b. Using $f(x, y) = x^3 y^2$ as a potential function for **F**.

32. Integral along different paths Evaluate $\int_C 2x \cos y \, dx - x^2 \sin y \, dy$ along the following paths C in the xy-plane.

a. The parabola $y = (x - 1)^2$ from $(1, 0)$ to $(0, 1)$

b. The line segment from $(-1, \pi)$ to $(1, 0)$

c. The x-axis from $(-1, 0)$ to $(1, 0)$

d. The astroid $\mathbf{r}(t) = (\cos^3 t)\mathbf{i} + (\sin^3 t)\mathbf{j}, 0 \le t \le 2\pi$, counterclockwise from $(1, 0)$ back to $(1, 0)$

33. a. Exact differential form How are the constants a, b, and c related if the following differential form is exact?

$$(ay^2 + 2czx)\,dx + y(bx + cz)\,dy + (ay^2 + cx^2)\,dz$$

b. **Gradient field** For what values of b and c will

$$\mathbf{F} = (y^2 + 2czx)\mathbf{i} + y(bx + cz)\mathbf{j} + (y^2 + cx^2)\mathbf{k}$$

be a gradient field?

34. Gradient of a line integral Suppose that $\mathbf{F} = \nabla f$ is a conservative vector field and

$$g(x, y, z) = \int_{(0,0,0)}^{(x,y,z)} \mathbf{F} \cdot d\mathbf{r}.$$

Show that $\nabla g = \mathbf{F}$.

35. Path of least work You have been asked to find the path along which a force field **F** will perform the least work in moving a particle between two locations. A quick calculation on your part shows **F** to be conservative. How should you respond? Give reasons for your answer.

36. A revealing experiment By experiment, you find that a force field **F** performs only half as much work in moving an object along path C_1 from A to B as it does in moving the object along path C_2 from A to B. What can you conclude about **F**? Give reasons for your answer.

37. Work by a constant force Show that the work done by a constant force field $\mathbf{F} = a\mathbf{i} + b\mathbf{j} + c\mathbf{k}$ in moving a particle along any path from A to B is $W = \mathbf{F} \cdot \overrightarrow{AB}$.

38. Gravitational field

a. Find a potential function for the gravitational field

$$\mathbf{F} = -GmM\frac{x\mathbf{i} + y\mathbf{j} + z\mathbf{k}}{(x^2 + y^2 + z^2)^{3/2}} \qquad (G, m, \text{ and } M \text{ are constants}).$$

b. Let P_1 and P_2 be points at distance s_1 and s_2 from the origin. Show that the work done by the gravitational field in part (a) in moving a particle from P_1 to P_2 is

$$GmM\left(\frac{1}{s_2} - \frac{1}{s_1}\right).$$

14.4 Green's Theorem in the Plane

If **F** is a conservative field, then we know $\mathbf{F} = \nabla f$ for a differentiable function f, and we can calculate the work integral over any path C joining point A to B as $\int_C \mathbf{F} \cdot d\mathbf{r} = f(B) - f(A)$. In this section we derive a method for computing a work or flux integral over a *closed* curve C in the plane when the field **F** is *not* conservative. This method, known as Green's Theorem, allows us to convert the line integral into a double integral over the region enclosed by C.

The discussion is given in terms of velocity fields of fluid flows because they are easy to visualize. However, Green's Theorem applies to any vector field, independent of any particular interpretation of the field, provided the assumptions of the theorem are satisfied.

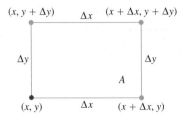

FIGURE 14.24 The rectangle for defining the divergence (flux density) of a vector field at a point (x, y).

Divergence

We need two new ideas for Green's Theorem. The first is the idea of the *divergence* of a vector field at a point, sometimes called the *flux density* of the vector field by physicists and engineers. We obtain it in the following way.

Suppose that $\mathbf{F}(x, y) = M(x, y)\mathbf{i} + N(x, y)\mathbf{j}$ is the velocity field of a fluid flow in the plane and that the first partial derivatives of M and N are continuous at each point of a region R. Let (x, y) be a point in R and let A be a small rectangle with one corner at (x, y) that, along with its interior, lies entirely in R (Figure 14.24). The sides of the rectangle, parallel to the coordinate axes, have lengths of Δx and Δy. The rate at which fluid leaves the rectangle across the bottom edge is approximately

$$\mathbf{F}(x, y) \cdot (-\mathbf{j}) \, \Delta x = -N(x, y)\Delta x.$$

This is the scalar component of the velocity at (x, y) in the direction of the outward normal times the length of the segment. If the velocity is in meters per second, for example, the exit rate will be in meters per second times meters or square meters per second. The rates at which the fluid crosses the other three sides in the directions of their outward normals can be estimated in a similar way. All told, we have

Exit Rates: Top: $\quad \mathbf{F}(x, y + \Delta y) \cdot \mathbf{j} \, \Delta x = N(x, y + \Delta y)\Delta x$

Bottom: $\quad \mathbf{F}(x, y) \cdot (-\mathbf{j}) \, \Delta x = -N(x, y)\Delta x$

Right: $\quad \mathbf{F}(x + \Delta x, y) \cdot \mathbf{i} \, \Delta y = M(x + \Delta x, y)\Delta y$

Left: $\quad \mathbf{F}(x, y) \cdot (-\mathbf{i}) \, \Delta y = -M(x, y)\Delta y.$

Combining opposite pairs gives

Top and bottom: $\quad (N(x, y + \Delta y) - N(x, y))\Delta x \approx \left(\dfrac{\partial N}{\partial y} \Delta y \right)\Delta x$

Right and left: $\quad (M(x + \Delta x, y) - M(x, y))\Delta y \approx \left(\dfrac{\partial M}{\partial x} \Delta x \right)\Delta y.$

Adding these last two equations gives

Flux across rectangle boundary $\approx \left(\dfrac{\partial M}{\partial x} + \dfrac{\partial N}{\partial y} \right)\Delta x \Delta y.$

We now divide by $\Delta x \Delta y$ to estimate the total flux per unit area or flux density for the rectangle:

$$\frac{\text{Flux across rectangle boundary}}{\text{rectangle area}} \approx \left(\frac{\partial M}{\partial x} + \frac{\partial N}{\partial y} \right).$$

Finally, we let Δx and Δy approach zero to define the *flux density* of $\mathbf{F}$ at the point (x, y). In mathematics, we call the flux density the *divergence* of $\mathbf{F}$. The symbol for it is div $\mathbf{F}$, pronounced "divergence of $\mathbf{F}$" or "div $\mathbf{F}$."

DEFINITION The **divergence (flux density)** of a vector field $\mathbf{F} = M\mathbf{i} + N\mathbf{j}$ at the point (x, y) is

$$\text{div } \mathbf{F} = \frac{\partial M}{\partial x} + \frac{\partial N}{\partial y}. \tag{1}$$

Source:

$$\text{div } \mathbf{F}\,(x_0, y_0) > 0$$

A gas expanding at the point (x_0, y_0).

Sink:

$$\text{div } \mathbf{F}\,(x_0, y_0) < 0$$

A gas compressing at the point (x_0, y_0).

FIGURE 14.25 If a gas is expanding at a point (x_0, y_0), the lines of flow have positive divergence; if the gas is compressing, the divergence is negative.

Intuitively, if a gas is expanding at the point (x_0, y_0), the lines of flow would diverge there (hence the name) and, since the gas would be flowing out of a small rectangle about (x_0, y_0) the divergence of $\mathbf{F}$ at (x_0, y_0) would be positive. If the gas were compressing instead of expanding, the divergence would be negative (Figure 14.25).

EXAMPLE 1 Find the divergence of $\mathbf{F}(x, y) = (x^2 - y)\mathbf{i} + (xy - y^2)\mathbf{j}$.

Solution We use the formula in Equation (1):

$$\text{div } \mathbf{F} = \frac{\partial M}{\partial x} + \frac{\partial N}{\partial y} = \frac{\partial}{\partial x}(x^2 - y) + \frac{\partial}{\partial y}(xy - y^2)$$

$$= 2x + x - 2y = 3x - 2y. \qquad \blacksquare$$

Spin Around an Axis: The k-Component of Curl

The second idea we need for Green's Theorem has to do with measuring how a paddle wheel spins at a point in a fluid flowing in a plane region. This idea gives some sense of how the fluid is circulating around axes located at different points and perpendicular to the region. Physicists sometimes refer to this as the *circulation density* of a vector field $\mathbf{F}$ at a point. To obtain it, we return to the velocity field

$$\mathbf{F}(x, y) = M(x, y)\mathbf{i} + N(x, y)\mathbf{j}$$

and the rectangle A. The rectangle is redrawn here as Figure 14.26.

The counterclockwise circulation of $\mathbf{F}$ around the boundary of A is the sum of flow rates along the sides. For the bottom edge, the flow rate is approximately

$$\mathbf{F}(x, y) \cdot \mathbf{i}\,\Delta x = M(x, y)\Delta x.$$

This is the scalar component of the velocity $\mathbf{F}(x, y)$ in the direction of the tangent vector $\mathbf{i}$ times the length of the segment. The rates of flow along the other sides in the counterclockwise direction are expressed in a similar way. In all, we have

Top:	$\mathbf{F}(x, y + \Delta y) \cdot (-\mathbf{i})\,\Delta x = -M(x, y + \Delta y)\Delta x$
Bottom:	$\mathbf{F}(x, y) \cdot \mathbf{i}\,\Delta x = M(x, y)\Delta x$
Right:	$\mathbf{F}(x + \Delta x, y) \cdot \mathbf{j}\,\Delta y = N(x + \Delta x, y)\Delta y$
Left:	$\mathbf{F}(x, y) \cdot (-\mathbf{j})\,\Delta y = -N(x, y)\Delta y.$

We add opposite pairs to get

Top and bottom:

$$-(M(x, y + \Delta y) - M(x, y))\Delta x \approx -\left(\frac{\partial M}{\partial y}\Delta y\right)\Delta x$$

Right and left:

$$(N(x + \Delta x, y) - N(x, y))\Delta y \approx \left(\frac{\partial N}{\partial x}\Delta x\right)\Delta y.$$

Adding these last two equations and dividing by $\Delta x \Delta y$ gives an estimate of the circulation density for the rectangle:

$$\frac{\text{Circulation around rectangle}}{\text{rectangle area}} \approx \frac{\partial N}{\partial x} - \frac{\partial M}{\partial y}.$$

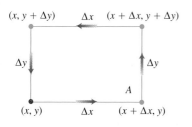

$(x, y + \Delta y)$ Δx $(x + \Delta x, y + \Delta y)$

Δy Δy

A

(x, y) Δx $(x + \Delta x, y)$

FIGURE 14.26 The rectangle for defining the curl (circulation density) of a vector field at a point (x, y).

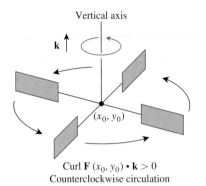

Curl $\mathbf{F}(x_0, y_0) \cdot \mathbf{k} > 0$
Counterclockwise circulation

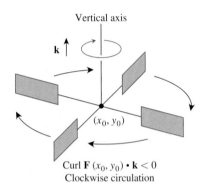

Curl $\mathbf{F}(x_0, y_0) \cdot \mathbf{k} < 0$
Clockwise circulation

FIGURE 14.27 In the flow of an incompressible fluid over a plane region, the **k**-component of the curl measures the rate of the fluid's rotation at a point. The **k**-component of the curl is positive at points where the rotation is counterclockwise and negative where the rotation is clockwise.

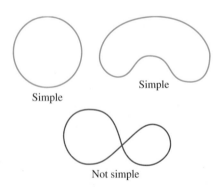

FIGURE 14.28 In proving Green's Theorem, we distinguish between two kinds of closed curves, simple and not simple. Simple curves do not cross themselves. A circle is simple but a figure 8 is not.

We let Δx and Δy approach zero to define the *circulation density* of **F** at the point (x, y).

The positive orientation of the circulation density for the plane is the *counterclockwise* rotation around the vertical axis, looking downward on the *xy*-plane from the tip of the (vertical) unit vector **k** (Figure 14.27). The circulation value is actually the **k**-component of a more general circulation vector field we define in Section 14.7, called the *curl* of the vector field **F**. For Green's Theorem, we need only this **k**-component.

> **DEFINITION** The **k-component of the curl (circulation density)** of a vector field $\mathbf{F} = M\mathbf{i} + N\mathbf{j}$ at the point (x, y) is the scalar
>
> $$(\operatorname{curl} \mathbf{F}) \cdot \mathbf{k} = \frac{\partial N}{\partial x} - \frac{\partial M}{\partial y}. \qquad (2)$$

If water is moving about a region in the *xy*-plane in a thin layer, then the **k**-component of the circulation, or curl, at a point (x_0, y_0) gives a way to measure how fast and in what direction a small paddle wheel will spin if it is put into the water at (x_0, y_0) with its axis perpendicular to the plane, parallel to **k** (Figure 14.27).

EXAMPLE 2 Find the **k**-component of the curl for the vector field

$$\mathbf{F}(x, y) = (x^2 - y)\mathbf{i} + (xy - y^2)\mathbf{j}.$$

Solution We use the formula in Equation (2):

$$(\operatorname{curl} \mathbf{F}) \cdot \mathbf{k} = \frac{\partial N}{\partial x} - \frac{\partial M}{\partial y} = \frac{\partial}{\partial x}(xy - y^2) - \frac{\partial}{\partial y}(x^2 - y) = y + 1. \qquad \blacksquare$$

Two Forms for Green's Theorem

In one form, Green's Theorem says that under suitable conditions the outward flux of a vector field across a simple closed curve in the plane (Figure 14.28) equals the double integral of the divergence of the field over the region enclosed by the curve. Recall the formulas for flux in Equations (3) and (4) in Section 14.2.

> **THEOREM 4—Green's Theorem (Flux-Divergence or Normal Form)** Let C be a piecewise smooth, simple closed curve enclosing a region R in the plane. Let $\mathbf{F} = M\mathbf{i} + N\mathbf{j}$ be a vector field with M and N having continuous first partial derivatives in an open region containing R. Then the outward flux of **F** across C equals the double integral of div **F** over the region R enclosed by C.
>
> $$\underbrace{\oint_C \mathbf{F} \cdot \mathbf{n} \, ds = \oint_C M \, dy - N \, dx}_{\text{Outward flux}} = \underbrace{\iint_R \left(\frac{\partial M}{\partial x} + \frac{\partial N}{\partial y} \right) dx \, dy}_{\text{Divergence integral}} \qquad (3)$$

In another form, Green's Theorem says that the counterclockwise circulation of a vector field around a simple closed curve is the double integral of the **k**-component of the curl of the field over the region enclosed by the curve. Recall the defining Equation (2) for circulation in Section 14.2.

THEOREM 5—Green's Theorem (Circulation-Curl or Tangential Form)   Let C be a piecewise smooth, simple closed curve enclosing a region R in the plane. Let $\mathbf{F} = M\mathbf{i} + N\mathbf{j}$ be a vector field with M and N having continuous first partial derivatives in an open region containing R. Then the counterclockwise circulation of $\mathbf{F}$ around C equals the double integral of $(\text{curl } \mathbf{F}) \cdot \mathbf{k}$ over R.

$$\underbrace{\oint_C \mathbf{F} \cdot \mathbf{T} \, ds = \oint_C M \, dx + N \, dy}_{\text{Counterclockwise circulation}} = \underbrace{\iint_R \left(\frac{\partial N}{\partial x} - \frac{\partial M}{\partial y} \right) dx \, dy}_{\text{Curl integral}} \qquad (4)$$

The two forms of Green's Theorem are equivalent. Applying Equation (3) to the field $\mathbf{G}_1 = N\mathbf{i} - M\mathbf{j}$ gives Equation (4), and applying Equation (4) to $\mathbf{G}_2 = -N\mathbf{i} + M\mathbf{j}$ gives Equation (3).

EXAMPLE 3   Verify both forms of Green's Theorem for the field

$$\mathbf{F}(x, y) = (x - y)\mathbf{i} + x\mathbf{j}$$

and the region R bounded by the unit circle

$$C: \quad \mathbf{r}(t) = (\cos t)\mathbf{i} + (\sin t)\mathbf{j}, \qquad 0 \le t \le 2\pi.$$

Solution   We have

$$M = \cos t - \sin t, \qquad dx = d(\cos t) = -\sin t \, dt,$$

$$N = \cos t, \qquad dy = d(\sin t) = \cos t \, dt,$$

$$\frac{\partial M}{\partial x} = 1, \qquad \frac{\partial M}{\partial y} = -1, \qquad \frac{\partial N}{\partial x} = 1, \qquad \frac{\partial N}{\partial y} = 0.$$

The two sides of Equation (3) are

$$\oint_C M \, dy - N \, dx = \int_{t=0}^{t=2\pi} (\cos t - \sin t)(\cos t \, dt) - (\cos t)(-\sin t \, dt)$$

$$= \int_0^{2\pi} \cos^2 t \, dt = \pi$$

$$\iint_R \left(\frac{\partial M}{\partial x} + \frac{\partial N}{\partial y} \right) dx \, dy = \iint_R (1 + 0) \, dx \, dy$$

$$= \iint_R dx \, dy = \text{area inside the unit circle} = \pi.$$

The two sides of Equation (4) are

$$\oint_C M \, dx + N \, dy = \int_{t=0}^{t=2\pi} (\cos t - \sin t)(-\sin t \, dt) + (\cos t)(\cos t \, dt)$$

$$= \int_0^{2\pi} (-\sin t \cos t + 1) \, dt = 2\pi$$

$$\iint_R \left(\frac{\partial N}{\partial x} - \frac{\partial M}{\partial y} \right) dx \, dy = \iint_R (1 - (-1)) \, dx \, dy = 2 \iint_R dx \, dy = 2\pi. \qquad \blacksquare$$

Using Green's Theorem to Evaluate Line Integrals

If we construct a closed curve C by piecing a number of different curves end to end, the process of evaluating a line integral over C can be lengthy because there are so many different integrals to evaluate. If C bounds a region R to which Green's Theorem applies, however, we can use Green's Theorem to change the line integral around C into one double integral over R.

EXAMPLE 4   Evaluate the integral

$$\oint_C xy \, dy - y^2 \, dx,$$

where C is the square cut from the first quadrant by the lines $x = 1$ and $y = 1$.

Solution   We can use either form of Green's Theorem to change the line integral into a double integral over the square.

1. *With the Normal Form Equation* (3): Taking $M = xy, N = y^2$, and C and R as the square's boundary and interior gives

$$\oint_C xy \, dy - y^2 \, dx = \iint_R (y + 2y) \, dx \, dy = \int_0^1 \int_0^1 3y \, dx \, dy$$

$$= \int_0^1 \left[3xy \right]_{x=0}^{x=1} dy = \int_0^1 3y \, dy = \frac{3}{2} y^2 \Big]_0^1 = \frac{3}{2}.$$

2. *With the Tangential Form Equation* (4): Taking $M = -y^2$ and $N = xy$ gives the same result:

$$\oint_C -y^2 \, dx + xy \, dy = \iint_R (y - (-2y)) \, dx \, dy = \frac{3}{2}. \qquad \blacksquare$$

EXAMPLE 5   Calculate the outward flux of the field $\mathbf{F}(x, y) = x\mathbf{i} + y^2\mathbf{j}$ across the square bounded by the lines $x = \pm 1$ and $y = \pm 1$.

Solution   Calculating the flux with a line integral would take four integrations, one for each side of the square. With Green's Theorem, we can change the line integral to one double integral. With $M = x$, $N = y^2$, C the square, and R the square's interior, we have

$$\text{Flux} = \oint_C \mathbf{F} \cdot \mathbf{n}\, ds = \oint_C M\, dy - N\, dx$$

$$= \iint_R \left(\frac{\partial M}{\partial x} + \frac{\partial N}{\partial y} \right) dx\, dy \qquad \text{Green's Theorem}$$

$$= \int_{-1}^{1} \int_{-1}^{1} (1 + 2y)\, dx\, dy = \int_{-1}^{1} \left[x + 2xy \right]_{x=-1}^{x=1} dy$$

$$= \int_{-1}^{1} (2 + 4y)\, dy = \left[2y + 2y^2 \right]_{-1}^{1} = 4. \qquad \blacksquare$$

Proof of Green's Theorem for Special Regions

Let C be a smooth simple closed curve in the xy-plane with the property that lines parallel to the axes cut it in no more than two points. Let R be the region enclosed by C and suppose that M, N, and their first partial derivatives are continuous at every point of some open region containing C and R. We want to prove the circulation-curl form of Green's Theorem,

$$\oint_C M\, dx + N\, dy = \iint_R \left(\frac{\partial N}{\partial x} - \frac{\partial M}{\partial y} \right) dx\, dy. \tag{5}$$

Figure 14.29 shows C made up of two directed parts:

$$C_1: \quad y = f_1(x), \quad a \le x \le b, \qquad C_2: \quad y = f_2(x), \quad b \ge x \ge a.$$

For any x between a and b, we can integrate $\partial M/\partial y$ with respect to y from $y = f_1(x)$ to $y = f_2(x)$ and obtain

$$\int_{f_1(x)}^{f_2(x)} \frac{\partial M}{\partial y}\, dy = M(x, y) \Big]_{y=f_1(x)}^{y=f_2(x)} = M(x, f_2(x)) - M(x, f_1(x)).$$

We can then integrate this with respect to x from a to b:

$$\int_a^b \int_{f_1(x)}^{f_2(x)} \frac{\partial M}{\partial y}\, dy\, dx = \int_a^b [M(x, f_2(x)) - M(x, f_1(x))]\, dx$$

$$= -\int_b^a M(x, f_2(x))\, dx - \int_a^b M(x, f_1(x))\, dx$$

$$= -\int_{C_2} M\, dx - \int_{C_1} M\, dx$$

$$= -\oint_C M\, dx.$$

Therefore

$$\oint_C M\, dx = \iint_R \left(-\frac{\partial M}{\partial y} \right) dx\, dy. \tag{6}$$

Equation (6) is half the result we need for Equation (5). We derive the other half by integrating $\partial N/\partial x$ first with respect to x and then with respect to y, as suggested by Figure 14.30. This shows the curve C of Figure 14.29 decomposed into the two directed

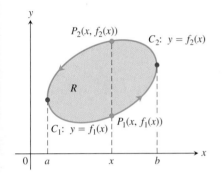

FIGURE 14.29 The boundary curve C is made up of C_1, the graph of $y = f_1(x)$, and C_2, the graph of $y = f_2(x)$.

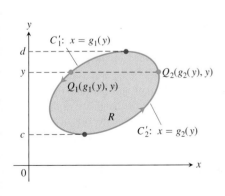

FIGURE 14.30 The boundary curve C is made up of C'_1, the graph of $x = g_1(y)$, and C'_2, the graph of $x = g_2(y)$.

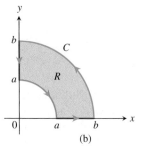

FIGURE 14.31 Other regions to which Green's Theorem applies.

parts $C_1': x = g_1(y), \; d \geq y \geq c$ and $C_2': x = g_2(y), c \leq y \leq d$. The result of this double integration is

$$\oint_C N \, dy = \iint_R \frac{\partial N}{\partial x} \, dx \, dy. \tag{7}$$

Summing Equations (6) and (7) gives Equation (5). This concludes the proof. ∎

Green's Theorem also holds for more general regions, such as those shown in Figures 14.31 and 14.32, but we will not prove this result here. Notice that the region in Figure 14.32 is not simply connected. The curves on its boundary are oriented so that the region R is always on the left-hand side as the curves are traversed in the directions shown. With this convention, Green's Theorem is valid for regions that are not simply connected.

While we stated the theorem in the xy-plane, Green's Theorem does apply for any region R contained in a plane bounded by a curve C in space. We will see how to express the double integral over R for this more general form of Green's Theorem in Section 14.7.

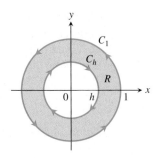

FIGURE 14.32 Green's Theorem may be applied to the annular region R by integrating along the boundaries as shown.

EXERCISES 14.4

In Exercises 1–4, verify the conclusion of Green's Theorem by evaluating both sides of Equations (3) and (4) for the field $\mathbf{F} = M\mathbf{i} + N\mathbf{j}$. Take the domains of integration in each case to be the disk $R: x^2 + y^2 \leq a^2$ and its bounding circle $C: \mathbf{r} = (a \cos t)\mathbf{i} + (a \sin t)\mathbf{j}, 0 \leq t \leq 2\pi$.

1. $\mathbf{F} = -y\mathbf{i} + x\mathbf{j}$ **2.** $\mathbf{F} = y\mathbf{i}$

3. $\mathbf{F} = 2x\mathbf{i} - 3y\mathbf{j}$ **4.** $\mathbf{F} = -x^2y\mathbf{i} + xy^2\mathbf{j}$

In Exercises 5–10, use Green's Theorem to find the counterclockwise circulation and outward flux for the field $\mathbf{F}$ and curve C.

5. $\mathbf{F} = (x - y)\mathbf{i} + (y - x)\mathbf{j}$
   C: The square bounded by $x = 0, x = 1, y = 0, y = 1$

6. $\mathbf{F} = (x^2 + 4y)\mathbf{i} + (x + y^2)\mathbf{j}$
   C: The square bounded by $x = 0, x = 1, y = 0, y = 1$

7. $\mathbf{F} = (y^2 - x^2)\mathbf{i} + (x^2 + y^2)\mathbf{j}$
   C: The triangle bounded by $y = 0, x = 3$, and $y = x$

8. $\mathbf{F} = (x + y)\mathbf{i} - (x^2 + y^2)\mathbf{j}$
   C: The triangle bounded by $y = 0, x = 1$, and $y = x$

9. $\mathbf{F} = (x + e^x \sin y)\mathbf{i} + (x + e^x \cos y)\mathbf{j}$
   C: The right-hand loop of the lemniscate $r^2 = \cos 2\theta$

10. $\mathbf{F} = \left(\tan^{-1} \dfrac{y}{x} \right)\mathbf{i} + \ln (x^2 + y^2)\mathbf{j}$
   C: The boundary of the region defined by the polar coordinate inequalities $1 \leq r \leq 2, 0 \leq \theta \leq \pi$

11. Find the counterclockwise circulation and outward flux of the field $\mathbf{F} = xy\mathbf{i} + y^2\mathbf{j}$ around and over the boundary of the region enclosed by the curves $y = x^2$ and $y = x$ in the first quadrant.

12. Find the counterclockwise circulation and the outward flux of the field $\mathbf{F} = (-\sin y)\mathbf{i} + (x \cos y)\mathbf{j}$ around and over the square cut from the first quadrant by the lines $x = \pi/2$ and $y = \pi/2$.

13. Find the outward flux of the field

$$\mathbf{F} = \left(3xy - \frac{x}{1 + y^2} \right)\mathbf{i} + (e^x + \tan^{-1} y)\mathbf{j}$$

across the cardioid $r = a(1 + \cos \theta), a > 0$.

14. Find the counterclockwise circulation of $\mathbf{F} = (y + e^x \ln y)\mathbf{i} + (e^x/y)\mathbf{j}$ around the boundary of the region that is bounded above by the curve $y = 3 - x^2$ and below by the curve $y = x^4 + 1$.

In Exercises 15 and 16, find the work done by $\mathbf{F}$ in moving a particle once counterclockwise around the given curve.

15. $\mathbf{F} = 2xy^3\mathbf{i} + 4x^2y^2\mathbf{j}$

    C: The boundary of the "triangular" region in the first quadrant enclosed by the x-axis, the line $x = 1$, and the curve $y = x^3$

16. $\mathbf{F} = (4x - 2y)\mathbf{i} + (2x - 4y)\mathbf{j}$

    C: The circle $(x - 2)^2 + (y - 2)^2 = 4$

Apply Green's Theorem to evaluate the integrals in Exercises 17–20.

17. $\oint_C (y^2\, dx + x^2\, dy)$

    C: The triangle bounded by $x = 0, x + y = 1, y = 0$

18. $\oint_C (3y\, dx + 2x\, dy)$

    C: The boundary of $0 \le x \le \pi, 0 \le y \le \sin x$

19. $\oint_C (6y + x)\, dx + (y + 2x)\, dy$

    C: The circle $(x - 2)^2 + (y - 3)^2 = 4$

20. $\oint_C (2x + y^2)\, dx + (2xy + 3y)\, dy$

    C: Any simple closed curve in the plane for which Green's Theorem holds

Calculating Area with Green's Theorem If a simple closed curve C in the plane and the region R it encloses satisfy the hypotheses of Green's Theorem, the area of R is given by

---
Green's Theorem Area Formula

$$\text{Area of } R = \frac{1}{2} \oint_C x\, dy - y\, dx$$

The reason is that by Equation (3), run backward,

$$\text{Area of } R = \iint_R dy\, dx = \iint_R \left(\frac{1}{2} + \frac{1}{2}\right) dy\, dx$$

$$= \oint_C \frac{1}{2}x\, dy - \frac{1}{2}y\, dx.$$

Use the Green's Theorem area formula given above to find the areas of the regions enclosed by the curves in Exercises 21–24.

21. The circle $\mathbf{r}(t) = (a \cos t)\mathbf{i} + (a \sin t)\mathbf{j}, \quad 0 \le t \le 2\pi$

22. The ellipse $\mathbf{r}(t) = (a \cos t)\mathbf{i} + (b \sin t)\mathbf{j}, \quad 0 \le t \le 2\pi$

23. The astroid $\mathbf{r}(t) = (\cos^3 t)\mathbf{i} + (\sin^3 t)\mathbf{j}, \quad 0 \le t \le 2\pi$

24. The curve $\mathbf{r}(t) = t^2\mathbf{i} + ((t^3/3) - t)\mathbf{j}, \quad -\sqrt{3} \le t \le \sqrt{3}$ (see accompanying figure).

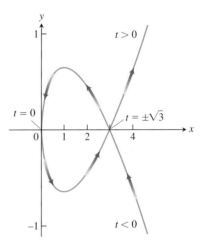

25. Let C be the boundary of a region on which Green's Theorem holds. Use Green's Theorem to calculate

    a. $\oint_C f(x)\, dx + g(y)\, dy$

    b. $\oint_C ky\, dx + hx\, dy \quad (k \text{ and } h \text{ constants}).$

26. Integral dependent only on area Show that the value of

$$\oint_C xy^2\, dx + (x^2y + 2x)\, dy$$

around any square depends only on the area of the square and not on its location in the plane.

27. What is special about the integral

$$\oint_C 4x^3y\, dx + x^4\, dy?$$

Give reasons for your answer.

28. What is special about the integral

$$\oint_C -y^3\, dy + x^3\, dx?$$

Give reasons for your answer.

29. Area as a line integral Show that if R is a region in the plane bounded by a piecewise smooth, simple closed curve C, then

$$\text{Area of } R = \oint_C x\, dy = -\oint_C y\, dx.$$

30. Definite integral as a line integral Suppose that a nonnegative function $y = f(x)$ has a continuous first derivative on $[a, b]$. Let C

be the boundary of the region in the xy-plane that is bounded below by the x-axis, above by the graph of f, and on the sides by the lines $x = a$ and $x = b$. Show that

$$\int_a^b f(x)\, dx = -\oint_C y\, dx.$$

31. Area and the centroid Let A be the area and $\bar{x}$ the x-coordinate of the centroid of a region R that is bounded by a piecewise smooth, simple closed curve C in the xy-plane. Show that

$$\frac{1}{2}\oint_C x^2\, dy = -\oint_C xy\, dx = \frac{1}{3}\oint_C x^2\, dy - xy\, dx = A\bar{x}.$$

32. Moment of inertia Let I_y be the moment of inertia about the y-axis of the region in Exercise 31. Show that

$$\frac{1}{3}\oint_C x^3\, dy = -\oint_C x^2 y\, dx = \frac{1}{4}\oint_C x^3\, dy - x^2 y\, dx = I_y.$$

33. Green's Theorem and Laplace's equation Assuming that all the necessary derivatives exist and are continuous, show that if $f(x, y)$ satisfies the Laplace equation

$$\frac{\partial^2 f}{\partial x^2} + \frac{\partial^2 f}{\partial y^2} = 0,$$

then

$$\oint_C \frac{\partial f}{\partial y}\, dx - \frac{\partial f}{\partial x}\, dy = 0$$

for all closed curves C to which Green's Theorem applies. (The converse is also true: If the line integral is always zero, then f satisfies the Laplace equation.)

34. Maximizing work Among all smooth, simple closed curves in the plane, oriented counterclockwise, find the one along which the work done by

$$\mathbf{F} = \left(\frac{1}{4}x^2 y + \frac{1}{3}y^3\right)\mathbf{i} + x\mathbf{j}$$

is greatest. (*Hint:* Where is $(\text{curl } \mathbf{F}) \cdot \mathbf{k}$ positive?)

35. Regions with many holes Green's Theorem holds for a region R with any finite number of holes as long as the bounding curves are smooth, simple, and closed and we integrate over each component of the boundary in the direction that keeps R on our immediate left as we go along (see accompanying figure).

a. Let $f(x, y) = \ln(x^2 + y^2)$ and let C be the circle $x^2 + y^2 = a^2$. Evaluate the flux integral

$$\oint_C \nabla f \cdot \mathbf{n}\, ds.$$

b. Let K be an arbitrary smooth, simple closed curve in the plane that does not pass through $(0, 0)$. Use Green's Theorem to show that

$$\oint_K \nabla f \cdot \mathbf{n}\, ds$$

has two possible values, depending on whether $(0, 0)$ lies inside K or outside K.

36. Bendixson's criterion The *streamlines* of a planar fluid flow are the smooth curves traced by the fluid's individual particles. The vectors $\mathbf{F} = M(x, y)\mathbf{i} + N(x, y)\mathbf{j}$ of the flow's velocity field are the tangent vectors of the streamlines. Show that if the flow takes place over a simply connected region R (no holes or missing points) and that if $M_x + N_y \neq 0$ throughout R, then none of the streamlines in R is closed. In other words, no particle of fluid ever has a closed trajectory in R. The criterion $M_x + N_y \neq 0$ is called **Bendixson's criterion** for the nonexistence of closed trajectories.

37. Establish Equation (7) to finish the proof of the special case of Green's Theorem.

38. Curl component of conservative fields Can anything be said about the curl component of a conservative two-dimensional vector field? Give reasons for your answer.

COMPUTER EXPLORATIONS

In Exercises 39–42, use a CAS and Green's Theorem to find the counterclockwise circulation of the field $\mathbf{F}$ around the simple closed curve C. Perform the following CAS steps.

a. Plot C in the xy-plane.

b. Determine the integrand $(\partial N/\partial x) - (\partial M/\partial y)$ for the curl form of Green's Theorem.

c. Determine the (double integral) limits of integration from your plot in part (a) and evaluate the curl integral for the circulation.

39. $\mathbf{F} = (2x - y)\mathbf{i} + (x + 3y)\mathbf{j}$, C: The ellipse $x^2 + 4y^2 = 4$

40. $\mathbf{F} = (2x^3 - y^3)\mathbf{i} + (x^3 + y^3)\mathbf{j}$, C: The ellipse $\dfrac{x^2}{4} + \dfrac{y^2}{9} = 1$

41. $\mathbf{F} = x^{-1}e^y\mathbf{i} + (e^y \ln x + 2x)\mathbf{j}$,

   C: The boundary of the region defined by $y = 1 + x^4$ (below) and $y = 2$ (above)

42. $\mathbf{F} = xe^y\mathbf{i} + 4x^2 \ln y\,\mathbf{j}$,

   C: The triangle with vertices $(0, 0)$, $(2, 0)$, and $(0, 4)$

14.5 Surfaces and Area

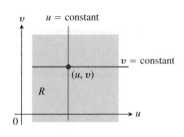

We have defined curves in the plane in three different ways:

Explicit form: $\qquad y = f(x)$

Implicit form: $\qquad F(x, y) = 0$

Parametric vector form: $\quad \mathbf{r}(t) = f(t)\mathbf{i} + g(t)\mathbf{j}, \quad a \le t \le b.$

We have analogous definitions of surfaces in space:

Explicit form: $\quad z = f(x, y)$

Implicit form: $\quad F(x, y, z) = 0.$

There is also a parametric form that gives the position of a point on the surface as a vector function of two variables. We discuss this new form in this section and apply the form to obtain the area of a surface as a double integral.

Parametrizations of Surfaces

Let

$$\mathbf{r}(u, v) = f(u, v)\mathbf{i} + g(u, v)\mathbf{j} + h(u, v)\mathbf{k} \qquad (1)$$

be a continuous vector function that is defined on a region R in the uv-plane and one-to-one on the interior of R (Figure 14.33). We call the range of $\mathbf{r}$ the **surface** S defined or traced by $\mathbf{r}$. Equation (1) together with the domain R constitute a **parametrization** of the surface. The variables u and v are the **parameters**, and R is the **parameter domain**. To simplify our discussion, we take R to be a rectangle defined by inequalities of the form $a \le u \le b, c \le v \le d$. The requirement that $\mathbf{r}$ be one-to-one on the interior of R ensures that S does not cross itself. Notice that Equation (1) is the vector equivalent of *three* parametric equations:

$$x = f(u, v), \qquad y = g(u, v), \qquad z = h(u, v).$$

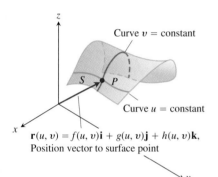

FIGURE 14.33 A parametrized surface S expressed as a vector function of two variables defined on a region R.

EXAMPLE 1 Find a parametrization of the cone

$$z = \sqrt{x^2 + y^2}, \qquad 0 \le z \le 1.$$

Solution Here, cylindrical coordinates provide a parametrization. A typical point (x, y, z) on the cone (Figure 14.34) has $x = r \cos \theta, y = r \sin \theta$, and $z = \sqrt{x^2 + y^2} = r$, with $0 \le r \le 1$ and $0 \le \theta \le 2\pi$. Taking $u = r$ and $v = \theta$ in Equation (1) gives the parametrization

$$\mathbf{r}(r, \theta) = (r \cos \theta)\mathbf{i} + (r \sin \theta)\mathbf{j} + r\mathbf{k}, \qquad 0 \le r \le 1, \quad 0 \le \theta \le 2\pi.$$

The parameterization is one-to-one on the interior of the domain R, though not on its boundary where $r = 0$. ∎

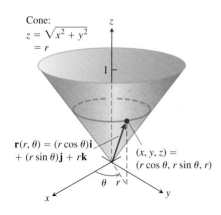

FIGURE 14.34 The cone in Example 1 can be parametrized using cylindrical coordinates.

EXAMPLE 2 Find a parametrization of the sphere $x^2 + y^2 + z^2 = a^2$.

Solution Spherical coordinates provide what we need. A typical point (x, y, z) on the sphere (Figure 14.35) has $x = a \sin \phi \cos \theta$, $y = a \sin \phi \sin \theta$, and $z = a \cos \phi$,

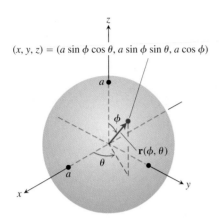

FIGURE 14.35 The sphere in Example 2 can be parametrized using spherical coordinates.

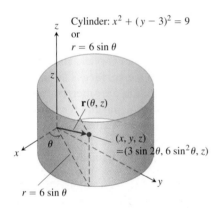

FIGURE 14.36 The cylinder in Example 3 can be parametrized using cylindrical coordinates.

$0 \leq \phi \leq \pi$, $0 \leq \theta \leq 2\pi$. Taking $u = \phi$ and $v = \theta$ in Equation (1) gives the parametrization

$$\mathbf{r}(\phi, \theta) = (a \sin \phi \cos \theta)\mathbf{i} + (a \sin \phi \sin \theta)\mathbf{j} + (a \cos \phi)\mathbf{k},$$
$$0 \leq \phi \leq \pi, \quad 0 \leq \theta \leq 2\pi. \qquad \blacksquare$$

EXAMPLE 3   Find a parametrization of the cylinder

$$x^2 + (y - 3)^2 = 9, \qquad 0 \leq z \leq 5.$$

Solution   In cylindrical coordinates, a point (x, y, z) has $x = r \cos \theta$, $y = r \sin \theta$, and $z = z$. For points on the cylinder $x^2 + (y - 3)^2 = 9$ (Figure 14.36), the equation is the same as the polar equation for the cylinder's base in the xy-plane:

$$x^2 + (y^2 - 6y + 9) = 9$$
$$r^2 - 6r \sin \theta = 0 \qquad \begin{array}{l} x^2 + y^2 = r^2, \\ y = r \sin \theta \end{array}$$

or

$$r = 6 \sin \theta, \qquad 0 \leq \theta \leq \pi.$$

A typical point on the cylinder therefore has

$$x = r \cos \theta = 6 \sin \theta \cos \theta = 3 \sin 2\theta$$
$$y = r \sin \theta = 6 \sin^2 \theta$$
$$z = z.$$

Taking $u = \theta$ and $v = z$ in Equation (1) gives the parametrization

$$\mathbf{r}(\theta, z) = (3 \sin 2\theta)\mathbf{i} + (6 \sin^2 \theta)\mathbf{j} + z\mathbf{k}, \qquad 0 \leq \theta \leq \pi, \quad 0 \leq z \leq 5. \qquad \blacksquare$$

Surface Area

Our goal is to find a double integral for calculating the area of a curved surface S based on the parametrization

$$\mathbf{r}(u, v) = f(u, v)\mathbf{i} + g(u, v)\mathbf{j} + h(u, v)\mathbf{k}, \qquad a \leq u \leq b, \quad c \leq v \leq d.$$

We need S to be smooth for the construction we are about to carry out. The definition of smoothness involves the partial derivatives of $\mathbf{r}$ with respect to u and v:

$$\mathbf{r}_u = \frac{\partial \mathbf{r}}{\partial u} = \frac{\partial f}{\partial u}\mathbf{i} + \frac{\partial g}{\partial u}\mathbf{j} + \frac{\partial h}{\partial u}\mathbf{k}$$

$$\mathbf{r}_v = \frac{\partial \mathbf{r}}{\partial v} = \frac{\partial f}{\partial v}\mathbf{i} + \frac{\partial g}{\partial v}\mathbf{j} + \frac{\partial h}{\partial v}\mathbf{k}.$$

> **DEFINITION**   A parametrized surface $\mathbf{r}(u, v) = f(u, v)\mathbf{i} + g(u, v)\mathbf{j} + h(u, v)\mathbf{k}$ is **smooth** if $\mathbf{r}_u$ and $\mathbf{r}_v$ are continuous and $\mathbf{r}_u \times \mathbf{r}_v$ is never zero on the interior of the parameter domain.

The condition that $\mathbf{r}_u \times \mathbf{r}_v$ is never the zero vector in the definition of smoothness means that the two vectors $\mathbf{r}_u$ and $\mathbf{r}_v$ are nonzero and never lie along the same line, so they always determine a plane tangent to the surface. We relax this condition on the boundary of the domain, but this does not affect the area computations.

Now consider a small rectangle ΔA_{uv} in R with sides on the lines $u = u_0, u = u_0 + \Delta u$, $v = v_0$ and $v = v_0 + \Delta v$ (Figure 14.37). Each side of ΔA_{uv} maps to a curve on the surface S, and together these four curves bound a "curved area element" $\Delta \sigma_{uv}$. In the notation of the figure, the side $v = v_0$ maps to curve C_1, the side $u = u_0$ maps to C_2, and their common vertex (u_0, v_0) maps to P_0.

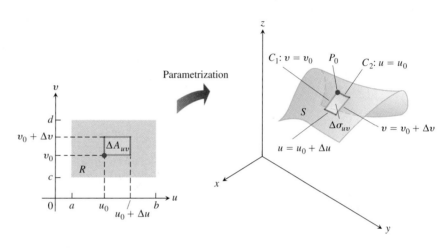

FIGURE 14.37 A rectangular area element ΔA_{uv} in the uv-plane maps onto a curved area element $\Delta \sigma_{uv}$ on S.

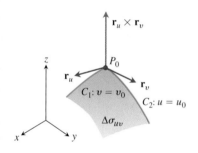

FIGURE 14.38 A magnified view of a surface area element $\Delta \sigma_{uv}$.

Figure 14.38 shows an enlarged view of $\Delta \sigma_{uv}$. The vector $\mathbf{r}_u(u_0, v_0)$ is tangent to C_1 at P_0. Likewise, $\mathbf{r}_v(u_0, v_0)$ is tangent to C_2 at P_0. The cross product $\mathbf{r}_u \times \mathbf{r}_v$ is normal to the surface at P_0. (Here is where we begin to use the assumption that S is smooth. We want to be sure that $\mathbf{r}_u \times \mathbf{r}_v \neq \mathbf{0}$.)

We next approximate the surface element $\Delta \sigma_{uv}$ by the parallelogram on the tangent plane whose sides are determined by the vectors $\Delta u \mathbf{r}_u$ and $\Delta v \mathbf{r}_v$ (Figure 14.39). The area of this parallelogram is

$$|\Delta u \mathbf{r}_u \times \Delta v \mathbf{r}_v| = |\mathbf{r}_u \times \mathbf{r}_v| \, \Delta u \, \Delta v. \qquad (2)$$

A partition of the region R in the uv-plane by rectangular regions ΔA_{uv} induces a partition of the surface S into surface area elements $\Delta \sigma_{uv}$. We approximate the area of each surface element $\Delta \sigma_{uv}$ by the parallelogram area in Equation (2) and sum these areas together to obtain an approximation of the area of S:

$$\sum_n |\mathbf{r}_u \times \mathbf{r}_v| \, \Delta u \, \Delta v. \qquad (3)$$

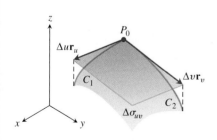

FIGURE 14.39 The parallelogram determined by the vectors $\Delta u \mathbf{r}_u$ and $\Delta v \mathbf{r}_v$ approximates the surface area element $\Delta \sigma_{uv}$.

As Δu and Δv approach zero independently, the number of area elements n tends to ∞ and the continuity of $\mathbf{r}_u$ and $\mathbf{r}_v$ guarantees that the sum in Equation (3) approaches the double integral $\int_c^d \int_a^b |\mathbf{r}_u \times \mathbf{r}_v| \, du \, dv$. This double integral over the region R defines the area of the surface S.

DEFINITION The **area** of the smooth surface

$$\mathbf{r}(u, v) = f(u, v)\mathbf{i} + g(u, v)\mathbf{j} + h(u, v)\mathbf{k}, \qquad a \le u \le b, \quad c \le v \le d$$

is

$$A = \iint\limits_{R} |\mathbf{r}_u \times \mathbf{r}_v| \, dA = \int_c^d \int_a^b |\mathbf{r}_u \times \mathbf{r}_v| \, du \, dv. \tag{4}$$

We can abbreviate the integral in Equation (4) by writing $d\sigma$ for $|\mathbf{r}_u \times \mathbf{r}_v| \, du \, dv$. The surface area differential $d\sigma$ is analogous to the arc length differential ds in Section 11.3.

Surface Area Differential for a Parametrized Surface

$$d\sigma = |\mathbf{r}_u \times \mathbf{r}_v| \, du \, dv \qquad\qquad \iint\limits_{S} d\sigma \tag{5}$$

Surface area   Differential formula
differential   for surface area

EXAMPLE 4 Find the surface area of the cone in Example 1 (Figure 14.34).

Solution In Example 1, we found the parametrization

$$\mathbf{r}(r, \theta) = (r \cos \theta)\mathbf{i} + (r \sin \theta)\mathbf{j} + r\mathbf{k}, \qquad 0 \le r \le 1, \quad 0 \le \theta \le 2\pi.$$

To apply Equation (4), we first find $\mathbf{r}_r \times \mathbf{r}_\theta$:

$$\mathbf{r}_r \times \mathbf{r}_\theta = \begin{vmatrix} \mathbf{i} & \mathbf{j} & \mathbf{k} \\ \cos \theta & \sin \theta & 1 \\ -r \sin \theta & r \cos \theta & 0 \end{vmatrix}$$

$$= -(r \cos \theta)\mathbf{i} - (r \sin \theta)\mathbf{j} + \underbrace{(r \cos^2 \theta + r \sin^2 \theta)}_{r}\mathbf{k}.$$

Thus, $|\mathbf{r}_r \times \mathbf{r}_\theta| = \sqrt{r^2 \cos^2 \theta + r^2 \sin^2 \theta + r^2} = \sqrt{2r^2} = \sqrt{2}\,r$. The area of the cone is

$$A = \int_0^{2\pi} \int_0^1 |\mathbf{r}_r \times \mathbf{r}_\theta| \, dr \, d\theta \qquad \text{Equation (4) with } u = r, v = \theta$$

$$= \int_0^{2\pi} \int_0^1 \sqrt{2}\,r \, dr \, d\theta = \int_0^{2\pi} \frac{\sqrt{2}}{2} \, d\theta = \frac{\sqrt{2}}{2}(2\pi) = \pi \sqrt{2} \text{ units squared.} \quad \blacksquare$$

EXAMPLE 5 Find the surface area of a sphere of radius a.

Solution We use the parametrization from Example 2:

$$\mathbf{r}(\phi, \theta) = (a \sin \phi \cos \theta)\mathbf{i} + (a \sin \phi \sin \theta)\mathbf{j} + (a \cos \phi)\mathbf{k},$$
$$0 \le \phi \le \pi, \quad 0 \le \theta \le 2\pi.$$

For $\mathbf{r}_\phi \times \mathbf{r}_\theta$, we get

$$\mathbf{r}_\phi \times \mathbf{r}_\theta = \begin{vmatrix} \mathbf{i} & \mathbf{j} & \mathbf{k} \\ a \cos \phi \cos \theta & a \cos \phi \sin \theta & -a \sin \phi \\ -a \sin \phi \sin \theta & a \sin \phi \cos \theta & 0 \end{vmatrix}$$

$$= (a^2 \sin^2 \phi \cos \theta)\mathbf{i} + (a^2 \sin^2 \phi \sin \theta)\mathbf{j} + (a^2 \sin \phi \cos \phi)\mathbf{k}.$$

Thus,

$$|\mathbf{r}_\phi \times \mathbf{r}_\theta| = \sqrt{a^4 \sin^4 \phi \cos^2 \theta + a^4 \sin^4 \phi \sin^2 \theta + a^4 \sin^2 \phi \cos^2 \phi}$$

$$= \sqrt{a^4 \sin^4 \phi + a^4 \sin^2 \phi \cos^2 \phi} = \sqrt{a^4 \sin^2 \phi (\sin^2 \phi + \cos^2 \phi)}$$

$$= a^2 \sqrt{\sin^2 \phi} = a^2 \sin \phi,$$

since $\sin \phi \ge 0$ for $0 \le \phi \le \pi$. Therefore, the area of the sphere is

$$A = \int_0^{2\pi} \int_0^\pi a^2 \sin \phi \, d\phi \, d\theta$$

$$= \int_0^{2\pi} \left[-a^2 \cos \phi \right]_0^\pi d\theta = \int_0^{2\pi} 2a^2 \, d\theta = 4\pi a^2 \text{ units squared.}$$

This agrees with the well-known formula for the surface area of a sphere. ∎

Implicit Surfaces

Surfaces are often presented as level sets of a function, described by an equation such as

$$F(x, y, z) = c,$$

for some constant c. Such a level surface does not come with an explicit parametrization, and is called an *implicitly defined surface*. Implicit surfaces arise, for example, as equipotential surfaces in electric or gravitational fields. Figure 14.40 shows a piece of such a surface. It may be difficult to find explicit formulas for the functions f, g, and h that describe the surface in the form $\mathbf{r}(u, v) = f(u, v)\mathbf{i} + g(u, v)\mathbf{j} + h(u, v)\mathbf{k}$. We now show how to compute the surface area differential $d\sigma$ for implicit surfaces.

Figure 14.40 shows a piece of an implicit surface S that lies above its "shadow" region R in the plane beneath it. The surface is defined by the equation $F(x, y, z) = c$ and $\mathbf{p}$ is a unit vector normal to the plane region R. We assume that the surface is **smooth** (F is differentiable and ∇F is nonzero and continuous on S) and that $\nabla F \cdot \mathbf{p} \neq 0$, so the surface never folds back over itself.

Assume that the normal vector $\mathbf{p}$ is the unit vector $\mathbf{k}$, so the region R in Figure 14.40 lies in the xy-plane. By assumption, we then have $\nabla F \cdot \mathbf{p} = \nabla F \cdot \mathbf{k} = F_z \neq 0$ on S. An advanced calculus theorem called the Implicit Function Theorem implies that S is then the graph of a differentiable function $z = h(x, y)$, although the function $h(x, y)$ is not explicitly known. Define the parameters u and v by $u = x$ and $v = y$. Then $z = h(u, v)$ and

$$\mathbf{r}(u, v) = u\mathbf{i} + v\mathbf{j} + h(u, v)\mathbf{k} \qquad (6)$$

Surface $F(x, y, z) = c$

S

$\mathbf{p}$

R

The vertical projection or "shadow" of S on a coordinate plane

FIGURE 14.40 As we soon see, the area of a surface S in space can be calculated by evaluating a related double integral over the vertical projection or "shadow" of S on a coordinate plane. The unit vector $\mathbf{p}$ is normal to the plane.

gives a parametrization of the surface S. We use Equation (4) to find the area of S. Calculating the partial derivatives of $\mathbf{r}$, we find

$$\mathbf{r}_u = \mathbf{i} + \frac{\partial h}{\partial u}\mathbf{k} \qquad \text{and} \qquad \mathbf{r}_v = \mathbf{j} + \frac{\partial h}{\partial v}\mathbf{k}.$$

Applying the chain rule for implicit differentiation (Theorem 8, Section 12.4) to $F(x, y, z) = c$, where $x = u, y = v$, and $z = h(u, v)$, we obtain the partial derivatives

$$\frac{\partial h}{\partial u} = -\frac{F_x}{F_z} \qquad \text{and} \qquad \frac{\partial h}{\partial v} = -\frac{F_y}{F_z}.$$

Substitution of these derivatives into the derivatives of $\mathbf{r}$ gives

$$\mathbf{r}_u = \mathbf{i} - \frac{F_x}{F_z}\mathbf{k} \qquad \text{and} \qquad \mathbf{r}_v = \mathbf{j} - \frac{F_y}{F_z}\mathbf{k}.$$

From a routine calculation of the cross product we find

$$\mathbf{r}_u \times \mathbf{r}_v = \frac{F_x}{F_z}\mathbf{i} + \frac{F_y}{F_z}\mathbf{j} + \mathbf{k} \qquad\qquad F_z \neq 0$$

$$= \frac{1}{F_z}(F_x\mathbf{i} + F_y\mathbf{j} + F_z\mathbf{k})$$

$$= \frac{\nabla F}{F_z} = \frac{\nabla F}{\nabla F \cdot \mathbf{k}}$$

$$= \frac{\nabla F}{\nabla F \cdot \mathbf{p}}. \qquad\qquad \mathbf{p} = \mathbf{k}$$

Therefore, the surface area differential is given by

$$d\sigma = |\mathbf{r}_u \times \mathbf{r}_v|\, du\, dv = \frac{|\nabla F|}{|\nabla F \cdot \mathbf{p}|} dx\, dy. \qquad u = x \text{ and } v = y$$

We obtain similar calculations if instead the vector $\mathbf{p} = \mathbf{j}$ is normal to the xz-plane when $F_y \neq 0$ on S, or if $\mathbf{p} = \mathbf{i}$ is normal to the yz-plane when $F_x \neq 0$ on S. Combining these results with Equation (4) then gives the following general formula.

Formula for Surface Area of an Implicit Surface
The area of the surface $F(x, y, z) = c$ over a closed and bounded plane region R is

$$\text{Surface area} = \iint\limits_R \frac{|\nabla F|}{|\nabla F \cdot \mathbf{p}|}\, dA, \tag{7}$$

where $\mathbf{p}$ is a unit vector normal to R and $\nabla F \cdot \mathbf{p} \neq 0$.

Thus, the area is the double integral over R of the magnitude of ∇F divided by the magnitude of the scalar component of ∇F normal to R.

We reached Equation (7) under the assumption that $\nabla F \cdot \mathbf{p} \neq 0$ throughout R and that ∇F is continuous. Whenever the integral exists, however, we define its value to be the area of the portion of the surface $F(x, y, z) = c$ that lies over R. (Recall that the projection is assumed to be one-to-one.)

EXAMPLE 6 Find the area of the surface cut from the bottom of the paraboloid $x^2 + y^2 - z = 0$ by the plane $z = 4$.

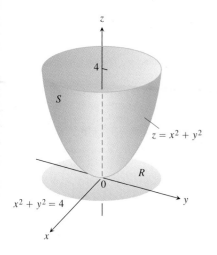

FIGURE 14.41 The area of this parabolic surface is calculated in Example 6.

Solution We sketch the surface S and the region R below it in the xy-plane (Figure 14.41). The surface S is part of the level surface $F(x, y, z) = x^2 + y^2 - z = 0$, and R is the disk $x^2 + y^2 \leq 4$ in the xy-plane. To get a unit vector normal to the plane of R, we can take $\mathbf{p} = \mathbf{k}$.

At any point (x, y, z) on the surface, we have

$$F(x, y, z) = x^2 + y^2 - z$$

$$\nabla F = 2x\mathbf{i} + 2y\mathbf{j} - \mathbf{k}$$

$$|\nabla F| = \sqrt{(2x)^2 + (2y)^2 + (-1)^2}$$

$$= \sqrt{4x^2 + 4y^2 + 1}$$

$$|\nabla F \cdot \mathbf{p}| = |\nabla F \cdot \mathbf{k}| = |-1| = 1.$$

In the region R, $dA = dx\,dy$. Therefore,

$$\text{Surface area} = \iint_R \frac{|\nabla F|}{|\nabla F \cdot \mathbf{p}|} \, dA \qquad \text{Equation (7)}$$

$$= \iint_{x^2+y^2\leq 4} \sqrt{4x^2 + 4y^2 + 1} \, dx\,dy$$

$$= \int_0^{2\pi} \int_0^2 \sqrt{4r^2 + 1}\, r\, dr\, d\theta \qquad \text{Polar coordinates}$$

$$= \int_0^{2\pi} \left[\frac{1}{12}(4r^2 + 1)^{3/2} \right]_0^2 d\theta$$

$$= \int_0^{2\pi} \frac{1}{12}(17^{3/2} - 1)\, d\theta = \frac{\pi}{6}\left(17\sqrt{17} - 1\right). \qquad \blacksquare$$

For a graph $z = f(x, y)$ over a region R in the xy-plane, the surface area formula is

$$A = \iint_R \sqrt{f_x^2 + f_y^2 + 1} \, dx\,dy.$$

(See Exercises 49–55.)

EXERCISES 14.5

In Exercises 1–16, find a parametrization of the surface. (There are many correct ways to do these, so your answers may not be the same as those in the back of the book.)

1. The paraboloid $z = x^2 + y^2, z \leq 4$

2. The paraboloid $z = 9 - x^2 - y^2, z \geq 0$

3. Cone frustum The first-octant portion of the cone $z = \sqrt{x^2 + y^2}/2$ between the planes $z = 0$ and $z = 3$

4. Cone frustum The portion of the cone $z = 2\sqrt{x^2 + y^2}$ between the planes $z = 2$ and $z = 4$

5. Spherical cap The cap cut from the sphere $x^2 + y^2 + z^2 = 9$ by the cone $z = \sqrt{x^2 + y^2}$

6. Spherical cap The portion of the sphere $x^2 + y^2 + z^2 = 4$ in the first octant between the xy-plane and the cone $z = \sqrt{x^2 + y^2}$

7. **Spherical band** The portion of the sphere $x^2 + y^2 + z^2 = 3$ between the planes $z = \sqrt{3}/2$ and $z = -\sqrt{3}/2$

8. **Spherical cap** The upper portion cut from the sphere $x^2 + y^2 + z^2 = 8$ by the plane $z = -2$

9. **Parabolic cylinder between planes** The surface cut from the parabolic cylinder $z = 4 - y^2$ by the planes $x = 0$, $x = 2$, and $z = 0$

10. **Parabolic cylinder between planes** The surface cut from the parabolic cylinder $y = x^2$ by the planes $z = 0, z = 3$ and $y = 2$

11. **Circular cylinder band** The portion of the cylinder $y^2 + z^2 = 9$ between the planes $x = 0$ and $x = 3$

12. **Circular cylinder band** The portion of the cylinder $x^2 + z^2 = 4$ above the xy-plane between the planes $y = -2$ and $y = 2$

13. **Tilted plane inside cylinder** The portion of the plane $x + y + z = 1$

 a. Inside the cylinder $x^2 + y^2 = 9$

 b. Inside the cylinder $y^2 + z^2 = 9$

14. **Tilted plane inside cylinder** The portion of the plane $x - y + 2z = 2$

 a. Inside the cylinder $x^2 + z^2 = 3$

 b. Inside the cylinder $y^2 + z^2 = 2$

15. **Circular cylinder band** The portion of the cylinder $(x - 2)^2 + z^2 = 4$ between the planes $y = 0$ and $y = 3$

16. **Circular cylinder band** The portion of the cylinder $y^2 + (z - 5)^2 = 25$ between the planes $x = 0$ and $x = 10$

In Exercises 17–26, use a parametrization to express the area of the surface as a double integral. Then evaluate the integral. (There are many correct ways to set up the integrals, so your integrals may not be the same as those in the back of the book. They should have the same values, however.)

17. **Tilted plane inside cylinder** The portion of the plane $y + 2z = 2$ inside the cylinder $x^2 + y^2 = 1$

18. **Plane inside cylinder** The portion of the plane $z = -x$ inside the cylinder $x^2 + y^2 = 4$

19. **Cone frustum** The portion of the cone $z = 2\sqrt{x^2 + y^2}$ between the planes $z = 2$ and $z = 6$

20. **Cone frustum** The portion of the cone $z = \sqrt{x^2 + y^2}/3$ between the planes $z = 1$ and $z = 4/3$

21. **Circular cylinder band** The portion of the cylinder $x^2 + y^2 = 1$ between the planes $z = 1$ and $z = 4$

22. **Circular cylinder band** The portion of the cylinder $x^2 + z^2 = 10$ between the planes $y = -1$ and $y = 1$

23. **Parabolic cap** The cap cut from the paraboloid $z = 2 - x^2 - y^2$ by the cone $z = \sqrt{x^2 + y^2}$

24. **Parabolic band** The portion of the paraboloid $z = x^2 + y^2$ between the planes $z = 1$ and $z = 4$

25. **Sawed-off sphere** The lower portion cut from the sphere $x^2 + y^2 + z^2 = 2$ by the cone $z = \sqrt{x^2 + y^2}$

26. **Spherical band** The portion of the sphere $x^2 + y^2 + z^2 = 4$ between the planes $z = -1$ and $z = \sqrt{3}$

Planes Tangent to Parametrized Surfaces The tangent plane at a point $P_0(f(u_0, v_0), g(u_0, v_0), h(u_0, v_0))$ on a parametrized surface $\mathbf{r}(u, v) = f(u, v)\mathbf{i} + g(u, v)\mathbf{j} + h(u, v)\mathbf{k}$ is the plane through P_0 normal to the vector $\mathbf{r}_u(u_0, v_0) \times \mathbf{r}_v(u_0, v_0)$, the cross product of the tangent vectors $\mathbf{r}_u(u_0, v_0)$ and $\mathbf{r}_v(u_0, v_0)$ at P_0. In Exercises 27–30, find an equation for the plane tangent to the surface at P_0. Then find a Cartesian equation for the surface and sketch the surface and tangent plane together.

27. **Cone** The cone $\mathbf{r}(r, \theta) = (r \cos \theta)\mathbf{i} + (r \sin \theta)\mathbf{j} + r\mathbf{k}, r \geq 0$, $0 \leq \theta \leq 2\pi$ at the point $P_0(\sqrt{2}, \sqrt{2}, 2)$ corresponding to $(r, \theta) = (2, \pi/4)$

28. **Hemisphere** The hemisphere surface $\mathbf{r}(\phi, \theta) = (4 \sin \phi \cos \theta)\mathbf{i} + (4 \sin \phi \sin \theta)\mathbf{j} + (4 \cos \phi)\mathbf{k}, 0 \leq \phi \leq \pi/2, 0 \leq \theta \leq 2\pi$, at the point $P_0(\sqrt{2}, \sqrt{2}, 2\sqrt{3})$ corresponding to $(\phi, \theta) = (\pi/6, \pi/4)$

29. **Circular cylinder** The circular cylinder $\mathbf{r}(\theta, z) = (3 \sin 2\theta)\mathbf{i} + (6 \sin^2 \theta)\mathbf{j} + z\mathbf{k}, 0 \leq \theta \leq \pi$, at the point $P_0(3\sqrt{3}/2, 9/2, 0)$ corresponding to $(\theta, z) = (\pi/3, 0)$ (See Example 3.)

30. **Parabolic cylinder** The parabolic cylinder surface $\mathbf{r}(x, y) = x\mathbf{i} + y\mathbf{j} - x^2\mathbf{k}, -\infty < x < \infty, -\infty < y < \infty$, at the point $P_0(1, 2, -1)$ corresponding to $(x, y) = (1, 2)$

31. a. A *torus of revolution* (doughnut) is obtained by rotating a circle C in the xz-plane about the z-axis in space. (See the accompanying figure.) If C has radius $r > 0$ and center $(R, 0, 0)$, show that a parametrization of the torus is

$$\mathbf{r}(u, v) = ((R + r \cos u)\cos v)\mathbf{i}$$
$$+ ((R + r \cos u)\sin v)\mathbf{j} + (r \sin u)\mathbf{k},$$

where $0 \leq u \leq 2\pi$ and $0 \leq v \leq 2\pi$ are the angles in the figure.

 b. Show that the surface area of the torus is $A = 4\pi^2 Rr$.

32. **Parametrization of a surface of revolution** Suppose that the parametrized curve C: $(f(u), g(u))$ is revolved about the x-axis, where $g(u) > 0$ for $a \le u \le b$.

   a. Show that

   $$\mathbf{r}(u, v) = f(u)\mathbf{i} + (g(u)\cos v)\mathbf{j} + (g(u)\sin v)\mathbf{k}$$

   is a parametrization of the resulting surface of revolution, where $0 \le v \le 2\pi$ is the angle from the xy-plane to the point $\mathbf{r}(u, v)$ on the surface. (See the accompanying figure.) Notice that $f(u)$ measures distance *along* the axis of revolution and $g(u)$ measures distance *from* the axis of revolution.

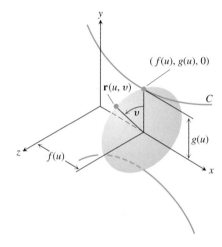

   b. Find a parametrization for the surface obtained by revolving the curve $x = y^2, y \ge 0$, about the x-axis.

33. a. **Parametrization of an ellipsoid** Recall the parametrization $x = a \cos \theta, y = b \sin \theta, 0 \le \theta \le 2\pi$ for the ellipse $(x^2/a^2) + (y^2/b^2) = 1$ (Section 3.5, Example 11). Using the angles θ and ϕ in spherical coordinates, show that

   $$\mathbf{r}(\theta, \phi) = (a \cos \theta \cos \phi)\mathbf{i} + (b \sin \theta \cos \phi)\mathbf{j} + (c \sin \phi)\mathbf{k}$$

   is a parametrization of the ellipsoid $(x^2/a^2) + (y^2/b^2) + (z^2/c^2) = 1$.

   b. Write an integral for the surface area of the ellipsoid, but do not evaluate the integral.

34. **Hyperboloid of one sheet**

   a. Find a parametrization for the hyperboloid of one sheet $x^2 + y^2 - z^2 = 1$ in terms of the angle θ associated with the circle $x^2 + y^2 = r^2$ and the hyperbolic parameter u associated with the hyperbolic function $r^2 - z^2 = 1$. (*Hint:* $\cosh^2 u - \sinh^2 u = 1$)

   b. Generalize the result in part (a) to the hyperboloid $(x^2/a^2) + (y^2/b^2) - (z^2/c^2) = 1$.

35. (*Continuation of Exercise 34.*) Find a Cartesian equation for the plane tangent to the hyperboloid $x^2 + y^2 - z^2 = 25$ at the point $(x_0, y_0, 0)$, where $x_0^2 + y_0^2 = 25$.

36. **Hyperboloid of two sheets** Find a parametrization of the hyperboloid of two sheets $(z^2/c^2) - (x^2/a^2) - (y^2/b^2) = 1$.

37. Find the area of the surface cut from the paraboloid $x^2 + y^2 - z = 0$ by the plane $z = 2$.

38. Find the area of the band cut from the paraboloid $x^2 + y^2 - z = 0$ by the planes $z = 2$ and $z = 6$.

39. Find the area of the region cut from the plane $x + 2y + 2z = 5$ by the cylinder whose walls are $x = y^2$ and $x = 2 - y^2$.

40. Find the area of the portion of the surface $x^2 - 2z = 0$ that lies above the triangle bounded by the lines $x = \sqrt{3}, y = 0$, and $y = x$ in the xy-plane.

41. Find the area of the surface $x^2 - 2y - 2z = 0$ that lies above the triangle bounded by the lines $x = 2, y = 0$, and $y = 3x$ in the xy-plane.

42. Find the area of the cap cut from the sphere $x^2 + y^2 + z^2 = 2$ by the cone $z = \sqrt{x^2 + y^2}$.

43. Find the area of the ellipse cut from the plane $z = cx$ (c a constant) by the cylinder $x^2 + y^2 = 1$.

44. Find the area of the upper portion of the cylinder $x^2 + z^2 = 1$ that lies between the planes $x = \pm 1/2$ and $y = \pm 1/2$.

45. Find the area of the portion of the paraboloid $x = 4 - y^2 - z^2$ that lies above the ring $1 \le y^2 + z^2 \le 4$ in the yz-plane.

46. Find the area of the surface cut from the paraboloid $x^2 + y + z^2 = 2$ by the plane $y = 0$.

47. Find the area of the surface $x^2 - 2 \ln x + \sqrt{15}y - z = 0$ above the square R: $1 \le x \le 2, 0 \le y \le 1$, in the xy-plane.

48. Find the area of the surface $2x^{3/2} + 2y^{3/2} - 3z = 0$ above the square R: $0 \le x \le 1, 0 \le y \le 1$, in the xy-plane.

Surface Area for Explicit Forms If S is the surface defined by a function $z = f(x, y)$ that has continuous first partial derivatives throughout a region R_{xy} in the xy-plane (Figure 14.42), then S is also the level surface $F(x, y, z) = 0$ of the function $F(x, y, z) = f(x, y) - z$. Taking the unit normal to R_{xy} to be $\mathbf{p} = \mathbf{k}$ then gives

$$|\nabla F| = |f_x\mathbf{i} + f_y\mathbf{j} - \mathbf{k}| = \sqrt{f_x^2 + f_y^2 + 1}$$

$$|\nabla F \cdot \mathbf{p}| = |(f_x\mathbf{i} + f_y\mathbf{j} - \mathbf{k}) \cdot \mathbf{k}| = |-1| = 1$$

and

$$\iint_{R_{xy}} \frac{|\nabla F|}{|\nabla F \cdot \mathbf{p}|} \, dA = \iint_{R_{xy}} \sqrt{f_x^2 + f_y^2 + 1} \, dx \, dy, \qquad (8)$$

Similarly, the area of a smooth surface $x = f(y, z)$ over a region R_{yz} in the yz-plane is

$$A = \iint_{R_{yz}} \sqrt{f_y^2 + f_z^2 + 1} \, dy \, dz, \qquad (9)$$

and the area of a smooth $y = f(x, z)$ over a region R_{xz} in the xz-plane is

$$A = \iint_{R_{xz}} \sqrt{f_x^2 + f_z^2 + 1} \; dx \, dz. \qquad (10)$$

FIGURE 14.42 For a surface $z = f(x, y)$, the surface area formula in Equation (7) takes the form

$$A = \iint_{R_{xy}} \sqrt{f_x^2 + f_y^2 + 1} \; dx \, dy.$$

Use Equations (8)–(10) to find the area of the surfaces in Exercises 49–54.

49. The surface cut from the bottom of the paraboloid $z = x^2 + y^2$ by the plane $z = 3$

50. The surface cut from the "nose" of the paraboloid $x = 1 - y^2 - z^2$ by the yz-plane

51. The portion of the cone $z = \sqrt{x^2 + y^2}$ that lies over the region between the circle $x^2 + y^2 = 1$ and the ellipse $9x^2 + 4y^2 = 36$ in the xy-plane. (*Hint:* Use formulas from geometry to find the area of the region.)

52. The triangle cut from the plane $2x + 6y + 3z = 6$ by the bounding planes of the first octant. Calculate the area three ways, once with each area formula.

53. The surface in the first octant cut from the cylinder $y = (2/3)z^{3/2}$ by the planes $x = 1$ and $y = 16/3$

54. The portion of the plane $y + z = 4$ that lies above the region cut from the first quadrant of the xz-plane by the parabola $x = 4 - z^2$

55. Use the parametrization

$$\mathbf{r}(x, y) = x\mathbf{i} + y\mathbf{j} + f(x, y)\mathbf{k}$$

and Equation (5) to derive the double integral formula for $d\sigma$ in Equation (8).

56. Let S be the surface obtained by rotating the smooth curve $y = f(x)$, $a \le x \le b$, about the x-axis, where $f(x) \ge 0$.

   a. Show that the vector function

   $$\mathbf{r}(x, \theta) = x\mathbf{i} + f(x)\cos\theta\,\mathbf{j} + f(x)\sin\theta\,\mathbf{k}$$

   is a parametrization of S, where θ is the angle of rotation around the x-axis (see the accompanying figure).

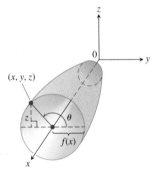

   b. Use Equation (4) to show that the surface area of this surface of revolution is given by

   $$A = \int_a^b 2\pi f(x)\sqrt{1 + [f'(x)]^2} \; dx.$$

14.6 Surface Integrals and Flux

To compute quantities such as the flow of liquid across a curved membrane or the upward force on a falling parachute, we need to integrate a function over a curved surface in space. This concept of a *surface integral* is an extension of the idea of a line integral when integrating over a curve.

Surface Integrals

Suppose, for example, that we have an electrical charge distributed over a surface S, and that the function $G(x, y, z)$ gives the charge per unit area (charge density) at each point on S. Then we may calculate the total charge on S as an integral in the following way.

Assume, as in Section 14.5, that the surface S is defined parametrically on a region R in the uv-plane,

$$\mathbf{r}(u, v) = f(u, v)\mathbf{i} + g(u, v)\mathbf{j} + h(u, v)\mathbf{k}, \qquad (u, v) \in R.$$

In Figure 14.37, we saw how a subdivision of R (considered as a rectangle for simplicity) divided the surface S into corresponding curved surface elements, or patches, of area

$$\Delta\sigma_{uv} \approx |\mathbf{r}_u \times \mathbf{r}_v| \, du \, dv.$$

As we did for the subdivisions when defining double integrals in Section 13.1, we number the surface element patches in some order with their areas given by $\Delta\sigma_1, \Delta\sigma_2, \ldots, \Delta\sigma_n$. To form a Riemann sum over S, we choose a point (x_k, y_k, z_k) in the kth patch, multiply the value of the function G at that point by the area $\Delta\sigma_k$, and add together the products:

$$\sum_{k=1}^{n} G(x_k, y_k, z_k) \, \Delta\sigma_k.$$

Depending on how we pick (x_k, y_k, z_k) in the kth patch, we may get different values for this Riemann sum. Then we take the limit as the number of surface patches increases and their areas shrink to zero. This limit, whenever it exists, defines the **surface integral of G over the surface S** as

$$\iint_S G(x, y, z) \, d\sigma = \lim_{n \to \infty} \sum_{k=1}^{n} G(x_k, y_k, z_k) \, \Delta\sigma_k. \tag{1}$$

Notice the analogy with the definition of the double integral (Section 13.1) and with the line integral (Section 14.1). If S is a piecewise smooth surface, and G is continuous over S, then the surface integral defined by Equation (1) can be shown to exist.

The formula for evaluating the surface integral depends on the manner in which S is described, parametrically or implicitly, as discussed in Section 14.5.

Computing a Surface Integral

1. For a smooth surface S defined parametrically as $\mathbf{r}(u, v) = f(u, v)\mathbf{i} + g(u, v)\mathbf{j} + h(u, v)\mathbf{k}$, $(u, v) \in R$, and a continuous function $G(x, y, z)$ defined on S, the surface integral of G over S is given by the double integral over R

$$\iint_S G(x, y, z) \, d\sigma = \iint_R G(f(u, v), g(u, v), h(u, v)) \, |\mathbf{r}_u \times \mathbf{r}_v| \, du \, dv. \tag{2}$$

2. For a surface S given implicitly by $F(x, y, z) = c$, where F is a continuously differentiable function, with S lying above its closed and bounded shadow region R in the coordinate plane beneath it, the surface integral of the continuous function G over S is given by the double integral over R

$$\iint_S G(x, y, z) \, d\sigma = \iint_R G(x, y, z) \frac{|\nabla F|}{|\nabla F \cdot \mathbf{p}|} dA, \tag{3}$$

where $\mathbf{p}$ is a unit vector normal to R and $\nabla F \cdot \mathbf{p} \neq 0$.

The surface integral in Equation (1) takes on different meanings in different applications. If G has the constant value 1, the integral gives the area of S. If G gives the mass density of a thin shell of material modeled by S, the integral gives the mass of the shell.

EXAMPLE 1   Integrate $G(x, y, z) = x^2$ over the cone $z = \sqrt{x^2 + y^2}, 0 \le z \le 1$.

Solution   Using Equation (2) and the calculations from Example 4 in Section 14.5, we have $|\mathbf{r}_r \times \mathbf{r}_\theta| = \sqrt{2}r$ and

$$\iint\limits_{S} x^2 \, d\sigma = \int_0^{2\pi} \int_0^1 \left(r^2 \cos^2\theta\right)\left(\sqrt{2}r\right) dr \, d\theta \qquad x = r\cos\theta$$

$$= \sqrt{2} \int_0^{2\pi} \int_0^1 r^3 \cos^2\theta \, dr \, d\theta$$

$$= \frac{\sqrt{2}}{4} \int_0^{2\pi} \cos^2\theta \, d\theta = \frac{\sqrt{2}}{4}\left[\frac{\theta}{2} + \frac{1}{4}\sin 2\theta\right]_0^{2\pi} = \frac{\pi\sqrt{2}}{4}. \qquad \blacksquare$$

Surface integrals behave like other double integrals, the integral of the sum of two functions being the sum of their integrals and so on. The domain Additivity Property takes the form

$$\iint\limits_{S} G \, d\sigma = \iint\limits_{S_1} G \, d\sigma + \iint\limits_{S_2} G \, d\sigma + \cdots + \iint\limits_{S_n} G \, d\sigma.$$

The idea is that if S is partitioned by smooth curves into a finite number of nonoverlapping smooth patches (i.e., if S is **piecewise smooth**), then the integral over S is the sum of the integrals over the patches. Thus, the integral of a function over the surface of a cube is the sum of the integrals over the faces of the cube. We integrate over a turtle shell of welded plates by integrating over one plate at a time and adding the results.

EXAMPLE 2   Integrate $G(x, y, z) = xyz$ over the surface of the cube cut from the first octant by the planes $x = 1, y = 1$, and $z = 1$ (Figure 14.43).

Solution   We integrate xyz over each of the six sides and add the results. Since $xyz = 0$ on the sides that lie in the coordinate planes, the integral over the surface of the cube reduces to

$$\iint\limits_{\substack{\text{Cube} \\ \text{surface}}} xyz \, d\sigma = \iint\limits_{\text{Side } A} xyz \, d\sigma + \iint\limits_{\text{Side } B} xyz \, d\sigma + \iint\limits_{\text{Side } C} xyz \, d\sigma.$$

Side A is the surface $f(x, y, z) = z = 1$ over the square region $R_{xy}: 0 \le x \le 1$, $0 \le y \le 1$, in the xy-plane. For this surface and region,

$$\mathbf{p} = \mathbf{k}, \qquad \nabla f = \mathbf{k}, \qquad |\nabla f| = 1, \qquad |\nabla f \cdot \mathbf{p}| = |\mathbf{k} \cdot \mathbf{k}| = 1$$

$$d\sigma = \frac{|\nabla f|}{|\nabla f \cdot \mathbf{p}|} dA = \frac{1}{1} dx \, dy = dx \, dy$$

$$xyz = xy(1) = xy$$

and

$$\iint\limits_{\text{Side } A} xyz \, d\sigma = \iint\limits_{R_{xy}} xy \, dx \, dy = \int_0^1 \int_0^1 xy \, dx \, dy = \int_0^1 \frac{y}{2} dy = \frac{1}{4}.$$

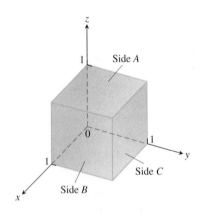

FIGURE 14.43   The cube in Example 2.

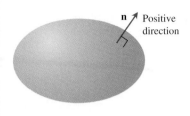

FIGURE 14.44 Smooth closed surfaces in space are orientable. The outward unit normal vector defines the positive direction at each point.

FIGURE 14.45 To make a Möbius band, take a rectangular strip of paper *abcd*, give the end *bc* a single twist, and paste the ends of the strip together to match *a* with *c* and *b* with *d*. The Möbius band is a nonorientable or one-sided surface.

Symmetry tells us that the integrals of *xyz* over sides *B* and *C* are also 1/4. Hence,

$$\iint_{\substack{Cube\\surface}} xyz\, d\sigma = \frac{1}{4} + \frac{1}{4} + \frac{1}{4} = \frac{3}{4}.$$ ■

Orientation

We call a smooth surface *S* **orientable** or **two-sided** if it is possible to define a field **n** of unit normal vectors on *S* that varies continuously with position. Any patch or subportion of an orientable surface is orientable. Spheres and other smooth closed surfaces in space (smooth surfaces that enclose solids) are orientable. By convention, we choose **n** on a closed surface to point outward.

Once **n** has been chosen, we say that we have **oriented** the surface, and we call the surface together with its normal field an **oriented surface**. The vector **n** at any point is called the **positive direction** at that point (Figure 14.44).

The Möbius band in Figure 14.45 is not orientable. No matter where you start to construct a continuous-unit normal field (shown as the shaft of a thumbtack in the figure), moving the vector continuously around the surface in the manner shown will return it to the starting point with a direction opposite to the one it had when it started out. The vector at that point cannot point both ways and yet it must if the field is to be continuous. We conclude that no such field exists.

Surface Integral for Flux

Suppose that **F** is a continuous vector field defined over an oriented surface *S* and that **n** is the chosen unit normal field on the surface. We call the integral of **F** · **n** over *S* the flux of **F** across *S* in the positive direction. Thus, the flux is the integral over *S* of the scalar component of **F** in the direction of **n**.

> **DEFINITION** The **flux** of a three-dimensional vector field **F** across an oriented surface *S* in the direction of **n** is
>
> $$\text{Flux} = \iint_S \mathbf{F} \cdot \mathbf{n}\, d\sigma. \qquad (4)$$

The definition is analogous to the flux of a two-dimensional field **F** across a plane curve *C*. In the plane (Section 14.2), the flux is

$$\int_C \mathbf{F} \cdot \mathbf{n}\, ds,$$

the integral of the scalar component of **F** normal to the curve.

If **F** is the velocity field of a three-dimensional fluid flow, the flux of **F** across *S* is the net rate at which fluid is crossing *S* in the chosen positive direction. We discuss such flows in more detail in Section 14.7.

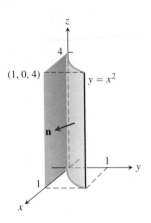

FIGURE 14.46 Finding the flux through the surface of a parabolic cylinder (Example 3).

EXAMPLE 3 Find the flux of $\mathbf{F} = yz\mathbf{i} + x\mathbf{j} - z^2\mathbf{k}$ outward through the parabolic cylinder $y = x^2$, $0 \le x \le 1$, $0 \le z \le 4$ (Figure 14.46).

Solution On the surface we have $x = x$, $y = x^2$, and $z = z$, so we automatically have the parametrization $\mathbf{r}(x, z) = x\mathbf{i} + x^2\mathbf{j} + z\mathbf{k}$, $0 \le x \le 1$, $0 \le z \le 4$. The cross product of tangent vectors is

$$\mathbf{r}_x \times \mathbf{r}_z = \begin{vmatrix} \mathbf{i} & \mathbf{j} & \mathbf{k} \\ 1 & 2x & 0 \\ 0 & 0 & 1 \end{vmatrix} = 2x\mathbf{i} - \mathbf{j}.$$

The unit normal pointing outward from the surface is

$$\mathbf{n} = \frac{\mathbf{r}_x \times \mathbf{r}_z}{|\mathbf{r}_x \times \mathbf{r}_z|} = \frac{2x\mathbf{i} - \mathbf{j}}{\sqrt{4x^2 + 1}}.$$

On the surface, $y = x^2$, so the vector field there is

$$\mathbf{F} = yz\mathbf{i} + x\mathbf{j} - z^2\mathbf{k} = x^2z\mathbf{i} + x\mathbf{j} - z^2\mathbf{k}.$$

Thus,

$$\mathbf{F} \cdot \mathbf{n} = \frac{1}{\sqrt{4x^2 + 1}}((x^2z)(2x) + (x)(-1) + (-z^2)(0))$$

$$= \frac{2x^3z - x}{\sqrt{4x^2 + 1}}.$$

The flux of $\mathbf{F}$ outward through the surface is

$$\iint\limits_S \mathbf{F} \cdot \mathbf{n} \, d\sigma = \int_0^4 \int_0^1 \frac{2x^3z - x}{\sqrt{4x^2 + 1}} |\mathbf{r}_x \times \mathbf{r}_z| \, dx \, dz$$

$$= \int_0^4 \int_0^1 \frac{2x^3z - x}{\sqrt{4x^2 + 1}} \sqrt{4x^2 + 1} \, dx \, dz$$

$$= \int_0^4 \int_0^1 (2x^3z - x) \, dx \, dz = \int_0^4 \left[\frac{1}{2}x^4z - \frac{1}{2}x^2 \right]_{x=0}^{x=1} dz$$

$$= \int_0^4 \frac{1}{2}(z - 1) \, dz = \frac{1}{4}(z - 1)^2 \Big]_0^4$$

$$= \frac{1}{4}(9) - \frac{1}{4}(1) = 2. \qquad \blacksquare$$

If S is part of a level surface $g(x, y, z) = c$, then $\mathbf{n}$ may be taken to be one of the two fields

$$\mathbf{n} = \pm \frac{\nabla g}{|\nabla g|}, \tag{5}$$

depending on which one gives the preferred direction. The corresponding flux is

$$\text{Flux} = \iint\limits_S \mathbf{F} \cdot \mathbf{n} \, d\sigma$$

$$= \iint\limits_R \left(\mathbf{F} \cdot \frac{\pm \nabla g}{|\nabla g|} \right) \frac{|\nabla g|}{|\nabla g \cdot \mathbf{p}|} \, dA \qquad \text{Equations (5) and (3)}$$

$$= \iint\limits_R \mathbf{F} \cdot \frac{\pm \nabla g}{|\nabla g \cdot \mathbf{p}|} \, dA. \tag{6}$$

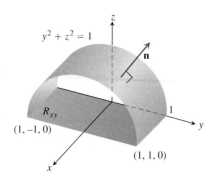

FIGURE 14.47 Calculating the flux of a vector field outward through this surface. The area of the shadow region R_{xy} is 2 (Example 4).

EXAMPLE 4 Find the flux of $\mathbf{F} = yz\mathbf{j} + z^2\mathbf{k}$ outward through the surface S cut from the cylinder $y^2 + z^2 = 1$, $z \geq 0$, by the planes $x = 0$ and $x = 1$.

Solution The outward normal field on S (Figure 14.47) may be calculated from the gradient of $g(x, y, z) = y^2 + z^2$ to be

$$\mathbf{n} = +\frac{\nabla g}{|\nabla g|} = \frac{2y\mathbf{j} + 2z\mathbf{k}}{\sqrt{4y^2 + 4z^2}} = \frac{2y\mathbf{j} + 2z\mathbf{k}}{2\sqrt{1}} = y\mathbf{j} + z\mathbf{k}.$$

With $\mathbf{p} = \mathbf{k}$, we also have

$$d\sigma = \frac{|\nabla g|}{|\nabla g \cdot \mathbf{k}|} dA = \frac{2}{|2z|} dA = \frac{1}{z} dA.$$

We can drop the absolute value bars because $z \geq 0$ on S.

The value of $\mathbf{F} \cdot \mathbf{n}$ on the surface is

$$\begin{aligned}\mathbf{F} \cdot \mathbf{n} &= (yz\mathbf{j} + z^2\mathbf{k}) \cdot (y\mathbf{j} + z\mathbf{k}) \\ &= y^2 z + z^3 = z(y^2 + z^2) \qquad {\scriptstyle y^2 + z^2 = 1 \text{ on } S}\\ &= z.\end{aligned}$$

Therefore, the flux of $\mathbf{F}$ outward through S is

$$\iint_S \mathbf{F} \cdot \mathbf{n}\, d\sigma = \iint_S (z)\left(\frac{1}{z} dA\right) = \iint_{R_{xy}} dA = \text{area}(R_{xy}) = 2. \qquad \blacksquare$$

Moments and Masses of Thin Shells

Thin shells of material like bowls, metal drums, and domes are modeled with surfaces. Their moments and masses are calculated with the formulas in Table 14.3. The derivations are similar to those in Section 6.7.

TABLE 14.3 Mass and moment formulas for very thin shells

Mass: $M = \iint_S \delta(x, y, z)\, d\sigma$ $\delta(x, y, z)$ = density at (x, y, z), mass per unit area

First moments about the coordinate planes:

$$M_{yz} = \iint_S x\,\delta\, d\sigma, \qquad M_{xz} = \iint_S y\,\delta\, d\sigma, \qquad M_{xy} = \iint_S z\,\delta\, d\sigma$$

Coordinates of center of mass:

$$\bar{x} = M_{yz}/M, \qquad \bar{y} = M_{xz}/M, \qquad \bar{z} = M_{xy}/M$$

Moments of inertia about coordinate axes:

$$I_x = \iint_S (y^2 + z^2)\,\delta\, d\sigma, \quad I_y = \iint_S (x^2 + z^2)\,\delta\, d\sigma, \quad I_z = \iint_S (x^2 + y^2)\,\delta\, d\sigma,$$

$$I_L = \iint_S r^2\delta\, d\sigma, \qquad r(x, y, z) = \text{distance from point } (x, y, z) \text{ to line } L$$

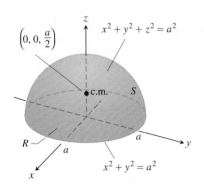

FIGURE 14.48 The center of mass of a thin hemispherical shell of constant density lies on the axis of symmetry halfway from the base to the top (Example 5).

EXAMPLE 5 Find the center of mass of a thin hemispherical shell of radius a and constant density δ.

Solution We model the shell with the hemisphere

$$f(x, y, z) = x^2 + y^2 + z^2 = a^2, \qquad z \geq 0$$

(Figure 14.48). The symmetry of the surface about the z-axis tells us that $\bar{x} = \bar{y} = 0$. It remains only to find $\bar{z}$ from the formula $\bar{z} = M_{xy}/M$.

The mass of the shell is

$$M = \iint_S \delta \, d\sigma = \delta \iint_S d\sigma = (\delta)(\text{area of } S) = 2\pi a^2 \delta.$$

To evaluate the integral for M_{xy}, we take $\mathbf{p} = \mathbf{k}$ and calculate

$$|\nabla f| = |2x\mathbf{i} + 2y\mathbf{j} + 2z\mathbf{k}| = 2\sqrt{x^2 + y^2 + z^2} = 2a$$

$$|\nabla f \cdot \mathbf{p}| = |\nabla f \cdot \mathbf{k}| = |2z| = 2z$$

$$d\sigma = \frac{|\nabla f|}{|\nabla f \cdot \mathbf{p}|} \, dA = \frac{a}{z} \, dA.$$

Then

$$M_{xy} = \iint_S z\delta \, d\sigma = \delta \iint_R z \frac{a}{z} \, dA = \delta a \iint_R dA = \delta a (\pi a^2) = \delta \pi a^3$$

$$\bar{z} = \frac{M_{xy}}{M} = \frac{\pi a^3 \delta}{2\pi a^2 \delta} = \frac{a}{2}.$$

The shell's center of mass is the point $(0, 0, a/2)$. ∎

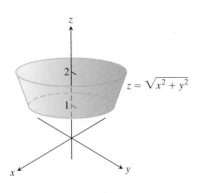

FIGURE 14.49 The cone frustum formed when the cone $z = \sqrt{x^2 + y^2}$ is cut by the planes $z = 1$ and $z = 2$ (Example 6).

EXAMPLE 6 Find the center of mass of a thin shell of constant density δ cut from the cone $z = \sqrt{x^2 + y^2}$ by the planes $z = 1$ and $z = 2$ (Figure 14.49).

Solution The symmetry of the surface about the z-axis tells us that $\bar{x} = \bar{y} = 0$. We find $\bar{z} = M_{xy}/M$. Working as in Example 4 of Section 14.5, we have

$$\mathbf{r}(r, \theta) = r\cos\theta\,\mathbf{i} + r\sin\theta\,\mathbf{j} + r\mathbf{k}, \qquad 1 \leq r \leq 2, \quad 0 \leq \theta \leq 2\pi,$$

and

$$|\mathbf{r}_r \times \mathbf{r}_\theta| = \sqrt{2}\,r.$$

Therefore,

$$M = \iint_S \delta \, d\sigma = \int_0^{2\pi} \int_1^2 \delta\sqrt{2}\,r \, dr \, d\theta$$

$$= \delta\sqrt{2} \int_0^{2\pi} \left[\frac{r^2}{2}\right]_1^2 d\theta = \delta\sqrt{2} \int_0^{2\pi} \left(2 - \frac{1}{2}\right) d\theta$$

$$= \delta\sqrt{2} \left[\frac{3\theta}{2}\right]_0^{2\pi} = 3\pi\delta\sqrt{2}$$

$$M_{xy} = \iint_S \delta z \, d\sigma = \int_0^{2\pi} \int_1^2 \delta r \sqrt{2} \, r \, dr \, d\theta$$

$$= \delta\sqrt{2} \int_0^{2\pi} \int_1^2 r^2 \, dr \, d\theta = \delta\sqrt{2} \int_0^{2\pi} \left[\frac{r^3}{3}\right]_1^2 d\theta$$

$$= \delta\sqrt{2} \int_0^{2\pi} \frac{7}{3} \, d\theta = \frac{14}{3} \pi\delta\sqrt{2}$$

$$\bar{z} = \frac{M_{xy}}{M} = \frac{14\pi\delta\sqrt{2}}{3\left(3\pi\delta\sqrt{2}\right)} = \frac{14}{9}.$$

The shell's center of mass is the point $(0, 0, 14/9)$. ∎

EXERCISES 14.6

In Exercises 1–8, integrate the given function over the given surface.

1. **Parabolic cylinder** $G(x, y, z) = x$, over the parabolic cylinder $y = x^2$, $0 \le x \le 2$, $0 \le z \le 3$

2. **Circular cylinder** $G(x, y, z) = z$, over the cylindrical surface $y^2 + z^2 = 4$, $z \ge 0$, $1 \le x \le 4$

3. **Sphere** $G(x, y, z) = x^2$, over the unit sphere $x^2 + y^2 + z^2 = 1$

4. **Hemisphere** $G(x, y, z) = z^2$, over the hemisphere $x^2 + y^2 + z^2 = a^2$, $z \ge 0$

5. **Portion of plane** $F(x, y, z) = z$, over the portion of the plane $x + y + z = 4$ that lies above the square $0 \le x \le 1$, $0 \le y \le 1$, in the xy-plane

6. **Cone** $F(x, y, z) = z - x$, over the cone $z = \sqrt{x^2 + y^2}$, $0 \le z \le 1$

7. **Parabolic dome** $H(x, y, z) = x^2\sqrt{5 - 4z}$, over the parabolic dome $z = 1 - x^2 - y^2$, $z \ge 0$

8. **Spherical cap** $H(x, y, z) = yz$, over the part of the sphere $x^2 + y^2 + z^2 = 4$ that lies above the cone $z = \sqrt{x^2 + y^2}$

9. Integrate $G(x, y, z) = x + y + z$ over the surface of the cube cut from the first octant by the planes $x = a$, $y = a$, $z = a$.

10. Integrate $G(x, y, z) = y + z$ over the surface of the wedge in the first octant bounded by the coordinate planes and the planes $x = 2$ and $y + z = 1$.

11. Integrate $G(x, y, z) = xyz$ over the surface of the rectangular solid cut from the first octant by the planes $x = a$, $y = b$, and $z = c$.

12. Integrate $G(x, y, z) = xyz$ over the surface of the rectangular solid bounded by the planes $x = \pm a$, $y = \pm b$, and $z = \pm c$.

13. Integrate $G(x, y, z) = x + y + z$ over the portion of the plane $2x + 2y + z = 2$ that lies in the first octant.

14. Integrate $G(x, y, z) = x\sqrt{y^2 + 4}$ over the surface cut from the parabolic cylinder $y^2 + 4z = 16$ by the planes $x = 0$, $x = 1$, and $z = 0$.

In Exercises 15–24, use a parametrization to find the flux $\iint_S \mathbf{F} \cdot \mathbf{n} \, d\sigma$ across the surface in the given direction.

15. **Parabolic cylinder** $\mathbf{F} = z^2\mathbf{i} + x\mathbf{j} - 3z\mathbf{k}$ outward (normal away from the x-axis) through the surface cut from the parabolic cylinder $z = 4 - y^2$ by the planes $x = 0$, $x = 1$, and $z = 0$

16. **Parabolic cylinder** $\mathbf{F} = x^2\mathbf{j} - xz\mathbf{k}$ outward (normal away from the yz-plane) through the surface cut from the parabolic cylinder $y = x^2$, $-1 \le x \le 1$, by the planes $z = 0$ and $z = 2$

17. **Sphere** $\mathbf{F} = z\mathbf{k}$ across the portion of the sphere $x^2 + y^2 + z^2 = a^2$ in the first octant in the direction away from the origin

18. **Sphere** $\mathbf{F} = x\mathbf{i} + y\mathbf{j} + z\mathbf{k}$ across the sphere $x^2 + y^2 + z^2 = a^2$ in the direction away from the origin

19. **Plane** $\mathbf{F} = 2xy\mathbf{i} + 2yz\mathbf{j} + 2xz\mathbf{k}$ upward across the portion of the plane $x + y + z = 2a$ that lies above the square $0 \le x \le a$, $0 \le y \le a$, in the xy-plane

20. **Cylinder** $\mathbf{F} = x\mathbf{i} + y\mathbf{j} + z\mathbf{k}$ outward through the portion of the cylinder $x^2 + y^2 = 1$ cut by the planes $z = 0$ and $z = a$

21. **Cone** $\mathbf{F} = xy\mathbf{i} - z\mathbf{k}$ outward (normal away from the z-axis) through the cone $z = \sqrt{x^2 + y^2}$, $0 \le z \le 1$

22. **Cone** $\mathbf{F} = y^2\mathbf{i} + xz\mathbf{j} - \mathbf{k}$ outward (normal away from the z-axis) through the cone $z = 2\sqrt{x^2 + y^2}$, $0 \le z \le 2$

23. **Cone frustum** $\mathbf{F} = -x\mathbf{i} - y\mathbf{j} + z^2\mathbf{k}$ outward (normal away from the z-axis) through the portion of the cone $z = \sqrt{x^2 + y^2}$ between the planes $z = 1$ and $z = 2$

24. **Paraboloid** $\mathbf{F} = 4x\mathbf{i} + 4y\mathbf{j} + 2\mathbf{k}$ outward (normal away from the z-axis) through the surface cut from the bottom of the paraboloid $z = x^2 + y^2$ by the plane $z = 1$

In Exercises 25 and 26, find the flux of the field $\mathbf{F}$ across the portion of the given surface in the specified direction.

25. $\mathbf{F}(x, y, z) = -\mathbf{i} + 2\mathbf{j} + 3\mathbf{k}$

   S: rectangular surface $z = 0$, $0 \le x \le 2$, $0 \le y \le 3$, direction $\mathbf{k}$

26. $\mathbf{F}(x, y, z) = yx^2\mathbf{i} - 2\mathbf{j} + xz\mathbf{k}$

   S: rectangular surface $y = 0$, $-1 \le x \le 2$, $2 \le z \le 7$, direction $-\mathbf{j}$

In Exercises 27–32, find the flux of the field $\mathbf{F}$ across the portion of the sphere $x^2 + y^2 + z^2 = a^2$ in the first octant in the direction away from the origin.

27. $\mathbf{F}(x, y, z) = z\mathbf{k}$ 28. $\mathbf{F}(x, y, z) = -y\mathbf{i} + x\mathbf{j}$

29. $\mathbf{F}(x, y, z) = y\mathbf{i} - x\mathbf{j} + \mathbf{k}$ 30. $\mathbf{F}(x, y, z) = zx\mathbf{i} + zy\mathbf{j} + z^2\mathbf{k}$

31. $\mathbf{F}(x, y, z) = x\mathbf{i} + y\mathbf{j} + z\mathbf{k}$

32. $\mathbf{F}(x, y, z) = \dfrac{x\mathbf{i} + y\mathbf{j} + z\mathbf{k}}{\sqrt{x^2 + y^2 + z^2}}$

33. Find the flux of the field $\mathbf{F}(x, y, z) = z^2\mathbf{i} + x\mathbf{j} - 3z\mathbf{k}$ outward through the surface cut from the parabolic cylinder $z = 4 - y^2$ by the planes $x = 0$, $x = 1$, and $z = 0$.

34. Find the flux of the field $\mathbf{F}(x, y, z) = 4x\mathbf{i} + 4y\mathbf{j} + 2\mathbf{k}$ outward (away from the z-axis) through the surface cut from the bottom of the paraboloid $z = x^2 + y^2$ by the plane $z = 1$.

35. Let S be the portion of the cylinder $y = e^x$ in the first octant that projects parallel to the x-axis onto the rectangle R_{yz}: $1 \le y \le 2$, $0 \le z \le 1$ in the yz-plane (see the accompanying figure). Let $\mathbf{n}$ be the unit vector normal to S that points away from the yz-plane. Find the flux of the field $\mathbf{F}(x, y, z) = -2\mathbf{i} + 2y\mathbf{j} + z\mathbf{k}$ across S in the direction of $\mathbf{n}$.

36. Let S be the portion of the cylinder $y = \ln x$ in the first octant whose projection parallel to the y-axis onto the xz-plane is the rectangle R_{xz}: $1 \le x \le e$, $0 \le z \le 1$. Let $\mathbf{n}$ be the unit vector normal to S that points away from the xz-plane. Find the flux of $\mathbf{F} = 2y\mathbf{j} + z\mathbf{k}$ through S in the direction of $\mathbf{n}$.

37. Find the outward flux of the field $\mathbf{F} = 2xy\mathbf{i} + 2yz\mathbf{j} + 2xz\mathbf{k}$ across the surface of the cube cut from the first octant by the planes $x = a, y = a, z = a$.

38. Find the outward flux of the field $\mathbf{F} = xz\mathbf{i} + yz\mathbf{j} + \mathbf{k}$ across the surface of the upper cap cut from the solid sphere $x^2 + y^2 + z^2 \le 25$ by the plane $z = 3$.

39. **Centroid** Find the centroid of the portion of the sphere $x^2 + y^2 + z^2 = a^2$ that lies in the first octant.

40. **Centroid** Find the centroid of the surface cut from the cylinder $y^2 + z^2 = 9, z \ge 0$, by the planes $x = 0$ and $x = 3$ (resembles the surface in Example 4).

41. **Thin shell of constant density** Find the center of mass and the moment of inertia about the z-axis of a thin shell of constant density δ cut from the cone $x^2 + y^2 - z^2 = 0$ by the planes $z = 1$ and $z = 2$.

42. **Conical surface of constant density** Find the moment of inertia about the z-axis of a thin shell of constant density δ cut from the cone $4x^2 + 4y^2 - z^2 = 0, z \ge 0$, by the circular cylinder $x^2 + y^2 = 2x$ (see the accompanying figure).

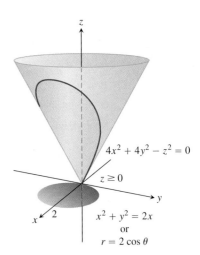

43. **Spherical shells**

   a. Find the moment of inertia about a diameter of a thin spherical shell of radius a and constant density δ. (Work with a hemispherical shell and double the result.)

   b. Use the Parallel Axis Theorem (Exercises 13.6) and the result in part (a) to find the moment of inertia about a line tangent to the shell.

44. **Conical Surface** Find the centroid of the lateral surface of a solid cone of base radius a and height h (cone surface minus the base).

14.7 Stokes' Theorem

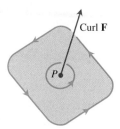

FIGURE 14.50 The circulation vector at a point P in a plane in a three-dimensional fluid flow. Notice its right-hand relation to the circulation line.

As we saw in Section 14.4, the circulation density or curl component of a two-dimensional field $\mathbf{F} = M\mathbf{i} + N\mathbf{j}$ at a point (x, y) is described by the scalar quantity $(\partial N/\partial x - \partial M/\partial y)$. In three dimensions, the circulation around a point P in a plane is described with a vector. This vector is normal to the plane of the circulation (Figure 14.50) and points in the direction that gives it a right-hand relation to the circulation line. The length of the vector gives the rate of the fluid's rotation, which usually varies as the circulation plane is tilted about P. It turns out that the vector of greatest circulation in a flow with velocity field $\mathbf{F} = M\mathbf{i} + N\mathbf{j} + P\mathbf{k}$ is the **curl vector**

$$\text{curl } \mathbf{F} = \left(\frac{\partial P}{\partial y} - \frac{\partial N}{\partial z}\right)\mathbf{i} + \left(\frac{\partial M}{\partial z} - \frac{\partial P}{\partial x}\right)\mathbf{j} + \left(\frac{\partial N}{\partial x} - \frac{\partial M}{\partial y}\right)\mathbf{k}. \qquad (1)$$

We get this information from Stokes' Theorem, the generalization to space of the circulation-curl form of Green's Theorem.

Notice that $(\text{curl } \mathbf{F}) \cdot \mathbf{k} = (\partial N/\partial x - \partial M/\partial y)$ is consistent with our definition in Section 14.4 when $\mathbf{F} = M(x, y)\mathbf{i} + N(x, y)\mathbf{j}$. The formula for curl $\mathbf{F}$ in Equation (1) is often written using the symbolic operator

$$\nabla = \mathbf{i}\frac{\partial}{\partial x} + \mathbf{j}\frac{\partial}{\partial y} + \mathbf{k}\frac{\partial}{\partial z}. \qquad (2)$$

(The symbol ∇ is pronounced "del.") The curl of $\mathbf{F}$ is $\nabla \times \mathbf{F}$:

$$\nabla \times \mathbf{F} = \begin{vmatrix} \mathbf{i} & \mathbf{j} & \mathbf{k} \\ \dfrac{\partial}{\partial x} & \dfrac{\partial}{\partial y} & \dfrac{\partial}{\partial z} \\ M & N & P \end{vmatrix}$$

$$= \left(\frac{\partial P}{\partial y} - \frac{\partial N}{\partial z}\right)\mathbf{i} + \left(\frac{\partial M}{\partial z} - \frac{\partial P}{\partial x}\right)\mathbf{j} + \left(\frac{\partial N}{\partial x} - \frac{\partial M}{\partial y}\right)\mathbf{k}$$

$$= \text{curl } \mathbf{F}.$$

$$\text{curl } \mathbf{F} = \nabla \times \mathbf{F} \qquad (3)$$

EXAMPLE 1 Find the curl of $\mathbf{F} = (x^2 - y)\mathbf{i} + 4z\mathbf{j} + x^2\mathbf{k}$.

Solution

$$\text{curl } \mathbf{F} = \nabla \times \mathbf{F} \qquad \text{Equation (3)}$$

$$= \begin{vmatrix} \mathbf{i} & \mathbf{j} & \mathbf{k} \\ \dfrac{\partial}{\partial x} & \dfrac{\partial}{\partial y} & \dfrac{\partial}{\partial z} \\ x^2 - y & 4z & x^2 \end{vmatrix}$$

$$= \left(\frac{\partial}{\partial y}(x^2) - \frac{\partial}{\partial z}(4z)\right)\mathbf{i} - \left(\frac{\partial}{\partial x}(x^2) - \frac{\partial}{\partial z}(x^2 - y)\right)\mathbf{j}$$

$$+ \left(\frac{\partial}{\partial x}(4z) - \frac{\partial}{\partial y}(x^2 - y)\right)\mathbf{k}$$

$$= (0 - 4)\mathbf{i} - (2x - 0)\mathbf{j} + (0 + 1)\mathbf{k}$$

$$= -4\mathbf{i} - 2x\mathbf{j} + \mathbf{k} \qquad \blacksquare$$

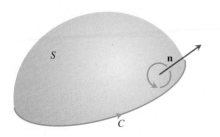

FIGURE 14.51 The orientation of the bounding curve C gives it a right-handed relation to the normal field $\mathbf{n}$. If the thumb of a right hand points along $\mathbf{n}$, the fingers curl in the direction of C.

As we will see, the operator ∇ has a number of other applications. For instance, when applied to a scalar function $f(x, y, z)$, it gives the gradient of f:

$$\nabla f = \frac{\partial f}{\partial x}\mathbf{i} + \frac{\partial f}{\partial y}\mathbf{j} + \frac{\partial f}{\partial z}\mathbf{k}.$$

It is sometimes read as "del f" as well as "grad f."

Stokes' Theorem

Stokes' Theorem says that the circulation of a vector field around the boundary of an oriented surface in space in the direction counterclockwise with respect to the surface's unit normal vector $\mathbf{n}$ (Figure 14.51) equals the integral of the normal component of the curl of the field over the surface. We require that the surface be **piecewise smooth**, which means that it is a finite union of smooth surfaces joining along smooth curves.

THEOREM 6—Stokes' Theorem   Let S be a piecewise smooth oriented surface having a piecewise smooth boundary curve C. Let $\mathbf{F} = M\mathbf{i} + N\mathbf{j} + P\mathbf{k}$ be a vector field whose components have continuous first partial derivatives on an open region containing S. Then the circulation of $\mathbf{F}$ around C in the direction counterclockwise with respect to the surface's unit normal vector $\mathbf{n}$ equals the integral of $\nabla \times \mathbf{F} \cdot \mathbf{n}$ over S.

$$\underset{\substack{C}}{\oint} \mathbf{F} \cdot d\mathbf{r} = \underset{S}{\iint} \nabla \times \mathbf{F} \cdot \mathbf{n}\, d\sigma \qquad (4)$$

$$\underbrace{}_{\substack{\text{Counterclockwise} \\ \text{circulation}}} \qquad \underbrace{}_{\text{Curl integral}}$$

Notice from Equation (4) that if two different oriented surfaces S_1 and S_2 have the same boundary C, their curl integrals are equal:

$$\underset{S_1}{\iint} \nabla \times \mathbf{F} \cdot \mathbf{n}_1\, d\sigma = \underset{S_2}{\iint} \nabla \times \mathbf{F} \cdot \mathbf{n}_2\, d\sigma.$$

Both curl integrals equal the counterclockwise circulation integral on the left side of Equation (4) as long as the unit normal vectors $\mathbf{n}_1$ and $\mathbf{n}_2$ correctly orient the surfaces.

If C is a curve in the xy-plane, oriented counterclockwise, and R is the region in the xy-plane bounded by C, then $d\sigma = dx\, dy$ and

$$(\nabla \times \mathbf{F}) \cdot \mathbf{n} = (\nabla \times \mathbf{F}) \cdot \mathbf{k} = \left(\frac{\partial N}{\partial x} - \frac{\partial M}{\partial y} \right).$$

Under these conditions, Stokes' equation becomes

$$\underset{\substack{C}}{\oint} \mathbf{F} \cdot d\mathbf{r} = \underset{R}{\iint} \left(\frac{\partial N}{\partial x} - \frac{\partial M}{\partial y} \right) dx\, dy,$$

which is the circulation-curl form of the equation in Green's Theorem. Conversely, by reversing these steps we can rewrite the circulation-curl form of Green's Theorem for

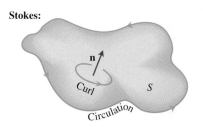

FIGURE 14.52 Comparison of Green's Theorem and Stokes' Theorem.

two-dimensional fields in del notation as

$$\oint_C \mathbf{F} \cdot d\mathbf{r} = \iint_R \nabla \times \mathbf{F} \cdot \mathbf{k} \, dA. \tag{5}$$

See Figure 14.52.

EXAMPLE 2 Evaluate Equation (4) for the hemisphere S: $x^2 + y^2 + z^2 = 9, z \geq 0$, its bounding circle C: $x^2 + y^2 = 9, z = 0$, and the field $\mathbf{F} = y\mathbf{i} - x\mathbf{j}$.

Solution We calculate the counterclockwise circulation around C (as viewed from above) using the parametrization $\mathbf{r}(\theta) = (3 \cos \theta)\mathbf{i} + (3 \sin \theta)\mathbf{j}, 0 \leq \theta \leq 2\pi$:

$$d\mathbf{r} = (-3 \sin \theta \, d\theta)\mathbf{i} + (3 \cos \theta \, d\theta)\mathbf{j}$$

$$\mathbf{F} = y\mathbf{i} - x\mathbf{j} = (3 \sin \theta)\mathbf{i} - (3 \cos \theta)\mathbf{j}$$

$$\mathbf{F} \cdot d\mathbf{r} = -9 \sin^2 \theta \, d\theta - 9 \cos^2 \theta \, d\theta = -9 \, d\theta$$

$$\oint_C \mathbf{F} \cdot d\mathbf{r} = \int_0^{2\pi} -9 \, d\theta = -18\pi.$$

For the curl integral of $\mathbf{F}$, we have

$$\nabla \times \mathbf{F} = \left(\frac{\partial P}{\partial y} - \frac{\partial N}{\partial z}\right)\mathbf{i} + \left(\frac{\partial M}{\partial z} - \frac{\partial P}{\partial x}\right)\mathbf{j} + \left(\frac{\partial N}{\partial x} - \frac{\partial M}{\partial y}\right)\mathbf{k}$$

$$= (0 - 0)\mathbf{i} + (0 - 0)\mathbf{j} + (-1 - 1)\mathbf{k} = -2\mathbf{k}$$

$$\mathbf{n} = \frac{x\mathbf{i} + y\mathbf{j} + z\mathbf{k}}{\sqrt{x^2 + y^2 + z^2}} = \frac{x\mathbf{i} + y\mathbf{j} + z\mathbf{k}}{3} \qquad \text{Outer unit normal}$$

$$d\sigma = \frac{3}{z} \, dA \qquad \text{Section 14.6, Example 5, with } a = 3$$

$$\nabla \times \mathbf{F} \cdot \mathbf{n} \, d\sigma = -\frac{2z}{3} \frac{3}{z} \, dA = -2 \, dA$$

and

$$\iint_S \nabla \times \mathbf{F} \cdot \mathbf{n} \, d\sigma = \iint_{x^2+y^2 \leq 9} -2 \, dA = -18\pi.$$

The circulation around the circle equals the integral of the curl over the hemisphere, as it should. ∎

EXAMPLE 3 Find the circulation of the field $\mathbf{F} = (x^2 - y)\mathbf{i} + 4z\mathbf{j} + x^2\mathbf{k}$ around the curve C in which the plane $z = 2$ meets the cone $z = \sqrt{x^2 + y^2}$, counterclockwise as viewed from above (Figure 14.53).

Solution Stokes' Theorem enables us to find the circulation by integrating over the surface of the cone. Traversing C in the counterclockwise direction viewed from above corresponds to taking the *inner* normal $\mathbf{n}$ to the cone, the normal with a positive z-component.

We parametrize the cone as

$$\mathbf{r}(r, \theta) = (r \cos \theta)\mathbf{i} + (r \sin \theta)\mathbf{j} + r\mathbf{k}, \qquad 0 \leq r \leq 2, \quad 0 \leq \theta \leq 2\pi.$$

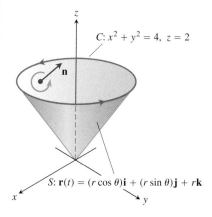

FIGURE 14.53 The curve C and cone S in Example 3.

We then have

$$\mathbf{n} = \frac{\mathbf{r}_r \times \mathbf{r}_\theta}{|\mathbf{r}_r \times \mathbf{r}_\theta|} = \frac{-(r\cos\theta)\mathbf{i} - (r\sin\theta)\mathbf{j} + r\mathbf{k}}{r\sqrt{2}} \qquad \text{Section 14.5, Example 4}$$

$$= \frac{1}{\sqrt{2}}\left(-(\cos\theta)\mathbf{i} - (\sin\theta)\mathbf{j} + \mathbf{k}\right)$$

$$d\sigma = r\sqrt{2}\,dr\,d\theta \qquad\qquad\qquad \text{Section 14.5, Example 4}$$

$$\nabla \times \mathbf{F} = -4\mathbf{i} - 2x\mathbf{j} + \mathbf{k} \qquad\qquad \text{Example 1}$$

$$= -4\mathbf{i} - 2r\cos\theta\mathbf{j} + \mathbf{k}. \qquad\quad x = r\cos\theta$$

Accordingly,

$$\nabla \times \mathbf{F}\cdot\mathbf{n} = \frac{1}{\sqrt{2}}\left(4\cos\theta + 2r\cos\theta\sin\theta + 1\right)$$

$$= \frac{1}{\sqrt{2}}\left(4\cos\theta + r\sin 2\theta + 1\right)$$

and the circulation is

$$\oint_C \mathbf{F}\cdot d\mathbf{r} = \iint_S \nabla \times \mathbf{F}\cdot\mathbf{n}\,d\sigma \qquad \text{Stokes' Theorem, Equation (4)}$$

$$= \int_0^{2\pi}\int_0^2 \frac{1}{\sqrt{2}}\left(4\cos\theta + r\sin 2\theta + 1\right)\left(r\sqrt{2}\,dr\,d\theta\right) = 4\pi. \qquad \blacksquare$$

Paddle Wheel Interpretation of $\nabla \times \mathbf{F}$

Suppose that $\mathbf{v}(x, y, z)$ is the velocity of a moving fluid whose density at (x, y, z) is $\delta(x, y, z)$ and let $\mathbf{F} = \delta\mathbf{v}$. Then

$$\oint_C \mathbf{F}\cdot d\mathbf{r}$$

is the circulation of the fluid around the closed curve C. By Stokes' Theorem, the circulation is equal to the flux of $\nabla \times \mathbf{F}$ through a surface S spanning C:

$$\oint_C \mathbf{F}\cdot d\mathbf{r} = \iint_S \nabla \times \mathbf{F}\cdot\mathbf{n}\,d\sigma.$$

Suppose we fix a point Q in the domain of $\mathbf{F}$ and a direction $\mathbf{u}$ at Q. Let C be a circle of radius ρ, with center at Q, whose plane is normal to $\mathbf{u}$. If $\nabla \times \mathbf{F}$ is continuous at Q, the average value of the $\mathbf{u}$-component of $\nabla \times \mathbf{F}$ over the circular disk S bounded by C approaches the $\mathbf{u}$-component of $\nabla \times \mathbf{F}$ at Q as $\rho \to 0$:

$$(\nabla \times \mathbf{F}\cdot\mathbf{u})_Q = \lim_{\rho\to 0}\frac{1}{\pi\rho^2}\iint_S \nabla \times \mathbf{F}\cdot\mathbf{u}\,d\sigma.$$

If we replace the surface integral in this last equation by the circulation, we get

$$(\nabla \times \mathbf{F}\cdot\mathbf{u})_Q = \lim_{\rho\to 0}\frac{1}{\pi\rho^2}\oint_C \mathbf{F}\cdot d\mathbf{r}. \qquad (6)$$

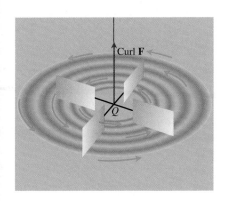

FIGURE 14.54 The paddle wheel interpretation of curl **F**.

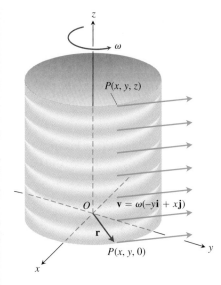

FIGURE 14.55 A steady rotational flow parallel to the xy-plane, with constant angular velocity ω in the positive (counterclockwise) direction (Example 4).

The left-hand side of Equation (6) has its maximum value when **u** is the direction of $\nabla \times \mathbf{F}$. When ρ is small, the limit on the right-hand side of Equation (6) is approximately

$$\frac{1}{\pi\rho^2} \oint_C \mathbf{F} \cdot d\mathbf{r},$$

which is the circulation around C divided by the area of the disk (circulation density). Suppose that a small paddle wheel of radius ρ is introduced into the fluid at Q, with its axle directed along **u**. The circulation of the fluid around C will affect the rate of spin of the paddle wheel. The wheel will spin fastest when the circulation integral is maximized; therefore it will spin fastest when the axle of the paddle wheel points in the direction of $\nabla \times \mathbf{F}$ (Figure 14.54).

EXAMPLE 4 A fluid of constant density rotates around the z-axis with velocity $\mathbf{v} = \omega(-y\mathbf{i} + x\mathbf{j})$, where ω is a positive constant called the *angular velocity* of the rotation (Figure 14.55). If $\mathbf{F} = \mathbf{v}$, find $\nabla \times \mathbf{F}$ and relate it to the circulation density.

Solution With $\mathbf{F} = \mathbf{v} = -\omega y\mathbf{i} + \omega x\mathbf{j}$,

$$\nabla \times \mathbf{F} = \left(\frac{\partial P}{\partial y} - \frac{\partial N}{\partial z}\right)\mathbf{i} + \left(\frac{\partial M}{\partial z} - \frac{\partial P}{\partial x}\right)\mathbf{j} + \left(\frac{\partial N}{\partial x} - \frac{\partial M}{\partial y}\right)\mathbf{k}$$

$$= (0 - 0)\mathbf{i} + (0 - 0)\mathbf{j} + (\omega - (-\omega))\mathbf{k} = 2\omega\mathbf{k}.$$

By Stokes' Theorem, the circulation of **F** around a circle C of radius ρ bounding a disk S in a plane normal to $\nabla \times \mathbf{F}$, say the xy-plane, is

$$\oint_C \mathbf{F} \cdot d\mathbf{r} = \iint_S \nabla \times \mathbf{F} \cdot \mathbf{n}\, d\sigma = \iint_S 2\omega\mathbf{k} \cdot \mathbf{k}\, dx\, dy = (2\omega)(\pi\rho^2).$$

Thus,

$$(\nabla \times \mathbf{F}) \cdot \mathbf{k} = 2\omega = \frac{1}{\pi\rho^2}\oint_C \mathbf{F} \cdot d\mathbf{r},$$

consistent with Equation (6) when $\mathbf{u} = \mathbf{k}$. ∎

EXAMPLE 5 Use Stokes' Theorem to evaluate $\int_C \mathbf{F} \cdot d\mathbf{r}$, if $\mathbf{F} = xz\mathbf{i} + xy\mathbf{j} + 3xz\mathbf{k}$ and C is the boundary of the portion of the plane $2x + y + z = 2$ in the first octant, traversed counterclockwise as viewed from above (Figure 14.56).

Solution The plane is the level surface $f(x, y, z) = 2$ of the function $f(x, y, z) = 2x + y + z$. The unit normal vector

$$\mathbf{n} = \frac{\nabla f}{|\nabla f|} = \frac{(2\mathbf{i} + \mathbf{j} + \mathbf{k})}{|2\mathbf{i} + \mathbf{j} + \mathbf{k}|} = \frac{1}{\sqrt{6}}\left(2\mathbf{i} + \mathbf{j} + \mathbf{k}\right)$$

is consistent with the counterclockwise motion around C. To apply Stokes' Theorem, we find

$$\text{curl } \mathbf{F} = \nabla \times \mathbf{F} = \begin{vmatrix} \mathbf{i} & \mathbf{j} & \mathbf{k} \\ \dfrac{\partial}{\partial x} & \dfrac{\partial}{\partial y} & \dfrac{\partial}{\partial z} \\ xz & xy & 3xz \end{vmatrix} = (x - 3z)\mathbf{j} + y\mathbf{k}.$$

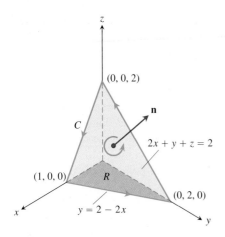

FIGURE 14.56   The planar surface in Example 5.

On the plane, z equals $2 - 2x - y$, so

$$\nabla \times \mathbf{F} = (x - 3(2 - 2x - y))\mathbf{j} + y\mathbf{k} = (7x + 3y - 6)\mathbf{j} + y\mathbf{k}$$

and

$$\nabla \times \mathbf{F} \cdot \mathbf{n} = \frac{1}{\sqrt{6}}\left(7x + 3y - 6 + y\right) = \frac{1}{\sqrt{6}}\left(7x + 4y - 6\right).$$

The surface area element is

$$d\sigma = \frac{|\nabla f|}{|\nabla f \cdot \mathbf{k}|}\, dA = \frac{\sqrt{6}}{1}\, dx\, dy.$$

The circulation is

$$\oint_C \mathbf{F} \cdot d\mathbf{r} = \iint_S \nabla \times \mathbf{F} \cdot \mathbf{n}\, d\sigma \qquad \text{Stokes' Theorem, Equation (4)}$$

$$= \int_0^1 \int_0^{2-2x} \frac{1}{\sqrt{6}}\left(7x + 4y - 6\right)\sqrt{6}\, dy\, dx$$

$$= \int_0^1 \int_0^{2-2x} (7x + 4y - 6)\, dy\, dx = -1. \qquad \blacksquare$$

Proof of Stokes' Theorem for Polyhedral Surfaces

Let S be a polyhedral surface consisting of a finite number of plane regions. (See Figure 14.57 for an example.) We apply Green's Theorem to each separate panel of S. There are two types of panels:

1. Those that are surrounded on all sides by other panels

2. Those that have one or more edges that are not adjacent to other panels.

FIGURE 14.57   Part of a polyhedral surface.

The boundary Δ of S consists of those edges of the type 2 panels that are not adjacent to other panels. In Figure 14.57, the triangles EAB, BCE, and CDE represent a part of S, with $ABCD$ part of the boundary Δ. We apply a generalized tangential form of Green's Theorem to the three triangles in turn and add the results to get

$$\left(\oint_{EAB} + \oint_{BCE} + \oint_{CDE}\right)\mathbf{F} \cdot d\mathbf{r} = \left(\iint_{EAB} + \iint_{BCE} + \iint_{CDE}\right)\nabla \times \mathbf{F} \cdot \mathbf{n}\, d\sigma. \qquad (7)$$

In the generalized form, the line integral of $\mathbf{F}$ around the curve enclosing the plane region R normal to $\mathbf{n}$ equals the double integral of (curl $\mathbf{F}$) $\cdot$ $\mathbf{n}$ over R.

The three line integrals on the left-hand side of Equation (7) combine into a single line integral taken around the periphery $ABCDE$ because the integrals along interior segments cancel in pairs. For example, the integral along segment BE in triangle ABE is opposite in sign to the integral along the same segment in triangle EBC. The same holds for segment CE. Hence, Equation (7) reduces to

$$\oint_{ABCDE} \mathbf{F} \cdot d\mathbf{r} = \iint_{ABCDE} \nabla \times \mathbf{F} \cdot \mathbf{n}\, d\sigma.$$

FIGURE 14.58 Stokes' Theorem also holds for oriented surfaces with holes.

When we apply the generalized form of Green's Theorem to all the panels and add the results, we get

$$\oint_\Delta \mathbf{F} \cdot d\mathbf{r} = \iint_S \nabla \times \mathbf{F} \cdot \mathbf{n} \, d\sigma.$$

This is Stokes' Theorem for a polyhedral surface S. You can find proofs for more general surfaces in advanced calculus texts. (More general surfaces are limits of polyhedral surfaces and the proof can be extended.)

Stokes' Theorem for Surfaces with Holes

Stokes' Theorem holds for an oriented surface S that has one or more holes (Figure 14.58). The surface integral over S of the normal component of $\nabla \times \mathbf{F}$ equals the sum of the line integrals around all the boundary curves of the tangential component of $\mathbf{F}$, where the curves are to be traced in the direction induced by the orientation of S.

An Important Identity

The following identity arises frequently in mathematics and the physical sciences.

$$\text{curl grad } f = \mathbf{0} \qquad \text{or} \qquad \nabla \times \nabla f = \mathbf{0} \qquad (8)$$

This identity holds for any function $f(x, y, z)$ whose second partial derivatives are continuous. The proof goes like this:

$$\nabla \times \nabla f = \begin{vmatrix} \mathbf{i} & \mathbf{j} & \mathbf{k} \\ \dfrac{\partial}{\partial x} & \dfrac{\partial}{\partial y} & \dfrac{\partial}{\partial z} \\ \dfrac{\partial f}{\partial x} & \dfrac{\partial f}{\partial y} & \dfrac{\partial f}{\partial z} \end{vmatrix} = (f_{zy} - f_{yz})\mathbf{i} - (f_{zx} - f_{xz})\mathbf{j} + (f_{yx} - f_{xy})\mathbf{k}.$$

If the second partial derivatives are continuous, the mixed second derivatives in parentheses are equal (Theorem 2, Section 12.3) and the vector is zero.

Conservative Fields and Stokes' Theorem

In Section 14.3, we found that a field $\mathbf{F}$ being conservative in an open region D in space is equivalent to the integral of $\mathbf{F}$ around every closed loop in D being zero. This, in turn, is equivalent in *simply connected* open regions to saying that $\nabla \times \mathbf{F} = \mathbf{0}$.

> **THEOREM 7—Curl F = 0 Related to the Closed-Loop Property** If $\nabla \times \mathbf{F} = \mathbf{0}$ at every point of a simply connected open region D in space, then on any piecewise-smooth closed path C in D,
>
> $$\oint_C \mathbf{F} \cdot d\mathbf{r} = 0.$$

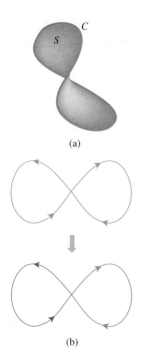

FIGURE 14.59 (a) In a simply connected open region in space, a simple closed curve C is the boundary of a smooth surface S. (b) Smooth curves that cross themselves can be divided into loops to which Stokes' Theorem applies.

Sketch of a Proof Theorem 7 is usually proved in two steps. The first step is for simple closed curves (loops that do not cross themselves), like the one in Figure 14.59a. A theorem from topology, a branch of advanced mathematics, states that every smooth simple closed curve C in a simply connected open region D is the boundary of a smooth two-sided surface S that also lies in D. Hence, by Stokes' Theorem,

$$\oint_C \mathbf{F} \cdot d\mathbf{r} = \iint_S \nabla \times \mathbf{F} \cdot \mathbf{n} \, d\sigma = 0.$$

The second step is for curves that cross themselves, like the one in Figure 14.59b. The idea is to break these into simple loops spanned by orientable surfaces, apply Stokes' Theorem one loop at a time, and add the results. ∎

The following diagram summarizes the results for conservative fields defined on connected, simply connected open regions.

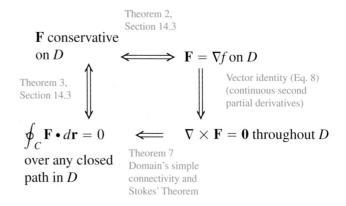

EXERCISES 14.7

In Exercises 1–6, use the surface integral in Stokes' Theorem to calculate the circulation of the field $\mathbf{F}$ around the curve C in the indicated direction.

1. $\mathbf{F} = x^2\mathbf{i} + 2x\mathbf{j} + z^2\mathbf{k}$
   C: The ellipse $4x^2 + y^2 = 4$ in the xy-plane, counterclockwise when viewed from above

2. $\mathbf{F} = 2y\mathbf{i} + 3x\mathbf{j} - z^2\mathbf{k}$
   C: The circle $x^2 + y^2 = 9$ in the xy-plane, counterclockwise when viewed from above

3. $\mathbf{F} = y\mathbf{i} + xz\mathbf{j} + x^2\mathbf{k}$
   C: The boundary of the triangle cut from the plane $x + y + z = 1$ by the first octant, counterclockwise when viewed from above

4. $\mathbf{F} = (y^2 + z^2)\mathbf{i} + (x^2 + z^2)\mathbf{j} + (x^2 + y^2)\mathbf{k}$
   C: The boundary of the triangle cut from the plane $x + y + z = 1$ by the first octant, counterclockwise when viewed from above

5. $\mathbf{F} = (y^2 + z^2)\mathbf{i} + (x^2 + y^2)\mathbf{j} + (x^2 + y^2)\mathbf{k}$
   C: The square bounded by the lines $x = \pm 1$ and $y = \pm 1$ in the xy-plane, counterclockwise when viewed from above

6. $\mathbf{F} = x^2y^3\mathbf{i} + \mathbf{j} + z\mathbf{k}$
   C: The intersection of the cylinder $x^2 + y^2 = 4$ and the hemisphere $x^2 + y^2 + z^2 = 16, z \geq 0$, counterclockwise when viewed from above.

7. Let $\mathbf{n}$ be the outer unit normal of the elliptical shell

   $$S: \quad 4x^2 + 9y^2 + 36z^2 = 36, \qquad z \geq 0,$$

   and let

   $$\mathbf{F} = y\mathbf{i} + x^2\mathbf{j} + (x^2 + y^4)^{3/2} \sin e^{\sqrt{xyz}}\,\mathbf{k}.$$

   Find the value of

   $$\iint_S \nabla \times \mathbf{F} \cdot \mathbf{n} \, d\sigma.$$

   (*Hint:* One parametrization of the ellipse at the base of the shell is $x = 3\cos t, y = 2\sin t, 0 \leq t \leq 2\pi$.)

8. Let $\mathbf{n}$ be the outer unit normal (normal away from the origin) of the parabolic shell

   $$S: \quad 4x^2 + y + z^2 = 4, \qquad y \geq 0,$$

and let

$$\mathbf{F} = \left(-z + \frac{1}{2+x}\right)\mathbf{i} + (\tan^{-1}y)\mathbf{j} + \left(x + \frac{1}{4+z}\right)\mathbf{k}.$$

Find the value of

$$\iint_S \nabla \times \mathbf{F} \cdot \mathbf{n} \, d\sigma.$$

9. Let S be the cylinder $x^2 + y^2 = a^2$, $0 \le z \le h$, together with its top, $x^2 + y^2 \le a^2$, $z = h$. Let $\mathbf{F} = -y\mathbf{i} + x\mathbf{j} + x^2\mathbf{k}$. Use Stokes' Theorem to find the flux of $\nabla \times \mathbf{F}$ outward through S.

10. Evaluate

$$\iint_S \nabla \times (y\mathbf{i}) \cdot \mathbf{n} \, d\sigma,$$

where S is the hemisphere $x^2 + y^2 + z^2 = 1$, $z \ge 0$.

11. Flux of curl F   Show that

$$\iint_S \nabla \times \mathbf{F} \cdot \mathbf{n} \, d\sigma$$

has the same value for all oriented surfaces S that span C and that induce the same positive direction on C.

12. Let $\mathbf{F}$ be a differentiable vector field defined on a region containing a smooth closed oriented surface S and its interior. Let $\mathbf{n}$ be the unit normal vector field on S. Suppose that S is the union of two surfaces S_1 and S_2 joined along a smooth simple closed curve C. Can anything be said about

$$\iint_S \nabla \times \mathbf{F} \cdot \mathbf{n} \, d\sigma?$$

Give reasons for your answer.

In Exercises 13–18, use the surface integral in Stokes' Theorem to calculate the flux of the curl of the field $\mathbf{F}$ across the surface S in the direction of the outward unit normal $\mathbf{n}$.

13. $\mathbf{F} = 2z\mathbf{i} + 3x\mathbf{j} + 5y\mathbf{k}$
  S: $\mathbf{r}(r, \theta) = (r\cos\theta)\mathbf{i} + (r\sin\theta)\mathbf{j} + (4 - r^2)\mathbf{k}$,
  $0 \le r \le 2$, $0 \le \theta \le 2\pi$

14. $\mathbf{F} = (y - z)\mathbf{i} + (z - x)\mathbf{j} + (x + z)\mathbf{k}$
  S: $\mathbf{r}(r, \theta) = (r\cos\theta)\mathbf{i} + (r\sin\theta)\mathbf{j} + (9 - r^2)\mathbf{k}$,
  $0 \le r \le 3$, $0 \le \theta \le 2\pi$

15. $\mathbf{F} = x^2y\mathbf{i} + 2y^3z\mathbf{j} + 3z\mathbf{k}$
  S: $\mathbf{r}(r, \theta) = (r\cos\theta)\mathbf{i} + (r\sin\theta)\mathbf{j} + r\mathbf{k}$,
  $0 \le r \le 1$, $0 \le \theta \le 2\pi$

16. $\mathbf{F} = (x - y)\mathbf{i} + (y - z)\mathbf{j} + (z - x)\mathbf{k}$
  S: $\mathbf{r}(r, \theta) = (r\cos\theta)\mathbf{i} + (r\sin\theta)\mathbf{j} + (5 - r)\mathbf{k}$,
  $0 \le r \le 5$, $0 \le \theta \le 2\pi$

17. $\mathbf{F} = 3y\mathbf{i} + (5 - 2x)\mathbf{j} + (z^2 - 2)\mathbf{k}$
  S: $\mathbf{r}(\phi, \theta) = \left(\sqrt{3}\sin\phi\cos\theta\right)\mathbf{i} + \left(\sqrt{3}\sin\phi\sin\theta\right)\mathbf{j} + \left(\sqrt{3}\cos\phi\right)\mathbf{k}$, $0 \le \phi \le \pi/2$, $0 \le \theta \le 2\pi$

18. $\mathbf{F} = y^2\mathbf{i} + z^2\mathbf{j} + x\mathbf{k}$
  S: $\mathbf{r}(\phi, \theta) = (2\sin\phi\cos\theta)\mathbf{i} + (2\sin\phi\sin\theta)\mathbf{j} + (2\cos\phi)\mathbf{k}$,
  $0 \le \phi \le \pi/2$, $0 \le \theta \le 2\pi$

19. Zero circulation   Use the identity $\nabla \times \nabla f = \mathbf{0}$ (Equation (8) in the text) and Stokes' Theorem to show that the circulations of the following fields around the boundary of any smooth orientable surface in space are zero.
  a. $\mathbf{F} = 2x\mathbf{i} + 2y\mathbf{j} + 2z\mathbf{k}$ **b.** $\mathbf{F} = \nabla(xy^2z^3)$
  c. $\mathbf{F} = \nabla \times (x\mathbf{i} + y\mathbf{j} + z\mathbf{k})$ **d.** $\mathbf{F} = \nabla f$

20. Zero circulation   Let $f(x, y, z) = (x^2 + y^2 + z^2)^{-1/2}$. Show that the clockwise circulation of the field $\mathbf{F} = \nabla f$ around the circle $x^2 + y^2 = a^2$ in the xy-plane is zero
  a. by taking $\mathbf{r} = (a\cos t)\mathbf{i} + (a\sin t)\mathbf{j}$, $0 \le t \le 2\pi$, and integrating $\mathbf{F} \cdot d\mathbf{r}$ over the circle.
  b. by applying Stokes' Theorem.

21. Let C be a simple closed smooth curve in the plane $2x + 2y + z = 2$, oriented as shown here. Show that

$$\oint_C 2y \, dx + 3z \, dy - x \, dz$$

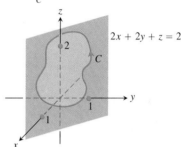

depends only on the area of the region enclosed by C and not on the position or shape of C.

22. Show that if $\mathbf{F} = x\mathbf{i} + y\mathbf{j} + z\mathbf{k}$, then $\nabla \times \mathbf{F} = \mathbf{0}$.

23. Find a vector field with twice-differentiable components whose curl is $x\mathbf{i} + y\mathbf{j} + z\mathbf{k}$ or prove that no such field exists.

24. Does Stokes' Theorem say anything special about circulation in a field whose curl is zero? Give reasons for your answer.

25. Let R be a region in the xy-plane that is bounded by a piecewise-smooth simple closed curve C and suppose that the moments of inertia of R about the x- and y-axes are known to be I_x and I_y. Evaluate the integral

$$\oint_C \nabla(r^4) \cdot \mathbf{n} \, ds,$$

where $r = \sqrt{x^2 + y^2}$, in terms of I_x and I_y.

26. Zero curl, yet field not conservative   Show that the curl of

$$\mathbf{F} = \frac{-y}{x^2 + y^2}\mathbf{i} + \frac{x}{x^2 + y^2}\mathbf{j} + z\mathbf{k}$$

is zero but that

$$\oint_C \mathbf{F} \cdot d\mathbf{r}$$

is not zero if C is the circle $x^2 + y^2 = 1$ in the xy-plane. (Theorem 7 does not apply here because the domain of $\mathbf{F}$ is not simply connected. The field $\mathbf{F}$ is not defined along the z-axis so there is no way to contract C to a point without leaving the domain of $\mathbf{F}$.)

The divergence form of Green's Theorem in the plane states that the net outward flux of a vector field across a simple closed curve can be calculated by integrating the divergence of the field over the region enclosed by the curve. The corresponding theorem in three dimensions, called the Divergence Theorem, states that the net outward flux of a vector field across a closed surface in space can be calculated by integrating the divergence of the field over the region enclosed by the surface. In this section, we prove the Divergence Theorem and show how it simplifies the calculation of flux. We also derive Gauss's law for flux in an electric field and the continuity equation of hydrodynamics. Finally, we unify the chapter's vector integral theorems into a single fundamental theorem.

Divergence in Three Dimensions

The **divergence** of a vector field $\mathbf{F} = M(x, y, z)\mathbf{i} + N(x, y, z)\mathbf{j} + P(x, y, z)\mathbf{k}$ is the scalar function

$$\text{div } \mathbf{F} = \nabla \cdot \mathbf{F} = \frac{\partial M}{\partial x} + \frac{\partial N}{\partial y} + \frac{\partial P}{\partial z}. \tag{1}$$

The symbol "div $\mathbf{F}$" is read as "divergence of $\mathbf{F}$" or "div $\mathbf{F}$." The notation $\nabla \cdot \mathbf{F}$ is read "del dot $\mathbf{F}$."

Div $\mathbf{F}$ has the same physical interpretation in three dimensions that it does in two. If $\mathbf{F}$ is the velocity field of a flowing gas, the value of div $\mathbf{F}$ at a point (x, y, z) is the rate at which the gas is compressing or expanding at (x, y, z). The divergence is the flux per unit volume or flux density at the point.

EXAMPLE 1   Find the divergence of $\mathbf{F} = 2xz\mathbf{i} - xy\mathbf{j} - z\mathbf{k}$.

Solution   The divergence of $\mathbf{F}$ is

$$\nabla \cdot \mathbf{F} = \frac{\partial}{\partial x}(2xz) + \frac{\partial}{\partial y}(-xy) + \frac{\partial}{\partial z}(-z) = 2z - x - 1. \qquad \blacksquare$$

Divergence Theorem

The Divergence Theorem says that under suitable conditions, the outward flux of a vector field across a closed surface (oriented outward) equals the triple integral of the divergence of the field over the region enclosed by the surface.

THEOREM 8—Divergence Theorem   Let $\mathbf{F}$ be a vector field whose components have continuous first partial derivatives, and let S be a piecewise smooth oriented closed surface. The flux of $\mathbf{F}$ across S in the direction of the surface's outward unit normal field $\mathbf{n}$ equals the integral of $\nabla \cdot \mathbf{F}$ over the region D enclosed by the surface:

$$\underset{S}{\iint} \mathbf{F} \cdot \mathbf{n} \, d\sigma = \underset{D}{\iiint} \nabla \cdot \mathbf{F} \, dV. \tag{2}$$

$$\underset{\substack{\text{Outward} \\ \text{flux}}}{} \qquad \underset{\substack{\text{Divergence} \\ \text{integral}}}{}$$

EXAMPLE 2   Evaluate both sides of Equation (2) for the field $\mathbf{F} = x\mathbf{i} + y\mathbf{j} + z\mathbf{k}$ over the sphere $x^2 + y^2 + z^2 = a^2$.

Solution   The outer unit normal to S, calculated from the gradient of $f(x, y, z) = x^2 + y^2 + z^2 - a^2$, is

$$\mathbf{n} = \frac{2(x\mathbf{i} + y\mathbf{j} + z\mathbf{k})}{\sqrt{4(x^2 + y^2 + z^2)}} = \frac{x\mathbf{i} + y\mathbf{j} + z\mathbf{k}}{a}.$$

Hence,

$$\mathbf{F} \cdot \mathbf{n} \, d\sigma = \frac{x^2 + y^2 + z^2}{a} \, d\sigma = \frac{a^2}{a} \, d\sigma = a \, d\sigma$$

because $x^2 + y^2 + z^2 = a^2$ on the surface. Therefore,

$$\iint_S \mathbf{F} \cdot \mathbf{n} \, d\sigma = \iint_S a \, d\sigma = a \iint_S d\sigma = a(4\pi a^2) = 4\pi a^3.$$

The divergence of $\mathbf{F}$ is

$$\nabla \cdot \mathbf{F} = \frac{\partial}{\partial x}(x) + \frac{\partial}{\partial y}(y) + \frac{\partial}{\partial z}(z) = 3,$$

so

$$\iiint_D \nabla \cdot \mathbf{F} \, dV = \iiint_D 3 \, dV = 3\left(\frac{4}{3}\pi a^3\right) = 4\pi a^3. \qquad \blacksquare$$

EXAMPLE 3   Find the flux of $\mathbf{F} = xy\mathbf{i} + yz\mathbf{j} + xz\mathbf{k}$ outward through the surface of the cube cut from the first octant by the planes $x = 1, y = 1$, and $z = 1$.

Solution   Instead of calculating the flux as a sum of six separate integrals, one for each face of the cube, we can calculate the flux by integrating the divergence

$$\nabla \cdot \mathbf{F} = \frac{\partial}{\partial x}(xy) + \frac{\partial}{\partial y}(yz) + \frac{\partial}{\partial z}(xz) = y + z + x$$

over the cube's interior:

$$\text{Flux} = \underbrace{\iint \mathbf{F} \cdot \mathbf{n} \, d\sigma}_{\substack{\text{Cube} \\ \text{surface}}} = \underbrace{\iiint \nabla \cdot \mathbf{F} \, dV}_{\substack{\text{Cube} \\ \text{interior}}} \qquad \text{The Divergence Theorem}$$

$$= \int_0^1 \int_0^1 \int_0^1 (x + y + z) \, dx \, dy \, dz = \frac{3}{2}. \qquad \text{Routine integration} \qquad \blacksquare$$

Proof of the Divergence Theorem for Special Regions

To prove the Divergence Theorem, we assume that the components of $\mathbf{F}$ have continuous first partial derivatives. We also assume that D is a convex region with no holes or bubbles, such as a solid ball, cube, or ellipsoid, and that S is a piecewise smooth surface. In addition, we assume that any line perpendicular to the xy-plane at an interior point of the

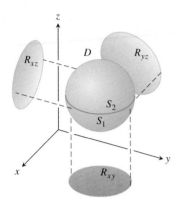

FIGURE 14.60 We prove the Divergence Theorem for the kind of three-dimensional region shown here.

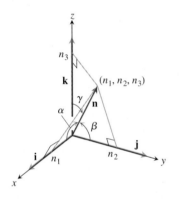

FIGURE 14.61 The components of **n** are the cosines of the angles α, β, and γ that it makes with **i**, **j**, and **k**.

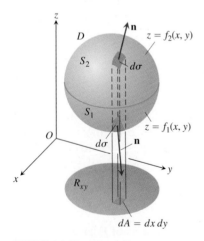

FIGURE 14.62 The region D enclosed by the surfaces S_1 and S_2 projects vertically onto R_{xy} in the xy-plane.

region R_{xy} that is the projection of D on the xy-plane intersects the surface S in exactly two points, producing surfaces

$$S_1: \quad z = f_1(x, y), \qquad (x, y) \text{ in } R_{xy}$$

$$S_2: \quad z = f_2(x, y), \qquad (x, y) \text{ in } R_{xy},$$

with $f_1 \leq f_2$. We make similar assumptions about the projection of D onto the other coordinate planes. See Figure 14.60.

The components of the unit normal vector $\mathbf{n} = n_1\mathbf{i} + n_2\mathbf{j} + n_3\mathbf{k}$ are the cosines of the angles α, β, and γ that $\mathbf{n}$ makes with $\mathbf{i}$, $\mathbf{j}$, and $\mathbf{k}$ (Figure 14.61). This is true because all the vectors involved are unit vectors. We have

$$n_1 = \mathbf{n} \cdot \mathbf{i} = |\mathbf{n}|\,|\mathbf{i}| \cos \alpha = \cos \alpha$$

$$n_2 = \mathbf{n} \cdot \mathbf{j} = |\mathbf{n}|\,|\mathbf{j}| \cos \beta = \cos \beta$$

$$n_3 = \mathbf{n} \cdot \mathbf{k} = |\mathbf{n}|\,|\mathbf{k}| \cos \gamma = \cos \gamma$$

Thus,

$$\mathbf{n} = (\cos \alpha)\mathbf{i} + (\cos \beta)\mathbf{j} + (\cos \gamma)\mathbf{k}$$

and

$$\mathbf{F} \cdot \mathbf{n} = M \cos \alpha + N \cos \beta + P \cos \gamma.$$

In component form, the Divergence Theorem states that

$$\iint_S (M \cos \alpha + N \cos \beta + P \cos \gamma)\, d\sigma = \iiint_D \left(\frac{\partial M}{\partial x} + \frac{\partial N}{\partial y} + \frac{\partial P}{\partial z} \right) dx\, dy\, dz.$$

We prove the theorem by proving the three following equalities:

$$\iint_S M \cos \alpha \, d\sigma = \iiint_D \frac{\partial M}{\partial x} \, dx\, dy\, dz \tag{3}$$

$$\iint_S N \cos \beta \, d\sigma = \iiint_D \frac{\partial N}{\partial y} \, dx\, dy\, dz \tag{4}$$

$$\iint_S P \cos \gamma \, d\sigma = \iiint_D \frac{\partial P}{\partial z} \, dx\, dy\, dz \tag{5}$$

Proof of Equation (5) We prove Equation (5) by converting the surface integral on the left to a double integral over the projection R_{xy} of D on the xy-plane (Figure 14.62). The surface S consists of an upper part S_2 whose equation is $z = f_2(x, y)$ and a lower part S_1 whose equation is $z = f_1(x, y)$. On S_2, the outer normal $\mathbf{n}$ has a positive **k**-component and

$$\cos \gamma \, d\sigma = dx\, dy \qquad \text{because} \qquad d\sigma = \frac{dA}{|\cos \gamma|} = \frac{dx\, dy}{\cos \gamma}.$$

See Figure 14.63. On S_1, the outer normal $\mathbf{n}$ has a negative **k**-component and

$$\cos \gamma \, d\sigma = -dx\, dy.$$

Therefore,

$$\iint_S P \cos \gamma \, d\sigma = \iint_{S_2} P \cos \gamma \, d\sigma + \iint_{S_1} P \cos \gamma \, d\sigma$$

$$= \iint_{R_{xy}} P(x, y, f_2(x, y)) \, dx \, dy - \iint_{R_{xy}} P(x, y, f_1(x, y)) \, dx \, dy$$

$$= \iint_{R_{xy}} [P(x, y, f_2(x, y)) - P(x, y, f_1(x, y))] \, dx \, dy$$

$$= \iint_{R_{xy}} \left[\int_{f_1(x,y)}^{f_2(x,y)} \frac{\partial P}{\partial z} \, dz \right] dx \, dy = \iiint_{D} \frac{\partial P}{\partial z} \, dz \, dx \, dy.$$

This proves Equation (5). ∎

The proofs for Equations (3) and (4) follow the same pattern; or just permute x, y, z; M, N, P; α, β, γ, in order, and get those results from Equation (5).

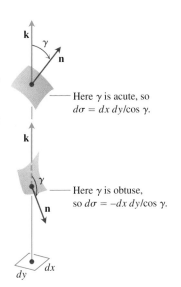

FIGURE 14.63 An enlarged view of the area patches in Figure 14.62. The relations $d\sigma = \pm dx \, dy / \cos \gamma$ come from Eq. (7) in Section 14.5.

In the figure:
- Here γ is acute, so $d\sigma = dx \, dy / \cos \gamma.$
- Here γ is obtuse, so $d\sigma = -dx \, dy / \cos \gamma.$

Divergence Theorem for Other Regions

The Divergence Theorem can be extended to regions that can be partitioned into a finite number of simple regions of the type just discussed and to regions that can be defined as limits of simpler regions in certain ways. For an example of one step in such a splitting process, suppose that D is the region between two concentric spheres and that $\mathbf{F}$ has continuously differentiable components throughout D and on the bounding surfaces. Split D by an equatorial plane and apply the Divergence Theorem to each half separately. The bottom half, D_1, is shown in Figure 14.64. The surface S_1 that bounds D_1 consists of an outer hemisphere, a plane washer-shaped base, and an inner hemisphere. The Divergence Theorem says that

$$\iint_{S_1} \mathbf{F} \cdot \mathbf{n}_1 \, d\sigma_1 = \iiint_{D_1} \nabla \cdot \mathbf{F} \, dV_1. \tag{6}$$

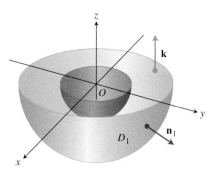

FIGURE 14.64 The lower half of the solid region between two concentric spheres.

The unit normal $\mathbf{n}_1$ that points outward from D_1 points away from the origin along the outer surface, equals $\mathbf{k}$ along the flat base, and points toward the origin along the inner surface. Next apply the Divergence Theorem to D_2, and its surface S_2 (Figure 14.65):

$$\iint_{S_2} \mathbf{F} \cdot \mathbf{n}_2 \, d\sigma_2 = \iiint_{D_2} \nabla \cdot \mathbf{F} \, dV_2. \tag{7}$$

As we follow $\mathbf{n}_2$ over S_2, pointing outward from D_2, we see that $\mathbf{n}_2$ equals $-\mathbf{k}$ along the washer-shaped base in the xy-plane, points away from the origin on the outer sphere, and points toward the origin on the inner sphere. When we add Equations (6) and (7), the integrals over the flat base cancel because of the opposite signs of $\mathbf{n}_1$ and $\mathbf{n}_2$. We thus arrive at the result

$$\iint_{S} \mathbf{F} \cdot \mathbf{n} \, d\sigma = \iiint_{D} \nabla \cdot \mathbf{F} \, dV,$$

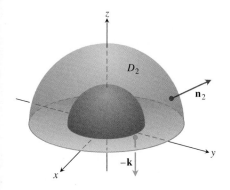

FIGURE 14.65 The upper half of the solid region between two concentric spheres.

with D the region between the spheres, S the boundary of D consisting of two spheres, and $\mathbf{n}$ the unit normal to S directed outward from D.

EXAMPLE 4   Find the net outward flux of the field

$$\mathbf{F} = \frac{x\mathbf{i} + y\mathbf{j} + z\mathbf{k}}{\rho^3}, \qquad \rho = \sqrt{x^2 + y^2 + z^2}$$

across the boundary of the region D: $0 < a^2 \leq x^2 + y^2 + z^2 \leq b^2$.

Solution   The flux can be calculated by integrating $\nabla \cdot \mathbf{F}$ over D. We have

$$\frac{\partial \rho}{\partial x} = \frac{1}{2}(x^2 + y^2 + z^2)^{-1/2}(2x) = \frac{x}{\rho}$$

and

$$\frac{\partial M}{\partial x} = \frac{\partial}{\partial x}(x\rho^{-3}) = \rho^{-3} - 3x\rho^{-4}\frac{\partial \rho}{\partial x} = \frac{1}{\rho^3} - \frac{3x^2}{\rho^5}.$$

Similarly,

$$\frac{\partial N}{\partial y} = \frac{1}{\rho^3} - \frac{3y^2}{\rho^5} \qquad \text{and} \qquad \frac{\partial P}{\partial z} = \frac{1}{\rho^3} - \frac{3z^2}{\rho^5}.$$

Hence,

$$\text{div } \mathbf{F} = \frac{3}{\rho^3} - \frac{3}{\rho^5}(x^2 + y^2 + z^2) = \frac{3}{\rho^3} - \frac{3\rho^2}{\rho^5} = 0$$

and

$$\iiint\limits_{D} \nabla \cdot \mathbf{F} \, dV = 0. \qquad \nabla \cdot \mathbf{F} = \text{div } \mathbf{F}$$

So the integral of $\nabla \cdot \mathbf{F}$ over D is zero and the net outward flux across the boundary of D is zero. There is more to learn from this example, though. The flux leaving D across the inner sphere S_a is the negative of the flux leaving D across the outer sphere S_b (because the sum of these fluxes is zero). Hence, the flux of $\mathbf{F}$ across S_a in the direction away from the origin equals the flux of $\mathbf{F}$ across S_b in the direction away from the origin. Thus, the flux of $\mathbf{F}$ across a sphere centered at the origin is independent of the radius of the sphere. What is this flux?

To find it, we evaluate the flux integral directly. The outward unit normal on the sphere of radius a is

$$\mathbf{n} = \frac{x\mathbf{i} + y\mathbf{j} + z\mathbf{k}}{\sqrt{x^2 + y^2 + z^2}} = \frac{x\mathbf{i} + y\mathbf{j} + z\mathbf{k}}{a}.$$

Hence, on the sphere,

$$\mathbf{F} \cdot \mathbf{n} = \frac{x\mathbf{i} + y\mathbf{j} + z\mathbf{k}}{a^3} \cdot \frac{x\mathbf{i} + y\mathbf{j} + z\mathbf{k}}{a} = \frac{x^2 + y^2 + z^2}{a^4} = \frac{a^2}{a^4} = \frac{1}{a^2}$$

and

$$\iint\limits_{S_a} \mathbf{F} \cdot \mathbf{n} \, d\sigma = \frac{1}{a^2}\iint\limits_{S_a} d\sigma = \frac{1}{a^2}(4\pi a^2) = 4\pi.$$

The outward flux of $\mathbf{F}$ across any sphere centered at the origin is 4π.   ∎

Gauss's Law: One of the Four Great Laws of Electromagnetic Theory

There is still more to be learned from Example 4. In electromagnetic theory, the electric field created by a point charge q located at the origin is

$$\mathbf{E}(x, y, z) = \frac{1}{4\pi\epsilon_0} \frac{q}{|\mathbf{r}|^2} \left(\frac{\mathbf{r}}{|\mathbf{r}|} \right) = \frac{q}{4\pi\epsilon_0} \frac{\mathbf{r}}{|\mathbf{r}|^3} = \frac{q}{4\pi\epsilon_0} \frac{x\mathbf{i} + y\mathbf{j} + z\mathbf{k}}{\rho^3},$$

where ϵ_0 is a physical constant, $\mathbf{r}$ is the position vector of the point (x, y, z), and $\rho = |\mathbf{r}| = \sqrt{x^2 + y^2 + z^2}$. In the notation of Example 4,

$$\mathbf{E} = \frac{q}{4\pi\epsilon_0} \mathbf{F}.$$

The calculations in Example 4 show that the outward flux of $\mathbf{E}$ across any sphere centered at the origin is q/ϵ_0, but this result is not confined to spheres. The outward flux of $\mathbf{E}$ across any closed surface S that encloses the origin (and to which the Divergence Theorem applies) is also q/ϵ_0. To see why, we have only to imagine a large sphere S_a centered at the origin and enclosing the surface S. Since

$$\nabla \cdot \mathbf{E} = \nabla \cdot \frac{q}{4\pi\epsilon_0} \mathbf{F} = \frac{q}{4\pi\epsilon_0} \nabla \cdot \mathbf{F} = 0$$

when $\rho > 0$, the integral of $\nabla \cdot \mathbf{E}$ over the region D between S and S_a is zero. Hence, by the Divergence Theorem,

$$\iint\limits_{\substack{\text{Boundary} \\ \text{of } D}} \mathbf{E} \cdot \mathbf{n} \, d\sigma = 0,$$

and the flux of $\mathbf{E}$ across S in the direction away from the origin must be the same as the flux of $\mathbf{E}$ across S_a in the direction away from the origin, which is q/ϵ_0. This statement, called *Gauss's Law*, also applies to charge distributions that are more general than the one assumed here, as you will see in nearly any physics text.

$$\text{Gauss's law:} \quad \iint\limits_{S} \mathbf{E} \cdot \mathbf{n} \, d\sigma = \frac{q}{\epsilon_0}$$

Continuity Equation of Hydrodynamics

Let D be a region in space bounded by a closed oriented surface S. If $\mathbf{v}(x, y, z)$ is the velocity field of a fluid flowing smoothly through D, $\delta = \delta(t, x, y, z)$ is the fluid's density at (x, y, z) at time t, and $\mathbf{F} = \delta\mathbf{v}$, then the **continuity equation** of hydrodynamics states that

$$\nabla \cdot \mathbf{F} + \frac{\partial \delta}{\partial t} = 0.$$

If the functions involved have continuous first partial derivatives, the equation evolves naturally from the Divergence Theorem, as we now see.

First, the integral

$$\iint\limits_{S} \mathbf{F} \cdot \mathbf{n} \, d\sigma$$

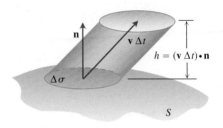

FIGURE 14.66 The fluid that flows upward through the patch $\Delta\sigma$ in a short time Δt fills a "cylinder" whose volume is approximately base $\times$ height = $\mathbf{v} \cdot \mathbf{n} \, \Delta\sigma \, \Delta t$.

is the rate at which mass leaves D across S (leaves because $\mathbf{n}$ is the outer normal). To see why, consider a patch of area $\Delta\sigma$ on the surface (Figure 14.66). In a short time interval Δt, the volume ΔV of fluid that flows across the patch is approximately equal to the volume of a cylinder with base area $\Delta\sigma$ and height $(\mathbf{v}\Delta t) \cdot \mathbf{n}$, where $\mathbf{v}$ is a velocity vector rooted at a point of the patch:

$$\Delta V \approx \mathbf{v} \cdot \mathbf{n} \, \Delta\sigma \, \Delta t.$$

The mass of this volume of fluid is about

$$\Delta m \approx \delta\mathbf{v} \cdot \mathbf{n} \, \Delta\sigma \, \Delta t,$$

so the rate at which mass is flowing out of D across the patch is about

$$\frac{\Delta m}{\Delta t} \approx \delta\mathbf{v} \cdot \mathbf{n} \, \Delta\sigma.$$

This leads to the approximation

$$\sum \frac{\Delta m}{\Delta t} \approx \sum \delta\mathbf{v} \cdot \mathbf{n} \, \Delta\sigma$$

as an estimate of the average rate at which mass flows across S. Finally, letting $\Delta\sigma \to 0$ and $\Delta t \to 0$ gives the instantaneous rate at which mass leaves D across S as

$$\frac{dm}{dt} = \iint\limits_{S} \delta\mathbf{v} \cdot \mathbf{n} \, d\sigma,$$

which for our particular flow is

$$\frac{dm}{dt} = \iint\limits_{S} \mathbf{F} \cdot \mathbf{n} \, d\sigma.$$

Now let B be a solid sphere centered at a point Q in the flow. The average value of $\nabla \cdot \mathbf{F}$ over B is

$$\frac{1}{\text{volume of } B} \iiint\limits_{B} \nabla \cdot \mathbf{F} \, dV.$$

It is a consequence of the continuity of the divergence that $\nabla \cdot \mathbf{F}$ actually takes on this value at some point P in B. Thus,

$$(\nabla \cdot \mathbf{F})_P = \frac{1}{\text{volume of } B} \iiint\limits_{B} \nabla \cdot \mathbf{F} \, dV = \frac{\displaystyle\iint\limits_{S} \mathbf{F} \cdot \mathbf{n} \, d\sigma}{\text{volume of } B}$$

$$= \frac{\text{rate at which mass leaves } B \text{ across its surface } S}{\text{volume of } B} \tag{8}$$

The fraction on the right describes decrease in mass per unit volume.

Now let the radius of B approach zero while the center Q stays fixed. The left side of Equation (8) converges to $(\nabla \cdot \mathbf{F})_Q$, the right side to $(-\partial\delta/\partial t)_Q$. The equality of these two limits is the continuity equation

$$\nabla \cdot \mathbf{F} = -\frac{\partial\delta}{\partial t}.$$

The continuity equation "explains" $\nabla \cdot \mathbf{F}$: The divergence of $\mathbf{F}$ at a point is the rate at which the density of the fluid is decreasing there.

The Divergence Theorem

$$\iint\limits_{S} \mathbf{F} \cdot \mathbf{n} \, d\sigma = \iiint\limits_{D} \nabla \cdot \mathbf{F} \, dV$$

now says that the net decrease in density of the fluid in region D is accounted for by the mass transported across the surface S. So, the theorem is a statement about conservation of mass (Exercise 31).

Unifying the Integral Theorems

If we think of a two-dimensional field $\mathbf{F} = M(x, y)\mathbf{i} + N(x, y)\mathbf{j}$ as a three-dimensional field whose $\mathbf{k}$-component is zero, then $\nabla \cdot \mathbf{F} = (\partial M/\partial x) + (\partial N/\partial y)$ and the normal form of Green's Theorem can be written as

$$\oint_{C} \mathbf{F} \cdot \mathbf{n} \, ds = \iint\limits_{R} \left(\frac{\partial M}{\partial x} + \frac{\partial N}{\partial y} \right) dx \, dy = \iint\limits_{R} \nabla \cdot \mathbf{F} \, dA.$$

Similarly, $\nabla \times \mathbf{F} \cdot \mathbf{k} = (\partial N/\partial x) - (\partial M/\partial y)$, so the tangential form of Green's Theorem can be written as

$$\oint_{C} \mathbf{F} \cdot d\mathbf{r} = \iint\limits_{R} \left(\frac{\partial N}{\partial x} - \frac{\partial M}{\partial y} \right) dx \, dy = \iint\limits_{R} \nabla \times \mathbf{F} \cdot \mathbf{k} \, dA.$$

With the equations of Green's Theorem now in del notation, we can see their relationships to the equations in Stokes' Theorem and the Divergence Theorem.

Green's Theorem and Its Generalization to Three Dimensions

Normal form of Green's Theorem: $\displaystyle\oint_{C} \mathbf{F} \cdot \mathbf{n} \, ds = \iint\limits_{R} \nabla \cdot \mathbf{F} \, dA$

Divergence Theorem: $\displaystyle\iint\limits_{S} \mathbf{F} \cdot \mathbf{n} \, d\sigma = \iiint\limits_{D} \nabla \cdot \mathbf{F} \, dV$

Tangential form of Green's Theorem: $\displaystyle\oint_{C} \mathbf{F} \cdot d\mathbf{r} = \iint\limits_{R} \nabla \times \mathbf{F} \cdot \mathbf{k} \, dA$

Stokes' Theorem: $\displaystyle\oint_{C} \mathbf{F} \cdot d\mathbf{r} = \iint\limits_{S} \nabla \times \mathbf{F} \cdot \mathbf{n} \, d\sigma$

Notice how Stokes' Theorem generalizes the tangential (curl) form of Green's Theorem from a flat surface in the plane to a surface in three-dimensional space. In each case, the integral of the normal component of curl $\mathbf{F}$ over the interior of the surface equals the circulation of $\mathbf{F}$ around the boundary.

Likewise, the Divergence Theorem generalizes the normal (flux) form of Green's Theorem from a two-dimensional region in the plane to a three-dimensional region in space.

In each case, the integral of $\nabla \cdot \mathbf{F}$ over the interior of the region equals the total flux of the field across the boundary.

There is still more to be learned here. All these results can be thought of as forms of a *single fundamental theorem*. Think back to the Fundamental Theorem of Calculus in Section 5.4. It says that if $f(x)$ is differentiable on (a, b) and continuous on $[a, b]$, then

$$\int_a^b \frac{df}{dx}\,dx = f(b) - f(a).$$

$\mathbf{n} = -\mathbf{i}$ $\mathbf{n} = \mathbf{i}$

a b x

FIGURE 14.67 The outward unit normals at the boundary of $[a, b]$ in one-dimensional space.

If we let $\mathbf{F} = f(x)\mathbf{i}$ throughout $[a, b]$, then $(df/dx) = \nabla \cdot \mathbf{F}$. If we define the unit vector field $\mathbf{n}$ normal to the boundary of $[a, b]$ to be $\mathbf{i}$ at b and $-\mathbf{i}$ at a (Figure 14.67), then

$$f(b) - f(a) = f(b)\mathbf{i} \cdot (\mathbf{i}) + f(a)\mathbf{i} \cdot (-\mathbf{i})$$
$$= \mathbf{F}(b) \cdot \mathbf{n} + \mathbf{F}(a) \cdot \mathbf{n}$$
$$= \text{total outward flux of } \mathbf{F} \text{ across the boundary of } [a, b].$$

The Fundamental Theorem now says that

$$\mathbf{F}(b) \cdot \mathbf{n} + \mathbf{F}(a) \cdot \mathbf{n} = \int_{[a,b]} \nabla \cdot \mathbf{F}\,dx.$$

The Fundamental Theorem of Calculus, the normal form of Green's Theorem, and the Divergence Theorem all say that the integral of the differential operator $\nabla \cdot$ operating on a field $\mathbf{F}$ over a region equals the sum of the normal field components over the boundary of the region. (Here we are interpreting the line integral in Green's Theorem and the surface integral in the Divergence Theorem as "sums" over the boundary.)

Stokes' Theorem and the tangential form of Green's Theorem say that, when things are properly oriented, the integral of the normal component of the curl operating on a field equals the sum of the tangential field components on the boundary of the surface.

The beauty of these interpretations is the observance of a single unifying principle, which we might state as follows.

> The integral of a differential operator acting on a field over a region equals the sum of the field components appropriate to the operator over the boundary of the region.

EXERCISES 14.8

In Exercises 1–4, find the divergence of the field.

1. The spin field in Figure 14.13.

2. The radial field in Figure 14.12.

3. The gravitational field in Figure 14.8 and Exercise 38a in Section 14.3.

4. The velocity field in Figure 14.11.

In Exercises 5–16, use the Divergence Theorem to find the outward flux of $\mathbf{F}$ across the boundary of the region D.

5. Cube $\mathbf{F} = (y - x)\mathbf{i} + (z - y)\mathbf{j} + (y - x)\mathbf{k}$

D: The cube bounded by the planes $x = \pm 1$, $y = \pm 1$, and $z = \pm 1$

6. $\mathbf{F} = x^2\mathbf{i} + y^2\mathbf{j} + z^2\mathbf{k}$

  a. Cube D: The cube cut from the first octant by the planes $x = 1, y = 1$, and $z = 1$

  b. Cube D: The cube bounded by the planes $x = \pm 1$, $y = \pm 1$, and $z = \pm 1$

  c. Cylindrical can D: The region cut from the solid cylinder $x^2 + y^2 \le 4$ by the planes $z = 0$ and $z = 1$

7. Cylinder and paraboloid $\mathbf{F} = y\mathbf{i} + xy\mathbf{j} - z\mathbf{k}$

D: The region inside the solid cylinder $x^2 + y^2 \leq 4$ between the plane $z = 0$ and the paraboloid $z = x^2 + y^2$

8. Sphere $\mathbf{F} = x^2\mathbf{i} + xz\mathbf{j} + 3z\mathbf{k}$

D: The solid sphere $x^2 + y^2 + z^2 \leq 4$

9. Portion of sphere $\mathbf{F} = x^2\mathbf{i} - 2xy\mathbf{j} + 3xz\mathbf{k}$

D: The region cut from the first octant by the sphere $x^2 + y^2 + z^2 = 4$

10. Cylindrical can $\mathbf{F} = (6x^2 + 2xy)\mathbf{i} + (2y + x^2z)\mathbf{j} + 4x^2y^3\mathbf{k}$

D: The region cut from the first octant by the cylinder $x^2 + y^2 = 4$ and the plane $z = 3$

11. Wedge $\mathbf{F} = 2xz\mathbf{i} - xy\mathbf{j} - z^2\mathbf{k}$

D: The wedge cut from the first octant by the plane $y + z = 4$ and the elliptical cylinder $4x^2 + y^2 = 16$

12. Sphere $\mathbf{F} = x^3\mathbf{i} + y^3\mathbf{j} + z^3\mathbf{k}$

D: The solid sphere $x^2 + y^2 + z^2 \leq a^2$

13. Thick sphere $\mathbf{F} = \sqrt{x^2 + y^2 + z^2}\,(x\mathbf{i} + y\mathbf{j} + z\mathbf{k})$

D: The region $1 \leq x^2 + y^2 + z^2 \leq 2$

14. Thick sphere $\mathbf{F} = (x\mathbf{i} + y\mathbf{j} + z\mathbf{k})/\sqrt{x^2 + y^2 + z^2}$

D: The region $1 \leq x^2 + y^2 + z^2 \leq 4$

15. Thick sphere $\mathbf{F} = (5x^3 + 12xy^2)\mathbf{i} + (y^3 + e^y\sin z)\mathbf{j} + (5z^3 + e^y\cos z)\mathbf{k}$

D: The solid region between the spheres $x^2 + y^2 + z^2 = 1$ and $x^2 + y^2 + z^2 = 2$

16. Thick cylinder $\mathbf{F} = \ln(x^2 + y^2)\mathbf{i} - \left(\dfrac{2z}{x}\tan^{-1}\dfrac{y}{x}\right)\mathbf{j} + z\sqrt{x^2 + y^2}\,\mathbf{k}$

D: The thick-walled cylinder $1 \leq x^2 + y^2 \leq 2$, $-1 \leq z \leq 2$

17. div (curl G) is zero

a. Show that if the necessary partial derivatives of the components of the field $\mathbf{G} = M\mathbf{i} + N\mathbf{j} + P\mathbf{k}$ are continuous, then $\nabla \cdot \nabla \times \mathbf{G} = 0$.

b. What, if anything, can you conclude about the flux of the field $\nabla \times \mathbf{G}$ across a closed surface? Give reasons for your answer.

18. Let $\mathbf{F}_1$ and $\mathbf{F}_2$ be differentiable vector fields and let a and b be arbitrary real constants. Verify the following identities.

a. $\nabla \cdot (a\mathbf{F}_1 + b\mathbf{F}_2) = a\nabla \cdot \mathbf{F}_1 + b\nabla \cdot \mathbf{F}_2$

b. $\nabla \times (a\mathbf{F}_1 + b\mathbf{F}_2) = a\nabla \times \mathbf{F}_1 + b\nabla \times \mathbf{F}_2$

c. $\nabla \cdot (\mathbf{F}_1 \times \mathbf{F}_2) = \mathbf{F}_2 \cdot \nabla \times \mathbf{F}_1 - \mathbf{F}_1 \cdot \nabla \times \mathbf{F}_2$

19. Let $\mathbf{F}$ be a differentiable vector field and let $g(x, y, z)$ be a differentiable scalar function. Verify the following identities.

a. $\nabla \cdot (g\mathbf{F}) = g\nabla \cdot \mathbf{F} + \nabla g \cdot \mathbf{F}$

b. $\nabla \times (g\mathbf{F}) = g\nabla \times \mathbf{F} + \nabla g \times \mathbf{F}$

20. If $\mathbf{F} = M\mathbf{i} + N\mathbf{j} + P\mathbf{k}$ is a differentiable vector field, we define the notation $\mathbf{F} \cdot \nabla$ to mean

$$M\frac{\partial}{\partial x} + N\frac{\partial}{\partial y} + P\frac{\partial}{\partial z}.$$

For differentiable vector fields $\mathbf{F}_1$ and $\mathbf{F}_2$, verify the following identities.

a. $\nabla \times (\mathbf{F}_1 \times \mathbf{F}_2) = (\mathbf{F}_2 \cdot \nabla)\mathbf{F}_1 - (\mathbf{F}_1 \cdot \nabla)\mathbf{F}_2 + (\nabla \cdot \mathbf{F}_2)\mathbf{F}_1 - (\nabla \cdot \mathbf{F}_1)\mathbf{F}_2$

b. $\nabla(\mathbf{F}_1 \cdot \mathbf{F}_2) = (\mathbf{F}_1 \cdot \nabla)\mathbf{F}_2 + (\mathbf{F}_2 \cdot \nabla)\mathbf{F}_1 + \mathbf{F}_1 \times (\nabla \times \mathbf{F}_2) + \mathbf{F}_2 \times (\nabla \times \mathbf{F}_1)$

21. Let $\mathbf{F}$ be a field whose components have continuous first partial derivatives throughout a portion of space containing a region D bounded by a smooth closed surface S. If $|\mathbf{F}| \leq 1$, can any bound be placed on the size of

$$\iiint\limits_D \nabla \cdot \mathbf{F}\, dV\,?$$

Give reasons for your answer.

22. The base of the closed cubelike surface shown here is the unit square in the xy-plane. The four sides lie in the planes $x = 0, x = 1, y = 0$, and $y = 1$. The top is an arbitrary smooth surface whose identity is unknown. Let $\mathbf{F} = x\mathbf{i} - 2y\mathbf{j} + (z + 3)\mathbf{k}$ and suppose the outward flux of $\mathbf{F}$ through side A is 1 and through side B is -3. Can you conclude anything about the outward flux through the top? Give reasons for your answer.

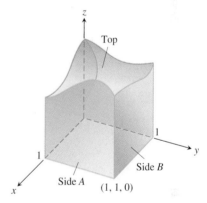

23. a. Show that the flux of the position vector field $\mathbf{F} = x\mathbf{i} + y\mathbf{j} + z\mathbf{k}$ outward through a smooth closed surface S is three times the volume of the region enclosed by the surface.

b. Let $\mathbf{n}$ be the outward unit normal vector field on S. Show that it is not possible for $\mathbf{F}$ to be orthogonal to $\mathbf{n}$ at every point of S.

24. Maximum flux Among all rectangular solids defined by the inequalities $0 \leq x \leq a, 0 \leq y \leq b, 0 \leq z \leq 1$, find the one for which the total flux of $\mathbf{F} = (-x^2 - 4xy)\mathbf{i} - 6yz\mathbf{j} + 12z\mathbf{k}$ outward through the six sides is greatest. What *is* the greatest flux?

25. Volume of a solid region Let $\mathbf{F} = x\mathbf{i} + y\mathbf{j} + z\mathbf{k}$ and suppose that the surface S and region D satisfy the hypotheses of the Divergence Theorem. Show that the volume of D is given by the formula

$$\text{Volume of } D = \frac{1}{3}\iint\limits_S \mathbf{F} \cdot \mathbf{n}\, d\sigma.$$

26. Flux of a constant field Show that the outward flux of a constant vector field $\mathbf{F} = \mathbf{C}$ across any closed surface to which the Divergence Theorem applies is zero.

27. Harmonic functions A function $f(x, y, z)$ is said to be *harmonic* in a region D in space if it satisfies the Laplace equation

$$\nabla^2 f = \nabla \cdot \nabla f = \frac{\partial^2 f}{\partial x^2} + \frac{\partial^2 f}{\partial y^2} + \frac{\partial^2 f}{\partial z^2} = 0$$

throughout D.

  a. Suppose that f is harmonic throughout a bounded region D enclosed by a smooth surface S and that $\mathbf{n}$ is the chosen unit normal vector on S. Show that the integral over S of $\nabla f \cdot \mathbf{n}$, the derivative of f in the direction of $\mathbf{n}$, is zero.

  b. Show that if f is harmonic on D, then

$$\iint_S f \, \nabla f \cdot \mathbf{n} \, d\sigma = \iiint_D |\nabla f|^2 \, dV.$$

28. Flux of a gradient field Let S be the surface of the portion of the solid sphere $x^2 + y^2 + z^2 \leq a^2$ that lies in the first octant and let $f(x, y, z) = \ln\sqrt{x^2 + y^2 + z^2}$. Calculate

$$\iint_S \nabla f \cdot \mathbf{n} \, d\sigma.$$

($\nabla f \cdot \mathbf{n}$ is the derivative of f in the direction of $\mathbf{n}$.)

29. Green's first formula Suppose that f and g are scalar functions with continuous first- and second-order partial derivatives throughout a region D that is bounded by a closed piecewise-smooth surface S. Show that

$$\iint_S f \, \nabla g \cdot \mathbf{n} \, d\sigma = \iiint_D (f \, \nabla^2 g + \nabla f \cdot \nabla g) \, dV. \qquad (9)$$

Equation (9) is **Green's first formula**. (*Hint:* Apply the Divergence Theorem to the field $\mathbf{F} = f \, \nabla g$.)

30. Green's second formula (*Continuation of Exercise 29.*) Interchange f and g in Equation (9) to obtain a similar formula. Then subtract this formula from Equation (9) to show that

$$\iint_S (f \, \nabla g - g \nabla f) \cdot \mathbf{n} \, d\sigma = \iiint_D (f \, \nabla^2 g - g \nabla^2 f) \, dV. \qquad (10)$$

This equation is **Green's second formula**.

31. Conservation of mass Let $\mathbf{v}(t, x, y, z)$ be a continuously differentiable vector field over the region D in space and let $p(t, x, y, z)$ be a continuously differentiable scalar function. The variable t represents the time domain. The Law of Conservation of Mass asserts that

$$\frac{d}{dt} \iiint_D p(t, x, y, z) \, dV = -\iint_S p\mathbf{v} \cdot \mathbf{n} \, d\sigma,$$

where S is the surface enclosing D.

  a. Give a physical interpretation of the conservation of mass law if $\mathbf{v}$ is a velocity flow field and p represents the density of the fluid at point (x, y, z) at time t.

  b. Use the Divergence Theorem and Leibniz's Rule,

$$\frac{d}{dt} \iiint_D p(t, x, y, z) \, dV = \iiint_D \frac{\partial p}{\partial t} \, dV,$$

to show that the Law of Conservation of Mass is equivalent to the continuity equation,

$$\nabla \cdot p\mathbf{v} + \frac{\partial p}{\partial t} = 0.$$

(In the first term $\nabla \cdot p\mathbf{v}$, the variable t is held fixed, and in the second term $\partial p/\partial t$, it is assumed that the point (x, y, z) in D is held fixed.)

32. The heat diffusion equation Let $T(t, x, y, z)$ be a function with continuous second derivatives giving the temperature at time t at the point (x, y, z) of a solid occupying a region D in space. If the solid's heat capacity and mass density are denoted by the constants c and ρ, respectively, the quantity $c\rho T$ is called the solid's **heat energy per unit volume**.

  a. Explain why $-\nabla T$ points in the direction of heat flow.

  b. Let $-k\nabla T$ denote the **energy flux vector**. (Here the constant k is called the **conductivity**.) Assuming the Law of Conservation of Mass with $-k\nabla T = \mathbf{v}$ and $c\rho T = p$ in Exercise 31, derive the diffusion (heat) equation

$$\frac{\partial T}{\partial t} = K\nabla^2 T,$$

where $K = k/(c\rho) > 0$ is the *diffusivity* constant. (Notice that if $T(t, x)$ represents the temperature at time t at position x in a uniform conducting rod with perfectly insulated sides, then $\nabla^2 T = \partial^2 T/\partial x^2$ and the diffusion equation reduces to the one-dimensional heat equation in Chapter 12's Additional Exercises.)

Chapter **14**   **Questions to Guide Your Review**

1. What are line integrals? How are they evaluated? Give examples.

2. How can you use line integrals to find the centers of mass of springs? Explain.

3. What is a vector field? A gradient field? Give examples.

4. How do you calculate the work done by a force in moving a particle along a curve? Give an example.

5. What are flow, circulation, and flux?

6. What is special about path independent fields?

7. How can you tell when a field is conservative?

8. What is a potential function? Show by example how to find a potential function for a conservative field.

9. What is a differential form? What does it mean for such a form to be exact? How do you test for exactness? Give examples.

10. What is the divergence of a vector field? How can you interpret it?

11. What is the curl of a vector field? How can you interpret it?

12. What is Green's theorem? How can you interpret it?

13. How do you calculate the area of a curved surface in space? Give an example.

14. What is an oriented surface? How do you calculate the flux of a three-dimensional vector field across an oriented surface? Give an example.

15. What are surface integrals? What can you calculate with them? Give an example.

16. What is a parametrized surface? How do you find the area of such a surface? Give examples.

17. How do you integrate a function over a parametrized surface? Give an example.

18. What is Stokes' Theorem? How can you interpret it?

19. Summarize the chapter's results on conservative fields.

20. What is the Divergence Theorem? How can you interpret it?

21. How does the Divergence Theorem generalize Green's Theorem?

22. How does Stokes' Theorem generalize Green's Theorem?

23. How can Green's Theorem, Stokes' Theorem, and the Divergence Theorem be thought of as forms of a single fundamental theorem?

Chapter **14**   **Practice Exercises**

1. The accompanying figure shows two polygonal paths in space joining the origin to the point $(1, 1, 1)$. Integrate $f(x, y, z) = 2x - 3y^2 - 2z + 3$ over each path.

Path 1

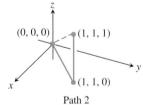

Path 2

2. The accompanying figure shows three polygonal paths joining the origin to the point $(1, 1, 1)$. Integrate $f(x, y, z) = x^2 + y - z$ over each path.

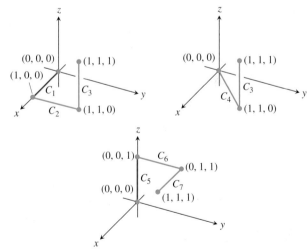

3. Integrate $f(x, y, z) = \sqrt{x^2 + z^2}$ over the circle
$$\mathbf{r}(t) = (a \cos t)\mathbf{j} + (a \sin t)\mathbf{k}, \qquad 0 \le t \le 2\pi.$$

4. Integrate $f(x, y, z) = \sqrt{x^2 + y^2}$ over the involute curve
$$\mathbf{r}(t) = (\cos t + t \sin t)\mathbf{i} + (\sin t - t \cos t)\mathbf{j}, \qquad 0 \le t \le \sqrt{3}.$$

Evaluate the integrals in Exercises 5 and 6.

5. $\displaystyle\int_{(-1,1,1)}^{(4,-3,0)} \frac{dx + dy + dz}{\sqrt{x + y + z}}$

6. $\displaystyle\int_{(1,1,1)}^{(10,3,3)} dx - \sqrt{\frac{z}{y}}\, dy - \sqrt{\frac{y}{z}}\, dz$

7. Integrate $\mathbf{F} = -(y\sin z)\mathbf{i} + (x\sin z)\mathbf{j} + (xy\cos z)\mathbf{k}$ around the circle cut from the sphere $x^2 + y^2 + z^2 = 5$ by the plane $z = -1$, clockwise as viewed from above.

8. Integrate $\mathbf{F} = 3x^2y\mathbf{i} + (x^3 + 1)\mathbf{j} + 9z^2\mathbf{k}$ around the circle cut from the sphere $x^2 + y^2 + z^2 = 9$ by the plane $x = 2$.

Evaluate the integrals in Exercises 9 and 10.

9. $\displaystyle\int_C 8x\sin y\, dx - 8y\cos x\, dy$

C is the square cut from the first quadrant by the lines $x = \pi/2$ and $y = \pi/2$.

10. $\displaystyle\int_C y^2\, dx + x^2\, dy$

C is the circle $x^2 + y^2 = 4$.

11. Area of an elliptical region Find the area of the elliptical region cut from the plane $x + y + z = 1$ by the cylinder $x^2 + y^2 = 1$.

12. Area of a parabolic cap Find the area of the cap cut from the paraboloid $y^2 + z^2 = 3x$ by the plane $x = 1$.

13. Area of a spherical cap Find the area of the cap cut from the top of the sphere $x^2 + y^2 + z^2 = 1$ by the plane $z = \sqrt{2}/2$.

14. a. Hemisphere cut by cylinder Find the area of the surface cut from the hemisphere $x^2 + y^2 + z^2 = 4, z \geq 0$, by the cylinder $x^2 + y^2 = 2x$.

b. Find the area of the portion of the cylinder that lies inside the hemisphere. (*Hint:* Project onto the xz-plane. Or evaluate the integral $\int h\, ds$, where h is the altitude of the cylinder and ds is the element of arc length on the circle $x^2 + y^2 = 2x$ in the xy-plane.)

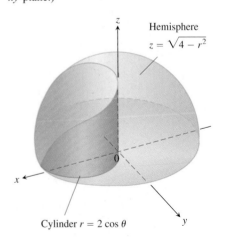

Hemisphere
$z = \sqrt{4 - r^2}$

Cylinder $r = 2\cos\theta$

15. Area of a triangle Find the area of the triangle in which the plane $(x/a) + (y/b) + (z/c) = 1\ (a, b, c > 0)$ intersects the first octant. Check your answer with an appropriate vector calculation.

16. Parabolic cylinder cut by planes Integrate

a. $g(x, y, z) = \dfrac{yz}{\sqrt{4y^2 + 1}}$

b. $g(x, y, z) = \dfrac{z}{\sqrt{4y^2 + 1}}$

over the surface cut from the parabolic cylinder $y^2 - z = 1$ by the planes $x = 0, x = 3$, and $z = 0$.

17. Circular cylinder cut by planes Integrate $g(x, y, z) = x^4y(y^2 + z^2)$ over the portion of the cylinder $y^2 + z^2 = 25$ that lies in the first octant between the planes $x = 0$ and $x = 1$ and above the plane $z = 3$.

18. Area of Wyoming The state of Wyoming is bounded by the meridians $111°3'$ and $104°3'$ west longitude and by the circles $41°$ and $45°$ north latitude. Assuming that Earth is a sphere of radius $R = 3959$ mi, find the area of Wyoming.

Find parametrizations for the surfaces in Exercises 19–24. (There are many ways to do these, so your answers may not be the same as those in the back of the book.)

19. Spherical band The portion of the sphere $x^2 + y^2 + z^2 = 36$ between the planes $z = -3$ and $z = 3\sqrt{3}$

20. Parabolic cap The portion of the paraboloid $z = -(x^2 + y^2)/2$ above the plane $z = -2$

21. Cone The cone $z = 1 + \sqrt{x^2 + y^2}, z \leq 3$

22. Plane above square The portion of the plane $4x + 2y + 4z = 12$ that lies above the square $0 \leq x \leq 2, 0 \leq y \leq 2$ in the first quadrant

23. Portion of paraboloid The portion of the paraboloid $y = 2(x^2 + z^2), y \leq 2$, that lies above the xy-plane

24. Portion of hemisphere The portion of the hemisphere $x^2 + y^2 + z^2 = 10, y \geq 0$, in the first octant

25. Surface area Find the area of the surface

$$\mathbf{r}(u, v) = (u + v)\mathbf{i} + (u - v)\mathbf{j} + v\mathbf{k},$$

$$0 \leq u \leq 1,\ 0 \leq v \leq 1.$$

26. Surface integral Integrate $f(x, y, z) = xy - z^2$ over the surface in Exercise 25.

27. Area of a helicoid Find the surface area of the helicoid

$\mathbf{r}(r, \theta) = (r\cos\theta)\mathbf{i} + (r\sin\theta)\mathbf{j} + \theta\mathbf{k},\quad 0 \leq \theta \leq 2\pi,\ 0 \leq r \leq 1,$

in the accompanying figure.

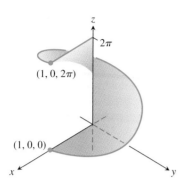

(1, 0, 2π)

2π

(1, 0, 0)

28. Surface integral Evaluate the integral $\iint_S \sqrt{x^2 + y^2 + 1}\, d\sigma$, where S is the helicoid in Exercise 27.

Which of the fields in Exercises 29–32 are conservative, and which are not?

29. $\mathbf{F} = x\mathbf{i} + y\mathbf{j} + z\mathbf{k}$

30. $\mathbf{F} = (x\mathbf{i} + y\mathbf{j} + z\mathbf{k})/(x^2 + y^2 + z^2)^{3/2}$

31. $\mathbf{F} = xe^y\mathbf{i} + ye^z\mathbf{j} + ze^x\mathbf{k}$

32. $\mathbf{F} = (\mathbf{i} + z\mathbf{j} + y\mathbf{k})/(x + yz)$

Find potential functions for the fields in Exercises 33 and 34.

33. $\mathbf{F} = 2\mathbf{i} + (2y + z)\mathbf{j} + (y + 1)\mathbf{k}$

34. $\mathbf{F} = (z\cos xz)\mathbf{i} + e^y\mathbf{j} + (x\cos xz)\mathbf{k}$

In Exercises 35 and 36, find the work done by each field along the paths from (0, 0, 0) to (1, 1, 1) in Exercise 1.

35. $\mathbf{F} = 2xy\mathbf{i} + \mathbf{j} + x^2\mathbf{k}$ **36.** $\mathbf{F} = 2xy\mathbf{i} + x^2\mathbf{j} + \mathbf{k}$

37. Finding work in two ways Find the work done by

$$\mathbf{F} = \frac{x\mathbf{i} + y\mathbf{j}}{(x^2 + y^2)^{3/2}}$$

over the plane curve $\mathbf{r}(t) = (e^t\cos t)\mathbf{i} + (e^t\sin t)\mathbf{j}$ from the point (1, 0) to the point $(e^{2\pi}, 0)$ in two ways:

a. By using the parametrization of the curve to evaluate the work integral

b. By evaluating a potential function for $\mathbf{F}$.

38. Flow along different paths Find the flow of the field $\mathbf{F} = \nabla(x^2ze^y)$

a. Once around the ellipse C in which the plane $x + y + z = 1$ intersects the cylinder $x^2 + z^2 = 25$, clockwise as viewed from the positive y-axis

b. Along the curved boundary of the helicoid in Exercise 27 from (1, 0, 0) to (1, 0, 2π).

In Exercises 39 and 40, use the surface integral in Stokes' Theorem to find the circulation of the field $\mathbf{F}$ around the curve C in the indicated direction.

39. Circulation around an ellipse $\mathbf{F} = y^2\mathbf{i} - y\mathbf{j} + 3z^2\mathbf{k}$

C: The ellipse in which the plane $2x + 6y - 3z = 6$ meets the cylinder $x^2 + y^2 = 1$, counterclockwise as viewed from above

40. Circulation around a circle $\mathbf{F} = (x^2 + y)\mathbf{i} + (x + y)\mathbf{j} + (4y^2 - z)\mathbf{k}$

C: The circle in which the plane $z = -y$ meets the sphere $x^2 + y^2 + z^2 = 4$, counterclockwise as viewed from above

41. Wire with different densities Find the mass of a thin wire lying along the curve $\mathbf{r}(t) = \sqrt{2}t\mathbf{i} + \sqrt{2}t\mathbf{j} + (4 - t^2)\mathbf{k}$, $0 \le t \le 1$, if the density at t is **(a)** $\delta = 3t$ and **(b)** $\delta = 1$.

42. Wire with variable density Find the center of mass of a thin wire lying along the curve $\mathbf{r}(t) = t\mathbf{i} + 2t\mathbf{j} + (2/3)t^{3/2}\mathbf{k}$, $0 \le t \le 2$, if the density at t is $\delta = 3\sqrt{5 + t}$.

43. Wire with variable density Find the center of mass and the moments of inertia about the coordinate axes of a thin wire lying along the curve

$$\mathbf{r}(t) = t\mathbf{i} + \frac{2\sqrt{2}}{3}t^{3/2}\mathbf{j} + \frac{t^2}{2}\mathbf{k}, \qquad 0 \le t \le 2,$$

if the density at t is $\delta = 1/(t + 1)$.

44. Center of mass of an arch A slender metal arch lies along the semicircle $y = \sqrt{a^2 - x^2}$ in the xy-plane. The density at the point (x, y) on the arch is $\delta(x, y) = 2a - y$. Find the center of mass.

45. Wire with constant density A wire of constant density $\delta = 1$ lies along the curve $\mathbf{r}(t) = (e^t\cos t)\mathbf{i} + (e^t\sin t)\mathbf{j} + e^t\mathbf{k}$, $0 \le t \le \ln 2$. Find $\bar{z}$ and I_z.

46. Helical wire with constant density Find the mass and center of mass of a wire of constant density δ that lies along the helix $\mathbf{r}(t) = (2\sin t)\mathbf{i} + (2\cos t)\mathbf{j} + 3t\mathbf{k}$, $0 \le t \le 2\pi$.

47. Inertia and center of mass of a shell Find I_z and the center of mass of a thin shell of density $\delta(x, y, z) = z$ cut from the upper portion of the sphere $x^2 + y^2 + z^2 = 25$ by the plane $z = 3$.

48. Moment of inertia of a cube Find the moment of inertia about the z-axis of the surface of the cube cut from the first octant by the planes $x = 1$, $y = 1$, and $z = 1$ if the density is $\delta = 1$.

Use Green's Theorem to find the counterclockwise circulation and outward flux for the fields and curves in Exercises 49 and 50.

49. Square $\mathbf{F} = (2xy + x)\mathbf{i} + (xy - y)\mathbf{j}$

C: The square bounded by $x = 0$, $x = 1$, $y = 0$, $y = 1$

50. Triangle $\mathbf{F} = (y - 6x^2)\mathbf{i} + (x + y^2)\mathbf{j}$

C: The triangle made by the lines $y = 0$, $y = x$, and $x = 1$

51. Zero line integral Show that

$$\oint_C \ln x \sin y \, dy - \frac{\cos y}{x}\, dx = 0$$

for any closed curve C to which Green's Theorem applies.

52. a. Outward flux and area   Show that the outward flux of the position vector field $\mathbf{F} = x\mathbf{i} + y\mathbf{j}$ across any closed curve to which Green's Theorem applies is twice the area of the region enclosed by the curve.

  b. Let $\mathbf{n}$ be the outward unit normal vector to a closed curve to which Green's Theorem applies. Show that it is not possible for $\mathbf{F} = x\mathbf{i} + y\mathbf{j}$ to be orthogonal to $\mathbf{n}$ at every point of C.

In Exercises 53–56, find the outward flux of $\mathbf{F}$ across the boundary of D.

53. Cube $\mathbf{F} = 2xy\mathbf{i} + 2yz\mathbf{j} + 2xz\mathbf{k}$

  D: The cube cut from the first octant by the planes $x = 1, y = 1,$ $z = 1$

54. Spherical cap $\mathbf{F} = xz\mathbf{i} + yz\mathbf{j} + \mathbf{k}$

  D: The entire surface of the upper cap cut from the solid sphere $x^2 + y^2 + z^2 \leq 25$ by the plane $z = 3$

55. Spherical cap $\mathbf{F} = -2x\mathbf{i} - 3y\mathbf{j} + z\mathbf{k}$

  D: The upper region cut from the solid sphere $x^2 + y^2 + z^2 \leq 2$ by the paraboloid $z = x^2 + y^2$

56. Cone and cylinder $\mathbf{F} = (6x + y)\mathbf{i} - (x + z)\mathbf{j} + 4yz\mathbf{k}$

  D: The region in the first octant bounded by the cone $z = \sqrt{x^2 + y^2}$, the cylinder $x^2 + y^2 = 1$, and the coordinate planes

57. Hemisphere, cylinder, and plane   Let S be the surface that is bounded on the left by the hemisphere $x^2 + y^2 + z^2 = a^2, y \leq 0$, in the middle by the cylinder $x^2 + z^2 = a^2, 0 \leq y \leq a$, and on the right by the plane $y = a$. Find the flux of $\mathbf{F} = y\mathbf{i} + z\mathbf{j} + x\mathbf{k}$ outward across S.

58. Cylinder and planes   Find the outward flux of the field $\mathbf{F} = 3xz^2\mathbf{i} + y\mathbf{j} - z^3\mathbf{k}$ across the surface of the solid in the first octant that is bounded by the cylinder $x^2 + 4y^2 = 16$ and the planes $y = 2z, x = 0$, and $z = 0$.

59. Cylindrical can   Use the Divergence Theorem to find the flux of $\mathbf{F} = xy^2\mathbf{i} + x^2y\mathbf{j} + y\mathbf{k}$ outward through the surface of the region enclosed by the cylinder $x^2 + y^2 = 1$ and the planes $z = 1$ and $z = -1$.

60. Hemisphere   Find the flux of $\mathbf{F} = (3z + 1)\mathbf{k}$ upward across the hemisphere $x^2 + y^2 + z^2 = a^2, z \geq 0$ **(a)** with the Divergence Theorem and **(b)** by evaluating the flux integral directly.

Chapter 14   Additional and Advanced Exercises

Use the Green's Theorem area formula, Equation (13) in Exercises 14.4, to find the areas of the regions enclosed by the curves in Exercises 1–4.

1. The limaçon $x = 2\cos t - \cos 2t, y = 2\sin t - \sin 2t,$ $0 \leq t \leq 2\pi$

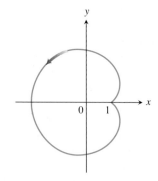

2. The deltoid $x = 2\cos t + \cos 2t, y = 2\sin t - \sin 2t,$ $0 \leq t \leq 2\pi$

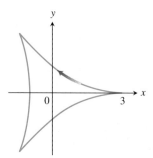

3. The eight curve $x = (1/2)\sin 2t, y = \sin t, 0 \leq t \leq \pi$ (one loop)

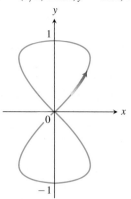

4. The teardrop $x = 2a \cos t - a \sin 2t$, $y = b \sin t$, $0 \le t \le 2\pi$

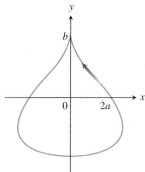

5. a. Give an example of a vector field $\mathbf{F}(x, y, z)$ that has value $\mathbf{0}$ at only one point and such that curl $\mathbf{F}$ is nonzero everywhere. Be sure to identify the point and compute the curl.

b. Give an example of a vector field $\mathbf{F}(x, y, z)$ that has value $\mathbf{0}$ on precisely one line and such that curl $\mathbf{F}$ is nonzero everywhere. Be sure to identify the line and compute the curl.

c. Give an example of a vector field $\mathbf{F}(x, y, z)$ that has value $\mathbf{0}$ on a surface and such that curl $\mathbf{F}$ is nonzero everywhere. Be sure to identify the surface and compute the curl.

6. Find all points (a, b, c) on the sphere $x^2 + y^2 + z^2 = R^2$ where the vector field $\mathbf{F} = yz^2\mathbf{i} + xz^2\mathbf{j} + 2xyz\mathbf{k}$ is normal to the surface and $\mathbf{F}(a, b, c) \ne \mathbf{0}$.

7. Find the mass of a spherical shell of radius R such that at each point (x, y, z) on the surface the mass density $\delta(x, y, z)$ is its distance to some fixed point (a, b, c) of the surface.

8. Find the mass of a helicoid

$$\mathbf{r}(r, \theta) = (r \cos \theta)\mathbf{i} + (r \sin \theta)\mathbf{j} + \theta\mathbf{k},$$

$0 \le r \le 1, 0 \le \theta \le 2\pi$, if the density function is $\delta(x, y, z) = 2\sqrt{x^2 + y^2}$. See Practice Exercise 27 for a figure.

9. Among all rectangular regions $0 \le x \le a, 0 \le y \le b$, find the one for which the total outward flux of $\mathbf{F} = (x^2 + 4xy)\mathbf{i} - 6y\mathbf{j}$ across the four sides is least. What *is* the least flux?

10. Find an equation for the plane through the origin such that the circulation of the flow field $\mathbf{F} = z\mathbf{i} + x\mathbf{j} + y\mathbf{k}$ around the circle of intersection of the plane with the sphere $x^2 + y^2 + z^2 = 4$ is a maximum.

11. A string lies along the circle $x^2 + y^2 = 4$ from $(2, 0)$ to $(0, 2)$ in the first quadrant. The density of the string is $\rho(x, y) = xy$.

a. Partition the string into a finite number of subarcs to show that the work done by gravity to move the string straight down to the x-axis is given by

$$\text{Work} = \lim_{n \to \infty} \sum_{k=1}^{n} g\, x_k y_k^2 \Delta s_k = \int_C g\, xy^2 \, ds,$$

where g is the gravitational constant.

b. Find the total work done by evaluating the line integral in part (a).

c. Show that the total work done equals the work required to move the string's center of mass $(\bar{x}, \bar{y})$ straight down to the x-axis.

12. A thin sheet lies along the portion of the plane $x + y + z = 1$ in the first octant. The density of the sheet is $\delta(x, y, z) = xy$.

a. Partition the sheet into a finite number of subpieces to show that the work done by gravity to move the sheet straight down to the xy-plane is given by

$$\text{Work} = \lim_{n \to \infty} \sum_{k=1}^{n} g\, x_k y_k z_k \Delta\sigma_k = \iint_S g\, xyz \, d\sigma,$$

where g is the gravitational constant.

b. Find the total work done by evaluating the surface integral in part (a).

c. Show that the total work done equals the work required to move the sheet's center of mass $(\bar{x}, \bar{y}, \bar{z})$ straight down to the xy-plane.

13. Archimedes' principle If an object such as a ball is placed in a liquid, it will either sink to the bottom, float, or sink a certain distance and remain suspended in the liquid. Suppose a fluid has constant weight density w and that the fluid's surface coincides with the plane $z = 4$. A spherical ball remains suspended in the fluid and occupies the region $x^2 + y^2 + (z - 2)^2 \le 1$.

a. Show that the surface integral giving the magnitude of the total force on the ball due to the fluid's pressure is

$$\text{Force} = \lim_{n \to \infty} \sum_{k=1}^{n} w(4 - z_k) \Delta\sigma_k = \iint_S w(4 - z) \, d\sigma.$$

b. Since the ball is not moving, it is being held up by the buoyant force of the liquid. Show that the magnitude of the buoyant force on the sphere is

$$\text{Buoyant force} = \iint_S w(z - 4)\mathbf{k} \cdot \mathbf{n} \, d\sigma,$$

where $\mathbf{n}$ is the outer unit normal at (x, y, z). This illustrates Archimedes' principle that the magnitude of the buoyant force on a submerged solid equals the weight of the displaced fluid.

c. Use the Divergence Theorem to find the magnitude of the buoyant force in part (b).

14. Fluid force on a curved surface A cone in the shape of the surface $z = \sqrt{x^2 + y^2}, 0 \le z \le 2$ is filled with a liquid of constant weight density w. Assuming the xy-plane is "ground level," show that the total force on the portion of the cone from $z = 1$ to $z = 2$ due to liquid pressure is the surface integral

$$F = \iint_S w(2 - z) \, d\sigma.$$

Evaluate the integral.

15. Faraday's Law If $\mathbf{E}(t, x, y, z)$ and $\mathbf{B}(t, x, y, z)$ represent the electric and magnetic fields at point (x, y, z) at time t, a basic principle of electromagnetic theory says that $\nabla \times \mathbf{E} = -\partial\mathbf{B}/\partial t$. In this expression $\nabla \times \mathbf{E}$ is computed with t held fixed and $\partial\mathbf{B}/\partial t$ is calculated with (x, y, z) fixed. Use Stokes' Theorem to derive Faraday's Law

$$\oint_C \mathbf{E} \cdot d\mathbf{r} = -\frac{\partial}{\partial t} \iint_S \mathbf{B} \cdot \mathbf{n} \, d\sigma,$$

where C represents a wire loop through which current flows counterclockwise with respect to the surface's unit normal $\mathbf{n}$, giving rise to the voltage

$$\oint_C \mathbf{E} \cdot d\mathbf{r}$$

around C. The surface integral on the right side of the equation is called the *magnetic flux*, and S is any oriented surface with boundary C.

16. Let

$$\mathbf{F} = -\frac{GmM}{|\mathbf{r}|^3} \mathbf{r}$$

be the gravitational force field defined for $\mathbf{r} \neq \mathbf{0}$. Use Gauss's Law in Section 14.8 to show that there is no continuously differentiable vector field $\mathbf{H}$ satisfying $\mathbf{F} = \nabla \times \mathbf{H}$.

17. If $f(x, y, z)$ and $g(x, y, z)$ are continuously differentiable scalar functions defined over the oriented surface S with boundary curve C, prove that

$$\iint_S (\nabla f \times \nabla g) \cdot \mathbf{n} \, d\sigma = \oint_C f \, \nabla g \cdot d\mathbf{r}.$$

18. Suppose that $\nabla \cdot \mathbf{F}_1 = \nabla \cdot \mathbf{F}_2$ and $\nabla \times \mathbf{F}_1 = \nabla \times \mathbf{F}_2$ over a region D enclosed by the oriented surface S with outward unit normal $\mathbf{n}$ and that $\mathbf{F}_1 \cdot \mathbf{n} = \mathbf{F}_2 \cdot \mathbf{n}$ on S. Prove that $\mathbf{F}_1 = \mathbf{F}_2$ throughout D.

19. Prove or disprove that if $\nabla \cdot \mathbf{F} = 0$ and $\nabla \times \mathbf{F} = \mathbf{0}$, then $\mathbf{F} = \mathbf{0}$.

20. Let S be an oriented surface parametrized by $\mathbf{r}(u, v)$. Define the notation $d\boldsymbol{\sigma} = \mathbf{r}_u \, du \times \mathbf{r}_v \, dv$ so that $d\boldsymbol{\sigma}$ is a vector normal to the surface. Also, the magnitude $d\sigma = |d\boldsymbol{\sigma}|$ is the element of surface area (by Equation 5 in Section 14.5). Derive the identity

$$d\sigma = (EG - F^2)^{1/2} \, du \, dv$$

where

$$E = |\mathbf{r}_u|^2, \quad F = \mathbf{r}_u \cdot \mathbf{r}_v, \quad \text{and} \quad G = |\mathbf{r}_v|^2.$$

21. Show that the volume V of a region D in space enclosed by the oriented surface S with outward normal $\mathbf{n}$ satisfies the identity

$$V = \frac{1}{3} \iint_S \mathbf{r} \cdot \mathbf{n} \, d\sigma,$$

where $\mathbf{r}$ is the position vector of the point (x, y, z) in D.

APPENDICES

Real Numbers and the Real Line

This section reviews real numbers, inequalities, intervals, and absolute values.

Real Numbers

Much of calculus is based on properties of the real number system. **Real numbers** are numbers that can be expressed as decimals, such as

$$-\frac{3}{4} = -0.75000\ldots$$

$$\frac{1}{3} = 0.33333\ldots$$

$$\sqrt{2} = 1.4142\ldots$$

The dots $\ldots$ in each case indicate that the sequence of decimal digits goes on forever. Every conceivable decimal expansion represents a real number, although some numbers have two representations. For instance, the infinite decimals $.999\ldots$ and $1.000\ldots$ represent the same real number 1. A similar statement holds for any number with an infinite tail of 9's.

The real numbers can be represented geometrically as points on a number line called the **real line**.

The symbol $\mathbb{R}$ denotes either the real number system or, equivalently, the real line.

The properties of the real number system fall into three categories: algebraic properties, order properties, and completeness. The **algebraic properties** say that the real numbers can be added, subtracted, multiplied, and divided (except by 0) to produce more real numbers under the usual rules of arithmetic. *You can never divide by* 0.

The **order properties** of real numbers are given in Appendix 7. The useful rules at the left can be derived from them, where the symbol $\Rightarrow$ means "implies."

Notice the rules for multiplying an inequality by a number. Multiplying by a positive number preserves the inequality; multiplying by a negative number reverses the inequality. Also, reciprocation reverses the inequality for numbers of the same sign. For example, $2 < 5$ but $-2 > -5$ and $1/2 > 1/5$.

Rules for inequalities

If a, b, and c are real numbers, then:

1. $a < b \Rightarrow a + c < b + c$

2. $a < b \Rightarrow a - c < b - c$

3. $a < b$ and $c > 0 \Rightarrow ac < bc$

4. $a < b$ and $c < 0 \Rightarrow bc < ac$
   Special case: $a < b \Rightarrow -b < -a$

5. $a > 0 \Rightarrow \frac{1}{a} > 0$

6. If a and b are both positive or both negative, then $a < b \Rightarrow \frac{1}{b} < \frac{1}{a}$

The **completeness property** of the real number system is deeper and harder to define precisely. However, the property is essential to the idea of a limit (Chapter 2). Roughly speaking, it says that there are enough real numbers to "complete" the real number line, in the sense that there are no "holes" or "gaps" in it. Many theorems of calculus would fail if the real number system were not complete. The topic is best saved for a more advanced course, but Appendix 7 hints about what is involved and how the real numbers are constructed.

We distinguish three special subsets of real numbers.

1. The **natural numbers**, namely $1, 2, 3, 4, \ldots$

2. The **integers**, namely $0, \pm 1, \pm 2, \pm 3, \ldots$

3. The **rational numbers**, namely the numbers that can be expressed in the form of a fraction m/n, where m and n are integers and $n \neq 0$. Examples are

$$\frac{1}{3}, \quad -\frac{4}{9} = \frac{-4}{9} = \frac{4}{-9}, \quad \frac{200}{13}, \quad \text{and} \quad 57 = \frac{57}{1}.$$

The rational numbers are precisely the real numbers with decimal expansions that are either

(a) terminating (ending in an infinite string of zeros), for example,

$$\frac{3}{4} = 0.75000\ldots = 0.75 \qquad \text{or}$$

(b) eventually repeating (ending with a block of digits that repeats over and over), for example

$$\frac{23}{11} = 2.090909\ldots = 2.\overline{09} \qquad \text{The bar indicates the block of repeating digits.}$$

A terminating decimal expansion is a special type of repeating decimal, since the ending zeros repeat.

The set of rational numbers has all the algebraic and order properties of the real numbers but lacks the completeness property. For example, there is no rational number whose square is 2; there is a "hole" in the rational line where $\sqrt{2}$ should be.

Real numbers that are not rational are called **irrational numbers**. They are characterized by having nonterminating and nonrepeating decimal expansions. Examples are π, $\sqrt{2}$, $\sqrt[3]{5}$, and $\log_{10} 3$. Since every decimal expansion represents a real number, it should be clear that there are infinitely many irrational numbers. Both rational and irrational numbers are found arbitrarily close to any point on the real line.

Set notation is very useful for specifying a particular subset of real numbers. A **set** is a collection of objects, and these objects are the **elements** of the set. If S is a set, the notation $a \in S$ means that a is an element of S, and $a \notin S$ means that a is not an element of S. If S and T are sets, then $S \cup T$ is their **union** and consists of all elements belonging either to S or T (or to both S and T). The **intersection** $S \cap T$ consists of all elements belonging to both S and T. The **empty set** $\varnothing$ is the set that contains no elements. For example, the intersection of the rational numbers and the irrational numbers is the empty set.

Some sets can be described by *listing* their elements in braces. For instance, the set A consisting of the natural numbers (or positive integers) less than 6 can be expressed as

$$A = \{1, 2, 3, 4, 5\}.$$

The entire set of integers is written as

$$\{0, \pm 1, \pm 2, \pm 3, \ldots\}.$$

Another way to describe a set is to enclose in braces a rule that generates all the elements of the set. For instance, the set

$$A = \{x \mid x \text{ is an integer and } 0 < x < 6\}$$

is the set of positive integers less than 6.

Intervals

A subset of the real line is called an **interval** if it contains at least two numbers and contains all the real numbers lying between any two of its elements. For example, the set of all real numbers x such that $x > 6$ is an interval, as is the set of all x such that $-2 \le x \le 5$. The set of all nonzero real numbers is not an interval; since 0 is absent, the set fails to contain every real number between -1 and 1 (for example).

Geometrically, intervals correspond to rays and line segments on the real line, along with the real line itself. Intervals of numbers corresponding to line segments are **finite intervals**; intervals corresponding to rays and the real line are **infinite intervals**.

A finite interval is said to be **closed** if it contains both of its endpoints, **half-open** if it contains one endpoint but not the other, and **open** if it contains neither endpoint. The endpoints are also called **boundary points**; they make up the interval's **boundary**. The remaining points of the interval are **interior points** and together comprise the interval's **interior**. Infinite intervals are closed if they contain a finite endpoint, and open otherwise. The entire real line $\mathbb{R}$ is an infinite interval that is both open and closed. Table A.1 summarizes the various types of intervals.

TABLE A.1 Types of intervals

Notation	Set description	Type	Picture
(a, b)	$\{x \mid a < x < b\}$	Open	
$[a, b]$	$\{x \mid a \le x \le b\}$	Closed	
$[a, b)$	$\{x \mid a \le x < b\}$	Half-open	
$(a, b]$	$\{x \mid a < x \le b\}$	Half-open	
(a, ∞)	$\{x \mid x > a\}$	Open	
$[a, \infty)$	$\{x \mid x \ge a\}$	Closed	
$(-\infty, b)$	$\{x \mid x < b\}$	Open	
$(-\infty, b]$	$\{x \mid x \le b\}$	Closed	
$(-\infty, \infty)$	$\mathbb{R}$ (set of all real numbers)	Both open and closed	

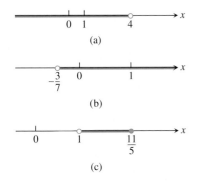

FIGURE A.1   Solution sets for the inequalities in Example 1.

Solving Inequalities

The process of finding the interval or intervals of numbers that satisfy an inequality in x is called **solving** the inequality.

EXAMPLE 1   Solve the following inequalities and show their solution sets on the real line.

(a) $2x - 1 < x + 3$ **(b)** $-\dfrac{x}{3} < 2x + 1$ **(c)** $\dfrac{6}{x - 1} \geq 5$

Solution

(a)
$$2x - 1 < x + 3$$
$$2x < x + 4 \qquad \text{Add 1 to both sides.}$$
$$x < 4 \qquad \text{Subtract } x \text{ from both sides.}$$

The solution set is the open interval $(-\infty, 4)$ (Figure A.1a).

(b)
$$-\frac{x}{3} < 2x + 1$$
$$-x < 6x + 3 \qquad \text{Multiply both sides by 3.}$$
$$0 < 7x + 3 \qquad \text{Add } x \text{ to both sides.}$$
$$-3 < 7x \qquad \text{Subtract 3 from both sides.}$$
$$-\frac{3}{7} < x \qquad \text{Divide by 7.}$$

The solution set is the open interval $(-3/7, \infty)$ (Figure A.1b).

(c) The inequality $6/(x - 1) \geq 5$ can hold only if $x > 1$, because otherwise $6/(x - 1)$ is undefined or negative. Therefore, $(x - 1)$ is positive and the inequality will be preserved if we multiply both sides by $(x - 1)$, and we have

$$\frac{6}{x - 1} \geq 5$$
$$6 \geq 5x - 5 \qquad \text{Multiply both sides by } (x - 1).$$
$$11 \geq 5x \qquad \text{Add 5 to both sides.}$$
$$\frac{11}{5} \geq x. \qquad \text{Or } x \leq \frac{11}{5}.$$

The solution set is the half-open interval $(1, 11/5\,]$ (Figure A.1c).   ∎

Absolute Value

The **absolute value** of a number x, denoted by $|x|$, is defined by the formula

$$|x| = \begin{cases} x, & x \geq 0 \\ -x, & x < 0. \end{cases}$$

EXAMPLE 2 $|3| = 3, \quad |0| = 0, \quad |-5| = -(-5) = 5, \quad |-|a|| = |a|$ ∎

Geometrically, the absolute value of x is the distance from x to 0 on the real number line. Since distances are always positive or 0, we see that $|x| \geq 0$ for every real number x,

FIGURE A.2 Absolute values give distances between points on the number line.

and $|x| = 0$ if and only if $x = 0$. Also,

$$|x - y| = \text{the distance between } x \text{ and } y$$

on the real line (Figure A.2).

Since the symbol $\sqrt{a}$ always denotes the *nonnegative* square root of a, an alternate definition of $|x|$ is

$$|x| = \sqrt{x^2}.$$

It is important to remember that $\sqrt{a^2} = |a|$. Do not write $\sqrt{a^2} = a$ unless you already know that $a \geq 0$.

The absolute value has the following properties. (You are asked to prove these properties in the exercises.)

Absolute Value Properties

1. $|-a| = |a|$ A number and its additive inverse or negative have the same absolute value.

2. $|ab| = |a||b|$ The absolute value of a product is the product of the absolute values.

3. $\left|\dfrac{a}{b}\right| = \dfrac{|a|}{|b|}$ The absolute value of a quotient is the quotient of the absolute values.

4. $|a + b| \leq |a| + |b|$ The **triangle inequality**. The absolute value of the sum of two numbers is less than or equal to the sum of their absolute values.

FIGURE A.3 $|x| < a$ means x lies between $-a$ and a.

Absolute values and intervals

If a is any positive number, then

5. $|x| = a \iff x = \pm a$

6. $|x| < a \iff -a < x < a$

7. $|x| > a \iff x > a \text{ or } x < -a$

8. $|x| \leq a \iff -a \leq x \leq a$

9. $|x| \geq a \iff x \geq a \text{ or } x \leq -a$

Note that $|-a| \neq -|a|$. For example, $|-3| = 3$, whereas $-|3| = -3$. If a and b differ in sign, then $|a + b|$ is less than $|a| + |b|$. In all other cases, $|a + b|$ equals $|a| + |b|$. Absolute value bars in expressions like $|-3 + 5|$ work like parentheses: We do the arithmetic inside *before* taking the absolute value.

EXAMPLE 3

$$|-3 + 5| = |2| = 2 < |-3| + |5| = 8$$
$$|3 + 5| = |8| = |3| + |5|$$
$$|-3 - 5| = |-8| = 8 = |-3| + |-5|$$ ∎

The inequality $|x| < a$ says that the distance from x to 0 is less than the positive number a. This means that x must lie between $-a$ and a, as we can see from Figure A.3.

The statements in the table are all consequences of the definition of absolute value and are often helpful when solving equations or inequalities involving absolute values.

The symbol $\iff$ is often used by mathematicians to denote the "if and only if" logical relationship. It also means "implies and is implied by."

EXAMPLE 4   Solve the equation $|2x - 3| = 7$.

Solution   By Property 5, $2x - 3 = \pm 7$, so there are two possibilities:

$$2x - 3 = 7 \qquad 2x - 3 = -7 \qquad \text{Equivalent equations without absolute values}$$
$$2x = 10 \qquad 2x = -4 \qquad \text{Solve as usual.}$$
$$x = 5 \qquad x = -2$$

The solutions of $|2x - 3| = 7$ are $x = 5$ and $x = -2$.   ∎

EXAMPLE 5   Solve the inequality $\left|5 - \dfrac{2}{x}\right| < 1$.

Solution   We have

$$\left|5 - \frac{2}{x}\right| < 1 \Leftrightarrow -1 < 5 - \frac{2}{x} < 1 \qquad \text{Property 6}$$

$$\Leftrightarrow -6 < -\frac{2}{x} < -4 \qquad \text{Subtract 5.}$$

$$\Leftrightarrow 3 > \frac{1}{x} > 2 \qquad \text{Multiply by } -\frac{1}{2}.$$

$$\Leftrightarrow \frac{1}{3} < x < \frac{1}{2}. \qquad \text{Take reciprocals.}$$

Notice how the various rules for inequalities were used here. Multiplying by a negative number reverses the inequality. So does taking reciprocals in an inequality in which both sides are positive. The original inequality holds if and only if $(1/3) < x < (1/2)$. The solution set is the open interval $(1/3, 1/2)$.   ∎

EXERCISES A.1

1. Express $1/9$ as a repeating decimal, using a bar to indicate the repeating digits. What are the decimal representations of $2/9$? $3/9$? $8/9$? $9/9$?

2. If $2 < x < 6$, which of the following statements about x are necessarily true, and which are not necessarily true?
   - **a.** $0 < x < 4$
   - **b.** $0 < x - 2 < 4$
   - **c.** $1 < \dfrac{x}{2} < 3$
   - **d.** $\dfrac{1}{6} < \dfrac{1}{x} < \dfrac{1}{2}$
   - **e.** $1 < \dfrac{6}{x} < 3$
   - **f.** $|x - 4| < 2$
   - **g.** $-6 < -x < 2$
   - **h.** $-6 < -x < -2$

In Exercises 3–6, solve the inequalities and show the solution sets on the real line.
3. $-2x > 4$
4. $5x - 3 \le 7 - 3x$
5. $2x - \dfrac{1}{2} \ge 7x + \dfrac{7}{6}$
6. $\dfrac{4}{5}(x - 2) < \dfrac{1}{3}(x - 6)$

Solve the equations in Exercises 7–9.
7. $|y| = 3$
8. $|2t + 5| = 4$
9. $|8 - 3s| = \dfrac{9}{2}$

Solve the inequalities in Exercises 10–17, expressing the solution sets as intervals or unions of intervals. Also, show each solution set on the real line.
10. $|x| < 2$
11. $|t - 1| \le 3$
12. $|3y - 7| < 4$
13. $\left|\dfrac{z}{5} - 1\right| \le 1$
14. $\left|3 - \dfrac{1}{x}\right| < \dfrac{1}{2}$
15. $|2s| \ge 4$
16. $|1 - x| > 1$
17. $\left|\dfrac{r + 1}{2}\right| \ge 1$

Solve the inequalities in Exercises 18–21. Express the solution sets as intervals or unions of intervals and show them on the real line. Use the result $\sqrt{a^2} = |a|$ as appropriate.
18. $x^2 < 2$
19. $4 < x^2 < 9$
20. $(x - 1)^2 < 4$
21. $x^2 - x < 0$

22. Do not fall into the trap $|-a| = a$. For what real numbers a is this equation true? For what real numbers is it false?

23. Solve the equation $|x - 1| = 1 - x$.

24. A proof of the triangle inequality Give the reason justifying each of the numbered steps in the following proof of the triangle inequality.

$$|a + b|^2 = (a + b)^2 \tag{1}$$
$$= a^2 + 2ab + b^2$$
$$\leq a^2 + 2|a||b| + b^2 \tag{2}$$
$$= |a|^2 + 2|a||b| + |b|^2 \tag{3}$$
$$= (|a| + |b|)^2$$
$$|a + b| \leq |a| + |b| \tag{4}$$

25. Prove that $|ab| = |a||b|$ for any numbers a and b.

26. If $|x| \leq 3$ and $x > -1/2$, what can you say about x?

27. Graph the inequality $|x| + |y| \leq 1$.

28. For any number a, prove that $|-a| = |a|$.

29. Let a be any positive number. Prove that $|x| > a$ if and only if $x > a$ or $x < -a$.

30. a. If b is any nonzero real number, prove that $|1/b| = 1/|b|$.

b. Prove that $\left|\dfrac{a}{b}\right| = \dfrac{|a|}{|b|}$ for any numbers a and $b \neq 0$.

A.2 Mathematical Induction

Many formulas, like

$$1 + 2 + \cdots + n = \frac{n(n + 1)}{2},$$

can be shown to hold for every positive integer n by applying an axiom called the *mathematical induction principle*. A proof that uses this axiom is called a *proof by mathematical induction* or a *proof by induction*.

The steps in proving a formula by induction are the following:

1. Check that the formula holds for $n = 1$.

2. Prove that if the formula holds for any positive integer $n = k$, then it also holds for the next integer, $n = k + 1$.

The induction axiom says that once these steps are completed, the formula holds for all positive integers n. By Step 1 it holds for $n = 1$. By Step 2 it holds for $n = 2$, and therefore by Step 2 also for $n = 3$, and by Step 2 again for $n = 4$, and so on. If the first domino falls, and the kth domino always knocks over the $(k + 1)$st when it falls, all the dominoes fall.

From another point of view, suppose we have a sequence of statements S_1, $S_2, \ldots, S_n, \ldots$, one for each positive integer. Suppose we can show that assuming any one of the statements to be true implies that the next statement in line is true. Suppose that we can also show that S_1 is true. Then we may conclude that the statements are true from S_1 on.

EXAMPLE 1 Use mathematical induction to prove that for every positive integer n,

$$1 + 2 + \cdots + n = \frac{n(n + 1)}{2}.$$

Solution We accomplish the proof by carrying out the two steps above.

1. The formula holds for $n = 1$ because

$$1 = \frac{1(1 + 1)}{2}.$$

2.   If the formula holds for $n = k$, does it also hold for $n = k + 1$? The answer is yes, as we now show. If

$$1 + 2 + \cdots + k = \frac{k(k + 1)}{2},$$

then

$$1 + 2 + \cdots + k + (k + 1) = \frac{k(k + 1)}{2} + (k + 1) = \frac{k^2 + k + 2k + 2}{2}$$

$$= \frac{(k + 1)(k + 2)}{2} = \frac{(k + 1)((k + 1) + 1)}{2}.$$

The last expression in this string of equalities is the expression $n(n + 1)/2$ for $n = (k + 1)$.

The mathematical induction principle now guarantees the original formula for all positive integers n.   ■

In Example 4 of Section 5.2 we gave another proof for the formula giving the sum of the first n integers. However, proof by mathematical induction is more general. It can be used to find the sums of the squares and cubes of the first n integers (Exercises 9 and 10). Here is another example.

EXAMPLE 2   Show by mathematical induction that for all positive integers n,

$$\frac{1}{2^1} + \frac{1}{2^2} + \cdots + \frac{1}{2^n} = 1 - \frac{1}{2^n}.$$

Solution   We accomplish the proof by carrying out the two steps of mathematical induction.

1.   The formula holds for $n = 1$ because

$$\frac{1}{2^1} = 1 - \frac{1}{2^1}.$$

2.   If

$$\frac{1}{2^1} + \frac{1}{2^2} + \cdots + \frac{1}{2^k} = 1 - \frac{1}{2^k},$$

then

$$\frac{1}{2^1} + \frac{1}{2^2} + \cdots + \frac{1}{2^k} + \frac{1}{2^{k+1}} = 1 - \frac{1}{2^k} + \frac{1}{2^{k+1}} = 1 - \frac{1 \cdot 2}{2^k \cdot 2} + \frac{1}{2^{k+1}}$$

$$= 1 - \frac{2}{2^{k+1}} + \frac{1}{2^{k+1}} = 1 - \frac{1}{2^{k+1}}.$$

Thus, the original formula holds for $n = (k + 1)$ whenever it holds for $n = k$.

With these steps verified, the mathematical induction principle now guarantees the formula for every positive integer n.   ■

Other Starting Integers

Instead of starting at $n = 1$ some induction arguments start at another integer. The steps for such an argument are as follows.

1.   Check that the formula holds for $n = n_1$ (the first appropriate integer).
2.   Prove that if the formula holds for any integer $n = k \geq n_1$, then it also holds for $n = (k + 1)$.

Once these steps are completed, the mathematical induction principle guarantees the formula for all $n \geq n_1$.

EXAMPLE 3   Show that $n! > 3^n$ if n is large enough.

Solution   How large is large enough? We experiment:

n	1	2	3	4	5	6	7
$n!$	1	2	6	24	120	720	5040
3^n	3	9	27	81	243	729	2187

It looks as if $n! > 3^n$ for $n \geq 7$. To be sure, we apply mathematical induction. We take $n_1 = 7$ in Step 1 and complete Step 2.

Suppose $k! > 3^k$ for some $k \geq 7$. Then

$$(k + 1)! = (k + 1)(k!) > (k + 1)3^k > 7 \cdot 3^k > 3^{k+1}.$$

Thus, for $k \geq 7$,

$$k! > 3^k \quad \text{implies} \quad (k + 1)! > 3^{k+1}.$$

The mathematical induction principle now guarantees $n! \geq 3^n$ for all $n \geq 7$.   ∎

EXERCISES A.2

1. Assuming that the triangle inequality $|a + b| \leq |a| + |b|$ holds for any two numbers a and b. Show that

$$|x_1 + x_2 + \cdots + x_n| \leq |x_1| + |x_2| + \cdots + |x_n|$$

for any n numbers.

2. Show that if $r \neq 1$, then

$$1 + r + r^2 + \cdots + r^n = \frac{1 - r^{n+1}}{1 - r}$$

for every positive integer n.

3. Use the Product Rule, $\dfrac{d}{dx}(uv) = u\dfrac{dv}{dx} + v\dfrac{du}{dx}$, and the fact that $\dfrac{d}{dx}(x) = 1$ to show that $\dfrac{d}{dx}(x^n) = nx^{n-1}$ for every positive integer n.

4. Suppose that a function $f(x)$ has the property that $f(x_1 x_2) = f(x_1) + f(x_2)$ for any two positive numbers x_1 and x_2. Show that

$$f(x_1 x_2 \cdots x_n) = f(x_1) + f(x_2) + \cdots + f(x_n)$$

for the product of any n positive numbers $x_1, x_2, \ldots, x_n$.

5. Show that

$$\frac{2}{3^1} + \frac{2}{3^2} + \cdots + \frac{2}{3^n} = 1 - \frac{1}{3^n}$$

for all positive integers n.

6. Show that $n! > n^3$ if n is large enough.

7. Show that $2^n > n^2$ if n is large enough.

8. Show that $2^n \geq 1/8$ for $n \geq -3$.

9. Sums of squares Show that the sum of the squares of the first n positive integers is

$$\frac{n\left(n + \dfrac{1}{2}\right)(n + 1)}{3}.$$

10. Sums of cubes Show that the sum of the cubes of the first n positive integers is $(n(n + 1)/2)^2$.

11. Rules for finite sums Show that the following finite sum rules hold for every positive integer n. (See Section 5.2.)

a. $\displaystyle\sum_{k=1}^{n} (a_k + b_k) = \sum_{k=1}^{n} a_k + \sum_{k=1}^{n} b_k$

b. $\displaystyle\sum_{k=1}^{n} (a_k - b_k) = \sum_{k=1}^{n} a_k - \sum_{k=1}^{n} b_k$

c. $\displaystyle\sum_{k=1}^{n} ca_k = c \cdot \sum_{k=1}^{n} a_k$ (Any number c)

d. $\displaystyle\sum_{k=1}^{n} a_k = n \cdot c$ (if a_k has the constant value c)

12. Show that $|x^n| = |x|^n$ for every positive integer n and every real number x.

<table>
<tr><td>**A.3**</td><td>**Lines, Circles, and Parabolas**</td></tr>
</table>

This section reviews coordinates, lines, distance, circles, and parabolas in the plane. The notion of increment is also discussed.

Cartesian Coordinates in the Plane

FIGURE A.4 Cartesian coordinates in the plane are based on two perpendicular axes intersecting at the origin.

In Appendix 1 we identified the points on the line with real numbers by assigning them coordinates. Points in the plane can be identified with ordered pairs of real numbers. To begin, we draw two perpendicular coordinate lines that intersect at the 0-point of each line. These lines are called **coordinate axes** in the plane. On the horizontal x-axis, numbers are denoted by x and increase to the right. On the vertical y-axis, numbers are denoted by y and increase upward (Figure A.4). Thus "upward" and "to the right" are positive directions, whereas "downward" and "to the left" are considered as negative. The **origin** O, also labeled 0, of the coordinate system is the point in the plane where x and y are both zero.

If P is any point in the plane, it can be located by exactly one ordered pair of real numbers in the following way. Draw lines through P perpendicular to the two coordinate axes. These lines intersect the axes at points with coordinates a and b (Figure A.4). The ordered pair (a, b) is assigned to the point P and is called its **coordinate pair**. The first number a is the **x-coordinate** (or **abscissa**) of P; the second number b is the **y-coordinate** (or **ordinate**) of P. The x-coordinate of every point on the y-axis is 0. The y-coordinate of every point on the x-axis is 0. The origin is the point $(0, 0)$.

Starting with an ordered pair (a, b), we can reverse the process and arrive at a corresponding point P in the plane. Often we identify P with the ordered pair and write $P(a, b)$. We sometimes also refer to "the point (a, b)" and it will be clear from the context when (a, b) refers to a point in the plane and not to an open interval on the real line. Several points labeled by their coordinates are shown in Figure A.5.

This coordinate system is called the **rectangular coordinate system** or **Cartesian coordinate system** (after the sixteenth century French mathematician René Descartes).

HISTORICAL BIOGRAPHY

René Descartes
(1596–1650)

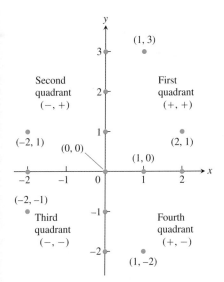

FIGURE A.5   Points labeled in the xy-coordinate or Cartesian plane. The points on the axes all have coordinate pairs but are usually labeled with single real numbers, (so $(1, 0)$ on the x-axis is labeled as 1). Notice the coordinate sign patterns of the quadrants.

The coordinate axes of this coordinate or Cartesian plane divide the plane into four regions called **quadrants**, numbered counterclockwise as shown in Figure A.5.

The **graph** of an equation or inequality in the variables x and y is the set of all points $P(x, y)$ in the plane whose coordinates satisfy the equation or inequality. When we plot data in the coordinate plane or graph formulas whose variables have different units of measure, we do not need to use the same scale on the two axes. If we plot time vs. thrust for a rocket motor, for example, there is no reason to place the mark that shows 1 sec on the time axis the same distance from the origin as the mark that shows 1 lb on the thrust axis.

Usually when we graph functions whose variables do not represent physical measurements and when we draw figures in the coordinate plane to study their geometry and trigonometry, we try to make the scales on the axes identical. A vertical unit of distance then looks the same as a horizontal unit. As on a surveyor's map or a scale drawing, line segments that are supposed to have the same length will look as if they do and angles that are supposed to be congruent will look congruent.

Computer displays and calculator displays are another matter. The vertical and horizontal scales on machine-generated graphs usually differ, and there are corresponding distortions in distances, slopes, and angles. Circles may look like ellipses, rectangles may look like squares, right angles may appear to be acute or obtuse, and so on. We discuss these displays and distortions in greater detail in Section 1.6.

Increments and Straight Lines

When a particle moves from one point in the plane to another, the net changes in its coordinates are called *increments*. They are calculated by subtracting the coordinates of the starting point from the coordinates of the ending point. If x changes from x_1 to x_2, the **increment** in x is

$$\Delta x = x_2 - x_1.$$

EXAMPLE 1   In going from the point $A(4, -3)$ to the point $B(2, 5)$ the increments in the x- and y-coordinates are

$$\Delta x = 2 - 4 = -2, \qquad \Delta y = 5 - (-3) = 8.$$

From $C(5, 6)$ to $D(5, 1)$ the coordinate increments are

$$\Delta x = 5 - 5 = 0, \qquad \Delta y = 1 - 6 = -5.$$

See Figure A.6.   ∎

Given two points $P_1(x_1, y_1)$ and $P_2(x_2, y_2)$ in the plane, we call the increments $\Delta x = x_2 - x_1$ and $\Delta y = y_2 - y_1$ the **run** and the **rise**, respectively, between P_1 and P_2. Two such points always determine a unique straight line (usually called simply a line) passing through them both. We call the line P_1P_2.

Any nonvertical line in the plane has the property that the ratio

$$m = \frac{\text{rise}}{\text{run}} = \frac{\Delta y}{\Delta x} = \frac{y_2 - y_1}{x_2 - x_1}$$

has the same value for every choice of the two points $P_1(x_1, y_1)$ and $P_2(x_2, y_2)$ on the line (Figure A.7). This is because the ratios of corresponding sides for similar triangles are equal.

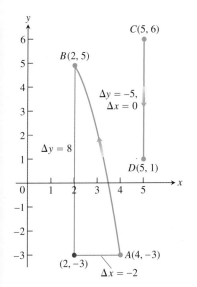

FIGURE A.6   Coordinate increments may be positive, negative, or zero (Example 1).

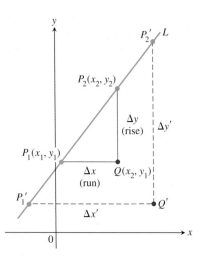

FIGURE A.7 Triangles P_1QP_2 and $P_1'Q'P_2'$ are similar, so the ratio of their sides has the same value for any two points on the line. This common value is the line's slope.

> **DEFINITION**   Slope
>
> The constant
>
> $$m = \frac{\text{rise}}{\text{run}} = \frac{\Delta y}{\Delta x} = \frac{y_2 - y_1}{x_2 - x_1}$$
>
> is the **slope** of the nonvertical line P_1P_2.

The slope tells us the direction (uphill, downhill) and steepness of a line. A line with positive slope rises uphill to the right; one with negative slope falls downhill to the right (Figure A.8). The greater the absolute value of the slope, the more rapid the rise or fall. The slope of a vertical line is *undefined*. Since the run Δx is zero for a vertical line, we cannot form the slope ratio m.

The direction and steepness of a line can also be measured with an angle. The **angle of inclination** of a line that crosses the x-axis is the smallest counterclockwise angle from the x-axis to the line (Figure A.9). The inclination of a horizontal line is 0°. The inclination of a vertical line is 90°. If ϕ (the Greek letter phi) is the inclination of a line, then $0 \le \phi < 180°$.

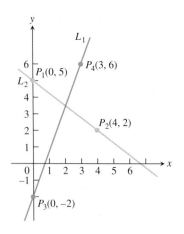

FIGURE A.8 The slope of L_1 is

$$m = \frac{\Delta y}{\Delta x} = \frac{6 - (-2)}{3 - 0} = \frac{8}{3}.$$

That is, y increases 8 units every time x increases 3 units. The slope of L_2 is

$$m = \frac{\Delta y}{\Delta x} = \frac{2 - 5}{4 - 0} = \frac{-3}{4}.$$

That is, y decreases 3 units every time x increases 4 units.

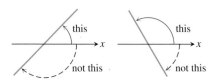

FIGURE A.9 Angles of inclination are measured counterclockwise from the x-axis.

The relationship between the slope m of a nonvertical line and the line's angle of inclination ϕ is shown in Figure A.10:

$$m = \tan \phi.$$

Straight lines have relatively simple equations. All points on the *vertical line* through the point a on the x-axis have x-coordinates equal to a. Thus, $x = a$ is an equation for the vertical line. Similarly, $y = b$ is an equation for the *horizontal line* meeting the y-axis at b. (See Figure A.11.)

We can write an equation for a nonvertical straight line L if we know its slope m and the coordinates of one point $P_1(x_1, y_1)$ on it. If $P(x, y)$ is *any* other point on L, then we can use the two points P_1 and P to compute the slope,

$$m = \frac{y - y_1}{x - x_1}$$

so that

$$y - y_1 = m(x - x_1) \qquad \text{or} \qquad y = y_1 + m(x - x_1).$$

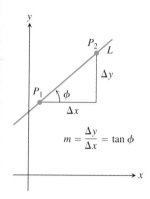

FIGURE A.10   The slope of a nonvertical line is the tangent of its angle of inclination.

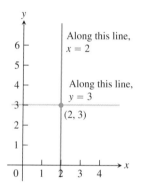

FIGURE A.11   The standard equations for the vertical and horizontal lines through (2, 3) are $x = 2$ and $y = 3$.

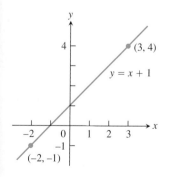

FIGURE A.12   The line in Example 3.

The equation

$$y = y_1 + m(x - x_1)$$

is the **point-slope equation** of the line that passes through the point (x_1, y_1) and has slope m.

EXAMPLE 2   Write an equation for the line through the point (2, 3) with slope $-3/2$.

Solution   We substitute $x_1 = 2$, $y_1 = 3$, and $m = -3/2$ into the point-slope equation and obtain

$$y = 3 - \frac{3}{2}\left(x - 2\right), \quad \text{or} \quad y = -\frac{3}{2}x + 6.$$

When $x = 0$, $y = 6$ so the line intersects the y-axis at $y = 6$.   ∎

EXAMPLE 3   Write an equation for the line through $(-2, -1)$ and $(3, 4)$.

Solution   The line's slope is

$$m = \frac{-1 - 4}{-2 - 3} = \frac{-5}{-5} = 1.$$

We can use this slope with either of the two given points in the point-slope equation:

With $(x_1, y_1) = (-2, -1)$

$y = -1 + 1 \cdot (x - (-2))$

$y = -1 + x + 2$

$y = x + 1$

With $(x_1, y_1) = (3, 4)$

$y = 4 + 1 \cdot (x - 3)$

$y = 4 + x - 3$

$y = x + 1$

Same result

Either way, $y = x + 1$ is an equation for the line (Figure A.12).   ∎

The y-coordinate of the point where a nonvertical line intersects the y-axis is called the **y-intercept** of the line. Similarly, the **x-intercept** of a nonhorizontal line is the x-coordinate of the point where it crosses the x-axis (Figure A.13). A line with slope m and y-intercept b passes through the point $(0, b)$, so it has equation

$$y = b + m(x - 0), \quad \text{or, more simply,} \quad y = mx + b.$$

The equation

$$y = mx + b$$

is called the **slope-intercept equation** of the line with slope m and y-intercept b.

Lines with equations of the form $y = mx$ have y-intercept 0 and so pass through the origin. Equations of lines are called **linear** equations.

The equation

$$Ax + By = C \quad (A \text{ and } B \text{ not both } 0)$$

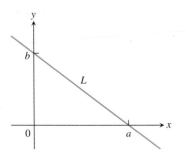

FIGURE A.13 Line L has x-intercept a and y-intercept b.

is called the **general linear equation** in x and y because its graph always represents a line and every line has an equation in this form (including lines with undefined slope).

Parallel and Perpendicular Lines

Lines that are parallel have equal angles of inclination, so they have the same slope (if they are not vertical). Conversely, lines with equal slopes have equal angles of inclination and so are parallel.

If two nonvertical lines L_1 and L_2 are perpendicular, their slopes m_1 and m_2 satisfy $m_1 m_2 = -1$, so each slope is the *negative reciprocal* of the other:

$$m_1 = -\frac{1}{m_2}, \qquad m_2 = -\frac{1}{m_1}.$$

To see this, notice by inspecting similar triangles in Figure A.14 that $m_1 = a/h$, and $m_2 = -h/a$. Hence, $m_1 m_2 = (a/h)(-h/a) = -1$.

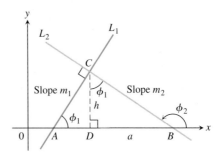

FIGURE A.14 $\triangle ADC$ is similar to $\triangle CDB$. Hence ϕ_1 is also the upper angle in $\triangle CDB$. From the sides of $\triangle CDB$, we read $\tan \phi_1 = a/h$.

Distance and Circles in the Plane

The distance between points in the plane is calculated with a formula that comes from the Pythagorean theorem (Figure A.15).

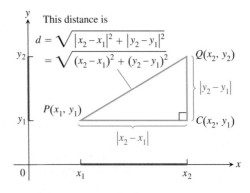

This distance is
$$d = \sqrt{|x_2 - x_1|^2 + |y_2 - y_1|^2}$$
$$= \sqrt{(x_2 - x_1)^2 + (y_2 - y_1)^2}$$

FIGURE A.15 To calculate the distance between $P(x_1, y_1)$ and $Q(x_2, y_2)$, apply the Pythagorean theorem to triangle PCQ.

> **Distance Formula for Points in the Plane**
>
> The distance between $P(x_1, y_1)$ and $Q(x_2, y_2)$ is
>
> $$d = \sqrt{(\Delta x)^2 + (\Delta y)^2} = \sqrt{(x_2 - x_1)^2 + (y_2 - y_1)^2}.$$

EXAMPLE 4

(a) The distance between $P(-1, 2)$ and $Q(3, 4)$ is

$$\sqrt{(3 - (-1))^2 + (4 - 2)^2} = \sqrt{(4)^2 + (2)^2} = \sqrt{20} = \sqrt{4 \cdot 5} = 2\sqrt{5}.$$

(b) The distance from the origin to $P(x, y)$ is

$$\sqrt{(x - 0)^2 + (y - 0)^2} = \sqrt{x^2 + y^2}.$$ ∎

By definition, a **circle** of radius a is the set of all points $P(x, y)$ whose distance from some center $C(h, k)$ equals a (Figure A.16). From the distance formula, P lies on the circle if and only if

$$\sqrt{(x - h)^2 + (y - k)^2} = a,$$

so

$P(x, y)$

a

$C(h, k)$

$(x - h)^2 + (y - k)^2 = a^2$

FIGURE A.16   A circle of radius a in the xy-plane, with center at (h, k).

> $$(x - h)^2 + (y - k)^2 = a^2. \tag{1}$$

Equation (1) is the **standard equation** of a circle with center (h, k) and radius a. The circle of radius $a = 1$ and centered at the origin is the **unit circle** with equation

$$x^2 + y^2 = 1.$$

EXAMPLE 5

(a) The standard equation for the circle of radius 2 centered at $(3, 4)$ is

$$(x - 3)^2 + (y - 4)^2 = 2^2 = 4.$$

(b) The circle

$$(x - 1)^2 + (y + 5)^2 = 3$$

has $h = 1$, $k = -5$, and $a = \sqrt{3}$. The center is the point $(h, k) = (1, -5)$ and the radius is $a = \sqrt{3}$.   ∎

If an equation for a circle is not in standard form, we can find the circle's center and radius by first converting the equation to standard form. The algebraic technique for doing so is *completing the square*.

EXAMPLE 6   Find the center and radius of the circle

$$x^2 + y^2 + 4x - 6y - 3 = 0.$$

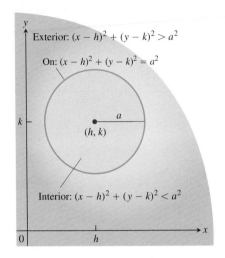

FIGURE A.17 The interior and exterior of the circle $(x - h)^2 + (y - k)^2 = a^2$.

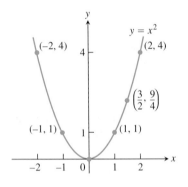

FIGURE A.18 The parabola $y = x^2$ (Example 7).

Solution We convert the equation to standard form by completing the squares in x and y:

$$x^2 + y^2 + 4x - 6y - 3 = 0$$ Start with the given equation.

$$(x^2 + 4x) + (y^2 - 6y) = 3$$ Gather terms. Move the constant to the right-hand side.

$$\left(x^2 + 4x + \left(\frac{4}{2}\right)^2\right) + \left(y^2 - 6y + \left(\frac{-6}{2}\right)^2\right) =$$
$$3 + \left(\frac{4}{2}\right)^2 + \left(\frac{-6}{2}\right)^2$$
Add the square of half the coefficient of x to each side of the equation. Do the same for y. The parenthetical expressions on the left-hand side are now perfect squares.

$$(x^2 + 4x + 4) + (y^2 - 6y + 9) = 3 + 4 + 9$$

$$(x + 2)^2 + (y - 3)^2 = 16$$
Write each quadratic as a squared linear expression.

The center is $(-2, 3)$ and the radius is $a = 4$. ∎

The points (x, y) satisfying the inequality

$$(x - h)^2 + (y - k)^2 < a^2$$

make up the **interior** region of the circle with center (h, k) and radius a (Figure A.17). The circle's **exterior** consists of the points (x, y) satisfying

$$(x - h)^2 + (y - k)^2 > a^2.$$

Parabolas

The geometric definition and properties of general parabolas are reviewed in Section 9.4. Here we look at parabolas arising as the graphs of equations of the form $y = ax^2 + bx + c$.

EXAMPLE 7 Consider the equation $y = x^2$. Some points whose coordinates satisfy this equation are $(0, 0)$, $(1, 1)$, $\left(\frac{3}{2}, \frac{9}{4}\right)$, $(-1, 1)$, $(2, 4)$, and $(-2, 4)$. These points (and all others satisfying the equation) make up a smooth curve called a parabola (Figure A.18). ∎

The graph of an equation of the form

$$y = ax^2$$

is a **parabola** whose **axis** (axis of symmetry) is the y-axis. The parabola's **vertex** (point where the parabola and axis cross) lies at the origin. The parabola opens upward if $a > 0$ and downward if $a < 0$. The larger the value of $|a|$, the narrower the parabola (Figure A.19).

Generally, the graph of $y = ax^2 + bx + c$ is a shifted and scaled version of the parabola $y = x^2$. We discuss shifting and scaling of graphs in more detail in Section 1.2.

The Graph of $y = ax^2 + bx + c$, $a \neq 0$

The graph of the equation $y = ax^2 + bx + c$, $a \neq 0$, is a parabola. The parabola opens upward if $a > 0$ and downward if $a < 0$. The **axis** is the line

$$x = -\frac{b}{2a}. \tag{2}$$

The **vertex** of the parabola is the point where the axis and parabola intersect. Its x-coordinate is $x = -b/2a$; its y-coordinate is found by substituting $x = -b/2a$ in the parabola's equation.

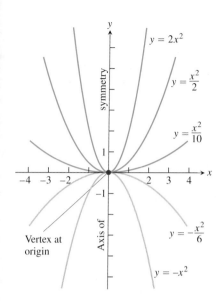

FIGURE A.19   Besides determining the direction in which the parabola $y = ax^2$ opens, the number a is a scaling factor. The parabola widens as a approaches zero and narrows as $|a|$ becomes large.

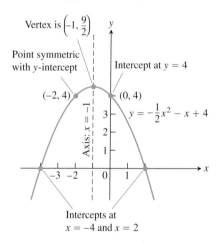

FIGURE A.20   The parabola in Example 8.

Notice that if $a = 0$, then we have $y = bx + c$ which is an equation for a line. The axis, given by Equation (2), can be found by completing the square or by using a technique we study in Section 4.1.

EXAMPLE 8   Graph the equation $y = -\dfrac{1}{2}x^2 - x + 4$.

Solution   Comparing the equation with $y = ax^2 + bx + c$ we see that

$$a = -\frac{1}{2}, \qquad b = -1, \qquad c = 4.$$

Since $a < 0$, the parabola opens downward. From Equation (2) the axis is the vertical line

$$x = -\frac{b}{2a} = -\frac{(-1)}{2(-1/2)} = -1.$$

When $x = -1$, we have

$$y = -\frac{1}{2}(-1)^2 - (-1) + 4 = \frac{9}{2}.$$

The vertex is $(-1, 9/2)$.

The x-intercepts are where $y = 0$:

$$-\frac{1}{2}x^2 - x + 4 = 0$$
$$x^2 + 2x - 8 = 0$$
$$(x - 2)(x + 4) = 0$$
$$x = 2, \qquad x = -4$$

We plot some points, sketch the axis, and use the direction of opening to complete the graph in Figure A.20. ∎

EXERCISES A.3

In Exercises 1 and 2, a particle moves from A to B in the coordinate plane. Find the increments Δx and Δy in the particle's coordinates. Also find the distance from A to B.

1. $A(-3, 2), \quad B(-1, -2)$ **2.** $A(-3.2, -2), B(-8.1, -2)$

Describe the graphs of the equations in Exercises 3 and 4.

3. $x^2 + y^2 = 1$ **4.** $x^2 + y^2 \le 3$

Plot the points in Exercises 5 and 6 and find the slope (if any) of the line they determine. Also find the common slope (if any) of the lines perpendicular to line AB.

5. $A(-1, 2), \quad B(-2, -1)$ **6.** $A(2, 3), \quad B(-1, 3)$

In Exercises 7 and 8, find an equation for (a) the vertical line and (b) the horizontal line through the given point.

7. $(-1, 4/3)$ **8.** $\left(0, -\sqrt{2}\right)$

In Exercises 9–15, write an equation for each line described.

9. Passes through $(-1, 1)$ with slope -1

10. Passes through $(3, 4)$ and $(-2, 5)$

11. Has slope $-5/4$ and y-intercept 6

12. Passes through $(-12, -9)$ and has slope 0

13. Has y-intercept 4 and x-intercept -1

14. Passes through $(5, -1)$ and is parallel to the line $2x + 5y = 15$

15. Passes through $(4, 10)$ and is perpendicular to the line $6x - 3y = 5$.

In Exercises 16 and 17, find the line's x- and y-intercepts and use this information to graph the line.

16. $3x + 4y = 12$ **17.** $\sqrt{2}x - \sqrt{3}y = \sqrt{6}$

18. Is there anything special about the relationship between the lines $Ax + By = C_1$ and $Bx - Ay = C_2 \, (A \neq 0, B \neq 0)$? Give reasons for your answer.

19. A particle starts at $A(-2, 3)$ and its coordinates change by increments $\Delta x = 5$, $\Delta y = -6$. Find its new position.

20. The coordinates of a particle change by $\Delta x = 5$ and $\Delta y = 6$ as it moves from $A(x, y)$ to $B(3, -3)$. Find x and y.

In Exercises 21–23, find an equation for the circle with the given center $C(h, k)$ and radius a. Then sketch the circle in the xy-plane. Include the circle's center in your sketch. Also, label the circle's x- and y-intercepts, if any, with their coordinate pairs.

21. $C(0, 2), \quad a = 2$ **22.** $C(-1, 5), \quad a = \sqrt{10}$

23. $C\left(-\sqrt{3}, -2\right), \quad a = 2$

Graph the circles whose equations are given in Exercises 24–26. Label each circle's center and intercepts (if any) with their coordinate pairs.

24. $x^2 + y^2 + 4x - 4y + 4 = 0$

25. $x^2 + y^2 - 3y - 4 = 0$ **26.** $x^2 + y^2 - 4x + 4y = 0$

Graph the parabolas in Exercises 27–30. Label the vertex, axis, and intercepts in each case.

27. $y = x^2 - 2x - 3$ **28.** $y = -x^2 + 4x$

29. $y = -x^2 - 6x - 5$ **30.** $y = \frac{1}{2}x^2 + x + 4$

Describe the regions defined by the inequalities and pairs of inequalities in Exercises 31–36.

31. $x^2 + y^2 > 7$ **32.** $(x - 1)^2 + y^2 \leq 4$

33. $x^2 + y^2 > 1, \quad x^2 + y^2 < 4$

34. $x^2 + y^2 + 6y < 0, \quad y > -3$

35. Write an inequality that describes the points that lie inside the circle with center $(-2, 1)$ and radius $\sqrt{6}$.

36. Write a pair of inequalities that describe the points that lie inside or on the circle with center $(0, 0)$ and radius $\sqrt{2}$, and on or to the right of the vertical line through $(1, 0)$.

In Exercises 37–40, graph the two equations and find the points in which the graphs intersect.

37. $y = 2x, \quad x^2 + y^2 = 1$ **38.** $y - x = 1, \quad y = x^2$

39. $y = -x^2, \quad y = 2x^2 - 1$

40. $x^2 + y^2 = 1, \quad (x - 1)^2 + y^2 = 1$

41. Insulation By measuring slopes in the accompanying figure, estimate the temperature change in degrees per inch for (a) the gypsum wallboard; (b) the fiberglass insulation; (c) the wood sheathing.

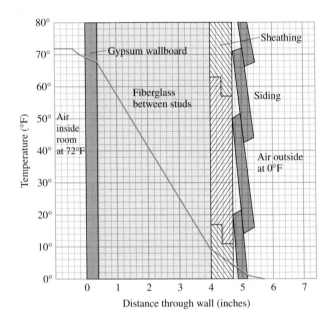

The temperature changes in the wall in Exercises 41 and 42.

42. Insulation According to the figure in Exercise 41, which of the materials is the best insulator? The poorest? Explain.

43. Pressure under water The pressure p experienced by a diver under water is related to the diver's depth d by an equation of the form $p = kd + 1$ (k a constant). At the surface, the pressure is 1 atmosphere. The pressure at 100 meters is about 10.94 atmospheres. Find the pressure at 50 meters.

44. Reflected light A ray of light comes in along the line $x + y = 1$ from the second quadrant and reflects off the x-axis (see the accompanying figure). The angle of incidence is equal to the angle of reflection. Write an equation for the line along which the departing light travels.

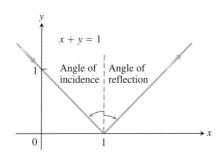

The path of the light ray in Exercise 44. Angles of incidence and reflection are measured from the perpendicular.

45. Fahrenheit vs. Celsius In the FC-plane, sketch the graph of the equation

$$C = \frac{5}{9}(F - 32)$$

linking Fahrenheit and Celsius temperatures. On the same graph sketch the line $C = F$. Is there a temperature at which a Celsius thermometer gives the same numerical reading as a Fahrenheit thermometer? If so, find it.

46. The Mt. Washington Cog Railway Civil engineers calculate the slope of roadbed as the ratio of the distance it rises or falls to the distance it runs horizontally. They call this ratio the **grade** of the roadbed,

usually written as a percentage. Along the coast, commercial railroad grades are usually less than 2%. In the mountains, they may go as high as 4%. Highway grades are usually less than 5%.

The steepest part of the Mt. Washington Cog Railway in New Hampshire has an exceptional 37.1% grade. Along this part of the track, the seats in the front of the car are 14 ft above those in the rear. About how far apart are the front and rear rows of seats?

47. By calculating the lengths of its sides, show that the triangle with vertices at the points $A(1, 2)$, $B(5, 5)$, and $C(4, -2)$ is isosceles but not equilateral.

48. Show that the triangle with vertices $A(0, 0)$, $B\left(1, \sqrt{3}\right)$, and $C(2, 0)$ is equilateral.

49. Show that the points $A(2, -1)$, $B(1, 3)$, and $C(-3, 2)$ are vertices of a square, and find the fourth vertex.

50. Three different parallelograms have vertices at $(-1, 1)$, $(2, 0)$, and $(2, 3)$. Sketch them and find the coordinates of the fourth vertex of each.

51. For what value of k is the line $2x + ky = 3$ perpendicular to the line $4x + y = 1$? For what value of k are the lines parallel?

52. Midpoint of a line segment Show that the point with coordinates

$$\left(\frac{x_1 + x_2}{2}, \frac{y_1 + y_2}{2}\right)$$

is the midpoint of the line segment joining $P(x_1, y_1)$ to $Q(x_2, y_2)$.

A.4 Trigonometry Formulas

Trigonometric Functions

Radian Measure

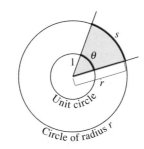

$$\frac{s}{r} = \frac{\theta}{1} = \theta \qquad \text{or} \qquad \theta = \frac{s}{r},$$

$$180° = \pi \text{ radians}.$$

The angles of two common triangles, in degrees and radians.

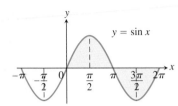

Domain: $(-\infty, \infty)$
Range: $[-1, 1]$

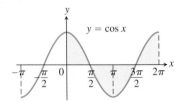

Domain: $(-\infty, \infty)$
Range: $[-1, 1]$

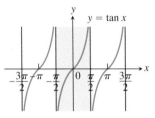

Domain: All real numbers except odd
   integer multiples of $\pi/2$
Range: $(-\infty, \infty)$

Domain: $x \neq \pm\dfrac{\pi}{2}, \pm\dfrac{3\pi}{2}, \dots$
Range: $(-\infty, -1] \cup [1, \infty)$

Domain: $x \neq 0, \pm\pi, \pm2\pi, \dots$
Range: $(-\infty, -1] \cup [1, \infty)$

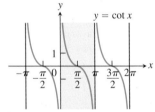

Domain: $x \neq 0, \pm\pi, \pm2\pi, \dots$
Range: $(-\infty, \infty)$

Trigonometry Formulas

Definitions and Fundamental Identities

Sine: $\quad \sin \theta = \dfrac{y}{r} = \dfrac{1}{\csc \theta}$

Cosine: $\quad \cos \theta = \dfrac{x}{r} = \dfrac{1}{\sec \theta}$

Tangent: $\quad \tan \theta = \dfrac{y}{x} = \dfrac{1}{\cot \theta}$

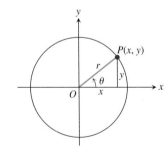

Identities

$\sin (-\theta) = -\sin \theta, \quad \cos (-\theta) = \cos \theta$

$\sin^2 \theta + \cos^2 \theta = 1, \quad \sec^2 \theta = 1 + \tan^2 \theta, \quad \csc^2 \theta = 1 + \cot^2 \theta$

$\sin 2\theta = 2 \sin \theta \cos \theta, \quad \cos 2\theta = \cos^2 \theta - \sin^2 \theta$

$\cos^2 \theta = \dfrac{1 + \cos 2\theta}{2}, \quad \sin^2 \theta = \dfrac{1 - \cos 2\theta}{2}$

$\sin (A + B) = \sin A \cos B + \cos A \sin B$

$\sin (A - B) = \sin A \cos B - \cos A \sin B$

$$\cos(A + B) = \cos A \cos B - \sin A \sin B$$

$$\cos(A - B) = \cos A \cos B + \sin A \sin B$$

$$\tan(A + B) = \frac{\tan A + \tan B}{1 - \tan A \tan B}, \qquad \tan(A - B) = \frac{\tan A - \tan B}{1 + \tan A \tan B}$$

$$\sin\left(A - \frac{\pi}{2}\right) = -\cos A, \qquad \cos\left(A - \frac{\pi}{2}\right) = \sin A$$

$$\sin\left(A + \frac{\pi}{2}\right) = \cos A, \qquad \cos\left(A + \frac{\pi}{2}\right) = -\sin A$$

$$\sin A \sin B = \frac{1}{2}\cos(A - B) - \frac{1}{2}\cos(A + B)$$

$$\cos A \cos B = \frac{1}{2}\cos(A - B) + \frac{1}{2}\cos(A + B)$$

$$\sin A \cos B = \frac{1}{2}\sin(A - B) + \frac{1}{2}\sin(A + B)$$

$$\sin A + \sin B = 2 \sin\frac{1}{2}(A + B)\cos\frac{1}{2}(A - B)$$

$$\sin A - \sin B = 2 \cos\frac{1}{2}(A + B)\sin\frac{1}{2}(A - B)$$

$$\cos A + \cos B = 2 \cos\frac{1}{2}(A + B)\cos\frac{1}{2}(A - B)$$

$$\cos A - \cos B = -2 \sin\frac{1}{2}(A + B)\sin\frac{1}{2}(A - B)$$

A.5   Proofs of Limit Theorems

This appendix proves Theorem 1, Parts 2–5, and Theorem 4 from Section 2.2.

THEOREM 1—Limit Laws   If L, M, c, and k are real numbers and

$$\lim_{x \to c} f(x) = L \quad \text{and} \quad \lim_{x \to c} g(x) = M, \quad \text{then}$$

1. *Sum Rule:* $\displaystyle\lim_{x \to c} \left(f(x) + g(x)\right) = L + M$

2. *Difference Rule:* $\displaystyle\lim_{x \to c} \left(f(x) - g(x)\right) = L - M$

3. *Product Rule:* $\displaystyle\lim_{x \to c} \left(f(x) \cdot g(x)\right) = L \cdot M$

4. *Constant Multiple Rule:* $\displaystyle\lim_{x \to c} \left(kf(x)\right) = kL$ (any number k)

5. *Quotient Rule:* $\displaystyle\lim_{x \to c} \frac{f(x)}{g(x)} = \frac{L}{M}$,   if $M \neq 0$

6. *Power Rule:*   If r and s are integers with no common factors and $s \neq 0$, then

$$\lim_{x \to c} (f(x))^{r/s} = L^{r/s}$$

provided that $L^{r/s}$ is a real number. (If s is even, we assume that $L > 0$.)

We proved the Sum Rule in Section 2.3 and the Power Rule is proved in more advanced texts. We obtain the Difference Rule by replacing $g(x)$ by $-g(x)$ and M by $-M$ in the Sum Rule. The Constant Multiple Rule is the special case $g(x) = k$ of the Product Rule. This leaves only the Product and Quotient Rules.

Proof of the Limit Product Rule   We show that for any $\epsilon > 0$ there exists a $\delta > 0$ such that for all x in the intersection D of the domains of f and g,

$$0 < |x - c| < \delta \quad \Rightarrow \quad |f(x)g(x) - LM| < \epsilon.$$

Suppose then that ϵ is a positive number, and write $f(x)$ and $g(x)$ as

$$f(x) = L + (f(x) - L), \quad g(x) = M + (g(x) - M).$$

Multiply these expressions together and subtract LM:

$$\begin{aligned}
f(x) \cdot g(x) - LM &= (L + (f(x) - L))(M + (g(x) - M)) - LM \\
&= LM + L(g(x) - M) + M(f(x) - L) \\
&\quad + (f(x) - L)(g(x) - M) - LM \\
&= L(g(x) - M) + M(f(x) - L) + (f(x) - L)(g(x) - M). \quad (1)
\end{aligned}$$

Since f and g have limits L and M as $x \to c$, there exist positive numbers $\delta_1, \delta_2, \delta_3,$ and δ_4 such that for all x in D

$$\begin{aligned}
0 < |x - c| < \delta_1 &\quad \Rightarrow \quad |f(x) - L| < \sqrt{\epsilon/3} \\
0 < |x - c| < \delta_2 &\quad \Rightarrow \quad |g(x) - M| < \sqrt{\epsilon/3} \\
0 < |x - c| < \delta_3 &\quad \Rightarrow \quad |f(x) - L| < \epsilon/(3(1 + |M|)) \\
0 < |x - c| < \delta_4 &\quad \Rightarrow \quad |g(x) - M| < \epsilon/(3(1 + |L|))
\end{aligned} \quad (2)$$

If we take δ to be the smallest numbers δ_1 through δ_4, the inequalities on the right-hand side of the Implications (2) will hold simultaneously for $0 < |x - c| < \delta$. Therefore, for all x in D, $0 < |x - c| < \delta$ implies

$$\begin{aligned}
|f(x) \cdot g(x) - LM| & \quad \text{\small Triangle inequality applied} \\
& \quad \text{\small to Equation (1)} \\
&\leq |L||g(x) - M| + |M||f(x) - L| + |f(x) - L||g(x) - M| \\
&\leq (1 + |L|)|g(x) - M| + (1 + |M|)|f(x) - L| + |f(x) - L||g(x) - M| \\
&< \frac{\epsilon}{3} + \frac{\epsilon}{3} + \sqrt{\frac{\epsilon}{3}}\sqrt{\frac{\epsilon}{3}} = \epsilon. \quad \text{\small Values from (2)}
\end{aligned}$$

This completes the proof of the Limit Product Rule.   ∎

Proof of the Limit Quotient Rule   We show that $\lim_{x \to c}(1/g(x)) = 1/M$. We can then conclude that

$$\lim_{x \to c} \frac{f(x)}{g(x)} = \lim_{x \to c}\left(f(x) \cdot \frac{1}{g(x)}\right) = \lim_{x \to c} f(x) \cdot \lim_{x \to c} \frac{1}{g(x)} = L \cdot \frac{1}{M} = \frac{L}{M}$$

by the Limit Product Rule.

Let $\epsilon > 0$ be given. To show that $\lim_{x \to c}(1/g(x)) = 1/M$, we need to show that there exists a $\delta > 0$ such that for all x

$$0 < |x - c| < \delta \quad \Rightarrow \quad \left|\frac{1}{g(x)} - \frac{1}{M}\right| < \epsilon.$$

Since $|M| > 0$, there exists a positive number δ_1 such that for all x

$$0 < |x - c| < \delta_1 \quad \Rightarrow \quad |g(x) - M| < \frac{M}{2}. \tag{3}$$

For any numbers A and B it can be shown that $|A| - |B| \leq |A - B|$ and $|B| - |A| \leq |A - B|$, from which it follows that $||A| - |B|| \leq |A - B|$. With $A = g(x)$ and $B = M$, this becomes

$$||g(x)| - |M|| \leq |g(x) - M|,$$

which can be combined with the inequality on the right in Implication (3) to get, in turn,

$$||g(x)| - |M|| < \frac{|M|}{2}$$

$$-\frac{|M|}{2} < |g(x)| - |M| < \frac{|M|}{2}$$

$$\frac{|M|}{2} < |g(x)| < \frac{3|M|}{2}$$

$$|M| < 2|g(x)| < 3|M|$$

$$\frac{1}{|g(x)|} < \frac{2}{|M|} < \frac{3}{|g(x)|} \tag{4}$$

Therefore, $0 < |x - c| < \delta_1$ implies that

$$\left| \frac{1}{g(x)} - \frac{1}{M} \right| = \left| \frac{M - g(x)}{Mg(x)} \right| \leq \frac{1}{|M|} \cdot \frac{1}{|g(x)|} \cdot |M - g(x)|$$

$$< \frac{1}{|M|} \cdot \frac{2}{|M|} \cdot |M - g(x)|. \qquad \text{Inequality (4)} \tag{5}$$

Since $(1/2)|M|^2 \epsilon > 0$, there exists a number $\delta_2 > 0$ such that for all x

$$0 < |x - c| < \delta_2 \quad \Rightarrow \quad |M - g(x)| < \frac{\epsilon}{2}|M|^2. \tag{6}$$

If we take δ to be the smaller of δ_1 and δ_2, the conclusions in (5) and (6) both hold for all x such that $0 < |x - c| < \delta$. Combining these conclusions gives

$$0 < |x - c| < \delta \quad \Rightarrow \quad \left| \frac{1}{g(x)} - \frac{1}{M} \right| < \epsilon.$$

This concludes the proof of the Limit Quotient Rule.   ∎

THEOREM 4—The Sandwich Theorem   Suppose that $g(x) \leq f(x) \leq h(x)$ for all x in some open interval I containing c, except possibly at $x = c$ itself. Suppose also that $\lim_{x \to c} g(x) = \lim_{x \to c} h(x) = L$. Then $\lim_{x \to c} f(x) = L$.

Proof for Right-Hand Limits   Suppose $\lim_{x \to c^+} g(x) = \lim_{x \to c^+} h(x) = L$. Then for any $\epsilon > 0$ there exists a $\delta > 0$ such that for all x the interval $c < x < c + \delta$ is contained in I and the inequality implies

$$L - \epsilon < g(x) < L + \epsilon \quad \text{and} \quad L - \epsilon < h(x) < L + \epsilon.$$

These inequalities combine with the inequality $g(x) \le f(x) \le h(x)$ to give

$$L - \epsilon < g(x) \le f(x) \le h(x) < L + \epsilon,$$
$$L - \epsilon < f(x) < L + \epsilon,$$
$$- \epsilon < f(x) - L < \epsilon.$$

Therefore, for all x, the inequality $c < x < c + \delta$ implies $|f(x) - L| < \epsilon$.

Proof for Left-Hand Limits   Suppose $\lim_{x \to c^-} g(x) = \lim_{x \to c^-} h(x) = L$. Then for any $\epsilon > 0$ there exists a $\delta > 0$ such that for all x the interval $c - \delta < x < c$ is contained in I and the inequality implies

$$L - \epsilon < g(x) < L + \epsilon \quad \text{and} \quad L - \epsilon < h(x) < L + \epsilon.$$

We conclude as before that for all x, $c - \delta < x < c$ implies $|f(x) - L| < \epsilon$.

Proof for Two-Sided Limits   If $\lim_{x \to c} g(x) = \lim_{x \to c} h(x) = L$, then $g(x)$ and $h(x)$ both approach L as $x \to c^+$ and as $x \to c^-$; so $\lim_{x \to c^+} f(x) = L$ and $\lim_{x \to c^-} f(x) = L$. Hence $\lim_{x \to c} f(x)$ exists and equals L.   ∎

EXERCISES A.5

1. Suppose that functions $f_1(x)$, $f_2(x)$, and $f_3(x)$ have limits L_1, L_2, and L_3, respectively, as $x \to c$. Show that their sum has limit $L_1 + L_2 + L_3$. Use mathematical induction (Appendix 2) to generalize this result to the sum of any finite number of functions.

2. Use mathematical induction and the Limit Product Rule in Theorem 1 to show that if functions $f_1(x)$, $f_2(x)$, ..., $f_n(x)$ have limits $L_1, L_2, \ldots, L_n$ as $x \to c$, then

$$\lim_{x \to c} f_1(x) f_2(x) \cdot \cdots \cdot f_n(x) = L_1 \cdot L_2 \cdot \cdots \cdot L_n.$$

3. Use the fact that $\lim_{x \to c} x = c$ and the result of Exercise 2 to show that $\lim_{x \to c} x^n = c^n$ for any integer $n > 1$.

4. **Limits of polynomials**   Use the fact that $\lim_{x \to c}(k) = k$ for any number k together with the results of Exercises 1 and 3 to show that $\lim_{x \to c} f(x) = f(c)$ for any polynomial function

$$f(x) = a_n x^n + a_{n-1} x^{n-1} + \cdots + a_1 x + a_0.$$

5. **Limits of rational functions**   Use Theorem 1 and the result of Exercise 4 to show that if $f(x)$ and $g(x)$ are polynomial functions and $g(c) \ne 0$, then

$$\lim_{x \to c} \frac{f(x)}{g(x)} = \frac{f(c)}{g(c)}.$$

6. **Composites of continuous functions**   Figure A.21 gives the diagram for a proof that the composite of two continuous functions is continuous. Reconstruct the proof from the diagram. The statement to be proved is this: If f is continuous at $x = c$ and g is continuous at $f(c)$, then $g \circ f$ is continuous at c.

   Assume that c is an interior point of the domain of f and that $f(c)$ is an interior point of the domain of g. This will make the limits involved two-sided. (The arguments for the cases that involve one-sided limits are similar.)

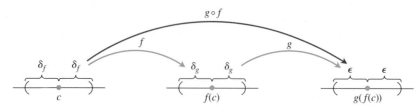

FIGURE A.21   The diagram for a proof that the composite of two continuous functions is continuous.

A.6   Commonly Occurring Limits

This appendix verifies limits (4)–(6) in Theorem 5 of Section 8.1.

Limit 4: If $|x| < 1$, $\lim\limits_{n \to \infty} x^n = 0$ We need to show that to each $\epsilon > 0$ there corresponds an integer N so large that $|x^n| < \epsilon$ for all n greater than N. Since $\epsilon^{1/n} \to 1$, while $|x| < 1$, there exists an integer N for which $\epsilon^{1/N} > |x|$. In other words,

$$|x^N| = |x|^N < \epsilon. \tag{1}$$

This is the integer we seek because, if $|x| < 1$, then

$$|x^n| < |x^N| \quad \text{for all } n > N. \tag{2}$$

Combining (1) and (2) produces $|x^n| < \epsilon$ for all $n > N$, concluding the proof.   ∎

Limit 5: For any number x, $\lim\limits_{n \to \infty} \left(1 + \dfrac{x}{n} \right)^n = e^x$ Let

$$a_n = \left(1 + \frac{x}{n} \right)^n.$$

Then

$$\ln a_n = \ln \left(1 + \frac{x}{n} \right)^n = n \ln \left(1 + \frac{x}{n} \right) \to x,$$

as we can see by the following application of l'Hôpital's Rule, in which we differentiate with respect to n:

$$\lim_{n \to \infty} n \ln \left(1 + \frac{x}{n} \right) = \lim_{n \to \infty} \frac{\ln(1 + x/n)}{1/n}$$

$$= \lim_{n \to \infty} \frac{\left(\dfrac{1}{1 + x/n} \right) \cdot \left(-\dfrac{x}{n^2} \right)}{-1/n^2} = \lim_{n \to \infty} \frac{x}{1 + x/n} = x.$$

Apply Theorem 3, Section 8.1, with $f(x) = e^x$ to conclude that

$$\left(1 + \frac{x}{n} \right)^n = a_n = e^{\ln a_n} \to e^x.$$ ∎

Limit 6: For any number x, $\lim\limits_{n \to \infty} \dfrac{x^n}{n!} = 0$ Since

$$-\frac{|x|^n}{n!} \le \frac{x^n}{n!} \le \frac{|x|^n}{n!},$$

all we need to show is that $|x|^n/n! \to 0$. We can then apply the Sandwich Theorem for Sequences (Section 8.1, Theorem 2) to conclude that $x^n/n! \to 0$.

The first step in showing that $|x|^n/n! \to 0$ is to choose an integer $M > |x|$, so that $(|x|/M) < 1$. By Limit 4, just proved, we then have $(|x|/M)^n \to 0$. We then restrict our attention to values of $n > M$. For these values of n, we can write

$$\frac{|x|^n}{n!} = \frac{|x|^n}{1 \cdot 2 \cdot \cdots \cdot M \cdot \underbrace{(M+1)(M+2) \cdot \cdots \cdot n}_{(n-M) \text{ factors}}}$$

$$\leq \frac{|x|^n}{M! M^{n-M}} = \frac{|x|^n M^M}{M! M^n} = \frac{M^M}{M!} \left(\frac{|x|}{M}\right)^n.$$

Thus,

$$0 \leq \frac{|x|^n}{n!} \leq \frac{M^M}{M!} \left(\frac{|x|}{M}\right)^n.$$

Now, the constant $M^M/M!$ does not change as n increases. Thus the Sandwich Theorem tells us that $|x|^n/n! \to 0$ because $(|x|/M)^n \to 0$. ∎

A.7 Theory of the Real Numbers

A rigorous development of calculus is based on properties of the real numbers. Many results about functions, derivatives, and integrals would be false if stated for functions defined only on the rational numbers. In this appendix we briefly examine some basic concepts of the theory of the reals that hint at what might be learned in a deeper, more theoretical study of calculus.

Three types of properties make the real numbers what they are. These are the **algebraic, order**, and **completeness** properties. The algebraic properties involve addition and multiplication, subtraction and division. They apply to rational or complex numbers as well as to the reals.

The structure of numbers is built around a set with addition and multiplication operations. The following properties are required of addition and multiplication.

A1 $a + (b + c) = (a + b) + c$ for all a, b, c.

A2 $a + b = b + a$ for all a, b, c.

A3   There is a number called "0" such that $a + 0 = a$ for all a.

A4   For each number a, there is a b such that $a + b = 0$.

M1 $a(bc) = (ab)c$ for all a, b, c.

M2 $ab = ba$ for all a, b.

M3   There is a number called "1" such that $a \cdot 1 = a$ for all a.

M4   For each nonzero a, there is a b such that $ab = 1$.

D $a(b + c) = ab + bc$ for all a, b, c.

A1 and M1 are *associative laws*, A2 and M2 are *commutativity laws*, A3 and M3 are *identity laws*, and D is the *distributive law*. Sets that have these algebraic properties are examples of **fields**, and are studied in depth in the area of theoretical mathematics called abstract algebra.

The **order** properties allow us to compare the size of any two numbers. The order properties are

O1   For any a and b, either $a \le b$ or $b \le a$ or both.

O2   If $a \le b$ and $b \le a$ then $a = b$.

O3   If $a \le b$ and $b \le c$ then $a \le c$.

O4   If $a \le b$ then $a + c \le b + c$.

O5   If $a \le b$ and $0 \le c$ then $ac \le bc$.

O3 is the *transitivity law*, and O4 and O5 relate ordering to addition and multiplication.

We can order the reals, the integers, and the rational numbers, but we cannot order the complex numbers. There is no reasonable way to decide whether a number like $i = \sqrt{-1}$ is bigger or smaller than zero. A field in which the size of any two elements can be compared as above is called an **ordered field**. Both the rational numbers and the real numbers are ordered fields, and there are many others.

We can think of real numbers geometrically, lining them up as points on a line. The **completeness property** says that the real numbers correspond to all points on the line, with no "holes" or "gaps." The rationals, in contrast, omit points such as $\sqrt{2}$ and π, and the integers even leave out fractions like $1/2$. The reals, having the completeness property, omit no points.

What exactly do we mean by this vague idea of missing holes? To answer this we must give a more precise description of completeness. A number M is an **upper bound** for a set of numbers if all numbers in the set are smaller than or equal to M. M is a **least upper bound** if it is the smallest upper bound. For example, $M = 2$ is an upper bound for the negative numbers. So is $M = 1$, showing that 2 is not a least upper bound. The least upper bound for the set of negative numbers is $M = 0$. We define a **complete** ordered field to be one in which every nonempty set bounded above has a least upper bound.

If we work with just the rational numbers, the set of numbers less than $\sqrt{2}$ is bounded, but it does not have a rational least upper bound, since any rational upper bound M can be replaced by a slightly smaller rational number that is still larger than $\sqrt{2}$. So the rationals are not complete. In the real numbers, a set that is bounded above always has a least upper bound. The reals are a complete ordered field.

The completeness property is at the heart of many results in calculus. One example occurs when searching for a maximum value for a function on a closed interval $[a, b]$, as in Section 4.1. The function $y = x - x^3$ has a maximum value on $[0, 1]$ at the point x satisfying $1 - 3x^2 = 0$, or $x = \sqrt{1/3}$. If we limited our consideration to functions defined only on rational numbers, we would have to conclude that the function has no maximum, since $\sqrt{1/3}$ is irrational (Figure A.22). The Extreme Value Theorem (Section 4.1), which implies that continuous functions on closed intervals $[a, b]$ have a maximum value, is not true for functions defined only on the rationals.

The Intermediate Value Theorem implies that a continuous function f on an interval $[a, b]$ with $f(a) < 0$ and $f(b) > 0$ must be zero somewhere in $[a, b]$. The function values cannot jump from negative to positive without there being some point x in $[a, b]$ where $f(x) = 0$. The Intermediate Value Theorem also relies on the completeness of the real numbers and is false for continuous functions defined only on the rationals. The function $f(x) = 3x^2 - 1$ has $f(0) = -1$ and $f(1) = 2$, but if we consider f only on the rational numbers, it never equals zero. The only value of x for which $f(x) = 0$ is $x = \sqrt{1/3}$, an irrational number.

We have captured the desired properties of the reals by saying that the real numbers are a complete ordered field. But we're not quite finished. Greek mathematicians in the

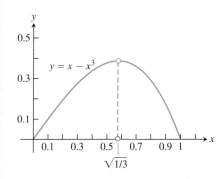

FIGURE A.22   The maximum value of $y = x - x^3$ on $[0, 1]$ occurs at the irrational number $x = \sqrt{1/3}$.

school of Pythagoras tried to impose another property on the numbers of the real line, the condition that all numbers are ratios of integers. They learned that their effort was doomed when they discovered irrational numbers such as $\sqrt{2}$. How do we know that our efforts to specify the real numbers are not also flawed, for some unseen reason? The artist Escher drew optical illusions of spiral staircases that went up and up until they rejoined themselves at the bottom. An engineer trying to build such a staircase would find that no structure realized the plans the architect had drawn. Could it be that our design for the reals contains some subtle contradiction, and that no construction of such a number system can be made?

We resolve this issue by giving a specific description of the real numbers and verifying that the algebraic, order, and completeness properties are satisfied in this model. This is called a **construction** of the reals, and just as stairs can be built with wood, stone, or steel, there are several approaches to constructing the reals. One construction treats the reals as all the infinite decimals,

$$a.d_1 d_2 d_3 d_4 \ldots$$

In this approach a real number is an integer a followed by a sequence of decimal digits $d_1, d_2, d_3, \ldots$, each between 0 and 9. This sequence may stop, or repeat in a periodic pattern, or keep going forever with no pattern. In this form, $2.00, 0.3333333\ldots$ and $3.1415926535898\ldots$ represent three familiar real numbers. The real meaning of the dots "$\ldots$" following these digits requires development of the theory of sequences and series, as in Chapter 8. Each real number is constructed as the limit of a sequence of rational numbers given by its finite decimal approximations. An infinite decimal is then the same as a series

$$a + \frac{d_1}{10} + \frac{d_2}{100} + \cdots.$$

This decimal construction of the real numbers is not entirely straightforward. It's easy enough to check that it gives numbers that satisfy the completeness and order properties, but verifying the algebraic properties is rather involved. Even adding or multiplying two numbers requires an infinite number of operations. Making sense of division requires a careful argument involving limits of rational approximations to infinite decimals.

A different approach was taken by Richard Dedekind (1831–1916), a German mathematician, who gave the first rigorous construction of the real numbers in 1872. Given any real number x, we can divide the rational numbers into two sets: those less than or equal to x and those greater. Dedekind cleverly reversed this reasoning and defined a real number to be a division of the rational numbers into two such sets. This seems like a strange approach, but such indirect methods of constructing new structures from old are common in theoretical mathematics.

These and other approaches can be used to construct a system of numbers having the desired algebraic, order, and completeness properties. A final issue that arises is whether all the constructions give the same thing. Is it possible that different constructions result in different number systems satisfying all the required properties? If yes, which of these is the real numbers? Fortunately, the answer turns out to be no. The reals are the only number system satisfying the algebraic, order, and completeness properties.

Confusion about the nature of the numbers and about limits caused considerable controversy in the early development of calculus. Calculus pioneers such as Newton, Leibniz, and their successors, when looking at what happens to the difference quotient

$$\frac{\Delta y}{\Delta x} = \frac{f(x + \Delta x) - f(x)}{\Delta x}$$

as each of Δy and Δx approach zero, talked about the resulting derivative being a quotient of two infinitely small quantities. These "infinitesimals," written dx and dy, were thought to be some new kind of number, smaller than any fixed number but not zero. Similarly, a definite integral was thought of as a sum of an infinite number of infinitesimals

$$f(x) \cdot dx$$

as x varied over a closed interval. While the approximating difference quotients $\Delta y/\Delta x$ were understood much as today, it was the quotient of infinitesimal quantities, rather than a limit, that was thought to encapsulate the meaning of the derivative. This way of thinking led to logical difficulties, as attempted definitions and manipulations of infinitesimals ran into contradictions and inconsistencies. The more concrete and computable difference quotients did not cause such trouble, but they were thought of merely as useful calculation tools. Difference quotients were used to work out the numerical value of the derivative and to derive general formulas for calculation, but were not considered to be at the heart of the question of what the derivative actually was. Today we realize that the logical problems associated with infinitesimals can be avoided by *defining* the derivative to be the limit of its approximating difference quotients. The ambiguities of the old approach are no longer present, and in the standard theory of calculus, infinitesimals are neither needed nor used.

A.8   The Distributive Law for Vector Cross Products

In this appendix, we prove the Distributive Law

$$\mathbf{u} \times (\mathbf{v} + \mathbf{w}) = \mathbf{u} \times \mathbf{v} + \mathbf{u} \times \mathbf{w}$$

which is Property 2 in Section 10.4.

Proof   To derive the Distributive Law, we construct $\mathbf{u} \times \mathbf{v}$ a new way. We draw $\mathbf{u}$ and $\mathbf{v}$ from the common point O and construct a plane M perpendicular to $\mathbf{u}$ at O (Figure A.23). We then project $\mathbf{v}$ orthogonally onto M, yielding a vector $\mathbf{v}'$ with length $|\mathbf{v}|\sin\theta$. We rotate $\mathbf{v}'$ 90° about $\mathbf{u}$ in the positive sense to produce a vector $\mathbf{v}''$. Finally, we multiply $\mathbf{v}''$ by the length of $\mathbf{u}$. The resulting vector $|\mathbf{u}|\mathbf{v}''$ is equal to $\mathbf{u} \times \mathbf{v}$ since $\mathbf{v}''$ has the same direction as $\mathbf{u} \times \mathbf{v}$ by its construction (Figure A.23) and

$$|\mathbf{u}||\mathbf{v}''| = |\mathbf{u}||\mathbf{v}'| = |\mathbf{u}||\mathbf{v}|\sin\theta = |\mathbf{u} \times \mathbf{v}|.$$

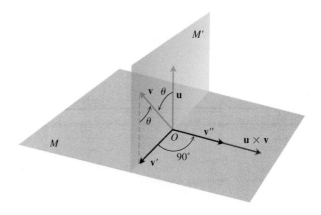

FIGURE A.23   As explained in the text, $\mathbf{u} \times \mathbf{v} = |\mathbf{u}|\mathbf{v}''$.

Now each of these three operations, namely,

1. projection onto M
2. rotation about **u** through $90°$
3. multiplication by the scalar $|\mathbf{u}|$

when applied to a triangle whose plane is not parallel to **u**, will produce another triangle. If we start with the triangle whose sides are **v**, **w**, and $\mathbf{v} + \mathbf{w}$ (Figure A.24) and apply these three steps, we successively obtain the following:

1. A triangle whose sides are $\mathbf{v}'$, $\mathbf{w}'$, and $(\mathbf{v} + \mathbf{w})'$ satisfying the vector equation

$$\mathbf{v}' + \mathbf{w}' = (\mathbf{v} + \mathbf{w})'$$

2. A triangle whose sides are $\mathbf{v}''$, $\mathbf{w}''$, and $(\mathbf{v} + \mathbf{w})''$ satisfying the vector equation

$$\mathbf{v}'' + \mathbf{w}'' = (\mathbf{v} + \mathbf{w})''$$

(the double prime on each vector has the same meaning as in Figure A.23)

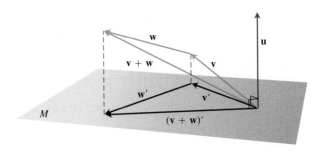

FIGURE A.24   The vectors, **v**, **w**, $\mathbf{v} + \mathbf{w}$, and their projections onto a plane perpendicular to **u**.

3. A triangle whose sides are $|\mathbf{u}|\mathbf{v}''$, $|\mathbf{u}|\mathbf{w}''$, and $|\mathbf{u}|(\mathbf{v} + \mathbf{w})''$ satisfying the vector equation

$$|\mathbf{u}|\mathbf{v}'' + |\mathbf{u}|\mathbf{w}'' = |\mathbf{u}|(\mathbf{v} + \mathbf{w})''.$$

Substituting $|\mathbf{u}|\mathbf{v}'' = \mathbf{u} \times \mathbf{v}$, $|\mathbf{u}|\mathbf{w}'' = \mathbf{u} \times \mathbf{w}$, and $|\mathbf{u}|(\mathbf{v} + \mathbf{w})'' = \mathbf{u} \times (\mathbf{v} + \mathbf{w})$ from our discussion above into this last equation gives

$$\mathbf{u} \times \mathbf{v} + \mathbf{u} \times \mathbf{w} = \mathbf{u} \times (\mathbf{v} + \mathbf{w}),$$

which is the law we wanted to establish. ∎

A.9   The Mixed Derivative Theorem and the Increment Theorem

This appendix derives the Mixed Derivative Theorem (Theorem 2, Section 12.3) and the Increment Theorem for Functions of Two Variables (Theorem 3, Section 12.3). Euler first published the Mixed Derivative Theorem in 1734, in a series of papers he wrote on hydrodynamics.

> **THEOREM 2—The Mixed Derivative Theorem**   If $f(x, y)$ and its partial derivatives $f_x, f_y, f_{xy},$ and f_{yx} are defined throughout an open region containing a point (a, b) and are all continuous at (a, b), then $f_{xy}(a, b) = f_{yx}(a, b)$.

Proof   The equality of $f_{xy}(a, b)$ and $f_{yx}(a, b)$ can be established by four applications of the Mean Value Theorem (Theorem 4, Section 4.2). By hypothesis, the point (a, b) lies in the interior of a rectangle R in the xy-plane on which $f, f_x, f_y, f_{xy},$ and f_{yx} are all defined. We let h and k be the numbers such that the point $(a + h, b + k)$ also lies in R, and we consider the difference

$$\Delta = F(a + h) - F(a), \tag{1}$$

where

$$F(x) = f(x, b + k) - f(x, b). \tag{2}$$

We apply the Mean Value Theorem to F, which is continuous because it is differentiable. Then Equation (1) becomes

$$\Delta = hF'(c_1), \tag{3}$$

where c_1 lies between a and $a + h$. From Equation (2).

$$F'(x) = f_x(x, b + k) - f_x(x, b),$$

so Equation (3) becomes

$$\Delta = h[f_x(c_1, b + k) - f_x(c_1, b)]. \tag{4}$$

Now we apply the Mean Value Theorem to the function $g(y) = f_x(c_1, y)$ and have

$$g(b + k) - g(b) = kg'(d_1),$$

or

$$f_x(c_1, b + k) - f_x(c_1, b) = kf_{xy}(c_1, d_1)$$

for some d_1 between b and $b + k$. By substituting this into Equation (4), we get

$$\Delta = hkf_{xy}(c_1, d_1) \tag{5}$$

for some point (c_1, d_1) in the rectangle R' whose vertices are the four points (a, b), $(a + h, b), (a + h, b + k),$ and $(a, b + k)$. (See Figure A.25.)

By substituting from Equation (2) into Equation (1), we may also write

$$\begin{aligned} \Delta &= f(a + h, b + k) - f(a + h, b) - f(a, b + k) + f(a, b) \\ &= [f(a + h, b + k) - f(a, b + k)] - [f(a + h, b) - f(a, b)] \\ &= \phi(b + k) - \phi(b), \end{aligned} \tag{6}$$

where

$$\phi(y) = f(a + h, y) - f(a, y). \tag{7}$$

The Mean Value Theorem applied to Equation (6) now gives

$$\Delta = k\phi'(d_2) \tag{8}$$

for some d_2 between b and $b + k$. By Equation (7),

$$\phi'(y) = f_y(a + h, y) - f_y(a, y). \tag{9}$$

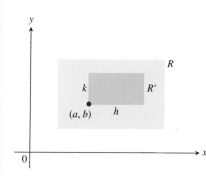

FIGURE A.25   The key to proving $f_{xy}(a, b) = f_{yx}(a, b)$ is that no matter how small R' is, f_{xy} and f_{yx} take on equal values somewhere inside R' (although not necessarily at the same point).

Substituting from Equation (9) into Equation (8) gives

$$\Delta = k[f_y(a + h, d_2) - f_y(a, d_2)].$$

Finally, we apply the Mean Value Theorem to the expression in brackets and get

$$\Delta = khf_{yx}(c_2, d_2) \tag{10}$$

for some c_2 between a and $a + h$.

Together, Equations (5) and (10) show that

$$f_{xy}(c_1, d_1) = f_{yx}(c_2, d_2), \tag{11}$$

where (c_1, d_1) and (c_2, d_2) both lie in the rectangle R' (Figure A.25). Equation (11) is not quite the result we want, since it says only that f_{xy} has the same value at (c_1, d_1) that f_{yx} has at (c_2, d_2). The numbers h and k in our discussion, however, may be made as small as we wish. The hypothesis that f_{xy} and f_{yx} are both continuous at (a, b) means that $f_{xy}(c_1, d_1) = f_{xy}(a, b) + \epsilon_1$ and $f_{yx}(c_2, d_2) = f_{yx}(a, b) + \epsilon_2$, where each of $\epsilon_1, \epsilon_2 \to 0$ as both $h, k \to 0$. Hence, if we let h and $k \to 0$, we have $f_{xy}(a, b) = f_{yx}(a, b)$.   ∎

The equality of $f_{xy}(a, b)$ and $f_{yx}(a, b)$ can be proved with hypotheses weaker than the ones we assumed. For example, it is enough for f, f_x, and f_y to exist in R and for f_{xy} to be continuous at (a, b). Then f_{yx} will exist at (a, b) and equal f_{xy} at that point.

THEOREM 3—The Increment Theorem for Functions of Two Variables

Suppose that the first partial derivatives of $z = f(x, y)$ are defined throughout an open region R containing the point (x_0, y_0) and that f_x and f_y are continuous at (x_0, y_0). Then the change $\Delta z = f(x_0 + \Delta x, y_0 + \Delta y) - f(x_0, y_0)$ in the value of f that results from moving from (x_0, y_0) to another point $(x_0 + \Delta x, y_0 + \Delta y)$ in R satisfies an equation of the form

$$\Delta z = f_x(x_0, y_0)\Delta x + f_y(x_0, y_0)\Delta y + \epsilon_1 \Delta x + \epsilon_2 \Delta y,$$

in which each of $\epsilon_1, \epsilon_2 \to 0$ as both $\Delta x, \Delta y \to 0$.

Proof   We work within a rectangle T centered at $A(x_0, y_0)$ and lying within R, and we assume that Δx and Δy are already so small that the line segment joining A to $B(x_0 + \Delta x, y_0)$ and the line segment joining B to $C(x_0 + \Delta x, y_0 + \Delta y)$ lie in the interior of T (Figure A.26).

We may think of Δz as the sum $\Delta z = \Delta z_1 + \Delta z_2$ of two increments, where

$$\Delta z_1 = f(x_0 + \Delta x, y_0) - f(x_0, y_0)$$

is the change in the value of f from A to B and

$$\Delta z_2 = f(x_0 + \Delta x, y_0 + \Delta y) - f(x_0 + \Delta x, y_0)$$

is the change in the value of f from B to C (Figure A.27).

On the closed interval of x-values joining x_0 to $x_0 + \Delta x$, the function $F(x) = f(x, y_0)$ is a differentiable (and hence continuous) function of x, with derivative

$$F'(x) = f_x(x, y_0).$$

By the Mean Value Theorem (Theorem 4, Section 4.2), there is an x-value c between x_0 and $x_0 + \Delta x$ at which

$$F(x_0 + \Delta x) - F(x_0) = F'(c)\,\Delta x$$

FIGURE A.26   The rectangular region T in the proof of the Increment Theorem. The figure is drawn for Δx and Δy positive, but either increment might be zero or negative.

In the figure:
$C(x_0 + \Delta x, y_0 + \Delta y)$
$A(x_0, y_0)$
$B(x_0 + \Delta x, y_0)$
T

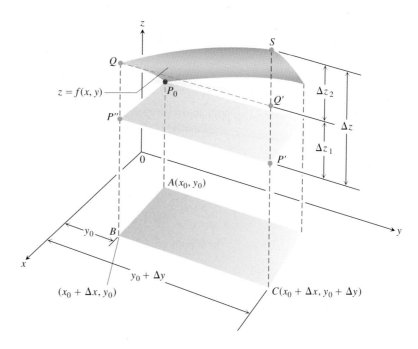

FIGURE A.27 Part of the surface $z = f(x, y)$ near $P_0(x_0, y_0, f(x_0, y_0))$. The points P_0, P', and P'' have the same height $z_0 = f(x_0, y_0)$ above the xy-plane. The change in z is $\Delta z = P'S$. The change

$$\Delta z_1 = f(x_0 + \Delta x, y_0) - f(x_0, y_0),$$

shown as $P''Q = P'Q'$, is caused by changing x from x_0 to $x_0 + \Delta x$ while holding y equal to y_0. Then, with x held equal to $x_0 + \Delta x$,

$$\Delta z_2 = f(x_0 + \Delta x, y_0 + \Delta y) - f(x_0 + \Delta x, y_0)$$

is the change in z caused by changing y_0 from $y_0 + \Delta y$, which is represented by $Q'S$? The total change in z is the sum of Δz_1 and Δz_2.

or

$$f(x_0 + \Delta x, y_0) - f(x_0, y_0) = f_x(c, y_0) \, \Delta x$$

or

$$\Delta z_1 = f_x(c, y_0) \, \Delta x. \tag{12}$$

Similarly, $G(y) = f(x_0 + \Delta x, y)$ is a differentiable (and hence continuous) function of y on the closed y-interval joining y_0 and $y_0 + \Delta y$, with derivative

$$G'(y) = f_y(x_0 + \Delta x, y).$$

Hence, there is a y-value d between y_0 and $y_0 + \Delta y$ at which

$$G(y_0 + \Delta y) - G(y_0) = G'(d) \, \Delta y$$

or

$$f(x_0 + \Delta x, y_0 + \Delta y) - f(x_0 + \Delta x, y) = f_y(x_0 + \Delta x, d) \, \Delta y$$

or

$$\Delta z_2 = f_y(x_0 + \Delta x, d) \, \Delta y. \tag{13}$$

Now, as both Δx and $\Delta y \to 0$, we know that $c \to x_0$ and $d \to y_0$. Therefore, since f_x and f_y are continuous at (x_0, y_0), the quantities

$$\begin{aligned}
\epsilon_1 &= f_x(c, y_0) - f_x(x_0, y_0), \\
\epsilon_2 &= f_y(x_0 + \Delta x, d) - f_y(x_0, y_0)
\end{aligned} \tag{14}$$

both approach zero as both Δx and $\Delta y \to 0$.

Finally,

$$\begin{aligned}
\Delta z &= \Delta z_1 + \Delta z_2 \\
&= f_x(c, y_0)\Delta x + f_y(x_0 + \Delta x, d)\Delta y && \text{From Equations} \\
& && \text{(12) and (13)} \\
&= [f_x(x_0, y_0) + \epsilon_1]\Delta x + [f_y(x_0, y_0) + \epsilon_2]\Delta y && \text{From Equation (14)} \\
&= f_x(x_0, y_0)\Delta x + f_y(x_0, y_0)\Delta y + \epsilon_1\Delta x + \epsilon_2\Delta y,
\end{aligned}$$

where both ϵ_1 and $\epsilon_2 \to 0$ as both Δx and $\Delta y \to 0$, which is what we set out to prove. ∎

Analogous results hold for functions of any finite number of independent variables. Suppose that the first partial derivatives of $w = f(x, y, z)$ are defined throughout an open region containing the point (x_0, y_0, z_0) and that f_x, f_y, and f_z are continuous at (x_0, y_0, z_0). Then

$$\begin{aligned}
\Delta w &= f(x_0 + \Delta x, y_0 + \Delta y, z_0 + \Delta z) - f(x_0, y_0, z_0) \\
&= f_x\Delta x + f_y\Delta y + f_z\Delta z + \epsilon_1\Delta x + \epsilon_2\Delta y + \epsilon_3\Delta z, \tag{15}
\end{aligned}$$

where $\epsilon_1, \epsilon_2, \epsilon_3 \to 0$ as $\Delta x, \Delta y$, and $\Delta z \to 0$.

The partial derivatives f_x, f_y, f_z in Equation (15) are to be evaluated at the point (x_0, y_0, z_0).

Equation (15) can be proved by treating Δw as the sum of three increments,

$$\Delta w_1 = f(x_0 + \Delta x, y_0, z_0) - f(x_0, y_0, z_0) \tag{16}$$

$$\Delta w_2 = f(x_0 + \Delta x, y_0 + \Delta y, z_0) - f(x_0 + \Delta x, y_0, z_0) \tag{17}$$

$$\Delta w_3 = f(x_0 + \Delta x, y_0 + \Delta y, z_0 + \Delta z) - f(x_0 + \Delta x, y_0 + \Delta y, z_0), \tag{18}$$

and applying the Mean Value Theorem to each of these separately. Two coordinates remain constant and only one varies in each of these partial increments $\Delta w_1, \Delta w_2, \Delta w_3$. In Equation (17), for example, only y varies, since x is held equal to $x_0 + \Delta x$ and z is held equal to z_0. Since $f(x_0 + \Delta x, y, z_0)$ is a continuous function of y with a derivative f_y, it is subject to the Mean Value Theorem, and we have

$$\Delta w_2 = f_y(x_0 + \Delta x, y_1, z_0)\,\Delta y$$

for some y_1 between y_0 and $y_0 + \Delta y$.

ANSWERS

CHAPTER 8
Section 8.1, pp. 511–514

1. $a_1 = 0, a_2 = -1/4, a_3 = -2/9, a_4 = -3/16$

3. $a_1 = 1, a_2 = -1/3, a_3 = 1/5, a_4 = -1/7$

5. $a_1 = 1/2, a_2 = 1/2, a_3 = 1/2, a_4 = 1/2$

7. $1, \dfrac{3}{2}, \dfrac{7}{4}, \dfrac{15}{8}, \dfrac{31}{16}, \dfrac{63}{32}, \dfrac{127}{64}, \dfrac{255}{128}, \dfrac{511}{256}, \dfrac{1023}{512}$

9. $2, 1, -\dfrac{1}{2}, -\dfrac{1}{4}, \dfrac{1}{8}, \dfrac{1}{16}, -\dfrac{1}{32}, -\dfrac{1}{64}, \dfrac{1}{128}, \dfrac{1}{256}$

11. $1, 1, 2, 3, 5, 8, 13, 21, 34, 55$ **13.** $a_n = (-1)^{n+1}, n \geq 1$

15. $a_n = (-1)^{n+1}(n)^2, n \geq 1$ **17.** $a_n = n^2 - 1, n \geq 1$

19. $a_n = 4n - 3, n \geq 1$ **21.** $a_n = \dfrac{1 + (-1)^{n+1}}{2}, n \geq 1$

23. Converges, 2   **25.** Converges, -1 **27.** Converges, -5

29. Diverges   **31.** Diverges   **33.** Converges, $1/2$

35. Converges, 0   **37.** Converges, $\sqrt{2}$ **39.** Converges, 1

41. Converges, 0   **43.** Converges, 0   **45.** Converges, 0

47. Converges, 1   **49.** Converges, e^7 **51.** Converges, 1

53. Converges, 1   **55.** Diverges   **57.** Converges, 4

59. Converges, 0   **61.** Diverges   **63.** Converges, e^{-1}

65. Converges, $e^{2/3}$ **67.** Converges, $x \, (x > 0)$

69. Converges, 0   **71.** Converges, 1   **73.** Converges, $1/2$

75. Converges, $\pi/2$ **77.** Converges, 0   **79.** Converges, 0

81. Converges, $1/2$ **83.** Converges, 0   **85.** $x_n = 2^{n-2}$

87. (a) $f(x) = x^2 - 2, 1.414213562 \approx \sqrt{2}$

    (b) $f(x) = \tan(x) - 1, 0.7853981635 \approx \pi/4$

    (c) $f(x) = e^x$, diverges

89. (b) 1   **97.** Nondecreasing, bounded

99. Not nondecreasing, bounded

101. Converges, nondecreasing sequence theorem

103. Converges, nondecreasing sequence theorem

105. Diverges, definition of divergence   **109.** Converges

111. Converges   **121.** $N = 692, a_n = \sqrt[n]{0.5}, L = 1$

123. $N = 65, a_n = (0.9)^n, L = 0$ **125. (b)** $\sqrt{3}$

Section 8.2, pp. 522–523

1. $s_n = \dfrac{2(1 - (1/3)^n)}{1 - (1/3)}, 3$ **3.** $s_n = \dfrac{1 - (-1/2)^n}{1 - (-1/2)}, 2/3$

5. $s_n = \dfrac{1}{2} - \dfrac{1}{n + 2}, \dfrac{1}{2}$ **7.** $1 - \dfrac{1}{4} + \dfrac{1}{16} - \dfrac{1}{64} + \cdots, \dfrac{4}{5}$

9. $\dfrac{7}{4} + \dfrac{7}{16} + \dfrac{7}{64} + \cdots, \dfrac{7}{3}$

11. $(5 + 1) + \left(\dfrac{5}{2} + \dfrac{1}{3}\right) + \left(\dfrac{5}{4} + \dfrac{1}{9}\right) + \left(\dfrac{5}{8} + \dfrac{1}{27}\right) + \cdots, \dfrac{23}{2}$

13. $(1 + 1) + \left(\dfrac{1}{2} - \dfrac{1}{5}\right) + \left(\dfrac{1}{4} + \dfrac{1}{25}\right) + \left(\dfrac{1}{8} - \dfrac{1}{125}\right) + \cdots, \dfrac{17}{6}$

15. 1   **17.** 5   **19.** 1   **21.** $-\dfrac{1}{\ln 2}$ **23.** Converges, $2 + \sqrt{2}$

25. Converges, 1   **27.** Diverges   **29.** Converges, $\dfrac{e^2}{e^2 - 1}$

31. Converges, 2/9   **33.** Converges, 3/2   **35.** Diverges

37. Diverges   **39.** Converges, $\dfrac{\pi}{\pi - e}$

41. $a = 1, r = -x$; converges to $1/(1 + x)$ for $|x| < 1$

43. $a = 3, r = (x - 1)/2$; converges to $6/(3 - x)$ for x in $(-1, 3)$

45. $|x| < \dfrac{1}{2}, \dfrac{1}{1 - 2x}$ **47.** $-2 < x < 0, \dfrac{1}{2 + x}$

49. $x \neq (2k + 1)\dfrac{\pi}{2}, k$ an integer; $\dfrac{1}{1 - \sin x}$

51. 23/99   **53.** 7/9   **55.** 1/15   **57.** 41333/33300

59. (a) $\displaystyle\sum_{n=-2}^{\infty} \dfrac{1}{(n + 4)(n + 5)}$ **(b)** $\displaystyle\sum_{n=0}^{\infty} \dfrac{1}{(n + 2)(n + 3)}$

(c) $\displaystyle\sum_{n=5}^{\infty} \dfrac{1}{(n - 3)(n - 2)}$

69. (a) $r = 3/5$ **(b)** $r = -3/10$ **71.** $|r| < 1, \dfrac{1 + 2r}{1 - r^2}$

Section 8.3, pp. 527–528

1. Converges; geometric series, $r = \dfrac{1}{10} < 1$

3. Diverges; $\displaystyle\lim_{n\to\infty} \dfrac{n}{n + 1} = 1 \neq 0$ **5.** Diverges; p-series, $p < 1$

7. Converges; geometric series, $r = \dfrac{1}{8} < 1$

9. Diverges; Integral Test
11. Converges; geometric series, $r = 2/3 < 1$

13. Diverges; Integral Test   **15.** Diverges; $\displaystyle\lim_{n\to\infty} \dfrac{2^n}{n + 1} \neq 0$

17. Diverges; $\lim_{n\to\infty} \left(\sqrt{n}/\ln n\right) \neq 0$

19. Diverges; geometric series, $r = \dfrac{1}{\ln 2} > 1$

21. Converges; Integral Test   **23.** Diverges; nth-Term Test
25. Converges; Integral Test   **27.** Converges; Integral Test
29. Converges; Integral Test   **31.** $a = 1$ **33. (b)** About 41.55
35. True
43. (a) $1.20166 \leq S \leq 1.20253$ **(b)** $S \approx 1.2021$, error < 0.0005

Section 8.4, p. 532

1. Diverges; limit comparison with $\sum\left(1/\sqrt{n}\right)$
3. Converges; compare with $\sum\left(1/2^n\right)$
5. Diverges; nth-Term Test

7. Converges; $\left(\dfrac{n}{3n + 1}\right)^n < \left(\dfrac{n}{3n}\right)^n = \left(\dfrac{1}{3}\right)^n$

9. Diverges; direct comparison with $\sum(1/n)$
11. Converges; limit comparison with $\sum(1/n^2)$
13. Diverges; limit comparison with $\sum(1/n)$
15. Diverges; limit comparison with $\sum(1/n)$
17. Diverges; Integral Test
19. Converges; compare with $\sum(1/n^{3/2})$

21. Converges; $\dfrac{1}{n2^n} \leq \dfrac{1}{2^n}$ **23.** Converges; $\dfrac{1}{3^{n-1} + 1} < \dfrac{1}{3^{n-1}}$

25. Diverges; limit comparison with $\sum(1/n)$
27. Converges; compare with $\sum(1/n^2)$

29. Converges; $\dfrac{\tan^{-1} n}{n^{1.1}} < \dfrac{\pi/2}{n^{1.1}}$

31. Converges; compare with $\sum(1/n^2)$
33. Diverges; limit comparison with $\sum(1/n)$
35. Converges; limit comparison with $\sum(1/n^2)$

Section 8.5, pp. 536–537

1. Converges; Ratio Test   **3.** Diverges; Ratio Test
5. Converges; Ratio Test
7. Converges; compare with $\sum(3/(1.25)^n)$

9. Diverges; $\displaystyle\lim_{n\to\infty}\left(1 - \dfrac{3}{n}\right)^n = e^{-3} \neq 0$

11. Converges; compare with $\sum(1/n^2)$
13. Diverges; compare with $\sum(1/(2n))$
15. Diverges; compare with $\sum(1/n)$ **17.** Converges; Ratio Test
19. Converges; Ratio Test   **21.** Converges; Ratio Test
23. Converges; Root Test
25. Converges; compare with $\sum(1/n^2)$
27. Converges; Ratio Test   **29.** Diverges; Ratio Test
31. Converges; Ratio Test   **33.** Converges; Ratio Test

35. Diverges; $a_n = \left(\dfrac{1}{3}\right)^{(1/n!)} \to 1$ **37.** Converges; Ratio Test

39. Diverges; Root Test   **41.** Converges; Root Test
43. Converges; Ratio Test   **47.** Yes

Section 8.6, pp. 542–543

1. Converges by Theorem 16   **3.** Diverges; $a_n \nrightarrow 0$
5. Converges by Theorem 16   **7.** Diverges; $a_n \to 1/2$
9. Converges by Theorem 16
11. Converges absolutely. Series of absolute values is a convergent geometric series.

13. Converges conditionally. $1/\sqrt{n} \to 0$ but $\sum_{n=1}^{\infty} \dfrac{1}{\sqrt{n}}$ diverges.

15. Converges absolutely. Compare with $\sum_{n=1}^{\infty}(1/n^2)$.

17. Converges conditionally. $1/(n + 3) \to 0$ but $\sum_{n=1}^{\infty} \dfrac{1}{n + 3}$ diverges (compare with $\sum_{n=1}^{\infty}(1/n)$).

19. Diverges; $\dfrac{3 + n}{5 + n} \to 1$

21. Converges conditionally; $\left(\dfrac{1}{n^2} + \dfrac{1}{n}\right) \to 0$ but $(1 + n)/n^2 > 1/n$

23. Converges absolutely; Ratio Test

25. Converges absolutely by Integral Test   **27.** Diverges; $a_n \nrightarrow 0$

29. Converges absolutely by the Ratio Test

31. Converges absolutely; $\dfrac{1}{n^2 + 2n + 1} < \dfrac{1}{n^2}$

33. Converges absolutely since $\left|\dfrac{\cos n\pi}{n\sqrt{n}}\right| = \left|\dfrac{(-1)^{n+1}}{n^{3/2}}\right| = \dfrac{1}{n^{3/2}}$
(convergent p-series)

35. Converges absolutely by Root Test   **37.** Diverges; $a_n \to \infty$

39. Converges conditionally; $\sqrt{n + 1} - \sqrt{n} = 1/\left(\sqrt{n} + \sqrt{n + 1}\right) \to 0$, but series of absolute values diverges (compare with $\Sigma\left(1/\sqrt{n}\right)$)

41. Diverges, $a_n \to 1/2 \neq 0$

43. Converges absolutely; sech $n = \dfrac{2}{e^n + e^{-n}} = \dfrac{2e^n}{e^{2n} + 1} < \dfrac{2e^n}{e^{2n}} = \dfrac{2}{e^n}$, a term from a convergent geometric series.

45. $|\text{Error}| < 0.2$ **47.** $|\text{Error}| < 2 \times 10^{-11}$ **49.** 0.54030

51. **(a)** $a_n \geq a_{n+1}$ **(b)** $-1/2$

Section 8.7, pp. 552–553

1. **(a)** $1, -1 < x < 1$ **(b)** $-1 < x < 1$ **(c)** none

3. **(a)** $1/4, -1/2 < x < 0$ **(b)** $-1/2 < x < 0$ **(c)** none

5. **(a)** $10, -8 < x < 12$ **(b)** $-8 < x < 12$ **(c)** none

7. **(a)** $1, -1 < x < 1$ **(b)** $-1 < x < 1$ **(c)** none

9. **(a)** $3, -3 \leq x \leq 3$ **(b)** $-3 \leq x \leq 3$ **(c)** none

11. **(a)** ∞, for all x **(b)** for all x **(c)** none

13. **(a)** ∞, for all x **(b)** for all x **(c)** none

15. **(a)** $1, -1 \leq x < 1$ **(b)** $-1 < x < 1$ **(c)** $x = -1$

17. **(a)** $5, -8 < x < 2$ **(b)** $-8 < x < 2$ **(c)** none

19. **(a)** $3, -3 < x < 3$ **(b)** $-3 < x < 3$ **(c)** none

21. **(a)** $1, -1 < x < 1$ **(b)** $-1 < x < 1$ **(c)** none

23. **(a)** $0, x = 0$ **(b)** $x = 0$ **(c)** none

25. **(a)** $2, -4 < x \leq 0$ **(b)** $-4 < x < 0$ **(c)** $x = 0$

27. **(a)** $1, -1 \leq x \leq 1$ **(b)** $-1 \leq x \leq 1$ **(c)** none

29. **(a)** $1/4, 1 \leq x \leq 3/2$ **(b)** $1 \leq x \leq 3/2$ **(c)** none

31. **(a)** $1, (-1 - \pi) \leq x < (1 - \pi)$
    (b) $(-1 - \pi) < x < (1 - \pi)$ **(c)** $x = -1 - \pi$

33. $-1 < x < 3$, $4/(3 + 2x - x^2)$

35. $0 < x < 16$, $2/\left(4 - \sqrt{x}\right)$

37. $-\sqrt{2} < x < \sqrt{2}$, $3/(2 - x^2)$

39. $1 < x < 5$, $2/(x - 1)$, $1 < x < 5$, $-2/(x - 1)^2$

41. **(a)** $\cos x = 1 - \dfrac{x^2}{2!} + \dfrac{x^4}{4!} - \dfrac{x^6}{6!} + \dfrac{x^8}{8!} - \dfrac{x^{10}}{10!} + \cdots$; converges for all x

(b) Same answer as part (c)

(c) $2x - \dfrac{2^3x^3}{3!} + \dfrac{2^5x^5}{5!} - \dfrac{2^7x^7}{7!} + \dfrac{2^9x^9}{9!} - \dfrac{2^{11}x^{11}}{11!} + \cdots$

43. **(a)** $\dfrac{x^2}{2} + \dfrac{x^4}{12} + \dfrac{x^6}{45} + \dfrac{17x^8}{2520} + \dfrac{31x^{10}}{14175}, -\dfrac{\pi}{2} < x < \dfrac{\pi}{2}$

(b) $1 + x^2 + \dfrac{2x^4}{3} + \dfrac{17x^6}{45} + \dfrac{62x^8}{315} + \cdots, -\dfrac{\pi}{2} < x \leq \dfrac{\pi}{2}$

Section 8.8, pp. 558–559

1. $P_0(x) = 0, P_1(x) = x - 1, P_2(x) = (x - 1) - \dfrac{1}{2}(x - 1)^2$,

$P_3(x) = (x - 1) - \dfrac{1}{2}(x - 1)^2 + \dfrac{1}{3}(x - 1)^3$

3. $P_0(x) = \dfrac{1}{2}, P_1(x) = \dfrac{1}{2} - \dfrac{1}{4}(x - 2)$,

$P_2(x) = \dfrac{1}{2} - \dfrac{1}{4}(x - 2) + \dfrac{1}{8}(x - 2)^2$,

$P_3(x) = \dfrac{1}{2} - \dfrac{1}{4}(x - 2) + \dfrac{1}{8}(x - 2)^2 - \dfrac{1}{16}(x - 2)^3$

5. $P_0(x) = \dfrac{\sqrt{2}}{2}, P_1(x) = \dfrac{\sqrt{2}}{2} + \dfrac{\sqrt{2}}{2}\left(x - \dfrac{\pi}{4}\right)$,

$P_2(x) = \dfrac{\sqrt{2}}{2} + \dfrac{\sqrt{2}}{2}\left(x - \dfrac{\pi}{4}\right) - \dfrac{\sqrt{2}}{4}\left(x - \dfrac{\pi}{4}\right)^2$,

$P_3(x) = \dfrac{\sqrt{2}}{2} + \dfrac{\sqrt{2}}{2}\left(x - \dfrac{\pi}{4}\right) - \dfrac{\sqrt{2}}{4}\left(x - \dfrac{\pi}{4}\right)^2$

$\qquad - \dfrac{\sqrt{2}}{12}\left(x - \dfrac{\pi}{4}\right)^3$

7. $P_0(x) = 2, P_1(x) = 2 + \dfrac{1}{4}(x - 4)$,

$P_2(x) = 2 + \dfrac{1}{4}(x - 4) - \dfrac{1}{64}(x - 4)^2$,

$P_3(x) = 2 + \dfrac{1}{4}(x - 4) - \dfrac{1}{64}(x - 4)^2 + \dfrac{1}{512}(x - 4)^3$

9. $\displaystyle\sum_{n=0}^{\infty} \dfrac{(-x)^n}{n!} = 1 - x + \dfrac{x^2}{2!} - \dfrac{x^3}{3!} + \dfrac{x^4}{4!} - \cdots$

11. $\displaystyle\sum_{n=0}^{\infty}(-1)^n x^n = 1 - x + x^2 - x^3 + \cdots$

13. $\displaystyle\sum_{n=0}^{\infty} \dfrac{(-1)^n 3^{2n+1} x^{2n+1}}{(2n + 1)!}$ **15.** $7\displaystyle\sum_{n=0}^{\infty} \dfrac{(-1)^n x^{2n}}{(2n)!}$ **17.** $\displaystyle\sum_{n=0}^{\infty} \dfrac{x^{2n}}{(2n)!}$

19. $x^4 - 2x^3 - 5x + 4$

21. $8 + 10(x - 2) + 6(x - 2)^2 + (x - 2)^3$

23. $21 - 36(x + 2) + 25(x + 2)^2 - 8(x + 2)^3 + (x + 2)^4$

25. $\displaystyle\sum_{n=0}^{\infty}(-1)^n(n + 1)(x - 1)^n$ **27.** $\displaystyle\sum_{n=0}^{\infty} \dfrac{e^2}{n!}(x - 2)^n$

33. $L(x) = 0, Q(x) = -x^2/2$ **35.** $L(x) = 1, Q(x) = 1 + x^2/2$

37. $L(x) = x, Q(x) = x$

Section 8.9, pp. 567–569

1. $\displaystyle\sum_{n=0}^{\infty} \dfrac{(-5x)^n}{n!} = 1 - 5x + \dfrac{5^2 x^2}{2!} - \dfrac{5^3 x^3}{3!} + \cdots$

3. $\displaystyle\sum_{n=0}^{\infty} \dfrac{5(-1)^n(-x)^{2n+1}}{(2n + 1)!} = \displaystyle\sum_{n=0}^{\infty} \dfrac{5(-1)^{n+1} x^{2n+1}}{(2n + 1)!}$

$\qquad = -5x + \dfrac{5x^3}{3!} - \dfrac{5x^5}{5!} + \dfrac{5x^7}{7!} + \cdots$

5. $\displaystyle\sum_{n=0}^{\infty} \frac{(-1)^n (x+1)^n}{(2n)!}$

7. $\displaystyle\sum_{n=0}^{\infty} \frac{x^{n+1}}{n!} = x + x^2 + \frac{x^3}{2!} + \frac{x^4}{3!} + \frac{x^5}{4!} + \cdots$

9. $\displaystyle\sum_{n=2}^{\infty} \frac{(-1)^n x^{2n}}{(2n)!} = \frac{x^4}{4!} - \frac{x^6}{6!} + \frac{x^8}{8!} - \frac{x^{10}}{10!} + \cdots$

11. $x - \dfrac{\pi^2 x^3}{2!} + \dfrac{\pi^4 x^5}{4!} - \dfrac{\pi^6 x^7}{6!} + \cdots = \displaystyle\sum_{n=0}^{\infty} \frac{(-1)^n \pi^{2n} x^{2n+1}}{(2n)!}$

13. $1 + \displaystyle\sum_{n=1}^{\infty} \frac{(-1)^n (2x)^{2n}}{2 \cdot (2n)!} =$

$1 - \dfrac{(2x)^2}{2 \cdot 2!} + \dfrac{(2x)^4}{2 \cdot 4!} - \dfrac{(2x)^6}{2 \cdot 6!} + \dfrac{(2x)^8}{2 \cdot 8!} - \cdots$

15. $x^2 \displaystyle\sum_{n=0}^{\infty} (2x)^n = x^2 + 2x^3 + 4x^4 + \cdots$

17. $\displaystyle\sum_{n=1}^{\infty} n x^{n-1} = 1 + 2x + 3x^2 + 4x^3 + \cdots$

19. $|x| < (0.06)^{1/5} < 0.56968$

21. $|\text{Error}| < (10^{-3})^3/6 < 1.67 \times 10^{-10},\ -10^{-3} < x < 0$

23. $|\text{Error}| < (3^{0.1})(0.1)^3/6 < 1.87 \times 10^{-4}$

25. $\dfrac{x^3}{3} - \dfrac{x^7}{7 \cdot 3!} + \dfrac{x^{11}}{11 \cdot 5!}$

27. (a) $\dfrac{x^2}{2} - \dfrac{x^4}{12}$

(b) $\dfrac{x^2}{2} - \dfrac{x^4}{3 \cdot 4} + \dfrac{x^6}{5 \cdot 6} - \dfrac{x^8}{7 \cdot 8} + \cdots + (-1)^{15} \dfrac{x^{32}}{31 \cdot 32}$

29. $1/2$ **31.** $-1/24$ **33.** $1/3$

35. $e^x \sin x = x + x^2 + \dfrac{x^3}{3} - \dfrac{x^5}{30} - \dfrac{x^6}{90} \cdots$

43. (a) $Q(x) = 1 + kx + \dfrac{k(k-1)}{2} x^2$ (b) for $0 \le x < 100^{-1/3}$

49. (a) -1 (b) $\left(1/\sqrt{2}\right)(1+i)$ (c) $-i$

53. $x + x^2 + \dfrac{1}{3} x^3 - \dfrac{1}{30} x^5 \cdots$; will converge for all x

Section 8.10, p. 572

1. $1 + \dfrac{x}{2} - \dfrac{x^2}{8} + \dfrac{x^3}{16}$ **3.** $1 + \dfrac{1}{2}x + \dfrac{3}{8}x^2 + \dfrac{5}{16}x^3 + \cdots$

5. $1 - x + \dfrac{3x^2}{4} - \dfrac{x^3}{2}$ **7.** $1 - \dfrac{x^3}{2} + \dfrac{3x^6}{8} - \dfrac{5x^9}{16}$

9. $1 + \dfrac{1}{2x} - \dfrac{1}{8x^2} + \dfrac{1}{16x^3}$

11. $(1+x)^4 = 1 + 4x + 6x^2 + 4x^3 + x^4$

13. $(1-2x)^3 = 1 - 6x + 12x^2 - 8x^3$

15. (a) $x + \dfrac{x^3}{6} + \dfrac{3x^5}{40} + \dfrac{5x^7}{112}$, radius of convergence $= 1$

(b) $\dfrac{\pi}{2} - x - \dfrac{x^3}{6} - \dfrac{3x^5}{40} - \dfrac{5x^7}{112}$

17. $1 - 2x + 3x^2 - 4x^3 + \cdots$

Practice Exercises, pp. 573–575

1. Converges to 1   **3.** Converges to -1 **5.** Diverges
7. Converges to 0   **9.** Converges to 1   **11.** Converges to e^{-5}
13. Converges to 3   **15.** Converges to $\ln 2$ **17.** Diverges
19. $1/6$ **21.** $3/2$ **23.** $e/(e-1)$ **25.** Diverges
27. Converges conditionally   **29.** Converges conditionally
31. Converges absolutely   **33.** Converges absolutely
35. Converges absolutely   **37.** Converges absolutely
39. Converges absolutely
41. (a) $3, -7 \le x < -1$ (b) $-7 < x < -1$ (c) $x = -7$
43. (a) $1/3, 0 \le x \le 2/3$ (b) $0 \le x \le 2/3$ (c) none
45. (a) ∞, for all x (b) for all x (c) none
47. (a) $\sqrt{3}, -\sqrt{3} < x < \sqrt{3}$ (b) $-\sqrt{3} < x < \sqrt{3}$
(c) none
49. (a) e, $-e < x < e$ (b) $-e < x < e$ (c) empty set
51. $\dfrac{1}{1+x}, \dfrac{1}{4}, \dfrac{4}{5}$ **53.** $\sin x, \pi, 0$ **55.** $e^x, \ln 2, 2$ **57.** $\displaystyle\sum_{n=0}^{\infty} 2^n x^n$
59. $\displaystyle\sum_{n=0}^{\infty} \frac{(-1)^n \pi^{2n+1} x^{2n+1}}{(2n+1)!}$ **61.** $\displaystyle\sum_{n=0}^{\infty} \frac{(-1)^n x^{5n}}{(2n)!}$ **63.** $\displaystyle\sum_{n=0}^{\infty} \frac{((\pi x)/2)^n}{n!}$
65. $2 - \dfrac{(x+1)}{2 \cdot 1!} + \dfrac{3(x+1)^2}{2^3 \cdot 2!} + \dfrac{9(x+1)^3}{2^5 \cdot 3!} + \cdots$
67. $\dfrac{1}{4} - \dfrac{1}{4^2}(x-3) + \dfrac{1}{4^3}(x-3)^2 - \dfrac{1}{4^4}(x-3)^3$
69. 0.4849171431 **71.** ≈ 0.4872223583 **73.** $7/2$ **75.** $1/12$
77. -2 **79.** $r = -3, s = 9/2$ **81.** $2/3$
83. $\ln\left(\dfrac{n+1}{2n}\right)$; the series converges to $\ln\left(\dfrac{1}{2}\right)$.
85. (a) ∞ (b) $a = 1,\ b = 0$ **87.** It converges.

Additional and Advanced Exercises, pp. 575–576

1. Converges; Comparison Test   **3.** Diverges; nth-Term Test
5. Converges; Comparison Test   **7.** Diverges; nth-Term Test
9. With $a = \pi/3$, $\cos x = \dfrac{1}{2} - \dfrac{\sqrt{3}}{2}(x - \pi/3) - \dfrac{1}{4}(x - \pi/3)^2$
$\qquad + \dfrac{\sqrt{3}}{12}(x - \pi/3)^3 + \cdots$
11. With $a = 0$, $e^x = 1 + x + \dfrac{x^2}{2!} + \dfrac{x^3}{3!} + \cdots$
13. With $a = 22\pi$, $\cos x = 1 - \dfrac{1}{2}(x - 22\pi)^2 + \dfrac{1}{4!}(x - 22\pi)^4$
$\qquad - \dfrac{1}{6!}(x - 22\pi)^6 + \cdots$
15. Converges, limit $= b$ **17.** $\pi/2$ **21.** $b = \pm\dfrac{1}{5}$
23. $a = 2, L = -7/6$ **27.** (b) Yes
31. (a) $\displaystyle\sum_{n=1}^{\infty} n x^{n-1}$ (b) 6   (c) $1/q$

CHAPTER 9

Section 9.1, pp. 581–582

1. a, e; b, g; c, h; d, f   **3.**

(a) $\left(2, \dfrac{\pi}{2} + 2n\pi\right)$ and $\left(-2, \dfrac{\pi}{2} + (2n+1)\pi\right)$, n an integer

(b) $(2, 2n\pi)$ and $(-2, (2n+1)\pi)$, n an integer

(c) $\left(2, \dfrac{3\pi}{2} + 2n\pi\right)$ and $\left(-2, \dfrac{3\pi}{2} + (2n+1)\pi\right)$,
n an integer

(d) $(2, (2n+1)\pi)$ and $(-2, 2n\pi)$, n an integer

5. (a) $(3, 0)$ **(b)** $(-3, 0)$ **(c)** $\left(-1, \sqrt{3}\right)$ **(d)** $\left(1, \sqrt{3}\right)$
(e) $(3, 0)$ **(f)** $\left(1, \sqrt{3}\right)$ **(g)** $(-3, 0)$ **(h)** $\left(-1, \sqrt{3}\right)$

7.

9.

11.

13.

15.

17.

19.

21.

23. $x = 2$, vertical line through $(2, 0)$ **25.** $y = 0$, the x-axis
27. $y = 4$, horizontal line through $(0, 4)$
29. $x + y = 1$, line, $m = -1$, $b = 1$
31. $x^2 + y^2 = 1$, circle, $C(0, 0)$, radius 1
33. $y - 2x = 5$, line: $m = 2$, $b = 5$
35. $y^2 = x$, parabola, vertex $(0, 0)$, opens right
37. $y = e^x$, graph of natural exponential function
39. $x + y = \pm1$, two straight lines of slope -1, y-intercepts
   $b = \pm1$
41. $(x + 2)^2 + y^2 = 4$, circle, $C(-2, 0)$, radius 2
43. $x^2 + (y - 4)^2 = 16$, circle, $C(0, 4)$, radius 4
45. $(x - 1)^2 + (y - 1)^2 = 2$, circle, $C(1, 1)$, radius $\sqrt{2}$
47. $\sqrt{3}y + x = 4$ **49.** $r\cos\theta = 7$ **51.** $\theta = \pi/4$
53. $r = 2$ or $r = -2$ **55.** $4r^2\cos^2\theta + 9r^2\sin^2\theta = 36$
57. $r\sin^2\theta = 4\cos\theta$ **59.** $r = 4\sin\theta$
61. $r^2 = 6r\cos\theta - 2r\sin\theta - 6$ **63.** $(0, \theta)$, where θ is any angle

Section 9.2, pp. 585–586

1. x-axis   **3.** y-axis

5. y-axis   **7.** x-axis

9. x-axis, y-axis, origin   **11.** y-axis, x-axis, origin

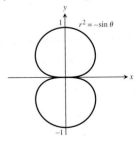

13. x-axis, y-axis, origin   **15.** Origin

17. The slope at $(-1, \pi/2)$ is -1, at $(-1, -\pi/2)$ is 1.

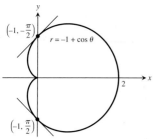

19. The slope at $(1, \pi/4)$ is -1, at $(-1, -\pi/4)$ is 1, at $(-1, 3\pi/4)$ is 1, at $(1, -3\pi/4)$ is -1.

21. (a)

(b)

23. (a)

(b)

25.

27.

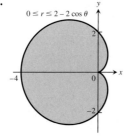

29. (a)

Section 9.3, pp. 589–590

1. 18π **3.** $\pi/8$ **5.** 2   **7.** $\dfrac{\pi}{2} - 1$ **9.** $5\pi - 8$

11. $3\sqrt{3} - \pi$ **13.** $\dfrac{\pi}{3} + \dfrac{\sqrt{3}}{2}$ **15. (a)** $\dfrac{3}{2} - \dfrac{\pi}{4}$ **17.** $19/3$

19. 8   **21.** $3\left(\sqrt{2} + \ln\left(1 + \sqrt{2}\right)\right)$ **23.** $\dfrac{\pi}{8} + \dfrac{3}{8}$ **25.** 2π

Section 9.4, pp. 597–599

1. $y^2 = 8x$, $F(2, 0)$, directrix: $x = -2$

3. $x^2 = -6y$, $F(0, -3/2)$, directrix: $y = 3/2$

5. $\dfrac{x^2}{4} - \dfrac{y^2}{9} = 1$, $F\left(\pm\sqrt{13}, 0\right)$, $V(\pm 2, 0)$,
asymptotes: $y = \pm\dfrac{3}{2}x$

7. $\dfrac{x^2}{2} + y^2 = 1$, $F(\pm 1, 0)$, $V\left(\pm\sqrt{2}, 0\right)$

9.

11.

13.

15.

17.

19.

21.

23.

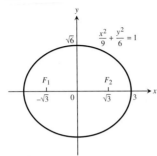

41. **(a)** Foci: $\left(4 \pm \sqrt{7}, 3\right)$; vertices: $(8, 3)$ and $(0, 3)$; center: $(4, 3)$
(b)

25. $\dfrac{x^2}{4} + \dfrac{y^2}{2} = 1$

27. Asymptotes: $y = \pm x$ **29.** Asymptotes: $y = \pm x$

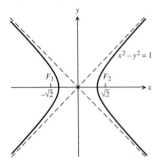

43. **(a)** Center: $(2, 0)$; foci: $(7, 0)$ and $(-3, 0)$; vertices: $(6, 0)$ and $(-2, 0)$; asymptotes: $y = \pm\dfrac{3}{4}(x - 2)$

(b)

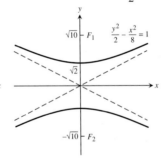

31. Asymptotes: $y = \pm 2x$ **33.** Asymptotes: $y = \pm\dfrac{x}{2}$

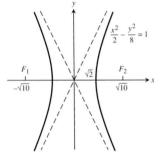

35. $y^2 - x^2 = 1$ **37.** $\dfrac{x^2}{9} - \dfrac{y^2}{16} = 1$

39. **(a)** Vertex: $(1, -2)$; focus: $(3, -2)$; directrix: $x = -1$
(b)

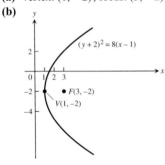

45. $(y + 3)^2 = 4(x + 2)$, $V(-2, -3)$, $F(-1, -3)$, directrix: $x = -3$

47. $(x - 1)^2 = 8(y + 7)$, $V(1, -7)$, $F(1, -5)$, directrix: $y = -9$

49. $\dfrac{(x + 2)^2}{6} + \dfrac{(y + 1)^2}{9} = 1$, $F\left(-2, \pm\sqrt{3} - 1\right)$, $V(-2, \pm 3 - 1)$, $C(-2, -1)$

51. $\dfrac{(x - 2)^2}{3} + \dfrac{(y - 3)^2}{2} = 1$, $F(3, 3)$ and $F(1, 3)$, $V\left(\pm\sqrt{3} + 2, 3\right)$, $C(2, 3)$

53. $\dfrac{(x - 2)^2}{4} - \dfrac{(y - 2)^2}{5} = 1$, $C(2, 2)$, $F(5, 2)$ and $F(-1, 2)$, $V(4, 2)$ and $V(0, 2)$;   asymptotes: $(y - 2) = \pm\dfrac{\sqrt{5}}{2}(x - 2)$

55. $(y + 1)^2 - (x + 1)^2 = 1$, $C(-1, -1)$, $F\left(-1, \sqrt{2} - 1\right)$ and $F\left(-1, -\sqrt{2} - 1\right)$, $V(-1, 0)$ and $V(-1, -2)$; asymptotes $(y + 1) = \pm(x + 1)$

57. $C(-2, 0)$, $a = 4$ **59.** $V(-1, 1)$, $F(-1, 0)$

61. Ellipse: $\dfrac{(x + 2)^2}{5} + y^2 = 1$, $C(-2, 0)$, $F(0, 0)$ and $F(-4, 0)$, $V\left(\sqrt{5} - 2, 0\right)$ and $V\left(-\sqrt{5} - 2, 0\right)$

63. Ellipse: $\dfrac{(x - 1)^2}{2} + (y - 1)^2 = 1$, $C(1, 1)$, $F(2, 1)$ and $F(0, 1)$, $V\left(\sqrt{2} + 1, 1\right)$ and $V\left(-\sqrt{2} + 1, 1\right)$

65. Hyperbola: $(x - 1)^2 - (y - 2)^2 = 1$, $C(1, 2)$,
$F\left(1 + \sqrt{2}, 2\right)$ and $F\left(1 - \sqrt{2}, 2\right)$, $V(2, 2)$ and
$V(0, 2)$; asymptotes: $(y - 2) = \pm(x - 1)$

67. Hyperbola: $\dfrac{(y - 3)^2}{6} - \dfrac{x^2}{3} = 1$, $C(0, 3)$, $F(0, 6)$ and
$F(0, 0)$, $V\left(0, \sqrt{6} + 3\right)$ and $V\left(0, -\sqrt{6} + 3\right)$;
asymptotes: $y = \sqrt{2}x + 3$ or $y = -\sqrt{2}x + 3$

69. (b) $1{:}1$ **73.** Length $= 2\sqrt{2}$, width $= \sqrt{2}$, area $= 4$ **75.** 24π

Section 9.5, pp. 604–606

1. $e = 3/5$, $F(\pm 3, 0)$, $x = \pm 25/3$

3. $e = 1/\sqrt{2}$, $F(0, \pm 1)$, $y = \pm 2$

5. $e = 1/\sqrt{3}$, $F(0, \pm 1)$, $y = \pm 3$

7. $e = \sqrt{3}/3$, $F\left(\pm\sqrt{3}, 0\right)$, $x = \pm 3\sqrt{3}$ **9.** $\dfrac{x^2}{27} + \dfrac{y^2}{36} = 1$

11. $\dfrac{x^2}{4851} + \dfrac{y^2}{4900} = 1$ **13.** $e = \dfrac{\sqrt{5}}{3}, \dfrac{x^2}{9} + \dfrac{y^2}{4} = 1$

15. $e = 1/2, \dfrac{x^2}{64} + \dfrac{y^2}{48} = 1$

17. $e = \sqrt{2}$, $F\left(\pm\sqrt{2}, 0\right)$, $x = \pm 1/\sqrt{2}$

19. $e = \sqrt{2}$, $F(0, \pm 4)$, $y = \pm 2$

21. $e = \sqrt{5}$, $F\left(\pm\sqrt{10}, 0\right)$, $x = \pm 2/\sqrt{10}$

23. $e = \sqrt{5}$, $F\left(0, \pm\sqrt{10}\right)$, $y = \pm 2/\sqrt{10}$

25. $y^2 - \dfrac{x^2}{8} = 1$ **27.** $x^2 - \dfrac{y^2}{8} = 1$ **29.** $r = 2/(1 + \cos\theta)$

31. $r = 30/(1 - 5\sin\theta)$ **33.** $r = 1/(2 + \cos\theta)$

35. $r = 10/(5 - \sin\theta)$

37.

39.

41.

43.

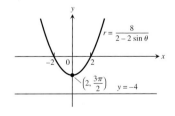

45. $y = 2 - x$ **47.** $y = \left(\sqrt{3}/3\right)x + 2\sqrt{3}$

49. $r\cos\left(\theta - \dfrac{\pi}{4}\right) = 3$ **51.** $r\cos\left(\theta + \dfrac{\pi}{2}\right) = 5$

53. $C(2, 0)$, radius $= 2$ **55.** $C(1, \pi)$, radius $= 1$

57. $(x - 6)^2 + y^2 = 36, r = 12\cos\theta$

59. $x^2 + (y - 5)^2 = 25, r = 10\sin\theta$

61. $(x + 1)^2 + y^2 = 1, r = -2\cos\theta$

63. $x^2 + (y + 1/2)^2 = 1/4, r = -\sin\theta$

75. (b)

Planet	Perihelion	Aphelion
Mercury	0.3075 AU	0.4667 AU
Venus	0.7184 AU	0.7282 AU
Earth	0.9833 AU	1.0167 AU
Mars	1.3817 AU	1.6663 AU
Jupiter	4.9512 AU	5.4548 AU
Saturn	9.0210 AU	10.0570 AU
Uranus	18.2977 AU	20.0623 AU
Neptune	29.8135 AU	30.3065 AU
Pluto	29.6549 AU	49.2251 AU

77. $r = 2a\sin\theta$ (a circle)   **79.** $r\cos(\theta - \alpha) = p$ (a line)

Section 9.6, pp. 608–609

1.

3.

5.

7.

9.

11.

13. $x = (a - b)\cos\theta + b\cos\left(\dfrac{a - b}{b}\theta\right)$,

$\quad y = (a - b)\sin\theta - b\sin\left(\dfrac{a - b}{b}\theta\right)$

15. $x = a\sin^2 t\tan t, y = a\sin^2 t$ **17.** $(1, 1)$

Practice Exercises, pp. 610–612

1. $y = \left(\sqrt{3}/3\right)x - 4$ **3.** $x = 2$ **5.** $y = -3/2$
7. $x^2 + (y + 2)^2 = 4$ **9.** $\left(x - \sqrt{2}\right)^2 + y^2 = 2$
11. $r = -5\sin\theta$ **13.** $r = 3\cos\theta$
15.   **17.** (d)   **19.** (l)   **21.** (k)

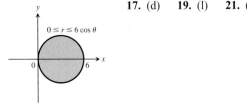

23. (i)   **25.** $9\pi/2$ **27.** $2 + \pi/4$ **29.** 8   **31.** $\pi - 3$
33.   **35.**

   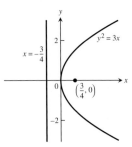

37. $e = 3/4$ **39.** $e = 2$

   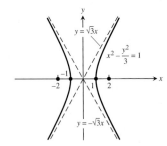

41. $(x - 2)^2 = -12(y - 3)$, $V(2, 3)$, $F(2, 0)$;   directrix: $y = 6$
43. $\dfrac{(x + 3)^2}{9} + \dfrac{(y + 5)^2}{25} = 1$, $C(-3, -5)$, $V(-3, 0)$ and
   $V(-3, -10)$, $F(-3, -1)$ and $F(-3, -9)$
45. $\dfrac{\left(y - 2\sqrt{2}\right)^2}{8} - \dfrac{(x - 2)^2}{2} = 1$, $C\left(2, 2\sqrt{2}\right)$, $V\left(2, 4\sqrt{2}\right)$
   and $V(2, 0)$, $F\left(2, \sqrt{10} + 2\sqrt{2}\right)$ and $F\left(2, -\sqrt{10} + 2\sqrt{2}\right)$;
   asymptotes: $y = 2x - 4 + 2\sqrt{2}$ and $y = -2x + 4 + 2\sqrt{2}$
47. Hyperbola: $\dfrac{(x - 2)^2}{4} - y^2 = 1$, $F\left(2 \pm \sqrt{5}, 0\right)$,
   $V(2 \pm 2, 0)$, $C(2, 0)$;   asymptotes: $y = \pm\dfrac{1}{2}(x - 2)$
49. Parabola: $(y - 1)^2 = -16(x + 3)$, $V(-3, 1)$,
   $F(-7, 1)$;   directrix: $x = 1$
51. Ellipse: $\dfrac{(x + 3)^2}{16} + \dfrac{(y - 2)^2}{9} = 1$, $F\left(\pm\sqrt{7} - 3, 2\right)$,
   $V(\pm 4 - 3, 2)$, $C(-3, 2)$

53. Circle: $(x - 1)^2 + (y - 1)^2 = 2$, $C(1, 1)$, radius $= \sqrt{2}$
55.   **57.**

59. $r = \dfrac{4}{1 + 2\cos\theta}$ **61.** $r = \dfrac{2}{2 + \sin\theta}$
63.   **65.**

   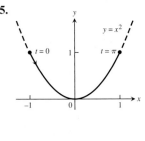

67. (a) 24π (b) 16π

Additional and Advanced Exercises, pp. 612–613

1.

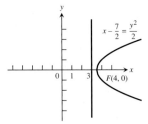

3. $3x^2 + 3y^2 - 8y + 4 = 0$ **5.** $(0, \pm 1)$

7. (a) $\dfrac{(y - 1)^2}{16} - \dfrac{x^2}{48} = 1$ (b) $\dfrac{16\left(y + \dfrac{3}{4}\right)^2}{25} - \dfrac{2x^2}{75} = 1$

11.   **13.**

15.

$9x^2 + 4y^2 - 36 = 0$ $4x^2 + 9y^2 = 16$

17. (a) $r = e^{2\theta}$ **(b)** $\dfrac{\sqrt{5}}{2}(e^{4\pi} - 1)$

19. $r = \dfrac{4}{1 + 2\cos\theta}$ **21.** $r = \dfrac{2}{2 + \sin\theta}$

23. $x = (a + b)\cos\theta - b\cos\left(\dfrac{a + b}{b}\theta\right),$

$y = (a + b)\sin\theta - b\sin\left(\dfrac{a + b}{b}\theta\right)$

27. $\pi/2$

CHAPTER 10

Section 10.1, pp. 617–618

1. The line through the point $(2, 3, 0)$ parallel to the z-axis
3. The x-axis   **5.** The circle $x^2 + y^2 = 4$ in the xy-plane
7. The circle $x^2 + z^2 = 4$ in the xz-plane
9. The circle $y^2 + z^2 = 1$ in the yz-plane
11. The circle $x^2 + y^2 = 16$ in the xy-plane
13. (a) The first quadrant of the xy-plane
   (b) The fourth quadrant of the xy-plane
15. (a) The ball of radius 1 centered at the origin
   (b) All points greater than 1 unit from the origin
17. (a) The upper hemisphere of radius 1 centered at the origin
   (b) The solid upper hemisphere of radius 1 centered at the origin
19. (a) $x = 3$ **(b)** $y = -1$ **(c)** $z = -2$
21. (a) $z = 1$ **(b)** $x = 3$ **(c)** $y = -1$
23. (a) $x^2 + (y - 2)^2 = 4, z = 0$
   (b) $(y - 2)^2 + z^2 = 4, x = 0$ **(c)** $x^2 + z^2 = 4, y = 2$
25. (a) $y = 3, z = -1$ **(b)** $x = 1, z = -1$ **(c)** $x = 1, y = 3$
27. $x^2 + y^2 + z^2 = 25, z = 3$ **29.** $0 \le z \le 1$ **31.** $z \le 0$
33. (a) $(x - 1)^2 + (y - 1)^2 + (z - 1)^2 < 1$
   (b) $(x - 1)^2 + (y - 1)^2 + (z - 1)^2 > 1$
35. 3   **37.** 7   **39.** $2\sqrt{3}$ **41.** $C(-2, 0, 2), a = 2\sqrt{2}$
43. $C(\sqrt{2}, \sqrt{2}, -\sqrt{2}), a = \sqrt{2}$
45. $(x - 1)^2 + (y - 2)^2 + (z - 3)^2 = 14$
47. $(x + 2)^2 + y^2 + z^2 = 3$ **49.** $C(-2, 0, 2), a = \sqrt{8}$
51. $C\left(-\dfrac{1}{4}, -\dfrac{1}{4}, -\dfrac{1}{4}\right), a = \dfrac{5\sqrt{3}}{4}$
53. (a) $\sqrt{y^2 + z^2}$ **(b)** $\sqrt{x^2 + z^2}$ **(c)** $\sqrt{x^2 + y^2}$
55. $\sqrt{17} + \sqrt{33} + 6$

Section 10.2, pp. 626–627

1. (a) $\langle 9, -6 \rangle$ **(b)** $3\sqrt{13}$ **3. (a)** $\langle 1, 3 \rangle$ **(b)** $\sqrt{10}$
5. (a) $\langle 12, -19 \rangle$ **(b)** $\sqrt{505}$ **7. (a)** $\left\langle \dfrac{1}{5}, \dfrac{14}{5} \right\rangle$ **(b)** $\dfrac{\sqrt{197}}{5}$
9. $\langle 1, -4 \rangle$ **11.** $\langle -2, -3 \rangle$ **13.** $\left\langle -\dfrac{1}{2}, \dfrac{\sqrt{3}}{2} \right\rangle$
15. $\left\langle -\dfrac{\sqrt{3}}{2}, -\dfrac{1}{2} \right\rangle$ **17.** $\langle -3\mathbf{i} + 2\mathbf{j} - \mathbf{k} \rangle$ **19.** $\langle -3\mathbf{i} + 16\mathbf{j} \rangle$
21. $\langle 3\mathbf{i} + 5\mathbf{j} - 8\mathbf{k} \rangle$
23. The vector $\mathbf{v}$ is horizontal and 1 in. long. The vectors $\mathbf{u}$ and $\mathbf{w}$ are $\dfrac{11}{16}$ in. long. $\mathbf{w}$ is vertical and $\mathbf{u}$ makes a $45°$ angle with the horizontal. All vectors must be drawn to scale.

(a)

(b)

(c)

(d)

25. $3\left(\dfrac{2}{3}\mathbf{i} + \dfrac{1}{3}\mathbf{j} - \dfrac{2}{3}\mathbf{k}\right)$ **27.** $5(\mathbf{k})$
29. $\sqrt{\dfrac{1}{2}}\left(\dfrac{1}{\sqrt{3}}\mathbf{i} - \dfrac{1}{\sqrt{3}}\mathbf{j} - \dfrac{1}{\sqrt{3}}\mathbf{k}\right)$
31. (a) $2\mathbf{i}$ **(b)** $-\sqrt{3}\mathbf{k}$ **(c)** $\dfrac{3}{10}\mathbf{j} + \dfrac{2}{5}\mathbf{k}$ **(d)** $6\mathbf{i} - 2\mathbf{j} + 3\mathbf{k}$
33. $\dfrac{7}{13}(12\mathbf{i} - 5\mathbf{k})$
35. (a) $\dfrac{3}{5\sqrt{2}}\mathbf{i} + \dfrac{4}{5\sqrt{2}}\mathbf{j} - \dfrac{1}{\sqrt{2}}\mathbf{k}$ **(b)** $(1/2, 3, 5/2)$
37. (a) $-\dfrac{1}{\sqrt{3}}\mathbf{i} - \dfrac{1}{\sqrt{3}}\mathbf{j} - \dfrac{1}{\sqrt{3}}\mathbf{k}$ **(b)** $\left(\dfrac{5}{2}, \dfrac{7}{2}, \dfrac{9}{2}\right)$
39. $A(4, -3, 5)$ **41.** $a = \dfrac{3}{2}, b = \dfrac{1}{2}$ **43.** $5\sqrt{3}\mathbf{i}, 5\mathbf{j}$
45. $\approx \langle -338.095, 725.046 \rangle$
47. (a) $(5\cos 60°, 5\sin 60°) = \left(\dfrac{5}{2}, \dfrac{5\sqrt{3}}{2}\right)$
   (b) $(5\cos 60° + 10\cos 315°, 5\sin 60° + 10\sin 315°) =$
   $\left(\dfrac{5 + 10\sqrt{2}}{2}, \dfrac{5\sqrt{3} - 10\sqrt{2}}{2}\right)$
49. (a) $\dfrac{3}{2}\mathbf{i} + \dfrac{3}{2}\mathbf{j} - 3\mathbf{k}$ **(b)** $\mathbf{i} + \mathbf{j} - 2\mathbf{k}$ **(c)** $(2, 2, 1)$

Section 10.3, pp. 634–636

1. (a) $-25, 5, 5$ **(b)** -1 **(c)** -5 **(d)** $-2\mathbf{i} + 4\mathbf{j} - \sqrt{5}\mathbf{k}$
3. (a) $25, 15, 5$ **(b)** $\dfrac{1}{3}$ **(c)** $\dfrac{5}{3}$ **(d)** $\dfrac{1}{9}(10\mathbf{i} + 11\mathbf{j} - 2\mathbf{k})$

5. (a) $2, \sqrt{34}, \sqrt{3}$ **(b)** $\dfrac{2}{\sqrt{3}\sqrt{34}}$ **(c)** $\dfrac{2}{\sqrt{34}}$

(d) $\dfrac{1}{17}(5\mathbf{j} - 3\mathbf{k})$

7. (a) $10 + \sqrt{17}, \sqrt{26}, \sqrt{21}$ **(b)** $\dfrac{10 + \sqrt{17}}{\sqrt{546}}$

(c) $\dfrac{10 + \sqrt{17}}{\sqrt{26}}$ **(d)** $\dfrac{10 + \sqrt{17}}{26}(5\mathbf{i} + \mathbf{j})$

9. 0.75 rad   **11.** 1.77 rad

13. Angle at $A = \cos^{-1}\left(\dfrac{1}{\sqrt{5}}\right) \approx 63.435$ degrees, angle at

$B = \cos^{-1}\left(\dfrac{3}{5}\right) \approx 53.130$ degrees, angle at

$C = \cos^{-1}\left(\dfrac{1}{\sqrt{5}}\right) \approx 63.435$ degrees.

17. The sum of two vectors of equal length is *always* orthogonal to their difference, as we can see from the equation
$(\mathbf{v}_1 + \mathbf{v}_2) \cdot (\mathbf{v}_1 - \mathbf{v}_2) = \mathbf{v}_1 \cdot \mathbf{v}_1 + \mathbf{v}_2 \cdot \mathbf{v}_1 - \mathbf{v}_1 \cdot \mathbf{v}_2 - \mathbf{v}_2 \cdot \mathbf{v}_2 = |\mathbf{v}_1|^2 - |\mathbf{v}_2|^2 = 0.$

23. Horizontal component: ≈ 1188 ft/sec, vertical component: ≈ 167 ft/sec

25. (a) Since $|\cos\theta| \le 1$, we have $|\mathbf{u} \cdot \mathbf{v}| = |\mathbf{u}||\mathbf{v}||\cos\theta| \le |\mathbf{u}||\mathbf{v}|(1) = |\mathbf{u}||\mathbf{v}|.$

(b) We have equality precisely when $|\cos\theta| = 1$ or when one or both of $\mathbf{u}$ and $\mathbf{v}$ are $\mathbf{0}$. In the case of nonzero vectors, we have equality when $\theta = 0$ or π, that is, when the vectors are parallel.

27. a

31. $x + 2y = 4$ **33.** $-2x + y = -3$

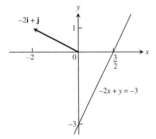

35. $x + y = -1$ **37.** $2x - y = 0$

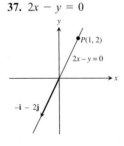

39. 5 J   **41.** 3464 J   **43.** $\dfrac{\pi}{4}$ **45.** $\dfrac{\pi}{6}$ **47.** 0.14

Section 10.4, pp. 641–642

1. $|\mathbf{u} \times \mathbf{v}| = 3$, direction is $\dfrac{2}{3}\mathbf{i} + \dfrac{1}{3}\mathbf{j} + \dfrac{2}{3}\mathbf{k}$; $|\mathbf{v} \times \mathbf{u}| = 3$, direction is $-\dfrac{2}{3}\mathbf{i} - \dfrac{1}{3}\mathbf{j} - \dfrac{2}{3}\mathbf{k}$

3. $|\mathbf{u} \times \mathbf{v}| = 0$, no direction; $|\mathbf{v} \times \mathbf{u}| = 0$, no direction

5. $|\mathbf{u} \times \mathbf{v}| = 6$, direction is $-\mathbf{k}$; $|\mathbf{v} \times \mathbf{u}| = 6$, direction is $\mathbf{k}$

7. $|\mathbf{u} \times \mathbf{v}| = 6\sqrt{5}$, direction is $\dfrac{1}{\sqrt{5}}\mathbf{i} - \dfrac{2}{\sqrt{5}}\mathbf{k}$; $|\mathbf{v} \times \mathbf{u}| = 6\sqrt{5}$, direction is $-\dfrac{1}{\sqrt{5}}\mathbf{i} + \dfrac{2}{\sqrt{5}}\mathbf{k}$

9.

11.

13.

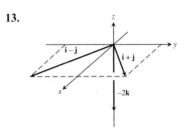

15. (a) $2\sqrt{6}$ **(b)** $\pm\dfrac{1}{\sqrt{6}}(2\mathbf{i} + \mathbf{j} + \mathbf{k})$

17. (a) $\dfrac{\sqrt{2}}{2}$ **(b)** $\pm\dfrac{1}{\sqrt{2}}(\mathbf{i} - \mathbf{j})$

19. 8 **21.** 7 **23. (a)** None   **(b)** $\mathbf{u}$ and $\mathbf{w}$ **25.** $10\sqrt{3}$ ft-lb

27. (a) True   **(b)** Not always true   **(c)** True   **(d)** True
(e) Not always true   **(f)** True   **(g)** True   **(h)** True

29. (a) $\text{proj}_\mathbf{v}\,\mathbf{u} = \dfrac{\mathbf{u} \cdot \mathbf{v}}{\mathbf{v} \cdot \mathbf{v}}\mathbf{v}$ **(b)** $\pm\mathbf{u} \times \mathbf{v}$ **(c)** $\pm(\mathbf{u} \times \mathbf{v}) \times \mathbf{w}$
(d) $|(\mathbf{u} \times \mathbf{v}) \cdot \mathbf{w}|$

31. (a) Yes   **(b)** No   **(c)** Yes   **(d)** No

33. No, $\mathbf{v}$ need not equal $\mathbf{w}$. For example, $\mathbf{i} + \mathbf{j} \ne -\mathbf{i} + \mathbf{j}$, but
$\mathbf{i} \times (\mathbf{i} + \mathbf{j}) = \mathbf{i} \times \mathbf{i} + \mathbf{i} \times \mathbf{j} = \mathbf{0} + \mathbf{k} = \mathbf{k}$ and
$\mathbf{i} \times (-\mathbf{i} + \mathbf{j}) = -\mathbf{i} \times \mathbf{i} + \mathbf{i} \times \mathbf{j} = \mathbf{0} + \mathbf{k} = \mathbf{k}.$

35. 2 **37.** 13 **39.** $11/2$ **41.** $25/2$

43. If $\mathbf{A} = a_1\mathbf{i} + a_2\mathbf{j}$ and $\mathbf{B} = b_1\mathbf{i} + b_2\mathbf{j}$, then

$$\mathbf{A} \times \mathbf{B} = \begin{vmatrix} \mathbf{i} & \mathbf{j} & \mathbf{k} \\ a_1 & a_2 & 0 \\ b_1 & b_2 & 0 \end{vmatrix} = \begin{vmatrix} a_1 & a_2 \\ b_1 & b_2 \end{vmatrix}\mathbf{k}$$

and the triangle's area is

$$\frac{1}{2}\left|\mathbf{A} \times \mathbf{B}\right| = \pm\frac{1}{2}\begin{vmatrix} a_1 & a_2 \\ b_1 & b_2 \end{vmatrix}.$$

The applicable sign is $(+)$ if the acute angle from $\mathbf{A}$ to $\mathbf{B}$ runs counterclockwise in the xy-plane, and $(-)$ if it runs clockwise.

Section 10.5, pp. 650–651

1. $x = 3 + t, y = -4 + t, z = -1 + t$

3. $x = -2 + 5t, y = 5t, z = 3 - 5t$ **5.** $x = 0, y = 2t, z = t$

7. $x = 1, y = 1, z = 1 + t$ **9.** $x = t, y = -7 + 2t, z = 2t$

11. $x = t, y = 0, z = 0$

13. $x = t, y = t, z = \dfrac{3}{2}t, 0 \le t \le 1$

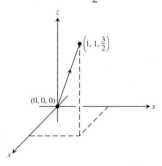

15. $x = 1, y = 1 + t, z = 0, -1 \le t \le 0$

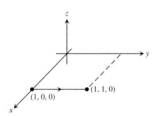

17. $x = 0, y = 1 - 2t, z = 1, 0 \le t \le 1$

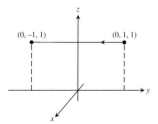

19. $x = 2 - 2t, y = 2t, z = 2 - 2t, 0 \le t \le 1$

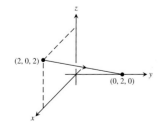

21. $3x - 2y - z = -3$ **23.** $7x - 5y - 4z = 6$

25. $x + 3y + 4z = 34$ **27.** $(1, 2, 3), -20x + 12y + z = 7$

29. $y + z = 3$ **31.** $x - y + z = 0$ **33.** $2\sqrt{30}$ **35.** 0

37. $\dfrac{9\sqrt{42}}{7}$ **39.** 3 **41.** $19/5$ **43.** $5/3$ **45.** $9/\sqrt{41}$

47. $\pi/4$ **49.** 1.76 rad   **51.** 0.82 rad   **53.** $\left(\dfrac{3}{2}, -\dfrac{3}{2}, \dfrac{1}{2}\right)$

55. $(1, 1, 0)$ **57.** $x = 1 - t, y = 1 + t, z = -1$

59. $x = 4, y = 3 + 6t, z = 1 + 3t$

61. $L1$ intersects $L2$; $L2$ is parallel to $L3$; $L1$ and $L3$ are skew.

63. $x = 2 + 2t, y = -4 - t, z = 7 + 3t; x = -2 - t,$
$y = -2 + (1/2)t, z = 1 - (3/2)t$

65. $\left(0, -\dfrac{1}{2}, -\dfrac{3}{2}\right), (-1, 0, -3), (1, -1, 0)$

69. Many possible answers. One possibility: $x + y = 3$ and $2y + z = 7$

71. $(x/a) + (y/b) + (z/c) = 1$ describes all planes *except* those through the origin or parallel to a coordinate axis.

Section 10.6, pp. 656–657

1. (d), ellipsoid   **3. (a),** cylinder   **5. (l),** hyperbolic paraboloid

7. (b), cylinder   **9. (k),** hyperbolic paraboloid   **11. (h),** cone

13.   **15.**

17.   **19.**

21.   **23.**

25.

$x^2 + y^2 = z^2$

27.

$x^2 + y^2 - z^2 = 1$

29.

$z^2 - x^2 - y^2 = 1$

31.

$y^2 - x^2 = z$

33.

$z^2 = 1 + y^2 - x^2$

35.

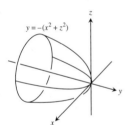

$y = -(x^2 + z^2)$

37.

$x^2 + y^2 - z^2 = 4$

39.

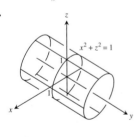

$x^2 + z^2 = 1$

41.

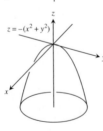

$z = -(x^2 + y^2)$

43.

$4y^2 + z^2 - 4x^2 = 4$

45. (a) $\dfrac{2\pi(9 - c^2)}{9}$ (b) 8π (c) $\dfrac{4\pi abc}{3}$

Practice Exercises, pp. 658–660

1. (a) $\langle -17, 32 \rangle$ (b) $\sqrt{1313}$

3. (a) $\langle 6, -8 \rangle$ (b) 10

5. $\left\langle -\dfrac{\sqrt{3}}{2}, -\dfrac{1}{2} \right\rangle$ [assuming counterclockwise]

7. $\left\langle \dfrac{8}{\sqrt{17}}, -\dfrac{2}{\sqrt{17}} \right\rangle$

9. Length $= 2$, direction is $\dfrac{1}{\sqrt{2}}\mathbf{i} + \dfrac{1}{\sqrt{2}}\mathbf{j}$.

11. $\mathbf{v}(\pi/2) = 2(-\mathbf{i})$

13. Length $= 7$, direction is $\dfrac{2}{7}\mathbf{i} - \dfrac{3}{7}\mathbf{j} + \dfrac{6}{7}\mathbf{k}$.

15. $\dfrac{8}{\sqrt{33}}\mathbf{i} - \dfrac{2}{\sqrt{33}}\mathbf{j} + \dfrac{8}{\sqrt{33}}\mathbf{k}$

17. $|\mathbf{v}| = \sqrt{2}, |\mathbf{u}| = 3, \mathbf{v} \cdot \mathbf{u} = \mathbf{u} \cdot \mathbf{v} = 3, \mathbf{v} \times \mathbf{u} = -2\mathbf{i} + 2\mathbf{j} - \mathbf{k},$

$\mathbf{u} \times \mathbf{v} = 2\mathbf{i} - 2\mathbf{j} + \mathbf{k}, |\mathbf{v} \times \mathbf{u}| = 3, \theta = \cos^{-1}\left(\dfrac{1}{\sqrt{2}}\right) = \dfrac{\pi}{4},$

$|\mathbf{u}|\cos\theta = \dfrac{3}{\sqrt{2}}, \text{proj}_{\mathbf{v}}\,\mathbf{u} = \dfrac{3}{2}(\mathbf{i} + \mathbf{j})$

19. $\dfrac{4}{3}(2\mathbf{i} + \mathbf{j} - \mathbf{k})$

21. $\mathbf{u} \times \mathbf{v} = \mathbf{k}$

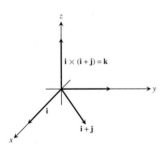

23. $2\sqrt{7}$ **25.** (a) $\sqrt{14}$ (b) 1 **29.** $\sqrt{78}/3$

31. $x = 1 - 3t, y = 2, z = 3 + 7t$ **33.** $\sqrt{2}$

35. $2x + y + z = 5$

37. $-9x + y + 7z = 4$

39. $\left(0, -\dfrac{1}{2}, -\dfrac{3}{2}\right), (-1, 0, -3), (1, -1, 0)$ **41.** $\pi/3$

43. $x = -5 + 5t, y = 3 - t, z = -3t$

45. (b) $x = -12t, y = 19/12 + 15t, z = 1/6 + 6t$

47. Yes; $\mathbf{v}$ is parallel to the plane.

49. 3 **51.** $-3\mathbf{j} + 3\mathbf{k}$

53. $\dfrac{2}{\sqrt{35}}\left(5\mathbf{i} - \mathbf{j} - 3\mathbf{k}\right)$

55. $\left(\dfrac{11}{9}, \dfrac{26}{9}, \dfrac{7}{9}\right)$

57. $(1, -2, -1); x = 1 - 5t, y = -2 + 3t, z = -1 + 4t$

59. $2x + 7y + 2z + 10 = 0$

61. (a) no (b) no (c) no (d) no (e) yes

63. $11/\sqrt{107}$

65. $x^2 + y^2 + z^2 = 4$

67. $4x^2 + 4y^2 + z^2 = 4$

69. $z = -(x^2 + y^2)$

71. $x^2 + y^2 = z^2$

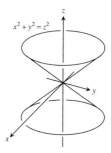

73. $x^2 + y^2 - z^2 = 4$

75. $y^2 - x^2 - z^2 = 1$

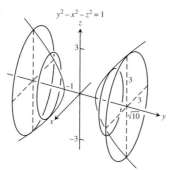

Additional and Advanced Exercises, pp. 660–662

1. $(26, 23, -1/3)$ **3.** $|\mathbf{F}| = 20$ lb

7. (a) $\vec{BD} = \vec{AD} - \vec{AB}$ **(b)** $\vec{AP} = \frac{1}{2}\vec{AB} + \frac{1}{2}\vec{AD}$

13. $\frac{32}{41}\mathbf{i} + \frac{23}{41}\mathbf{j} - \frac{13}{41}\mathbf{k}$

15. (a) $0, 0$ **(b)** $-10\mathbf{i} - 2\mathbf{j} + 6\mathbf{k}, -9\mathbf{i} - 2\mathbf{j} + 7\mathbf{k}$

  (c) $-4\mathbf{i} - 6\mathbf{j} + 2\mathbf{k}, \mathbf{i} - 2\mathbf{j} - 4\mathbf{k}$

  (d) $-10\mathbf{i} - 10\mathbf{k}, -12\mathbf{i} - 4\mathbf{j} - 8\mathbf{k}$

CHAPTER 11

Section 11.1, pp. 670–672

1. $y = x^2 - 2x$, $\mathbf{v} = \mathbf{i} + 2\mathbf{j}$, $\mathbf{a} = 2\mathbf{j}$

3. $y = \frac{2}{9}x^2$, $\mathbf{v} = 3\mathbf{i} + 4\mathbf{j}$, $\mathbf{a} = 3\mathbf{i} + 8\mathbf{j}$

5. $t = \frac{\pi}{4}$: $\mathbf{v} = \frac{\sqrt{2}}{2}\mathbf{i} - \frac{\sqrt{2}}{2}\mathbf{j}$, $\mathbf{a} = \frac{-\sqrt{2}}{2}\mathbf{i} - \frac{\sqrt{2}}{2}\mathbf{j}$;

  $t = \pi/2$: $\mathbf{v} = -\mathbf{j}$, $\mathbf{a} = -\mathbf{i}$

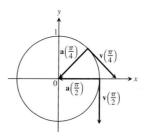

7. $t = \pi$: $\mathbf{v} = 2\mathbf{i}$, $\mathbf{a} = -\mathbf{j}$; $t = \frac{3\pi}{2}$: $\mathbf{v} = \mathbf{i} - \mathbf{j}$, $\mathbf{a} = -\mathbf{i}$

9. $\mathbf{v} = \mathbf{i} + 2t\mathbf{j} + 2\mathbf{k}$; $\mathbf{a} = 2\mathbf{j}$; speed: 3; direction: $\frac{1}{3}\mathbf{i} + \frac{2}{3}\mathbf{j} + \frac{2}{3}\mathbf{k}$; $\mathbf{v}(1) = 3\left(\frac{1}{3}\mathbf{i} + \frac{2}{3}\mathbf{j} + \frac{2}{3}\mathbf{k}\right)$

11. $\mathbf{v} = (-2\sin t)\mathbf{i} + (3\cos t)\mathbf{j} + 4\mathbf{k}$;

  $\mathbf{a} = (-2\cos t)\mathbf{i} - (3\sin t)\mathbf{j}$; speed: $2\sqrt{5}$;

  direction: $\left(-1/\sqrt{5}\right)\mathbf{i} + \left(2/\sqrt{5}\right)\mathbf{k}$;

  $\mathbf{v}(\pi/2) = 2\sqrt{5}\left[\left(-1/\sqrt{5}\right)\mathbf{i} + \left(2/\sqrt{5}\right)\mathbf{k}\right]$

13. $\mathbf{v} = \left(\frac{2}{t+1}\right)\mathbf{i} + 2t\mathbf{j} + t\mathbf{k}$; $\mathbf{a} = \left(\frac{-2}{(t+1)^2}\right)\mathbf{i} + 2\mathbf{j} + \mathbf{k}$;

  speed: $\sqrt{6}$; direction: $\frac{1}{\sqrt{6}}\mathbf{i} + \frac{2}{\sqrt{6}}\mathbf{j} + \frac{1}{\sqrt{6}}\mathbf{k}$;

  $\mathbf{v}(1) = \sqrt{6}\left(\frac{1}{\sqrt{6}}\mathbf{i} + \frac{2}{\sqrt{6}}\mathbf{j} + \frac{1}{\sqrt{6}}\mathbf{k}\right)$

15. $\pi/2$ **17.** $\pi/2$

19. $x = t, y = -1, z = 1 + t$ **21.** $x = at, y = a, z = 2\pi b + bt$

23. (a) (i): It has constant speed 1   (ii): Yes

  (iii): Counterclockwise   (iv): Yes

  (b) (i): It has constant speed 2   (ii): Yes

  (iii): Counterclockwise   (iv): Yes

  (c) (i): It has constant speed 1   (ii): Yes

  (iii): Counterclockwise

  (iv): It starts at $(0, -1)$ instead of $(1, 0)$

  (d) (i): It has constant speed 1   (ii): Yes

  (iii): Clockwise   (iv): Yes

  (e) (i): It has variable speed   (ii): No

  (iii): Counterclockwise   (iv): Yes

25. $\mathbf{v} = 2\sqrt{5}\mathbf{i} + \sqrt{5}\mathbf{j}$

Section 11.2, pp. 676–678

1. $(1/4)\mathbf{i} + 7\mathbf{j} + (3/2)\mathbf{k}$ **3.** $\left(\frac{\pi + 2\sqrt{2}}{2}\right)\mathbf{j} + 2\mathbf{k}$

5. $(\ln 4)\mathbf{i} + (\ln 4)\mathbf{j} + (\ln 2)\mathbf{k}$

7. $\mathbf{r}(t) = \left(\dfrac{-t^2}{2} + 1\right)\mathbf{i} + \left(\dfrac{-t^2}{2} + 2\right)\mathbf{j} + \left(\dfrac{-t^2}{2} + 3\right)\mathbf{k}$

9. $\mathbf{r}(t) = ((t + 1)^{3/2} - 1)\mathbf{i} + (-e^{-t} + 1)\mathbf{j} + (\ln(t + 1) + 1)\mathbf{k}$

11. $\mathbf{r}(t) = 8t\mathbf{i} + 8t\mathbf{j} + (-16t^2 + 100)\mathbf{k}$

13. $\mathbf{r}(t) = \left(\dfrac{3}{2}t^2 + \dfrac{6}{\sqrt{11}}t + 1\right)\mathbf{i} - \left(\dfrac{1}{2}t^2 + \dfrac{2}{\sqrt{11}}t - 2\right)\mathbf{j} +$

$\left(\dfrac{1}{2}t^2 + \dfrac{2}{\sqrt{11}}t + 3\right)\mathbf{k} = \left(\dfrac{1}{2}t^2 + \dfrac{2t}{\sqrt{11}}\right)(3\mathbf{i} - \mathbf{j} + \mathbf{k}) +$

$(\mathbf{i} + 2\mathbf{j} + 3\mathbf{k})$

15. 50 sec

17. (a) 72.2 sec; 25,510 m   (b) 4020 m   (c) 6378 m

19. (a) $v_0 \approx 9.9$ m/sec   (b) $\alpha \approx 18.4°$ or $71.6°$

21. $39.3°$ or $50.7°$

31. (a) (Assuming that "x" is zero at the point of impact.)
$\mathbf{r}(t) = (x(t))\mathbf{i} + (y(t))\mathbf{j}$, where $x(t) = (35 \cos 27°)t$ and
$y(t) = 4 + (35 \sin 27°)t - 16t^2$.
   (b) At $t \approx 0.497$ sec, it reaches its maximum height of about
   7.945 ft.
   (c) Range ≈ 37.45 ft; flight time ≈ 1.201 sec
   (d) At $t \approx 0.254$ and $t \approx 0.740$ sec, when it is ≈ 29.532 and
   ≈ 14.376 ft from where it will land.
   (e) Yes. It changes things because the ball won't clear the net.

Section 11.3, pp. 681–682

1. $\mathbf{T} = \left(-\dfrac{2}{3}\sin t\right)\mathbf{i} + \left(\dfrac{2}{3}\cos t\right)\mathbf{j} + \dfrac{\sqrt{5}}{3}\mathbf{k}, 3\pi$

3. $\mathbf{T} = \dfrac{1}{\sqrt{1 + t}}\mathbf{i} + \dfrac{\sqrt{t}}{\sqrt{1 + t}}\mathbf{k}, \dfrac{52}{3}$

5. $\mathbf{T} = -\cos t\mathbf{j} + \sin t\mathbf{k}, \dfrac{3}{2}$

7. $\mathbf{T} = \left(\dfrac{\cos t - t\sin t}{t + 1}\right)\mathbf{i} + \left(\dfrac{\sin t + t\cos t}{t + 1}\right)\mathbf{j} +$

$\left(\dfrac{\sqrt{2}t^{1/2}}{t + 1}\right)\mathbf{k}, \dfrac{\pi^2}{2} + \pi$

9. $(0, 5, 24\pi)$ **11.** $s(t) = 5t, L = \dfrac{5\pi}{2}$

13. $s(t) = \sqrt{3}e^t - \sqrt{3}, L = \dfrac{3\sqrt{3}}{4}$ **15.** $\sqrt{2} + \ln\left(1 + \sqrt{2}\right)$

17. (a) Cylinder is $x^2 + y^2 = 1$, plane is $x + z = 1$.
   (b) and (c)

(d) $L = \displaystyle\int_0^{2\pi} \sqrt{1 + \sin^2 t}\, dt$ **(e)** $L \approx 7.64$

Section 11.4, pp. 687–689

1. $\mathbf{T} = (\cos t)\mathbf{i} - (\sin t)\mathbf{j}, \mathbf{N} = (-\sin t)\mathbf{i} - (\cos t)\mathbf{j}, \kappa = \cos t$

3. $\mathbf{T} = \dfrac{1}{\sqrt{1 + t^2}}\mathbf{i} - \dfrac{t}{\sqrt{1 + t^2}}\mathbf{j}, \mathbf{N} = \dfrac{-t}{\sqrt{1 + t^2}}\mathbf{i} - \dfrac{1}{\sqrt{1 + t^2}}\mathbf{j},$

$\kappa = \dfrac{1}{2\left(\sqrt{1 + t^2}\right)^3}$

5. (b) $\cos x$

7. (b) $\mathbf{N} = \dfrac{-2e^{2t}}{\sqrt{1 + 4e^{4t}}}\mathbf{i} + \dfrac{1}{\sqrt{1 + 4e^{4t}}}\mathbf{j}$

   (c) $\mathbf{N} = -\dfrac{1}{2}\left(\sqrt{4 - t^2}\mathbf{i} + t\mathbf{j}\right)$

9. $\mathbf{T} = \dfrac{3\cos t}{5}\mathbf{i} - \dfrac{3\sin t}{5}\mathbf{j} + \dfrac{4}{5}\mathbf{k}, \mathbf{N} = (-\sin t)\mathbf{i} - (\cos t)\mathbf{j},$

$\kappa = \dfrac{3}{25}$

11. $\mathbf{T} = \left(\dfrac{\cos t - \sin t}{\sqrt{2}}\right)\mathbf{i} + \left(\dfrac{\cos t + \sin t}{\sqrt{2}}\right)\mathbf{j},$

$\mathbf{N} = \left(\dfrac{-\cos t - \sin t}{\sqrt{2}}\right)\mathbf{i} + \left(\dfrac{-\sin t + \cos t}{\sqrt{2}}\right)\mathbf{j},$

$\kappa = \dfrac{1}{e^t\sqrt{2}}$

13. $\mathbf{T} = \dfrac{t}{\sqrt{t^2 + 1}}\mathbf{i} + \dfrac{1}{\sqrt{t^2 + 1}}\mathbf{j}, \mathbf{N} = \dfrac{\mathbf{i}}{\sqrt{t^2 + 1}} - \dfrac{t\mathbf{j}}{\sqrt{t^2 + 1}},$

$\kappa = \dfrac{1}{t(t^2 + 1)^{3/2}}$

15. $\mathbf{T} = \left(\text{sech}\,\dfrac{t}{a}\right)\mathbf{i} + \left(\tanh\dfrac{t}{a}\right)\mathbf{j},$

$\mathbf{N} = \left(-\tanh\dfrac{t}{a}\right)\mathbf{i} + \left(\text{sech}\,\dfrac{t}{a}\right)\mathbf{j},$

$\kappa = \dfrac{1}{a}\text{sech}^2\dfrac{t}{a}$

19. $1/(2b)$ **21.** $\left(x - \dfrac{\pi}{2}\right)^2 + y^2 = 1$

23. $\kappa(x) = 2/(1 + 4x^2)^{3/2}$ **25.** $\kappa(x) = |\sin x|/(1 + \cos^2 x)^{3/2}$

Section 11.5, pp. 693–694

1. $\mathbf{a} = |a|\mathbf{N}$ **3.** $\mathbf{a}(1) = \dfrac{4}{3}\mathbf{T} + \dfrac{2\sqrt{5}}{3}\mathbf{N}$ **5.** $\mathbf{a}(0) = 2\mathbf{N}$

7. $\mathbf{r}\left(\dfrac{\pi}{4}\right) = \dfrac{\sqrt{2}}{2}\mathbf{i} + \dfrac{\sqrt{2}}{2}\mathbf{j} - \mathbf{k}, \mathbf{T}\left(\dfrac{\pi}{4}\right) = -\dfrac{\sqrt{2}}{2}\mathbf{i} + \dfrac{\sqrt{2}}{2}\mathbf{j},$

$\mathbf{N}\left(\dfrac{\pi}{4}\right) = -\dfrac{\sqrt{2}}{2}\mathbf{i} - \dfrac{\sqrt{2}}{2}\mathbf{j}, \mathbf{B}\left(\dfrac{\pi}{4}\right) = \mathbf{k}$; osculating plane:
$z = -1$; normal plane: $-x + y = 0$; rectifying plane:
$x + y = \sqrt{2}$

9. $\mathbf{B} = \left(\dfrac{4}{5}\cos t\right)\mathbf{i} - \left(\dfrac{4}{5}\sin t\right)\mathbf{j} - \dfrac{3}{5}\mathbf{k}, \tau = -\dfrac{4}{25}$

11. $\mathbf{B} = \mathbf{k}, \tau = 0$ **13.** $\mathbf{B} = -\mathbf{k}, \tau = 0$ **15.** $\mathbf{B} = \mathbf{k}, \tau = 0$

17. Yes. If the car is moving on a curved path ($\kappa \neq 0$), then $a_N = \kappa|\mathbf{v}|^2 \neq 0$ and $\mathbf{a} \neq \mathbf{0}$.

23. $\kappa = \dfrac{1}{t}, \rho = t$

29. Components of $\mathbf{v}$: $-1.8701, 0.7089, 1.0000$
Components of $\mathbf{a}$: $-1.6960, -2.0307, 0$
Speed: 2.2361; Components of $\mathbf{T}$: $-0.8364, 0.3170, 0.4472$
Components of $\mathbf{N}$: $-0.4143, -0.8998, -0.1369$
Components of $\mathbf{B}$: $0.3590, -0.2998, 0.8839$; Curvature: 0.5060
Torsion: 0.2813; Tangential component of acceleration: 0.7746
Normal component of acceleration: 2.5298

31. Components of $\mathbf{v}$: $2.0000, 0, 0.1629$
Components of $\mathbf{a}$: $0, -1.0000, 0.0086$; Speed: 2.0066
Components of $\mathbf{T}$: $0.9967, 0, 0.0812$
Components of $\mathbf{N}$: $-0.0007, -1.0000, 0.0086$
Components of $\mathbf{B}$: $0.0812, -0.0086, -0.9967$;
Curvature: 0.2484
Torsion: -0.0411; Tangential component of
acceleration: 0.0007
Normal component of acceleration: 1.0000

Section 11.6, pp. 697–698

1. $\mathbf{v} = (3a\sin\theta)\mathbf{u}_r + 3a(1 - \cos\theta)\mathbf{u}_\theta$
$\mathbf{a} = 9a(2\cos\theta - 1)\mathbf{u}_r + (18a\sin\theta)\mathbf{u}_\theta$

3. $\mathbf{v} = 2ae^{a\theta}\mathbf{u}_r + 2e^{a\theta}\mathbf{u}_\theta$
$\mathbf{a} = 4e^{a\theta}(a^2 - 2)\mathbf{u}_r + 8ae^{a\theta}\mathbf{u}_\theta$

5. $\mathbf{v} = (-8\sin 4t)\mathbf{u}_r + (4\cos 4t)\mathbf{u}_\theta$
$\mathbf{a} = (-40\cos 4t)\mathbf{u}_r - (32\sin 4t)\mathbf{u}_\theta$

Practice Exercises, pp. 698–700

1. $\dfrac{x^2}{16} + \dfrac{y^2}{2} = 1$

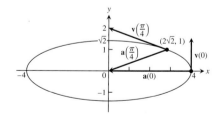

At $t = 0$: $a_T = 0, a_N = 4, \kappa = 2$;

At $t = \dfrac{\pi}{4}$: $a_T = \dfrac{7}{3}, a_N = \dfrac{4\sqrt{2}}{3}, \kappa = \dfrac{4\sqrt{2}}{27}$

3. $|\mathbf{v}|_{max} = 1$ **5.** $\kappa = 1/5$ **7.** $dy/dt = -x$; clockwise

11. Shot put is on the ground, about 66 ft, 3 in. from the stopboard.

15. Length $= \dfrac{\pi}{4}\sqrt{1 + \dfrac{\pi^2}{16}} + \ln\left(\dfrac{\pi}{4} + \sqrt{1 + \dfrac{\pi^2}{16}}\right)$

17. $\mathbf{T}(0) = \dfrac{2}{3}\mathbf{i} - \dfrac{2}{3}\mathbf{j} + \dfrac{1}{3}\mathbf{k}$; $\mathbf{N}(0) = \dfrac{1}{\sqrt{2}}\mathbf{i} + \dfrac{1}{\sqrt{2}}\mathbf{j}$;

$\mathbf{B}(0) = -\dfrac{1}{3\sqrt{2}}\mathbf{i} + \dfrac{1}{3\sqrt{2}}\mathbf{j} + \dfrac{4}{3\sqrt{2}}\mathbf{k}; \kappa = \dfrac{\sqrt{2}}{3}; \tau = \dfrac{1}{6}$

19. $\mathbf{T}(\ln 2) = \dfrac{1}{\sqrt{17}}\mathbf{i} + \dfrac{4}{\sqrt{17}}\mathbf{j}$; $\mathbf{N}(\ln 2) = -\dfrac{4}{\sqrt{17}}\mathbf{i} + \dfrac{1}{\sqrt{17}}\mathbf{j}$;

$\mathbf{B}(\ln 2) = \mathbf{k}; \kappa = \dfrac{8}{17\sqrt{17}}; \tau = 0$

21. $\mathbf{a}(0) = 10\mathbf{T} + 6\mathbf{N}$

23. $\mathbf{T} = \left(\dfrac{1}{\sqrt{2}}\cos t\right)\mathbf{i} - (\sin t)\mathbf{j} + \left(\dfrac{1}{\sqrt{2}}\cos t\right)\mathbf{k}$;

$\mathbf{N} = \left(-\dfrac{1}{\sqrt{2}}\sin t\right)\mathbf{i} - (\cos t)\mathbf{j} - \left(\dfrac{1}{\sqrt{2}}\sin t\right)\mathbf{k}$;

$\mathbf{B} = \dfrac{1}{\sqrt{2}}\mathbf{i} - \dfrac{1}{\sqrt{2}}\mathbf{k}; \kappa = \dfrac{1}{\sqrt{2}}; \tau = 0$

25. $\pi/3$ **27.** $x = 1 + t, y = t, z = -t$ **31.** $\kappa = 1/a$

Additional and Advanced Exercises, pp. 700–701

1. (a) $\dfrac{d\theta}{dt}\bigg|_{\theta=2\pi} = 2\sqrt{\dfrac{\pi gb}{a^2 + b^2}}$

(b) $\theta = \dfrac{gbt^2}{2(a^2 + b^2)}, z = \dfrac{gb^2 t^2}{2(a^2 + b^2)}$

(c) $\mathbf{v}(t) = \dfrac{gbt}{\sqrt{a^2 + b^2}}\mathbf{T}; \dfrac{d^2\mathbf{r}}{dt^2} = \dfrac{bg}{\sqrt{a^2 + b^2}}\mathbf{T} +$

$a\left(\dfrac{bgt}{a^2 + b^2}\right)^2\mathbf{N}$

There is no component in the direction of $\mathbf{B}$.

5. (a) $\dfrac{dx}{dt} = \dot{r}\cos\theta - r\dot{\theta}\sin\theta, \dfrac{dy}{dt} = \dot{r}\sin\theta + r\dot{\theta}\cos\theta$

(b) $\dfrac{dr}{dt} = \dot{x}\cos\theta + \dot{y}\sin\theta, r\dfrac{d\theta}{dt} = -\dot{x}\sin\theta + \dot{y}\cos\theta$

7. (a) $\mathbf{a}(1) = -9\mathbf{u}_r - 6\mathbf{u}_\theta, \mathbf{v}(1) = -\mathbf{u}_r + 3\mathbf{u}_\theta$ **(b)** 6.5 in.

9. (c) $\mathbf{v} = \dot{r}\mathbf{u}_r + r\dot{\theta}\mathbf{u}_\theta + \dot{z}\mathbf{k}, \mathbf{a} = (\ddot{r} - r\dot{\theta}^2)\mathbf{u}_r + (r\ddot{\theta} + 2\dot{r}\dot{\theta})\mathbf{u}_\theta + \ddot{z}\mathbf{k}$

CHAPTER 12

Section 12.1, pp. 708–710

1. (a) All points in the xy-plane **(b)** All reals
(c) The lines $y - x = c$ **(d)** No boundary points
(e) Both open and closed **(f)** Unbounded

3. (a) All points in the xy-plane **(b)** $z \geq 0$
(c) For $f(x, y) = 0$, the origin; for $f(x, y) \neq 0$, ellipses with the center $(0, 0)$, and major and minor axes, along the x- and y-axes, respectively

(d) No boundary points   **(e)** Both open and closed
(f) Unbounded

5. (a) All points in the xy-plane   **(b)** All reals
(c) For $f(x, y) = 0$, the x- and y-axes; for $f(x, y) \neq 0$, hyperbolas with the x- and y-axes as asymptotes
(d) No boundary points   **(e)** Both open and closed
(f) Unbounded

7. (a) All (x, y) satisfying $x^2 + y^2 < 16$ **(b)** $z \geq 1/4$
(c) Circles centered at the origin with radii $r < 4$
(d) Boundary is the circle $x^2 + y^2 = 16$
(e) Open   **(f)** Bounded

9. (a) $(x, y) \neq (0, 0)$ **(b)** All reals
(c) The circles with center $(0, 0)$ and radii $r > 0$
(d) Boundary is the single point $(0, 0)$
(e) Open   **(f)** Unbounded

11. (a) All (x, y) satisfying $-1 \leq y - x \leq 1$
(b) $-\pi/2 \leq z \leq \pi/2$
(c) Straight lines of the form $y - x = c$ where $-1 \leq c \leq 1$
(d) Boundary is two straight lines $y = 1 + x$ and $y = -1 + x$
(e) Closed   **(f)** Unbounded

13. (f)   **15.** (a)   **17.** (d)

19. (a)

(b)

21. (a)

(b)

23. (a)

(b)

25. (a)

(b)

27. (a)

(b)

29. $x^2 + y^2 = 10$ **31.** $\tan^{-1} y - \tan^{-1} x = 2 \tan^{-1} \sqrt{2}$

33.

$f(x, y, z) = x^2 + y^2 + z^2 = 1$

35.

$f(x, y, z) = x + z = 1$

37.

$f(x, y, z) = x^2 + y^2 = 1$

39.

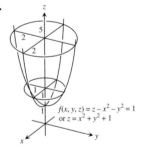

$f(x, y, z) = z - x^2 - y^2 = 1$
or $z = x^2 + y^2 + 1$

41. $\sqrt{x - y} - \ln z = 2$ **43.** $\dfrac{x + y}{z} = \ln 2$

Section 12.2, pp. 717–719

1. $5/2$ **3.** $2\sqrt{6}$ **5.** 1   **7.** $1/2$ **9.** 1   **11.** 0   **13.** 0
15. -1 **17.** 2   **19.** $1/4$ **21.** $19/12$ **23.** 2   **25.** 3
27. (a) All (x, y) **(b)** All (x, y) except $(0, 0)$
29. (a) All (x, y) except where $x = 0$ or $y = 0$ **(b)** All (x, y)

31. (a) All (x, y, z) **(b)** All (x, y, z) except the interior of the cylinder $x^2 + y^2 = 1$

33. (a) All (x, y, z) with $z \neq 0$ **(b)** All (x, y, z) with $x^2 + z^2 \neq 1$

35. Consider paths along $y = x, x > 0$, and along $y = x, x < 0$

37. Consider the paths $y = kx^2$, k a constant

39. Consider the paths $y = mx$, m a constant, $m \neq -1$

41. Consider the paths $y = kx^2$, k a constant, $k \neq 0$

45. The limit is 1   **47.** The limit is 0

49. (a) $f(x, y)|_{y=mx} = \sin 2\theta$ where $\tan \theta = m$ **51.** 0

53. Does not exist   **55.** $\pi/2$ **57.** $f(0, 0) = \ln 3$

59. $\delta = 0.1$ **61.** $\delta = 0.005$ **63.** $\delta = \sqrt{0.015}$

65. $\delta = 0.005$

Section 12.3, pp. 728–730

1. $\dfrac{\partial f}{\partial x} = 4x, \dfrac{\partial f}{\partial y} = -3$ **3.** $\dfrac{\partial f}{\partial x} = 2x(y + 2), \dfrac{\partial f}{\partial y} = x^2 - 1$

5. $\dfrac{\partial f}{\partial x} = 2y(xy - 1), \dfrac{\partial f}{\partial y} = 2x(xy - 1)$

7. $\dfrac{\partial f}{\partial x} = \dfrac{x}{\sqrt{x^2 + y^2}}, \dfrac{\partial f}{\partial y} = \dfrac{y}{\sqrt{x^2 + y^2}}$

9. $\dfrac{\partial f}{\partial x} = \dfrac{-1}{(x + y)^2}, \dfrac{\partial f}{\partial y} = \dfrac{-1}{(x + y)^2}$

11. $\dfrac{\partial f}{\partial x} = \dfrac{-y^2 - 1}{(xy - 1)^2}, \dfrac{\partial f}{\partial y} = \dfrac{-x^2 - 1}{(xy - 1)^2}$

13. $\dfrac{\partial f}{\partial x} = e^{x+y+1}, \dfrac{\partial f}{\partial y} = e^{x+y+1}$ **15.** $\dfrac{\partial f}{\partial x} = \dfrac{1}{x + y}, \dfrac{\partial f}{\partial y} = \dfrac{1}{x + y}$

17. $\dfrac{\partial f}{\partial x} = 2 \sin (x - 3y) \cos (x - 3y),$

$\dfrac{\partial f}{\partial y} = -6 \sin (x - 3y) \cos (x - 3y)$

19. $\dfrac{\partial f}{\partial x} = yx^{y-1}, \dfrac{\partial f}{\partial y} = x^y \ln x$ **21.** $\dfrac{\partial f}{\partial x} = -g(x), \dfrac{\partial f}{\partial y} = g(y)$

23. $f_x = y^2, f_y = 2xy, f_z = -4z$

25. $f_x = 1, f_y = -y(y^2 + z^2)^{-1/2}, f_z = -z(y^2 + z^2)^{-1/2}$

27. $f_x = \dfrac{yz}{\sqrt{1 - x^2y^2z^2}}, f_y = \dfrac{xz}{\sqrt{1 - x^2y^2z^2}}, f_z = \dfrac{xy}{\sqrt{1 - x^2y^2z^2}}$

29. $f_x = \dfrac{1}{x + 2y + 3z}, f_y = \dfrac{2}{x + 2y + 3z}, f_z = \dfrac{3}{x + 2y + 3z}$

31. $f_x = -2xe^{-(x^2+y^2+z^2)}, f_y = -2ye^{-(x^2+y^2+z^2)}, f_z = -2ze^{-(x^2+y^2+z^2)}$

33. $f_x = \text{sech}^2(x + 2y + 3z), f_y = 2 \text{sech}^2(x + 2y + 3z),$
$f_z = 3 \text{sech}^2(x + 2y + 3z)$

35. $\dfrac{\partial f}{\partial t} = -2\pi \sin (2\pi t - \alpha), \dfrac{\partial f}{\partial \alpha} = \sin (2\pi t - \alpha)$

37. $\dfrac{\partial h}{\partial \rho} = \sin \phi \cos \theta, \dfrac{\partial h}{\partial \phi} = \rho \cos \phi \cos \theta, \dfrac{\partial h}{\partial \theta} = -\rho \sin \phi \sin \theta$

39. $W_P(P, V, \delta, v, g) = V, W_V(P, V, \delta, v, g) = P + \dfrac{\delta v^2}{2g},$

$W_\delta(P, V, \delta, v, g) = \dfrac{Vv^2}{2g}, W_v(P, V, \delta, v, g) = \dfrac{V\delta v}{g},$

$W_g(P, V, \delta, v, g) = -\dfrac{V\delta v^2}{2g^2}$

41. $\dfrac{\partial f}{\partial x} = 1 + y, \dfrac{\partial f}{\partial y} = 1 + x, \dfrac{\partial^2 f}{\partial x^2} = 0, \dfrac{\partial^2 f}{\partial y^2} = 0,$

$\dfrac{\partial^2 f}{\partial y \, \partial x} = \dfrac{\partial^2 f}{\partial x \, \partial y} = 1$

43. $\dfrac{\partial g}{\partial x} = 2xy + y \cos x, \dfrac{\partial g}{\partial y} = x^2 - \sin y + \sin x,$

$\dfrac{\partial^2 g}{\partial x^2} = 2y - y \sin x, \dfrac{\partial^2 g}{\partial y^2} = -\cos y,$

$\dfrac{\partial^2 g}{\partial y \, \partial x} = \dfrac{\partial^2 g}{\partial x \, \partial y} = 2x + \cos x$

45. $\dfrac{\partial r}{\partial x} = \dfrac{1}{x + y}, \dfrac{\partial r}{\partial y} = \dfrac{1}{x + y}, \dfrac{\partial^2 r}{\partial x^2} = \dfrac{-1}{(x + y)^2}, \dfrac{\partial^2 r}{\partial y^2} = \dfrac{-1}{(x + y)^2},$

$\dfrac{\partial^2 r}{\partial y \, \partial x} = \dfrac{\partial^2 r}{\partial x \, \partial y} = \dfrac{-1}{(x + y)^2}$

47. $\dfrac{\partial w}{\partial x} = \dfrac{2}{2x + 3y}, \dfrac{\partial w}{\partial y} = \dfrac{3}{2x + 3y}, \dfrac{\partial^2 w}{\partial y \, \partial x} = \dfrac{\partial^2 w}{\partial x \, \partial y} = \dfrac{-6}{(2x + 3y)^2}$

49. $\dfrac{\partial w}{\partial x} = y^2 + 2xy^3 + 3x^2y^4, \dfrac{\partial w}{\partial y} = 2xy + 3x^2y^2 + 4x^3y^3,$

$\dfrac{\partial^2 w}{\partial y \, \partial x} = \dfrac{\partial^2 w}{\partial x \, \partial y} = 2y + 6xy^2 + 12x^2y^3$

51. (a) x first **(b)** y first **(c)** x first
  (d) x first **(e)** y first **(f)** y first

53. $f_x(1, 2) = -13, f_y(1, 2) = -2$ **55.** 12   **57.** -2

59. $\dfrac{\partial A}{\partial a} = \dfrac{a}{bc \sin A}, \dfrac{\partial A}{\partial b} = \dfrac{c \cos A - b}{bc \sin A}$ **61.** $v_x = \dfrac{\ln v}{(\ln u)(\ln v) - 1}$

77. Yes

Section 12.4, pp. 737–739

1. (a) $\dfrac{dw}{dt} = 0,$ **(b)** $\dfrac{dw}{dt}(\pi) = 0$

3. (a) $\dfrac{dw}{dt} = 1,$ **(b)** $\dfrac{dw}{dt}(3) = 1$

5. (a) $\dfrac{dw}{dt} = 4t \tan^{-1} t + 1,$ **(b)** $\dfrac{dw}{dt}(1) = \pi + 1$

7. (a) $\dfrac{\partial z}{\partial u} = 4 \cos v \ln (u \sin v) + 4 \cos v,$

$\dfrac{\partial z}{\partial v} = -4u \sin v \ln (u \sin v) + \dfrac{4u \cos^2 v}{\sin v}$

  (b) $\dfrac{\partial z}{\partial u} = \sqrt{2} (\ln 2 + 2), \dfrac{\partial z}{\partial v} = -2\sqrt{2} (\ln 2 - 2)$

9. (a) $\dfrac{\partial w}{\partial u} = 2u + 4uv, \dfrac{\partial w}{\partial v} = -2v + 2u^2$

  (b) $\dfrac{\partial w}{\partial u} = 3, \dfrac{\partial w}{\partial v} = -\dfrac{3}{2}$

11. (a) $\dfrac{\partial u}{\partial x} = 0, \dfrac{\partial u}{\partial y} = \dfrac{z}{(z - y)^2}, \dfrac{\partial u}{\partial z} = \dfrac{-y}{(z - y)^2}$

  (b) $\dfrac{\partial u}{\partial x} = 0, \dfrac{\partial u}{\partial y} = 1, \dfrac{\partial u}{\partial z} = -2$

13. $\dfrac{dz}{dt} = \dfrac{\partial z}{\partial x}\dfrac{dx}{dt} + \dfrac{\partial z}{\partial y}\dfrac{dy}{dt}$

15. $\dfrac{\partial w}{\partial u} = \dfrac{\partial w}{\partial x}\dfrac{\partial x}{\partial u} + \dfrac{\partial w}{\partial y}\dfrac{\partial y}{\partial u} + \dfrac{\partial w}{\partial z}\dfrac{\partial z}{\partial u},$
$\dfrac{\partial w}{\partial v} = \dfrac{\partial w}{\partial x}\dfrac{\partial x}{\partial v} + \dfrac{\partial w}{\partial y}\dfrac{\partial y}{\partial v} + \dfrac{\partial w}{\partial z}\dfrac{\partial z}{\partial v}$

17. $\dfrac{\partial w}{\partial u} = \dfrac{\partial w}{\partial x}\dfrac{\partial x}{\partial u} + \dfrac{\partial w}{\partial y}\dfrac{\partial y}{\partial u}, \dfrac{\partial w}{\partial v} = \dfrac{\partial w}{\partial x}\dfrac{\partial x}{\partial v} + \dfrac{\partial w}{\partial y}\dfrac{\partial y}{\partial v}$

19. $\dfrac{\partial z}{\partial t} = \dfrac{\partial z}{\partial x}\dfrac{\partial x}{\partial t} + \dfrac{\partial z}{\partial y}\dfrac{\partial y}{\partial t}, \dfrac{\partial z}{\partial s} = \dfrac{\partial z}{\partial x}\dfrac{\partial x}{\partial s} + \dfrac{\partial z}{\partial y}\dfrac{\partial y}{\partial s}$

21. $\dfrac{\partial w}{\partial s} = \dfrac{dw}{du}\dfrac{\partial u}{\partial s}, \dfrac{\partial w}{\partial t} = \dfrac{dw}{du}\dfrac{\partial u}{\partial t}$

23. $\dfrac{\partial w}{\partial r} = \dfrac{\partial w}{\partial x}\dfrac{dx}{dr} + \dfrac{\partial w}{\partial y}\dfrac{dy}{dr} = \dfrac{\partial w}{\partial x}\dfrac{dx}{dr}$ since $\dfrac{dy}{dr} = 0,$

$\dfrac{\partial w}{\partial s} = \dfrac{\partial w}{\partial x}\dfrac{dx}{ds} + \dfrac{\partial w}{\partial y}\dfrac{dy}{ds} = \dfrac{\partial w}{\partial y}\dfrac{dy}{ds}$ since $\dfrac{dx}{ds} = 0$

25. $4/3$ **27.** $-4/5$ **29.** $\dfrac{\partial z}{\partial x} = \dfrac{1}{4}, \dfrac{\partial z}{\partial y} = -\dfrac{3}{4}$

31. $\dfrac{\partial z}{\partial x} = -1, \dfrac{\partial z}{\partial y} = -1$ **33.** 12 **35.** -7

37. $\dfrac{\partial z}{\partial u} = 2, \dfrac{\partial z}{\partial v} = 1$ **39.** -0.00005 amps/sec

45. $(\cos 1, \sin 1, 1)$ and $(\cos(-2), \sin(-2), -2)$

47. (a) Maximum at $\left(-\dfrac{\sqrt{2}}{2}, \dfrac{\sqrt{2}}{2}\right)$ and $\left(\dfrac{\sqrt{2}}{2}, -\dfrac{\sqrt{2}}{2}\right)$; minimum

at $\left(\dfrac{\sqrt{2}}{2}, \dfrac{\sqrt{2}}{2}\right)$ and $\left(-\dfrac{\sqrt{2}}{2}, -\dfrac{\sqrt{2}}{2}\right)$

(b) Max $= 6$, min $= 2$

49. $2x\sqrt{x^8 + x^3} + \displaystyle\int_0^{x^2} \dfrac{3x^2}{2\sqrt{t^4 + x^3}}\, dt$

Section 12.5, pp. 746–747

1.   **3.**

5. $\nabla f = 3\mathbf{i} + 2\mathbf{j} - 4\mathbf{k}$ **7.** $\nabla f = -\dfrac{26}{27}\mathbf{i} + \dfrac{23}{54}\mathbf{j} - \dfrac{23}{54}\mathbf{k}$

9. -4 **11.** $31/13$ **13.** 3 **15.** 2

17. $\mathbf{u} = -\dfrac{1}{\sqrt{2}}\mathbf{i} + \dfrac{1}{\sqrt{2}}\mathbf{j}, (D_{\mathbf{u}}f)_{P_0} = \sqrt{2}; -\mathbf{u} = \dfrac{1}{\sqrt{2}}\mathbf{i} - \dfrac{1}{\sqrt{2}}\mathbf{j},$
$(D_{-\mathbf{u}}f)_{P_0} = -\sqrt{2}$

19. $\mathbf{u} = \dfrac{1}{3\sqrt{3}}\mathbf{i} - \dfrac{5}{3\sqrt{3}}\mathbf{j} - \dfrac{1}{3\sqrt{3}}\mathbf{k}, (D_{\mathbf{u}}f)_{P_0} = 3\sqrt{3};$
$-\mathbf{u} = -\dfrac{1}{3\sqrt{3}}\mathbf{i} + \dfrac{5}{3\sqrt{3}}\mathbf{j} + \dfrac{1}{3\sqrt{3}}\mathbf{k}, (D_{-\mathbf{u}}f)_{P_0} = -3\sqrt{3}$

21. $\mathbf{u} = \dfrac{1}{\sqrt{3}}(\mathbf{i} + \mathbf{j} + \mathbf{k}), (D_{\mathbf{u}}f)_{P_0} = 2\sqrt{3};$
$-\mathbf{u} = -\dfrac{1}{\sqrt{3}}(\mathbf{i} + \mathbf{j} + \mathbf{k}), (D_{-\mathbf{u}}f)_{P_0} = -2\sqrt{3}$

23.   **25.**

27. $u = \dfrac{7}{\sqrt{53}}i - \dfrac{2}{\sqrt{53}}j, \quad -u = -\dfrac{7}{\sqrt{53}}i + \dfrac{2}{\sqrt{53}}j$

29. No, the maximum rate of change is $\sqrt{185} < 14$.

31. $-\dfrac{7}{\sqrt{5}}$

Section 12.6, pp. 754–756

1. (a) $x + y + z = 3$
  (b) $x = 1 + 2t, y = 1 + 2t, z = 1 + 2t$
3. (a) $2x - z - 2 = 0$ **(b)** $x = 2 - 4t, y = 0, z = 2 + 2t$
5. (a) $2x + 2y + z - 4 = 0$
  (b) $x = 2t, y = 1 + 2t, z = 2 + t$
7. (a) $x + y + z - 1 = 0$ **(b)** $x = t, y = 1 + t, z = t$
9. $2x - z - 2 = 0$ **11.** $x - y + 2z - 1 = 0$
13. $x = 1, y = 1 + 2t, z = 1 - 2t$
15. $x = 1 - 2t, y = 1, z = \dfrac{1}{2} + 2t$
17. $x = 1 + 90t, y = 1 - 90t, z = 3$
19. $df = \dfrac{9}{11{,}830} \approx 0.0008$ **21.** $dg = 0$
23. (a) $\dfrac{\sqrt{3}}{2}\sin\sqrt{3} - \dfrac{1}{2}\cos\sqrt{3} \approx 0.935°C/ft$
  (b) $\sqrt{3}\sin\sqrt{3} - \cos\sqrt{3} \approx 1.87°C/sec$
25. (a) $L(x, y) = 1$ **(b)** $L(x, y) = 2x + 2y - 1$
27. (a) $L(x, y) = 3x - 4y + 5$ **(b)** $L(x, y) = 3x - 4y + 5$
29. (a) $L(x, y) = 1 + x$ **(b)** $L(x, y) = -y + \dfrac{\pi}{2}$
31. $L(x, y) = 7 + x - 6y; 0.06$ **33.** $L(x, y) = x + y + 1; 0.08$
35. $L(x, y) = 1 + x; 0.0222$
37. (a) $L(x, y, z) = 2x + 2y + 2z - 3$ **(b)** $L(x, y, z) = y + z$
  (c) $L(x, y, z) = 0$
39. (a) $L(x, y, z) = x$ **(b)** $L(x, y, z) = \dfrac{1}{\sqrt{2}}x + \dfrac{1}{\sqrt{2}}y$
  (c) $L(x, y, z) = \dfrac{1}{3}x + \dfrac{2}{3}y + \dfrac{2}{3}z$
41. (a) $L(x, y, z) = 2 + x$
  (b) $L(x, y, z) = x - y - z + \dfrac{\pi}{2} + 1$
  (c) $L(x, y, z) = x - y - z + \dfrac{\pi}{2} + 1$

43. $L(x, y, z) = 2x - 6y - 2z + 6, 0.0024$
45. $L(x, y, z) = x + y - z - 1, 0.00135$
47. Maximum error (estimate) ≤ 0.31 in magnitude
49. Pay more attention to the smaller of the two dimensions. It will generate the larger partial derivative.
51. f is most sensitive to a change in d.
55. At $-\dfrac{\pi}{4}, -\dfrac{\pi}{2\sqrt{2}}$; at $0, 0$; at $\dfrac{\pi}{4}, \dfrac{\pi}{2\sqrt{2}}$

Section 12.7, pp. 763–765

1. $f(-3, 3) = -5$, local minimum
3. $f\left(\dfrac{2}{3}, \dfrac{4}{3}\right) = 0$, local maximum   **5.** $f(-2, 1)$, saddle point
7. $f\left(\dfrac{6}{5}, \dfrac{69}{25}\right)$, saddle point   **9.** $f(2, 1)$, saddle point
11. $f(2, -1) = -6$, local minimum   **13.** $f(1, 2)$, saddle point
15. $f(0, 0)$, saddle point
17. $f(0, 0)$, saddle point; $f\left(-\dfrac{2}{3}, \dfrac{2}{3}\right) = \dfrac{170}{27}$, local maximum
19. $f(0, 0) = 0$, local minimum; $f(1, -1)$, saddle point
21. $f(0, 0)$, saddle point; $f\left(\dfrac{4}{9}, \dfrac{4}{3}\right) = -\dfrac{64}{81}$, local minimum
23. $f(0, 0)$, saddle point; $f(0, 2) = -12$, local minimum; $f(-2, 0) = -4$, local maximum; $f(-2, 2)$, saddle point
25. $f(0, 0)$, saddle point; $f(1, 1) = 2, f(-1, -1) = 2$, local maxima
27. $f(0, 0) = -1$, local maximum
29. $f(n\pi, 0)$, saddle point; $f(n\pi, 0) = 0$ for every n
31. Absolute maximum: 1 at $(0, 0)$; absolute minimum: -5 at $(1, 2)$
33. Absolute maximum: 4 at $(0, 2)$; absolute minimum: 0 at $(0, 0)$
35. Absolute maximum: 11 at $(0, -3)$; absolute minimum: -10 at $(4, -2)$
37. Absolute maximum: 4 at $(2, 0)$; absolute minimum: $\dfrac{3\sqrt{2}}{2}$ at $\left(3, -\dfrac{\pi}{4}\right), \left(3, \dfrac{\pi}{4}\right), \left(1, -\dfrac{\pi}{4}\right)$, and $\left(1, \dfrac{\pi}{4}\right)$
39. $a = -3, b = 2$
41. (a) $f(0, 0)$, saddle point   **(b)** $f(1, 2)$, local minimum
  (c) $f(1, -2)$, local minimum; $f(-1, -2)$, saddle point
47. $\left(\dfrac{1}{6}, \dfrac{1}{3}, \dfrac{355}{36}\right)$
51. (a) On the semicircle, max $f = 2\sqrt{2}$ at $t = \pi/4$, min $f = -2$ at $t = \pi$. On the quarter circle, max $f = 2\sqrt{2}$ at $t = \pi/4$, min $f = 2$ at $t = 0, \pi/2$.
  (b) On the semicircle, max $g = 2$ at $t = \pi/4$, min $g = -2$ at $t = 3\pi/4$. On the quarter circle, max $g = 2$ at $t = \pi/4$, min $g = 0$ at $t = 0, \pi/2$.
  (c) On the semicircle, max $h = 8$ at $t = 0, \pi$; min $h = 4$ at $t = \pi/2$. On the quarter circle, max $h = 8$ at $t = 0$, min $h = 4$ at $t = \pi/2$.

53. i) min $f = -1/2$ at $t = -1/2$; no max   ii) max $f = 0$ at
$t = -1, 0$; min $f = -1/2$ at $t = -1/2$ iii) max $f = 4$ at
$t = 1$; min $f = 0$ at $t = 0$

57. $y = -\dfrac{20}{13}x + \dfrac{9}{13}$, $y|_{x=4} = -\dfrac{71}{13}$

Section 12.8, pp. 773–775

1. $\left(\pm\dfrac{1}{\sqrt{2}}, \dfrac{1}{2}\right), \left(\pm\dfrac{1}{\sqrt{2}}, -\dfrac{1}{2}\right)$ **3.** 39   **5.** $\left(3, \pm3\sqrt{2}\right)$

7. (a) 8   **(b)** 64

9. $r = 2$ cm, $h = 4$ cm   **11.** Length $= 4\sqrt{2}$, width $= 3\sqrt{2}$

13. $f(0, 0) = 0$ is minimum, $f(2, 4) = 20$ is maximum.

15. Lowest $= 0°$, highest $= 125°$

17. $\left(\dfrac{3}{2}, 2, \dfrac{5}{2}\right)$ **19.** 1   **21.** $(0, 0, 2), (0, 0, -2)$

23. $f(1, -2, 5) = 30$ is maximum, $f(-1, 2, -5) = -30$ is
minimum.

25. 3, 3, 3   **27.** $\dfrac{2}{\sqrt{3}}$ by $\dfrac{2}{\sqrt{3}}$ by $\dfrac{2}{\sqrt{3}}$ units

29. $(\pm4/3, -4/3, -4/3)$ **31.** $U(8, 14) = \$128$

33. $f(2/3, 4/3, -4/3) = \dfrac{4}{3}$ **35.** $(2, 4, 4)$

37. Maximum is $1 + 6\sqrt{3}$ at $\left(\pm\sqrt{6}, \sqrt{3}, 1\right)$, minimum is
$1 - 6\sqrt{3}$ at $\left(\pm\sqrt{6}, -\sqrt{3}, 1\right)$.

39. Maximum is 4 at $(0, 0, \pm2)$, minimum is 2 at $\left(\pm\sqrt{2}, \pm\sqrt{2}, 0\right)$.

Section 12.9, p. 779

1. Quadratic: $x + xy$; cubic: $x + xy + \dfrac{1}{2}xy^2$

3. Quadratic: xy; cubic: xy

5. Quadratic: $y + \dfrac{1}{2}(2xy - y^2)$;

cubic: $y + \dfrac{1}{2}(2xy - y^2) + \dfrac{1}{6}(3x^2y - 3xy^2 + 2y^3)$

7. Quadratic: $\dfrac{1}{2}(2x^2 + 2y^2) = x^2 + y^2$; cubic: $x^2 + y^2$

9. Quadratic: $1 + (x + y) + (x + y)^2$;
cubic: $1 + (x + y) + (x + y)^2 + (x + y)^3$

11. Quadratic: $1 - \dfrac{1}{2}x^2 - \dfrac{1}{2}y^2$; $E(x, y) \leq 0.00134$

Practice Exercises, pp. 780–783

1. Domain: all points in the xy-plane; range: $z \geq 0$. Level curves
are ellipses with major axis along the y-axis and minor axis
along the x-axis.

3. Domain: all (x, y) such that $x \neq 0$ and $y \neq 0$; range: $z \neq 0$.
Level curves are hyperbolas with the x- and y-axes as asymptotes.

5. Domain: all points in xyz-space; range: all real numbers. Level
surfaces are paraboloids of revolution with the z-axis as axis.

7. Domain: all (x, y, z) such that $(x, y, z) \neq (0, 0, 0)$; range: positive real numbers. Level surfaces are spheres with center $(0, 0, 0)$
and radius $r > 0$.

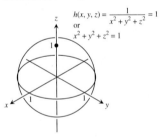

9. -2 **11.** $1/2$ **13.** 1   **15.** Let $y = kx^2, k \neq 1$

17. No; $\lim_{(x,y)\to(0,0)} f(x, y)$ does not exist.

19. $\dfrac{\partial g}{\partial r} = \cos\theta + \sin\theta, \dfrac{\partial g}{\partial \theta} = -r\sin\theta + r\cos\theta$

21. $\dfrac{\partial f}{\partial R_1} = -\dfrac{1}{R_1{}^2}, \dfrac{\partial f}{\partial R_2} = -\dfrac{1}{R_2{}^2}, \dfrac{\partial f}{\partial R_3} = -\dfrac{1}{R_3{}^2}$

23. $\dfrac{\partial P}{\partial n} = \dfrac{RT}{V}, \dfrac{\partial P}{\partial R} = \dfrac{nT}{V}, \dfrac{\partial P}{\partial T} = \dfrac{nR}{V}, \dfrac{\partial P}{\partial V} = -\dfrac{nRT}{V^2}$

25. $\dfrac{\partial^2 g}{\partial x^2} = 0, \dfrac{\partial^2 g}{\partial y^2} = \dfrac{2x}{y^3}, \dfrac{\partial^2 g}{\partial y\,\partial x} = \dfrac{\partial^2 g}{\partial x\,\partial y} = -\dfrac{1}{y^2}$

27. $\dfrac{\partial^2 f}{\partial x^2} = -30x + \dfrac{2 - 2x^2}{(x^2 + 1)^2}, \dfrac{\partial^2 f}{\partial y^2} = 0, \dfrac{\partial^2 f}{\partial y\,\partial x} = \dfrac{\partial^2 f}{\partial x\,\partial y} = 1$

29. $\dfrac{dw}{dt}\bigg|_{t=0} = -1$

31. $\dfrac{\partial w}{\partial r}\bigg|_{(r, s)=(\pi, 0)} = 2, \dfrac{\partial w}{\partial s}\bigg|_{(r, s)=(\pi, 0)} = 2 - \pi$

33. $\dfrac{df}{dt}\Big|_{t=1} = -(\sin 1 + \cos 2)(\sin 1) + (\cos 1 + \cos 2)(\cos 1)$
$\qquad\qquad - 2(\sin 1 + \cos 1)(\sin 2)$

35. $\dfrac{dy}{dx}\Big|_{(x,y)=(0,1)} = -1$

37. Increases most rapidly in the direction $\mathbf{u} = -\dfrac{\sqrt{2}}{2}\mathbf{i} - \dfrac{\sqrt{2}}{2}\mathbf{j}$;

decreases most rapidly in the direction $-\mathbf{u} = \dfrac{\sqrt{2}}{2}\mathbf{i} + \dfrac{\sqrt{2}}{2}\mathbf{j}$;

$D_{\mathbf{u}}f = \dfrac{\sqrt{2}}{2}; D_{-\mathbf{u}}f = -\dfrac{\sqrt{2}}{2}; D_{\mathbf{u}_1}f = -\dfrac{7}{10}$ where $\mathbf{u}_1 = \dfrac{\mathbf{v}}{|\mathbf{v}|}$

39. Increases most rapidly in the direction $\mathbf{u} = \dfrac{2}{7}\mathbf{i} + \dfrac{3}{7}\mathbf{j} + \dfrac{6}{7}\mathbf{k}$;

decreases most rapidly in the direction $-\mathbf{u} = -\dfrac{2}{7}\mathbf{i} - \dfrac{3}{7}\mathbf{j} - \dfrac{6}{7}\mathbf{k}$;

$D_{\mathbf{u}}f = 7; D_{-\mathbf{u}}f = -7; D_{\mathbf{u}_1}f = 7$ where $\mathbf{u}_1 = \dfrac{\mathbf{v}}{|\mathbf{v}|}$

41. $\pi/\sqrt{2}$

43. **(a)** $f_x(1,2) = f_y(1,2) = 2$ **(b)** $14/5$

45.

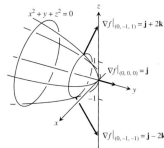

47. Tangent: $4x - y - 5z = 4$; normal line:
$x = 2 + 4t, y = -1 - t, z = 1 - 5t$

49. $2y - z - 2 = 0$

51. Tangent: $x + y = \pi + 1$; normal line: $y = x - \pi + 1$

53. $x = 1 - 2t, y = 1, z = 1/2 + 2t$

55. Answers will depend on the upper bound used for
$|f_{xx}|, |f_{xy}|, |f_{yy}|$. With $M = \sqrt{2}/2, |E| \leq 0.0142$. With
$M = 1, |E| \leq 0.02$.

57. $L(x, y, z) = y - 3z, L(x, y, z) = x + y - z - 1$

59. Local minimum of -8 at $(-2, -2)$

61. Saddle point at $(0, 0), f(0, 0) = 0$; local maximum of $1/4$ at
$(-1/2, -1/2)$

63. Saddle point at $(0, 0), f(0, 0) = 0$; local minimum of -4 at
$(0, 2)$; local maximum of 4 at $(-2, 0)$; saddle point at $(-2, 2)$,
$f(-2, 2) = 0$

65. Absolute maximum: 28 at $(0, 4)$; absolute minimum: $-9/4$ at
$(3/2, 0)$

67. Absolute maximum: 18 at $(2, -2)$; absolute minimum: $-17/4$ at
$(-2, 1/2)$

69. Absolute maximum: 8 at $(-2, 0)$; absolute minimum: -1 at $(1, 0)$

71. Absolute maximum: 4 at $(1, 0)$; absolute minimum: -4 at $(0, -1)$

73. Absolute maximum: 1 at $(0, \pm 1)$ and $(1, 0)$; absolute minimum:
-1 at $(-1, 0)$

75. Maximum: 5 at $(0, 1)$; minimum: $-1/3$ at $(0, -1/3)$

77. Maximum: $\sqrt{3}$ at $\left(\dfrac{1}{\sqrt{3}}, -\dfrac{1}{\sqrt{3}}, \dfrac{1}{\sqrt{3}}\right)$; minimum: $-\sqrt{3}$ at

$\left(-\dfrac{1}{\sqrt{3}}, \dfrac{1}{\sqrt{3}}, -\dfrac{1}{\sqrt{3}}\right)$

79. Width $= \left(\dfrac{c^2V}{ab}\right)^{1/3}$, depth $= \left(\dfrac{b^2V}{ac}\right)^{1/3}$, height $= \left(\dfrac{a^2V}{bc}\right)^{1/3}$

81. Maximum: $\dfrac{3}{2}$ at $\left(\dfrac{1}{\sqrt{2}}, \dfrac{1}{\sqrt{2}}, \sqrt{2}\right)$ and $\left(-\dfrac{1}{\sqrt{2}}, -\dfrac{1}{\sqrt{2}}, -\sqrt{2}\right)$;

minimum: $\dfrac{1}{2}$ at $\left(-\dfrac{1}{\sqrt{2}}, \dfrac{1}{\sqrt{2}}, -\sqrt{2}\right)$ and $\left(\dfrac{1}{\sqrt{2}}, -\dfrac{1}{\sqrt{2}}, \sqrt{2}\right)$

83. $\dfrac{\partial w}{\partial x} = \cos\theta\dfrac{\partial w}{\partial r} - \dfrac{\sin\theta}{r}\dfrac{\partial w}{\partial\theta}, \dfrac{\partial w}{\partial y} = \sin\theta\dfrac{\partial w}{\partial r} + \dfrac{\cos\theta}{r}\dfrac{\partial w}{\partial\theta}$

89. $(t, -t \pm 4, t)$, t a real number

Additional and Advanced Exercises, pp. 783–784

1. $f_{xy}(0, 0) = -1, f_{yx}(0, 0) = 1$

7. **(c)** $\dfrac{r^2}{2} = \dfrac{1}{2}(x^2 + y^2 + z^2)$ **13.** $V = \dfrac{\sqrt{3}abc}{2}$

17. $f(x, y) = \dfrac{y}{2} + 4, g(x, y) = \dfrac{x}{2} + \dfrac{9}{2}$

19. $y = 2\ln|\sin x| + \ln 2$

21. **(a)** $\dfrac{1}{\sqrt{53}}(2\mathbf{i} + 7\mathbf{j})$ **(b)** $\dfrac{-1}{\sqrt{29,097}}(98\mathbf{i} - 127\mathbf{j} + 58\mathbf{k})$

CHAPTER 13

Section 13.1, pp. 789–790

1. 24 **3.** 1 **5.** 16 **7.** $2\ln 2 - 1$ **9.** $(3/2)(5 - e)$ **11.** $3/2$
13. 14 **15.** 0 **17.** $1/2$ **19.** $2\ln 2$ **21.** $(\ln 2)^2$ **23.** $8/3$
25. 1 **27.** $\sqrt{2}$

Section 13.2, pp. 797–798

1. $\dfrac{\pi^2}{2} + 2$

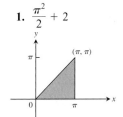

3. $8\ln 8 - 16 + e$

5. $e - 2$

7. $\dfrac{3}{2}\ln 2$ **9.** $-1/10$

11. 8

13. 2π

15. $\displaystyle\int_2^4\int_0^{(4-y)/2} dx\, dy$

17. $\displaystyle\int_0^1\int_{x^2}^{x} dy\, dx$

19. $\displaystyle\int_1^e\int_{\ln y}^{1} dx\, dy$

21. $\displaystyle\int_0^9\int_0^{(\sqrt{9-y})/2} 16x\, dx\, dy$

23. $\displaystyle\int_{-1}^1\int_0^{\sqrt{1-x^2}} 3y\, dy\, dx$

25. 2

27. $\dfrac{e-2}{2}$

29. 2

31. $1/(80\pi)$

33. $-2/3$

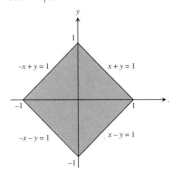

35. $4/3$ **37.** $625/12$ **39.** 16   **41.** 20   **43.** $2(1 + \ln 2)$

45. 1   **47.** π^2 **49.** $-\dfrac{3}{32}$ **51.** $\dfrac{20\sqrt{3}}{9}$

53. $\displaystyle\int_0^1\int_x^{2-x} (x^2 + y^2)\, dy\, dx = \dfrac{4}{3}$

55. R is the set of points (x, y) such that $x^2 + 2y^2 < 4$

57. No, by Fubini's Theorem, the two orders of integration must give the same result.

61. 0.603   **63.** 0.233

Section 13.3, pp. 801–802

1. $\displaystyle\int_0^2\int_0^{2-x} dy\, dx = 2$ or $\displaystyle\int_0^2\int_0^{2-y} dx\, dy = 2$

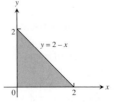

3. $\displaystyle\int_{-2}^{1}\int_{y-2}^{-y^2} dx\,dy = \frac{9}{2}$

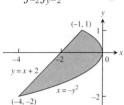

5. $\displaystyle\int_{0}^{\ln 2}\int_{0}^{e^x} dy\,dx = 1$

7. $\displaystyle\int_{0}^{1}\int_{y^2}^{2y-y^2} dx\,dy = \frac{1}{3}$

9. 12

11. $\sqrt{2} - 1$

13. $\dfrac{3}{2}$

15. (a) 0   (b) $4/\pi^2$ **17.** 8/3
19. $40{,}000(1 - e^{-2})\ln(7/2) \approx 43{,}329$

Section 13.4, pp. 806–807

1. $\pi/2$ **3.** $\pi/8$ **5.** πa^2 **7.** 36   **9.** $(1 - \ln 2)\pi$
11. $(2\ln 2 - 1)(\pi/2)$ **13.** $(\pi/2) + 1$ **15.** $\pi(\ln 4 - 1)$
17. $2(\pi - 1)$ **19.** 12π **21.** $(3\pi/8) + 1$ **23.** $\dfrac{2a}{3}$ **25.** $\dfrac{2a}{3}$
27. $2\pi(2 - \sqrt{e})$ **29.** $\dfrac{4}{3} + \dfrac{5\pi}{8}$ **31.** (a) $\dfrac{\sqrt{\pi}}{2}$ (b) 1
33. $\pi\ln 4$, no   **35.** $\dfrac{1}{2}(a^2 + 2h^2)$

Section 13.5, pp. 813–816

1. 1/6

3. $\displaystyle\int_{0}^{1}\int_{0}^{2-2x}\int_{0}^{3-3x-3y/2} dz\,dy\,dx,\quad \int_{0}^{2}\int_{0}^{1-y/2}\int_{0}^{3-3x-3y/2} dz\,dx\,dy,$
$\displaystyle\int_{0}^{1}\int_{0}^{3-3x}\int_{0}^{2-2x-2z/3} dy\,dz\,dx,\quad \int_{0}^{3}\int_{0}^{1-z/3}\int_{0}^{2-2x-2z/3} dy\,dx\,dz,$

$\displaystyle\int_{0}^{2}\int_{0}^{3-3y/2}\int_{0}^{1-y/2-z/3} dx\,dz\,dy,$
$\displaystyle\int_{0}^{3}\int_{0}^{2-2z/3}\int_{0}^{1-y/2-z/3} dx\,dy\,dz.$

The value of all six integrals is 1.

5. $\displaystyle\int_{-2}^{2}\int_{-\sqrt{4-x^2}}^{\sqrt{4-x^2}}\int_{x^2+y^2}^{8-x^2-y^2} 1\,dz\,dx\,dy,\quad \int_{-2}^{2}\int_{-\sqrt{4-y^2}}^{\sqrt{4-y^2}}\int_{x^2+y^2}^{8-x^2-y^2} 1\,dz\,dx\,dy,$

$\displaystyle\int_{-2}^{2}\int_{4}^{8-y^2}\int_{-\sqrt{8-z-y^2}}^{\sqrt{8-z-y^2}} 1\,dx\,dz\,dy + \int_{-2}^{2}\int_{y^2}^{4}\int_{-\sqrt{z-y^2}}^{\sqrt{z-y^2}} 1\,dx\,dz\,dy,$

$\displaystyle\int_{4}^{8}\int_{-\sqrt{8-z}}^{\sqrt{8-z}}\int_{-\sqrt{8-z-y^2}}^{\sqrt{8-z-y^2}} 1\,dx\,dy\,dz + \int_{0}^{4}\int_{-\sqrt{z}}^{\sqrt{z}}\int_{-\sqrt{z-y^2}}^{\sqrt{z-y^2}} 1\,dx\,dy\,dz,$

$\displaystyle\int_{-2}^{2}\int_{4}^{8-x^2}\int_{-\sqrt{8-z-x^2}}^{\sqrt{8-z-x^2}} 1\,dy\,dz\,dx + \int_{-2}^{2}\int_{x^2}^{4}\int_{-\sqrt{z-x^2}}^{\sqrt{z-x^2}} 1\,dy\,dz\,dx,$

$\displaystyle\int_{4}^{8}\int_{-\sqrt{8-z}}^{\sqrt{8-z}}\int_{-\sqrt{8-z-x^2}}^{\sqrt{8-z-x^2}} 1\,dy\,dx\,dz + \int_{0}^{4}\int_{-\sqrt{z}}^{\sqrt{z}}\int_{-\sqrt{z-x^2}}^{\sqrt{z-x^2}} 1\,dy\,dx\,dz.$

The value of all six integrals is 16π.

7. 1   **9.** 1   **11.** $\dfrac{\pi^3}{2}(1 - \cos 1)$ **13.** 18   **15.** 7/6

17. 0   **19.** $\dfrac{1}{2} - \dfrac{\pi}{8}$

21. (a) $\displaystyle\int_{-1}^{1}\int_{0}^{1-x^2}\int_{x^2}^{1-z} dy\,dz\,dx$ (b) $\displaystyle\int_{0}^{1}\int_{-\sqrt{1-z}}^{\sqrt{1-z}}\int_{x^2}^{1-z} dy\,dx\,dz$

(c) $\displaystyle\int_{0}^{1}\int_{0}^{1-z}\int_{-\sqrt{y}}^{\sqrt{y}} dx\,dy\,dz$ (d) $\displaystyle\int_{0}^{1}\int_{0}^{1-y}\int_{-\sqrt{y}}^{\sqrt{y}} dx\,dz\,dy$

(e) $\displaystyle\int_{0}^{1}\int_{-\sqrt{y}}^{\sqrt{y}}\int_{0}^{1-y} dz\,dx\,dy$

23. 2/3   **25.** 20/3   **27.** 1   **29.** 16/3   **31.** $8\pi - \dfrac{32}{3}$
33. 2   **35.** 4π **37.** 31/3   **39.** 1   **41.** $2\sin 4$ **43.** 4
45. $a = 3$ or $a = 13/3$
47. The domain is the set of all points (x, y, z) such that $4x^2 + 4y^2 + z^2 \le 4$.

Section 13.6, pp. 822–824

1. $\bar{x} = 5/14, \bar{y} = 38/35$ **3.** $\bar{x} = 64/35, \bar{y} = 5/7$
5. $\bar{x} = \bar{y} = 4a/(3\pi)$ **7.** $I_x = I_y = 4\pi, I_0 = 8\pi$
9. $\bar{x} = -1, \bar{y} = 1/4$ **11.** $I_x = 64/105$
13. $\bar{x} = 3/8, \bar{y} = 17/16$
15. $\bar{x} = 11/3, \bar{y} = 14/27, I_y = 432$
17. $\bar{x} = 0, \bar{y} = 13/31, I_y = 7/5$
19. $\bar{x} = 0, \bar{y} = 7/10; I_x = 9/10, I_y = 3/10, I_0 = 6/5$
21. $I_x = \dfrac{M}{3}(b^2 + c^2), I_y = \dfrac{M}{3}(a^2 + c^2), I_z = \dfrac{M}{3}(a^2 + b^2)$
23. $\bar{x} = \bar{y} = 0, \bar{z} = 12/5, I_x = 7904/105 \approx 75.28,$
$I_y = 4832/63 \approx 76.70, I_z = 256/45 \approx 5.69$

25. (a) $\bar{x} = \bar{y} = 0, \bar{z} = 8/3$ **(b)** $c = 2\sqrt{2}$

27. $I_L = 1386$

29. (a) $4/3$ **(b)** $\bar{x} = 4/5, \bar{y} = \bar{z} = 2/5$

31. (a) $5/2$ **(b)** $\bar{x} = \bar{y} = \bar{z} = 8/15$ **(c)** $I_x = I_y = I_z = 11/6$

33. 3

37. (a) $I_{c.m.} = \dfrac{abc(a^2 + b^2)}{12}, R_{c.m.} = \sqrt{\dfrac{a^2 + b^2}{12}}$

(b) $I_L = \dfrac{abc(a^2 + 7b^2)}{3}, R_L = \sqrt{\dfrac{a^2 + 7b^2}{3}}$

Section 13.7, pp. 833–837

1. $\dfrac{4\pi(\sqrt{2} - 1)}{3}$ **3.** $\dfrac{17\pi}{5}$ **5.** $\pi(6\sqrt{2} - 8)$ **7.** $\dfrac{3\pi}{10}$

9. $\pi/3$

11. (a) $\displaystyle\int_0^{2\pi}\int_0^1\int_0^{\sqrt{4-r^2}} r\, dz\, dr\, d\theta$

(b) $\displaystyle\int_0^{2\pi}\int_0^{\sqrt{3}}\int_0^1 r\, dr\, dz\, d\theta + \int_0^{2\pi}\int_{\sqrt{3}}^2\int_0^{\sqrt{4-z^2}} r\, dr\, dz\, d\theta$

(c) $\displaystyle\int_0^1\int_0^{\sqrt{4-r^2}}\int_0^{2\pi} r\, d\theta\, dz\, dr$

13. $\displaystyle\int_{-\pi/2}^{\pi/2}\int_0^{\cos\theta}\int_0^{3r^2} f(r, \theta, z)\, dz\, r\, dr\, d\theta$

15. $\displaystyle\int_0^{\pi}\int_0^{2\sin\theta}\int_0^{4-r\sin\theta} f(r, \theta, z)\, dz\, r\, dr\, d\theta$

17. $\displaystyle\int_{-\pi/2}^{\pi/2}\int_1^{1+\cos\theta}\int_0^4 f(r, \theta, z)\, dz\, r\, dr\, d\theta$

19. $\displaystyle\int_0^{\pi/4}\int_0^{\sec\theta}\int_0^{2-r\sin\theta} f(r, \theta, z)\, dz\, r\, dr\, d\theta$ **21.** π^2 **23.** $\pi/3$

25. 5π **27.** 2π **29.** $\left(\dfrac{8 - 5\sqrt{2}}{2}\right)\pi$

31. (a) $\displaystyle\int_0^{2\pi}\int_0^{\pi/6}\int_0^2 \rho^2\sin\phi\, d\rho\, d\phi\, d\theta +$

$\displaystyle\int_0^{2\pi}\int_{\pi/6}^{\pi/2}\int_0^{\csc\phi} \rho^2\sin\phi\, d\rho\, d\phi\, d\theta$

(b) $\displaystyle\int_0^{2\pi}\int_1^2\int_{\pi/6}^{\sin^{-1}(1/\rho)} \rho^2\sin\phi\, d\phi\, d\rho\, d\theta +$

$\displaystyle\int_0^{2\pi}\int_0^2\int_0^{\pi/6} \rho^2\sin\phi\, d\phi\, d\rho\, d\theta$

33. $\displaystyle\int_0^{2\pi}\int_0^{\pi/2}\int_{\cos\phi}^2 \rho^2\sin\phi\, d\rho\, d\phi\, d\theta = \dfrac{31\pi}{6}$

35. $\displaystyle\int_0^{2\pi}\int_0^{\pi}\int_0^{1-\cos\phi} \rho^2\sin\phi\, d\rho\, d\phi\, d\theta = \dfrac{8\pi}{3}$

37. $\displaystyle\int_0^{2\pi}\int_{\pi/4}^{\pi/2}\int_0^{2\cos\phi} \rho^2\sin\phi\, d\rho\, d\phi\, d\theta = \dfrac{\pi}{3}$

39. (a) $8\displaystyle\int_0^{\pi/2}\int_0^{\pi/2}\int_0^2 \rho^2\sin\phi\, d\rho\, d\phi\, d\theta$

(b) $8\displaystyle\int_0^{\pi/2}\int_0^2\int_0^{\sqrt{4-r^2}} r\, dz\, dr\, d\theta$

(c) $8\displaystyle\int_0^2\int_0^{\sqrt{4-x^2}}\int_0^{\sqrt{4-x^2-y^2}} dz\, dy\, dx$

41. (a) $\displaystyle\int_0^{2\pi}\int_0^{\pi/3}\int_{\sec\phi}^2 \rho^2\sin\phi\, d\rho\, d\phi\, d\theta$

(b) $\displaystyle\int_0^{2\pi}\int_0^{\sqrt{3}}\int_1^{\sqrt{4-r^2}} r\, dz\, dr\, d\theta$

(c) $\displaystyle\int_{-\sqrt{3}}^{\sqrt{3}}\int_{-\sqrt{3-x^2}}^{\sqrt{3-x^2}}\int_1^{\sqrt{4-x^2-y^2}} dz\, dy\, dx$ **(d)** $5\pi/3$

43. $8\pi/3$ **45.** $9/4$ **47.** $\dfrac{3\pi - 4}{18}$ **49.** $\dfrac{2\pi a^3}{3}$ **51.** $5\pi/3$

53. $\pi/2$ **55.** $\dfrac{4(2\sqrt{2} - 1)\pi}{3}$ **57.** 16π **59.** $5\pi/2$

61. $\dfrac{4\pi(8 - 3\sqrt{3})}{3}$ **63.** $2/3$ **65.** $3/4$

67. $\bar{x} = \bar{y} = 0, \bar{z} = 3/8$ **69.** $(\bar{x}, \bar{y}, \bar{z}) = (0, 0, 3/8)$

71. $\bar{x} = \bar{y} = 0, \bar{z} = 5/6$ **73.** $I_x = \pi/4$ **75.** $\dfrac{a^4 h\pi}{10}$

77. (a) $(\bar{x}, \bar{y}, \bar{z}) = \left(0, 0, \dfrac{4}{5}\right), I_z = \dfrac{\pi}{12}$

(b) $(\bar{x}, \bar{y}, \bar{z}) = \left(0, 0, \dfrac{5}{6}\right), I_z = \dfrac{\pi}{14}$

81. $\dfrac{3M}{\pi R^3}$

85. The surface's equation $r = f(z)$ tells us that the point $(r, \theta, z) = (f(z), \theta, z)$ will lie on the surface for all θ. In particular, $(f(z), \theta + \pi, z)$ lies on the surface whenever $(f(z), \theta, z)$ lies on the surface, so the surface is symmetric with respect to the z-axis.

Section 13.8, pp. 844–845

1. (a) $x = \dfrac{u + v}{3}, y = \dfrac{v - 2u}{3}; \dfrac{1}{3}$

(b) Triangular region with boundaries $u = 0$, $v = 0$, and $u + v = 3$

3. (a) $x = \dfrac{1}{5}(2u - v), y = \dfrac{1}{10}(3v - u); \dfrac{1}{10}$

(b) Triangular region with boundaries $3v = u$, $v = 2u$, and $3u + v = 10$

7. $64/5$ **9.** $\displaystyle\int_1^2 \int_1^3 (u + v)\dfrac{2u}{v}\, du\, dv = 8 + \dfrac{52}{3}\ln 2$

11. $\dfrac{\pi ab(a^2 + b^2)}{4}$ **13.** $\dfrac{1}{3}\left(1 + \dfrac{3}{e^2}\right) \approx 0.4687$

15. (a) $\begin{vmatrix} \cos v & -u\sin v \\ \sin v & u\cos v \end{vmatrix} = u\cos^2 v + u\sin^2 v = u$

(b) $\begin{vmatrix} \sin v & u\cos v \\ \cos v & -u\sin v \end{vmatrix} = -u\sin^2 v - u\cos^2 v = -u$

19. 12 **21.** $\dfrac{a^2 b^2 c^2}{6}$

Practice Exercises, pp. 846–848

1. $9e - 9$ **3.** $9/2$

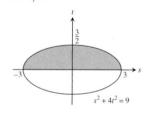

5. $\displaystyle\int_{-2}^0 \int_{2x+4}^{4-x^2} dy\, dx = \dfrac{4}{3}$ **7.** $\displaystyle\int_{-3}^3 \int_0^{(1/2)\sqrt{9-x^2}} y\, dy\, dx = \dfrac{9}{2}$

9. $\sin 4$ **11.** $\dfrac{\ln 17}{4}$ **13.** $4/3$ **15.** $4/3$ **17.** $1/4$ **19.** π

21. $\dfrac{\pi - 2}{4}$ **23.** 0 **25.** $8/35$ **27.** $\pi/2$ **29.** $\dfrac{2(31 - 3^{5/2})}{3}$

31. (a) $\displaystyle\int_{-\sqrt{2}}^{\sqrt{2}} \int_{-\sqrt{2-y^2}}^{\sqrt{2-y^2}} \int_{\sqrt{x^2+y^2}}^{\sqrt{4-x^2-y^2}} 3\, dz\, dx\, dy$

(b) $\displaystyle\int_0^{2\pi} \int_0^{\pi/4} \int_0^2 3\,\rho^2 \sin\phi\, d\rho\, d\phi\, d\theta$ **(c)** $2\pi(8 - 4\sqrt{2})$

33. $\displaystyle\int_0^{2\pi} \int_0^{\pi/4} \int_0^{\sec\phi} \rho^2 \sin\phi\, d\rho\, d\phi\, d\theta = \dfrac{\pi}{3}$

35. $\displaystyle\int_0^1 \int_{\sqrt{1-x^2}}^{\sqrt{3-x^2}} \int_1^{\sqrt{4-x^2-y^2}} z^2 xy\, dz\, dy\, dx$

$+ \displaystyle\int_1^{\sqrt{3}} \int_0^{\sqrt{3-x^2}} \int_1^{\sqrt{4-x^2-y^2}} z^2 xy\, dz\, dy\, dx$

37. (a) $\dfrac{8\pi(4\sqrt{2} - 5)}{3}$ **(b)** $\dfrac{8\pi(4\sqrt{2} - 5)}{3}$

39. $I_z = \dfrac{8\pi\delta(b^5 - a^5)}{15}$

41. $\bar{x} = \bar{y} = \dfrac{1}{2 - \ln 4}$ **43.** $I_0 = 104$ **45.** $I_x = 2\delta$

47. $M = 4, M_x = 0, M_y = 0$ **49.** $\bar{x} = \dfrac{3\sqrt{3}}{\pi}, \bar{y} = 0$

51. (a) $\bar{x} = \dfrac{15\pi + 32}{6\pi + 48}, \bar{y} = 0$

(b)

Additional and Advanced Exercises, pp. 848–850

1. (a) $\displaystyle\int_{-3}^2 \int_x^{6-x^2} x^2\, dy\, dx$ **(b)** $\displaystyle\int_{-3}^2 \int_x^{6-x^2} \int_0^{x^2} dz\, dy\, dx$

(c) $125/4$

3. 2π **5.** $3\pi/2$ **7. (a)** Hole radius $= 1$, sphere radius $= 2$

(b) $4\sqrt{3}\pi$ **9.** $\pi/4$ **11.** $\ln\left(\dfrac{b}{a}\right)$ **15.** $1/\sqrt[4]{3}$

17. Mass $= a^2 \cos^{-1}\left(\dfrac{b}{a}\right) - b\sqrt{a^2 - b^2}$,

$I_0 = \dfrac{a^4}{2}\cos^{-1}\left(\dfrac{b}{a}\right) - \dfrac{b^3}{2}\sqrt{a^2 - b^2} - \dfrac{b^3}{6}(a^2 - b^2)^{3/2}$

19. $\dfrac{1}{ab}(e^{a^2b^2} - 1)$ **21. (b)** 1 **(c)** 0

25. $h = \sqrt{20}$ in., $h = \sqrt{60}$ in.   **27.** $2\pi\left[\dfrac{1}{3} - \left(\dfrac{1}{3}\right)\dfrac{\sqrt{2}}{2}\right]$

CHAPTER 14

Section 14.1, pp. 855–856

1. Graph (c)   **3.** Graph (g)   **5.** Graph (d)   **7.** Graph (f)

9. $\sqrt{2}$ **11.** $\dfrac{13}{2}$ **13.** $3\sqrt{14}$ **15.** $\dfrac{1}{6}\left(5\sqrt{5}+9\right)$

17. $\sqrt{3}\ln\left(\dfrac{b}{a}\right)$ **19.** $\dfrac{10\sqrt{5}-2}{3}$ **21.** 8 **23.** $2\sqrt{2}-1$

25. (a) $4\sqrt{2}-2$ **(b)** $\sqrt{2}+\ln\left(1+\sqrt{2}\right)$ **27.** $I_z=2\pi\delta a^3$

29. (a) $I_z=2\pi\sqrt{2}\delta$ **(b)** $I_z=4\pi\sqrt{2}\delta$ **31.** $I_x=2\pi-2$

Section 14.2, pp. 865–867

1. $\nabla f = -(x\mathbf{i}+y\mathbf{j}+z\mathbf{k})(x^2+y^2+z^2)^{-3/2}$

3. $\nabla g = -\left(\dfrac{2x}{x^2+y^2}\right)\mathbf{i}-\left(\dfrac{2y}{x^2+y^2}\right)\mathbf{j}+e^z\mathbf{k}$

5. $\mathbf{F}=-\dfrac{kx}{(x^2+y^2)^{3/2}}\mathbf{i}-\dfrac{ky}{(x^2+y^2)^{3/2}}\mathbf{j}$, any $k>0$

7. (a) $9/2$ **(b)** $13/3$ **(c)** $9/2$

9. (a) $1/3$ **(b)** $-1/5$ **(c)** 0

11. (a) 2 **(b)** $3/2$ **(c)** $1/2$

13. $1/2$ **15.** $-\pi$ **17.** $69/4$ **19.** $-39/2$ **21.** $25/6$

23. (a) $\text{Circ}_1=0$, $\text{circ}_2=2\pi$, $\text{flux}_1=2\pi$, $\text{flux}_2=0$

    (b) $\text{Circ}_1=0$, $\text{circ}_2=8\pi$, $\text{flux}_1=8\pi$, $\text{flux}_2=0$

25. $\text{Circ}=0$, flux $=a^2\pi$ **27.** $\text{Circ}=a^2\pi$, flux $=0$

29. (a) $-\dfrac{\pi}{2}$ **(b)** 0 **(c)** 1

31.

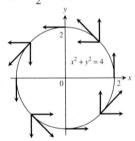

33. (a) $\mathbf{G}=-y\mathbf{i}+x\mathbf{j}$ **(b)** $\mathbf{G}=\sqrt{x^2+y^2}\,\mathbf{F}$

35. $\mathbf{F}=-\dfrac{x\mathbf{i}+y\mathbf{j}}{\sqrt{x^2+y^2}}$ **37.** 48 **39.** π **41.** 0 **43.** $\dfrac{1}{2}$

Section 14.3, pp. 876–877

1. Conservative **3.** Not conservative **5.** Not conservative

7. $f(x,y,z)=x^2+\dfrac{3y^2}{2}+2z^2+C$

9. $f(x,y,z)=xe^{y+2z}+C$

11. $f(x,y,z)=x\ln x-x+\tan(x+y)+\dfrac{1}{2}\ln(y^2+z^2)+C$

13. 49 **15.** -16 **17.** 1 **19.** $9\ln 2$ **21.** 0 **23.** -3

27. $\mathbf{F}=\nabla\left(\dfrac{x^2-1}{y}\right)$ **29. (a)** 1 **(b)** 1 **(c)** 1

31. (a) 2 **(b)** 2 **33. (a)** $c=b=2a$ **(b)** $c=b=2$

35. It does not matter what path you use. The work will be the same on any path because the field is conservative.

37. The force $\mathbf{F}$ is conservative because all partial derivatives of M, N, and P are zero. $f(x,y,z)=ax+by+cz+C$; $A=$

(xa, ya, za) and $B=(xb, yb, zb)$. Therefore, $\int \mathbf{F}\cdot d\mathbf{r}=f(B)-f(A)=a(xb-xa)+b(yb-ya)+c(zb-za)=\mathbf{F}\cdot\overrightarrow{AB}$.

Section 14.4, pp. 884–886

1. Flux $=0$, circ $=2\pi a^2$ **3.** Flux $=-\pi a^2$, circ $=0$

5. Flux $=2$, circ $=0$ **7.** Flux $=-9$, circ $=9$

9. Flux $=1/2$, circ $=1/2$ **11.** Flux $=1/5$, circ $=-1/12$

13. 0 **15.** $2/33$ **17.** 0 **19.** -16π **21.** πa^2 **23.** $\dfrac{3}{8}\pi$

25. (a) 0 if C is traversed counterclockwise

    (b) $(h-k)(\text{area of the region})$ **35. (a)** 0

Section 14.5, pp. 893–896

1. $\mathbf{r}(r,\theta)=(r\cos\theta)\mathbf{i}+(r\sin\theta)\mathbf{j}+r^2\mathbf{k}$, $0\le r\le 2$, $0\le\theta\le 2\pi$

3. $\mathbf{r}(r,\theta)=(r\cos\theta)\mathbf{i}+(r\sin\theta)\mathbf{j}+(r/2)\mathbf{k}$, $0\le r\le 6$, $0\le\theta\le\pi/2$

5. $\mathbf{r}(r,\theta)=(r\cos\theta)\mathbf{i}+(r\sin\theta)\mathbf{j}+\sqrt{9-r^2}\,\mathbf{k}$, $0\le r\le 3\sqrt{2}/2$, $0\le\theta\le 2\pi$; Also: $\mathbf{r}(\phi,\theta)=(3\sin\phi\cos\theta)\mathbf{i}+(3\sin\phi\sin\theta)\mathbf{j}+(3\cos\phi)\mathbf{k}$, $0\le\phi\le\pi/4$, $0\le\theta\le 2\pi$

7. $\mathbf{r}(\phi,\theta)=\left(\sqrt{3}\sin\phi\cos\theta\right)\mathbf{i}+\left(\sqrt{3}\sin\phi\sin\theta\right)\mathbf{j}+\left(\sqrt{3}\cos\phi\right)\mathbf{k}$, $\pi/3\le\phi\le 2\pi/3$, $0\le\theta\le 2\pi$

9. $\mathbf{r}(x,y)=x\mathbf{i}+y\mathbf{j}+(4-y^2)\mathbf{k}$, $0\le x\le 2$, $-2\le y\le 2$

11. $\mathbf{r}(u,v)=u\mathbf{i}+(3\cos v)\mathbf{j}+(3\sin v)\mathbf{k}$, $0\le u\le 3$, $0\le v\le 2\pi$

13. (a) $\mathbf{r}(r,\theta)=(r\cos\theta)\mathbf{i}+(r\sin\theta)\mathbf{j}+(1-r\cos\theta-r\sin\theta)\mathbf{k}$, $0\le r\le 3$, $0\le\theta\le 2\pi$

    (b) $\mathbf{r}(u,v)=(1-u\cos v-u\sin v)\mathbf{i}+(u\cos v)\mathbf{j}+(u\sin v)\mathbf{k}$, $0\le u\le 3$, $0\le v\le 2\pi$

15. $\mathbf{r}(u,v)=(4\cos^2 v)\mathbf{i}+u\mathbf{j}+(4\cos v\sin v)\mathbf{k}$, $0\le u\le 3$, $-(\pi/2)\le v\le(\pi/2)$; Another way: $\mathbf{r}(u,v)=(2+2\cos v)\mathbf{i}+u\mathbf{j}+(2\sin v)\mathbf{k}$, $0\le u\le 3$, $0\le v\le 2\pi$

17. $\displaystyle\int_0^{2\pi}\int_0^1\dfrac{\sqrt{5}}{2}r\,dr\,d\theta=\dfrac{\pi\sqrt{5}}{2}$

19. $\displaystyle\int_0^{2\pi}\int_1^3 r\sqrt{5}\,dr\,d\theta=8\pi\sqrt{5}$ **21.** $\displaystyle\int_0^{2\pi}\int_1^4 1\,du\,dv=6\pi$

23. $\displaystyle\int_0^{2\pi}\int_0^1 u\sqrt{4u^2+1}\,du\,dv=\dfrac{\left(5\sqrt{5}-1\right)}{6}\pi$

25. $\displaystyle\int_0^{2\pi}\int_{\pi/4}^{\pi}2\sin\phi\,d\phi\,d\theta=\left(4+2\sqrt{2}\right)\pi$

27.

29.

$x^2 + (y-3)^2 = 9$

$\sqrt{3}x + y = 9$

$\left(\frac{3\sqrt{3}}{2}, 9/2, 0\right)$

33. (b) $A = \int_0^{2\pi} \int_0^{\pi} [a^2 b^2 \sin^2 \phi \cos^2 \phi + b^2 c^2 \cos^4 \phi \cos^2 \theta + a^2 c^2 \cos^4 \phi \sin^2 \theta]^{1/2} \, d\phi \, d\theta$

35. $x_0 x + y_0 y = 25$ **37.** $\frac{13}{3}\pi$ **39.** 4   **41.** $6\sqrt{6} - 2\sqrt{2}$

43. $\pi\sqrt{c^2 + 1}$ **45.** $\frac{\pi}{6}\left(17\sqrt{17} - 5\sqrt{5}\right)$ **47.** $3 + 2\ln 2$

49. $\frac{\pi}{6}\left(13\sqrt{13} - 1\right)$ **51.** $5\pi\sqrt{2}$ **53.** $\frac{2}{3}\left(5\sqrt{5} - 1\right)$

Section 14.6, pp. 903–904

1. $\iint_S x \, d\sigma = \int_0^3 \int_0^2 u\sqrt{4u^2 + 1} \, du \, dv = \frac{17\sqrt{17} - 1}{4}$

3. $\iint_S x^2 \, d\sigma = \int_0^{2\pi} \int_0^{\pi} \sin^3 \phi \cos^2 \theta \, d\phi \, d\theta = \frac{4\pi}{3}$

5. $\iint_S z \, d\sigma = \int_0^1 \int_0^1 (4 - u - v)\sqrt{3} \, dv \, du = 3\sqrt{3}$
(for $x = u, y = v$)

7. $\iint_S x^2\sqrt{5 - 4z} \, d\sigma = \int_0^1 \int_0^{2\pi} u^2 \cos^2 v \cdot \sqrt{4u^2 + 1} \cdot u\sqrt{4u^2 + 1} \, dv \, du = \int_0^1 \int_0^{2\pi} u^3(4u^2 + 1)\cos^2 v \, dv \, du = \frac{11\pi}{12}$

9. $9a^3$ **11.** $\frac{abc}{4}(ab + ac + bc)$ **13.** 2   **15.** -32

17. $\pi a^3/6$ **19.** $13a^4/6$ **21.** $2\pi/3$ **23.** $-73\pi/6$ **25.** 18

27. $\frac{\pi a^3}{6}$ **29.** $\frac{\pi a^2}{4}$ **31.** $\frac{\pi a^3}{2}$ **33.** -32 **35.** -4 **37.** $3a^4$

39. $\left(\frac{a}{2}, \frac{a}{2}, \frac{a}{2}\right)$ **41.** $(\bar{x}, \bar{y}, \bar{z}) = \left(0, 0, \frac{14}{9}\right)$, $I_z = \frac{15\pi\sqrt{2}}{2}\delta$

43. (a) $\frac{8\pi}{3}a^4\delta$ **(b)** $\frac{20\pi}{3}a^4\delta$

Section 14.7, pp. 912–913

1. 4π **3.** $-5/6$ **5.** 0   **7.** -6π **9.** $2\pi a^2$ **13.** 12π

15. $-\pi/4$ **17.** -15π **25.** $16I_y + 16I_x$

Section 14.8, pp. 922–924

1. 0   **3.** 0   **5.** -16 **7.** -8π **9.** 3π **11.** $-40/3$

13. 12π **15.** $12\pi\left(4\sqrt{2} - 1\right)$

21. The integral's value never exceeds the surface area of S.

Practice Exercises, pp. 925–928

1. Path 1: $2\sqrt{3}$; path 2: $1 + 3\sqrt{2}$ **3.** $4a^2$ **5.** 0

7. $8\pi \sin(1)$ **9.** 0   **11.** $\pi\sqrt{3}$ **13.** $2\pi\left(1 - \frac{1}{\sqrt{2}}\right)$

15. $\frac{abc}{2}\sqrt{\frac{1}{a^2} + \frac{1}{b^2} + \frac{1}{c^2}}$ **17.** 50

19. $\mathbf{r}(\phi, \theta) = (6 \sin \phi \cos \theta)\mathbf{i} + (6 \sin \phi \sin \theta)\mathbf{j} + (6 \cos \phi)\mathbf{k}$, $\frac{\pi}{6} \le \phi \le \frac{2\pi}{3}, 0 \le \theta \le 2\pi$

21. $\mathbf{r}(r, \theta) = (r \cos \theta)\mathbf{i} + (r \sin \theta)\mathbf{j} + (1 + r)\mathbf{k}, 0 \le r \le 2$, $0 \le \theta \le 2\pi$

23. $\mathbf{r}(u, v) = (u \cos v)\mathbf{i} + 2u^2\mathbf{j} + (u \sin v)\mathbf{k}, 0 \le u \le 1$, $0 \le v \le \pi$

25. $\sqrt{6}$ **27.** $\pi\left[\sqrt{2} + \ln\left(1 + \sqrt{2}\right)\right]$ **29.** Conservative

31. Not conservative   **33.** $f(x, y, z) = y^2 + yz + 2x + z$

35. Path 1: 2; path 2: 8/3   **37. (a)** $1 - e^{-2\pi}$ **(b)** $1 - e^{-2\pi}$

39. 0   **41. (a)** $4\sqrt{2} - 2$ **(b)** $\sqrt{2} + \ln\left(1 + \sqrt{2}\right)$

43. $(\bar{x}, \bar{y}, \bar{z}) = \left(1, \frac{16}{15}, \frac{2}{3}\right)$; $I_x = \frac{232}{45}, I_y = \frac{64}{15}, I_z = \frac{56}{9}$

45. $\bar{z} = \frac{3}{2}, I_z = \frac{7\sqrt{3}}{3}$ **47.** $(\bar{x}, \bar{y}, \bar{z}) = (0, 0, 49/12), I_z = 640\pi$

49. Flux: 3/2; circ: $-1/2$ **53.** 3   **55.** $\frac{2\pi}{3}\left(7 - 8\sqrt{2}\right)$

57. 0   **59.** π

Additional and Advanced Exercises, pp. 928–930

1. 6π **3.** 2/3

5. (a) $\mathbf{F}(x, y, z) = z\mathbf{i} + x\mathbf{j} + y\mathbf{k}$ **(b)** $\mathbf{F}(x, y, z) = z\mathbf{i} + y\mathbf{k}$
(c) $\mathbf{F}(x, y, z) = z\mathbf{i}$

7. $\frac{16\pi R^3}{3}$ **9.** $a = 2, b = 1$. The minimum flux is -4.

11. (b) $\frac{16}{3}g$

(c) Work $= \left(\int_C gxy \, ds\right)\bar{y} = g\int_C xy^2 \, ds = \frac{16}{3}g$

13. (c) $\frac{4}{3}\pi w$ **19.** False if $\mathbf{F} = y\mathbf{i} + x\mathbf{j}$

APPENDICES

Appendix 1, pp. AP-6–AP-7

1. $0.\overline{1}, 0.\overline{2}, 0.\overline{3}, 0.\overline{8}, 0.\overline{9}$ or 1

3. $x < -2$

5. $x \le -\dfrac{1}{3}$

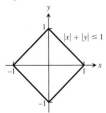

7. $3, -3$ **9.** $\dfrac{7}{6}, \dfrac{25}{6}$

11. $-2 \le t \le 4$ **13.** $0 \le z \le 10$

15. $(-\infty, -2] \cup [2, \infty)$ **17.** $(-\infty, -3] \cup [1, \infty)$

19. $(-3, -2) \cup (2, 3)$ **21.** $(0, 1)$ **23.** $(-\infty, 1]$

29. Graph of $|x| + |y| \le 1$ is the interior of "diamond-shaped" region.

Appendix 3, pp. AP-17–AP-19

1. $2, -4; 2\sqrt{5}$ **3.** Unit circle

5. $m_\perp = -\dfrac{1}{3}$

7. **(a)** $x = -1$ **(b)** $y = 4/3$ **9.** $y = -x$

11. $y = -\dfrac{5}{4}x + 6$ **13.** $y = 4x + 4$ **15.** $y = -\dfrac{x}{2} + 12$

17. x-intercept $= \sqrt{3}$, y-intercept $= -\sqrt{2}$

19. $(3, -3)$

21. $x^2 + (y - 2)^2 = 4$

23. $(x + \sqrt{3})^2 + (y + 2)^2 = 4$

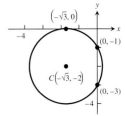

25. $x^2 + (y - 3/2)^2 = 25/4$ **27.**

29.

31. Exterior points of a circle of radius $\sqrt{7}$, centered at the origin

33. The washer between the circles $x^2 + y^2 = 1$ and $x^2 + y^2 = 4$ (points with distance from the origin between 1 and 2)

35. $(x + 2)^2 + (y - 1)^2 < 6$

37. $\left(\dfrac{1}{\sqrt{5}}, \dfrac{2}{\sqrt{5}}\right), \quad \left(-\dfrac{1}{\sqrt{5}}, -\dfrac{2}{\sqrt{5}}\right)$

39. $\left(-\dfrac{1}{\sqrt{3}}, -\dfrac{1}{3}\right), \quad \left(\dfrac{1}{\sqrt{3}}, -\dfrac{1}{3}\right)$

41. **(a)** ≈ -2.5 degrees/inch   **(b)** ≈ -16.1 degrees/inch
(c) ≈ -8.3 degrees/inch   **43.** 5.97 atm

45. Yes: $C = F = -40°$

51. $k = -8, \quad k = 1/2$

INDEX

A Brief Table of Integrals

1. $\int u \, dv = uv - \int v \, du$

2. $\int a^u \, du = \dfrac{a^u}{\ln a} + C, \quad a \neq 1, \quad a > 0$

3. $\int \cos u \, du = \sin u + C$

4. $\int \sin u \, du = -\cos u + C$

5. $\int (ax + b)^n \, dx = \dfrac{(ax + b)^{n+1}}{a(n + 1)} + C, \quad n \neq -1$

6. $\int (ax + b)^{-1} \, dx = \dfrac{1}{a} \ln |ax + b| + C$

7. $\int x(ax + b)^n \, dx = \dfrac{(ax + b)^{n+1}}{a^2} \left[\dfrac{ax + b}{n + 2} - \dfrac{b}{n + 1} \right] + C, \quad n \neq -1, -2$

8. $\int x(ax + b)^{-1} \, dx = \dfrac{x}{a} - \dfrac{b}{a^2} \ln |ax + b| + C$

9. $\int x(ax + b)^{-2} \, dx = \dfrac{1}{a^2} \left[\ln |ax + b| + \dfrac{b}{ax + b} \right] + C$

10. $\int \dfrac{dx}{x(ax + b)} = \dfrac{1}{b} \ln \left| \dfrac{x}{ax + b} \right| + C$

11. $\int \left(\sqrt{ax + b} \right)^n \, dx = \dfrac{2}{a} \dfrac{\left(\sqrt{ax + b} \right)^{n+2}}{n + 2} + C, \quad n \neq -2$

12. $\int \dfrac{\sqrt{ax + b}}{x} \, dx = 2\sqrt{ax + b} + b \int \dfrac{dx}{x\sqrt{ax + b}}$

13. (a) $\int \dfrac{dx}{x\sqrt{ax - b}} = \dfrac{2}{\sqrt{b}} \tan^{-1} \sqrt{\dfrac{ax - b}{b}} + C$ **(b)** $\int \dfrac{dx}{x\sqrt{ax + b}} = \dfrac{1}{\sqrt{b}} \ln \left| \dfrac{\sqrt{ax + b} - \sqrt{b}}{\sqrt{ax + b} + \sqrt{b}} \right| + C$

14. $\int \dfrac{\sqrt{ax + b}}{x^2} \, dx = -\dfrac{\sqrt{ax + b}}{x} + \dfrac{a}{2} \int \dfrac{dx}{x\sqrt{ax + b}} + C$

15. $\int \dfrac{dx}{x^2\sqrt{ax + b}} = -\dfrac{\sqrt{ax + b}}{bx} - \dfrac{a}{2b} \int \dfrac{dx}{x\sqrt{ax + b}} + C$

16. $\int \dfrac{dx}{a^2 + x^2} = \dfrac{1}{a} \tan^{-1} \dfrac{x}{a} + C$

17. $\int \dfrac{dx}{(a^2 + x^2)^2} = \dfrac{x}{2a^2(a^2 + x^2)} + \dfrac{1}{2a^3} \tan^{-1} \dfrac{x}{a} + C$

18. $\int \dfrac{dx}{a^2 - x^2} = \dfrac{1}{2a} \ln \left| \dfrac{x + a}{x - a} \right| + C$

19. $\int \dfrac{dx}{(a^2 - x^2)^2} = \dfrac{x}{2a^2(a^2 - x^2)} + \dfrac{1}{4a^3} \ln \left| \dfrac{x + a}{x - a} \right| + C$

20. $\int \dfrac{dx}{\sqrt{a^2 + x^2}} = \sinh^{-1} \dfrac{x}{a} + C = \ln \left(x + \sqrt{a^2 + x^2} \right) + C$

21. $\int \sqrt{a^2 + x^2} \, dx = \dfrac{x}{2} \sqrt{a^2 + x^2} + \dfrac{a^2}{2} \ln \left(x + \sqrt{a^2 + x^2} \right) + C$

22. $\int x^2 \sqrt{a^2 + x^2} \, dx = \dfrac{x}{8} (a^2 + 2x^2) \sqrt{a^2 + x^2} - \dfrac{a^4}{8} \ln \left(x + \sqrt{a^2 + x^2} \right) + C$

23. $\int \frac{\sqrt{a^2 + x^2}}{x} \, dx = \sqrt{a^2 + x^2} - a \ln \left| \frac{a + \sqrt{a^2 + x^2}}{x} \right| + C$

24. $\int \frac{\sqrt{a^2 + x^2}}{x^2} \, dx = \ln \left(x + \sqrt{a^2 + x^2} \right) - \frac{\sqrt{a^2 + x^2}}{x} + C$

25. $\int \frac{x^2}{\sqrt{a^2 + x^2}} \, dx = -\frac{a^2}{2} \ln \left(x + \sqrt{a^2 + x^2} \right) + \frac{x\sqrt{a^2 + x^2}}{2} + C$

26. $\int \frac{dx}{x\sqrt{a^2 + x^2}} = -\frac{1}{a} \ln \left| \frac{a + \sqrt{a^2 + x^2}}{x} \right| + C$ **27.** $\int \frac{dx}{x^2\sqrt{a^2 + x^2}} = -\frac{\sqrt{a^2 + x^2}}{a^2 x} + C$

28. $\int \frac{dx}{\sqrt{a^2 - x^2}} = \sin^{-1} \frac{x}{a} + C$ **29.** $\int \sqrt{a^2 - x^2} \, dx = \frac{x}{2}\sqrt{a^2 - x^2} + \frac{a^2}{2} \sin^{-1} \frac{x}{a} + C$

30. $\int x^2\sqrt{a^2 - x^2} \, dx = \frac{a^4}{8} \sin^{-1} \frac{x}{a} - \frac{1}{8} x\sqrt{a^2 - x^2} \, (a^2 - 2x^2) + C$

31. $\int \frac{\sqrt{a^2 - x^2}}{x} \, dx = \sqrt{a^2 - x^2} - a \ln \left| \frac{a + \sqrt{a^2 - x^2}}{x} \right| + C$ **32.** $\int \frac{\sqrt{a^2 - x^2}}{x^2} \, dx = -\sin^{-1} \frac{x}{a} - \frac{\sqrt{a^2 - x^2}}{x} + C$

33. $\int \frac{x^2}{\sqrt{a^2 - x^2}} \, dx = \frac{a^2}{2} \sin^{-1} \frac{x}{a} - \frac{1}{2} x\sqrt{a^2 - x^2} + C$ **34.** $\int \frac{dx}{x\sqrt{a^2 - x^2}} = -\frac{1}{a} \ln \left| \frac{a + \sqrt{a^2 - x^2}}{x} \right| + C$

35. $\int \frac{dx}{x^2\sqrt{a^2 - x^2}} = -\frac{\sqrt{a^2 - x^2}}{a^2 x} + C$ **36.** $\int \frac{dx}{\sqrt{x^2 - a^2}} = \cosh^{-1} \frac{x}{a} + C$

37. $\int \sqrt{x^2 - a^2} \, dx = \frac{x}{2}\sqrt{x^2 - a^2} - \frac{a^2}{2} \ln \left| x + \sqrt{x^2 - a^2} \right| + C$ $= \ln \left| x + \sqrt{x^2 - a^2} \right| + C$

38. $\int \left(\sqrt{x^2 - a^2} \right)^n \, dx = \frac{x \left(\sqrt{x^2 - a^2} \right)^n}{n + 1} - \frac{na^2}{n + 1} \int \left(\sqrt{x^2 - a^2} \right)^{n-2} \, dx, \quad n \neq -1$

39. $\int \frac{dx}{\left(\sqrt{x^2 - a^2} \right)^n} = \frac{x \left(\sqrt{x^2 - a^2} \right)^{2-n}}{(2 - n)a^2} - \frac{n - 3}{(n - 2)a^2} \int \frac{dx}{\left(\sqrt{x^2 - a^2} \right)^{n-2}}, \quad n \neq 2$

40. $\int x \left(\sqrt{x^2 - a^2} \right)^n \, dx = \frac{\left(\sqrt{x^2 - a^2} \right)^{n+2}}{n + 2} + C, \quad n \neq -2$

41. $\int x^2\sqrt{x^2 - a^2} \, dx = \frac{x}{8}(2x^2 - a^2)\sqrt{x^2 - a^2} - \frac{a^4}{8} \ln \left| x + \sqrt{x^2 - a^2} \right| + C$

42. $\int \frac{\sqrt{x^2 - a^2}}{x} \, dx = \sqrt{x^2 - a^2} - a \sec^{-1} \left| \frac{x}{a} \right| + C$

43. $\int \frac{\sqrt{x^2 - a^2}}{x^2} \, dx = \ln \left| x + \sqrt{x^2 - a^2} \right| - \frac{\sqrt{x^2 - a^2}}{x} + C$

44. $\int \frac{x^2}{\sqrt{x^2 - a^2}} \, dx = \frac{a^2}{2} \ln \left| x + \sqrt{x^2 - a^2} \right| + \frac{x}{2}\sqrt{x^2 - a^2} + C$

45. $\int \frac{dx}{x\sqrt{x^2 - a^2}} = \frac{1}{a} \sec^{-1} \left| \frac{x}{a} \right| + C = \frac{1}{a} \cos^{-1} \left| \frac{a}{x} \right| + C$ **46.** $\int \frac{dx}{x^2\sqrt{x^2 - a^2}} = \frac{\sqrt{x^2 - a^2}}{a^2 x} + C$

47. $\displaystyle\int \frac{dx}{\sqrt{2ax - x^2}} = \sin^{-1}\left(\frac{x - a}{a}\right) + C$

48. $\displaystyle\int \sqrt{2ax - x^2}\, dx = \frac{x - a}{2}\sqrt{2ax - x^2} + \frac{a^2}{2}\sin^{-1}\left(\frac{x - a}{a}\right) + C$

49. $\displaystyle\int \left(\sqrt{2ax - x^2}\right)^n dx = \frac{(x - a)\left(\sqrt{2ax - x^2}\right)^n}{n + 1} + \frac{na^2}{n + 1}\int \left(\sqrt{2ax - x^2}\right)^{n-2} dx$

50. $\displaystyle\int \frac{dx}{\left(\sqrt{2ax - x^2}\right)^n} = \frac{(x - a)\left(\sqrt{2ax - x^2}\right)^{2-n}}{(n - 2)a^2} + \frac{n - 3}{(n - 2)a^2}\int \frac{dx}{\left(\sqrt{2ax - x^2}\right)^{n-2}}$

51. $\displaystyle\int x\sqrt{2ax - x^2}\, dx = \frac{(x + a)(2x - 3a)\sqrt{2ax - x^2}}{6} + \frac{a^3}{2}\sin^{-1}\left(\frac{x - a}{a}\right) + C$

52. $\displaystyle\int \frac{\sqrt{2ax - x^2}}{x}\, dx = \sqrt{2ax - x^2} + a\sin^{-1}\left(\frac{x - a}{a}\right) + C$

53. $\displaystyle\int \frac{\sqrt{2ax - x^2}}{x^2}\, dx = -2\sqrt{\frac{2a - x}{x}} - \sin^{-1}\left(\frac{x - a}{a}\right) + C$

54. $\displaystyle\int \frac{x\, dx}{\sqrt{2ax - x^2}} = a\sin^{-1}\left(\frac{x - a}{a}\right) - \sqrt{2ax - x^2} + C$ **55.** $\displaystyle\int \frac{dx}{x\sqrt{2ax - x^2}} = -\frac{1}{a}\sqrt{\frac{2a - x}{x}} + C$

56. $\displaystyle\int \sin ax\, dx = -\frac{1}{a}\cos ax + C$ **57.** $\displaystyle\int \cos ax\, dx = \frac{1}{a}\sin ax + C$

58. $\displaystyle\int \sin^2 ax\, dx = \frac{x}{2} - \frac{\sin 2ax}{4a} + C$ **59.** $\displaystyle\int \cos^2 ax\, dx = \frac{x}{2} + \frac{\sin 2ax}{4a} + C$

60. $\displaystyle\int \sin^n ax\, dx = -\frac{\sin^{n-1} ax \cos ax}{na} + \frac{n - 1}{n}\int \sin^{n-2} ax\, dx$

61. $\displaystyle\int \cos^n ax\, dx = \frac{\cos^{n-1} ax \sin ax}{na} + \frac{n - 1}{n}\int \cos^{n-2} ax\, dx$

62. (a) $\displaystyle\int \sin ax \cos bx\, dx = -\frac{\cos(a + b)x}{2(a + b)} - \frac{\cos(a - b)x}{2(a - b)} + C, \quad a^2 \neq b^2$

(b) $\displaystyle\int \sin ax \sin bx\, dx = \frac{\sin(a - b)x}{2(a - b)} - \frac{\sin(a + b)x}{2(a + b)} + C, \quad a^2 \neq b^2$

(c) $\displaystyle\int \cos ax \cos bx\, dx = \frac{\sin(a - b)x}{2(a - b)} + \frac{\sin(a + b)x}{2(a + b)} + C, \quad a^2 \neq b^2$

63. $\displaystyle\int \sin ax \cos ax\, dx = -\frac{\cos 2ax}{4a} + C$ **64.** $\displaystyle\int \sin^n ax \cos ax\, dx = \frac{\sin^{n+1} ax}{(n + 1)a} + C, \quad n \neq -1$

65. $\displaystyle\int \frac{\cos ax}{\sin ax}\, dx = \frac{1}{a}\ln|\sin ax| + C$ **66.** $\displaystyle\int \cos^n ax \sin ax\, dx = -\frac{\cos^{n+1} ax}{(n + 1)a} + C, \quad n \neq -1$

67. $\displaystyle\int \frac{\sin ax}{\cos ax}\, dx = -\frac{1}{a}\ln|\cos ax| + C$

68. $\displaystyle\int \sin^n ax \cos^m ax\, dx = -\frac{\sin^{n-1} ax \cos^{m+1} ax}{a(m+n)} + \frac{n-1}{m+n}\int \sin^{n-2} ax \cos^m ax\, dx,\quad n \neq -m$ (reduces $\sin^n ax$)

69. $\displaystyle\int \sin^n ax \cos^m ax\, dx = \frac{\sin^{n+1} ax \cos^{m-1} ax}{a(m+n)} + \frac{m-1}{m+n}\int \sin^n ax \cos^{m-2} ax\, dx,\quad m \neq -n$ (reduces $\cos^m ax$)

70. $\displaystyle\int \frac{dx}{b + c\sin ax} = \frac{-2}{a\sqrt{b^2 - c^2}}\tan^{-1}\left[\sqrt{\frac{b-c}{b+c}}\tan\left(\frac{\pi}{4} - \frac{ax}{2}\right)\right] + C,\quad b^2 > c^2$

71. $\displaystyle\int \frac{dx}{b + c\sin ax} = \frac{-1}{a\sqrt{c^2 - b^2}}\ln\left|\frac{c + b\sin ax + \sqrt{c^2 - b^2}\cos ax}{b + c\sin ax}\right| + C,\quad b^2 < c^2$

72. $\displaystyle\int \frac{dx}{1 + \sin ax} = -\frac{1}{a}\tan\left(\frac{\pi}{4} - \frac{ax}{2}\right) + C$

73. $\displaystyle\int \frac{dx}{1 - \sin ax} = \frac{1}{a}\tan\left(\frac{\pi}{4} + \frac{ax}{2}\right) + C$

74. $\displaystyle\int \frac{dx}{b + c\cos ax} = \frac{2}{a\sqrt{b^2 - c^2}}\tan^{-1}\left[\sqrt{\frac{b-c}{b+c}}\tan\frac{ax}{2}\right] + C,\quad b^2 > c^2$

75. $\displaystyle\int \frac{dx}{b + c\cos ax} = \frac{1}{a\sqrt{c^2 - b^2}}\ln\left|\frac{c + b\cos ax + \sqrt{c^2 - b^2}\sin ax}{b + c\cos ax}\right| + C,\quad b^2 < c^2$

76. $\displaystyle\int \frac{dx}{1 + \cos ax} = \frac{1}{a}\tan\frac{ax}{2} + C$

77. $\displaystyle\int \frac{dx}{1 - \cos ax} = -\frac{1}{a}\cot\frac{ax}{2} + C$

78. $\displaystyle\int x\sin ax\, dx = \frac{1}{a^2}\sin ax - \frac{x}{a}\cos ax + C$

79. $\displaystyle\int x\cos ax\, dx = \frac{1}{a^2}\cos ax + \frac{x}{a}\sin ax + C$

80. $\displaystyle\int x^n \sin ax\, dx = -\frac{x^n}{a}\cos ax + \frac{n}{a}\int x^{n-1}\cos ax\, dx$

81. $\displaystyle\int x^n \cos ax\, dx = \frac{x^n}{a}\sin ax - \frac{n}{a}\int x^{n-1}\sin ax\, dx$

82. $\displaystyle\int \tan ax\, dx = \frac{1}{a}\ln|\sec ax| + C$

83. $\displaystyle\int \cot ax\, dx = \frac{1}{a}\ln|\sin ax| + C$

84. $\displaystyle\int \tan^2 ax\, dx = \frac{1}{a}\tan ax - x + C$

85. $\displaystyle\int \cot^2 ax\, dx = -\frac{1}{a}\cot ax - x + C$

86. $\displaystyle\int \tan^n ax\, dx = \frac{\tan^{n-1} ax}{a(n-1)} - \int \tan^{n-2} ax\, dx,\quad n \neq 1$

87. $\displaystyle\int \cot^n ax\, dx = -\frac{\cot^{n-1} ax}{a(n-1)} - \int \cot^{n-2} ax\, dx,\quad n \neq 1$

88. $\displaystyle\int \sec ax\, dx = \frac{1}{a}\ln|\sec ax + \tan ax| + C$

89. $\displaystyle\int \csc ax\, dx = -\frac{1}{a}\ln|\csc ax + \cot ax| + C$

90. $\displaystyle\int \sec^2 ax\, dx = \frac{1}{a}\tan ax + C$

91. $\displaystyle\int \csc^2 ax\, dx = -\frac{1}{a}\cot ax + C$

92. $\displaystyle\int \sec^n ax\, dx = \frac{\sec^{n-2} ax \tan ax}{a(n-1)} + \frac{n-2}{n-1}\int \sec^{n-2} ax\, dx,\quad n \neq 1$

93. $\displaystyle\int \csc^n ax\, dx = -\frac{\csc^{n-2} ax \cot ax}{a(n-1)} + \frac{n-2}{n-1}\int \csc^{n-2} ax\, dx,\quad n \neq 1$

94. $\displaystyle\int \sec^n ax \tan ax\, dx = \frac{\sec^n ax}{na} + C,\quad n \neq 0$

95. $\displaystyle\int \csc^n ax \cot ax\, dx = -\frac{\csc^n ax}{na} + C,\quad n \neq 0$

96. $\displaystyle\int \sin^{-1} ax \, dx = x \sin^{-1} ax + \frac{1}{a}\sqrt{1 - a^2x^2} + C$

97. $\displaystyle\int \cos^{-1} ax \, dx = x \cos^{-1} ax - \frac{1}{a}\sqrt{1 - a^2x^2} + C$

98. $\displaystyle\int \tan^{-1} ax \, dx = x \tan^{-1} ax - \frac{1}{2a} \ln(1 + a^2x^2) + C$

99. $\displaystyle\int x^n \sin^{-1} ax \, dx = \frac{x^{n+1}}{n+1} \sin^{-1} ax - \frac{a}{n+1} \int \frac{x^{n+1} \, dx}{\sqrt{1 - a^2x^2}}, \quad n \neq -1$

100. $\displaystyle\int x^n \cos^{-1} ax \, dx = \frac{x^{n+1}}{n+1} \cos^{-1} ax + \frac{a}{n+1} \int \frac{x^{n+1} \, dx}{\sqrt{1 - a^2x^2}}, \quad n \neq -1$

101. $\displaystyle\int x^n \tan^{-1} ax \, dx = \frac{x^{n+1}}{n+1} \tan^{-1} ax - \frac{a}{n+1} \int \frac{x^{n+1} \, dx}{1 + a^2x^2}, \quad n \neq -1$

102. $\displaystyle\int e^{ax} \, dx = \frac{1}{a} e^{ax} + C$

103. $\displaystyle\int b^{ax} \, dx = \frac{1}{a} \frac{b^{ax}}{\ln b} + C, \quad b > 0, b \neq 1$

104. $\displaystyle\int xe^{ax} \, dx = \frac{e^{ax}}{a^2}(ax - 1) + C$

105. $\displaystyle\int x^n e^{ax} \, dx = \frac{1}{a} x^n e^{ax} - \frac{n}{a} \int x^{n-1} e^{ax} \, dx$

106. $\displaystyle\int x^n b^{ax} \, dx = \frac{x^n b^{ax}}{a \ln b} - \frac{n}{a \ln b} \int x^{n-1} b^{ax} \, dx, \quad b > 0, b \neq 1$

107. $\displaystyle\int e^{ax} \sin bx \, dx = \frac{e^{ax}}{a^2 + b^2}(a \sin bx - b \cos bx) + C$

108. $\displaystyle\int e^{ax} \cos bx \, dx = \frac{e^{ax}}{a^2 + b^2}(a \cos bx + b \sin bx) + C$

109. $\displaystyle\int \ln ax \, dx = x \ln ax - x + C$

110. $\displaystyle\int x^n (\ln ax)^m \, dx = \frac{x^{n+1}(\ln ax)^m}{n+1} - \frac{m}{n+1} \int x^n (\ln ax)^{m-1} \, dx, \quad n \neq -1$

111. $\displaystyle\int x^{-1}(\ln ax)^m \, dx = \frac{(\ln ax)^{m+1}}{m+1} + C, \quad m \neq -1$

112. $\displaystyle\int \frac{dx}{x \ln ax} = \ln |\ln ax| + C$

113. $\displaystyle\int \sinh ax \, dx = \frac{1}{a} \cosh ax + C$

114. $\displaystyle\int \cosh ax \, dx = \frac{1}{a} \sinh ax + C$

115. $\displaystyle\int \sinh^2 ax \, dx = \frac{\sinh 2ax}{4a} - \frac{x}{2} + C$

116. $\displaystyle\int \cosh^2 ax \, dx = \frac{\sinh 2ax}{4a} + \frac{x}{2} + C$

117. $\displaystyle\int \sinh^n ax \, dx = \frac{\sinh^{n-1} ax \cosh ax}{na} - \frac{n-1}{n} \int \sinh^{n-2} ax \, dx, \quad n \neq 0$

118. $\displaystyle\int \cosh^n ax \, dx = \frac{\cosh^{n-1} ax \sinh ax}{na} + \frac{n-1}{n} \int \cosh^{n-2} ax \, dx, \quad n \neq 0$

119. $\displaystyle\int x \sinh ax \, dx = \frac{x}{a} \cosh ax - \frac{1}{a^2} \sinh ax + C$

120. $\displaystyle\int x \cosh ax \, dx = \frac{x}{a} \sinh ax - \frac{1}{a^2} \cosh ax + C$

121. $\displaystyle\int x^n \sinh ax \, dx = \frac{x^n}{a} \cosh ax - \frac{n}{a} \int x^{n-1} \cosh ax \, dx$

122. $\displaystyle\int x^n \cosh ax \, dx = \frac{x^n}{a} \sinh ax - \frac{n}{a} \int x^{n-1} \sinh ax \, dx$

123. $\displaystyle\int \tanh ax \, dx = \frac{1}{a} \ln(\cosh ax) + C$

124. $\displaystyle\int \coth ax \, dx = \frac{1}{a} \ln |\sinh ax| + C$

125. $\int \tanh^2 ax \, dx = x - \frac{1}{a} \tanh ax + C$

126. $\int \coth^2 ax \, dx = x - \frac{1}{a} \coth ax + C$

127. $\int \tanh^n ax \, dx = -\frac{\tanh^{n-1} ax}{(n-1)a} + \int \tanh^{n-2} ax \, dx, \quad n \neq 1$

128. $\int \coth^n ax \, dx = -\frac{\coth^{n-1} ax}{(n-1)a} + \int \coth^{n-2} ax \, dx, \quad n \neq 1$

129. $\int \operatorname{sech} ax \, dx = \frac{1}{a} \sin^{-1}(\tanh ax) + C$

130. $\int \operatorname{csch} ax \, dx = \frac{1}{a} \ln \left| \tanh \frac{ax}{2} \right| + C$

131. $\int \operatorname{sech}^2 ax \, dx = \frac{1}{a} \tanh ax + C$

132. $\int \operatorname{csch}^2 ax \, dx = -\frac{1}{a} \coth ax + C$

133. $\int \operatorname{sech}^n ax \, dx = \frac{\operatorname{sech}^{n-2} ax \tanh ax}{(n-1)a} + \frac{n-2}{n-1} \int \operatorname{sech}^{n-2} ax \, dx, \quad n \neq 1$

134. $\int \operatorname{csch}^n ax \, dx = -\frac{\operatorname{csch}^{n-2} ax \coth ax}{(n-1)a} - \frac{n-2}{n-1} \int \operatorname{csch}^{n-2} ax \, dx, \quad n \neq 1$

135. $\int \operatorname{sech}^n ax \tanh ax \, dx = -\frac{\operatorname{sech}^n ax}{na} + C, \quad n \neq 0$

136. $\int \operatorname{csch}^n ax \coth ax \, dx = -\frac{\operatorname{csch}^n ax}{na} + C, \quad n \neq 0$

137. $\int e^{ax} \sinh bx \, dx = \frac{e^{ax}}{2} \left[\frac{e^{bx}}{a+b} - \frac{e^{-bx}}{a-b} \right] + C, \quad a^2 \neq b^2$

138. $\int e^{ax} \cosh bx \, dx = \frac{e^{ax}}{2} \left[\frac{e^{bx}}{a+b} + \frac{e^{-bx}}{a-b} \right] + C, \quad a^2 \neq b^2$

139. $\int_0^\infty x^{n-1} e^{-x} \, dx = \Gamma(n) = (n-1)!, \quad n > 0$

140. $\int_0^\infty e^{-ax^2} \, dx = \frac{1}{2} \sqrt{\frac{\pi}{a}}, \quad a > 0$

141. $\int_0^{\pi/2} \sin^n x \, dx = \int_0^{\pi/2} \cos^n x \, dx = \begin{cases} \dfrac{1 \cdot 3 \cdot 5 \cdot \cdots \cdot (n-1)}{2 \cdot 4 \cdot 6 \cdot \cdots \cdot n} \cdot \dfrac{\pi}{2}, & \text{if } n \text{ is an even integer} \geq 2 \\[2ex] \dfrac{2 \cdot 4 \cdot 6 \cdot \cdots \cdot (n-1)}{3 \cdot 5 \cdot 7 \cdot \cdots \cdot n}, & \text{if } n \text{ is an odd integer} \geq 3 \end{cases}$

CREDITS

Page 155, Section 3.3, photo for Exercise 19, *PSSC Physics*, 2nd ed., DC Heath & Co. with Education Development Center, Inc.; **Page 231, Figure 3.58,** NCPMF "Differentiation" by W.U. Walton et al., Project CALC, Education Development Center, Inc.; **Page 235, Chapter 3 Additional and Advanced Exercises, photo for Exercise 9,** AP/Wide World Photos; **Page 437, Figure 6.41**, *PSSC Physics*, 2nd ed., DC Heath & Co. with Education Development Center, Inc.; **Page 517, Figure 8.6,** *PSSC Physics*, 2nd ed., DC Heath & Co. with Education Development Center, Inc.; **Page 706, Figure 12.6,** Reproduced by permission from Appalachian Mountain Club; **Page 740, Figure 12.23,** Department of History, U.S. Military Academy, West Point, New York; **Page 857, Figures 14.6 and 14.7,** *NCFMF Book of Film Notes*, 1974, MIT Press with Education Development Center, Inc.; **Page 858, Figure 14.14,** InterNetwork Media, Inc., and NASA/JPL.

Trigonometry Formulas

1. Definitions and Fundamental Identities

Sine: $\sin \theta = \dfrac{y}{r} = \dfrac{1}{\csc \theta}$

Cosine: $\cos \theta = \dfrac{x}{r} = \dfrac{1}{\sec \theta}$

Tangent: $\tan \theta = \dfrac{y}{x} = \dfrac{1}{\cot \theta}$

2. Identities

$\sin(-\theta) = -\sin \theta, \quad \cos(-\theta) = \cos \theta$

$\sin^2 \theta + \cos^2 \theta = 1, \quad \sec^2 \theta = 1 + \tan^2 \theta, \quad \csc^2 \theta = 1 + \cot^2 \theta$

$\sin 2\theta = 2 \sin \theta \cos \theta, \quad \cos 2\theta = \cos^2 \theta - \sin^2 \theta$

$\cos^2 \theta = \dfrac{1 + \cos 2\theta}{2}, \quad \sin^2 \theta = \dfrac{1 - \cos 2\theta}{2}$

$\sin(A + B) = \sin A \cos B + \cos A \sin B$

$\sin(A - B) = \sin A \cos B - \cos A \sin B$

$\cos(A + B) = \cos A \cos B - \sin A \sin B$

$\cos(A - B) = \cos A \cos B + \sin A \sin B$

$\tan(A + B) = \dfrac{\tan A + \tan B}{1 - \tan A \tan B}$

$\tan(A - B) = \dfrac{\tan A - \tan B}{1 + \tan A \tan B}$

$\sin\left(A - \dfrac{\pi}{2}\right) = -\cos A, \qquad \cos\left(A - \dfrac{\pi}{2}\right) = \sin$

$\sin\left(A + \dfrac{\pi}{2}\right) = \cos A, \qquad \cos\left(A + \dfrac{\pi}{2}\right) = -\sin$

$\sin A \sin B = \dfrac{1}{2} \cos(A - B) - \dfrac{1}{2} \cos(A + B)$

$\cos A \cos B = \dfrac{1}{2} \cos(A - B) + \dfrac{1}{2} \cos(A + B)$

$\sin A \cos B = \dfrac{1}{2} \sin(A - B) + \dfrac{1}{2} \sin(A + B)$

$\sin A + \sin B = 2 \sin \dfrac{1}{2}(A + B) \cos \dfrac{1}{2}(A - B)$

$\sin A - \sin B = 2 \cos \dfrac{1}{2}(A + B) \sin \dfrac{1}{2}(A - B)$

$\cos A + \cos B = 2 \cos \dfrac{1}{2}(A + B) \cos \dfrac{1}{2}(A - B)$

$\cos A - \cos B = -2 \sin \dfrac{1}{2}(A + B) \sin \dfrac{1}{2}(A - B)$

Trigonometric Functions

Radian Measure

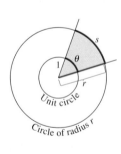

Unit circle

Circle of radius r

Degrees	Radians

The angles of two common triangles, in degrees and radians.

$\dfrac{s}{r} = \dfrac{\theta}{1} = \theta \quad$ or $\quad \theta = \dfrac{s}{r},$

$180° = \pi$ radians.

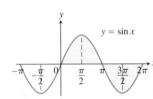

Domain: $(-\infty, \infty)$
Range: $[-1, 1]$

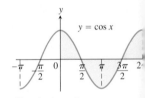

Domain: $(-\infty, \infty)$
Range: $[-1, 1]$

Domain: All real numbers except odd integer multiples of $\pi/2$
Range: $(-\infty, \infty)$

Domain: All real numbers except odd integer multiples of $\pi/2$
Range: $(-\infty, -1] \cup [1, \infty)$

Domain: $x \neq 0, \pm\pi, \pm2\pi, \ldots$
Range: $(-\infty, -1] \cup [1, \infty)$

Domain: $x \neq 0, \pm\pi, \pm2\pi, \ldots$
Range: $(-\infty, \infty)$

SERIES

Taylor Series

$$\frac{1}{1-x} = 1 + x + x^2 + \cdots + x^n + \cdots = \sum_{n=0}^{\infty} x^n, \qquad |x| < 1$$

$$\frac{1}{1+x} = 1 - x + x^2 - \cdots + (-x)^n + \cdots = \sum_{n=0}^{\infty} (-1)^n x^n, \qquad |x| < 1$$

$$e^x = 1 + x + \frac{x^2}{2!} + \cdots + \frac{x^n}{n!} + \cdots = \sum_{n=0}^{\infty} \frac{x^n}{n!}, \qquad |x| < \infty$$

$$\sin x = x - \frac{x^3}{3!} + \frac{x^5}{5!} - \cdots + (-1)^n \frac{x^{2n+1}}{(2n+1)!} + \cdots = \sum_{n=0}^{\infty} \frac{(-1)^n x^{2n+1}}{(2n+1)!}, \qquad |x| < \infty$$

$$\cos x = 1 - \frac{x^2}{2!} + \frac{x^4}{4!} - \cdots + (-1)^n \frac{x^{2n}}{(2n)!} + \cdots = \sum_{n=0}^{\infty} \frac{(-1)^n x^{2n}}{(2n)!}, \qquad |x| < \infty$$

$$\ln(1+x) = x - \frac{x^2}{2} + \frac{x^3}{3} - \cdots + (-1)^{n-1} \frac{x^n}{n} + \cdots = \sum_{n=1}^{\infty} \frac{(-1)^{n-1} x^n}{n}, \qquad -1 < x \leq 1$$

$$\ln \frac{1+x}{1-x} = 2 \tanh^{-1} x = 2\left(x + \frac{x^3}{3} + \frac{x^5}{5} + \cdots + \frac{x^{2n+1}}{2n+1} + \cdots \right) = 2 \sum_{n=0}^{\infty} \frac{x^{2n+1}}{2n+1}, \qquad |x| < 1$$

$$\tan^{-1} x = x - \frac{x^3}{3} + \frac{x^5}{5} - \cdots + (-1)^n \frac{x^{2n+1}}{2n+1} + \cdots = \sum_{n=0}^{\infty} \frac{(-1)^n x^{2n+1}}{2n+1}, \qquad |x| \leq 1$$

Binomial Series

$$(1+x)^m = 1 + mx + \frac{m(m-1)x^2}{2!} + \frac{m(m-1)(m-2)x^3}{3!} + \cdots + \frac{m(m-1)(m-2)\cdots(m-k+1)x^k}{k!} + \cdots$$

$$= 1 + \sum_{k=1}^{\infty} \binom{m}{k} x^k, \qquad |x| < 1,$$

where

$$\binom{m}{1} = m, \qquad \binom{m}{2} = \frac{m(m-1)}{2!}, \qquad \binom{m}{k} = \frac{m(m-1)\cdots(m-k+1)}{k!} \qquad \text{for } k \geq 3.$$

VECTOR OPERATOR FORMULAS (CARTESIAN FORM)

Formulas for Grad, Div, Curl, and the Laplacian

	Cartesian (x, y, z) $\mathbf{i}$, $\mathbf{j}$, and $\mathbf{k}$ are unit vectors in the directions of increasing x, y, and z. M, N, and P are the scalar components of $\mathbf{F}(x, y, z)$ in these directions.
Gradient	$\nabla f = \dfrac{\partial f}{\partial x}\mathbf{i} + \dfrac{\partial f}{\partial y}\mathbf{j} + \dfrac{\partial f}{\partial z}\mathbf{k}$
Divergence	$\nabla \cdot \mathbf{F} = \dfrac{\partial M}{\partial x} + \dfrac{\partial N}{\partial y} + \dfrac{\partial P}{\partial z}$
Curl	$\nabla \times \mathbf{F} = \begin{vmatrix} \mathbf{i} & \mathbf{j} & \mathbf{k} \\ \dfrac{\partial}{\partial x} & \dfrac{\partial}{\partial y} & \dfrac{\partial}{\partial z} \\ M & N & P \end{vmatrix}$
Laplacian	$\nabla^2 f = \dfrac{\partial^2 f}{\partial x^2} + \dfrac{\partial^2 f}{\partial y^2} + \dfrac{\partial^2 f}{\partial z^2}$

Vector Triple Products

$$(\mathbf{u} \times \mathbf{v}) \cdot \mathbf{w} = (\mathbf{v} \times \mathbf{w}) \cdot \mathbf{u} = (\mathbf{w} \times \mathbf{u}) \cdot \mathbf{v}$$

$$\mathbf{u} \times (\mathbf{v} \times \mathbf{w}) = (\mathbf{u} \cdot \mathbf{w})\mathbf{v} - (\mathbf{u} \cdot \mathbf{v})\mathbf{w}$$

The Fundamental Theorem of Line Integrals

1. Let $\mathbf{F} = M\mathbf{i} + N\mathbf{j} + P\mathbf{k}$ be a vector field whose components are continuous throughout an open connected region D in space. Then there exists a differentiable function f such that

$$\mathbf{F} = \nabla f = \frac{\partial f}{\partial x}\mathbf{i} + \frac{\partial f}{\partial y}\mathbf{j} + \frac{\partial f}{\partial z}\mathbf{k}$$

   if and only if for all points A and B in D the value of $\int_A^B \mathbf{F} \cdot d\mathbf{r}$ is independent of the path joining A to B in D.

2. If the integral is independent of the path from A to B, its value is

$$\int_A^B \mathbf{F} \cdot d\mathbf{r} = f(B) - f(A).$$

Green's Theorem and Its Generalization to Three Dimensions

Normal form of Green's Theorem:
$$\oint_C \mathbf{F} \cdot \mathbf{n}\, ds = \iint_R \nabla \cdot \mathbf{F}\, dA$$

Divergence Theorem:
$$\iint_S \mathbf{F} \cdot \mathbf{n}\, d\sigma = \iiint_D \nabla \cdot \mathbf{F}\, dV$$

Tangential form of Green's Theorem:
$$\oint_C \mathbf{F} \cdot d\mathbf{r} = \iint_R \nabla \times \mathbf{F} \cdot \mathbf{k}\, dA$$

Stokes' Theorem:
$$\oint_C \mathbf{F} \cdot d\mathbf{r} = \iint_S \nabla \times \mathbf{F} \cdot \mathbf{n}\, d\sigma$$

Vector Identities

In the identities here, f and g are differentiable scalar functions, $\mathbf{F}$, $\mathbf{F}_1$, and $\mathbf{F}_2$ are differentiable vector fields, and a and b are real constants.

$$\nabla \times (\nabla f) = \mathbf{0}$$

$$\nabla(fg) = f\nabla g + g\nabla f$$

$$\nabla \cdot (g\mathbf{F}) = g\nabla \cdot \mathbf{F} + \nabla g \cdot \mathbf{F}$$

$$\nabla \times (g\mathbf{F}) = g\nabla \times \mathbf{F} + \nabla g \times \mathbf{F}$$

$$\nabla \cdot (a\mathbf{F}_1 + b\mathbf{F}_2) = a\nabla \cdot \mathbf{F}_1 + b\nabla \cdot \mathbf{F}_2$$

$$\nabla \times (a\mathbf{F}_1 + b\mathbf{F}_2) = a\nabla \times \mathbf{F}_1 + b\nabla \times \mathbf{F}_2$$

$$\nabla(\mathbf{F}_1 \cdot \mathbf{F}_2) = (\mathbf{F}_1 \cdot \nabla)\mathbf{F}_2 + (\mathbf{F}_2 \cdot \nabla)\mathbf{F}_1 + \mathbf{F}_1 \times (\nabla \times \mathbf{F}_2) + \mathbf{F}_2 \times (\nabla \times \mathbf{F}_1)$$

$$\nabla \cdot (\mathbf{F}_1 \times \mathbf{F}_2) = \mathbf{F}_2 \cdot \nabla \times \mathbf{F}_1 - \mathbf{F}_1 \cdot \nabla \times \mathbf{F}_2$$

$$\nabla \times (\mathbf{F}_1 \times \mathbf{F}_2) = (\mathbf{F}_2 \cdot \nabla)\mathbf{F}_1 - (\mathbf{F}_1 \cdot \nabla)\mathbf{F}_2 + (\nabla \cdot \mathbf{F}_2)\mathbf{F}_1 - (\nabla \cdot \mathbf{F}_1)\mathbf{F}_2$$

$$\nabla \times (\nabla \times \mathbf{F}) = \nabla(\nabla \cdot \mathbf{F}) - (\nabla \cdot \nabla)\mathbf{F} = \nabla(\nabla \cdot \mathbf{F}) - \nabla^2 \mathbf{F}$$

$$(\nabla \times \mathbf{F}) \times \mathbf{F} = (\mathbf{F} \cdot \nabla)\mathbf{F} - \frac{1}{2}\nabla(\mathbf{F} \cdot \mathbf{F})$$

LIMITS

General Laws

If L, M, c, and k are real numbers and

$$\lim_{x \to c} f(x) = L \quad \text{and} \quad \lim_{x \to c} g(x) = M, \quad \text{then}$$

Sum Rule: $\quad\quad\quad\quad\quad \lim_{x \to c}(f(x) + g(x)) = L + M$

Difference Rule: $\quad\quad\quad \lim_{x \to c}(f(x) - g(x)) = L - M$

Product Rule: $\quad\quad\quad\quad \lim_{x \to c}(f(x) \cdot g(x)) = L \cdot M$

Constant Multiple Rule: $\quad \lim_{x \to c}(k \cdot f(x)) = k \cdot L$

Quotient Rule: $\quad\quad\quad\quad \lim_{x \to c} \dfrac{f(x)}{g(x)} = \dfrac{L}{M}, \quad M \neq 0$

The Sandwich Theorem

If $g(x) \leq f(x) \leq h(x)$ in an open interval containing c, except possibly at $x = c$, and if

$$\lim_{x \to c} g(x) = \lim_{x \to c} h(x) = L,$$

then $\lim_{x \to c} f(x) = L$.

Inequalities

If $f(x) \leq g(x)$ in an open interval containing c, except possibly at $x = c$, and both limits exist, then

$$\lim_{x \to c} f(x) \leq \lim_{x \to c} g(x).$$

Continuity

If g is continuous at L and $\lim_{x \to c} f(x) = L$, then

$$\lim_{x \to c} g(f(x)) = g(L).$$

Specific Formulas

If $P(x) = a_n x^n + a_{n-1} x^{n-1} + \cdots + a_0,$ then

$$\lim_{x \to c} P(x) = P(c) = a_n c^n + a_{n-1} c^{n-1} + \cdots + a_0.$$

If $P(x)$ and $Q(x)$ are polynomials and $Q(c) \neq 0$, then

$$\lim_{x \to c} \frac{P(x)}{Q(x)} = \frac{P(c)}{Q(c)}.$$

If $f(x)$ is continuous at $x = c$, then

$$\lim_{x \to c} f(x) = f(c).$$

$$\lim_{x \to 0} \frac{\sin x}{x} = 1 \quad \text{and} \quad \lim_{x \to 0} \frac{1 - \cos x}{x} = 0$$

L'Hôpital's Rule

If $f(a) = g(a) = 0$, both f' and g' exist in an open interval I containing a, and $g'(x) \neq 0$ on I if $x \neq a$, then

$$\lim_{x \to a} \frac{f(x)}{g(x)} = \lim_{x \to a} \frac{f'(x)}{g'(x)},$$

assuming the limit on the right side exists.

DIFFERENTIATION RULES

General Formulas

Assume u and v are differentiable functions of x.

Constant: $\dfrac{d}{dx}(c) = 0$

Sum: $\dfrac{d}{dx}(u + v) = \dfrac{du}{dx} + \dfrac{dv}{dx}$

Difference: $\dfrac{d}{dx}(u - v) = \dfrac{du}{dx} - \dfrac{dv}{dx}$

Constant Multiple: $\dfrac{d}{dx}(cu) = c\dfrac{du}{dx}$

Product: $\dfrac{d}{dx}(uv) = u\dfrac{dv}{dx} + v\dfrac{du}{dx}$

Quotient: $\dfrac{d}{dx}\left(\dfrac{u}{v}\right) = \dfrac{v\dfrac{du}{dx} - u\dfrac{dv}{dx}}{v^2}$

Power: $\dfrac{d}{dx}x^n = nx^{n-1}$

Chain Rule: $\dfrac{d}{dx}(f(g(x)) = f'(g(x)) \cdot g'(x)$

Trigonometric Functions

$\dfrac{d}{dx}(\sin x) = \cos x \qquad \dfrac{d}{dx}(\cos x) = -\sin x$

$\dfrac{d}{dx}(\tan x) = \sec^2 x \qquad \dfrac{d}{dx}(\sec x) = \sec x \tan x$

$\dfrac{d}{dx}(\cot x) = -\csc^2 x \qquad \dfrac{d}{dx}(\csc x) = -\csc x \cot x$

Exponential and Logarithmic Functions

$\dfrac{d}{dx}e^x = e^x \qquad \dfrac{d}{dx}\ln x = \dfrac{1}{x}$

$\dfrac{d}{dx}a^x = a^x \ln a \qquad \dfrac{d}{dx}(\log_a x) = \dfrac{1}{x \ln a}$

Inverse Trigonometric Functions

$\dfrac{d}{dx}(\sin^{-1} x) = \dfrac{1}{\sqrt{1 - x^2}} \qquad \dfrac{d}{dx}(\cos^{-1} x) = -\dfrac{1}{\sqrt{1 - x^2}}$

$\dfrac{d}{dx}(\tan^{-1} x) = \dfrac{1}{1 + x^2} \qquad \dfrac{d}{dx}(\sec^{-1} x) = \dfrac{1}{|x|\sqrt{x^2 - 1}}$

$\dfrac{d}{dx}(\cot^{-1} x) = -\dfrac{1}{1 + x^2} \qquad \dfrac{d}{dx}(\csc^{-1} x) = -\dfrac{1}{|x|\sqrt{x^2 - 1}}$

Hyperbolic Functions

$\dfrac{d}{dx}(\sinh x) = \cosh x \qquad \dfrac{d}{dx}(\cosh x) = \sinh x$

$\dfrac{d}{dx}(\tanh x) = \operatorname{sech}^2 x \qquad \dfrac{d}{dx}(\operatorname{sech} x) = -\operatorname{sech} x \tanh x$

$\dfrac{d}{dx}(\coth x) = -\operatorname{csch}^2 x \qquad \dfrac{d}{dx}(\operatorname{csch} x) = -\operatorname{csch} x \coth x$

Inverse Hyperbolic Functions

$\dfrac{d}{dx}(\sinh^{-1} x) = \dfrac{1}{\sqrt{1 + x^2}} \qquad \dfrac{d}{dx}(\cosh^{-1} x) = \dfrac{1}{\sqrt{x^2 - 1}}$

$\dfrac{d}{dx}(\tanh^{-1} x) = \dfrac{1}{1 - x^2} \qquad \dfrac{d}{dx}(\operatorname{sech}^{-1} x) = -\dfrac{1}{x\sqrt{1 - x^2}}$

$\dfrac{d}{dx}(\coth^{-1} x) = \dfrac{1}{1 - x^2} \qquad \dfrac{d}{dx}(\operatorname{csch}^{-1} x) = -\dfrac{1}{|x|\sqrt{1 + x^2}}$

Parametric Equations

If $x = f(t)$ and $y = g(t)$ are differentiable, then

$$y' = \frac{dy}{dx} = \frac{dy/dt}{dx/dt} \qquad \text{and} \qquad \frac{d^2y}{dx^2} = \frac{dy'/dt}{dx/dt}.$$

INTEGRATION RULES

General Formulas

Zero:
$$\int_a^a f(x)\, dx = 0$$

Order of Integration:
$$\int_b^a f(x)\, dx = -\int_a^b f(x)\, dx$$

Constant Multiples:
$$\int_a^b kf(x)\, dx = k\int_a^b f(x)\, dx \qquad (\text{Any number } k)$$

$$\int_a^b -f(x)\, dx = -\int_a^b f(x)\, dx \qquad (k = -1)$$

Sums and Differences:
$$\int_a^b (f(x) \pm g(x))\, dx = \int_a^b f(x)\, dx \pm \int_a^b g(x)\, dx$$

Additivity:
$$\int_a^b f(x)\, dx + \int_b^c f(x)\, dx = \int_a^c f(x)\, dx$$

Max-Min Inequality: If $\max f$ and $\min f$ are the maximum and minimum values of f on $[a, b]$, then

$$\min f \cdot (b - a) \le \int_a^b f(x)\, dx \le \max f \cdot (b - a).$$

Domination: $\quad f(x) \ge g(x) \quad$ on $\quad [a, b] \quad$ implies $\quad \displaystyle\int_a^b f(x)\, dx \ge \int_a^b g(x)\, dx$

$$f(x) \ge 0 \quad \text{on} \quad [a, b] \quad \text{implies} \quad \int_a^b f(x)\, dx \ge 0$$

The Fundamental Theorem of Calculus

Part 1 If f is continuous on $[a, b]$, then $F(x) = \int_a^x f(t)\, dt$ is continuous on $[a, b]$ and differentiable on (a, b) and its derivative is $f(x)$;

$$F'(x) = \frac{d}{dx}\int_a^x f(t)\, dt = f(x).$$

Part 2 If f is continuous at every point of $[a, b]$ and F is any antiderivative of f on $[a, b]$, then

$$\int_a^b f(x)\, dx = F(b) - F(a).$$

Substitution in Definite Integrals

$$\int_a^b f(g(x)) \cdot g'(x)\, dx = \int_{g(a)}^{g(b)} f(u)\, du$$

Integration by Parts

$$\int_a^b f(x)g'(x)\, dx = f(x)g(x)\Big]_a^b - \int_a^b f'(x)g(x)\, dx$$